D0893763

Viral Ecology

Viral Ecology

Edited by

Christon J. Hurst

Cincinnati, Ohio

ACADEMIC PRESS

San Diego London Boston New York Sydney Tokyo Toronto

Front cover photograph: Cinara research station in Puerto Mallarino, near Cali, Colombia. Photograph by Christon J. Hurst.

This book is printed on acid-free paper. (∞)

Copyright © 2000 by ACADEMIC PRESS

Academic Press
a Harcourt Science and Technology Company
525 B Street, Suite 1900, San Diego, California 92101-4495
http://www.academicpress.com

Academic Press Limited
Harcourt Place, 32 Jamestown Road, London NW1 7BY, UK
http://www.hbuk.co.uk/ap/

Library of Congress Catalog Card Number: 00-100242

International Standard Book Number: 0-12-362675-7

PRINTED IN UNITED STATES OF AMERICA
00   01   02   03   04   05   SB   9   8   7   6   5   4   3   2   1

To my children Allen and Rachel

Contents

Section I:
Structure and Behavior of Viruses: An Introduction

1 Defining the Ecology of Viruses

Christon J. Hurst and H. D. Alan Lindquist

2 An Introduction to Viral Taxonomy and the Proposal of Akamara, a Potential Domain for the Genomic Acellular Agents

Christon J. Hurst

Section II

Viruses of Other Microorganisms

Section III

Viruses of Macroscopic Plants

9   Viroid Diseases of Plants

Robert A. Owens

Section IV

Viruses of Macroscopic Animals

10   Ecology of Insect Viruses

Lorne D. Rothman and Judith H. Myers

11   Ecology of Viruses of Cold-Blooded Vertebrates

V. Gregory Chinchar

16   Prion Diseases of Humans and Animals

Glenn C. Telling

Contributors

Numbers in parentheses indicate the pages on which the authors' contributions begin.

Noreen J. Adcock (519), United States Environmental Protection Agency, 26 Martin Luther King Drive West, Cincinnati, Ohio 45268

Jeremy A. Bruenn (297), Department of Biological Sciences, State University of New York at Buffalo, Buffalo, New York 14260

Charles H. Calisher (493), Arthropod-Borne and Infectious Diseases Laboratory, Colorado State University, Fort Collins, Colorado 80523

Anju Chatterji (321), Department of Biology, Washington University, St. Louis, Missouri 63105

V. Gregory Chinchar (413), Department of Microbiology, University of Mississippi Medical Center, Jackson, Mississippi 39216

Victor R. DeFilippis (125), Department of Ecology and Evolutionary Biology, University of California at Irvine, Irvine, California 92697

Claude M. Fauquet (321), Department of Biology, Washington University, St. Louis, Missouri 63105

Frank J. Fenner (493), The John Curtin School of Medical Research, The Australian National University, Canberra, ACT 0200, Australia

Christon J. Hurst (3, 41, 519), United States Environmental Protection Agency, 26 Martin Luther King Drive West, Cincinnati, Ohio 45268

Christina A. Kellogg (211), Department of Marine Science, University of South Florida, St. Petersburg, Florida 33701

H. D. Alan Lindquist (3), United States Environmental Protection Agency, 26 Martin Luther King Drive West, Cincinnati, Ohio 45268

Judith H. Myers (385), Centre for Biodiversity Research, Departments of Zoology and Plant Science, University of British Columbia, Vancouver, British Columbia V6T 1Z4, Canada

Debi P. Nayak (63), Department of Microbiology and Immunology, Jonsson Comprehensive Cancer Center, School of Medicine, University of California at Los Angeles, Los Angeles, California 90095-1747

Robert A. Owens (353), Molecular Plant Pathology Laboratory, United States Department of Agriculture, Beltsville Agricultural Research Center, Beltsville, Maryland 20705

John H. Paul (211), Department of Marine Science, University of South Florida, St. Petersburg, Florida 33701

Michael L. Perdue (549), Southeast Poultry Research Laboratory, Agriculture Research Service, United States Department of Agriculture, Athens, Georgia 30605

Lorne D. Rothman (385), SAS Institute (Canada) Inc., Toronto, Ontario M5J 2T3, Canada

Bruce S. Seal (549), Southeast Poultry Research Laboratory, Agriculture Research Service, United States Department of Agriculture, Athens, Georgia 30605

Alvin W. Smith (447), College of Veterinary Medicine, Oregon State University, Corvallis, Oregon 97331

Curtis A. Suttle (247), Departments of Earth and Ocean Sciences (Oceanography), Botany, and Microbiology and Immunology, The University of British Columbia, Vancouver, British Columbia, Canada V6T 1Z4

Glenn C. Telling (593), Prion Pathogenesis Group, Sanders–Brown Center on Aging, University of Kentucky, Lexington, Kentucky 40536-0230

Luis P. Villarreal (125), Department of Molecular Biology and Biochemistry, University of California at Irvine, Irvine, California 92697

Preface

Virology is a field of study that has grown and expanded greatly since the viruses as a group first received their name in 1898. Many of the people presently learning virology have come to perceive these acellular biological entities as trinkets of nucleic acid to be cloned, probed, and spliced. However, viruses are much more than trinkets to be played with in molecular biology laboratories. Indeed, they are highly evolved biological entities with an organismal biology that is complex and interwoven with the biology of their hosting species.

The purpose of this book is to define and explain the ecology of viruses, that is, to examine what life might seem like from a "virocentric" point of view as opposed to our normal "anthropocentric" perspective. As we begin our examination of virocentric life, it is important to realize that in nature both the viruses of macroorganisms and the viruses of microorganisms exist in cycles with their respective hosts. Under normal conditions, the impact of viruses on their natural host populations may be barely apparent due to such factors as evolutionary coadaptation between the virus and those natural hosts. However, when viruses find access to new types of hosts and alternate transmission cycles, or when they encounter a concentrated population of susceptible genetically similar hosts such as occurs in densely populated human communities, communities of cultivated plants or animals, or algal blooms, then the impact of the virus on its host population can appear catastrophic. The key to understanding these types of cycles lies in understanding the viruses and how their ecology relates to the ecology of their hosts, their alternate hosts, and any vectors they utilize, as well as their relationship to the availability of suitable vehicles that can transport the different viral groups.

This book would have reached its audience a year earlier if one of the echoviruses had not invited itself into my brain to cause encephalitis. But, since these members of the genus *Enterovirus* usually just maim their hosts, rather than killing them outright, the book finally has seen completion. I wish to thank the many authors who contributed to this book, and I offer special appreciation to Noreen J. Adcock, who kindly assisted me by editing Chapters 1 and 2.

As required, I must hereby state that the United States of America's Environmental Protection Agency was not involved with the preparation of this book.

Christon J. Hurst

I

Structure and Behavior of Viruses: An Introduction

1

Defining the Ecology of Viruses

CHRISTON J. HURST and H. D. ALAN LINDQUIST

United States Environmental Protection Agency
26 Martin Luther King Drive West
Cincinnati, Ohio 45268

I. INTRODUCTION

The purpose of this book is to define the ecology of viruses and, in so doing, try
to approach the question of what life is like from a "virocentric" (as opposed to
our normal anthropocentric) point of view. Ecology is defined as the branch of

3

science that addresses the relationships between an organism of interest and the other organisms with which it interacts, the interactions between the organism of interest and its environment, and the geographic distribution of the organism of interest. The objective of this chapter is to introduce the main concepts of viral ecology. The remaining chapters of this book will then address those concepts in greater detail and illustrate the way in which those concepts apply to various host systems.

A. What Is a Virus?

Viruses are biological entities that possess a genome composed of either ribonucleic acid (RNA) or deoxyribonucleic acid (DNA) (Campbell *et al.*, 1997; Eigen, 1993; Strauss and Strauss, 1999). Viruses are infectious agents that do not possess a cellular structure of their own, and hence are "acellular infectious agents." Furthermore, the viruses are obligate intracellular parasites, meaning that they live (if we can say that about viruses) and replicate within living host cells at the expense of those host cells. Viruses accomplish their replication by usurping control of the host cell's biomolecular machinery. Those that are termed "classical viruses" will form a physical structure termed a "virion," which consists of their RNA or DNA genome surrounded by a layer of proteins (termed "capsid proteins"), which form a shell or "capsid" that protects the genomic material. Together, this capsid structure and its enclosed genomic material are often referred to as being a "nucleocapsid." The genetic coding for the capsid proteins generally is carried by the viral genome. Most of the presently known virus types code for their own capsid proteins. However, there are some viruses that are termed as being "satellite viruses." The satellite viruses encapsidate with proteins that are coded for by the genome of another virus that coinfects (simultaneously infects) that same host cell. The virus that loans its help by giving its capsid proteins to the satellite virus is termed as being a "helper virus." The capsid or nucleocapsid is, in the case of some virus groups, surrounded in turn by one or more concentric lipid bilayer membranes obtained from the host cell. There exist many other types of acellular infectious agents that have commonalities with the classical viruses in terms of their ecology. Two of these other types of acellular infectious agents — the viroids and prions — are included in this book and are addressed within their own respective chapters (9 and 16). Viroids are biological entities akin to the classical viruses and likewise can replicate only within host cells. The viroids possess RNA genomes but lack capsid proteins. The agents that we refer to as prions were once considered to be nonclassical viruses. However, we now know that the prions appear to be aberrant cellular protein products that, at least in the case of those afflicting mammals, have acquired the potential to be environmentally transmitted. The natural environmental acquisition of a prion infection occurs when a susceptible host mammal ingests the bodily material of an infected host mammal. The reproduction of prions is not a replication, but rather seems to result from a conversion of a normal host protein into an abnormal form (Prusiner, 1997; Vogel, 1997).

B. What Is Viral Ecology?

Ecology is the study of the relationships between organisms and their surroundings. Therefore, Viral ecology is the relationship between viruses, other organisms, and the environments that a virus must face as it attempts to comply with the basic biological imperatives of genetic survival and replication. As shown in Figure 1,

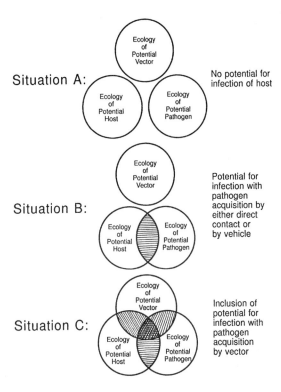

Fig. 1.   Interactions between organisms (biological entities) occur in areas where the physical and chemical ecologies of the involved organisms overlap. Infectious disease is a type of interaction in which a microorganism acts as a parasitic predator. The microorganism is referred to as a pathogen in these instances.

interactions between species and their constituent individual organisms (biological entities) occur in the areas where there exist overlaps in the temporal, physical, and biomolecular (or biochemical) aspects of the ecological zones of those different species. Many types of interactions can develop between species as they share an environment. One of the possible types of interactions is predation. When a microorganism is the predator, it is referred to as being a pathogen and the prey is referred to as being a host.

When we study viral ecology, we can view the two genetic imperatives that every biological entity must face — namely, that it survive and that it reproduce — in the perspective of a biological life cycle. A generalized biological life cycle is depicted in Figure 2. In its most basic form, this type of cycle exists at the level of the individual virus or individual cellular being. However, it must be understood that, in the case of a multicellular being, this biological life cycle exists not only at the level of each individual cell, but also at the tissue or tissue system level, as well as at the organ level. This biological life cycle likewise exists on even larger scales, where it operates at levels that describe the existence of each species as a whole, at the biological genus level, and also seems to operate further upward to at least the biological family level. Ecologically, the life cycles of those different individuals and respective species that affect one another will become interconnected temporally, geographically, and biologically. Thus, there will occur an evolution of the entire biological assemblage, and this process of biotic evolution will in turn be obliged to adapt to any abiotic changes that occur in the environment that those organisms share. While the physiologic capacities of a species establish the potential limits of the niche that it could occupy within this shared environment, the actual operational boundaries of its niche are more restricted and defined by its interspecies connections and biological competitions.

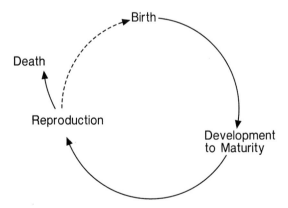

Fig. 2.   Generalized biological life cycle. Ecologically, the life cycles of different organisms that affect one another are temporally interconnected.

C. Why Study Viral Ecology?

The interplay that occurs between a virus and the living organisms that surround it, while all simultaneously pursue their own biological drive to achieve genetic survival and replication, creates an interest for studying the ecology of viruses (Doyle, 1985; Fuller, 1974; Larson, 1998; Morell, 1997; Packer, 1997; Preston, 1994; Zinkernagel, 1996). While examining this topic, we improve our understanding of the behavioral nature of viruses as predatory biological entities. It is important to realize that in nature the viruses of both macroorganisms and microorganisms normally exist in a cycle with their respective hosts. Under normal conditions, the impact of viruses on their natural hosts may be barely apparent due to factors such as evolutionary coadaptation between the virus and its host (evolutionary coadaptation is the process by which species try to achieve a mutually acceptable coexistence by evolving in ways that enable them to adapt to one another). However, when viruses find access to new types of hosts and alternate transmission cycles, or when they encounter a concentrated population of susceptible genetically similar hosts such as occurs in densely populated human communities, communities of cultivated plants or animals, or algal blooms, then the impact of the virus on its host population can appear catastrophic (Nathanson, 1997; Subbarao *et al.*, 1998).

As we study viral ecology, we come to understand not only those interconnections that exist between the entities of virus and host, but also the interconnections between these two entities and any vectors or vehicles that the virus may utilize (Cooper, 1995; Nathanson, 1996). As shown in Figure 3, this interplay can be represented by the four vertices of a tetrahedron. The possible routes by which a virus may move from one host organism to another can be illustrated as the

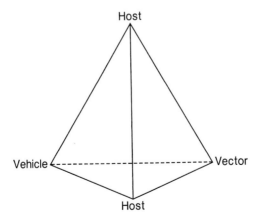

Fig. 3. The lines connecting the four vertices of this tetrahedron represent the possible routes by which a virus can move from one host organism to another.

8   Christon J. Hurst and H. D. Alan Lindquist

interconnecting lines between those vertices, which represent two hosts (present and proximate) plus one vertex apiece representing the concepts of vector and vehicle. Figure 4, which represents a flattened form of the tetrahedron shown in the Figure 3, can be considered our point of reference as we move forward in examining viral ecology. The virus must survive when in association with the present host and then successfully move from that (infected) host organism (center of Fig. 4) to another host organism (Cooper, 1995). This movement, or transmission, may occur via direct contact between the two host organisms or via routes that involve vectors and vehicles (Cooper, 1995; Hurst and Murphy, 1996). By

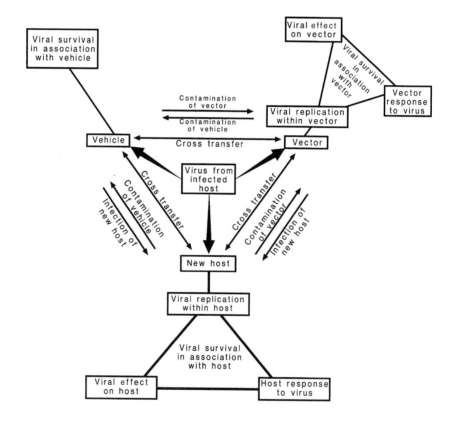

Fig. 4. Viral ecology can be represented by this diagram, which depicts a flattened form of the tetrahedron shown in Figure 3. The virus must successfully move from an infected host organism to another host organism. This movement, or transmission, may occur via direct transfer or via routes that involve vehicles and vectors. To sustain this cycle of transmission within a population of host organisms, the virus must survive when in association with the subsequently encountered hosts, vehicles, and vectors.

definition, vectors are animate (living) objects, while vehicles are, by definition, inanimate (nonliving) objects. Any virus that utilizes either vectors or vehicles must possess the means to survive when in association with those vectors and vehicles in order to sustain its cycle of transmission within a population of host organisms. If a virus replicates enough to increase its population while in association with a vector, then that vector is termed to be "biological" in nature. If the virus population does not increase while in association with a vector, then that vector is termed to be "mechanical" in nature. Because viruses are obligate intracellular parasites, and vehicles are by definition nonliving, then we must assume that the virus cannot increase its population while in association with a vehicle.

Environmentally, there are several organizational levels at which a virus must function. The first and most basic of these levels is the individual host cell. That one cell can comprise the entire host organism. Otherwise, that host cell may be part of a tissue. If within a tissue, then the tissue will be contained within a larger structure, termed either a tissue system (plant terminology) or an organ (plant and animal terminology) (Hickman and Roberts, 1995). That tissue system or organ will be contained within an organism. The host organism is but one part of a population of other organisms belonging to its same species. The members of that host species will be surrounded by populations of other types of organisms. Those populations of other types of organisms will be serving as hosts and vectors for either the same or other viruses. Each one of these organizational levels represents a different environment that the virus must successfully confront. The effects of a virus on its hosts and vectors will draw responses against which the virus must defend itself if it is to survive. Also, the virus must always be ready to do battle with its potential biological competitors. Contrariwise, the virus must be open to considering newly encountered (or reencountered) species as possible hosts or vectors. Because of their acellular nature, when viruses are viewed in ambient environments (air, soil, and water) they appear to exist in a form that essentially is biologically inert. However, they exhibit a very actively involved behavior when viewed in these many other organismal environments.

Considering the fact that viruses are obligate intracellular parasites, their ecology must be presented in terms that also include aspects of the ecology of their hosts and of any vectors they may utilize. Those factors or aspects of viral ecology that we study and thus will be considered in this book include the following:

Host-Related Issues

1. What are the principal and alternate hosts for the viruses?

2. What types of replication strategies do the viruses employ on a host cellular level, host tissue or tissue system level, host organ level, the level of the host as a whole being, and the host population level?

3. What types of survival strategies have the viruses evolved that protect them as they confront and biologically interact with the environments internal to their host (many of those internal environments are actively hostile, as the hosts have developed many powerful defensive mechanisms)?

4. What direct effects does a virus have on its hosts, that is, do the hosts get sick, and, if the hosts get sick, then how severe is the disease and does that disease directly threaten the life of the host?

5. What indirect effects does a virus have on its hosts, that is, if the virus does not directly cause the death of the hosts or if viral-induced death occurs in a temporally delayed manner, as is the case with slow or inapparent viral infections, then how might the virus affect the fitness of the host to compete for food resources or to avoid the host's predators?

General Transmission-Related Issue

6. What types of transmission strategies do viruses employ as they move between hosts, including their principal and alternate transmission routes, which may include vehicles and vectors?

Vector-Related Issues

7. In reference to biological vectors (during association with a biological vector, a virus will replicate and usually be carried within the body of the vector), what types of replication strategies do viruses employ on a vector cellular level, a vector tissue or tissue system level, a vector organ level, the level of the vector as a whole being, and also on a vector population level?

8. In reference to biological vectors, what types of survival strategies have viruses evolved that protect them as they confront and biologically interact with the environments internal to their vectors (those internal environments may be actively hostile, as vectors have developed many powerful defensive mechanisms)?

9. In reference to biological vectors, what direct effects does a virus have on its vectors, that is, do the vectors get sick, and, if the vectors get sick, then how severe is the disease and does that disease directly threaten the life of the vectors?

10. In reference to biological vectors, what indirect effects does a virus have on its vectors, that is, if the virus does not directly cause the death of the vectors or if viral-induced death occurs in a temporally delayed manner, as is the case with slow or inapparent viral infections, then how might that

virus affect the fitness of the vectors to compete for food resources or to avoid the vector's predators?

11. In reference to mechanical vectors, what types of survival strategies have been evolved by those viruses that are transmitted by (and during that event usually carried on the external surfaces of) mechanical vectors, since while in association with a mechanical vector the virus must successfully confront any compounds naturally present on the body surface of the vector as well as confront the passively hostile ambient environments of either air, water, or soil through which the vector will be moving?

Vehicle-Related Issue

12. What types of survival strategies have been evolved by viruses transmitted by way of vehicles and that thereby must successfully confront the passively hostile ambient environments of either air, water, or soil as the virus itself is transferred through those environments?

If biological curiosity alone were not a sufficient reason for studying viral ecology, then perhaps we would study the viruses out of a desire to both understand them as predators and to contemplate the ways in which we might enlist their aid as ecological tools.

II. SURVIVING THE GAME: THE VIRUS AND ITS HOST

Remember that *so long as a virus finds a new host, whether or not the current host survives is unimportant.* Although it may be beneficial to not kill a current host until that host has reproduced to help provide a new generation of potential host organisms, if the host-to-virus ratio is large enough, then even this latter point may be unimportant.

This section presents, in general terms, the relationship between a virus and its host. The generalities of relationships between viruses, vectors, and vehicles will be discussed in Section III. The specific subject of the practical limits of viral virulence in association with hosts and vectors will be addressed in Section IV.

While in association with a host, a virus has only one principal goal: to replicate itself in sufficient numbers that it can achieve transmission to another host. This goal can be attained by one of two basic strategies. The first would be a productive infection, for which three basic patterns can be defined. The second strategy would be a nonproductive infection. The goal of a productive infection is for the virus to produce infectious viral particles (those capable of infecting cells), which are termed "virions," during association of the virus with the current host. Subsequent spread of the infection to the next host occurs by transfer of these produced virions. In contrast, some of those agents that exhibit a nonproductive pattern may seldom or never produce actual virions. Thus, the usual goal of a nonproductive strategy of infection is to pass the infection to the next host by directly transferring only the viral genomic sequences (van der Kuyl et al., 1995). The patterns of productive infection are as follows:

> **"Short-term initial"** — in which viral production has only a short-term initial course, after which the viral infection ends, with the outcome depending on the virus type and historical exposure to that type within the host population, the situation being that in otherwise healthy members of a multicellular host population with which the virus has coevolved, these infections are usually mild and by themselves normally associated with a fairly low incidence of mortality.

> **"Recurrent"** — in which repeated episodes of viral production occur. This pattern often has a very pronounced initial period of viral production, after which the virus persists in a latent state within the body of the host with periodic reinitiations of viral production that usually are not life threatening.

> **"Increasing to end-stage"** — in which viral infection is normally associated with a slow, almost innocuous, start followed by a gradual progression associated with an increasing level of viral production and eventual death of the host. In these instances, death of the host may relate to destruction of the host's immunological defense systems, which then results in death by secondary infection.

There are two options with the "short-term initial" pattern. The first option is a very rapid and highly virulent approach, which is termed "fulminate" (seemingly explosive), and usually results in rapid death of the host organism. This option usually represents the product of an encounter between a virus and a host with which the virus has not coevolved. The second option is for the virus to be less virulent, causing an infection that often progresses more slowly, and appears more benign to the host. The "recurrent" and "increasing to end-stage" patterns incorporate latency into their scheme. Latency is the establishment of a condition in which the virus remains forever associated with that individual host organism and generally shows a slow and possibly only sporadic replication rate that, for some virus–host combinations, may never be life threatening to the host. The strategy of achieving a nonproductive, or virtually nonproductive, pattern of infection involves achieving an endogenous state. Endogeny implies that the genome of the virus is passed through the host's germ cells to all offspring of the infected host (van der Kuyl et al., 1995; Villareal, 1997).

The product of interspecies encounters between a virus and its natural host will usually lead to a relatively benign (mild, or not directly fatal) and statistically predictable outcome that results from adaptive coevolution between the two species. Still, these normal relationships do not represent a static coexistence between the virus and the natural host, but rather a tenuous equilibrium. Both the virus species and its evolved host species will be struggling to get the upper hand during each of their encounters (Moineau et al., 1994). The result will normally be some morbidity and even some mortality among the host population as a result of infection by that virus. Yet, because the virus as a species may not be able to survive without this natural host (Alexander, 1981), excessive viral-related mortality in the host population is not in the long-term best interest of the virus. Some endogenous viruses have evolved to offer a survival-related benefit to their natural host, and this can give an added measure of stability to their mutual relationship. Two examples of this type of relationship are the hypovirulence element associated with some strains of the chestnut blight fungus, and the endogenous retroviruses of placental mammals. The hypovirulence (reduced virulence) that the virus-derived genetic element affords to the fungus that causes the chestnut blight disease reduces the virulence of that fungus (Koonin et al., 1991; Shapira et al., 1991a,b). This reduced virulence allows the host tree and, in turn, the fungus to survive. Placental mammals, including humans, have permanently incorporated species of endogenous retroviruses into the chromosomes of their genomes. It has been hypothesized that the incorporation of these viruses has allowed the evolution of the placental mammals by suppressing maternal immunity during pregnancy (Villareal, 1997).

However, the impact of a virus on what either is, or could become, a natural host population can sometimes appear catastrophic. The most disastrous, from the host's perspective, are the biological invasions that occur when that host population encounters a virus that appears new to the host (Gao et al., 1999; Parrish, 1997). Three categories of events can lead to biological invasions of a virus into a host population. These categories are, first, that the virus species and host species (or subpopulation of the host species) may never have previously encountered one another (examples of this occurring in human populations would be the introduction of measles into the Pacific islands and the current introduction of HIV); second, if there have been previous encounters, the virus may have since changed to the point that antigenically it appears new to the host population (an example of this in humans would be the influenza pandemic of 1918–1919) (Pennisi, 1997b); and third, even if the two species may have had previous encounters, that this subpopulation of the host species subsequently may have been geographically isolated for such a length of time that most of the current host population represents a completely new generation of susceptible individuals (examples in humans are outbreaks of viral gastroenteritis found in remotely isolated communities on small islands as related to the occasional arrival of ill passengers by air- or watercraft).

Sadly, the biological invasion of the HIV viruses into human populations seems to be successful (Caldwell and Caldwell, 1996), and the extreme host death rate associated with this invasion can be assumed to indicate that the two species have not had time to coevolve. The sporadic, but limited, outbreaks in human populations of viruses such as those that cause the hemorrhagic fevers known as Ebola and Lassa represent examples of unsuccessful biological invasions. The limited chain of transmission for these latter two illnesses (Fuller, 1974; Preston, 1994), with their serial transfers often being limited to only two or three hosts in succession, represents what will occur when a virus species appears genetically unable to establish a stable relationship with a host species. The observation of extremely virulent and fulminate symptomatology, as associated with infections by Lassa and Ebola in humans, can generally be assumed to indicate either that the host in which these drastic symptoms are observed is not the natural host for those viruses or, at the very least, that these two species have not had time to coevolve. In fact, the extreme symptomatology and mortality that result in humans from Ebola and Lassa fevers seem to represent an overblown immune response on the part of the host (Spear, 1998). While having the death of a host individual occur as the product of an encounter with a pathogen may seem like a dire outcome, this outcome represents a mechanism of defense operating at the level of the host population. If a particular infectious agent is something against which members of the host population could not easily defend themselves, then it may be better to have that particular host individual die (and die very quickly!) to reduce possible spread of the contagion to other members of the host population.

A. Cell Sweet Cell, and Struggles at Home

As diagrammed in Figure 5, viruses can arrive at their new host either directly from the previously infected host, via an intermediate vehicle, or via an intermediate vector. Viral survival in association with the new host will first depend on the virus finding its appropriate receptor molecules on the host cell's surface (Spear, 1998). After this initial location, the virus must be capable of entering and modifying the host cell so that the virus can reproduce within that cell. If the host is multicellular, then the virus may first have to successfully navigate within the body of the host until it finds the particular host tissue that contains its correct host cells (Xoconostle-Cázares et al., 1999).

Within a multicellular host, the virus may face anatomically associated barriers including membranous tissues. The virus also may face nonspecific non-immune biological defenses (Baker et al., 1997; Moffat, 1994), including such chemical

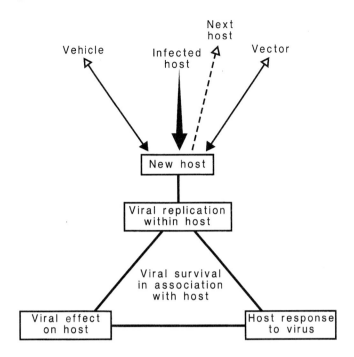

Fig. 5. Viruses can arrive at their new host (filled arrowheads) either directly from the previously infected host, by way of an intermediate vehicle, or via an intermediate vector. Viral survival in association with that new host depends on viral replication within that new host, the effects the virus has on that host, and the response of that host to the virus. Successful viral survival in association with this new host will allow possible subsequent transfer of the virus (open arrowheads) to its next host either directly, by way of a vehicle, or via a vector. This represents a segment from Figure 4.

factors as the enzymes found in both tears and saliva, and the acid found in gastric secretions. The types of anatomical and nonspecific non-immune defenses encountered can vary depending on the viral transmission route and the portal by which the virus gains entry into the host body. After a virus finds its initial host cell and succeeds at beginning its replication, the effects that the virus has on the host can elicit a defensive biological response. The category of nonspecific non-immune responses that a virus may encounter at this stage even include such things as changes in host body temperature. As if in a game of spy versus spy, most importantly, the virus must survive the host's specific immune defenses (Balter, 1996; Beck and Habicht, 1996; Gauntt, 1997; Gupta *et al.*, 1998; Levin *et al.*, 1999; Litman, 1996; Ploegh, 1998; Zinkernagel, 1996). Options for surviving the immune defenses of the host can include such techniques as:

"**You don't know me**" — a virus infecting an accidental host, in which case a very rapid proliferation may occur, an example being Lassa fever in humans.

"**Being very, very quiet**" — forming a pattern of latency in association with the virus's persistence within that host, an example being herpes viruses.

"**Virus of a thousand faces**" — antigen shifting, an example being the lentiviruses.

"**Keep to his left, that's his blind spot**" — maintaining low antigenicity, an approach used by viroids and prions.

"**Committing the perfect crime**" — infecting the immune system, an approach taken by many retroviruses and herpesviruses.

"**Finding a permanent home**" — taking up permanent genetic residence within the host and therefore automatically being transmitted to the host's progeny, an approach taken by viroids and endogenous retroviruses.

Each virus must successfully confront the host's responses while the virus tries to replicate in sufficient numbers that it has a realistic chance of being transmitted to another candidate host. Failure to successfully confront the host's responses will result in genetic termination of the virus, and, on a broader scale, such failure may eventually result in extinction for that viral species.

B. I Want a Niche Just Like the Niche That Nurtured Dear Old Mom and Dad

The initial tissue type in which a virus replicates may be inextricably linked with the initial transmission mode and portal (or site) of entry into the body of the host. For example, those viruses of mammals that are acquired by fecal–oral transmission tend to initiate their replication either in the nasopharyngeal tissues or else in the gastrointestinal tissues. There then are subsequent host tissue and organ types affected, some of which may be related to the efforts of the virus to reach its proper portal of exit. Others of the host tissues affected by the virus may be unrelated to interhost viral transmission, although the effect on those other tissues may play a strong role in the severity of illness that is associated with that viral infection. An example of the latter would be infection of brain neurons in association with echoviral conjunctivitis, an infection acquired from fomites and transmitted fecal–orally. In this case, the encephalitis causes nearly all of the associated morbidity but does not seem to benefit transmission of the virus (personal observation by author C. J. Hurst).

C. Being Societal

Successful viral survival in association with this new host will allow a possible subsequent transfer of the virus (open arrowheads, Fig. 5) to its next host either directly, via a vehicle, or via a vector. The movement of a viral infection through a population of host organisms can be examined and mathematically modeled. An epidemic transmission pattern characterized by a short-term higher-than-normal rate of infection within a host population is represented by the compartmental model shown in Figure 6 (Hurst and Murphy, 1996). An endemic transmission pattern characterized by a relatively constant long-term incidence rate of infection within a host population is represented by the compartmental model shown in Figure 7 (Hurst and Murphy, 1996).

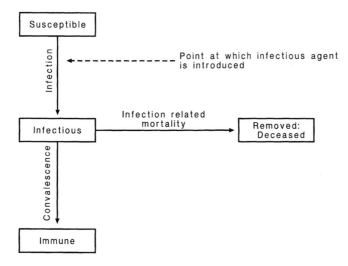

Fig. 6. Epidemic transmission of a virus within a host population is represented by this type of compartmental model (Hurst and Murphy, 1996). Each of the boxes, referred to as compartments, represents a decimal fraction of the host population, with the sum of those decimal fractions equal to 1.0. The compartments that represent actively included members of the host population are those labeled susceptible, infectious, and immune. This model incorporates only a single category of removed individuals, representing those whose demise was due to infection-related mortality. The solid arrows represent the rates at which individual members of the host species move between the different compartments during the course of an epidemic. Those rates of movement are often expressed in terms of individuals per day, as described by Hurst and Murphy (1996).

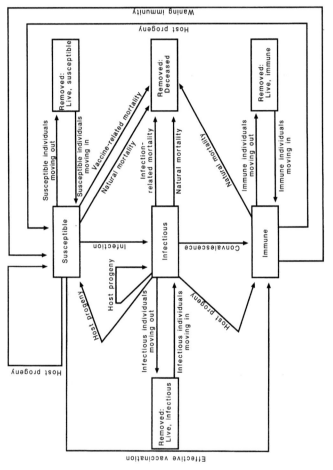

Fig. 7. Endemic transmission of a virus within a host population is represented by this type of compartmental model (Hurst and Murphy, 1996). This model is essentially an extension of the model presented in Figure 6. This model contains the same three compartments (susceptible, infectious, and immune), representing actively included individuals and the category of individuals removed by infection related mortality as described for Figure 6. This model differs in that it must also consider the various possible categories of removed live individuals that can move into and out from the compartments of actively included individuals. Their removal represents the fact that they do not interact with the actively included individuals in such a way that the virus can reach them, often due to spatial isolation. This model also includes the fact that the immune status of individuals can naturally wane or diminish with time such that immune individuals return to the compartment labeled susceptible; production of host progeny, representing reproductive success of the members of the host species; natural mortality, as a means of removing members of the population; and the possible use of vaccination to circumvent the infection process plus the associated vaccine-related mortality. Please notice that the progeny of infectious individuals may be susceptible, infectious, or immune at the time of their birth depending on the type of virus involved and whether or not that viral infection is passed to the progeny.

III. STEPPIN' OUT AND TAKING THE A TRAIN: REACHING OUT AND TOUCHING SOMEONE BY VECTOR OR VEHICLE

Remember that host–vector choices and cycles and vehicle utilizations as they exist today may (and probably do!) reflect evolutionary progression from prior species interactions and ecological relationships.

After a virus has successfully replicated within the body of its current (present) host, it must seek successful transmission to its next (proximate) host. The resulting chain of transmission is the end-all of viral reproduction. There are three basic approaches by which this can be attained: transmission by direct contact between the present and proximate hosts, transmission mediated by a vector (Brogdon and McAllister, 1998; Hurst and Murphy, 1996; Johnston and Peters, 1996; Mills and Childs, 1998; Monath and Heinz, 1996; Scott and Weaver, 1989; Weaver, 1997), and transmission mediated by a vehicle (Hurst and Murphy, 1996). While considering these approaches, it is important to keep in mind that the chains of transmission originate by random chance, followed by evolution.

A. "Down and Dirty" (Just between Us Hosts)

This heading is one that can be used to describe host-to-host transmission (transmission by host-to-host contact). While this is one of the most notorious, it is not the most common route of viral transmission between animals. This route only serves to a limited extent in microbes. Even worse, this route essentially does not seem to function in vascular plants due to the relative immobility of those hosts.

B. "The Hitchhiker" (Finding a Vector)

Transmission by vectors may be the most prevalent route by which plant viruses are spread among their hosts. Also, this route clearly exists for some viruses of animals (Weaver, 1997; Writer, 1997). However, this route has not yet been defined in terms of viruses that infect microbes. By definition, vectors are animate objects; more specifically, they are live organisms. Being a vector implies, although by definition does not require, that the entity serving as the vector has self-mobility. Thus, plants could serve as vectors, although when we consider the topic of viral vectors we usually tend to think of the vectors as being invertebrate animals. Vertebrate animals can also serve as vectors, as can some cellular microbes.

There are two categories of vectors: biological and mechanical. As stated earlier, if the virus increases in number while in association with a vector, then that vector is termed *biological*. Conversely, the vector is termed *mechanical* if the virus does not increase in number while in association with that vector. Beyond this there lie some deeper differences between mechanical and biological vectors. These differences include the fact that acquisition of a virus by a biological vector usually involves a feeding process. Phagic habits of the biological vector result in the virus being acquired from an infected host when the vector ingests virally contaminated host body materials acquired through a bite or sting. Subsequent transfer of the infection from the contaminated biological vector to the virus's next host occurs when the biological vector wounds and feeds on the next host. Actual transfer of the virus to the next host occurs incidentally when the vector contaminates the wound by discharging viruses contained either in the vector's saliva or regurgitated stomach or intestinal contents. Essentially, any animal is capable of serving as a biological vector, provided that the wound that it inflicts while feeding on a new, infected host plant or animal does not result in the death of that new host until the virus has had the chance to replicate and subsequently be transmitted onward to some proximate host. There are many issues surrounding the question of what makes a good biological vector. These issues include physical contact between the host and the potential vector during a feeding event, viral reproduction within that potential vector, and survival of the infected vector for long enough to transmit the virus to a new host. It also helps if there is some factor driving the vector to pass along the infection, such as the virus finding its way into the vector's saliva or the virus increasing the physical aggressiveness of the vector.

The fact that biological vectors usually acquire the viral contaminant while wounding and ingesting tissues from an infected host brings us to another distinguishing difference between biological and mechanical vectors: viral contamination of a biological vector is usually associated with the virus being carried internally, i.e., within the body of the vector. Replication of the virus then occurs within the body of the biological vector. In contrast, viral contamination of a mechanical vector usually occurs on the external surface of the vector, and the virus subsequently tends to remain on the external surface of the mechanical vector. One possible example of mechanical vectoring would be the acquisition of plant viruses by pollinating animals (e.g., bees, bats) during their feeding process. These pollinators can serve as mechanical vectors if they are subsequently able to passively transfer the virus from their body surface to the next plant from which they will feed. In the case of these pollinators, the acquired virus presumably is carried external to the pollinator's body. Conversely, it is possible that a plant being visited by a pollinator might become contaminated by viruses afflicting the pollinator, and the plant could then passively serve as a mechanical vector if subsequent pollinators should become infected when they visit that plant. Biting flies can serve as biological vectors if during feeding they ingest a pathogen that can replicate in association with that fly and then be passed onward when the fly bites its next victim (Hurst and Murphy, 1996). Nonbiting flies can passively serve as mechanical vectors if they feed on contaminated material and then subsequently transmit those microbial contaminants to the food of a new host without that pathogen having been able to replicate while in association with the nonbiting fly (Hurst and Murphy, 1996). Arthropods such as wasps, which repeatedly can sting multiple animals, could serve as mechanical vectors by transporting viruses on stinger surfaces. Also, passive surface contamination of pets that occurs unrelated to a feeding event can result in the pets serving as mechanical vectors (Hurst and Murphy, 1996).

When a virus is transported inside the body of a vector, then that transportation is referred to as being an "internal carriage." Contrastingly, transportation of a virus on the external body surfaces of a vector is referred to as being an "external carriage." As will be discussed in Chapter 8, there are some plant viruses that are transported through internal carriage by invertebrates that represent mechanical vectors (because the virus does not increase its population level when in association with those invertebrates). Thus, although the biological vectoring of a virus usually involves internal carriage, the fact of internal carriage does not alone always indicate that the vectoring is biological.

Because a virus must (by definition!) replicate in association with a biological vector, we can view the viral–vector association (Fig. 8) in the same manner as

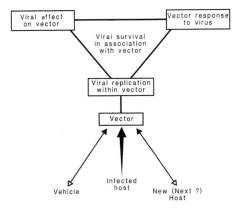

Fig. 8. This diagram addresses viral association with a biological vector and represents a segment of Figure 4. Vectors are, by definition, animate objects and are categorized either as *biological*, meaning that the virus increases in number during association with that vector, or *mechanical*, meaning that the virus does not increase in number during association with that vector. Biological vectors seem to have far greater importance than do mechanical vectors in terms of the spread of viral infections. Viruses can arrive at the biological vector (filled arrowheads) either directly from an infected host or via an intermediate vehicle. Transmission of the virus via this vector to a new host (or perhaps more accurately, the "next" host, since in the case of viruses biological vectors may be considered as alternate hosts) requires that the virus both survive and replicate while in association with that biological vector. Thus, examining viral survival in association with a biological vector also involves considering the effects that viral replication has on that vector and the response of that vector to the virus. Successful viral survival in association with the vector will allow possible subsequent transfer of the virus to its next host either directly or via a vehicle (open arrowheads).

was done for that of a virus and its host (Fig. 5). Indeed, it is often difficult to know which species is actually the viral host and which the viral vector, i.e., to distinguish the victim from the messenger. Traditionally, we have often taken the view that humans are a high form of life and that there is a descending hierarchy down to microbes. From this traditional, and sadly very anthropogenic, viewpoint, we might assume that any living thing that transmits a virus between humans must be the vector, as humans surely must be in the more respectable position of serving as the host. Another version of this philosophy would consider a vertebrate to be the host and any invertebrate the vector. Still a third version has been based on relative size, with the largest creature considered the host and the smaller the vector. Since we stated earlier that this chapter is intended to consider life from a virocentric perspective, we could easily accept the virocentric view that finds no clear distinction between host and vector. Rather, any biological vector can likewise be viewed as a host. The argument as to which one, the traditional host or traditional vector, really serves as the host would then become moot.

Because many types of viruses are capable of infecting more than a single host species, we are left to ponder how to distinguish the principal host and the alternate hosts. Settlement of this issue is usually done by examining the comparative virulence of the virus in different types of hosting species. That species for which the virus seems less virulent is assumed to be the more natural, most coevolved, host. Then it is assumed that those species for which the virus seems to have greater virulence are alternate hosts. While trying to appreciate this conundrum, it must be understood that, from a virocentric perspective, both the principal and alternate hosts, as well as any biological vectors utilized by a virus, will all represent hosting species, and thus we may never be able to sort out the answer. Any further discussion of this issue is best left to only the most insistent of philosophers! Perhaps what is left to be said here is that examples of the transmission of a virus by a biological vector are schematized in Figure 9, and that ecological interactions between a virus and its principal hosts, alternate hosts, and biological vectors can be illustrated by the example shown in Figure 10.

C. "In a Dirty Glass" (Going There by Vehicle)

Viruses also can be transmitted by vehicles. By definition, vehicles are inanimate objects. More specifically, the term *vehicle* applies to all objects other than living organisms. There are four general categories of vehicles, and these are foods, water, fomites (pronounced fō-mi′-tēz, defined as contaminated environmental surfaces that can serve in the transmission of pathogens), and aerosols. Figure 11 diagrams viral association with a vehicle. Transmission of the virus via a vehicle to a new host first requires contamination of that vehicle (shown by the

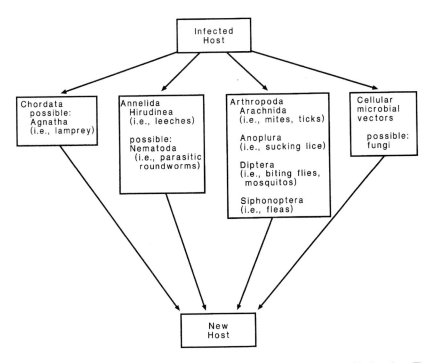

Fig. 9. Transmission of a virus via a biological vector can be represented by this flowchart. The virus is acquired as the biological vector feeds on natural bodily fluids or else enzymatically liquified bodily components of the infected host. Subsequent transmission of the virus to a new host results when the vector releases contaminated secretions while feeding on that new host.

filled arrowheads in Fig. 11). The virus must then survive while in association with the vehicle. Because viruses are by definition obligate intracellular parasites, and vehicles are by definition nonliving, a virus can neither replicate on nor within a vehicle. Likewise, because vehicles are by definition nonliving, we do not expect any specific antiviral response to be produced by the vehicle. Transfer of the virus to its next host can occur either directly or via a vector (shown as the open arrowheads in Fig. 11). One possible indication as to the difference between a vector and a vehicle is that, while a live mosquito can serve as a biological vector, after its death that same mosquito represents a vehicle. Transmission of a virus by way of a vehicle can be represented by the diagram shown in Figure 12. Acquisition of the virus by the next host or vector from that contaminated vehicle results from either ingestion of the vehicle (associated with foods and water), surface contact with either contaminated water or a contaminated solid object (a fomite), or inhalation (aerosols). The topic of vehicle-associated transmission of pathogens is discussed at length by Hurst and Murphy (1996).

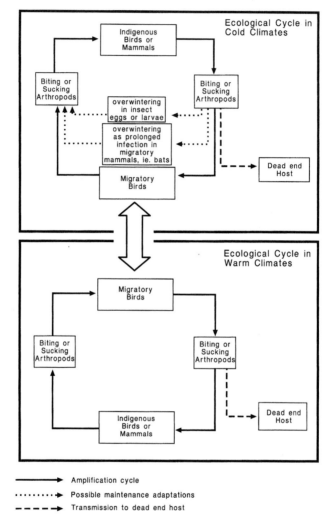

Fig. 10. This model represents a generalization of the ecological interactions that lead to insect-transmitted viral encephalitids. These infections generally are either enzootic or epizootic, meaning that their natural hosts are animals. Humans normally represent dead-end hosts for these viruses, meaning that the virus is not efficiently transmitted from infected humans to other hosts. The example shown here is of a virus that has evolved ecological cycles both in warm tropical climates and in cold temperate climates. The cycle that has evolved in the warm climates can utilize arthropod vectors that do not have to go through the process of overwintering, thus allowing for an active year-round transmission cycle. Migratory birds, which can travel thousands of miles during seasonal migrations, can shuttle the virus infection to the temperate zones, where the ecological cycle of the virus may need to include strategies for overwintering in insect eggs or larva and the possibility of survival as a prolonged infection in animals that may migrate lesser distances, such as bats.

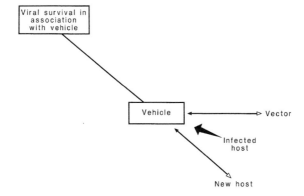

Fig. 11. This illustration addresses viral association with a vehicle and represents a segment of Figure 4. Viral transmission between hosts can occur by means of a vehicle. Vehicles are by definition inanimate. Viral contaminants can reach the vehicle (filled arrowhead) either directly from an infected host or via an intermediate vector. Transmission of the virus via this vehicle to a new host requires that the virus survive in association with the vehicle. Transfer of the virus to its next host can occur either directly or via a vector (open arrowheads).

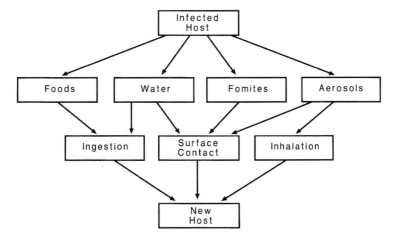

Fig. 12. The transmission of a virus via a vehicle can be represented by this flowchart. Food items can be contaminated by the action of an infected host. Alternatively, the food in question may actually be the body of an infected host that subsequently is consumed by a susceptible predatory new host. Viral contaminants present in water can be acquired by a new host either directly through ingestion of the water or following contact between the new host and an environmental surface (serving as a secondary intermediate vehicle) that has been contaminated by that water. Fomites are solid environmental (nonfood) objects whose surfaces may be involved in the transfer of infectious agents. Viral aerosols may result in infection of a new host either directly through inhalation of the aerosol or following contact between the new host and some other vehicle (either food, water, or fomite) contaminated by that aerosol.

D. Bringing Concepts Together

Biological entities exist over a spectrum of complexities, ranging from the viruses, viroids, and prions (yes, even the prions are biological entities!) to multicellular organisms. The process of maintaining the viability of even the largest of organisms is, and perhaps must be, organized at small levels. Biologically, this has been achieved by a highly evolved process of internal compartmentalization of functions with systemic coordination (Hickman and Roberts, 1995). If we consider for a moment one of the most enormous of the currently living multicellular organisms, the blue whale (*Balaenoptera musculus*), we notice that this kind of compartmentalization and coordination begins all the way down at the level of the subcellular structures and organelles within individual cells. The compartmentalization and coordination then continue upward through a number of levels, including the various individual types of cells, the tissues into which those cells are organized, the organs comprised by the tissues, and finally the total internal coordination of all of these through nerve signaling and hormonal regulation. At every one of these biological levels there is a "taking from" and a "leaving behind" exchange of material with respect to the immediate surrounding environment. This results in the existence of dramatic environmental differences at all levels, even down to the many microenvironments that exist within the organizational regions of a single cell.

Every virus must try to comply with the basic biological imperatives of genetic survival and replication. While complying with these imperatives, the viruses must, as obligate intracellular parasites, not only face but also survive within and successfully be transported through the various environments that are internal to the host. Those viruses transmitted by biological vectors must also have evolved the capability to survive and be transported through internal environments within the vector. Viruses transmitted by mechanical vectors generally must possess an additional evolved ability to survive on the surface of that vector. Likewise, both the viruses transmitted by mechanical vectors and those transmitted by vehicles must possess the ability to survive exposure to natural ambient environments encountered either in the atmosphere, hydrosphere, or lithosphere. These numerous environments are presented in Figure 13. Conditions confronted at the interface zones, as indicated by the dashed lines in the diagram, represent areas of still additional environmental complexity. While viruses appear biologically inert when viewed in ambient environments, they display their biology and interact with their surroundings when they reach environments internal to their hosts and biological vectors.

The adaptability of a species in terms of its biological cycle and biological needs will determine the potential distribution range of that species. This potential distribution range is limited in actuality to a smaller range based on interspecies

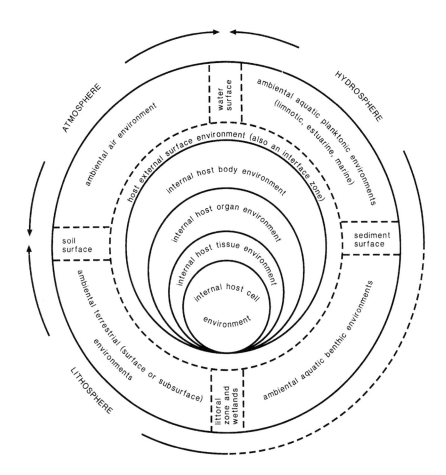

Fig. 13. This figure integrates the concepts of host, vehicle, and biological vector by representing the environments potentially faced by a virus. As obligate intracellular parasites, the viruses must face, survive within, and successfully be transported through environments that are internal to the host. Those viruses transmitted by biological vectors must also have evolved the capability to survive and be transported through internal environments faced within the vector. Viruses transmitted by vehicles and mechanical vectors must additionally possess an evolved ability to survive in natural ambient environments (atmosphere, hydrosphere, and lithosphere). Conditions confronted at the interface zones, as indicated by dashed lines, represent areas of additional environmental complexity.

relationships and competitions (Hickman and Roberts, 1995). Being large multicellular creatures ourselves, we humans normally think of a distribution range as being geographical in nature. As microbiologists, many of us have come to understand the concept of distribution range in finer detail, an example being the depth within a body of water where a particular species of microorganism normally will be found. At the level of viral ecology, the concept of species distribution range encompasses everything from tissue and organ tropisms (those tissues and organs that a virus seems to attack preferentially) upward to the geographical availability of host species and vector species, and the prevailing directional flow of appropriate vehicles such as air and water. The larger geographical end of this scale is depicted in Figure 14.

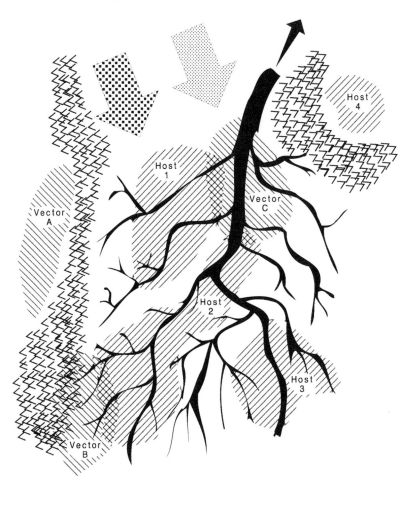

Mountains

Water Flow

Air Flow

Migratory Host (or Vector) Population

Indigenous Host Population

Indigenous Vector Population

While considering the factors addressed in Figure 14, it is important to keep in mind that, though the choice of hosts, vectors, and routes of transmission originates by random chance, the attainment of reliable continued viral success requires that such random selection events be followed and strengthened by evolution. This explains why viruses do not appear to develop the ability to use a different vehicle suddenly. Indeed, it is perhaps likely that, in order to use a vehicle such as air or water, the virus must preadapt itself to the conditions that it will encounter in association with that vehicle. Nearly every individual species of virus that achieves transmission by vehicles seems invariably to use only one type of vehicle. This trait seems to hold true for all species belonging to any given viral genus. Furthermore, this identification seems to nearly always hold true at the viral family level. In fact, this is one of the defining characteristics of the ecology of a viral group. The only virus that seems to have evolved the ability to utilize more than a single vehicle is hepatitis A (Hurst and Murphy, 1996), which has evolved a most remarkable ability to be effectively transmitted both by water and on fomites. Perhaps accordingly, the hepatitis A virus currently exists in a genus (*Hepatovirus*) it shares only with the simian hepatitis A virus. It then follows that we should not be surprised if we were to find that the simian hepatitis A virus likewise uses these same two vehicles. It is for these reasons that fears expressed in the public press that such viruses as Ebola will suddenly take flight and be transmitted over large distances via aerosol transmission amount to nothing more than over-hyped and

Fig. 14 (opposite).   This figure presents a hypothetical example of the way in which the ecology of a virus is delineated by the spatial relationships between its potential hosts, vectors, and vehicles. It represents a viral infection existing in a watershed basin whose area encompasses tens of millions of hectares. An assumption is made that the four potential indigenous host populations and three potential indigenous vector populations are terrestrial organisms whose ecological areas are delineated and that these organisms do not migrate outside their own respective ecological areas. Indigenous host populations 1, 2, and 3 reside in riverine ecological areas within the basin. Indigenous vector population B has a highland ecology, while vector population C has a lowland ecology, and both of these vector populations reside within the basin. Indigenous vector population A and indigenous host population 4 are excluded from participation in the viral infection cycle due to their geographical isolation, and, because of their geographical exclusion from the basin, we do not need to be concerned with the nature of their ecological zones. Vector population B is capable of interacting in a cycle of transmission involving host population 2. Vector population C is capable of interacting in a cycle of transmission involving host populations 1 and 2. None of the indigenous vector populations is capable of interacting in a cycle of transmission involving host population 3. A virus capable of being transmitted by surface waters could move from host population 3 to host population 2, since host population 2 is located downstream of host population 3. That same surface waterborne route could not spread the virus to host population 1, because host population 1 is not situated downstream of either host populations 2 or 3. Likewise, neither could the surface waterborne route spread the virus in an upstream direction from host population 1 to host population 2, nor from host population 2 to host population 3. Alternatively, a migratory host or vector population could carry the virus from host population 1 to host populations 2 and 3, as likewise could airflow if the virus is capable of being transmitted as an aerosol.

frightening speculation. Why is it just speculation? Because that route of transmission is not part of the virus's ecology. Invasive medical devices such as syringes and endoscopes, and transplanted tissues, including transfused blood and blood products, represent exceptions to this rule. These devices and tissues represent unnatural vehicles, which by their nature allow the virus abnormal access to the interior of a new host (Hurst and Murphy, 1996; Pennisi, 1997a). Any virus that would naturally be transmissible by direct contact with either an infected host or any type of vector can also be transmitted by one of these unnatural vehicular routes.

Viruses occasionally will appear in association with "apparently new" (unexpected) hosts and biological vectors (Brookesmith, 1997). These latter occurrences with unexpected hosts or vectors represent the identification of sporadic events that occur when geographical boundaries are breached by movement of those potential hosts and vectors for which the virus in question already has a preevolved disposition (Nettles, 1996). These preevolved dispositions may represent, at some basic level, the renewal of old acquaintances between virus, vector, and host. Alternatively, if these particular viral, host, and vector species truly never have met before, then an important aspect that can factor into these encounters is the biological relatedness between these "apparently new" hosts or vectors and those other hosts or vectors that the virus more normally would use.

E. Is There No Hope?

Many host-related factors do play a role in the transmission of viral-induced illnesses. These include the following:

> **"Finding the wrong host"** — the "whoops" or accidental occurrence factor, wherein viruses occasionally will encounter and successfully infect living beings other than their natural hosting species (Brookesmith, 1997), an event which represents a mistake not only for the host (often fated to die for want of having inherited an evolved capability to mount an effective defense against that virus) as well as a mistake for the virus (often not able to subsequently find one of its natural hosts and hence also ceases to exist).

> **"Only the good die young"** — culling the herd for communal protection can be of some advantage for the host population as a whole if those individuals that demonstrate a lesser ability to resist the virus are weakened enough by the infection that they are then more easily killed by predators (an act that reduces the likelihood that other members of the host population will become infected by that virus strain and also

may improve the gene pool of the host species by selectively eliminating the most susceptible members).

"Being your own worst enemy" — behavioral opportunities for disease transmission do exist, and ethnic or social customs often play a role in disease transmission (including the probable reality that a lack of male circumcision has spelled disaster for the human population of Africa by facilitating heterosexual transmission of HIV) (Caldwell and Caldwell, 1996), and in fact most of those vector-borne diseases that afflict humans can be avoided by changes in host behavior.

If we view this situation from the human perspective, there is a basis for hope in terms of the health of hosts. Our most important advantage lies in the use of barriers, which represent a very effective means for reducing transmission of all types of infectious agents. Barriers can be classified by their nature as physical (Table I), chemical (Table II), and biological (Table III). In many cases, these barriers already exist in nature. Natural examples of barriers include both high and low temperatures (thermal, a physical barrier), sunlight (radiation, a physical

TABLE I

Categories of Physical Barriers

Thermal

Acoustic (usually ultrasonic)

Pressure
    barometric
    hydrostatic
    osmotic

Radiation
    electronic
    neutronic
    photonic
    protonic

Impaction (includes gravitational)

Adhesion (adsorption)
    electrostatic
    van der Waals

Filtration (size exclusion)

Geographic features

Atmospheric factors
    (includes such meteorological aspects as humidity,
    precipitation, and prevailing winds)

TABLE II

Categories of Chemical Barriers

Ionic (includes pH and salinity)

Surfactant

Oxidant

Alkylant

Desiccant

Denaturant

TABLE III

Categories of Biological Barriers

Immunological (includes specific as well as nonspecific)
    naturally induced (intrinsic response)
    naturally transferred (lacteal, transovarian, transplacental, etc.)
    artificially induced (includes cytokine injection and vaccination)
    artificially transferred (includes injection with antiserum and
        tissue transfers such as transfusion and grafting)

Biomolecular resistance (not immune related)
    lack of receptor molecules
    molecular attack mechanisms (includes nucleotide-based restrictions)
    antibiotic compounds (metabolic inhibitors, either intrinsic or
        artificially supplied)

Competitive (other species in ecological competition with either the virus, its vectors,
    or its hosts)

barrier), the natural salinity of water (both osmotic, a physical barrier, and desiccant, a chemical barrier), and ecological competition (competitive, a biological barrier). The intentional use of barriers can involve both individual and combined applications. One example of a combined barrier application is the retorting of canned products, a process that employs a combination of elevated temperature and hydrostatic pressure to achieve either disinfection or sterilization (a process similar to autoclaving). Many of these barrier concepts, such as filtration, can be applied at different levels. For example, some particle exclusion filtration devices have pore sizes small enough that they can act as a filtration barrier against virus particles themselves: natural latex condoms and disposable gloves act as filtration barriers against a liquid vehicle (they contain pores larger than the virus particles yet smaller than the droplets of liquid in which the virus is contained); window screens and mosquito netting act as filtration barriers against flying vectors; and walls, fences, doors, and gates can act as filtration barriers against infected hosts. When viewed from the virocentric perspective, the use of barrier techniques to prevent viral transmission represents cause for despair instead of hope.

IV. WHY THINGS ARE THE WAY THEY ARE

The ability of a virus to pass on its genetic content is its primary consideration. We now understand how this gets done on a molecular level. What still remains to be understood is how this thing gets done and how it has come about at the species level (Eigen, 1993).

A. To Kill or Not to Kill: A Question of Virulence

One of the nagging questions that a virus must face is what the extent of its virulence should be, that is, whether or not it should kill its hosts and biological vectors as a consequence of their encounters (Ewald, 1993; Lederberg, 1997). When considered in purely evolutionary terms, virulence is the ability of the disease agent to reduce the reproductive fitness of its host. The relative virulence of a virus with respect to one of its hosts or biological vectors is generally presumed to be a marker of coevolution. More specifically stated, it seems that the less virulent a virus is for one of its hosts or vectors, the more greatly coevolved is the relationship. Why should this be so? It should be clear that, were a virus to infect an individual member of a host or biological vector population prior to that individual having reached reproductive age, it would be in the best interest of the virus to not kill that host or vector. Contrariwise, in a very strict sense, the death of that host or biological vector should not matter to the virus if the individual host or biological vector has passed the end of its normal reproductive life-span. The reason for this latter strategy is that, even if this particular host were to survive, it would not produce more susceptible offspring. Additionally, within each species of potential host or biological vector there would be a strong genetic drive to enable their infants to mount sufficient immunological defense so as to reach the age of reproductive maturity. That same genetic drive does not, by definition, act on the preservation of individuals who have lived past their reproductive years. One example of the result of interaction of these forces is the fact that infections caused by the hepatitis A virus can go nearly unnoticed in human infants, yet hepatitis A virus infections can be disastrous in human adults.

Figure 15 poses the question of how the success of a virus relates to its virulence. The virus will not be successful if the result of viral infection is too deleterious in terms of affecting the ability of the present host or biological vector to survive before that virus has been able to achieve transmission to the next host or biological vector.

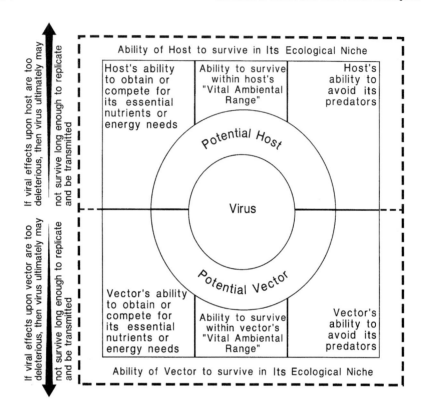

Fig. 15. This diagram illustrates how the success of a virus relates to its virulence. Success requires that the virus replicate within the body of its host and any biological vectors to concentrations high enough so that the virus has a reasonable chance of being passed on to infect either its next host or its next biological vector. The virus will not be successful if within this period of replication the result of viral infection is too deleterious in terms of affecting the ability of the present host or biological vector to survive within its own respective ecological niche. The survival requirements of those potential hosts and biological vectors include the respective ability of those hosts and biological vectors to compete for their essential needs; their ability to survive within their own vital ambient range as defined by factors such as temperature, plus either humidity and altitude (if terrestrial) or depth and salinity (if aquatic); and their ability to avoid being consumed by predatory individuals.

B. Genetic Equilibrium (Versus Disequilibrium)

One of the hallmarks of relationships between viral species and their host species is their apparent goal of reaching a mutually acceptable genetically based equilibrium (Lederberg, 1997; Zinkernagel, 1996). Some viruses also seem to have interchanged genetic material with their hosts while striving to evolutionarily reach a level of mutual coexistence.

There are many considerations associated with an apparent genetic equilibrium. In most instances of endemic viral infection in populations of a coevolved host or biological vector, the infections appear relatively unnoticed or relatively innocuous. This may change when the virus encounters a concentrated population of genetically similar susceptible hosts or biological vectors concentrated within a small radius, perhaps resulting in an epidemic. It also may change when the virus invades a population of novel hosts or vectors (hosts or vectors to which that virus appears to be new), which is termed a "biological invasion." Excessive virulence may represent reduced genetic fitness with respect to the virus, host, or biological vector. Limited virulence on the part of the virus seems to represent a state of coevolution, but with some remaining flux in the virus–host interaction. This state may have a beneficial effect by acting as genetic screening on both the host species and the viral species. In contrast, avirulence may represent a far more evolved steady state, although evolutionarily it may not be the final state, between the viral and host populations. Avirulence is normally acquired by repeated successive passage of the virus through members of a host or biological vector population.

What are the considerations associated with an apparent genetic disequilibrium? If the virus seems to make all members of a species extremely sick, then presumably it may not normally be hosted or vectored by that species. If a virus causes a reduction in the genetic fitness of the host (the ability of the host to pass on its genetic heritage), then the virus is viewed as being in disequilibrium with the host. Incompatible genetic differences may both fan the flames of virulence and allow a constant state of genetic disequilibrium to exist. Genetic equilibria need time to establish. Constant disequilibrium may be viewed as a competitive strategy effected via "Evolutionary Cheating" (included in loving memory of Dr. Alex Frasier, who taught me to understand evolution). Evolutionary cheating involves finding ways to change the rules of fair competition and thereby tilt the playing field in favor of your species. One good example of evolutionary cheating would be eating those species that compete against you. Viruses tend to steal genes from their hosts (Balter, 1998), and that would represent another example of evolutionary cheating.

C. Evolution

As we look at the relationships among viruses and their hosts and vectors, we might ask ourselves this age-old question: "Which came first, the virus or the cell?" It is perhaps more likely that viruses and cells arose simultaneously. We do not know either to what, or to where, the viruses are leading, although in a true biological sense it is not necessary for the viruses to "lead" anywhere. From a virocentric view, a perfectly organized virus reproducing from host to host (perhaps with a few vectors included for spice) and transmitting its genetic information over time is a sufficient trend. In considering the evolution of viruses, we must remember the wisdom of Niles Eldredge (1991), that no existing biological entity can be said to represent an end-product of evolution. Rather, it is only the extinct biological life forms that clearly can be said to have represented end-products of evolution. Likewise, we do not and perhaps never may know if viruses arose only once or else have arisen at many times, with their evolutionary arrival bounded only by the practical limits of some definable adaptive zone. Understanding this comes from the realization that sabre-toothed cats evolved at three different points in history and that they evolved from different lineages (Eldredge, 1991). Each evolution would have corresponded to the opening of an appropriate niche, and each extinction would have corresponded to the closing of that niche. Just as it is true that the availability of a niche can drive evolution, so too can the closure of a niche drive extinction.

Although the lack of viral fossils restricts our efforts at following the evolution of viruses, we can draw hypotheses by looking at parallels between a few of the virus groups and their hosts. To begin this process, we see that some of the presently existing viral families (we know nothing about those viral families that may be extinct) seem restricted to different host groups. It is likely that, as time has gone by, these viruses and their hosts have coevolved and perhaps even undergone phylogenation (the evolution of phylogenetic groupings) in parallel. For example, those viruses that we know as the Myoviridae seem restricted to infecting prokaryotic cells. This could suggest that the ancestors of the Myoviridae are relatively new or else relatively ancient. Members of the Siphoviridae, which also infect prokaryotes, have developed a relatively stable mechanism of endogeny (in their case referred to as lysogeny) (Campbell *et al.*, 1997) that may be suggestive of these viruses having had a long period of coevolutionary adaptation with their host cells. We can see that the viroids of plants, which genetically bear a link to the viruses (see Chapter 2 on viral taxonomy), seemingly have developed such a highly evolved endogenous state that they never produce anything resembling a virion and indeed may not use or even need a natural route of transmission. Additional examination on the existing viral groups and the establishment of parallels between these and the known evolution of animal phyla reveal that virus groups such as the Iridoviridae, which do produce virions, seem restricted to invertebrates and poikilothermic vertebrates. This latter examination could lead to the suggestion that ancestors of the iridoviruses followed the animal phylogenation pathway upward to a point just short of the evolution of euthermia. The retroviruses have gone on to infect euthermic animals, and it has been hypothesized that at least some retroviruses have coevolved with their hosts to the extent that they allowed development of the placental mammals (Villareal, 1997).

Why are the viruses still around? The viruses might serve as an evolutionary benefit to the cellular organisms by gradually transferring genetic information between different sources (Fenner et al., 1974; Williams, 1996). Perhaps this is why their hosting species continue allowing the viruses to exist. Perhaps the pure beauty of a virus, when viewed as an evolutionary element, is that it can break free from one host to enter another. Gradually, that virus could coevolve until it might settle at last on a permanent home as some endogenous genetic element within a single hosting species. Alternatively, the virus may play the role of eternally being a rebel in search of a cause. Oh, to be so free as a species!

What will the viruses become with time? As stated above, in a strictly evolutionary sense it is not necessary for the viruses to be leading anywhere. However, if we can draw parallels and make the assumption that the relationship between virus and host moves with time toward avirulence and an eventual genetic equilibrium, then we can make hypotheses. Perhaps some of the viruses will indeed continue on the way of being predatory outsiders. Others, however, seem destined for symbiosis and thus to become a part of us. We see at least two clues pointing to the latter destiny. One lies in Villareal's (1997) hypothesis that, by evolving to have the same biological agenda as their placental mammalian hosts, the endogenous retroviruses have symbiotically joined with their hosts to create a single species. The hypovirulence element of the fungus that causes chestnut blight disease is another clue (Koonin *et al.*, 1991; Shapira *et al.*, 1991a,b). This element apparently evolved from a virus and seemingly has achieved symbiosis. The hypovirulence element sustains its existence by reducing the virulence of its host fungus, so that, in turn, the host fungus does not kill the tree on which the fungus feeds.

Alas, it might also be true that the evolution of viruses represents a question that we cannot yet even attempt to answer.

V. SUMMARY (CAN THERE BE CONCLUSIONS?)

The ecology of a virus primarily consists of its interactions with the organisms that serve as its hosting species (principal hosts, alternate hosts, and vectors). The routes by which viruses achieve transmission between these other organisms represent a second aspect of the ecology of viruses. Furthermore, an examination of the interactions between a virus and its hosts and biological vectors brings up many questions. Foremost among them is the reason why the outcome of viral infections sometimes appears to be so disastrous, and yet at other times unnoticeable.

One of the founding principles in biology is that natural selection serves as the basis for the population dynamics that produce the many different outcomes we observe as scientists. When we use this principle as the lens through which to examine interactions between viruses and their host and vector species, we notice that many possible strategies exist, more than can be explained. The strategies that we do find in evidence began at random and exist because selection has not done away with them. While we do not know how the viruses arose, or what their destiny will be, we can assume that there may be viruses for as long as there are cells.

REFERENCES

Alexander, M. (1981). Why microbial predators and parasites do not eliminate their prey and hosts. *Ann. Rev. Microbiol.* **35**, 113–133.

Baker, B., Zambryski, P., Staskawicz, B., and Dinesh-Kumar, S. P. (1997). Signaling in plant–microbe interactions. *Science* **276**, 726–733.

Balter, M. (1996). HIV's other immune-system targets: Macrophages. *Science* **274**, 1464–1465.

Balter, M. (1998). Viruses have many ways to be unwelcome guests. *Science* **280**, 204–205.

Beck, G., and Habicht, G. S. (1996). Immunity and the invertebrates. *Scient. Am.* **275**(5), 60–66.

Brogdon, W. G., and McAllister, J. C. (1998). Insecticide resistance and vector control. *Emerg. Infect. Dis.* **4**, 605–613.

Brookesmith, P. (1997). "Biohazard, the Hot Zone and Beyond: Mankind's Battle Against Deadly Disease." Barnes & Noble, New York.

Caldwell, J. C., and Caldwell, P. (1996). The African AIDS epidemic. *Scient. Am.* **274**(3), 62–68.

Campbell, N. A., Mitchell, L. G., and Reece, J. B. (1997). *In* "Biology: Concepts & Connections," 2nd ed., pp. 171–195. Benjamin/Cummings, Menlo Park, CA.

Cooper, J. I. (1995). *In* "Viruses and the Environment," 2nd ed., pp. 34–60, 98–129, 140–167. Chapman and Hall, London.

Doyle, J. (1985). "Altered Harvest." Viking, New York.

Eigen, M. (1993). Viral quasispecies. *Scient. Am.* **269**(1), 42–49.

Eldredge, N. (1991). *In* "Fossils: The Evolution and Extinction of Species," pp. 4–30. Harry N. Abrams, New York.

Ewald, P. W. (1993). The evolution of virulence. *Scient. Am.* **268**(4), 86–93.

Fenner, F., McAuslan, B. R., Mims, C. A., Sambrook, J., and White, D. O. (1974). Evolutionary aspects of viral diseases. *In* "The Biology of Animal Viruses," 2nd ed., pp. 618–641. Academic Press, New York.

Fuller, J. G. (1974). "Fever!: The Hunt for a New Killer Virus." Reader's Digest, New York.

Gao, F., Bailes, E., Robertson, D. L., Chen, Y., Rodenburg, C. M., Michael, S. F., Cummins, L. B., Arthur, L. O., Peeters, M., Shaw, G. M., Sharp, P. M., and Hahn, B. H. (1999). Origin of HIV-1 in the chimpanzee *Pan troglodytes troglodytes*. *Nature* **397**, 436–441.

Gauntt, C. J. (1997). Nutrients often influence viral diseases. *ASM News* **63**, 133–135.

Gupta, S., Ferguson, N., and Anderson, R. (1998). Chaos, persistence, and evolution of strain structure in antigenically diverse infectious agents. *Science* **280**, 912–915.

Hickman Jr., C. P., and Roberts, L. S. (1995). *In* "Animal Diversity," pp. 33–62, 113–138, 163–176. McGraw-Hill, Boston.

Hurst, C. J., and Murphy, P. A. (1996). The transmission and prevention of infectious disease. *In* "Modeling Disease Transmission and Its Prevention by Disinfection" (C. J. Hurst, ed.), pp. 3–54. Cambridge University Press, Cambridge.

Johnston, R. E., and Peters, C. J. (1996). Alphaviruses. *In* "Field's Virology" (B. N. Fields, D. M. Knipe, P. M. Howley, R. M. Chanock, J. L. Melnick, T. P. Monath, B. Roizman, and S. E. Straus, eds.), 3rd ed., pp. 843–898. Lippincott-Raven, Philadelphia.

Koonin, E. V., Choi, G. H., Nuss, D. L., Shapira, R., and Carrington, J. C. (1991). Evidence for common ancestry of a chestnut blight hypovirulence-associated double-stranded RNA and a group of positive-strand RNA plant viruses. *Proc. Natl. Acad. Sci. U.S.A.* **88**, 10647–10651.

Larson, G. (1998). "There's a Hair in my Dirt! A Worm's Story." HarperCollins, New York.

Lederberg, J. (1997). Infectious disease as an evolutionary paradigm. *Emerg. Infect. Dis.* **3**, 417–423.

Levin, B. R., Lipsitch, M., and Bonhoeffer, S. (1999). Population biology, evolution, and infectious disease: Convergence and synthesis. *Science* **283**, 806–809.

Litman, G. W. (1996). Sharks and the origins of vertebrate immunity. *Scient. Am.* **275**(5), 67–71.

Mills, J. N., and Childs, J. E. (1998). Ecologic studies of rodent reservoirs: Their relevance for human health. *Emerg. Infect. Dis.* **4**, 529–537.

Moffat, A. S. (1994). Mapping the sequence of disease resistance. *Science* **265**, 1804–1805.

Moineau, S., Pandian, S, and Klaenhammer, T. R. (1994). Evolution of a lytic bacteriophage via DNA acquisition from the *Lactococcus lactis* chromosome. *Appl. Environ. Microbiol.* **60**, 1832–1841.

Monath, T. P., and Heinz, F. X. (1996). Flaviviruses. *In* "Field's Virology" (B. N. Fields, D. M. Knipe, P. M. Howley, R. M. Chanock, J. L. Melnick, T. P. Monath, B. Roizman, and S. E. Straus, eds.), 3rd ed., pp. 961–1034. Lippincott-Raven, Philadelphia.

Morell, V. (1997). Return of the forest. *Science* **278**, 2059.

Nathanson, N. (1996). Epidemiology. *In* "Field's Virology" (B. N. Fields, D. M. Knipe, P. M. Howley, R. M. Chanock, J. L. Melnick, T. P. Monath, B. Roizman, and S. E. Straus, eds.), 3rd ed., pp. 251–271. Lippincott-Raven, Philadelphia.

Nathanson, N. (1997). The emergence of infectious diseases: Societal causes and consequences. *ASM News* **63**, 83–88.

Nettles, V. F. (1996). Reemerging and emerging infectious diseases: Economic and other impacts on wildlife. *ASM News* **62**, 589–591.

Packer, C. (1997). Virus hunter. *Nat. Hist.* **106**(9), 36–41.

Parrish, C. R. (1997). How canine parvovirus suddenly shifted host range. *ASM News* **63**, 307–311.

Pennisi, E. (1997a). Monkey virus DNA found in rare human cancers. *Science* **275**, 748–749.

Pennisi, E. (1997b). First genes isolated from the deadly 1918 flu virus. *Science* **275**, 1739.

Ploegh, H. L. (1998). Viral strategies of immune evasion. *Science* **280**, 248–253.

Preston, R. (1994). "The Hot Zone." Random House, New York.

Prusiner, S. B. (1997). Prion diseases and the BSE crisis. *Science* **278**, 245–251.

Scott, T. W., and Weaver, S. C. (1989). Eastern equine encephalomyelitis virus: Epidemiology and evolution of mosquito transmission. *Adv. Virus Res.* **37**, 277–328.

Shapira, R., Choi, G. H., Hillman, B. I., and Nuss, D. L. (1991a). The contribution of defective RNAs to the complexity of viral-encoded double-stranded RNA populations present in hypovirulent strains of the chestnut blight fungus *Cryphonectria parasitica*. *EMBO J.* **10**, 741–746.

Shapira, R., Choi, G. H., and Nuss, D. L. (1991b). Virus-like genetic organization and expression strategy for a double-stranded RNA genetic element associated with biological control of chestnut blight. *EMBO J.* **10**, 731–739.

Spear, P. G. (1998). A welcome mat for leprosy and Lassa fever. *Science* **282**, 1999–2000.

Strauss, J. H., and Strauss, E. G. (1999). With a little help from the host. *Science* **283**, 802–804.

Subbarao, K., Klimov, A., Katz, J., Regnery, H., Lim, W., Hall, H., Perdue, M., Swayne, D., Bender, C., Huang, J., Hemphill, M., Rowe, T., Shaw, M., Xu, X., Fukuda, K., and Cox, N. (1998). Characterization of an avian influenza A (H5N1) virus isolated from a child with a fatal respiratory illness. *Science* **279**, 393–396.

van der Kuyl, A. C., Dekker, J. T., and Goudsmit, J. (1995). Distribution of baboon endogenous virus among species of African monkeys suggests multiple ancient cross-species transmissions in shared habitats. *J. Virol.* **69**, 7877–7887.

Villareal, L. P. (1997). On viruses, sex, and motherhood. *J. Virol.* **71**, 859–865.

Vogel, G. (1997). B cells may propagate prions. *Science* 278, 2050.

Weaver, S. C. (1997). Vector biology in arboviral pathogenesis. *In* "Viral Pathogenesis" (N. Nathanson, R. Ahmed, F. Gonzalez-Scarano, D. E. Griffin, K. V. Holmes, F. A. Murphy and H. L. Robinson, eds.), pp. 329–352. Lippincott-Raven, Philadelphia.

Williams, N. (1996). Phage transfer: A new player turns up in cholera infection. *Science* **272**, 1869–1870.

Writer, J. V. (1997). Did the mosquito do it? *Am. Hist.*, February, pp. 45–*51*.

Xoconostle-Cázares, B., Xiang, Y., Ruiz-Medrano, R., Wang, H.-L., Monzer, J., Yoo, B.-C., McFarland, K. C., Franceschi, V. R., and Lucas, W. J. (1999). Plant paralog to viral movement protein that potentiates transport of mRNA into the phloem. *Science* **283**, 94–98.

Zinkernagel, R. M. (1996). Immunology taught by viruses. *Science* **271**, 173–178.

2

An Introduction to Viral Taxonomy and the Proposal of Akamara, a Potential Domain for the Genomic Acellular Agents

CHRISTON J. HURST

United States Environmental Protection Agency
26 Martin Luther King Drive West
Cincinnati, Ohio 45268

I. INTRODUCTION

Taxonomy is literally the naming of taxons (plural also termed *taxa*), which by definition are groupings of items that have identifiable similarities. The viruses are a group of biological entities that have in common the fact that they possess a nucleic acid genome that is composed either of DNA or RNA. That nucleic acid genome is surrounded and protected by a shell of proteins that is termed either to be a *nucleocapsid* or, more simply, a *capsid*. The nucleocapsids of some viruses are, in turn, surrounded by a lipid membrane. The viruses have been grouped by many different methods. Those viral taxonomic groupings that presently are recognized by the International Committee on Taxonomy of Viruses (ICTV) divide these biological entities into families, genera, and species. There also currently exists one recognized viral family group.

The illustrations for this chapter were drawn by the author.

VIRAL ECOLOGY

This chapter also introduces the idea that the taxonomy of the viruses and their biological relatives could be extended to the domain level. There currently exist three biological domains — Archaea, Bacteria and Eukarya — that consist only of cellular organisms. The establishment of these three existing domains and the taxonomic placement of biological entities within them is based largely on the ribosomal RNA nucleotide sequence of those constituent organisms. This chapter proposes the creation of an additional biological domain that would represent the acellular infectious agents that possess nucleic acid genomes (termed "genomic acellular infectious agents" for the purpose of this proposal). The proposed constituents of this domain are the agents commonly termed to be either viruses, satellite viruses, virusoids, or viroids. The proposed domain title is *Akamara* (ακαμαρα), whose derivation from the Greek (*a* + *kamara*) would translate as "without chamber" or "without vault," and is suggested as describing the fact that these agents lack a cellular structure of their own. A possible organizational structure within this proposed new domain is also suggested, with its occupants shown as being divided into two kingdoms, plus phyla and classes premised on the basic characteristics of the genomic biochemistry of the organisms. The kingdom Euviria (true viruses) is suggested as containing the "conventional" viruses plus those viral-like agents that likewise possess genomes that code for their own structural "shell" or "capsid" proteins. The kingdom Viroidia would contain the genus *Deltavirus* plus the viroids and virusoids, whose members are RNA agents that have in common the trait that their genomic structure has endowed them with the capacity for evolutionary survival even though their genomes do not code for such structural proteins. The members of the kingdom Euviria are suggested as being subdivided into two phyla based on whether their genome is composed of RNA (phylum Ribovira) or DNA (phylum Deoxyribovira), and these are further subdivided into classes based on whether their genomes are "negative-sense" single-stranded RNA or "plus-sense" single-stranded RNA, double-stranded RNA, single-stranded DNA, or double-stranded DNA. The kingdom Viroidia is suggested to contain one phylum, Viroida, encompassing the viroids, virusoids, and genus *Deltavirus*, all of which possess RNA genomes.

II. THE EXISTING VIRAL FAMILIES

The viruses recognized by the ICTV have been assigned into genera, and nearly all of these genera have been grouped into families. Those genera that have not been incorporated into families are considered "floating genera." This concept is very fluid (Pun intended! Never accept taxonomy as though it is chiseled in stone and handed down by some deity or divine biological committee.) The viral family groupings and floating genera that existed at the time when the ICTV published its sixth report (ICTV, 1995) are listed in Table I. These viral families and floating genera are depicted, along with their basic morphological characteristics, in Figures 1–5. An examination of the drawings of the virus groups presented in these five figures reveals that the known viral capsid structures can be categorized as being either helical or icosahedral. The basic form of a helical capsid structure is illustrated in Figure 6, and the icosahedral capsid structure is shown in Figures 7 and 8.

TABLE I

Listing of Viral Taxonomic Groups

Viral group (name)	Taxonomic level (family vs. unassociated or "floating" genus)	Nature of genome	Host range as presently known	Refer to figure number
Adenoviridae	Family	DNA, double-stranded	Vertebrates	1
African swine fever-like viruses	Floating genus	DNA, double-stranded	Vertebrates	1
Arenaviridae	Family	RNA, single-stranded	Vertebrates	4
Arterivirus	Floating genus	RNA, single-stranded	Vertebrates	5
Astroviridae	Family	RNA, single-stranded	Vertebrates	5
Baculoviridae	Family	DNA, double-stranded	Invertebrates	1
Badnavirus	Floating genus	DNA, double-stranded	Plants	1
Barnaviridae	Family	RNA, single-stranded	Fungi	5
Birnaviridae	Family	RNA, double-stranded	Invertebrates, vertebrates	3
Bromoviridae	Family	RNA, single-stranded	Plants	5
Bunyaviridae	Family	RNA, single-stranded	Plants, invertebrates, vertebrates	4
Caliciviridae	Family	RNA, single-stranded	Vertebrates	5
Capillovirus	Floating genus	RNA, single-stranded	Plants	5
Carlavirus	Floating genus	RNA, single-stranded	Plants	5
Caulimovirus	Floating genus	DNA, double-stranded	Plants	1
Circoviridae	Family	DNA, single-stranded	Plants (possibly), vertebrates	2
Closterovirus	Floating genus	RNA, single-stranded	Plants	5
Comoviridae	Family	RNA, single-stranded	Plants	5
Coronaviridae	Family	RNA, single-stranded	Vertebrates	5
Corticoviridae	Family	DNA, double-stranded	Bacteria	1
Cystoviridae	Family	RNA, double-stranded	Bacteria	3
Deltavirus	Floating genus	RNA, single-stranded	Vertebrates	4
Dianthovirus	Floating genus	RNA, single-stranded	Plants	5
Enamovirus	Floating genus	RNA, single-stranded	Plants	5
Filoviridae	Family	RNA, single-stranded	Vertebrates	4
Flaviviridae	Family	RNA, single-stranded	Invertebrates, vertebrates	5
Furovirus	Floating genus	RNA, single-stranded	Plants	5
Fuselloviridae	Family	DNA, double-stranded	Bacteria	1
Geminiviridae	Family	DNA, single-stranded	Plants	2

cont'd

Viral group (name)	Taxonomic level	Nature of genome	Host range as presently known	Refer to figure number
Hepadnaviridae	Family	DNA, single-stranded	Vertebrates	2
Herpesviridae	Family	DNA, double-stranded	Vertebrates	1
Hordeivirus	Floating genus	RNA, single-stranded	Plants	5
Hypoviridae	Family	RNA, double-stranded	Fungi	3
Idaeovirus	Floating genus	RNA, single-stranded	Plants	5
Inoviridae	Family	DNA, single-stranded	Bacteria, mycoplasma	2
Iridoviridae	Family	DNA, double-stranded	Invertebrates, vertebrates	1
Leviviridae	Family	RNA, single-stranded	Bacteria	5
Lipothrixviridae	Family	DNA, double-stranded	Bacteria	1
Luteovirus	Floating genus	RNA, single-stranded	Plants	5
Machlomovirus	Floating genus	RNA, single-stranded	Plants	5
Marafivirus	Floating genus	RNA, single-stranded	Plants	5
Microviridae	Family	DNA, single-stranded	Bacteria	2
Myoviridae	Family	DNA, double-stranded	Bacteria	1
Necrovirus	Floating genus	RNA, single-stranded	Plants	5
Nodaviridae	Family	RNA, single-stranded	Invertebrates, vertebrates	5
Orthomyxoviridae	Family	RNA, single-stranded	Vertebrates	4
Papovaviridae	Family	DNA, single-stranded	Vertebrates	2
Paramyxoviridae	Family	RNA, single-stranded	Vertebrates	4
Partitiviridae	Family	RNA, double-stranded	Fungi, plants	3
Parvoviridae	Family	DNA, single-stranded	Invertebrates, vertebrates	2
Phycodnaviridae	Family	DNA, double-stranded	Algae	1
Picornaviridae	Family	RNA, single-stranded	Invertebrates, vertebrates	5
Plasmaviridae	Family	DNA, double-stranded	Mycoplasma	1
Podoviridae	Family	DNA, double-stranded	Bacteria	1
Polydnaviridae	Family	DNA, double-stranded	Invertebrates	1
Potexvirus	Floating genus	RNA, single-stranded	Plants	5
Potyviridae	Family	RNA, single-stranded	Plants	5
Poxviridae	Family	DNA, double-stranded	Vertebrates	1
Reoviridae	Family	RNA, double-stranded	Plants, invertebrates, vertebrates	3
Retroviridae	Family	RNA, single-stranded	Vertebrates	5
Rhabdoviridae	Family	RNA, single-stranded	Plants, invertebrates, vertebrates	4
Rhizidiovirus	Floating genus	DNA, double-stranded	Fungi	1
Sequiviridae	Family	RNA, single-stranded	Plants	5
Siphoviridae	Family	DNA, double-stranded	Bacteria	1
Sobemovirus	Floating genus	RNA, single-stranded	Plants	5
Tectiviridae	Family	DNA, double-stranded	Bacteria	1
Tenuivirus	Floating genus	RNA, single-stranded	Plants	4
Tetraviridae	Family	RNA, single-stranded	Invertebrates	5
Tobamovirus	Floating genus	RNA, single-stranded	Plants	5
Tobravirus	Floating genus	RNA, single-stranded	Plants	5
Togaviridae	Family	RNA, single-stranded	Invertebrates, vertebrates	5
Tombusviridae	Family	RNA, single-stranded	Plants	5
Totiviridae	Family	RNA, double-stranded	Protozoans, fungi	3
Trichovirus	Floating genus	RNA, single-stranded	Plants	5
Tymovirus	Floating genus	RNA, single-stranded	Plants	5
Umbravirus	Floating genus	RNA, single-stranded	Plants	5

100 nm

Plate 1.1   Family: Adenoviridae. Nucleic acid: DNA. Genome: Double-stranded, 1 linear segment. Morphology: Non-enveloped. Virion: Icosahedral (MW = 1.5–1.8 × 10⁸). Nucleocapsid: Icosahedral.

100 nm

Plate 1.2   Floating genus: *African swine fever-like viruses*. Nucleic acid: DNA. Genome: Double-stranded, 1 linear segment. Morphology: Enveloped. Virion: Spherical (MW not specified). Nucleocapsid: Icosahedral.

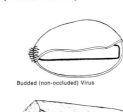

Budded (non-occluded) Virus

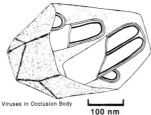

Viruses in Occlusion Body

100 nm

Plate 1.3   Family: Baculoviridae. Nucleic acid: DNA. Genome: Double-stranded, 1 circular segment. Morphology: Enveloped. Virion: Bacilliform (MW not specified). Nucleocapsid: Bacilliform. Distinguishing feature: In one phenotype the virions are found occluded inside a crystalline protein "polyhedra" or "occlusion body." Virions of the other phenotype are referred to as being "budded" or "non-occluded."

100 nm

Plate 1.4   Floating genus: *Badnavirus*. Nucleic acid: DNA. Genome: Double-stranded, 1 circular segment. Morphology: Non-enveloped. Virion: Bacilliform (MW not specified). Nucleocapsid: Tubular, comprised of repeating hexamer subunits. Distinguishing feature: Nucleocapsid structure based on an icosahedra cut across its 3-fold axis.

100 nm

Plate 1.5   Floating genus: *Caulimovirus*. Nucleic acid: DNA. Genome: Double-stranded, 1 circular segment. Morphology: Non-enveloped. Virion: Icosahedral (MW = 2.0 × 10⁷). Nucleocapsid: Icosahedral.

100 nm

Plate 1.6   Family: Corticoviridae. Nucleic acid: DNA. Genome: Double-stranded, 1 circular segment. Morphology: Non-enveloped. Virion: Icosahedral (MW = 5.8 × 10⁶). Nucleocapsid: Icosahedral. Distinguishing feature: Nucleocapsid consists of two concentric protein shells enclosing an internal lipid bilayer.

100 nm

Plate 1.7   Family: Fuselloviridae. Nucleic acid: DNA. Genome: Double-stranded, 1 circular segment. Morphology: Enveloped. Virion: Lemon-shaped (MW not specified). Nucleocapsid: Possibly helical.

Fig. 1.   Relative sizes and basic information for those viruses that possess double-stranded DNA genomes.

Plate 1.8 Family: Herpesviridae. Nucleic acid: DNA. Genome: Double-stranded, 1 linear segment. Morphology: Enveloped. Virion: Quasi-spherical [(MW = (approximate) 4.6×10^8)]. Nucleocapsid: Icosahedral.

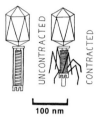

Plate 1.11 Family: Myoviridae. Nucleic acid: DNA. Genome: Double-stranded, 1 circular segment. Morphology: Non-enveloped. Virion: Tailed (MW = 2.1×10^8). Nucleocapsid: Elongated head with contractile tail and long tail fibers. Distinguishing feature: Contractile tail.

Plate 1.9 Family: Iridoviridae. Nucleic acid: DNA. Genome: Double-stranded, 1 linear segment. Morphology: Enveloped. Virion: Icosahedral (MW not specified). Nucleocapsid: Icosahedral.

Plate 1.12 Family: Phycodnaviridae. Nucleic acid: DNA. Genome: Double-stranded, 1 linear segment. Morphology: Non-enveloped. Virion: Icosahedral (MW = 1.0×10^9). Nucleocapsid: Icosahedral (multilaminate).

Plate 1.13 Family: Plasmaviridae. Nucleic acid: DNA. Genome: Double-stranded, 1 circular segment. Morphology: Enveloped. Virion: Spherical (Pleomorphic) (MW not specified). Nucleocapsid: None (possibly contains an asymmetric nucleoprotein condensate).

Plate 1.10 Family: Lipothrixviridae. Nucleic acid: DNA. Genome: Double-stranded, 1 linear segment. Morphology: Enveloped. Virion: Rodlike (rigid) (MW = 3.3×10^8). Nucleocapsid: Helical.

Fig. 1. continued

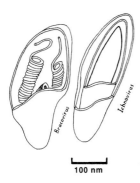

Plate 1.14 Family: Podoviridae. Nucleic acid: DNA. Genome: Double-stranded, 1 linear segment. Morphology: Non-enveloped. Virion: Tailed (MW = 4.8 × 10⁷). Nucleocapsid: Isometric head with short rigid tail and short tail fibers.

Plate 1.17 Floating genus: *Rhizidiovirus*. Nucleic acid: DNA. Genome: Double-stranded, 1 linear segment. Morphology: Non-enveloped. Virion: Icosahedral (MW not specified). Nucleocapsid: Icosahedral.

Plate 1.15 Family: Polydnaviridae. Nucleic acid: DNA. Genome: Double-stranded, Multiple circular segments (number of segments may vary by species). Morphology: Enveloped. Virion: Ellipsoidal (MW not specified). Nucleocapsid: Helical. Distinguishing features: The two genera grouped within this family differ greatly in morphology. Members of the genus *Bracovirus* contain only a single unit membrane envelope. Members of the genus *Ichnovirus* uniquely have a double envelope consisting of two concentric unit membranes surrounding the nucleocapsid.

Plate 1.18 Family: Siphoviridae. Nucleic acid: DNA. Genome: Double-stranded, 1 linear segment. Morphology: Non-enveloped. Virion: Tailed (MW = 6.0 × 10⁷). Nucleocapsid: Isometric head with long flexible, non-contractile tail and short tail fibers.

Plate 1.16 Family: Poxviridae. Nucleic acid: DNA. Genome: Double-stranded, 1 linear segment. Morphology: Enveloped. Virion: Ovoid (MW = 1.2 × 10⁸). Nucleocapsid: Cylindrical or biconcave (genus-specific). Distinguishing feature: Virions contain both an external envelope plus internal surface and core membranes.

Plate 1.19 Family: Tectiviridae. Nucleic acid: DNA. Genome: Double-stranded, 1 linear segment. Morphology: Non-enveloped. Virion: Icosahedral (MW = 7.0 × 10⁷). Nucleocapsid: Icosahedral.

Fig. 1.   continued

100 nm

Plate 2.1 Family: Circoviridae. Nucleic acid: DNA. Genome: Single-stranded (sense not specified), 1 circular segment. Morphology: Non-enveloped. Virion: Icosahedral (MW not specified). Nucleocapsid: Icosahedral.

100 nm

Plate 2.5 Family: Microviridae. Nucleic acid: DNA. Genome: Single-stranded, positive sense, 1 circular segment. Morphology: Non-enveloped. Virion: Icosahedral (MW = $6.0–7.0 \times 10^6$). Nucleocapsid: Icosahedral.

100 nm

Plate 2.2 Family: Geminiviridae. Nucleic acid: DNA. Genome: Single-stranded, ambisense, 1 circular segment (per virion). Morphology: Non-enveloped. Virion: Geminate (MW not specified). Nucleocapsid: Geminate. Distinguishing feature: Virion capsid consists of multiple (usually two) adjoined incomplete icosahedra.

100 nm

Plate 2.6 Family: Papovaviridae. Nucleic acid: DNA. Genome: Single-stranded, negative sense (presumably), 1 circular segment. Morphology: Non-enveloped. Virion: Icosahedral (MW = $2.5–4.7 \times 10^7$). Nucleocapsid: Icosahedral.

100 nm

Plate 2.3 Family: Hepadnaviridae. Nucleic acid: DNA. Genome: Single-stranded (partially double-stranded), negative sense, 1 circular segment. Morphology: Enveloped. Virion: Spherical (pleomorphic) (MW = $1.6–1.8 \times 10^6$). Nucleocapsid: Icosahedral.

Plate 2.4 Family: Inoviridae. Nucleic acid: DNA. Genome: Single-stranded, positive sense, 1 circular segment. Morphology: Non-enveloped. Virion: Rodlike (flexuous) (MW = $1.2–2.3 \times 10^7$). Nucleocapsid: Helical.

100 nm

Plate 2.7 Family: Parvoviridae. Nucleic acid: DNA. Genome: Single-stranded, strands of either sense can be encapsidated, 1 linear segment. Morphology: Non-enveloped. Virion: Icosahedral (MW = $5.5–6.2 \times 10^6$). Nucleocapsid: Icosahedral.

Fig. 2. Relative sizes and basic information for those viruses that possess single-stranded DNA genomes.

Plate 3.1 Family: Birnaviridae. Nucleic acid: RNA. Genome: Double-stranded, 2 linear segments. Morphology: Non-enveloped. Virion: Icosahedral (MW = 5.5 × 10^7). Nucleocapsid: Icosahedral.

Plate 3.4 Family: Partitiviridae. Nucleic acid: RNA. Genome: Double-stranded, 2 linear segments. Morphology: Non-enveloped. Virion: Icosahedral (MW = 6.0–9.0 × 10^6). Nucleocapsid: Icosahedral.

Plate 3.2 Family: Cystoviridae. Nucleic acid: RNA. Genome: Double-stranded, 3 linear segments. Morphology: Enveloped. Virion: Spherical (MW = 9.9 × 10^7). Nucleocapsid: Icosahedral.

Plate 3.5 Family: Reoviridae. Nucleic acid: RNA. Genome: Double-stranded, 10–12 linear segments. Morphology: Non-enveloped. Virion: Icosahedral (MW = 1.2 × 10^8). Nucleocapsid: Icosahedral. Distinguishing feature: Nucleocapsid contains several concentric protein layers.

Plate 3.3 Family: Hypoviridae. Nucleic acid: RNA. Genome: Double-stranded, 1 linear segment. Morphology: Encapsulating lipid vesicle. Virion: No true virion (MW not specified). Nucleocapsid: None.

Plate 3.6 Family: Totiviridae. Nucleic acid: RNA. Genome: Double-stranded, 1 linear segment. Morphology: Non-enveloped. Virion: Icosahedral (MW = 1.2 × 10^7). Nucleocapsid: Icosahedral.

Fig. 3. Relative sizes and basic information for those viruses that possess double-stranded RNA genomes.

100 nm

Plate 4.1 Family: Arenaviridae. Nucleic acid: RNA. Genome: Single-stranded, ambisense, 2 circular segments. Morphology: Enveloped. Virion: Spherical (pleomorphic) (MW not specified). Nucleocapsid: Helical. Distinguishing feature: Virions contain ribosomes from host cell.

100 nm

Plate 4.2 Family: Bunyaviridae. Nucleic acid: RNA. Genome: Single-stranded, ambisense, 3 linear segments. Morphology: Enveloped. Virion: Spherical (pleomorphic) (MW = 3.0–4.0 × 10⁸). Nucleocapsid: Helical.

100 nm

Plate 4.3 Floating genus: *Deltavirus*. Nucleic acid: RNA. Genome: Single-stranded, negative sense, 1 circular segment. Morphology: Enveloped. Virion: Spherical (MW not specified). Nucleocapsid: Presumably icosahedral.

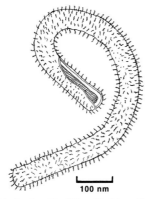

100 nm

Plate 4.4 Family: Filoviridae. Nucleic acid: RNA. Genome: Single-stranded, negative sense, 1 linear segment. Morphology: Enveloped. Virion: Filamental (pleomorphic) (MW = 4.2 × 10⁶). Nucleocapsid: Helical.

100 nm

Plate 4.5 Family: Orthomyxoviridae. Nucleic acid: RNA. Genome: Single-stranded, negative sense, 6 to 8 linear segments. Morphology: Enveloped. Virion: Spherical (pleomorphic) (MW = 2.5 × 10⁸). Nucleocapsid: Helical.

100 nm

Plate 4.6 Family: Paramyxoviridae. Nucleic acid: RNA. Genome: Single-stranded, negative sense, 1 linear segment. Morphology: Enveloped. Virion: Spherical (pleomorphic) (MW = 5.0 × 10⁸ and upwards). Nucleocapsid: Helical.

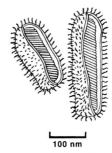

100 nm

Plate 4.7 Family: Rhabdoviridae. Nucleic acid: RNA. Genome: Single-stranded, negative sense, 1 linear segment. Morphology: Enveloped. Virion: Bullet as well as bacilliform (MW = 0.3–1.0 × 10⁹). Nucleocapsid: Helical.

100 nm

Plate 4.8 Floating genus: *Tenuivirus*. Nucleic acid: RNA. Genome: Single-stranded, ambisense, 4 to 5 different length categories. Morphology: Non-enveloped. Virion: Filamental (MW not specified). Nucleocapsid: Helical.

Fig. 4. Relative sizes and basic information for those viruses that possess single-stranded RNA genomes having either negative sense or ambisense coding.

100 nm

Plate 5.1 Floating genus: *Arterivirus*. Nucleic acid: RNA. Genome: Single-stranded, positive sense, 1 linear segment. Morphology: Enveloped. Virion: Spherical (MW not specified). Nucleocapsid: Icosahedral.

100 nm

Plate 5.2 Family: Astroviridae. Nucleic acid: RNA. Genome: Single-stranded, positive sense, 1 linear segment. Morphology: Non-enveloped. Virion: Icosahedral (MW = 8.0×10^6). Nucleocapsid: Icosahedral.

100 nm

Plate 5.3 Family: Barnaviridae. Nucleic acid: RNA. Genome: Single-stranded, positive sense, 1 linear segment. Morphology: Non-enveloped. Virion: Bacilliform (MW = 7.1×10^6). Nucleocapsid: Uncertain (possibly icosahedral).

100 nm

Plate 5.4 Family: Bromoviridae. Nucleic acid: RNA. Genome: Single-stranded, positive sense, 4 linear segments. Morphology: Non-enveloped. Virion: Icosahedral (the genomic RNA segments are encapsidated individually) (MW = $3.5-6.9 \times 10^6$). Nucleocapsid: Icosahedral.

100 nm

Plate 5.5 Family: Caliciviridae. Nucleic acid: RNA. Genome: Single-stranded, positive sense, 1 linear segment. Morphology: non-enveloped. Virion: Icosahedral (MW = 1.5×10^7). Nucleocapsid: Icosahedral.

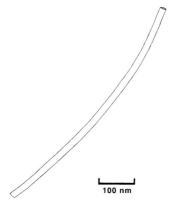

100 nm

Plate 5.6 Floating genus: *Capillovirus*. Nucleic acid: RNA. Genome: Single-stranded, positive sense, 1 linear segment. Morphology: Non-enveloped. Virion: Filamental (flexuous) [MW = (estimate) 4.2×10^6]. Nucleocapsid: Helical.

100 nm

Plate 5.7 Floating genus: *Carlavirus*. Nucleic acid: RNA. Genome: Single-stranded, positive sense, 1 linear segment. Morphology: Non-enveloped. Virion: Slightly flexuous (filamental) (MW = 6.0×10^7). Nucleocapsid: Helical.

Fig. 5. Relative sizes and basic information for those viruses that possess single-stranded RNA genomes having positive-sense coding.

100 nm

Plate 5.13 Family: Flaviviridae. Nucleic acid: RNA. Genome: Single-stranded, positive sense, 1 linear segment. Morphology: Enveloped. Virion: Spherical (MW = 6.0 × 10⁷). Nucleocapsid: Icosahedral.

100 nm

Plate 5.8 Floating genus: *Closterovirus*. Nucleic acid: RNA. Genome: Single-stranded, positive sense, 1 linear segment. Morphology: Non-enveloped. Virion: Filamental (flexuous) [MW = (estimate) 8.0–9.0× 10⁶]. Nucleocapsid: Helical.

100 nm

Plate 5.9 Family: Comoviridae. Nucleic acid: RNA. Genome: Single-stranded, positive sense, 2 linear segments. Morphology: Non-enveloped. Virion: Icosahedral (MW = 3.2–3.8 × 10⁶). Nucleocapsid: Icosahedral.

100 nm

Plate 5.14 Floating genus: *Furovirus*. Nucleic acid: RNA. Genome: Single-stranded, positive sense, 2 linear segments. Morphology: Non-enveloped. Virion: Rodlike (rigid) (MW not specified). Nucleocapsid: Helical.

100 nm

Plate 5.10 Family: Coronaviridae. Nucleic acid: RNA. Genome: Single-stranded, positive sense, 1 linear segment. Morphology: Enveloped. Virion: Spherical (pleomorphic) (MW = 4.0× 10⁸). Nucleocapsid: Helical.

100 nm

Plate 5.15 Floating genus: *Hordeivirus*. Nucleic acid: RNA. Genome: Single-stranded, positive sense, 3 linear segments. Morphology: Non-enveloped. Virion: Rodlike (rigid) (MW not specified). Nucleocapsid: Helical.

100 nm

Plate 5.11 Floating genus: *Dianthovirus*. Nucleic acid: RNA. Genome: Single-stranded, positive sense, 2 linear segments. Morphology: Non-enveloped. Virion: Icosahedral (MW = 8.6 × 10⁶). Nucleocapsid: Icosahedral.

100 nm

Plate 5.16 Floating genus: *Idaeovirus*. Nucleic acid: RNA. Genome: Single-stranded, positive sense, 3 linear segments. Morphology: Non-enveloped. Virion: Icosahedral (MW = 7.5 × 10⁶). Nucleocapsid: Icosahedral.

100 nm

Plate 5.12 Floating genus: *Enamovirus*. Nucleic acid: RNA. Genome: Single-stranded, positive sense, 2 linear segments. Morphology: Non-enveloped. Virion: Icosahedral (MW = 5.0–6.0 × 10⁶). Nucleocapsid: Icosahedral.

100 nm

Plate 5.17 Family: Leviviridae. Nucleic acid: RNA. Genome: Single-stranded, positive sense, 1 linear segment. Morphology: Non-enveloped. Virion: Icosahedral (MW = 3.6–4.2 × 10⁶). Nucleocapsid: Icosahedral.

Fig. 5. continued

100 nm

Plate 5.18 Floating genus: *Luteovirus*. Nucleic acid: RNA. Genome: Single-stranded, positive sense, 1 linear segment. Morphology: Non-enveloped. Virion: Icosahedral (MW = 6.5 × 10⁶). Nucleocapsid: Icosahedral.

100 nm

Plate 5.19 Floating genus: *Machlomovirus*. Nucleic acid: RNA. Genome: Single-stranded, positive sense, 1 linear segment. Morphology: Non-enveloped. Virion: Icosahedral (MW = 6.1 × 10⁶). Nucleocapsid: Icosahedral.

100 nm

Plate 5.20 Floating genus: *Marafivirus*. Nucleic acid: RNA. Genome: Single-stranded, positive sense, 1 linear segment. Morphology: Non-enveloped. Virion: Icosahedral [MW = (estimate) 6.0–9.6 × 10⁶]. Nucleocapsid: Icosahedral.

100 nm

Plate 5.21 Floating genus: *Necrovirus*. Nucleic acid: RNA. Genome: Single-stranded, positive sense, 1 linear segment. Morphology: Non-enveloped. Virion: Icosahedral (MW = 7.6 × 10⁶). Nucleocapsid: Icosahedral.

100 nm

Plate 5.22 Family: Nodaviridae. Nucleic acid: RNA. Genome: Single-stranded, positive sense, 2 linear segments. Morphology: Non-enveloped. Virion: Icosahedral (MW = 8.0 × 10⁶). Nucleocapsid: Icosahedral.

100 nm

Plate 5.23 Family: Picornaviridae. Nucleic acid: RNA. Genome: Single-stranded, positive sense, 1 linear segment. Morphology: Non-enveloped. Virion: Icosahedral (MW = 8.0–9.0 × 10⁶). Nucleocapsid: Icosahedral.

100 nm

Plate 5.24 Floating genus: *Potexvirus*. Nucleic acid: RNA. Genome: Single-stranded, positive sense, 1 linear segment. Morphology: Non-enveloped. Virion: Rodlike (flexuous) (MW = 3.5 × 10⁶). Nucleocapsid: Helical.

100 nm

Plate 5.25 Family: Potyviridae. Nucleic acid: RNA. Genome: Single-stranded, positive sense, 1 to 2 linear segments. Morphology: Non-enveloped. Virion: Filamental flexuous [MW = (estimate) 1.0 × 10⁷]. Nucleocapsid: Helical.

100 nm

Plate 5.26 Family: Retroviridae. Nucleic acid: RNA. Genome: Single-stranded, positive sense, dimer of a linear segment (utilizes reverse transcription during replication). Morphology: Enveloped. Virion: Spherical (MW not specified). Nucleocapsid: Icosahedral.

Fig. 5.   continued

100 nm

Plate 5.27 Family: Sequiviridae. Nucleic acid: RNA. Genome: Single-stranded, positive sense, 1 linear segment. Morphology: Non-enveloped. Virion: Icosahedral (MW = not specified). Nucleocapsid: Icosahedral.

100 nm

Plate 5.28 Floating genus: *Sobemovirus*. Nucleic acid: RNA. Genome: Single-stranded, positive sense, 1 linear segment. Morphology: Non-enveloped. Virion: Icosahedral (MW = 6.6×10^6). Nucleocapsid: Icosahedral.

100 nm

Plate 5.29 Family: Tetraviridae. Nucleic acid: RNA. Genome: Single-stranded, positive sense, 1 to 2 linear segments. Morphology: Non-enveloped. Virion: Icosahedral (MW = 1.6×10^7). Nucleocapsid: Icosahedral.

100 nm

Plate 5.30 Floating genus: *Tobamovirus*. Nucleic acid: RNA. Genome: Single-stranded, positive sense, 1 linear segment. Morphology: Non-enveloped. Virion: Cylindrical (rigid) (MW = 4.0×10^7). Nucleocapsid: Helical.

100 nm

Plate 5.31 Floating genus: *Tobravirus*. Nucleic acid: RNA. Genome: Single-stranded, positive sense, 2 linear segments. Morphology: Non-enveloped. Virion: Tubular [two distinct length categories: S (short) and L (long); (LMW = $4.8–5.0 \times 10^7$, SMW = $1.1–2.9 \times 10^7$)]. Nucleocapsid: Helical.

100 nm

Plate 5.32 Family: Togaviridae. Nucleic acid: RNA. Genome: Single-stranded, positive sense, 1 linear segment. Morphology: Enveloped. Virion: Spherical (MW = 5.2×10^7). Nucleocapsid: Icosahedral.

100 nm

Plate 5.33 Family: Tombusviridae. Nucleic acid: RNA. Genome: Single-stranded, positive sense, 1 linear segment. Morphology: Non-enveloped. Virion: Icosahedral (MW = $8.2–8.9 \times 10^6$). Nucleocapsid: Icosahedral.

100 nm

Plate 5.34 Floating genus: *Trichovirus*. Nucleic acid: RNA. Genome: Single-stranded, positive sense, 1 linear segment. Morphology: Non-enveloped. Virion: Filamental (flexuous) [MW = (estimate) $4.2–5.3 \times 10^6$]. Nucleocapsid: Helical.

100 nm

Plate 5.35 Floating genus: *Tymovirus*. Nucleic acid: RNA. Genome: Single-stranded, positive sense, 1 linear segment. Morphology: Non-enveloped. Virion: Icosahedral (MW = $3.6–5.6 \times 10^6$). Nucleocapsid: Icosahedral.

100 nm

Plate 5.36 Floating genus: *Umbravirus*. Nucleic acid: RNA. Genome: Single-stranded, unspecified sense, 1 segment. Morphology: Presumably enveloped. Virion: Spherical (MW not specified). Nucleocapsid: Unspecified.

Fig. 5.   continued

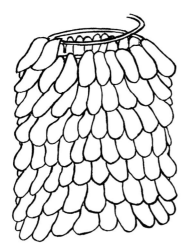

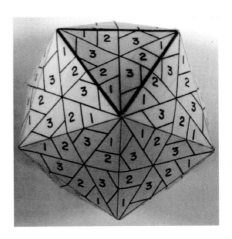

Fig. 6. Drawing of a helical capsid structure showing how the capsid proteins attach to the helical coil of the viral nucleic acid genome. Presumably, all of the capsid proteins are identical to one another in a helical structure.

Fig. 7. Photograph of the assembled model published by Hurst *et al.* (1987) showing the protein arrangement in an icosahedral capsid structure. This particular structure is a representation of the viral family Picornaviridae. The members of this family produce capsids that contain multiple copies of three major (larger-sized, numbered 1, 2, and 3) capsid proteins and one minor (smaller-sized) capsid protein. The relative positions of the three major capsid proteins are shown as trapezoids numbered 1, 2, and 3. The trapezoidal shape is used for illustrative purposes, as the true shapes of these proteins is more complex and not truly trapezoidal. The darkly outlined triangle represents one of the twenty sides of the viral capsid. Although these sides are often referred to as "faces," the term *icosahedron* literally interprets from the Greek as meaning that this structure has twenty surfaces upon which it could rest.

Fig. 8. Drawing of an icosahedral capsid structure showing what would be a mirror image of the shape of the capsid proteins for the viral family Bromoviridae. Unlike the picornaviral model, the bromoviral capsid seems to contain multiple copies of only one type of capsid protein. Presumably, those copies of the same protein would be rotated into different relative positions such that they can arrange into an icosahedron. This drawing shows how those capsid proteins combine to produce the twofold (left), threefold (center), and fivefold (right) axes of symmetry that define an icosahedral structure.

III. THE PROPOSED DOMAIN OF AKAMARA

It has been suggested (Morell, 1996) that, with the sequencing of an archaeon microbe (Bult *et al.*, 1996), the last of life's three domains has been elucidated, and that these domains are Archaea, Bacteria, and Eukarya (Woese *et al.*, 1990). That assessment leaves out something very important, namely, the viruses and the other acellular infectious agents. Indeed, it must be remembered that the first life form whose genome was sequenced in entirety was not cellular in nature, but rather was the virus SV40, and we have had knowledge of its full genome for more than 20 years (Fiers *et al.*, 1978).

Perhaps the time has come to suggest a fourth biological domain to give a higher taxonomic home to the viruses and their genomic relatives. One logical suggestion would be to include as a group the "conventional" viruses plus those satellite viruses whose genomes likewise code for their own structural "shell" or "capsid" proteins, and which are also commonly defined as "viruses" (Strauss *et al.*, 1996). A second group within this domain might consist of the viroids, virusoids, and the viral genus *Deltavirus*, which share strong commonalities with respect to their RNA genomic structure and the fact that they do not code for such structural proteins (Taylor, 1996; Strauss *et al.*, 1996). These infectious agents are excluded from the three existing domains for two reasons. First, their genomes do not code for ribosomal RNA, which is the defining characteristic for membership in the three domains. Second, they lack a cellular structure of their own, a fact that also kept them officially excluded from the older kingdom classifications.

The conventional viruses are very heterogenous with respect to their genomic structure and vary widely in the extent of genetic coding that they carry. Some of them, such as T4, a member of the family Myoviridae, carry a major amount of the genetic coding that is necessary to replicate themselves. Other viruses, such as the human polioviruses that belong to the family Picornaviridae, carry just barely more than the limited amount of genome needed to code for their structural proteins. In comparison, as a group, the RNA genomes of the viroids and virusoids are more homogeneous and uniquely seem to evidence an evolutionary stability as infectious agents despite the fact that their genomes do not code for any such structural proteins. The variety of agents known as virusoids "borrow" encapsulating proteins from a helper virus (Strauss *et al.*, 1996; Taylor, 1996). The viroids either have done away with the need for encapsulating proteins or perhaps never possessed them. The genome of the hepatitis D virus, which is the constituent species of the floating genus *Deltavirus* (ICTV, 1995; Taylor, 1996), represents what seems to be an interesting evolutionary anomaly. This agent of humans is essentially identical to that of the viroids, which are plant pathogens, with the exception that the genome of the hepatitis D virus carries the genetic coding for a protein that it apparently picked up from a cellular host (Brazas and Ganem, 1996; Robertson, 1996). Despite its very limited coding capacity, as with the virusoids, the hepatitis D virus needs to "borrow" enveloping structural proteins that are coded for by a helper virus.

The assignment of taxonomic levels for cellular organisms was initially based on their similarities at the level of physical traits and was aided by a trail of fossilized remains. This approach has since been superseded (Woese *et al.*, 1990) by the suggestion that such assignments could be based on molecular chemistry, specifically the nucleotide sequence of the organism's ribosomal RNA. These assignments based on RNA sequence are assumed to represent the phylogenetic origin and evolutionary history of the organisms, and they have largely confirmed the preexisting eukaryote classifications that had been based on physical traits. Similarly, defining the genetic relatedness of the viruses on the taxonomic levels of order, family, genus, and species as elaborated by the ICTV (Calisher *et al.*, 1995; ICTV, 1995) was initially based on the morphologic and antigenic characteristics of the viruses. These older viral classifications have subsequently been refined and largely confirmed based on the nucleotide sequence and organizational structure of the viral genomes. The proposed taxonomic structure for the genomic acellular infectious agents (Fig. 9) suggests a logical placement of the existing ICTV taxonomic classifications into a higher-level schematic by progressing upward using successively more basic attributes of the viral genomes. The domain name suggested here is Akamara (ακαμαρα), whose derivation from the Greek [α (without) + καμαρα (vault, chamber)] could represent the fact that these life forms do not possess a cellular structure of their own.

All of the groups of infectious agents shown in Figure 9 are depicted as belonging to a common domain, as it would seem perhaps improbable to premise an accurate grouping of these agents based on which of the three commonly suggested evolutionary sources represented their respective origins. In examining this point, we should remember the three possible theories about how viruses began (Strauss *et al.*, 1996): (1) that the viruses are remnants of the primordial soup, (2) that they represent degenerated cellular organisms, and (3) that they are regulatory cellular elements gone awry.

The suggested domainial classifications as shown in Figure 9 would group together the numerous agents whose genomes code for their own structural "shell" or "capsid" proteins as one kingdom (Euviria, signifying "true" viruses). The viroids, a group of agents that share a unique and very homogeneous single-stranded RNA genomic organization that somehow has enabled them to evolutionarily persist despite the fact that they do not code for proteins, have been suggested as constituting a second kingdom (Viroidia) along with other groups of related agents whose genomes likewise do not code for their structural "shell" or "capsid" proteins. This may be perceived as a key biological difference, since all of the cellular organisms as well as the Euviria completely code for their own structural proteins. The kingdom Euviria is suggested as being divided into two phyla, which separate the Euviria with respect to whether their genomes are composed of RNA

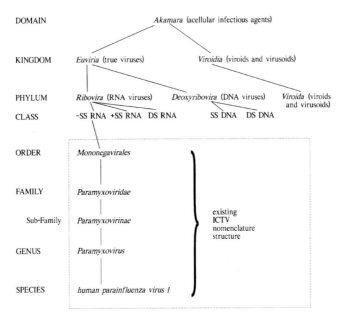

Fig. 9. Proposed new domain, Akamara, plus proposed taxonomic structure at the levels of kingdom, phylum, and class. The abbreviated designations at the class level represent: –SSRNA = "negative"-sense single-stranded RNA genome; +SSRNA = "positive"-sense single-stranded RNA genome; DSRNA = double-stranded RNA genome; SSDNA = single-stranded DNA genome; and DSDNA = double-stranded DNA genome. The existing ICTV nomenclature structure for viruses, an example of which is shown within the inset box, covers only taxonomic levels from order through species and could be adopted directly into this domain.

(Ribovira) or DNA (Deoxyribovira). It is suggested, in turn, that these two phyla of the Euviria could logically be subdivided into classes based on whether their genomes are double or single stranded. Those Euviria that possess single-stranded RNA genomes could logically be divided as to whether their genomes are "plus" sense, meaning they can be directly translated by ribosomes, or "negative" sense. This assignment of phyla and classes is premised on principal biochemical differences in the agents' genomes and also follows basic commonalities in terms of the molecular strategies of these infectious agents. As noted above, the viroids, virusoids, and floating genus *Deltavirus* are shown as being assigned a separate kingdom level, named Viroidia, and the suggested schematic carries their placement intact to the phylum level, as all of the agents grouped into that category possess RNA genomes. The current ICTV nomenclature structure groups these genomic acellular infectious agents at the levels of order through species and could be adopted directly into this proposed new domain. An example of the existing ICTV structure is shown within the box in Figure 9. The ICTV nomenclature structure is now considered to be based on viral nucleotide sequence commonalities at the lowest levels, progressing through commonalities in genome organization according to a "bottom-up" philosophy. The levels of taxonomy proposed here can be seen as logically continuing that trend by progressing upward to the trait of strandedness at the class level, where a distinction of single- versus double-stranded genome is used and aided by the designation of plus versus negative sense in the case of those viruses whose strategies are based on single-stranded RNA genomes, and to the still more basic distinction of RNA (Ribovira) versus DNA (Deoxyribovira) genome at the phylum level. Simultaneously, this proposed structure can also be perceived as progressing from the top downward to meet the current ICTV structure.

Perhaps it is necessary to ask whether these genomic acellular infectious agents indeed deserve to be considered among the living and therefore assigned a taxonomic structure. In examining this point, we should remember the three possible theories about how viruses began: that these agents as we know them represent the evolutionary product from either primordial remnants, degenerated cells, or rogue cellular elements (Strauss *et al.*, 1996). It is indeed possible and perhaps likely that the origin and evolutionary course that the various groups of genomic acellular infectious agents have followed to attain their present forms does in fact reflect a combination of contributions from all three sources. Many of the large viruses certainly have replicative traits that seem to mimic some of the molecular complexity that we associate with cellularity. In contrast, the idea that these groups of acellular infectious agents began as subcellular elements that gained independence is certainly suggested by some of the smaller viruses, which scarcely carry any more genetic coding than that required for their few capsid proteins, and by the viroids and virusoids that do not code for capsid proteins. However, since these groups of acellular infectious agents seem to have evolutionarily taken on an identity of their own, then perhaps we should recognize that identity as a life form. When considering the possibility that these groups of acellular infectious agents might in fact have begun as cellular organisms that have since lost biochemical complexity, if we should consider them now to be nonliving based on the application of complex definitions of living creatures, then we would be faced with deciding at which exact point in the process of evolutionary simplification the term "life" would cease to be applicable. If, alternatively, we could more simply define life by indicating that living things are naturally existing organic entities that are capable of catalyzing their biochemical self-replication, then yes, these genomic acellular infectious agents are a form of life.

It is not necessarily implied that the suggested taxonomic separation of these genomic acellular infectious agents into kingdoms, phyla, and classes represents a strict phylogeny. This is a departure from the current philosophy that the existing ICTV taxonomic nomenclature, which goes no higher than the level of order, does carry strong evidence for common phylogeny. Taxonomic systems change with the evolution of our scientific philosophy, and the proposed categories presented in this chapter are only a suggestion. However, it might be necessary to admit that we may never be able to use strict phylogeny to base an exact classification of these agents at the very highest taxonomic levels. This is due to the idea that, after presumably billions of years of coevolution and biochemical interactions within their host cells, the physical appearance of these modern descendants may be very different from that of their evolutionary ancestors, and they have left no identifiable fossilized remains to guide us. Instead, basic biochemistry has been used as the basis for this proposed division of the genomic acellular infectious agents into kingdoms, phyla, and classes. If there is a fault to be found with this proposal, it is perhaps that this organizational structure could be seen as relating to a cellular origin for the genomic acellular infectious agents. However, this proposal is not intended to imply that cellularity was an initiating condition for the evolution of these acellular agents. Rather, as indicated earlier, we do not and perhaps cannot know the conditions under which these life forms initially began, and the same present-day endpoint could have resulted from gradual evolution regardless of whether either the viruses or other genomic acellular infectious agents began as primordial components, cellular organisms, or rogue cellular elements. Likewise, it would not be possible to place these organisms into the three domains described by Woese *et al.* (1990), since those domains are defined by the nucleotide sequences of their constituent organism's genes that code for ribosomal RNA, whereas the genomes of the viruses, satellite viruses, virusoids, and viroids do not code for ribosomal RNA. There is a suggestion that single-point evolutionary connections exist at the branch junctures where the three domains used to describe the cellular organisms separate from each another. However, assigning a single point of evolutionary connection between the life forms contained in this newly proposed domain Akamara and the domains of Woese *et al.* (1990) may not be possible, as that might necessitate stating that these acellular organisms either evolved into or from the cellular organisms that are represented by the three existing domains.

IV. CONCLUSIONS

The purpose of this chapter is to help us take stock of what we as virologists now have available in terms of taxonomy for the viruses and their relatives. We have just passed the century mark for use of the term *virus* (Beijerinck, 1898) and for official biological recognition of the viruses by a scientific commission (Loeffler and Frosch, 1898). It would seem to be time for the scientific community to consider extending the existing viral taxonomy by recognizing the viruses on a domain level. The taxonomic schematic proposed in this chapter, of course, is only one possible suggestion, and no taxonomic scheme can be considered infallible. However, this proposal is logically based and represents an assessment that relies on successive generations of sound biochemical research that has been conducted and published by tens of thousands of virologists worldwide.

What does this proposal leave out? Classification of those RNA viruses that utilize reverse transcription to produce a DNA equivalent of their genome during the course of their replication could be perceived as requiring a separate class or even phylum within the proposed kingdom Euviria. Those viruses that possess single-stranded RNA genomes that are positive sense imitate messenger RNA molecules. Those viruses that possess single-stranded RNA genomes that are ambisense — meaning that their genomes are partially negative sense and partially positive sense — can be seen as representing a significant departure from those viruses that possess single-stranded RNA genomes that strictly are positive sense. The reason why this represents a molecular departure is because the negative-sense regions of those virus RNA genomes must be copied to form a matching positive-sense strand before they can be translated into protein. Perhaps the ambisense single-stranded RNA viruses could be grouped with the negative-sense single-stranded RNA viruses because the members of the latter viral group also have incorporated the same molecular departure into their biochemistry. This commonality is the reason why the ambisense and negative-sense single-stranded RNA viruses were combined into Figure 4.

What else is left in the field of biological entities? There are reasons for biologists to potentially consider at least some plasmids to represent a type of infectious agent, and perhaps therefore to be a form of life and worthy of an eventual home within some taxonomic scheme. A noteworthy example of this might be the conjugational plasmids, which carry coding for specific protein structures that are then expressed by their cellular host organisms and are important in facilitating transmission of that plasmid to a new cellular host. The prions of mammals are considered infectious agents and might also eventually be thought to represent a form of life, despite the fact that they apparently do not carry any genomic coding of their own (and thus would be defined as agenomic or nongenomic) as they move from one host to another. Indeed, we must consider that perhaps an infectious agent would not need to carry any genomic material with it if its new host cells already possessed all of the coding necessary for replicating that agent. In fact, while two of the traits that prions possess, a measure of resistance to acidic conditions and to proteolytic enzymes (Prusiner, 1996; Caughey et al., 1991), would seem to contribute to the pathogenic process associated with the prions (Prusiner, 1996), these same two traits could be seen as representing an evolutionary adaptation to the acidic conditions and enzymatic milieu encountered in the digestive tract of their host animals during the course of the natural transmission of these prions as an enterically acquired infection. These remaining questions represent fodder for future thought.

REFERENCES

Beijerinck, M. W. (1898). Ueber ein Contagium vivum fluidum als Ursache der Fleckenkrankheit der Tabaksblatter. *Verhandelingen der Koninklijke Akademie Wetenschappen te Amsterdam, II* **6**(5), 1–21.

Braza, R., and Ganem, D. (1996). A cellular homologue of hepatitis delta antigen: Implications for viral replication and evolution. *Science* **274**, 90–94.

Bult, C. J., White, O., Olsen, G. J., Zhou, L., Fleischmann, R. D., Sutton, G. G., Blake, J. A., FitzGerald, L. M., Clayton, R. A., Gocayne, J. D., Kerlavage, A. R., Dougherty, B. A., Tomb, J.-F., Adams, M. D., Reich, C. I., Overbeek, R., Kirkness, E. F., Weinstock, K. G., Merrick, J. M., Glodek, A., Scott, J. L., Geoghagen, N. S. M., Weidman, J. F., Fuhrmann, J. L., Nguyen, D., Utterback, T. R., Kelley, J. M., Peterson, J. D., Sadow, P. W., Hanna, M. C., Cotton, M. D., Roberts, K. M., Hurst, M. A., Kaine, B. P., Borodovsky, M., Klenk, H.-P., Fraser, C. M., Smith, H. O., Woese, C. R., and Venter, J. C. (1996). Complete genome sequence of the methanogenic archaeon *Methanococcus jannaschii*. *Science* **273**, 1058–1073.

Calisher, C. H., Horzinek, M. C., Mayo, M. A., Ackermann, H.-W., and Maniloff, J. (1995). Sequence analyses and a unifying system of virus taxonomy: Consensus via consent. *Arch. Virol.* **140**, 2093–2099.

Caughey, B., Raymond G. J., Ernst, D., and Race, R. E. (1991). N-terminal truncation of the scrapie-associated form of PrP by lysosomal protease(s): Implications regarding the site of conversion of PrP to the protease-resistant state. *J. Vir.* **65**, 6597–6603.

Fiers, W., Contreras, R., Haegeman, G., Rogiers, R., van de Voorde, A., Van Heuverswyn, H., Van Herreweghe, J., Volckaert, G., and Ysebaert, M. (1978). Complete nucleotide sequence of SV40 DNA. *Nature* **273**, 113–120.

Hurst, C. J., Benton, W. H., and Enneking, J. M. (1987). Three-dimensional model of human rhino-virus type 14. *Trends Biochem. Sci.* **12**, 460.

ICTV (International Committee on Taxonomy of Viruses) (1995). "Virus Taxonomy: Sixth Report of the International Committee on Taxonomy of Viruses" (F. A. Murphy, C. M. Fauquet, D. H. L. Bishop, S. A. Ghabrial, A. W. Jarvis, G. P. Martelli, M. A. Mayo, and M. D. Summers, eds.), pp. 1–507. Springer-Verlag, Vienna.

Loeffler and Frosch. (1898). Berichte der Kommission zur Erforschung der Maulund Klauenseuche bei dem Institut fur Infektionskrankheiten in Berlin. *Zentralblatt fur Bakteriologie, Parasitenkunde und Infektionskrankheiten. 1 Abt. Medizinsch-hygiensche Bakteriologie und tierische Parasitenkunde* **23**, 371–391.

Morell, V. (1996). Life's last domain. *Science* **273**, 1043–1045.

Prusiner, S. B. (1996). Prions. *In* "Fundamental Virology," 3rd ed. (B. N. Fields, D. M. Knipe, P. M. Howley, R. M. Chanock, J. L. Melnick, T. P. Monath, B. Roizman, and S. E. Straus, eds.), pp. 1245–1294. Lippincott-Raven, Philadelphia.

Robertson, H. D. (1996). How did replicating and coding RNAs first get together? *Science* **274**, 66–67.

Strauss, E. G., Strauss, J. H., and Levine, A. J. (1996). Virus evolution. *In* "Fundamental Virology," 3rd ed. (B. N. Fields, D. M. Knipe, P. M. Howley, R. M. Chanock, J. L. Melnick, T. P. Monath, B. Roizman, and S. E. Straus, eds.), pp. 141–159. Lippincott-Raven, Philadelphia.

Taylor, J. M. (1996). Hepatitis delta virus and its replication. *In* "Fundamental Virology," 3rd ed. (B. N. Fields, D. M. Knipe, P. M. Howley, R. M. Chanock, J. L. Melnick, T. P. Monath, B. Roizman, and S. E. Straus, eds.), pp. 1235–1244. Lippincott-Raven, Philadelphia.

Woese, C. R., Kandler, O., and Wheelis, M. L. (1990). Towards a natural system of organisms: Proposal for the domains Archaea, Bacteria, and Eucarya. *Proc. Natl. Acad. Sci U.S.A.* **87**, 4576–4579.

3

Virus Morphology, Replication, and Assembly

DEBI P. NAYAK

Department of Microbiology and Immunology
Jonsson Comprehensive Cancer Center
School of Medicine
University of California at Los Angeles
Los Angeles, California 90095-1747

VIRAL ECOLOGY

I. GLOSSARY OF ABBREVIATIONS AND DEFINITIONS

Ambisense RNA: These RNAs are of partly positive-sense and partly negative-sense polarity.

Capsid coat or shell: The protein shell in contact or directly surrounding the viral nucleic acid (genome).

Capsomeres: These are morphological units that form capsids. Capsomeres consist of oligomers of one or more viral proteins.

CPE: Cytopathic effect; could be due to apoptosis, necrosis, or syncytium formation.

cRNA: Full-length plus-strand template RNA complementary to the minus-strand genomic RNA.

CTL: Cytotoxic T lymphocytes.

DI: Defective interfering viruses. DI virus particles contain a smaller viral genome, are noninfectious, and need the help of infectious (wild-type) virus for replication but, in turn, interfere with replication of homologous infectious (standard) viruses.

Ectodomain: The portion of the transmembrane protein that remains exposed outside of the cell or virus particle.

EIS: These are the *cis* elements of the unsegmented minus-strand RNA genome (e.g., VSV). "E" denotes the end or transcription termination and polyadenylation sequence; "I" stands for intragenic sequence not transcribed in messenger RNA (mRNA); "S" indicates the start sequence for the next mRNA.

Endocytic pathways (endocytosis): The process of internalization of external macromolecules or viruses, which involves specific binding to cell surface receptors. Viruses use this mechanism to enter into host cells. In this process, clathrin-coated vesicles and subcellular organelles like endosomes and lysosomes are involved.

Envelope: The viral membrane containing the lipid bilayer and associated proteins, which surround the nucleocapsid of enveloped viruses and form the outermost barrier of the enveloped virus particle.

Enveloped viruses: Viruses that possess an envelope or membrane surrounding the nucleocapsid. For enveloped viruses, the naked nucleocapsid is not infectious.

Episomal, extrachromosomal: The state of existence of nucleic acid molecules that do not became integrated into host cell chromosomes. They exist and multiply independently within the cell nucleus or cell cytoplasm.

Escape mutants: Virus mutants that are not neutralized by antibodies. These viruses possess amino acid change in the epitope and therefore no longer bind to the neutralizing antibody.

Exocytic pathways (exocytosis): The mechanism for transporting intracellular transmembrane or secretory proteins from intracellular compartments to the cell surface or extracellular environment. In this process, various subcellular compartments, like endoplasmic reticulum (ER) and Golgi complexes, are involved in protein transport.

Genome: The complete genetic information (DNA or RNA) of an organism.

Glycosylation: In this process, one or several carbohydrate groups are attached to proteins during their transport through the exocytic pathways. Sugar residues are attached at specific sites to amino acids such as serine or threonine (for o-linked) or asparagine (for N-linked) carbohydrate moieties. These carbohydrate moieties are also called glycans.

GSL: Glycosphingolipid.

HBV: Hepatitis B virus.

Helical capsids: These structures are spiral, springlike, and flexible rods. The RNA genome in a helical capsid is either exposed (influenza viruses) or enclosed (paramyxoviruses, rhabdovirus) by the nucleoprotein molecules constituting the nucleocapsid.

H1N1, etc.: Subtype specificity of influenza type A viruses. H denotes hemagglutinin (H1–H14), and N stands for neuraminidase (N1–N9) subtypes.

HIV: Human immunodeficiency virus.

HSV: Herpes simplex virus.

HRV-14: Human rhinovirus strain 14.

ICAM-1: Intracellular adhesion molecule-1, the receptor for rhinoviruses.

Icosahedron, icosadeltahedron, icosahedral symmetry, icosahedral capsids: Icosahedron is a structure with a twofold, threefold, and fivefold rotational symmetry. It is a polyhedron with 20 faces, 12 vertices, and 30 edges. Most icosahedral viruses have 60 (multiple of 60) subunits (e.g., polioviruses, togaviruses).

Inclusion bodies: Microscopic structures, produced in some virus-infected cells consisting of viral proteins, nucleic acids, and cellular elements (particularly cytoskeletal elements). Inclusion bodies can be intranuclear (herpesviruses), intracytoplasmic (paramyxoviruses).

IRES: Internal ribosome entry site.

LB: Lateral bodies found in poxviruses.

LCMV: Lymphocytic choriomeningitis virus, a member of the Arenavirus group.

LT: Large T antigen (ST = small T antigen) of SV40.

MOI: Multiplicity of infection, that is, infectious units adsorbed per cell.

Naked or non-enveloped viruses: These viruses do not have any membrane and the nucleocapsids represent the infectious virus.

NP: Nucleoprotein.

NTS, NLS: Nuclear targeting (or localizing) signal.

Nucleocapsid: The complete nucleic acid–protein complex of a virus particle. Sometimes the term viral ribonucleoprotein (vRNP) is used to indicate nucleocapsid (e.g., vRNP of influenza viruses).

ORF: Open reading frame.

Panhandle: A circular nucleic acid structure of single-stranded (ss) DNA or RNA with a double-stranded stem at the end produced by intrastrand hybridization due to partial complementarity of the nucleic acid sequences at both the 5′ and 3′ termini of ssRNA or DNA. The panhandle structures function as the promoter and are important for transcription and replication.

pfu: Plaque-forming unit.

Phagocytosis, viropexis: Uptake of particles by cells not totally dependent on receptor-mediated endocytosis. The particle on the surface is engulfed by the cell membrane into a phagocytic vesicle. These phagocytic vesicles then undergo similar changes as the endosome. Poxviruses enter cells by phagocytosis.

Poly(A): Polyadenylation at the 3′ end of an RNA molecule.

Protomer: The term often used to indicate a structural unit containing one or more nonidentical protein subunits. Promoters are used as a building-block for virus capsid assembly.

RDRP: RNA-dependent RNA polymerase, also called RNA transcriptase and RNA replicase.

RNA of positive and negative polarity: The RNA strand of the same polarity as the mRNA-encoding proteins is called positive-, plus-, or + strand RNA. When the RNA is of polarity opposite to the mRNA (i.e., cannot code for a protein), it is called negative-, minus-, or – strand RNA.

RNP: Ribonucleoprotein.

RSV: Respiratory syncytial virus.

RT: Reverse transcriptase, RNA-dependent DNA polymerase.

ST: Small T antigen (LT = large T antigen) of SV40.

Structural and nonstructural proteins: Structural proteins are those proteins that are found in virions either as components of capsid or envelope. Nonstructural proteins are those virally encoded proteins produced in the infected cells but not found in virions. Nonstructural proteins are usually catalytic and regulatory in nature and are also involved in modifying host functions.

Synchronous infection: When all cells in the culture are infected simultaneously. Cells are infected at a high MOI (>5) and at low temperatures (4°C). Then the temperature is raised to 37°C to permit entry and uncoating of all cell-bound viruses at the same time.

Syncytium (multinucleated giant cells): Cells possessing multiple nuclei are formed due to fusion among a number of cells. Usually viruses that can undergo fusion at a neutral pH (paramyxoviruses, retroviruses) produce syncytium.

Temperature-sensitive (ts) mutant: A mutant virus that will replicate at a permissive (low) temperature but not at the nonpermissive or restrictive (high) temperature. This phenotype is usually caused by missense mutations of one or more nucleotides, causing alteration of amino acid(s) of a protein that cannot assume the functional configuration at the nonpermissive (restrictive) temperature.

TGN: Trans-Golgi network.

Transmembrane proteins: These are membrane proteins that are anchored to the membrane by spanning the lipid bilayer of the membrane via transmembrane domains. These proteins can be classified as type I (e.g., influenza virus HA), type II (e.g., influenza virus NA), type III (e.g., influenza virus M2), or complex (e.g., coronaviral E1) depending on the orientation of the NH_2 and COOH termini (type I, II, or III), cleavage of signal peptide (type I), and multiple transmembrane spanning domains (complex).

Virion: The entire virus particle. It usually refers to infectious or complete virus particle as opposed to noninfectious or defective virus particles.

VSV: Vesicular stomatitis virus.

WSN/33 (H1N1): A neurotropic variant of WS/33 (H1N1), a human influenza virus isolated in 1933 (Francis and Moore, 1940).

II. INTRODUCTION

Viruses are unique life forms different from all other living organisms, either eukaryotes or prokaryotes, for three fundamental reasons: (1) the nature of environment in which they grow and multiply, (2) the nature of their genome, and

(3) the mode of their multiplication. First, they can function and multiply only inside another living organism, which may be either a prokaryotic or eukaryotic cell depending on the virus. Viruses are acellular and metabolically inert outside the host cell and are obligatory parasites. Although there are other examples of obligatory parasites among the eukaryotes and prokaryotes, the nature of the intimate relationship between viruses and their host (i.e., environment) is much different. For example, some viruses extend their parasitic behavior to another level of mutual coexistence with their host, that is, they not only exist intracellularly but can, and do in some cases, integrate their genome into the genome of their host and thus tie their fate to the fate of the host. In fact, under these conditions, the integrated viral genome behaves like a host gene(s), undergoing similar regulatory control in transcription and replication and similar evolutionary changes as do the host gene(s). Second, whereas all other living forms can use only DNA (and not RNA) as their genetic material (genome) for information transmission from parent to progeny, viruses can use either DNA or RNA as their genome, that is, some viruses can use only RNA (and not DNA) as their genetic material. Therefore, these classes of RNA viruses have developed new sets of enzymes for replicating and transcribing RNA from an RNA template, as such enzymes (RNA-dependent RNA polymerase or RDRP) are not normally found in either eukaryotic or prokaryotic cells. Finally, all eukaryotic and prokaryotic cells divide and multiply as a whole unit, that is, $1 \rightarrow 2 \rightarrow 4 \rightarrow 8$ and so on. However, viruses do not multiply as a unit. In fact, they have developed a much more efficient way to multiply just as complex machines are made in a modern factory. Different viral components are made separately from independent templates, and then these components are assembled into the whole and infectious units, also called virus particles (virions), just as the complex machines are efficiently assembled from individual components. In this chapter, I will discuss aspects of viral morphology, the mode of viral replication, and viral morphogenesis.

Viruses are a heterogenous group of microorganisms that vary with respect to size, morphology, and chemical composition. The size of virions ranges from 20 nm (parvovirus) to ~300 nm (poxvirus) in diameter, as compared to the size of *Escherichia coli*, which is about 1000 nm in length. Viral shape also varies. Some viruses are round (spherical), others filamentous, and still others pleomorphic. Usually, naked (non-enveloped) viruses have specific shapes and sizes, whereas some enveloped viruses (particularly enveloped viruses possessing helical nucleocapsids) are highly pleomorphic (e.g., orthomyxoviruses), with shapes varying from spherical to filamentous (Table I).

III. CHEMICAL COMPOSITION

The chemical composition of a virus depends on the nature of that virus, that is, the nature of the viral genome (RNA or DNA), the composition of the protein shell

TABLE I

Properties of the Virions of the Major Genera of DNA and RNA Animal Viruses

Viruses	Genome nature	Envelope	Shape	Genome polarity	Size (nm)	Transcriptase in virion	Symmetry of nucleocapsid
RNA viruses							
Enterovirus	S,[b] 1[a]	–	Icosahedral	+	~20–30	–	Icosahedral
Rhinovirus	S, 1	–	Icosahedral	+	20–30	–	Icosahedral
Calicivirus	S, 1	–	Icosahedral	+	20–30	–	Icosahedral
Alphavirus	S, 1	+	Spheroidal	+	50–60	–	Icosahedral
Flavivirus	S, 1	+	Spheroidal	+	40–50	–	Icosahedral
Orthomyxovirus	S, 8	+	Spheroidal[h]	–	80–120	+	Helical
Paramyxovirus	S, 1	+	Spheroidal	–	100–150	+	Helical
Coronavirus	S, 1	+	Spheroidal	+	80–220	–	Helical
Arenavirus	S, 2	+	Spheroidal	±[c]	85–120	+	Helical[d]
Bunyavirus	S, 3	+	Spheroidal	±[c]	90–100	+	Helical[d]
Retrovirus	S, 1[g]	+	Spheroidal	+	100–120	+[f]	Icosahedral[i]
Rhabdovirus	S, 1	+	Bullet-shaped	–	175 × 70	+	Helical
Reovirus	D, 10	–	Icosahedral	±	70–80	+	Icosahedral
Orbivirus	D, 10	–	Icosahedral	±	50–60	+	Icosahedral
Filovirus	S, 1	–	filamentous	–	≥80 × 800	+	Helical

DNA viruses

	Genome		Capsid symmetry		Size (nm)		Capsid symmetry
Papillomavirus	D, circular	–	Icosahedral	±	55	–	Icosahedral
Polyomavirus	D, circular	–	Icosahedral	±	45	–	Icosahedral
Adenovirus	D, linear	–	Icosahedral	±	70–80	–	Icosahedral
Herpesvirus	D, linear	+	Spheroidal	±	150	–	Icosahedral
Iridovirus[e]	D, linear	+[e]	Spheroidal	±	125 × 300	+	Icosahedral
Poxvirus	D, linear	+	Brick-shaped	±	300 × 240 × 140[j]	+	Complex
Parvovirus	S, linear	–	Icosahedral	+, –[k]	20	–	Icosahedral

[a] Genome, the number indicates the segments of RNA present in the virus particle. All RNA genome is haploid except retrovirus (diploid).

[b] D = double-stranded; S = single-stranded.

[c] Ambisense (contains coding for protein on both genomic and complementary RNA strands).

[d] Circular helical nucleocapsid.

[e] Insect iridoviruses have no envelope; vertebrate members are enveloped.

[f] Reverse transcriptase (RT).

[g] Diploid, two molecules of the same RNA (+ strand) segment are present in one virus particle.

[h] Pleomorphic including filamentous forms.

[i] The capsid structure of mature retroviruses is not fully known, although it appears icosahedral.

[j] Length × width × thickness.

[k] Some virus particles contain (+)-strand and others contain (–)-strand DNAs.

called the viral "nucleocapsid" surrounding the genome, and the presence or absence of viral membrane depending on whether the virus is enveloped or naked. All viruses have nucleocapsids and therefore contain nucleic acids and proteins. The nucleic acid is the genome that contains the information necessary for viral function and multiplication, and this information is passed from the parent to progeny viruses. Some viruses contain extragenomic nucleic acid, for example, tRNA in retroviruses and ribosomal RNA in arenaviruses. Viral proteins have three primary functions. They (1) provide the shell to protect the nucleic acid from degradation by environmental nucleases, (2) facilitate transfer of the genome from one host to another, and (3) provide many of the enzymatic and regulatory functions needed for transcription and replication so that viruses can survive, multiply, and perpetuate. In addition to the capsid shell, many viruses also possess an envelope (or viral membrane) around the nucleocapsid. The envelope in these viruses is critical for transmission of those viruses from one host to another host. The naked nucleocapsids of enveloped viruses are noninfectious because they lack the receptor binding protein. The viral envelope contains lipids and carbohydrates in addition to "envelope- or membrane-associated" viral proteins. The viral genome codes for most, if not all, of the proteins associated with the viral envelope. Lipids of the viral membrane are synthesized by the host cell and derived from the host cell. Therefore, viral lipid composition varies depending on the host cell in which the virus grows and also on the type of the cellular membrane (e.g., ER, Golgi, plasma or nuclear membrane) from which the particular type of virus buds. The carbohydrate content of the viral envelope is usually determined by the nature of glycosylation (N-glycosylation, O-glycosylation, complex versus simple sugar addition) of the viral envelope proteins, which may in turn undergo other modifications such as myristylation, palmitoylation, sulfation, and phosphorylation.

A. Viral Nucleic Acid (Genome)

Genomes of different viruses are widely diverse in size and complexity. Some are comprised of DNA, others of RNA. As mentioned earlier, only in viruses is RNA known to function as a genome. Viral DNA genomes vary in complexity ranging from 5 kb containing 5–6 genes (parvoviruses, SV40) to 300 kb (avipoxviruses) containing more than 200 genes. Some DNA genomes are double-stranded (SV40), some are partially double-stranded (hepatitis B virus), and still others are single-stranded (parvoviruses) (Table I and II). The single-stranded viral DNAs can be of either plus or minus polarity. Some DNA genomes are circular (and supercoiled), while others are linear. Some linear DNA genomes become circular intermediates during replication. Many viral DNA genomes are terminally redundant in their nucleotide sequences.

RNA genomes of viruses also vary in complexity but not as widely as do DNA genomes. They range from ~7 kb (rhinoviruses) to ~30 kb (coronaviruses). Coro-

TABLE II

Replication of DNA Viruses

Virus	Form of DNA	Polym-erase	Activity	Presence in virion	Replication site in cell
Papovaviruses	ds[a]	Host	DNA *pol*	–	Nucleus
Adenoviruses	ds	Viral	DNA *pol*	–	Nucleus
Herpesviruses	ds	Viral	DNA *pol*	–	Nucleus
Poxviruses	ds	Viral	DNA *pol*	–[b]	Cytoplasm
Parvoviruses	ss	Host	DNA *pol*	–	Nucleus
Hepadnaviruses	Partially ds	Viral	Reverse transcriptase (RT)	+	Nucleus/ cytoplasm

[a]ds = double-stranded; ss = single-stranded.

[b]Virions contain DNA-dependent RNA transcriptase and many other enzymes, but not DNA-dependent DNA polymerase.

naviral RNA represents the largest stable single-stranded RNA found in nature. Viral RNA can be single- or double-stranded (Tables I and III).

The viral RNA genome may be unsegmented, consisting of a single RNA molecule, or segmented, consisting of multiple segments. Usually, viral genomes are haploid, but some are diploid (e.g., retroviruses; Fig. 1). Some viral RNA genomes may be linear, whereas others have partial terminal complementarity assuming panhandle structures (e.g., orthomyxoviruses). Some of the single-stranded RNA genomes are of plus or "positive" polarity, meaning that they can be translated directly into proteins, and others are of minus or "negative" polarity, meaning they must be used as a template to synthesize a translatable complementary strand (mRNA), and still others are ambisense (Table I). The plus-polarity naked viral genomes (except for retroviruses), completely free from all viral proteins, are infectious when introduced into a permissive cell, whereas minus-polarity naked genomes are noninfectious. Viruses possessing the minus-polarity genome therefore must carry an enzyme, RNA-dependent RNA polymerase (RDRP), inside the virus particle in order to initiate the infectious cycle. Similarly, retroviruses must possess reverse transcriptase (RT, RNA-dependent DNA polymerase) to initiate the infectious cycle inside host cells. However, using reverse genetics, many of the RNA genomes of both plus and minus polarity can be converted into infectious double-stranded DNA, thus permitting artificially induced mutational changes and genetic analysis of the viral genome, as well as use in DNA vaccination and gene therapy. Some of the DNA (adenoviruses, hepadnaviruses) as well as RNA (polioviruses) viral genomes possess a covalently linked terminal protein at the 5′ end of a genomic nucleic acid strand, which

TABLE III

Replication of RNA Viruses

Virus	Form of RNA	Source of nucleic polym-erase	Nature of polym-erase activity	Pres-ence of polym-erase in virion	Viral replica-tion site within host cell
A Paramyxovirus, Rhabdovirus	ss[a] (–), unseg-mented	Viral	RNA-dependent RNA polymerase (RDRP)	+	Cytoplasm
B. Bunyavirus, Arenavirus	ss[c](±), segmented	Viral	RDRP	+	Cytoplasm
C. Orthomyxovirus (Influenza virus)	ss (–), segmented	Viral	RDRP	+	Nucleus
D. Rotavirus, Reovirus, Orbivirus	ds[b](±), segmented	Viral	RDRP	+	Cytoplasm
E. Picornavirus (Poliovirus, Hepatitis A), Togavirus (Sindbis virus), Coronavirus	ss (+), unseg-mented	Viral	RDRP	–	Cytoplasm
F. Retrovirus, (HIV)	ss (+), unseg-mented, diploid	Viral	Reverse transcrip-tase	+	Nucleus

[a]ss = single-stranded.

[b]ds = double-stranded; + or – indicates positive or negative polarity.

[c]± = ambisense genome.

provides critical functions for initiating DNA or RNA replication. Some positive-strand RNA viral genomes are also capped at the 5′ end and polyadenylated at the 3′ end (togaviruses), while others are not capped at the 5′ end (polioviruses) but possess polyadenylation (poly(A)) at the 3′ end. The minus-strand RNA genomes do not possess either the cap at the 5′ end or poly(A) at the 3′ end. Usually, the 5′ and 3′ ends of the minus-strand RNA genome are partially complementary, often forming panhandles by intrastrand hybridization and function as their own promoters for transcription and replication.

Organization of genes in the RNA genome varies between different groups of viruses. For positive-strand naked RNA viruses (e.g., polioviruses), which are

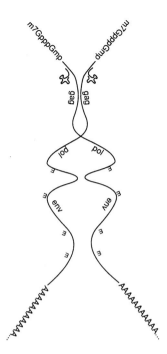

Fig. 1.   Features of the retrovirus genome. The diploid RNA genome includes the following from 5′ to 3′: the m⁷Gppp capping group, the primer tRNA, the coding regions, M_6A residues (m), and the 3′ poly(AAAAA) sequence. Reprinted with permission from Fields and Knipe (1990).

translated into a single large polyprotein, the 5′ end of the genome is not capped but is rather covalently linked to a small protein, VPg (Fig. 2). The 5′ end of these viral genomes contains an untranslated region possessing a highly ordered secondary structure for internal ribosome entry, followed next in sequence by the genes of capsid proteins (VP4, VP2, VP3, VP1). The genes for nonstructural proteins including proteases and viral replicase (an RNA-dependent RNA polymerase) are located in the 3′ half of the genome. However, for the plus-strand enveloped RNA viruses (e.g., Sindbis viruses), the genes for the nonstructural proteins are present at the 5′ end and structural proteins including capsid and envelope proteins are present in the 3′ half of the genome. Structural genes of this latter type of viruses are translated from a separate subgenomic mRNA, whereas their nonstructural proteins are translated from the genomic RNA. The large plus-strand coronavirus RNA genome possesses the nonstructural genes in the 5′ half and structural genes in the 3′ half of the genome. The gene for the highly abundant nucleoprotein (N protein) of coronaviruses is present at the 3′ end of the genome.

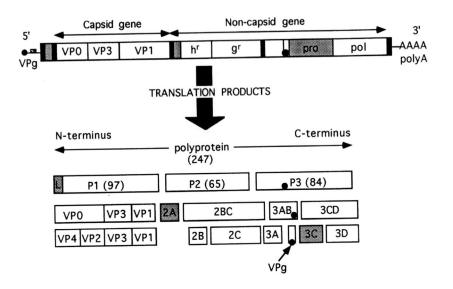

Fig. 2. Organization of picornaviral genome (RNA) and its translation products. P1, P2, and P3 indicate three intermediate precursor proteins cleaved from the polyprotein. These precursor proteins are further cleaved by virus-encoded proteases into mature functional proteins. Numbers in parentheses indicate molecular weights in thousands. h^r and g^r indicate host range and guanidine resistance determinants, respectively. 2A and 3C are proteinases involved in cleavage of the polyprotein and precursor proteins into mature viral proteins. Vpg, VP0, etc. indicate specific viral proteins.

For unsegmented minus-strand RNA genomes, the order of genes for both rhabdo- and paramyxoviruses are similar. Structural genes for capsid (N and P proteins) and envelope proteins are at the 3′ half, and the large polymerase (L) gene occupies the entire 5′ half of the minus-strand RNA genome (Fig. 3). The 3′ end of the template (minus-strand) RNA is transcribed into a leader (ℓ) sequence that is not present in the mRNA, and the region between two genes is separated by an element called the EIS. It consists of an "E" (end) sequence for transcription termination and polyadenylation of a gene, an "I" (intergenic) sequence that allows the viral transcriptase to escape (therefore the "I" sequence is not represented in the mRNA), and "S" (the start) sequences, which denote the start of the next gene. EIS sequences in the genome vary for different viruses in these groups.

B. Viral Proteins

Proteins are major constituents of the viral structure, and their main functions, as indicated earlier, are to protect the nucleic acid from nucleases and to provide

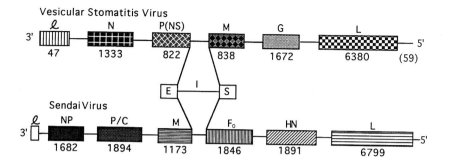

Fig. 3. Genome of unsegmented negative-strand RNA viruses (vesicular stomatitis virus [VSV] and Sendai viruses). Numbers underneath rectangles represent the number of nucleotides in each gene (shown above the line). ℓ = leader sequence; E = end (or transcription termination) sequence; I = intergenic sequence (not transcribed); S = start sequence of mRNA of the next gene; N, NP = nucleoproteins; P/C, P (NS) = phosphoprotein; M = matrix protein; G, F, HN = glycoproteins; L = polymerase protein.

receptor-binding site(s) for virus attachment, which is required for efficient transmission of virus from one host to another. Viral proteins can be classified as either nonstructural or structural. Nonstructural proteins are those encoded by the virion genome, and expressed inside the virus-infected host cells, but not found in the virion particles. These nonstructural proteins usually have regulatory or catalytic functions, which are involved either in viral replication or transcription processes or are involved in modifying host functions. Structural proteins are broadly defined as those proteins found in virus particles. The majority of these structural proteins constitute the viral capsid or core, and are intimately associated with the viral genome to form the nucleocapsid. The cores of some viruses also contain regulatory or catalytic proteins as minor structural proteins (e.g., proteins with enzymatic functions, such as transcriptase (RDRP) or reverse transcriptase (RT) (Tables I–III). In addition, some viruses include host proteins such as histones associated with the viral genome in virus particles (e.g., minichromosome in SV40) or ribosomes, as is the case in arenaviruses. Although these minor virus-coded and host-coded proteins are critically involved in virus replication and infectivity, they are not essential for formation of viral capsids.

In addition to having viral capsids, the enveloped viruses possess membranes (or envelopes) surrounding the viral capsids. These viral membranes, as noted earlier, contain lipids derived from the host membrane and proteins specified by the viral genome. Two types of proteins are found in the viral membrane: transmembrane proteins and matrix proteins.

1. Transmembrane Proteins

Transmembrane proteins, which can be either type I (such as influenza virus hemagglutinin), type II (such as influenza virus neuraminidase), and type III (such as influenza virus M2) depending on their molecular orientation; or complex proteins, containing multiple transmembrane domains (such as E1 glycoprotein of coronaviruses). Enveloped viruses may contain either only one (as in the G protein of VSV), two (as in the HN and F proteins in paramyxoviruses) or multiple transmembrane proteins (as in the influenza viruses, herpesviruses, poxviruses, etc.) on their envelope. These transmembrane proteins are often glycosylated via N- or O-glycosidic bonds, and their carbohydrate moieties can be comprised of simple sugars, usually consisting of mannose molecules, or complex sugars, including galactose, glucosamine, galactosamine, fucose, and mannose, as well as also sialic acid residues. Proper glycosylation of viral proteins is often important to provide for the necessary molecular stability, solubility, oligomer formation, and intracellular transport of viral proteins, as well as for modulating the host immune response, including epitope masking and unmasking. These glycans also may be important in apical sorting of proteins within polarized epithelial cells. It often is the case that one or more of these transmembrane proteins are involved in providing important functions in the processes of receptor binding, fusing of the viral envelope, uncoating of the viral genome, releasing of mature viruses from the infected cells, and spreading of viruses from cell to cell (e.g., function of NA, the neuraminidase protein, in releasing influenza viruses). These envelope proteins are also important for host defense, where they elicit neutralizing antibodies as well as CTL response against the virus infection in infected hosts and therefore play a critical role in vaccination and protection against viral infections.

2. Matrix Proteins

In addition to the transmembrane proteins, the majority of these enveloped viruses also contain another type of membrane protein called a *matrix protein* (e.g., M1 protein of influenza viruses), which forms a shell underneath the membrane, enclosing the capsid. The matrix proteins are therefore likely to interact with the lipid bilayer and transmembrane proteins of the viral envelope on the outer side and with the nucleocapsid on the inner side. Matrix proteins are also usually the most abundant proteins in enveloped virus particles and are critical for the budding of enveloped viruses. Some enveloped viruses containing icosahedral capsids do not possess typical matrix proteins around the nucleocapsids (e.g., togaviruses).

IV. MORPHOLOGY

Viruses vary greatly in size and shape. They can be either spherical or cylindrical (rod-shaped) (Fig. 4) or even pleomorphic (Figs. 5 and 6). Primarily, the virus

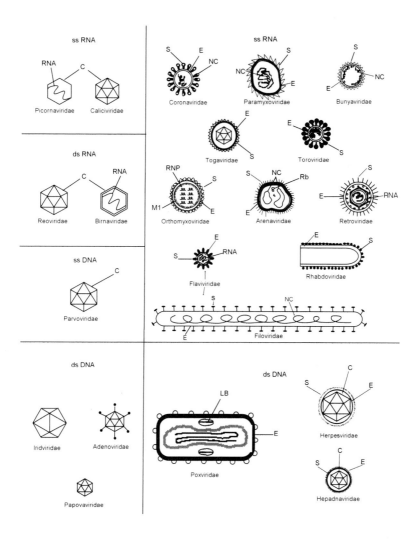

Fig. 4.   Schematic presentation of different forms of viral structures. C = capsid; S = spike on viral
envelope; E = viral lipid envelope; NC = nucleocapsid (i.e., capsid proteins in association with either
RNA or DNA); M1 = matrix protein of influenza virus; LB = lateral bodies present in poxviruses; ss
= single-stranded, ds = double-stranded RNA or DNA.

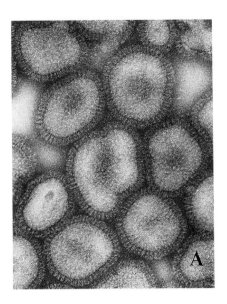

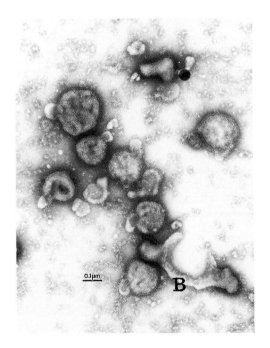

Fig. 5. Transmission electron micrographs of influenza virus (**A**), Sendai virus (**B**), and helical Sendai viral nucleocapsid (**C**). These electron micrographs were provided by and are reprinted with permission from K. G. Murti of St. Jude Children's Research Hospital of Memphis, Tennessee.

structure is determined by the nature of the capsid and whether the capsid is naked or surrounded by an envelope. The structure of the capsid is in part determined by the protein and nucleic acid (nucleocapsid) interactions, but principally by the protein–protein interactions of the capsid protein(s). In most cases, the nucleic acid is incorporated after the majority of the protein shell of the capsid has been formed, or capsids can remain empty, resulting in noninfectious virus particles. The capsids are composed of repeating protein subunits called capsomeres. Capsomeres are composed of multimeric units of either a single protein, or often heteromeric units of more than one protein.

The formation of the viral capsid and its shape is primarily determined by the three-dimensional structure of the capsid proteins, which in turn is determined by the specific amino-acid sequence encoded by the viral nucleic acid. The amino-acid sequence is considered the primary structure of the protein, whose three-dimensional structure is composed of secondary structures such as α helices, β sheets, and random coils. These secondary structures interact with each other, forming the tertiary and quaternary structures, which are usually stabilized by noncovalent interactions (sometimes by covalent disulfide linkages), and represent folding of the proteins into relatively stable structures of microdomains (e.g., globular heads). In addition, extended and flexible regions of the proteins, called hinges, are also present, and these hinges become important for interaction with other members of the protein subunits that form the capsomeres. In most viruses, contacts between capsomeres are repeated, exhibiting a symmetry. This is a process of self-assembly driven by the stability of interaction among the protein subunits forming the capsomeres and the capsomeres forming the capsid. Viral capsids have either a helical (springlike) or icosahedral-based (cuboidal or spherical) symmetry.

A. Helical Capsids

Helical capsids are usually flexible and rodlike. The length of the helical capsid is usually determined by the length of the nucleic acids, that is, some defective interfering (DI) viruses having shorter nucleic acids will have a shorter helical nucleocapsid (e.g., DI RNA of VSV). Helical capsids can be naked, that is, without an envelope (e.g., tobacco mosaic virus). However, there is no known example of an animal virus with a naked helical nucleocapsid. All animal viruses with helical capsids found to date are enveloped. However, such helical capsids when enclosed in an envelope can appear to be either rod-shaped (e.g., rhabdoviruses) or spherical (e.g., orthomyxo- or paramyxoviruses), indicating that the helical capsid in these viruses is flexible (Fig. 5). Some helical capsids can be further folded, forming supercoiled nucleocapsids (e.g., orthomyxoviruses). Usually, enveloped viruses with helical capsids are pleomorphic, forming either spherical or filamentous particles (Figs. 5 and 6). Helical capsids can package only single-stranded RNA, but not double-stranded DNA or RNA, possibly because of the rigidity of the double-stranded nucleic acids. However, some viruses with helical capsids may possess only one capsid containing one virion RNA (unsegmented) molecule (rhabdoviruses, paramyxoviruses) or multiple capsids containing multiple RNA segments (orthomyxoviruses). Viruses containing multiple RNA segments can undergo reassortment with other related viruses, thus exchanging different RNA segments and giving rise to new viruses with different antigenic and virulence determinants (e.g., the antigenic shift that occurs in influenza viruses). The genomic RNA is protected by the helical capsid in some viruses (e.g., paramyxo- and rhabdoviruses) but remains exposed in others (e.g., orthomyxoviruses). A single viral protein (e.g., NP protein of orthomyxoviruses) is usually involved in helical capsid formation.

B. Icosahedral Capsids

Viruses with icosahedral capsids possess a closed shell enclosing the nucleic acid inside (Fig. 4). An icosahedron has 20 triangular faces, 30 edges, and 12 vertices and is characterized by a 5:3:2-fold rotational symmetry. Unlike helical nucleocapsids that package only single-stranded nucleic acid, icosahedral capsids can be used to package either single- or double-stranded RNA and DNA molecules. However, whereas either plus- or minus-strand DNAs are found in the icosahedral capsids of parvoviruses, there are as yet no examples of an icosahedral virus with minus-strand RNA. An icosahedral virus can be either naked or enveloped; but, unlike the helical enveloped viruses, the enveloped icosahedral viruses are less pleomorphic in their shape because the icosahedron capsid structure is rather rigid, and in addition, with icosahedral capsids, the overall size is fixed for a particular virus. The virus particle's formation, stability, and size do not depend on the amount of nucleic acid in the capsid. Although the packaging of the nucleic inside the icosahedral capsid is relatively fixed and does not vary greatly, noninfectious viruses containing empty capsids (i.e., without nucleic acid), can often be seen in virus populations. In recent years, the complete three-dimensional structures of several icosahedral viruses have been determined at the atomic level using the powerful tools of cryoelectron microscopy and X-ray diffraction analysis. Such analyses have led to the rational design of a number of antiviral drugs. Some examples of such three-dimensional viral structures are presented in Figure 7.

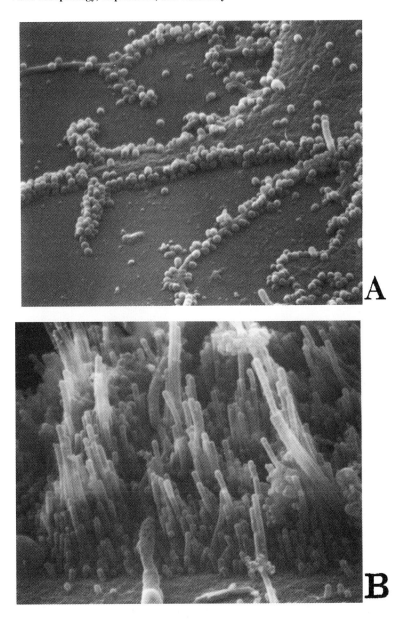

Fig. 6.   Scanning electron micrographs of influenza viruses budding from infected cells. Spherical (**A**) and filamentous (**B**) forms are seen (×40,000). These micrographs were provided by and are reprinted with permission from David Hockley of the National Institute for Biological Standards and Control at Hertfordshire, UK.

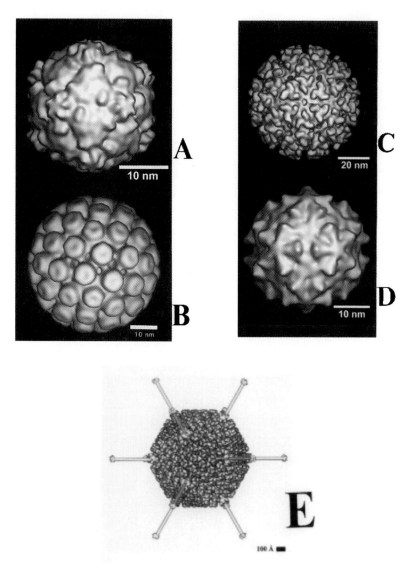

Fig. 7. Structure of representative RNA and DNA viruses as determined by cryoelectron micros-copy: (**A**) human rhinovirus 14, an ssRNA virus; (**B**) SV40, a dsDNA virus; (**C**) Sindbis viral capsid, an ssRNA virus; (**D**) Flockhouse virus, a positive-strand bipartite ssRNA insect virus; (**E**) adenovirus, a dsDNA virus. The micrographs of human rhinovirus 14, SV40, Sindbis virus, and Flockhouse virus were provided by and are reprinted with permission from Norm Olson and Jim Baker of Purdue University. The adenovirus micrograph was provided by and is reprinted with permission from Phoebe Stewart of UCLA.

V. VIRAL REPLICATION CYCLE

To survive, viruses must multiply. Since viruses cannot multiply outside the host cell, they must infect host cells and use cellular machinery and energy supplies to replicate and produce the progeny viruses, which must in turn infect other hosts, and the cycle continues. Host–virus interaction at the cellular level is therefore obligatory for virus replication. Specific host cells can be either nonsusceptible (i.e., resistant or nonpermissive) or susceptible (i.e., permissive) to a particular virus. Nonsusceptibility of cells can be either at the attachment and entry phase (e.g., lack of a suitable receptor for a virus at the cell surface), at the intracellular phase (i.e., a block in synthesis of viral macromolecules), or at the assembly and exit phase. Following infection, viruses can cause abortive (nonproductive) or productive infection. Only productive infection yields infectious progeny virus particles. Following either abortive or productive infection, the host cell may survive or die (i.e., the cytopathic effect [CPE]). CPE caused by a virus does not necessarily indicate the permissiveness of a cell to a virus leading to productive infection. The viral genome in abortive infection may be degraded or may integrate into the host DNA or exist as extrachromosomal (episomal) DNA in the surviving cell. The growth properties of such cells may be altered, including the possibility that they may become transformed and cancerous. Alternatively, cells containing the integrated viral DNA may behave normally, exhibiting little change in their normal properties. Malignant transformation of the infected cells often depends on the site of viral genomic integration leading to activation of cellular oncogenes, disruption or inhibition of tumor suppressor genes, or synthesis of the viral oncogene products that are encoded by the virus in its genome. In the infected cells, the virus genome may remain dormant, resulting in a latent infection, and it can be activated later, producing infectious viruses, as occurs with herpesviruses. Alternatively, infected cells may yield virus at a low level without affecting cell survival, resulting in persistent infection, as occurs with LCMV.

The effect of virus infection has been studied at both the cellular and organismic levels. At the organismic level, it is called "viral pathogenesis," while at the cellular level it is called the "cytopathic effect" (CPE). Under these conditions, cells may undergo morphological changes, including rounding, detachment, cell death and cell lysis (either apoptotic or necrotic), and syncytium (giant multinucleated cell) formation as well as inclusion body formation. Many of these changes are caused by the toxic effects of viral proteins affecting host macromolecular synthesis, including DNA replication, DNA fragmentation, mRNA transcription, translation, protein modification, and degradation, as well as other cellular synthetic and catalytic processes. Furthermore, since the same cellular machineries are directed toward viral macromolecular synthesis, the host is deprived of their functions. In addition to direct cell killing, virus infection can indirectly cause injury to tissues in a complex organism, as a result of complex host–viral immune interactions (i.e., immunopathology), as well as by cytokine production causing inflammatory reactions.

It is evident from the foregoing that, for successful replication of a virus, it must find susceptible host cells, and it must be able to attach itself to and penetrate into the host cell, and be uncoated, rendering the viral genome available for interaction of the viral and cellular machineries for transcription, translation, and replication of the viral genome. Finally, the newly synthesized viral components must be assembled into progeny viruses and released into the medium (outside environment) to infect other hosts. Whether with the cultured cells in laboratory or the complex organisms in nature, the virus–host interaction always occurs at the level of single cells. Thus, the viral infectious cycle (also known as the viral growth cycle, or replication cycle) can be divided into different phases, namely: (1) adsorption (attachment), penetration, and uncoating; (2) transcription, translation, and replication; and (3) assembly and release.

A. Adsorption

Viral adsorption is defined as the specific binding of a virus to a cellular (host) receptor. It is a receptor to ligand interaction in which viruses function as specific ligands and bind to the receptors present on the cell surface. Ligand functions of the virus are provided by the specific viral proteins present on the surface of the virus. For naked (i.e., non-enveloped) viruses, this function is performed by one of the capsid proteins, and for enveloped viruses, one of the membrane proteins functions as the ligand (variously also known as the receptor-binding protein, viral attachment protein or antireceptor) for the host receptor. Usually only one viral protein provides the receptor binding function, although one or more cellular proteins can function as receptor and coreceptor. For enveloped viruses, a classic example of a viral ligand (i.e., receptor-binding protein) is influenza virus hemagglutinin (HA), and its receptor-binding site is present on the globular head of the HA spike. For non-enveloped viruses, a classic example of a viral ligand is the VP1 of rhinoviruses. When five VP1 proteins are packed together within the viral capsid structure, the confluence of these grooves forms a depression called a canyon. The canyon has been shown to be the site for interaction between human rhinovirus-14 (HRV-14) and the cellular molecule ICAM-1 (receptor for rhinovirus). The amino acids lining the floor of these canyons are highly conserved, but residues on the surface of the canyon are variable (Fig. 8). Antibodies can bind to the surface epitopes around and in the proximity of the receptor-binding site and thus interfere with virus attachment by steric hindrance. Viruses can accept mutations in these surface epitopes and thereby escape (and are thus known as

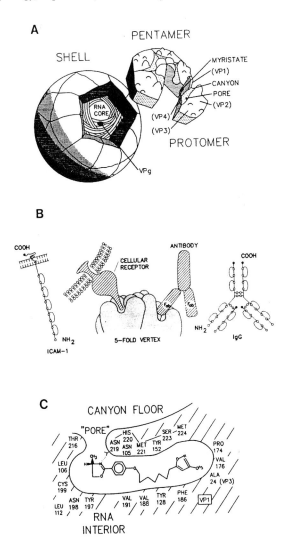

Fig. 8.   Key features in the function of cellular receptor interactions with an invading virus, like a typical picornavirus. (**A**) Exploded diagram showing internal location at the canyon-like center of the pentamer fivefold vertex with myristate residues on the NH$_2$ terminus of VP4. (**B**) Binding of cellular receptor (ICAM-1 molecule) to the floor of the canyon. Note that the binding site of the ICAM-1 molecule, identified as a major rhinovirus receptor, has a diameter roughly half that of an IgG antibody molecule. (**C**) Location of a drug-binding site in VP1 of HRV14 (human rhinovirus 14) and identity of amino-acid residues lining the wall. The drug depicted here, WIN 52084, prevents attachment of HRV14 by deforming part of the canyon floor. The pentamer vertex lies to the right. Reprinted with permission from Fields and Knipe (1990).

escape mutants) neutralization by specific antibodies, but the receptor-binding site usually does not undergo mutational changes and remains conserved. This also appears to be the case with influenza virus hemagglutinin and other viral receptor-binding sites that remain conserved despite the variation in the neutralizing epitopes of the same viral protein. Thus, the viral receptor-binding site is usually a depression or canyon and is therefore protected from the mutational pressure of antibodies.

The cellular receptors of many viruses have been recently identified. Cellular receptors must be present on the cell surface and are either carbohydrates, lipids, or proteins. Sialooligosaccharides present on either glycoproteins or glycolipids function as receptors for orthomyxoviruses, paramyxoviruses, or polyomaviruses; phosphatidylserine and phosphatidylinositol are the likely receptors for VSV; and immunoglobulin superfamily molecules (CD4 for HIV, ICAM-1 for rhinovirus) as well as hormone or neurotransmitter receptors function as receptors for a number of other viruses (e.g., epidermal growth factor for vaccinia viruses, β-adrenergic receptor for reovirus, acetylcholine receptor for rabies virus). Some viruses have more than one receptor, one being the primary receptor and the other a coreceptor. A classic example of this is the case of CD4 and chemokine receptors (CXCR4, CCR5, etc.) respectively functioning as the receptor and coreceptor for HIV. Both the receptor and coreceptor are needed for productive HIV infection, although only one viral protein (gp120) provides the receptor-binding sites for both receptor and coreceptors. Receptor–virus interaction is a major reason for the host and tissue tropism of viruses. Recent studies have shown that lack of a specific coreceptor on a cell's surface provides resistance to HIV infection in some persons. Receptor–virus interactions are specific, and the noncovalent binding is independent of energy or temperature. Thus, the kinetics of viral binding to cells can be determined at 4°C, which serves as a research aid since their interaction at that temperature prevents viral penetration and uncoating. Therefore, binding virus to cells at 4°C and subsequently raising the temperature to 37°C can be used to infect cells synchronously and to study the subsequent events such as uncoating and penetration of virus into host cells. The time course of viral adsorption follows first-order kinetics and is dependent on virus-to-cell concentration. Usually, cells contain a large number of receptors: in the range of 10^4–10^5 per cell.

B. Penetration and Uncoating

Following specific ligand-to-receptor interaction, the next steps in virus replication include penetration of virus into the host cell and uncoating of the viral genome, which are energy-dependent processes and can be prevented experimentally in the laboratory by subjecting the virus–cell complex to low temperatures (4°C). Penetration refers to entry of the surface-bound virus particles inside the cell, where they either exist free in the cytoplasm or inside the host cell vesicles (usually within endosomes). Quantitatively, penetration of virus particles is measured by the loss of the ability of antiviral antibodies to neutralize the cell-bound virus particles after adsorption, an effect that occurs because, after the viral particles have entered the cell, they are protected and no longer accessible to antibodies outside the cell. Uncoating, on the other hand, refers to disruption of virus particles, causing partial or complete separation of nucleic acid from the capsid, and is needed for initiation of transcription and translation of the viral genome. Uncoating can be assessed by, among other things, alterations in viral morphology or viral density, release of nucleocapsid and membrane proteins from enveloped virus particles, as well as by the accessibility of the viral genome to nucleases. For such viruses as orthomyxovirus and poliovirus these processes are separated temporally (i.e., penetration is followed by uncoating in the cytoplasm), but for others both penetration and uncoating occur simultaneously at the cell surface (e.g., paramyxoviruses, HIV). Uncoating refers to the step in which the viral genome becomes functional either transcriptionally or translationally. However, complete separation of nucleic acid from all capsid proteins is not required for most viruses. For naked viruses, uncoating is a postpenetration process that occurs either in the endosome or in the nucleus. Viruses that undergo uncoating in the cytoplasm following endocytosis require low pH (~5) in the endosome for uncoating, whereas viruses that undergo fusion at the cell surface can undergo uncoating in a pH-independent manner.

Naked viruses like the RNA-based picornaviruses enter into the cytoplasm of the infected cells via receptor-mediated endocytosis (Fig. 9) or by phagocytosis (viropexis). In the endosome, the virus particle undergoes alteration in structural and antigenic properties and becomes acid-labile and noninfectious. During uncoating, VP4 (a capsid protein) is released and the viral RNA is extruded from the capsid structure through the hole in the capsid caused by VP4 release into the cytoplasm. How the viral RNA gets through the endosomal membrane is not clear, but it is speculated that pore formation may occur by the interaction of the myristylated NH_2 terminus of VP4 with the endosomal membrane (Flint *et al.*, 1999). The viral RNA now becomes available for translation and replication (Fig. 9). However, only a small fraction of the viruses in the endosomes undergoes successful uncoating. The majority of the virus particles in the endosomes instead become noninfectious due to acid-induced structural alteration and are released outside the cell by the abortive pathway (Fig. 9). SV40, a naked DNA virus, also enters into the cytoplasm via receptor-mediated endocytosis. Some alteration in the SV40 virion structure occurs in the endosome as VP3, a viral capsid protein, becomes exposed. However, in the case of SV40, the virus is extruded essentially intact from the endosome into the cytoplasm and targeted to the nucleus. Therefore, the uncoating of the SV40 genome occurs in the nucleus and not in the cytoplasm. In the nucleus, the viral minichromosome (viral DNA-containing histones) is released from the capsid and becomes available for transcription and replication. Therefore, although entry of SV40 into the cell likewise occurs via an endosome, its uncoating takes place within the nucleus in a pH-independent

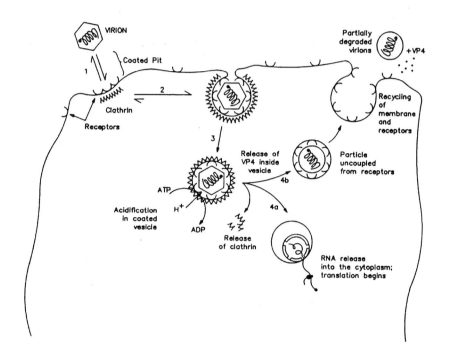

Fig. 9. Receptor-mediated endocytosis of viruses like polioviruses (steps 1 through 4a,b). The virus binds to *cell surface receptors*, usually glycoproteins, which undergo clustering at *clathrin-coated pits* (step 1) and is followed by invagination (step 2) and internalization (endocytosis) to form *clathrin*-coated vesicles (step 3). Acidification inside the coated vesicles, brought about by an energy-requiring ATPase-coupled proton pump, triggers the release of VP4 and unfolding of hydrophobic polypeptide patches previously buried inside the viral capsid. Fusion of the lipid bilayer with hydrophobic patches in the acid-unfolded capsid protein presumably triggers release and transfer of RNA from virion into the cytosol, where ribosomes can begin translating the plus-strand viral genome (step 4a). Fusion of uncoated vesicles with other kinds of intracellular lysosome-like vesicles may also be involved in the uncoating process. Some virus particles are not fully uncoated after acid-induced changes in the endosomes and are released into the extracellular medium via an abortive pathway (step 4b). These partially degraded extracellular virus particles are noninfectious. Reprinted with permission from Fields and Knipe (1990).

manner. However, how the SV40 virus is released form the endosome into the cytoplasm prior to nuclear entry remains unclear. Reovirus, a double-stranded naked RNA virus, uses the host proteolytic enzymes present in the lysosome to partially remove the outer capsid proteins and activate the core RNA transcriptase for initiation of viral mRNA synthesis.

For enveloped viruses, uncoating occurs through fusion of the viral membrane with the cellular membrane using either pH-independent or pH-dependent pathways. In the pH independent pathway, virus penetration and uncoating occur simultaneously and at the cell surface after virus–host interaction. This is best illustrated by the entry process of paramyxoviruses and retroviruses (e.g., HIV). In both cases, viruses bind to the cell surface receptors (i.e., sialic acid present either on the cell surface glycolipids or glycoproteins for paramyxoviruses and the receptor protein CD4 and coreceptors for HIV). Either one (gp160 for HIV) or two (F and HN for paramyxovirus) separate viral glycoproteins are involved in this binding and fusion processes. One of these proteins must be cleaved in the infecting virus for fusion to occur (examples being gp160 → gp120 and gp 41 for HIV and F → F1 and F2 for Sendai virus). For HIV, the gp120/gp41 complex undergoes conformational changes after binding to the cellular receptor and coreceptor, releasing the hydrophobic domain of gp41, which then functions as a fusion peptide and causes fusion of the viral membrane with the plasma membrane, thereby releasing the nucleocapsid containing the viral RNA and reverse transcriptase into the cytoplasm. Subsequently, cyclophilin A, present in HIV particles, aids in the uncoating process by destabilizing the capsid and initiating reverse transcription of the viral RNA. For paramyxoviruses, HN protein binds to the sialic acid on the cell surface receptor and induces, in some way, conformational changes in the other viral envelope protein, known as the F1/F2 complex, and thereby facilitates the fusion domain of F1 to cause fusion between the viral membrane and the plasma membrane, and release of the viral nucleocapsid containing the transcriptase (RDRP) into the cytoplasm. For paramyxovirus, the entire viral replication process takes place in the cytoplasm, whereas for retroviruses the proviral DNA is formed in the cytoplasm after reverse transcription of the viral RNA and is then transported into the nucleus for integration and transcription. How the receptor–protein interaction facilitates conformational changes leading to fusion of the viral and cellular membranes in a pH independent manner is not fully understood. Furthermore, fusion for these viruses occurs not only between viruses and host cells but also between virus-infected cells expressing the cleaved viral membrane proteins on the cell surface and uninfected cells containing the receptors (and coreceptors) present on the cell surface. These cell-to-cell interactions lead to formation of syncytium or multinucleated giant cells. Such multinucleated giant cells are important diagnostic markers for a number of viral infections (e.g., respiratory syncytial virus [RSV], mumps, measles viruses). The process of fusion of HIV-infected cells to uninfected $CD4^+$ T cells is implicated in AIDS pathogenesis, which causes depletion of $CD4^+$ T cells in HIV infected persons.

For other enveloped viruses like VSV and influenza viruses, penetration and uncoating are two separate events. Following receptor binding, these other viruses enter the cytoplasm by receptor-mediated endocytosis, and fusion and uncoating occur within the endosome in a pH-dependent (low pH of ~5) manner. The fusion and uncoating of these viruses can be blocked by agents like monensin, which increases endosomal pH. For VSV, the G protein binds to the receptor and becomes activated for fusion at low pH, even though it remains uncleaved. Although the VSV G protein contains a hydrophobic fusion region, the mechanism of its fusion process within the endosome is not well understood. The fusion and uncoating processes are best understood at the molecular level for influenza viruses. Again, for influenza viruses, although fusion and uncoating occur simultaneously, they are considered two separate events. Following binding to sialic acid on the cell surface receptor, influenza virus undergoes receptor-mediated endocytosis and the cleaved HA trimer (i.e., HA1/HA2 heterotrimer complex) present on the viral membrane undergoes conformational changes at the low pH of endosomes (~5).

Acidic pH specifically alters the structure of HA2, which attains the fusiogenic state. In conjunction with this process, HA1 becomes dissociated from the stem of the HA spike, and the fusion peptide present at the NH_2 terminus of HA2, which normally remains buried in the protein interior of the HA trimer, is released and the polypeptide structural loop becomes transformed into a helix to form an extended coiled coil structure that relocates the hydrophobic fusion peptide toward and into the target (endosomal) membrane (Fig. 10). This process leads first to hemifusion by mixing of the outer lipids of the bilayers of both viral and endosomal membranes and later to complete fusion of both lipid bilayers of the membranes, leading to formation of a pore between the two compartments. Subsequently, the pore dilates and leads to mixing of the cytosol and virion contents and delivery of the viral nucleocapsid into the cytoplasm (Fig. 11). In addition to causing fusion, low pH also aids in the uncoating of influenza viral nucleocapsid. Uncoating in this case is defined as the separation of a nucleocapsid from the virus matrix protein (M1). Therefore, with this type of virus, low pH (~5) is not only crucial to the outside of the virus particle (virion) for inducing conformational changes of HA1 and HA2 but is also needed inside the virus particle for separation of M1 from the nucleocapsid. The viral membrane also possesses a small number of M2 tetramers (16 to 20 per virus particle) formed by a small transmembrane type III M2 protein. These M2 tetramers are ion channels that remain closed at neutral pH and open at low pH (~5) to allow protons (H^+) to enter from the endosomes into the core of the virus particle. The resulting acidic pH inside the virus particles causes dissociation of M1 from the viral RNP (also known as the vRNP or nucleocapsid) containing vRNA, and so the M1-free viral RNP is released into the cytoplasm (Fig. 12). Both the opening of the M2 ion channel and the uncoating of some influenza A viruses can be blocked by amantadine (or rimantadine), a drug currently used in treating influenza infection. The dissociation of M1 from the vRNP is important since the released vRNP can now be

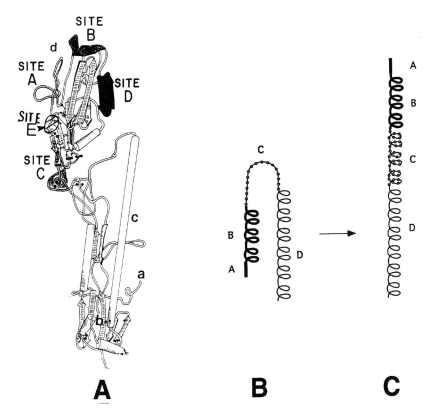

Fig. 10. A model for the fusiogenic state of HA of influenza virus. HA in its native state is a trimer. However, for demonstration of the conformational changes during fusion, only a monomer is depicted in this figure. Panel **A** shows an HA monomer in its native state, containing the cleaved HA1 and HA2. Five epitope sites (A–E) are shown on the globular head. "a" and "b," respectively, are the NH_2 terminus of HA2 and the COOH terminus of HA1 after cleavage. Panel **B** shows the native form of HA2 (again only a monomer is depicted) in which the NH_2 terminus (A) would be buried inside the core of the coiled coil of HA trimer. B and D represent two α helices, and C is the loop region. Panel **C** shows the fusiogenic state at pH 5, at which point the HA1 subunits (not shown) are dissociated from the stem and the fusion peptide (A) at the NH_2 terminus of HA2 is released from the protein interior, and the loop (C) has "sprung" into a helical conformation to form an extended coil that relocates the fusion peptide (A) 100 Å toward the target membrane to promote membrane fusion. Note the conversion of the native form of HA2 in panel **B** to the fusiogenic state in panel **C**.

translocated into the host nucleus where the transcription and replication of vRNA can occur. M1, on the other hand, interferes with transport of vRNP into the nucleus and also inhibits vRNP transcription.

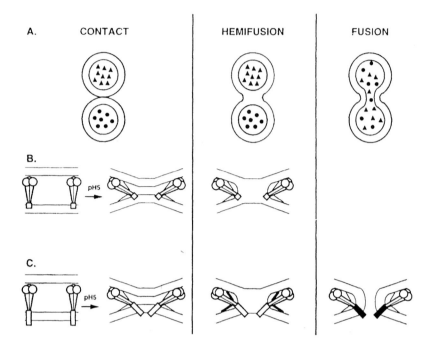

Fig. 11.   Membrane fusion via a hemifusion intermediate. (**A**) Hemifusion in model systems. Two membrane-bound vesicles contact (CONTACT). Next, they form a hemifusion intermediate (HEMI-FUSION). In this state, lipids of the outer leaflets, but not the inner leaflets, mix. At the point of hemifusion, the aqueous contents of the two vesicles (triangles and circles, representing host cell and virion) still remain separated. Next, the lipids of the inner leaflets mix and complete the fusion process (FUSION), resulting in mixing of the aqueous contents of both vesicles, that is, releasing the viral nucleocapsid into the cytoplasm of the infected cell. (**B**) and (**C**) represent enlargements of the respective contact areas from **A**, showing the proposed placement of HA (hemagglutinin protein) trimmers. After opening of the initial narrow fusion pore, the pore dilates (not shown). Reprinted with permission from Kemble *et al.* (1994).

C. Targeting Viral Nucleocapsids to the Replication Site

Viral replication occurs either in the nucleus or in the cytoplasm of infected cells. For those viruses that replicate in the cytoplasm, which customarily are those with RNA genomes, except for the DNA-containing poxviruses, the uncoating process releases the viral nucleocapsid directly into the cytoplasm, which is the site of transcription and replication. For viruses that replicate within the nucleus, which tend to be the ones having DNA genomes with notable exceptions such as the RNA-containing influenza viruses and retroviruses, the viral nucleocapsids

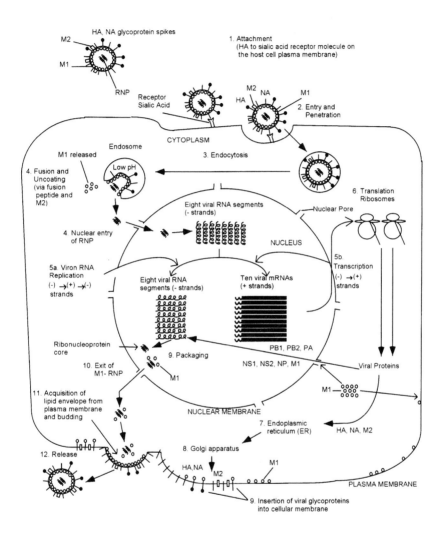

Fig. 12.   Schematic presentation of the infectious cycle of an influenza virus. The steps in the replication cycle are noted as 1 (attachment) through 12 (release). PB1, PB2, PA, NS1, NS2, NP, M1, HA, NA, and M2 are the virus-encoded proteins translated from 10 mRNAs, which are transcribed from eight vRNA segments of negative polarity.

that are released in the cytoplasm after uncoating must be targeted into the nucleus. Nuclear targeting requires that these viral nucleocapsids contain proteins possessing nuclear targeting signal(s) (NTSs or NLSs), which are recognized by the cellular nuclear targeting machinery and translocated into the nucleus via nuclear pores. However, the stage of uncoating at which nuclear targeting takes place varies with viruses. For SV40, essentially the entire virus particle that has been taken into the cytoplasm is transported into the nucleus, and it is only in the nucleus that uncoating of the capsid occurs concomitant with release of the viral minichromosome. For adenoviruses, uncoating occurs at the nuclear pore where the viral nucleocapsid docks and the viral DNA is delivered into the nucleus through the nuclear pore. For influenza viruses, uncoating occurs during introduction of the nucleocapsid into the cytoplasm by dissociation of M1 from the vRNP. This M1-free vRNP is then transported into the nucleus. For retroviruses, not only uncoating but also additional biosynthetic processes — including reverse transcription of the RNA genome and synthesis of the double-stranded proviral DNA — occur in the cytoplasm. Then the retroviral DNA along with integrase is translocated into the nucleus for integration of the proviral DNA into the host genome. Transcription of the retroviral genomic and subgenomic mRNAs occurs only from the integrated proviral DNA in the nucleus. For hepatitis B virus, the partially double-stranded DNA, the viral genome following uncoating in the cytoplasm, becomes fully double-stranded and circularized in the cytoplasm, and then it is translocated into the nucleus for subsequent transcription of genomic and subgenomic mRNAs.

D. Post-Uncoating Events

The "immediate events" in the viral replication cycle, those that occur following uncoating, vary with the nature of the viral genome. For plus-strand RNA viruses except retroviruses, translation of the viral RNA follows immediately after uncoating. The viral RNA extruded from the capsid is then used by the host translation machinery for directing protein synthesis (Fig. 9). For all other viruses, whether of DNA or RNA genome, the step immediately following uncoating is either transcription of the genome yielding functional mRNAs or reverse transcription of vRNA yielding proviral DNA (retroviruses).

E. Transcription of Viral Genes

From the transcription viewpoint, viruses can be classified into two major categories, that is, whether they possess a DNA genome or an RNA genome. Of the first group, the DNA genome of different viruses varies greatly in complexity between virus families, encoding from only 4 to 5 genes to more than 200 genes or open reading frames (ORFs). DNA viruses use DNA-dependent RNA polymerase, which can be either virus-specified (e.g., poxviral RNA polymerase) or host-specified (e.g., RNA *pol* II) to generate their mRNAs. RNA viruses, however, must use RNA-dependent RNA polymerase (RDRP), which is always virus-specified and is therefore different and specific for each virus group to generate their mRNAs.

1. Transcription of DNA Viruses

All DNA viruses except for the poxviruses transcribe and replicate their genomic material in the host cell nucleus. Poxviruses transcribe and replicate in the cytoplasm. In addition, all DNA viruses except poxviruses use host *pol* II for transcription of their DNA into mRNAs. Poxviruses use virus-specified polymerase for transcription of its genome. Viral DNA genomes like host DNA often possess the *cis*-acting elements, which are essential for successful transcription of their DNA. These elements are called the viral promoter and enhancer. The promoter is the RNA polymerase binding site on viral DNA (e.g., TATA box, CAT box, GC box) localized in the vicinity (usually upstream) of the transcription initiation point. The enhancer element, which enhances transcription of the viral mRNA over the basal level, is found either in the proximal or distal region of the promoter and may be located either upstream or downstream of the promoter element. Transcription of the viral DNA genome can broadly be divided into the early and late phases. Early genes are usually catalytic and regulatory in nature, involved in regulating transcription of mRNAs and replication of viral DNA. Late genes usually produce mRNAs for structural viral proteins, which are the major components of either the viral capsid or envelope. Early genes are usually transcribed prior to initiation of viral DNA synthesis, and late genes are transcribed only after viral DNA synthesis has been initiated. Thus, synthesis of the progeny viral DNA demarcates the dividing line between early and late gene transcription. However, for complex viruses such as herpes simplex virus (HSV), the different classes of regulatory genes, for example, immediate early (α), delayed early (β), and late (γ1, γ2) are transcribed at different phases of the viral replication cycle, each having different regulatory functions for either turning on or shutting off other viral genes. Viral genes can be transcribed from either of the two DNA strands, with the coding sequences thus running in a direction opposite to a duplex DNA. These viral genes usually possess the structural features of eukaryotic cellular genes, and the viral mRNAs similarly undergo posttranscriptional processing like cellular genes. The mRNAs are usually capped at the 5′ end, polyadenylated at the 3′ end, and may undergo posttranscriptional splicing prior to their exit from the host cell nucleus. However, poxviral mRNAs, which are also capped and polyadenylated, do not undergo splicing since they are made in the cytoplasm.

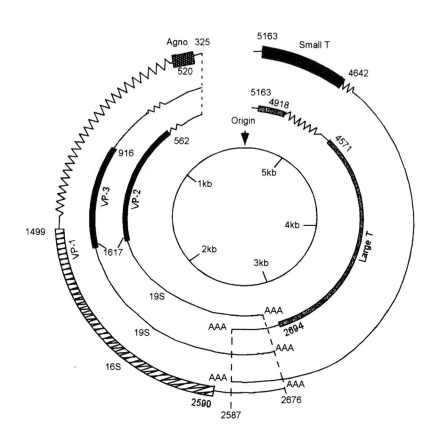

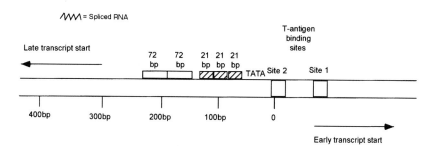

Fig. 13. Genome and transcription map of SV40 (top). The origin of replication is shown at the top of the inner circle. The numbers indicate the nucleotide position in the SV40 DNA, while zigzag markings indicate spliced introns. Different shaded regions indicate different protein-coding sequences. The bottom drawing shows the details of the transcription regulatory elements in the proximity of the "origin" region and the direction of the early and late transcription.

An example of transcription of a small double-stranded viral DNA genome (SV40) is shown in Figure 13. Transcription of the SV40 genome is carried out by the host cell's RNA polymerase II. Early mRNAs (large T and small T) are transcribed from the early promoter of the early DNA strand, whereas late mRNAs (i.e., the mRNAs for VP1, VP2, VP3 and agno proteins) are transcribed from the late promoter and the opposite DNA strand. Both early and late transcription in SV40 are initiated from the common control region in opposite directions at different phases of the replication cycle. This control region also regulates viral DNA replication. This region consists of a series of repeat elements with different functions: three 21-base repeats that together contain six copies of GC-containing hexamers and serve as the promoter for early transcription. Downstream of these repeats is a TATA-box and upstream are two 72-base repeats constituting the enhancer element (Fig. 13, bottom). These three regulatory elements bind specific cellular factors and are important in regulating early transcription. Of these, the 21-bp repeats and the enhancer elements are also important in regulating late transcription. The switch from early to late transcription is brought about by binding of large T antigen to specific sites in the control region and a change in the replicative state of the viral DNA. Large T (LT) and small T (ST) antigens are two early proteins, translated from two different mRNAs produced by differential splicing. The two late mRNAs have a common untranslated region and a common poly(A) addition site but are generated by differential splicing. Each of these late SV40 mRNAs are bicistronic, with alternative initiation codons. One of these late mRNAs is translated into VP2 and VP3, and the other into VP1 and the agno protein.

On entry into the cytoplasm of the infected cell, hepatitis B virus (HBV), a partially double-stranded DNA virus, uses virus-specified reverse transcriptase (P) to synthesize the complete circular DNA, which is then transported into the nucleus. Host cell *pol* II subsequently transcribes its genomic and subgenomic mRNAs from different initiation points (Fig. 14). They are all capped at the 5′ end, unspliced, and have a common termination and poly(A) addition site at the 3′ end. Different classes of genomic-length (3.5-kb) RNAs, possessing different 5′ but common 3′ termini, function as a template for making cDNA or are translated into the Pre-C and C proteins as well as the P protein. Subgenomic mRNAs are translated into the Pre-S1, Pre-S2, and S proteins as well as the X protein (Fig. 14).

2. Transcription of RNA Viruses

Among the different families of RNA viruses, the RNA viral genome appears to be much less complex as compared to the genomes of the highly complex DNA viruses. However, these RNA viruses use multiple strategies to encode different mRNAs and different proteins. Unlike DNA viruses, the majority of the RNA viruses (except for retro-, orthomyxo-, and related viruses) replicate in the cyto-

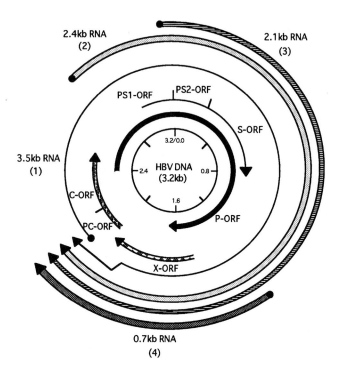

Fig. 14. Replication, transcription, and translation of hepatitis B virus (HBV) DNA. Four RNA classes: 3.5 kb (1), 2.4 kb (2), 2.1 kb (3), and 0.7 kb (4) are transcribed. The 3.5-kb product (#1) is used for full-length DNA (minus-strand) synthesis. Different classes of 3.5-kb product also function as mRNAs whose translation products are HBcAg, the polypeptide consisting of the PC-ORF (pre-core), and C-ORF (core) and P-ORF (P-protein, also called either polymerase or reverse transcriptase). The 2.4-kb mRNA (#2) makes a large protein consisting of the polypeptides PS1-ORF, PS2-ORF (presurface), and S-ORF (surface protein). The 2.1-kb mRNA (#3) makes the S-ORF (surface) protein, and the 0.7-kb mRNA (#4) encodes the X-ORF protein.

plasm, so that their mRNAs cannot undergo RNA splicing. RNA viruses also possess genes for regulatory and catalytic proteins as well as for structural proteins. However, transcription of mRNAs encoding these proteins is not as strictly demarcated with respect to the timing of their genomic nucleic acid replication, as is found for DNA viruses. On the other hand, with RNA viruses there is a great deal of variation in the level of transcription of different viral mRNAs. The mRNAs of the major structural proteins — like the nucleoprotein (NP) and matrix (M) glycoproteins — are usually made in larger amounts as compared to the lower amount of mRNAs synthesized for catalytic (e.g., polymerases) proteins. For unsegmented negative-strand RNA viruses, the level of mRNA transcription is regulated by the promoter-proximal position of a gene (e.g., for VSV or paramyxoviruses, see Fig. 3) or by an internal promoter that produces subgenomic mRNAs (e.g., togaviruses). The strategy used by different RNA viruses for mRNA transcription depends on the nature of the RNA genome (+ or − strand, segmented or unsegmented) and whether the nucleocapsid is icosahedral or helical.

a. For plus-strand icosahedral naked RNA viruses (e.g., poliovirus), the entire viral genomic RNA functions as the only mRNA and is translated from one ORF into a large polyprotein, which is then cleaved by specific proteases into different functional proteins representing the RNA polymerase and the capsid proteins (VP1, VP2, VP3, VP4), and so on (Fig. 2).

b. For some enveloped plus-strand RNA viruses (e.g., togaviruses), the 5′ half of the viral genomic RNA encodes and is translated into nonstructural (catalytic) proteins involved in RNA transcription and RNA replication, whereas a separate subgenomic 26-S mRNA (+), made from an internal promoter on the minus-strand RNA template, encodes the structural proteins (i.e., capsid and envelope proteins). This 26-S mRNA is synthesized in a larger quantity than is the genomic length RNA. However, another group (flaviviruses) of enveloped plus-strand RNA viruses possesses one large ORF in its genomic RNA encoding a single large polyprotein, which, as is the case with picornaviruses, is cleaved into specific proteins by a virus-encoded proteinases.

c. For coronaviruses, which contain a large plus-strand RNA genome of ~30 kb, multiple subgenomic mRNAs are found. However, each of these mRNAs possesses the same 5′ leader (i.e., leader-primed transcription) and the common 3′ end containing poly(A) sequences. These mRNAs therefore contain the nucleotide sequence of more than one ORF. Usually, however, only the first ORF at the 5′ end of mRNA is translated into protein.

d. Minus-strand RNA (-) viruses replicating in the cytoplasm may possess either one large genomic RNA molecule (unsegmented) or two or more different subgenomic RNAs (segmented). For those viruses that possess an unsegmented genomic RNA molecule (e.g., VSV), the viral genes are arranged sequentially in the genomic RNA (–) with stop, intergenic, and start (EIS) sequences (Fig. 3). RNA polymerase (RDRP) synthesizes the virus mRNAs by initiating transcription at the 3′ end (one entry) and then terminates at the stop sequence (E) of that gene, skips the intergenic sequence (I), and initiates at the start (S) sequence of the next gene, and so on. Therefore, the RNA polymerase sequentially transcribes the downstream genes and there is no independent internal entry of the RNA polymerase on the genome. Since RNA polymerase randomly falls off during transcription and cannot initiate internally, the mRNA level (and, consequently, the protein level) is determined by the location of a particular gene in the viral genome. For example, mRNA of the capsid protein (N or NP) is present at the extreme 3′ end of the minus-strand (i.e., proximal to the promoter) just after the leader (ℓ) sequence, and it is therefore made in the most abundant amount, since it is the first gene to be transcribed by RDRP into mRNA. On the other hand, the L (polymerase) gene, encompassing nearly half of the genome, is located at the 5′ end of the viral RNA (distal to the promoter), so that the L mRNAs and L proteins are made in the least amount (Fig. 3). Each mRNA is capped at the 5′ end and polyadenylated at the 3′ end by the virally encoded RDRP.

e. Orthomyxoviruses, which are segmented, minus-strand RNA viruses, possess 8 RNA segments, which in total encode 10 mRNAs and 10 proteins for type A and B viruses. Orthomyxoviruses are transcribed and replicated in the nucleus. Orthomyxoviruses use a unique strategy to initiate transcription. They cannot initiate *de novo* mRNA transcription without a primer and must use the host's capped RNA as the primer at the 5′ end for mRNA transcription. One of the three proteins (PB2) of the viral polymerase complex (PB1/PB2/PA) recognizes the newly synthesized capped host RNA and cleaves the capped host RNA around 12–15 nucleotides from its 5′ end. Then another protein (PB1) of the polymerase complex uses the capped primer for viral mRNA initiation and chain elongation. Therefore, each influenza viral mRNA possesses at its 5′ end a capped nonviral RNA sequence acquired from the host nuclear RNA (Fig. 15). In addition, two viral RNA segments (segments #7 and #8) generate both unspliced and spliced mRNAs, causing translational shift to a different reading frame. In this process, 8 influenza viral RNA segments of type A and B viruses give rise to 10 mRNAs and 10 proteins (Nayak, 1997).

Segmented ambisense RNA viruses (e.g., arenaviruses) on infection produce a subgenomic mRNA using the 3′ end of the genomic RNA as the template, and later on in the infectious cycle use the antigenomic RNA as the template to generate the mRNA with the same polarity as the 5′ end of genomic RNA.

f. Viruses that possess double-stranded (ds) RNA viral genomes, such as reoviruses, are segmented and replicate in the cytoplasm. Their viral transcriptase, which is also present within the virus particles, synthesizes single monocistronic mRNAs from each dsRNA segment.

g. Retroviruses, although possessing a plus-strand RNA genome, contain reverse transcriptase (RT) in the virion. Transcription of retroviral mRNAs occurs in the nucleus from the integrated proviral DNA template by the host RNA *pol* II. Usually, both the unspliced genomic-length mRNA as well as the subgenomic mRNA, the latter being produced by splicing in the nucleus, function in protein translation.

F. Translation

Virions have evolved to become very efficient organisms that package a relatively small amount of genomic DNA or RNA in their capsids but use this information efficiently to generate the maximum number of functional proteins

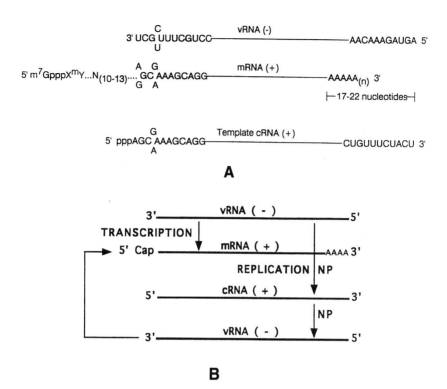

Fig. 15. Transcription and replication of the influenza virus RNA (vRNA). (**A**) The three classes of influenza virus-specific RNAs found in the virus-infected cells, vRNA of minus (–) polarity, and cRNA and mRNA, both of (+) polarity. Note that the viral mRNA (+) possesses nonviral (host) capped sequences at the 5′ end and lacks sequences of 17–22 nucleotides from the 3′ end but contains poly(A) sequences. The template cRNA (+ strand), on the other hand, is an exact copy from end to end of the vRNA (– strand) and does not possess either cap at the 5′ end or poly(A) at the 3′ end. (**B**) The transcriptive and replicative processes of influenza viral RNA.

required to produce infectious progeny virions. For some viruses like VSV, all of the viral proteins encoded by the genome and produced in the infected cells including the transcriptase are incorporated into the virion and become structural components of virus particles. For these viruses, there are by definition no nonstructural proteins, that is, there are no proteins that are encoded in the virion genome and produced in the infected cells but not incorporated into the virion. However, for the majority of viruses, one or more nonstructural proteins, either catalytic (enzymatic) or regulatory, are synthesized in virus-infected cells. These nonstructural proteins are required for the infectious cycle but are not incorporated into virion particles. Both structural and nonstructural proteins are translated from viral mRNAs, and the majority of viral mRNAs (except in the case of picornaviruses) possess structural features similar to that of the host mRNA (i.e., they possess a cap at the 5′ end, a translation initiation triplet [AUG] in the context of Kozak's rule, as well as translation termination triplets and poly(A) sequences at the 3′ end). These viral mRNAs undergo cap-dependent ribosome binding and ribosome scanning to locate the proper initiation triplet, a process that does not provide any advantage over the host mRNAs during translation. Therefore, after infection, the virus must overcome two major problems to achieve successful replication: (1) viruses must somehow overcome competition from host mRNAs for using translation machineries, and (2) viruses that possess only a limited amount of coding information must still be able to generate the considerable number of functional proteins needed for replication.

1. Viruses have developed a number of strategies to compete with host mRNAs for efficiently using the host translation machinery. These include the following. (a) Viral transcription machinery (especially in RNA viruses) are more efficient in generating high levels of mRNAs so that they can outcompete host mRNAs in translation. (b) Some viral proteins target and interfere with the host transcription machinery so that the host transcription level goes down or shuts off. Influenza viruses, however, use a novel system to their advantage. As mentioned earlier, one of the influenza polymerase proteins, PB2, recognizes, binds to, and cleaves the newly synthesized capped host hnRNAs around 13–15 nucleotides, and the capped oligonucleotide is used as the primer for mRNA synthesis. The cleavage of host hnRNAs, in turn, prevents host mRNA synthesis and processing. In addition, this virus interferes with nuclear export of the host mRNAs. (c) Some viruses modify the host translation machinery to use that machinery for its advantage while simultaneously shutting off host mRNA translation. This latter mechanistic approach is particularly evident for picornaviruses, which inactivate the cap-binding protein and modify the host translational factors (e.g., eIF2, eIF3/4B) and thus shut off cap-dependent host mRNA translation. However, picornaviral mRNA can still be translated efficiently because it does not have a cap at the 5′ end but rather possesses a unique RNA secondary structure known as an internal ribosome entry site (IRES) and is independent of Kozak's rule. The picornaviral mRNAs possessing an IRES can be translated efficiently in a cap-independent manner, while capped host mRNAs cannot be translated because of viral-mediated inactivation of some of the host translational factors.

2. Viruses have developed different strategies to produce a relatively large number of functional proteins from a small amount of genetic information using both transcriptional (or posttranscriptional) as well as translational (or posttranslational) processing.

a. Transcriptional (or posttranscriptional) generation of different mRNAs. Double-stranded DNA viruses can use both of their DNA strands to transcribe mRNAs, thereby increasing potential transfer of information into proteins. Some viruses that make mRNAs in the nucleus (either RNA or DNA viruses) can generate different mRNAs from the same genomic strand by using either unspliced mRNA or electing alternative splicing sites, thus even causing frame shifts in the subsequent translation. Influenza viral proteins M1, M2, NS1, NS2, and SV40 proteins (such as VP1, VP2, and large T and small T antigens) are classic examples of generating different mRNAs and proteins through splicing. Some viruses use RNA editing (i.e., nontemplated nucleotide addition in the mRNA) to shift the translation frame. This latter technique is frequently used by paramyxoviruses to generate their V and C proteins. Hepatitis delta virus uses adenosine deaminase for RNA editing as part of the transcription process to generate its δ Ag-L antigen. Other viruses selectively use different promoters to generate genomic and subgenomic mRNAs (e.g., HBV, togaviruses).

b. Translational (and posttranslational) generation of different viral proteins. The most common way to generate a number of functional proteins after translation is by proteolytic cleavage. These endoproteases, usually encoded by the virus, are sequence-specific and can generate a number of functional proteins from one large viral polypeptide. Classic examples of this type of cleavage activity are found with poliovirus (picornavirus) and flavivirus proteins. Poliovirus RNA is translated into a large polypeptide that sequentially undergoes endoproteolytic cleavage by different poliovirus proteases at specific amino-acid sites, generating 11 viral proteins (VP4, VP2, VP3, VP1, 2A, 2B, 2C, 3A, VPg, 3C, 3D) and other intermediate proteins (Fig. 2). The importance of virus-specific proteases was recently demonstrated in HIV infection, during which HIV protease inhibitors alone or in combination with RT (reverse transcriptase) inhibitors can be used in the treatment of AIDS to reduce patient virus load. Cleavage by host proteases is also sometimes critical to render viral proteins functional and viral particles infectious (e.g., conversion of influenza viral HA to HA1 and HA2, HIV gp160 to gp120 and gp 41).

Different initiation codons are also used in bicistronic mRNAs to translate different proteins. Depending on the initiation codon used, either one or the other protein can be translated (e.g., NB protein and NA protein from the same mRNA in influenza virus type B). Usually, one of the initiation codons is favored, thus regulating the levels of the two proteins produced from one bicistronic messenger RNA. Another strategy, often used by retroviruses, is either translational frameshift or translational suppression of termination codons. Translational frame-shift due to ribosomal slippage causes generation of the *gag*–*pro*–*pol* fusion protein in avian leukosis virus. This protein is then cleaved by a virus-specific protease (usually aspartic proteases) to generate individual functional proteins. Similarly, some retroviruses use translational termination suppression to continue translation in the same reading frame. In the *gag–UAG–pol* sequence, translation is normally terminated after the *gag* protein at the UAG codon. Occasionally, termination at UAG can be suppressed by a minor host tRNA capable of inserting glutamine and thereby generating a *gag–pol* fusion protein, which subsequently is cleaved by a viral protease to generate *gag* and *pol* proteins. Again, both frame-shift and/or in-frame suppression produces only a minority of fusion proteins with *pol*, thus regulating the amount of *pol* protein needed in small amounts in virus-infected cells.

G. Replication of Viral Genome

The replication pathway of different viral genomes varies depending on the nature of the viral genome. The overall strategy of viral genome replication can be grouped into seven pathways depending on the nature of the genome (Fig. 16). While all DNA viruses of eukaryotes except poxviruses replicate in the nucleus of their host cells, some use cellular DNA polymerase and others use DNA polymerase encoded by the virus genome (Table II). Poxviruses replicate their genome in the cytoplasm and use polymerase encoded by the viral genome (Table III). All RNA viruses except for retroviruses use an RNA-dependent RNA polymerase (RDRP) encoded by their own genome. Some of these (minus-stranded RNA viruses) carry RDRP in the virus particle to initiate transcription/replication of viral RNA following their entry and uncoating inside the cell. Retroviruses require reverse transcriptase (RT), an RNA-dependent DNA polymerase, in the virion particle to initiate replication. The majority of RNA viruses of eukaryotes replicate in the cytoplasm, except for the orthomyxo- and related viruses and the retroviruses. Orthomyxoviruses require cellular capped 5′ RNAs as primers for mRNA transcription, and retroviruses require production of proviral DNA and its integration into the host DNA as a prelude to both transcription and replication of the viral genome.

1. Replication of DNA Genome

Smaller DNA viruses (papova- and parvoviruses) rely on the host cell DNA polymerase, whereas more complex DNA viruses use their own virus-encoded DNA polymerase (Table II). The step for switching from transcription to replication of DNA viral genomes is primarily determined by the level of early viral proteins, which often are both regulatory and catalytic in nature. For SV40, when a sufficient amount of large T (LT) antigen has been synthesized, binding of the

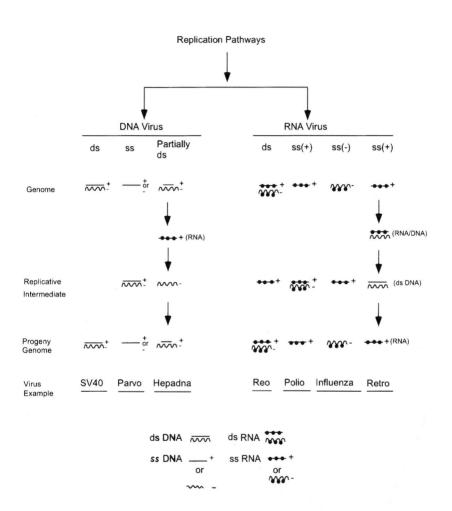

Fig. 16.   Seven replication pathways of the DNA and RNA genome of viruses. Examples of different viruses with DNA or RNA genomes are indicated. ds = double-stranded; ss = single-stranded; + and − indicate positive and negative polarity.

LT antigen-initiation complex to the transcription start site of early mRNAs (Fig. 13) causes suppression of early mRNA transcription. The helicase activity of the virus LT antigen then unwinds the DNA molecule, creating a replication bubble, whereupon the host DNA primase–polymerase α complex initiates DNA synthesis using an RNA primer, creating a replication fork. Synthesis of the SV40 DNA continues bidirectionally, creating circular intermediates (Fig. 17C). Adenoviruses use asymmetric DNA-replication, which initiates DNA synthesis at the 3′ end of one strand (template strand). At the 5′ end of that strand, a 55-kD protein, covalently linked to the DNA, is needed for initiation of DNA replication. The new growing opposite DNA strand then displaces the preexisting opposite strand. The displaced strand forms a panhandle structure by pairing the inverted terminal repeats before its own replication begins (Fig. 17A). In poxvirus DNA, two complementary forms are joined at the terminal repeat sections, forming palindromes. During replication, concatamers of two genomic length strands are formed. Unit length genomic molecules are then formed by separating the staggered ends and ligation (Fig. 17E). Linear herpes virus DNA becomes circularized inside the host cell nucleus and then replicates as a rolling circle, forming tandem concatamers. Finally, the unit length genomic DNA molecules are excised from concatamers (Fig. 17B). Single-stranded parvoviral linear DNA has terminal palindromes that form hairpin structures. These hairpins then serve to covalently link the plus and minus DNA strands and self-prime the replication. The progeny viral DNA genomes are then made by strand displacement (Fig. 17D).

Hepatitis B virus DNA uses reverse transcription for replication (Fig. 17F). The partially dsDNA in the virion contains a complete minus and a partial plus strand. After infection of the cell, the virion-associated reverse transcriptase renders the partially double-stranded viral DNA into a circular duplex DNA in the cytoplasm, which is then translocated into the nucleus and transcribed into a full-length plus-strand RNA by the host RNA *pol* II that already is present in the nucleus (Fig. 14). This full-length plus-strand viral RNA is encapsidated, transported into cytoplasm, and reverse transcribed into a full-length minus-strand and a partial plus-strand DNA before being released as infectious virion.

2. Replication of RNA Genome

Viral RNA genomes can be single-stranded and comprised either of a plus or minus strand, or double-stranded. Furthermore, while the genomes of some RNA viruses are segmented (multiple RNA molecules), others are nonsegmented (i.e., one RNA molecule) (Tables I and III). Switching from transcription to replication in the viral infectious cycle usually occurs after sufficient amounts of the capsid protein (e.g., the nucleoprotein) have been synthesized. The nucleoprotein functions as a regulator for switching from transcription to replication of the genome.

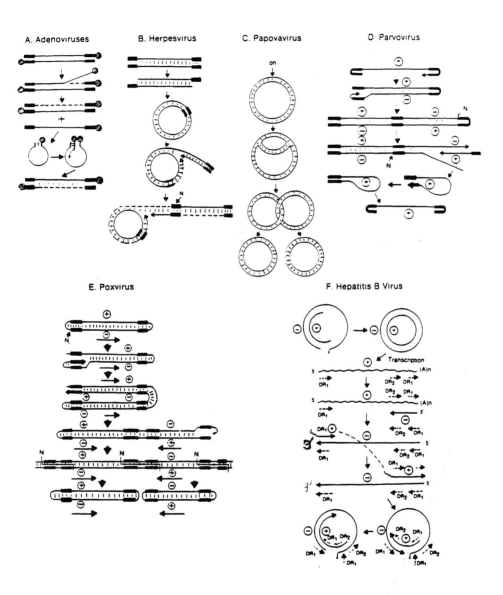

Fig. 17.   Replication pathways for viral DNA genomes. In panel **A**, dashed circles are terminal viral proteins attached to the 5′ end of the DNA strands. In panels **B** and **D**, N represents endonuclease cleavage site. The heavy lines shown in panel **D** are palindromes and self-priming steps, with + and − representing strand polarity. In panel **F**, the wavy lines represent RNA and the dashed lines represent DNA coding for direct repeats DR1 and DR2. Reprinted with permission from Davis *et al.* (1990).

Inhibition of nucleoprotein or protein synthesis will also inhibit vRNA replication without necessarily interfering with mRNA synthesis. The same core enzyme (i.e., RNA-dependent RNA polymerase, [RDRP]) is used for both transcription and replication, but the enzyme (and possibly the template RNA) becomes modified by viral nucleoprotein and cellular factors to effect the switch from transcription mode to replication mode. There are five different classes of RNA genomes, based upon the different strategies that these viruses use for genome replication (Table III).

a. *Replication of Single-Stranded Viral RNA.* Plus-strand RNA viruses are copied into a complete minus-strand RNA, which then serves as a template for synthesis of more plus strands via replicative RNA intermediates (Fig. 16). Minus-strand unsegmented RNA genomes are transcribed into two types of plus-strand RNAs: (a) subgenomic mRNAs (+ sense), which represent specific portions of the genome and are translated into proteins, and (b) full-length cRNA (+ sense), which represents a complete copy of the entire minus-strand genome and serves as the template for genomic RNA (– sense) synthesis. The synthesis of cRNA is regulated by a switch from transcription to replication that occurs after sufficient amounts of capsid proteins (e.g., NPs) have been synthesized. The capsid proteins provide the anti-termination factor required for cRNA synthesis. The cRNA is then copied back into minus-strand viral RNA, which is incorporated into virions. Orthomyxoviruses, which possess a segmented minus-strand RNA genome, likewise synthesize two classes of plus-strand RNA from the same minus-strand RNA template. However, the mRNAs and cRNAs of these viruses are different at both their 5′ and 3′ ends (Fig. 15). As indicated earlier, the orthomyxoviral mRNAs possess nonviral sequences from capped host mRNAs at their 5′ ends and terminate 18–22 nucleotides prior to the 3′ end with the addition of poly(A) sequences. However, the orthomyxoviral template cRNAs are complete copies of vRNA from end to end without any nonviral sequence at the 5′ end or poly(A) sequences at the 3′ end. Therefore, for orthomyxoviral cRNA synthesis to occur, the viral polymerase must be able to initiate RNA synthesis without any host capped primer at the 5′ end and the RNA synthesis must not terminate until the complete 3′ end is reached, thus fully copying the entire template vRNA from the 3′ end to the 5′ end and without any polyadenylation. Such complete cRNAs then function as templates for vRNA (minus-strand) synthesis.

b. *Replication of Double-Stranded Viral RNA.* Each segment of double-stranded viral RNA genome is replicated independently. First, the genome is transcribed to generate plus-strand mRNAs within the incoming virion core by the virion-associated RDRP. Next, the mRNA is used as a template by RDRP to synthesize the minus RNA strand, and thereby mRNAs become converted into double-stranded RNA, which is then packaged into progeny virion capsids.

c. ***Replication of RNA via a DNA Intermediate.*** Retroviruses contain a diploid genome consisting of two identical RNA molecules, a tRNA primer (Fig. 1), and a reverse transcriptase, an RNA-dependent DNA polymerase, which also possesses both RNase H and integrase activity. Conversion of the plus-strand viral RNA into double-stranded DNA is initiated by viral reverse transcriptase using the tRNA as a primer. The RNA-dependent DNA replication process is complex and requires strand switching twice (Fig. 18). Eventually, a double-stranded proviral DNA is made in the cytoplasm, which is translocated into the nucleus and becomes integrated into the host genome. The integrated proviral DNA is then transcribed by the host RNA *pol* II into full-length plus-strand RNA, and that full-length RNA is then transported into the cytoplasm and encapsidated into progeny virions.

VI. ASSEMBLY AND MORPHOGENESIS OF VIRUS PARTICLES

As indicated earlier, when compared to either eukaryotes or prokaryotes, viruses use a unique multiplication strategy to produce their progeny. All cells, either prokaryotic or eukaryotic, multiply as a whole unit from parent to progeny and in a geometric order, that is, from 1 to 2 to 4 to 8, and so duplicatively on. Viruses, on the other hand, do not multiply as units. Rather, they are assembled from component parts. Each component part of progeny virus particles is made separately, and they are often made in different amounts and at different locations and compartments within the host cell. These viral components are then put together to form the whole (infectious) virus particles (virions). In this assembly-line type of process, all individual viral components need not be assembled at the same time, and in fact, some components may be put together separately to form higher-ordered structures, that is, subviral particles (e.g., capsid), before they are assembled into a whole progeny virus particle. The number of steps involved and the complexity of the assembly process may vary greatly from one type virus to another. Some viruses, like the polioviruses, have only a few components to assemble, and yet others, like the pox or herpesviruses, have many components to assemble and their assembly compared to polioviruses is a far more complex process involving multiple steps.

With respect to the assembly processes, viruses can be classified into two major subclasses: naked viruses and enveloped viruses. Naked viruses consist of a nucleocapsid only (i.e., the capsid containing the genome and no envelope). The assembly of the protein capsid and incorporation of genomic nucleic acid into the capsid to create this nucleocapsid will render the virus particle infectious. For

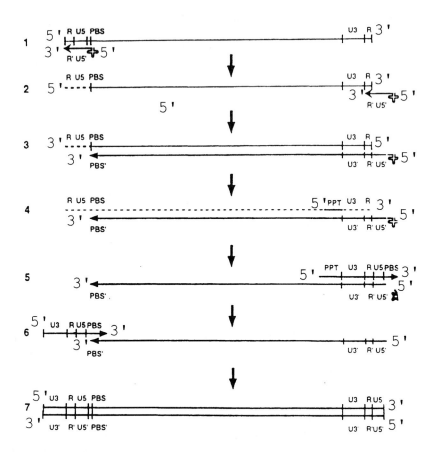

Fig. 18. Reverse transcription of retroviral genomic RNA into double-stranded proviral DNA. **Step 1**: Annealing of primer tRNA (shown as a cross-shaped symbol) to the primer-binding site (PBS), and synthesis of minus-strand strong-stop DNA. The R and U5 RNA is degraded by the RNase H activity of the reverse transcriptase, and the strong-stop DNA is released. **Step 2**: The first strand switch (or transfer). The minus-strand strong-stop DNA is annealed to the 3′ terminus of the genomic RNA via R–R′ hybridization (first strand jump). **Steps 3,4**: Further synthesis of minus-strand DNA, during which the genomic RNA is further degraded by RNase H. However, a small piece of RNA (the polypurine tract [PPT]) remains undegraded and serves as a primer for synthesis of plus-strand strong-stop DNA (step 5). **Step 5**: Termination of plus-strand DNA synthesis at 18 nucleotides into the primer tRNA, thus generating the new primer-binding site (PBS) sequence; the plus-strand DNA is then released from the minus-strand DNA. **Step 6**: The second jump (or the second transfer). The plus-strand strong-stop DNA is annealed to the 3′ terminus of the minus-strand DNA via PBS–PBS hybridization, completing the second jump. **Step 7**: Completion of the synthesis of the double-stranded proviral DNA. R = terminally redundant identical sequences at the 5′ and 3′ ends of viral RNA; U5 = unique nucleotide sequences near the 5′ end of the viral genome between the R and PBS (primer-binding site); U3 = the region near the 3′ end of the viral RNA between the initiation site of the plus-strand DNA synthesis and R sequences; PPT = polypurine tract that escapes RNase H digestion and serves as a primer for the second-strand DNA synthesis. Strong-stop DNA is the DNA copy of the region between the primer-binding site (PBS) and the 5′ end of the viral RNA genome. Reprinted with permission from Mak and Kleiman (1997).

these viruses, the virus receptor-binding proteins are part of the capsid proteins. Enveloped viruses, however, are those in which the nucleocapsid is surrounded by a lipid membrane containing the transmembrane viral proteins. In enveloped viruses, one of the transmembrane viral proteins (and not the capsid protein) contains the receptor-binding protein.

A. Assembly of Naked Viruses

The assembly of naked viruses occurs either in the cytoplasm (most RNA viruses) or nucleus (DNA viruses). For cytoplasmic viruses (e.g., the plus-sense RNA picornaviruses), the entire genomic RNA is translated into a single giant polyprotein (Fig. 2) that is cleaved by a virus-specific protease into P1 (a coat precursor protein); P1 is then cleaved by the protease 3C into VP0, VP3, and VP1 (5-S promoter). Five subunits (5-S promoter) — each containing one molecule of VP0, VP3, and VP1 — then assemble into pentamers (14 S). Twelve pentamers form the 60-subunit protein shell (capsid) for the picornaviruses. The viral RNA genome is then incorporated into the capsid, forming what is called the "provirion." Subsequently, the molecules of VP0 in the provirion are cleaved into molecules of VP4 and VP2 (Fig. 2), converting the provirion nucleocapsid into an infectious virion. This process of picornaviral capsid assembly is basically a self-assembly process whose rate is dependent on viral protein concentration.

For assembly of a naked virion to occur inside the nucleus, one of at least two distinct strategies can be used. The first of these would require that all capsid proteins, after their translation in the cytoplasm, must be transported into the nucleus either independently or cooperatively by forming a complex with other capsid proteins and that nucleocapsid assembly occurs around the viral genome in the host nucleus. This option is used by papovaviruses, whose DNA genomes, or minichromosomes, contain a single closed circular duplex DNA molecule complexed with cellular histone, which is organized into a nucleosome within the host nucleus. Papovaviral capsid assembly then proceeds in a stepwise fashion around the viral minichromosome. The capsid of SV40, which is a member of this virus group, contains 360 copies of its major viral protein (VP1) assembled into 72 pentamers plus 30 to 60 copies of internal proteins VP2 and VP3. VP2 contains the full VP3 sequence plus 100 extra amino acids at the NH_2 terminus, which are critical for interacting with the SV40 minichromosome. The papovaviral capsid proteins and minichromosomes assemble first into 200-S structures called provirions, which then mature into infectious virions. During this maturation, H1 histone protein is removed from the viral minichromosome and degraded.

Adenoviruses use a second type of strategy in which the capsid shell is first formed by the assembly of viral capsid proteins. Viral DNA, including core proteins, is then inserted into the empty capsid shells to form infectious virions. Both of these nuclear DNA viruses as well as the cytoplasmic naked RNA viruses are primarily released to the extracellular environment by cell lysis.

B. Assembly of Enveloped Viruses

The assembly of enveloped viruses is much more complex than that of naked viruses. It involves not only nucleocapsid formation but also envelopment of the nucleocapsid and budding of virions from different cellular organelles and membranes. Subsequently, the virus is released into the extracellular environment. The assembly and the budding site on the cellular membrane varies with different groups of viruses. Some viruses, like the poxviruses and rotaviruses, bud from the endoplasmic reticulum (ER), while others, like the bunyaviruses, bud from the Golgi complex, and still others bud from the nuclear membrane, like the herpesviruses. Still other viruses (e.g., orthomyxo-, paramyxo-, rhabdo-, and retroviruses) use the plasma membrane (apical or basolateral) as the budding site.

Since the assembly processes of orthomyxo- and paramyxoviruses have been well studied, the steps involved in morphogenesis of these viruses will be discussed below in some detail. For comparison, rhabdovirus and retrovirus assembly will also be included as needed. First, I will discuss the steps that are common to both orthomyxoviruses and paramyxoviruses and then later point out the differences between these two viruses in terms of assembly and morphogenesis. As noted earlier, orthomyxo- and paramyxoviruses are enveloped RNA viruses that contain single-stranded RNA genomes of negative (minus) polarity, and they are assembled into nucleocapsids that have helical symmetry (Table I). Electron-microscopic studies have demonstrated that these viruses bud from the plasma membrane into the outside environment and that complete virions are usually not observed inside the cell during the productive infectious cycle.

1. Steps Common for Enveloped Virus Morphogenesis

For elucidating viral assembly and budding processes, the viral structure can be separated into three major subviral components, each of which must be brought to the assembly site for morphogenesis. These subviral components are: (a) the viral nucleocapsid (or viral ribonucleoprotein [vRNP]) containing the vRNA, NP (nucleoprotein), and transcriptase complex, which together form the inner core of virus particle; (b) the matrix protein, which forms an outer protein shell around the nucleocapsid and constitutes the bridge between the envelope and nucleocapsid; and (c) the envelope (or membrane), which forms the outermost barrier of these enveloped virus particles, containing the virally coded transmembrane proteins and host cell lipids. Each of these virus groups (namely, orthomyxo-, paramyxo-, rhabdo-, and retroviruses) buds from the plasma membrane of infected cells. However, while the orthomyxo- and paramyxoviruses bud from the apical plasma membrane of polarized epithelial cells, both *in vivo* (e.g., in bronchial epithelium) and in cultured polarized epithelial cells (e.g., MDCK cells), the rhabdoviruses and retroviruses bud from the basolateral surface.

With respect to the processes involved in virus assembly, there are two major differences between orthomyxo- and paramyxoviruses: (a) Since the viral genome of orthomyxoviruses is segmented, multiple RNA segments (8 separate RNA segments for influenza types A and B, 7 RNA segments for influenza type C viruses) must be incorporated into infectious virions, whereas only one large RNA molecule is packaged in infectious paramyxovirus particles. (b) Since the transcription and replication of orthomyxoviral RNA and assembly of these viral nucleocapsids (vRNP) occur in the host nucleus, the viral nucleocapsids must be exported out of the nucleus into the cytoplasm for the final stages of viral assembly and for budding. In contrast, for paramyxoviruses, all of these steps, including assembly of viral nucleocapsids, take place in the cytoplasm.

Analysis of the steps involved in putting these subviral components together in an orderly fashion into an infectious virus particle is critical to an understanding of the assembly and budding processes. Two steps are obligatory for virus assembly and morphogenesis to occur. First, all of these viral components (or subviral particles) must be directed and brought to the assembly site, that is, the apical plasma membrane in polarized epithelial cells for assembly and budding of orthomyxo- and paramyxoviruses. Obviously, this step is the first obligatory requirement in virus assembly, since, if different viral components are misdirected to different locations or parts of the cell, virus assembly and morphogenesis cannot take place. Second, the viral components must interact with each other to form the proper virus structure during morphogenesis. It is possible that viral components may be directed to the assembly site but that defective interaction among these components will not yield infectious particles. However, although these two steps are obligatory, they alone may not be sufficient to form and release infectious virus particles. Therefore, virus components may be directed correctly to the assembly site and then interact with each other to form virus particles, yet infectious viruses may not be released into the medium. Such abortive virus morphogenesis in HeLa cells infected with influenza viruses has been observed where virus particles are formed on the plasma membrane but not released (Gujuluva *et al.*, 1994). Furthermore, each of these steps is indeed a complex multistep process.

Among the viral components, the greatest amount of information is available about the transport, sorting, and targeting of viral transmembrane proteins to the assembly site (prospective budding site) in the plasma membrane. It has been observed that the transmembrane viral proteins in virus-infected cells also preferentially accumulate at the virus assembly site, that is, the orthomyxo- and paramyxoviral proteins like HA, NA, F, and HN accumulate on the apical plasma membrane, whereas rhabdoviral and retroviral transmembrane proteins like the VSV G and the HIV gp160 proteins are targeted to the basolateral plasma membrane in virus-infected polarized epithelial cells (see Blau and Compans, 1996; Tucker and Compans, 1993). Furthermore, it was also observed that, when these transmembrane viral proteins were expressed alone from cloned cDNAs in the absence of any other viral protein, they also exhibited the same phenotype as in the virus-infected cells, that is, depending on the viral protein, they also accumulated on either the apical or basolateral plasma membrane in polarized epithelial cells. These results demonstrated that these transmembrane viral proteins do not need the presence of any other viral proteins for their successful targeting to the assembly site, and that they possess the necessary determinants in their protein structure for their sorting and targeting to the correct cell surface (apical or basolateral) in polarized epithelial cells. Furthermore, these studies indicated that the sorting and localization of transmembrane proteins may be a critical factor in selection of the assembly site of virus particles and also may be a critical determinant in viral pathogenesis. For example, the wild-type Sendai virus (a mouse pathogen), which buds from the cell's apical surface, is pneumotropic and less virulent as compared to a pantropic and highly virulent mutant Sendai virus that buds from both apical and basolateral surfaces (see §VI.C).

Although the nature of the cellular machineries that target either viral or cellular proteins to their appropriate apical or basolateral plasma membranes are not fully understood, it is now accepted that these viral proteins are targeted directly to apical or basolateral surfaces and that for apical versus basolateral targeting protein sorting occurs during the process of vesicular transport from the trans-Golgi network (TGN) to the plasma membrane. It has also been shown that separate sorting machineries (vesicles) are involved in targeting to the apical and basolateral surfaces. It is therefore expected that viral proteins must possess determinants that selectively interact with either the apical or basolateral sorting machineries of the host cells. Those proteins that are directed to the basolateral membrane, including the VSV (a member of family Rhabdoviridae) G protein, possess basolateral sorting determinants in their cytoplasmic tail region. However, the nature and location of the sorting determinant(s) of apical proteins are more complex. For example, type I apical viral proteins like influenza virus hemagglutinin (HA) and type II transmembrane proteins, like influenza virus neuraminidase (NA), have been shown to possess at least two apical sorting signals that function independent of each other (see Lin et al., 1998; Kundu et al., 1996). One of these apical determinants is in the ectodomain of the protein (the part of the membrane protein that is exposed to the outside). Glycans are the likely candidates for the apical sorting determinants present on protein ectodomains. In addition, another apical determinant is present in their transmembrane domain (the part of the protein that spans the distance between the inner and outer layers of the lipid membrane and anchors the protein to the lipid bilayer via lipid–protein interactions). The mechanisms by which these proteins are targeted to the apical plasma membrane are unclear, although the transmembrane domains of apical type I and

II proteins may interact with Triton X-100 (TX100) detergent-insoluble lipid rafts, enriched in cholesterol and glycosphingolipids (GSLs). These rafts may function as platforms for targeting lipids and proteins to the apical plasma membrane (Scheiffele *et al.*, 1997).

In addition to the transmembrane glycoproteins, other viral proteins and subviral complexes such as matrix protein and nucleocapsid must also reach the assembly site. How these viral components are also directed to the assembly site is not yet fully understood. The possibility exists that the matrix (M1) protein may be directed to the assembly site by its interaction with the transmembrane viral proteins using a piggyback mechanism. Similarly, viral nucleocapsids (or ribonucleoprotein) may be transported to the assembly site either independently or on the back of the M1 protein. A number of studies have shown that matrix proteins as well as matrix protein/nucleocapsid complexes interact with the glycoproteins in transit to the plasma membrane (see Nayak, 1996). Furthermore, it appears that cytoskeletal components are involved in these processes, by interacting with both the matrix proteins as well as the nucleocapsids (including the nucleoprotein [NP]) and thereby facilitating transport of these components to the assembly site (see Avalos *et al.*, 1997).

The viral matrix protein is a key component in virus assembly and morphogenesis. It is the most abundant protein in virus particles and is the rate-limiting component in particle formation since the virus particle is greatly reduced when matrix protein synthesis is defective or reduced. However, particles with reduced amounts of glycoproteins can be formed efficiently, although such viruses may be less infectious, or noninfectious. Freeze-fracture electron microscopy has shown that, during bud formation in virus-infected cells and in virus particles, the matrix protein is present as a sheet between the lipid bilayer and the viral nucleocapsid. The results of these and other studies imply that the matrix protein is likely to interact with both the lipid bilayer of the membrane and its associated viral glycoproteins on the outer side, and with the viral nucleocapsid on the inner side of the virus particles. In virus particles, the matrix protein remains bound strongly to the nucleocapsid under conditions where membrane glycoproteins can be dissociated using nonionic detergents. It is only after further treatment with either low-pH (influenza virus) or high-salt (paramyxovirus) buffer that matrix proteins can dissociate from the viral nucleocapsids. Therefore, we have strong evidence of interactions between the nucleocapsids and the matrix proteins. However, a detailed analysis of the exact nature of these associations in virus-infected cells at different stages of the assembly process has yet to be undertaken.

Likewise, the location of the matrix protein and glycoproteins in virus particles and in virus-infected cells predicts an interaction between the matrix protein and glycoproteins. However, the interaction of matrix proteins with membrane glycoproteins has been rather difficult to demonstrate. Studies using morphological as well as biochemical analyses have demonstrated that the Sendai virus (a member of family Paramyxoviridae) M (matrix) protein can bind independently to either Sendai virus membrane glycoproteins (F or HN), and that this interaction can take place on the plasma membrane as well as during exocytic transport of F and HN proteins through the Golgi complex. Biochemical and morphological studies have also shown that interaction of the Sendai virus M protein with viral glycoproteins can occur in the absence of nucleocapsid (or NP) protein (Sanderson et al., 1994). However, the nucleocapsid protein may facilitate further with membrane association of the M protein by interacting with viral glycoproteins. In Sendai virus-infected cells, the nucleocapsid possibly becomes membrane associated with the help of the M protein acting as a bridge between the nucleocapsid protein and the membrane glycoproteins. The fact that virus particles can be formed in the absence of one or more of their membrane glycoproteins suggests that interaction of the matrix proteins with all transmembrane viral proteins is not obligatory for viral morphogenesis to occur. Furthermore, influenza virus particles can be formed, albeit inefficiently, which contain HA and NA proteins that lack the cytoplasmic (inner membrane side) of the tail (Jin et al., 1997). However, these tail-minus influenza virus particles exhibited extremely high pleomorphism and bizarre morphological shape, suggesting an impairment of interaction between matrix proteins and glycoproteins. Similarly, in VSV-infected cells, virus particles are formed that possess VSV G protein with a foreign cytoplasmic tail, supporting the hypothesis that the interaction of matrix protein with the cytoplasmic glycoprotein tail is not an obligatory requirement for morphogenesis (Schnell et al., 1998).

Although the majority of host proteins present on the cell membranes are excluded during the budding process, two classes of host components (the lipids and cytoskeletal components) do appear to be involved in virus morphogenesis. Viral lipids are directly borrowed from the cellular lipids, and depend on the site of virus budding. For example, viruses budding from the apical or basolateral membrane will incorporate the lipids present on one specific side of the plasma membrane. It has also been specifically suggested that glycosphingolipids (GSLs) and cholesterol, which are present in the host cell apical membrane, may play a specific role in targeting viral transmembrane proteins to the apical membrane domains (budding sites). For example, both type I (influenza virus HA, Sendai virus F) and type II (influenza virus NA, Sendai virus HN) proteins associate with TX100 detergent-insoluble lipid rafts enriched in GSL and cholesterol during exocytic transport to the viral assembly sites on the plasma membrane.

Cytoskeletal components facilitate both the transport of viral proteins to the assembly site and the budding process. Subviral components such as the NP proteins of nucleocapsids, as well as the RNA of the influenza viral RNP and the matrix proteins, interact with the host cytoskeletal components during intracellular transport. Cytoskeletal components, particularly microtubules and microfilaments, are known to be involved in targeting proteins to apical and basolateral membrane domains. Actin filaments have been observed in budding paramyxoviruses, and are also present in released viral particles of both orthomyxo- and paramyxoviruses. These observations suggest that the host cytoskeletal elements are actively involved in the assembly and budding processes of virus particles. However, studies that attempted to examine these roles by using cytoskeletal disruptive agents like cytochalasins B and D have produced conflicting results. Some of these studies have reported drug-induced enhancement of virus assembly and release, while others found a decrease in or no effect on virus replication following treatment with these agents. It is likely that both the nature of host cells and viruses as well as the timing of the addition of these drugs in the infectious cycle may have contributed to the variable results obtained in these studies. For example, drugs added relatively early in the infectious cycle might interfere with transport of viral components and thereby inhibit virus assembly and release, whereas, if these drugs were added late in the infectious cycle, they might enhance virus release by disrupting microfilaments and facilitating closure of viral buds.

2. Assembly of Paramyxoviruses

For paramyxoviruses, three viral components — namely, the NP of viral nucleocapsids (RNP), the M protein, and two transmembrane glycoproteins (F and HN) — are critically important for the assembly and morphogenesis of virus particles. However, the function of some other viral proteins such as P or L in the assembly process is unknown. These latter two proteins are synthesized independently from different viral messengers, and in different cellular compartments than NP, M, F and HN, and their synthesis appears to be temporally regulated. NP, the major component of the viral capsid, is critically required for both genome replication and nucleocapsid formation. Both free (soluble) and nucleocapsid-bound NP are present in the cytoplasm of virus-infected cells. Viral genome (vRNA) synthesis is coordinated with NP synthesis such that vRNA synthesis will not take place in the absence of a sufficient level of free NP proteins. Both the M and NP proteins are synthesized on free polyribosomes in the cytoplasm. NP associates with vRNA to form the RNP complex, and M proteins interact with the vRNP to form the M/vRNP complex. These proteins and complexes are formed in the cytoplasm. These M/NP (or M/RNP) complexes as well as M and NP proteins become associated independently with cytoskeletal components.

These components then interact with the F and HN glycoproteins either individually or together during their transport through the exocytic pathway or following insertion of F and HN into the plasma membrane, or both. The site of interaction between the M/RNP complex and the viral glycoproteins is essentially regulated by the presence and availability of these viral membrane glycoproteins in specific membrane components. For example, late in the infectious cycle in virus-infected cells, the majority of the viral glycoproteins are already present on the plasma membrane; therefore, the M/RNP–glycoprotein interaction is likely to occur predominantly on the plasma membrane. In addition, Sendai virus M protein can interact with Sendai virus glycoproteins in the absence of NP protein or RNP during exocytic transport of glycoproteins (Sanderson et al., 1994). Although some M proteins can interact alone but inefficiently with lipid membranes, viral glycoproteins greatly facilitate membrane association of M protein, suggesting a direct interaction of M proteins with membrane-inserted viral glycoproteins. On the plasma membrane, M proteins will also interact with each other, forming a sheet of M proteins. This sheet of M proteins then acts to exclude, from viral budding sites, other cellular membrane proteins, except for viral glycoproteins, which are retained since they are bound to the M protein. This M protein sheet binds to the virus nucleocapsid and becomes the assembly site for budding. Nucleocapsids are directed to the assembly site by cytoskeletal components such as actin, which may also aid in bud formation. Eventually, the bud containing the vRNP (nucleocapsid) and matrix protein will be closed by fusion of membrane lipids. Experimental evidence supports a presumed interaction of M proteins with themselves and with viral RNP and glycoproteins, as well as interaction of M and NP (or RNP) with the cellular cytoskeleton elements (Sanderson *et al.*, 1995). However, the mechanism of bud formation and the forces causing membrane curvature for initiating bud formation remain unclear. It has been proposed that the interaction among the M proteins and pushed by actin filaments may be responsible for bud formation. In some cases, actin filaments are seen to extend into the viral bud. Therefore, closure of the bud seemingly would require disassembly of actin and might result in incorporation of some actin into the virus particles. Experimental evidence for the presence of actin within virus particles does support this proposal for the role of actin in bud formation and virus release. However, nothing is known about the specifics of actin disassembly during the process of budding and release of virus particles from the cell.

3. Assembly of Orthomyxoviruses

The assembly of orthomyxoviruses (influenza viruses, Fig. 12) differs from that of paramyxoviruses in two unique ways. First, unlike paramyxoviruses, influenza vRNP is synthesized in the nucleus. Therefore, vRNPs must exit from the nucleus into the cytoplasm. Second, eight separate vRNP segments must assemble into one infectious virus particle, requiring that multiple vRNP segments must either uniquely or randomly associate with each other in such a way that during virus assembly at least one of each of the eight separate vRNP segments is incorporated into every infectious budding progeny virus.

After uncoating, the infecting vRNPs that have entered the cell are transported into the nucleus for replication and transcription of viral RNAs, following which the progeny vRNPs are formed in the nucleus. In order for assembly of progeny influenza viruses to occur, the newly formed vRNPs must be transported out of the nucleus into the cytoplasm and then directed to the assembly site on the plasma membrane. However, since the NP protein possesses nuclear localization and nuclear retention signals, it is not clear what enables the vRNP to exit the nucleus. Although it appears that dissociation of the incoming M1 matrix protein in the acid pH of endosomes during the viral uncoating process is necessary in order for the vRNPs to subsequently enter nucleus, massive association of M1 with progeny vRNP in the nucleus is not required for exit of vRNPs from the nucleus into the cytoplasm. A study performed in ts51 (temperature-sensitive M1 mutant WSN of influenza virus) virus-infected cells has revealed that progeny vRNPs exit from the nucleus into cytoplasm at the restrictive temperature when the ts51 M1 protein is not functional, and the majority of this mutant's newly synthesized M1 proteins is retained in the nucleus. The ts defect is due to hyperphosphorylation of the M1 protein at the restrictive temperature. Recently, NS2 protein has been shown to provide the nuclear export signal for exit of viral RNP from the nucleus into the cytoplasm (O'Neill *et al.*, 1998). However, unlike such other nuclear export proteins as REV protein of HIV, NS2 does not have an RNA binding domain.

The following scenario is envisioned for the nuclear exit of vRNP (Nayak, 1996). In the nucleus there are likely two classes of vRNPs, of which one class is involved in active transcription and replication of viral RNA. This class of progeny vRNPs are not exported from the nucleus. A second class of progeny vRNPs that are to be exported from the nucleus are rendered transcriptionally inactive and sequestered from the transcriptionally active vRNPs. The support for incorporation of the transcriptionally inactive progeny vRNPs in virus particles comes from an observation that in released progeny virus particles there are polymerase molecules present only at one end of the vRNP that are not present all over of the vRNP, as would be expected if those progeny vRNPs had been undergoing active transcription within the nucleus during their exit from the nucleus and if such vRNPs were assembled into the progeny virions. Furthermore, the requirement of a 5′ primer for initiation of transcription and incorporation of that primer into the transcription product mRNA, as observed during *in vitro* studies, would also support the concept of initiation of transcription rather than chain elongation. On the other hand, simple chain elongation without a primer requirement would be expected to occur if instead the vRNPs had essentially been "frozen" during active transcription when they exited from the nucleus and budded into virus particles from the infected host cell. These results provide further evidence to support the hypothesis that only transcriptionally inactive vRNPs are exported from the nucleus. However, it is not clear how the progeny vRNPs are rendered transcriptionally inactive in the nucleus. Since M1 is known to inhibit transcription, it is possible that a few M1 molecules bind at the critical site on either the progeny vRNA or the viral polymerase, and that these M1 molecules then render those vRNPs transcriptionally inactive and sequester them from the transcriptionally active vRNP that are retained in the nucleus. NS2 then may bind to the transcriptionally inactive vRNPs and facilitate their export to the cytoplasm. Such a hypothesis can explain how in ts51-infected cells vRNPs are exported from the nucleus even though the majority of M1 is retained in the nucleus. Only a few M1 molecules will be able to block transcription of vRNPs and bind to NS2 for nuclear export of vRNPs.

Once vRNPs are exported from the nucleus into the cytoplasm, the subsequent sequence of events leading to assembly and budding have yet to be resolved for orthomyxoviruses. Late in the infectious cycle, the majority of the cytoplasmic M1 proteins in virus-infected cells become membrane bound immediately after their synthesis. Pulse-chase experiments have shown that late in the infectious cycle the newly synthesized M1 protein incorporated quickly into the progeny virus particles, suggesting that M1 is a rate-limiting factor in virus morphogenesis and release. This result also suggests that viral RNPs and glycoproteins are already present on the plasma membrane of virus-infected cells late in the replication cycle at the time when M1, soon after its synthesis, interacts with the membrane-bound nucleocapsid complexes. However, the interaction of M1 with membrane-associated components is not totally dependent on having viral glycoproteins associated with the membrane. This independence has been demonstrated by the fact that free M1 alone can become membrane bound and, possibly, even diffuse to the assembly site, where it becomes stably associated with the membrane after it interacts with other previously bound molecules of M1, membrane glycoproteins, and vRNPs. However, despite this evidence, the precise role of the viral components (namely, M1, glycoproteins, and nucleocapsids), as well as the sequence of steps involved in the association of M1 with the membrane during influenza virus assembly and budding, have yet to be determined. Since progeny virus particles can be formed that lack either the HA or NA proteins or lack the cytoplasmic tail of the HA or NA, it would seem that the association of M1 with any one of these glycoproteins is sufficient to allow for viral morphogenesis and that M1 may interact with the transmembrane domain as well as cytoplasmic tail of either of these glycoproteins. On the other hand, since only a few molecules of M2 protein are found in the virus particle, M2 is unlikely to be a major factor in membrane association of M1 or in assembly and morphogenesis of virus particles.

Finally, how the eight different RNA (or RNP) segments are incorporated into each infectious virus particle remains unclear. Two models have been proposed: (a) selective assembly of eight unique vRNP segments, and (b) random assembly of multiple vRNP segments into a virus particle. The latter model would propose that more than eight RNA segments are incorporated randomly into each particle, so that a fraction of virus particles will possess eight separate RNA segments and be infectious. The majority of data from genetic and biochemical experiments does not differentiate between these two models. Selective assembly predicts that each virus particle will possess only eight separate RNA segments. However, extra vRNA segments have been shown to be present in virus particles under forced selection, suggesting that eight vRNA segments can be randomly incorporated into virus particles. This, however, may represent only a minority of virus particles. On the other hand, an extrapolation of findings that describe the loss of a homologous vRNA segment in defective interfering (DI) virus particles would suggest that multiple segments of the same RNA segment are not favored for incorporation into the same virus particle and would therefore favor some selective process for achieving incorporation of specific vRNP segments into a virion during viral assembly. However, there are as yet neither definitive data nor a specific hypothesis to explain the mechanism by which selective incorporation of vRNPs in a virus particle is accomplished. Finally, the pleomorphism and plasticity of virus particles (Fig. 6) as well as a high particle-to-infectivity ratio (20 to 1000 particles/pfu) can accommodate either a selective or random mode for the strategy of vRNP assembly during viral morphogenesis (see Lamb and Choppin, 1983).

Last, after budding from the host cell, viruses must be released into the surrounding medium and spread outward to infect other cells. Some viral components are critically involved in the viral release process. The data from ts viruses at restrictive temperature clearly demonstrate that the viral neuraminidase (NA) protein is involved in virus release. The NA removes the sialic acid, which is the receptor for influenza virus, from membrane glycolipids and glycoproteins of both the virus and the virus-infected cells, and thus prevents self-aggregation among virus particles and reattachment to the virus-infected cell.

C. Role of Viral Budding in Viral Pathogenesis

The site and nature of budding can be an important contributory factor in viral pathogenesis, particularly for such respiratory viruses as the influenza and Sendai viruses. The influenza and Sendai viruses bud from the apical surface of polarized epithelial cells (e.g., bronchial epithelial cells) into the lumen of the lungs and are therefore usually pneumotropic, that is, restricted to the lungs, and do not cause viremia or invade other internal organs. However, occasionally, some influenza viruses — like fowl plague (H5 or H7) viruses (H5 or H7 indicates hemagglutinin subtype specificity of type A influenza viruses) and WSN (H1N1) viruses — are not restricted to the lungs and produce viremia infecting other internal organs (pantropism) and cause a high degree of mortality in infected animals. In humans, most of the influenza viruses are pneumotropic and do not spread to other internal organs. However, it is not clear if the Spanish Influenza of 1918, the most devastating influenza pandemic in recorded human history, which killed 20 to 40 million people worldwide, particularly affecting young healthy adults, was only pneumotropic. In addition to pneumonia, some people died due to massive pulmonary hemorrhage and edema (see Taubenberger, 1998). The 1918 flu virus, like fowl plague, may have caused viremia and infected other organs. Therefore, it is possible that highly virulent viruses may not be restricted to lungs in either chicken or humans. H5N1, the Hong Kong chicken virus, is extremely virulent and pantropic for chicken, causing viremia and spreading to other internal organs. In 1997, H5N1 was shown to have infected at least 16 peoples, including 5 who died of complications from the influenza infection. The majority of them had clinical pneumonia (or acute respiratory distress syndrome), gastrointestinal symptoms, and impaired hepatic and renal function. Although the virus did not spread from human to human or spread very inefficiently, the high morbidity and mortality in humans raises the question as to the cause for such high virulence in the infected individuals. The viruses that are restricted to lungs are called "pneumotropic," whereas the viruses that cause viremia and spread to other internal organs are called "pantropic." Why some influenza viruses are pneumotropic and others are pantropic is an important question for predicting the outcome of a major influenza epidemic or pandemic.

The severity of viral pathogenesis depends on both viral and host factors, including host immunity. The virulence determinants of influenza viruses are complex and multigenic. However, one factor that has been thought to be critical in viral growth and virulence is the cleavability of HA $\rightarrow$ HA1 and HA2. Influenza virus is normally restricted to the lungs because its HA can be cleaved by tryptase Clara, a serine protease restricted to the lungs. However, some HA that contains multiple basic amino acids at the HA1–HA2 junction, as is found only in H5 and H7 avian subtypes, can be cleaved by furin and subtilisin-type enzymes, which are present ubiquitously. Therefore, such viruses can grow in other organs. In addition, the NA of some influenza viruses (e.g., WSN viruses) binds and activates plasminogen into plasmin in the vicinity of HA, and the activated plasmin cleaves HA → HA1 and HA2, rendering the virus infectious. Therefore, WSN virus, which lacks multiple basic residues in its HA, can grow and multiply in tissues other than the lungs.

However, although the cleavage of HA → HA1 and HA2 is a major virulence factor, it is not the only factor contributing to the pantropism of a normally pneumotropic flu virus. For example, although WSN virus is pantropic and neurovirulent in the mouse, gene reassortment experiments demonstrated that the WSN NA gene responsible for the cleavage of HA was not sufficient for neurovirulence in chickens or mice. Other WSN genes, like the M and NS genes, in addition to the NA gene, were required for neurovirulence and, therefore, likely pantropism. The function of M and NS genes in neurovirulence is not known. The M gene in Sendai virus has been shown to affect apical versus basolateral budding and contribute to the pantropism of F1-R Sendai virus mutant (Tashiro and Seto, 1997). Therefore, it is possible that, in addition to increased cleavability of HA, the pantropic virus causes alteration in apical budding, releasing more virus basolaterally. Since blood vessels are proximal to the basolateral surface of cells, basolateral budding would facilitate more viruses entering into the blood, causing viremia, and invading other internal organs. Therefore pantropic influenza viruses like WSN/33 virus or highly virulent Hong Kong H5N1 and H7N1 fowl plague viruses may also cause altered budding from apical and basolateral surfaces. Thus, altered budding may be considered an important trait for the virulence of a specific strain of influenza virus. However, the role of altered budding in influenza virus pathogenesis remains to be determined.

Sendai virus, like influenza virus, is a pneumotropic mouse virus that buds apically. However, a Sendai virus mutant, F1-R, which exhibited pantropism possessed two characteristics (Tashiro and Seto, 1997): (1) cleavage mutation of F → F1 and F2 ubiquitously due to the presence of multiple basic residues, and (2) altered budding from both the apical and basolateral surfaces. On the other hand, the Sendai virus mutants that exhibited only cleavage of F → F1 and F2 or that exhibited only altered budding did not cause viremia or pantropism in the mouse. This would also support the argument that altered budding may be a factor that facilitates viremia and pantropism. Therefore, altered apical versus basolateral budding could be an important factor in dissemination of virus into the blood, invasion of internal organs, pantropism, and the higher virulence of a specific virus strain.

VII. CONCLUSIONS

The replication and morphogenesis processes of viruses are different from those of either prokaryotic or eukaryotic organisms. In this chapter, I have presented some of the general steps involved in the viral infectious cycle, including: entry, uncoating, transcription, translation, replication, and assembly processes, and the possible role of budding in viral pathogenesis. Of these, viral morphogenesis is the most obscure phase in the virus life cycle. Yet, knowledge of how the particles are formed during this morphogenetic stage is fundamental to understanding virus growth and multiplication, and therefore is crucial in defining viral infectivity, transmission, virulence, tissue tropism, host specificity, and pathogenesis, and contributes to an overall understanding of the disease process and progression of disease, including host morbidity and mortality. In addition, the site of budding can affect virus virulence and pathogenesis. Elucidation of the viral replicative and assembly processes is critical in terms of enabling us to find ways to block these steps and thereby intervene in the viral life cycle and disease process. Much remains to be done to achieve these goals, particularly in terms of elucidating those stages of the viral assembly process that relate to how viral components are brought to the assembly site, how those components interact with each other at the assembly site, and how viral budding actually occurs. A better understanding of viral replication and morphogenesis may lead us to develop novel therapeutic agents capable of interfering with these critical steps in viral multiplication and pathogenesis.

ACKNOWLEDGMENTS

Research in the author's laboratory was supported by grants from the National Institutes of Health. The author thanks Kiet Chi Tran for drawing some of these figures. The source for the other figures is gratefully acknowledged in the captions. The author is grateful to Dr. Felix Wettstein for helpful suggestions and Eleanor Berlin for typing the manuscript.

REFERENCES

Avalos, R. T., Yu, Z., and Nayak, D. P. (1997). Association of influenza virus NP and M1 proteins witH cellular cytoskeletal elements in influenza virus-infected cells. *J. Virol.* **71**, 2947–2958.

Blau, D. M., and Compans, R. W. (1996). Polarization of viral entry and release in epithelial cells. *Sem. Virol.* **7**, 245–253.

Davis, B. D., Dulbecco, R., Eisen, H. N., and Ginsberg, H. S. (1990). "Microbiology," 4th ed. J. B. Lippincott, Philadelphia.

Fields, B. N. and Knipe, D. M. (1990). "Fields' Virology," Vols. 1 and 2, 2nd ed. Raven, New York.

Fields, B. N., Knipe, D. M., and Howley, P. M. (1996). "Fundamental Virology," 3rd ed. Lippinicott-Raven, Philadelphia.

Flint, J., Enquist, L., Krug, R., Racaniello, V. R., and Skalka, A. M. (1999). "Principles of Virology: Molecular Biology, Pathogenesis, and Control." American Society of Microbiology, Washington, DC.

Francis, T., and Moore, H. E. (1940). A study of the neurotropic tendency in strains of virus of epidemic influenza. *J. Exp. Med.* **72**, 717–728.

Gujuluva, C. N., Kundu, A., Murti, K. G., and Nayak, D. P. (1994). Abortive replication of influenza virus A/WSN/33 in HeLa 229 cells: Defective membrane function during entry and budding processes. *Virology* **204**, 491–505.

Jin, H., Leser, G. P., Zhang, J., and Lamb, R. A. (1997). Influenza virus hemagglutinin and neuraminidase cytoplasmic tails control particle shape. *EMBO J.* **16**, 1236–1247.

Kemble, G. W., Daniell, T., and White, J. M. (1994). Lipid-anchored influenza hemagglutinin promotes hemifusion, not complete fusion. *Cell* **76**, 383–391.

Krug, R. M. (1989). "The Influenza Viruses." Plenum, New York.

Kundu, A., Avalos, R. T., Sanderson, C. M., and Nayak, D. P. (1996). Transmembrane domain of influenza virus neuraminidase, a type II protein, possesses an apical sorting signal in polarized MDCK cells. *J. Virol.* **70**, 6508–6515.

Lamb, R. A., and Choppin, P. W. (1983). Gene structure and replication of influenza virus. *Annu. Rev. Biochem.* **52**, 467–506.

Lin, S., Naim, H. Y., Rodrigues, A. C., and Roth, M. G. (1998). Mutations in the middle of the transmembrane domain reverse the polarity of transport of the influenza virus hemagglutinin in MDCK epithelial cells. *J. Cell. Biol.* **142**, 51–57.

Mak, J., and Kleiman, L. (1997). Primer tRNAs for reverse transcription. *J. Virol.* **71**, 8087–8095.

Nayak, D. P. (1994). Abortive replication of influenza virus A/WSN/33 in HeLa 229 cells: Defective viral entry and budding process. *Virology* **204**, 491–505.

Nayak, D. P. (1996). A look at assembly and morphogenesis of orthomyxo- and paramyxoviruses. *ASM News* **62**, 411–414.

Nayak, D. P. (1997). Influenza virus infection. *In* "Encyclopedia of Human Biology" (R. Dulbecco, ed.), Vol. 5, pp. 67–80. Academic Press, New York.

O'Neill, R. E., Talon, J., and Palese, P. (1998). The influenza virus NEP (NS2 protein) mediates nuclear export of viral ribonucleoprotein. *EMBO J.* **17**, 288–296.

Ray, R., Roux, L., and Compans, R. W. (1991). Intracellular targeting and assembly of paramyxovirus proteins. *In* "The Paramyxoviruses" (D. W. Kingsbury, ed.), pp. 457–479. Plenum, New York.

Sanderson, C. M., Wu, H.-H., and Nayak, D. P. (1994). Sendai virus M protein binds independently to either the F or the HN glycoprotein in vivo. *J. Virol.* **68**, 69–76.

Sanderson, C. M., Avalos, R., Kundu, A., and Nayak, D. P. (1995). Interaction of Sendai viral F, HN, and M protein cytoskeletal and lipid components in Sendai virus-infected BHK cells. *Virology* **209**, 701–707.

Scheiffele, P., Roth, M. G., and Simons, K. (1997). Interaction of influenza virus hemagglutinin with sphingolipid–cholesterol membrane via transmembrane domain. *EMBO J.* **16**, 5501–5508.

Schnell, M. J., Bunonocore, L., Boritz, E., Ghosh, H. P., Chemis, R., and Rose, J. K. (1998). Requirement for a nonspecific cytoplasmic domain sequence to drive efficient budding of vesicular stomatitis viruses. *EMBO J.* **17**, 1289–1296.

Tashiro, M., and Seto, J. T. (1997). Determinants of organ tropism of Sendai virus. *Frontiers Biosci.* **2**, 588–591.

Taubenberger, J. K. (1998). Influenza virus hemagglutinin cleavage into HA1, HA2: No laughing matter. *Proc. Natl. Acad. Sci. U.S.A.* **95**, 9713–9715.

Tucker, S. P., and Compans, R. W. (1993). Virus infection of polarized epithelial cells. *Adv. Virus Res.* **42**, 187–247.

Wolfgang, K., Joklik, H. P., and Willet, D. (1984). "Zinsser Microbiology," 17th ed. Appleton-Century-Crofts, New York.

4

An Introduction to the Evolutionary Ecology of Viruses

VICTOR R. DeFILIPPIS* and LUIS P. VILLARREAL[†]

*Department of Ecology and Evolutionary Biology
University of California, Irvine
Irvine, California 92697

[†]Department of Molecular Biology and Biochemistry
University of California, Irvine
Irvine, California 92697

I. INTRODUCTION: DEFINING VIRAL EVOLUTIONARY ECOLOGY

Distinguishing viruses, whether exogenous or endogenous, from other forms of life including parasitic genetic elements is an ambiguous but not impossible task. There are characteristics shared by all viruses that set them apart from other self-replicating systems. Viruses are obligate molecular parasites that rely on only one type of genomic material (either DNA or RNA) and use a proteinaceous genome package to facilitate dispersal. While viruses are not metabolically active organisms, their phenotypic variation results in differential replication success, thereby ensuring they are subject to the laws of evolution. Virus taxa, like organisms more frequently investigated evolutionarily, clearly change through time and to profound degrees. Unlike the cellular "macrobiota," however, our observations of viral transformation are very direct and never from examination of a large-scale past timeline such as the fossil record. If observed proportions of certain character state changes in viruses were applied to megafauna, hypotheses regarding the amount of time required to manufacture these changes would most likely be on the order of hundreds of thousands to millions of years. Yet critical changes in such factors as host range (habitat), capsid and envelope proteins (morphology), and life-history characters are common in some viruses within the time of a typical research project.

As with all forms of life, the pace and direction of viral evolution are determined by the selective environment that is occupied. This environment is both multidimensional and continually changing while constantly driving the increase in population fitness. The differences in viral as opposed to, say, multicellular selective environments are mostly in quantity, not quality. In other words, the fundamental nature of selection is not influenced by organism complexity (fitness must always move "uphill"), yet the character of selective pressure is determined by an organism's necessities for survival. It could also be argued that, because of the limited sequence space of viral genomes, the high probability of genetic drift during transmission bottlenecks, extremely large population sizes, and (at least for many RNA viruses) the high mutation and recombination rates, viruses exhibit greater mobility through the space of their selective or adaptive environments than do more complex organisms (Moya, 1997).

Evolutionary ecology is the study of how organisms respond and have responded evolutionarily to the selective environments in which they exist. The environment is defined as the physical, biological, and stochastic elements with which an organism interacts and uniquely determines the type and degree of natural selective pressure exerted.

Viruses have both general and specific requirements for replication and existence. All viruses require the use of a host cell's ATP, ribosome, nucleotide, tRNA, and amino acid resources. Many viruses also require use of other cell functions and components such as replication and splicing machinery. Often viruses will utilize more advanced host characteristics, especially when dealing with the demands of transmission. Here they may take advantage of host ecology, feeding habits, sexual behavior, social structure, etc. to facilitate host–host transfer.

No matter the level, one characteristic that is shared by all viruses' fundamental reproductive strategy is host exploitation. If a host's biology possesses a component that can be used to augment reproduction of some viral strain, then chances are it typically is. But as with all organisms, these requisite conditions are a double-edged sword in that, while they allow perpetuation of a given virus population, their variability selects for survival of only a subset of phenotypes from that population. Thus, while viruses are not typically subjected to the types of soft selection resulting from resource competition that many plant and animal species experience, their natural selection is often more commonly unpredictable and determined by probabilistic conditions.

Viruses' lack of relative complexity is clearly not paralleled in their adaptability, and nowhere is this more apparent than in their seeming omnipresence. Every extant species genome is thought to act as a viral host, and perhaps a thousand different virus types are known to infect *Homo sapiens* alone. Investigation of the evolutionary ecology of viruses seeks to explain observed patterns of viral diversity by invoking mechanisms that take into account evolutionary response to the environment(s) in which they exist. Such research relies (heavily) on very disparate areas of science. For instance, mere description of such fundamental virus characteristics as appearance, life cycle, and tissue tropism is only possible through the tools of biochemistry and molecular biology. Answers regarding virus population dynamics such as growth and size fluctuation are often provided by epidemiological research. And evolutionary interpretation and prediction are formulated using concepts of theoretical population biology. This area of inquiry is quite young and is destined to expand with increasing sophistication in data collection and in analytical methods and technology, the exponentially growing database of molecular information, and the continual emergence of viruses as pathogens.

Knowledge about the evolutionary ecology of viruses has valuable multidisciplinary applications as well. For example, it may aid investigation of emerging infectious diseases (of all types) by providing models of host switching (i.e., successful occupation of an additional host taxon for sustained transmission). It can make contributions to medical virology by giving researchers information about specific viral ecological requirements, thereby opening up possible routes of antiviral treatment or immunization. Such inquiry also assists theoreticians in many ways by providing real-life models for population genetic, phylogenetic, ecological, and even mathematical research.

Our discussion of viral evolutionary ecology will begin with examination of viral fitness and selection, with special emphasis on the many levels at which these concepts can be applied. We continue with discussion of ecological concepts such as niche, species (taxon) interactions, life-history evolution, and specialization. We will finish by looking at theoretical concepts such as sequence space and fitness peaks and relating such ideas to actual patterns of virus diversity and diversification.

II. FITNESS AND SELECTION

A. Defining Fitness and Selection

Evolution by natural selection is a temporal process resulting from differential reproductive success of phenotypes that allows persistence of corresponding genotypes through generations. Traditionally, this change is measured as a statistical difference in some population parameter(s) between separate generational cohorts of a given taxonomic group (usually population or species). The direction and extent of this change is determined by a combination of stochastic and environmental factors that are specific for a given time, space, and taxon. Therefore, the offspring of individuals or groups selected for reproduction in one biological circumstance are by no means guaranteed to thrive in the following generation since the selective environment may be slightly or radically different. An admittedly poorly defined and at times controversial evolutionary term, *adaptation* essentially refers to characters that arise and persist as a result of such selection and thus confer upon their bearers the ability to succeed genetically in a specific environment. Adaptation can also refer to the actual process of acquiring such traits. Many different levels of characters can thus be considered as adaptations: point mutations, coding region products, multigene assemblages, behavioral traits, and even populational characters may all endow their possessors with reproductive advantage and thus evolutionary persistence.

B. Measuring Fitness

Technically, the fitness of a unit under selection refers to the contribution it makes to the overall genetic makeup of the subsequent generation relative to other such units. In theory, selected units can be genes, cells, individuals, populations, and even phyla (Lewontin, 1970; Williams, 1966). Fitness is thus directly related to a unit's comparative degree of adaptedness. The concept of fitness as applied to viruses and other microparasites has been discussed at length elsewhere (May, 1995; Anderson and May, 1991; Domingo and Holland, 1997). Fitness and its measurement are central themes in any discussion of evolutionary ecology no matter the organism and thus warrant clearer elucidation.

Fitness can be examined in various ways. First, by using Fisher's (1930) "net reproductive rate" (R_0), which is merely the number of viable offspring produced per individual over their life-span (which for viruses is equal to the infection of one cell or one host, as will be discussed). R_0 is a conceptually convenient metric, especially when considering the *absolute* fitness of an organism (for instance, the potential for occupation of a host population by a specific virus) but may be misleading when interpreted as a measure of *relative* fitness since alone it provides no information about individual reproductive rate compared to the population reproductive rate over the same time-span. The competing virus with the highest R_0 can have the highest probability of long-term survival, but determining this probability for an individual would require comparison of all relevant R_0s in a population or group. Therefore, R_0 for an individual can be greater than 1 (implying "reproductive success"), but this provides no information about individual performance with respect to population growth as a whole. Therefore, an individual with $R_0 < 1$ may actually have high relative fitness under conditions of population decrease provided the individual's rate of decrease is correspondingly smaller.

R_0 also does not take into account variation in generation time and is therefore not always useful when examining fitness over absolute time. In other words, one phenotype may actually confer a lower R_0 than another yet still exhibit relatively higher reproductive output (i.e., higher fitness) if its generation time is appropriately shorter.

Because of these shortcomings in R_0, it is advantageous to use instantaneous growth rates for both the individual (ρ) and the population as a whole (r), since these are independent of generation length and thus measure reproductive output per unit (t) of absolute time. Giske and colleagues (1993) have described a particularly useful relative fitness term (Φ) that measures an individual's reproductive output relative to that of its population as the difference between ρ and r. Therefore, Φ is a measure of fitness that is both instantaneous in real time and relative. As such, any positive value of Φ indicates increasing proportional representation of a particular phenotype in a population even when $R_0 < 1$ or $\rho < 0$, provided r is sufficiently small. Relative fitness in this sense is both more accurate since it accounts for variation in generation length and perhaps more applicable since it describes an individual's reproductive performance in light of that for its population as a whole.

As will be discussed below, these concepts must be applied to viruses on at least two completely separate levels: within (intra-) hosts and between (inter-) hosts. Taking into consideration the purely endogenous nature of some agents does not influence the partitioned application of the two notions since new host inoculation (regardless of the route) is key to evolutionary persistence. The theme unifying both types of fitness, however, is competition between viral strains (taxa) for reproductive substrate. In the case of intrahost evolution, this substrate is host resources (cellular, molecular), while in the case of interhost evolution it is the actual occupation of the hosts themselves.

III. INTRAHOST EVOLUTION

Productive viral infection of a host need not be associated with pathogenicity, tissue destruction, morbidity, or mortality. It also need not result in full or partial mobilization of host immune/antiviral mechanisms. In fact, it is not uncommon to find viruses associated with hosts in mutualistic relationships (see section IV.B.6.c). A more universal expectation of host occupation is competition between different viral strains or individuals present at a given time. In this sense, selection is effectively acting on viruses as it would on nonparasitic organisms occupying some particular habitat. Intrahost evolutionary success is therefore largely independent from interhost success and subsequent evolutionary persistence. In other words, adaptive viral characters favored within the relatively closed system of one host individual arise and persist due to intrahost selection, the nature and strength of which is determined by the conditions and other virus strains contained therein. These characters may not necessarily be favored in the larger "open" arena that includes all viruses occupying a host population (discussed below).

Fundamentally, if two strains (or individuals) occupy a single host, the one with the highest net reproductive value (per-capita growth rate) will rise to the highest frequency. Assuming a positive linear association between probability of transmission and population size (viral titer), the long-term effect of this dominance may be more occupied hosts. This simplistic view of intrahost evolutionary dynamics is an unrealistic portrayal of viral life and death, however, and requires expansion.

A. The Host as an Ecosystem

Any ecological system is composed of species and individuals interacting both with each other and with abiotic conditions. Such interactions result in regulation of population sizes as individuals compete for limiting resources. This is similarly the case in viruses reproducing within a susceptible host or host tissue. Viral population sizes must obviously be regulated lest they grow to infinity, but the forces influencing such limits differ in their relative contributions between "viral host" ecosystems and species ecosystems.

Factors that influence species population change may include density-dependent interactions such as competition for resources (inter- and intraspecific) and predation, but also abiotic factors like physical environmental characteristics and probabilistic conditions. Such factors also exist in viral host ecosystems, but some have a much greater influence than others. On a smaller level, however, an individual host can be looked upon as an ecosystem: there are many biotic and abiotic components that interact in both predictable and unpredictable ways to influence the growth properties of viral populations.

B. Host Resources

Resources (i.e., reproductive substrate) required by viruses inside a host include living cells of given tissue type(s). Once an intrahost virus population has become established, the relative abundance of infectible cells is usually extremely high and, in a holistic sense, not a limiting factor. This is demonstrated by the observation that in most viral infections less than 1% of the susceptible host tissue is actually infected (Griffin, 1997). However, the ecological concept of metapopulation dynamics may have some application in this case. This notion assumes that populations are spatially structured so that they are actually aggregates of smaller, local subpopulations, and that migration between these subpopulations is significant enough to influence local dynamics (Hanski and Simberloff, 1997). While this is most commonly applied to sexually reproducing organisms and popularly utilized in topics of conservation biology, it is possible that the metapopulation concept could be useful in understanding viral intrahost evolutionary patterns.

Since cellular resource competition is in general not a primary force regulating viral population dynamics, it may have some significance at the microspatial level. For instance, very localized infection of tissue could produce a confined abundance of viral particles with limited access to mechanisms of dispersal, which, under optimal conditions, could carry them to other sources of infectible cells. This accumulation of both viral particles and non-infectible cells would result in isolated competition and subsequently produce selection for either quick and successful cell invasion or dispersal to more fertile areas. Successful dispersal may give rise to other local subpopulations and result in similar selective pressures. The concept of metapopulation dynamics has never been applied to viruses in such a context but may be worth further investigation when endeavoring to understand intrahost viral reproduction and adaptive evolution.

Resource competition may also be manifest under some circumstances of extreme viral induced cytopathicity, for instance, near the end of an infection that has resulted in high cytopathology (or even necrosis) of specific tissue required for virus replication. In such a situation, there may theoretically be a much larger population of infectious virus than available reproductive substrate can support. This gives rise to competitive conditions in which some may survive due to their superior ability to effectively occupy remaining cells and the remainder go extinct. It is conceivable that such may be the case in diseases like AIDS, in which CD4$^+$ cells required for HIV replication have been extensively "consumed," thus resulting in the reproduction of only a small portion of existing viral particles. In fact, up to 60% of CD4$^+$ lymphocytes may be infected with HIV during AIDS viremia (Bagasra et al., 1993; Hsia and Spector, 1991; Patterson et al., 1993; Schnittman et al., 1990).

C. Dynamics of Host Antiviral "Predation"

In nearly all cases, the most significant power guiding intrahost viral population dynamics is perhaps neutralization by host antiviral (immune) defenses. The relationship between viral growth and immune response has of course been studied quite extensively from a clinical standpoint, and most recently from the perspectives of evolution and population biology (Wassom, 1993; Garnett and Antia, 1994; Bonhoeffer and Nowak, 1994; Kilbourne, 1994; Bangham, 1995).

In many ways the virus-immune interaction mimics a typical predator–prey type relationship and can be understood fundamentally using the Lotka–Volterra ecological model of species interactions (Lotka, 1932; Volterra, 1926). The goal here is an interpretation of the temporal growth/decay dynamics of an intrahost virus population and cells and/or products manufactured by the host to neutralize virus infection. In the absence of any counteractive force(s), an intrahost population of viruses will grow exponentially as

$$\frac{dV}{dt} = r V \tag{1}$$

where V is the number of viruses (i.e., "prey") present at time t, and r is the per-capita rate of population growth per time unit, as discussed previously. Viruses cannot grow at this rate indefinitely, however, and population growth is diminished (in our model) at a rate proportional to the potency of host antiviral responses. Therefore, if the quantity of host antiviral components (immune cells, antiviral compounds, etc.) are collectively represented by I (i.e., "predatory influences"), which has an overall per-virus neutralization effectiveness coefficient of a (i.e., the probability of viral death on contact with these components) then the growth of virus populations becomes

$$\frac{dV}{dt} = rV - aIV \tag{2}$$

In many host systems, the presence of adaptive antiviral elements that react against specific viral types is related to the concentration of those viral particles. The population dynamics of this particular association has not been investigated from an "ecological" standpoint as thoroughly for viruses as it has for macroscopic organisms. Yet it was shown that B-cell populations in mice persist in relation to the amount of antigen present in the animal's periphery on which the B cells rely for continued stimulation, and that competition actually exists between B-cell subtypes for this limiting resource (McLean *et al.*, 1997). If such density dependence occurs for immune components as well, the "population" change therein may be described in terms of an equation similar to (2):

$$\frac{dI}{dt} = faIV - qI \tag{3}$$

where f is the rate of immune component increase (commonly due to stimulation by viral antigens), and q is the per-capita rate of immune component degradation (death). Therefore, it should be clear that $faIV$ represents the intrinsic rate of increase of the immune response and that this is proportional to V, the amount of virus present within a host.

D. Adaptation to the Intrahost Environment

The upshot of these relationships is their effect on viral fitness. An obvious way to increase population size is by modification of reproductive rates [i.e., r in Eq. (2)]. In terms of intrahost evolutionary dynamics, this rate should be maximized relative to that of other individuals (or strains). This may not translate into long-term evolutionary success due to the impact of host mortality on transmission, however (discussed below). Nevertheless, adaptations for increased individual reproductive rates are prevalent and frequently include effects on host cell molecular function. The list of such virus–host cell interactions is large and covered at length elsewhere (see chapters in Fields *et al.*, 1996a,b). Therefore, only some widespread viral techniques will be discussed here. To begin with, all viruses exhibit similar steps during their replication cycle, each of which is a likely target of optimization. These steps include host and/or cell entry, transport of viral materials (both intra- and intercellularly), transcription and translation of viral genes, genome replication, virion assembly, and virion release.

1. Cell Entry

The "first" potential virus–host cell interaction to be acted upon evolutionarily is cell surface receptor binding. Specificity of surface receptor usage is likely connected, at least to a certain extent, with the occurrence of particular molecules (proteins, carbohydrates, glycolipids) exposed on the surface of those candidate host cells that are likely to provide an optimal overall environment for virus replication. In other words, many receptor specificities may have arisen as methods of cell type identification in viruses, since viruses are continually under selective pressure to infect tissue with the highest potential for long-term virus propagation. This is an important variable in the course of viral infection since it is often the only barrier preventing replication within a given cell or tissue type.

For instance, human poliovirus receptor is found only on primate cells (McLaren *et al.*, 1959). The virus does not naturally replicate in other mammals but is fully capable of doing so in mice once its genome is artificially introduced into host cells (Holland *et al.*, 1959). More recently, Winkler and colleagues (1998) discovered a human polymorphism in the stromal-derived factor (SDF-1) gene that delays the onset of AIDS in patients homozygous for a variant allele. It is thought that the variant upregulates the expression of SDF-1, which is a ligand for the CXCR4 cell receptor. The CD4 molecule and this cell surface protein act together as coreceptors for T-tropic HIV strains that normally arise late in infection. Competitive interference from the increased presence of SDF-1 may actually be protecting susceptible cells from HIV binding and subsequent endocytosis, thereby postponing disease symptoms. Receptor specificity can thus have profound implications for the pathogenicity and evolutionary expansion of viruses into new tissue types and host ranges and will be discussed later.

Receptor specificity may also be related to the normal cellular function of the receptor molecule; that is, since the first viral replication objective is typically cellular internalization, selective pressure must exist to increase the efficiency and rapidity with which this occurs. Therefore, it would benefit the infecting virus to utilize surface receptors whose normal molecular purpose it is to transport objects across the cell membrane. Furthermore, it may be additionally advantageous to select among all the possible transport receptors those that can internalize the given virus type most effectively. In this way, selection can shape the use and specificity of the employed surface receptors to optimize infection. Along these same lines, it may be easy to see how many viruses could have adapted to employ as receptors those cell components that affect endocytotic mechanisms and conditions needed for viral entry and uncoating.

2. Viral Gene Expression and Replication

Another group of common targets for evolutionary modification includes those cellular components that enhance viral reproductive rate at the stages of transcription, translation, and replication of viral genes. A comprehensive listing of specific viral tactics used to optimize these activities would be quite large and beyond the scope of this chapter, yet some patterns definitely exist and warrant mentioning. For instance, the genomes of numerous viruses are configured so that those gene products that are translated early in infection are involved in and often direct the ensuing transcription and/or translation of late genes. Such early proteins may also act as *trans*-acting downregulators of their own expression on approach to threshold concentrations. In addition, many viruses have developed the ability to use host factors to control transcription.

3. Immunopathology as an Evolutionary Strategy

Aside from increasing the rate and efficiency of replication, Eq. (2) indicates that impairment of the host antiviral response will result in increased survival of an infecting viral population by decreasing losses caused by these components. Viruses within a host can subsequently increase their fitness in this context by decreasing their overall losses due to host antiviral responses. To this end, a number of potential mechanisms can be and are employed.

An extremely important technique used by viruses to decrease losses from the host antiviral response is impairment of the host's ability to mount an adequate viral attack. This phenomenon has perhaps no equivalent in species predator–prey ecological interactions. For example, by looking at Eq. (3) it can be seen how intensifying immune component death or degradation results in an increase in q. Likewise, interference with recognition and removal of non-self (i.e., particles either derived from a foreign non-host source or host material that is unrecognizable to the body as derived from a host gene) material will decrease a. Impeding immune component production (cell division, secretion, etc.) accordingly decreases f. All of these directly result in augmented virus survival by either paralyzing the "predators" or diminishing their population size. Fortunately for the viral evolutionist, virus-induced human immunopathology as well as immune suppression are extremely well researched topics for which there exist numerous specific examples of anti-immune strategies.

a. *Immune Cell (Tissue) Tropism.* Perhaps discussion of viral tactics of immune suppression should begin with examples of viruses that replicate within cells that are involved either directly or indirectly in the antiviral response. As mentioned, an obvious way to disable a host's antiviral capability is to destroy components of the response or the source of those components. For instance, the replication cycle of numerous viruses of vertebrate taxa requires cells of the host immune system such as macrophages, monocytes, lymphocytes, natural killer (NK) cells, antigen presenting cells (APCs), or stromal cells (see Table I and Griffin, 1997 for a review). This strategy provides two clear evolutionary benefits to such viruses: a source of viral progeny and a reduced host antiviral reaction.

TABLE I

Viruses Known to Infect Cells of the Mammalian Immune System and the Impact of that Infection on the Immune Response[a]

Host cell	Virus	Immuno-pathogenicity
B lymphocyte	Simian retrovirus	High
	Pancreatic necrosis virus	High
	Epstein–Barr virus	Moderate
	Murine leukemia virus	High
T lymphocyte	HTLV-I	Low
	Simian retrovirus	High
	Feline leukemia virus	High
	HIV	High
	SIV	High
	Feline immunodeficiency virus	High
	Mouse thymic virus	High
	Human herpes virus 6	Low
	Human herpes virus 7	Low
Monocyte	Venezuelan equine encephalitis virus	Unknown
	Rubella virus	Low
	Dengue virus	Unknown
	Lactic dehydrogenase elevating virus	Unknown
	Murine hepatitis virus	Unknown
	Lymphocytic choriomeningitis virus	Low
	Influenza virus	Low
	Sendai virus	Moderate
	Measles virus	Low
	HIV	High
	Visna-maedi virus	High
	Cytomegalovirus	Moderate
	VZV	Low

[a]After Griffin (1997).

Although potentially extreme and relatively rare, cytocidal replication within such tissues can obviously have a direct negative impact on the host's ability to destroy viral pathogens. The ability of viruses to endure host antiviral reactions is facilitated in infectious bursal disease virus of chickens and infectious pancreatic necrosis virus of trout by cytolytic infection of B lymphocytes. This lysis results in a significant loss of these cells and subsequent humoral immunosuppression due to decreased antibody production (Tate *et al.*, 1990; Saif, 1991; Muller, 1986).

More common among viruses is cytopathic infection of immune cells. Replication within the host immune cells by many viruses usually does not result in rapid and essential death of these cells, yet may impair cell function to such a degree that the antiviral response is weakened — all to the potential benefit of the resident virus population. A mammalian target immune cell of viral replication for which a great deal of research has been undertaken is the T lymphocyte. Lentiviruses are particularly known to infect T cells in humans (see Table I) as well as in other mammals, but other viruses such as some herpesviruses may also infect these cells. Such infection frequently results in abnormal proliferation of T cells. T-cell growth resulting from HTLV-1 infection, for example, is known to impair immunologic function, but the subsequent effect on viral population growth is unknown. HIV is also known to have profound effects on host immune function, mainly through its replication within $CD4^+$ T cells and monocytes, which is a likely contributor to the high viremia exhibited during the final stages of infection. In this case, the utilization of immune cells may not be as beneficial ecologically as it might appear, and for a very simple reason; that is, while the destruction of $CD4^+$ cells may promote viral replication by decreasing antiviral pressure, it simultaneously diminishes the amount of reproductive substrate available for replication (see above).

b. *Immunosuppressive Viral Factors*. A very prevalent viral technique of immunosuppression involves utilization of proteins, whether viral or cellular, to disrupt the host antiviral response. Many viruses have the ability to sabotage production or function of host cytokines using their own gene products. Such abilities of individual viruses give rise to obvious evolutionary benefits not only for that individual but also for other members of the infecting viral population. Many host molecules — including interferons, interleukins, and other cytokines — are susceptible to the action of immunosuppressive viruses (see Table II). The action of these molecules may be at the level of prevention of transcription (e.g., HBV terminal protein), or translation (e.g., adenovirus E1A), inhibition of cytokine activated enzymes (e.g., RNase-L inactivation in HSV infection), or sequestration of the molecules via complement binding (e.g., IFN-γ receptor decoy produced by myxoma virus), to name a few. Viral gene products may also interfere with other aspects of the host antiviral response such as antigen presentation (vaccinia virus, HSV) and downregulation of host cell genes required for proper

TABLE II

Viral Factors that Alter Host Immune Functions and Subsequently Result in Increased Virus Survival

Host defense	Virus(es)	Viral factor	Activity	Fitness[a]	References
Interleukins	Vaccinia/Cowpox	B15R	Binds to IL-1 & prevents normal function	Population	Spriggs et al. (1992), Alcami & Smith (1992)
	Cowpox	cmrA	Prevents IL-1β maturation	Population	Ray et al. (1992)
	EBV	BCRF-1	Blocks IL-10 synthesis	Population	Moore et al. (1993)
	Tanapox	38 kDa	Binds IL-2 and IL-5	Population	Essani et al. (1994)
Interferons	Adenoviruses	VA RNA	Block IFN-induced autophosphorylation of DAI	Individual	Matthews & Shenk (1991)
		E1A	Block IFN transcriptional signaling	Individual	Ackrill et al. (1991), Kalvakolanu et al. (1991), Gutch & Reich (1991)
	EBV	EBER RNA	DAI function inhibition	Individual	Bhat & Thimmappaya (1983)
		IL-10 homologue	Downregulates IFNγ synthesis	Individual	Moore et al. (1990)
	HIV-1	Tat	Decreases PKR expression	Individual	Roy et al. (1990)
	Reoviruses	σ3	Blocks activation of IFN-induced PKR	Individual	Imani & Jacobs (1988), Lloyd & Shatkin (1992)
	HBV	Terminal protein	Blocks IFN transcriptional signaling	Population	Foster et al. (1991)
	HSV-1	2-5(A) analogues	Blocks activation of IFN-induced PKR	Individual	Katze (1992), Sen & Ransohoff (1993)
	Myxoma	M-T7	Binds to and neutralizes IFNγ	Population	Upton et al. (1992)
	Influenza	Cellular P58	Inhibits IFN-induced PKR via cell protein activation	Individual	Lee et al. (1990)
	Vaccinia	E3L	Blocks activation of IFN induced PKR	Individual	Chang et al. (1992)
		K3L	PKR decoy substrate	Individual	Davies et al. (1993)
Complement components	Vaccinia	C21L	Blocks alternative & classical complement activation	Population	Kotwal et al. (1990), Isaacs et al. (1992)
	Herpesvirus saimiri	ORF 4	Blocks alternative & classical complement activation	Population	Albrecht & Fleckenstein (1992)
		ORF 15	Inhibits membrane attack complex	Individual	Albrecht et al. (1992)
	Herpes simplex	Glycoprotein C	Blocks alternative & classical complement activation	Individual	Fries et al. (1986), McNearney et al. (1987), Harris et al. (1990)
	EBV	Whole virion	Inhibits alternative pathway function	Population	Mold et al. (1988)

continued

	Virus	Gene	Function	Fitness[a]	References
TNF	Shope fibroma virus	T2	Binds to TNF and inhibits function	Population	Smith et al. (1991), Upton et al. (1991), Pickup et al. (1993)
	Adenoviruses	E1B	Increase resistance of cells to TNF-induced lysis & apoptosis	Individual	Gooding et al. (1991), White et al. (1992)
		E3	Increase resistance of cells to TNF-induced lysis	Individual	Gooding et al. (1988)
MHC-I	Adenoviruses	E3 (19 kDa gp)	Sequesters MHC proteins within ER	Individual	Anderson et al. (1987), Pääbo et al. (1987), Rawle et al. (1989)
	Adenovirus 12	E1A	Inhibits MHC-I transcription	Individual	Schrier et al. (1983), Ackrill & Blair (1988), Ge et al. (1992)
	MCMV	Early protein(s)	Prevents peptide-bound MHC-I transport to surface	Individual	del Val et al. (1992)
Other	HCMV	US28	Binds to and sequesters C—C chemokines	Population	Neote (1993)
	Myxoma	T2	Binds to & sequesters serpins	Population	Lomas et al. (1993), Upton et al. (1990), Macen et al. (1993)
	Poxviruses	EGF-like proteins	Stimulates regional cell growth to enhance progeny virus replication	Population	Brown et al. (1985), Chang et al. (1987)
	MMTV	LTR ORF	Stimulates T- & B-cell proliferation to enhance progeny virus replication	Population	Choi et al. (1991), Korman et al. (1992), Held et al. (1993), Golovkina et al. (1992)
	EBV	LMP-1, LMP-2	B-cell activation & latent-to-lytic infection switch	Population	Miller et al. (1993), Burkhardt et al. (1992)

[a]"Fitness" refers to whether the factor contributes to viral population or individual viral fitness.

antiviral function (e.g., MHC I and II, ICAM, LFA-3). This subject is discussed in greater detail as an evolutionary topic below.

   c. *Clonal Deletion and Immunologic Invisibility.*   Viruses may avoid contact with a functional immune response by becoming antigenically "invisible." For instance, some viruses that utilize vertical transmission do not elicit a specific antiviral response at all due to the inability of the immune system to recognize them as foreign. One such strategy is well exemplified by lymphocytic choriomeningitis virus (LCMV) infection in mice. In this system, viruses that infect adult mice are effectively cleared from the host principally by way of an LCMV-specific cytotoxic T-lymphocyte (CTL) response (Zinkernagel and Welsh, 1976; Byrne and Oldstone, 1984; Moskophidisy *et al.*, 1987; Ahmed *et al.*, 1988; Matloubian *et al.*, 1994). Mice that are infected *in utero* or perinatally, however, retain high lifelong titers of infectious LCMV throughout their bodies. This persistence is made possible by the lack of a specific CTL-mediated antiviral response due to clonal deletion of LCMV-specific T cells during mouse ontogeny (Buchmeier *et al.*, 1980; Ahmed *et al.*, 1984; Moskophidis *et al.*, 1987; Pircher *et al.*, 1989; Jamieson *et al.*, 1991). The early and pervasive presence of LCMV particles (antigen) results in immunologic "acceptance" of the virus by the host yet still allows for a completely functional immune response against other pathogens and their antigens.

4. Survival through Evasion of Antiviral Mechanisms

   a. *Spatial Evasion.*   One certain way to avoid destruction by an intact immune system is to avoid or limit direct interaction with its components altogether. This can be accomplished in a number of ways, among which is viral persistence, including the possibility of replication within immunologically privileged tissue that is inaccessible to the host antiviral response. For instance, many types of herpesviruses (varicella zoster, herpes simplex) lie latent in nonreplicating sensory neurons that are poorly penetrated by the immune system. These viruses may periodically move through axons to epithelial or ophthalmic sites for replication (where they may eventually be suppressed by the immune response) but never actually leave the host permanently. Some viruses will replicate within tissues that experience little or no exposure to antiviral host defenses. Human polyomaviruses BK and JC as well as cytomegalovirus benefit from limited T-cell effectiveness within kidney epithelial tissue by replicating there and thus avoid destruction quite efficiently (Ahmed *et al.*, 1997).

   b. *Transmission and Neutralization Thresholds: Temporal "Evasion."*   Other viruses may evade immune contact by completing sufficient replication and transmission before a significant host antiviral attack can be fully executed. Many viruses that replicate to high concentrations quickly employ this technique. Such viruses operate under conditions of three relevant thresholds: (1) the intrahost virus population size that triggers an immune response; (2) the intrahost population size that allows for interhost transmission; (3) the amount of time it takes an antiviral reaction to decrease the viral concentration below the transmission threshold.

Figure 1 provides a simple illustration of intrahost population growth for an imaginary acutely infecting virus. Under this model, a virus can only survive evolutionarily in a host population given the following specific conditions. The threshold viral concentration for effective interhost transmission (B) must be exceeded for sufficient time as to allow for occupation of a new susceptible host. This time will vary inversely with the quantity of infectious virus being shed by the host. A virus must therefore employ reproductive strategies that can include maximizing the amount of time it is being shed in adequate numbers (E–D) and/or minimizing B (or, conversely, maximizing A, the threshold concentration triggering a host antiviral response). This appears to be analogous to the strategies used by many well-examined viruses such as influenza, measles, rhinoviruses, and VZV, to name just a few.

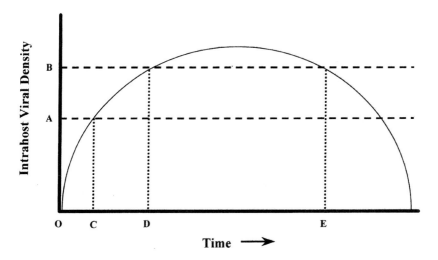

A = Threshold of host antiviral defense activation
B = Threshold of effective (or probable) interhost virus transmission
C = Time of delay between inoculation and activity of the triggered antiviral response
D = Time of delay between inoculation and attainment of probable transmission threshold
E - D = Probable transmission time

Fig. 1. Model of the interplay between intrahost population density, host antiviral response time, and interhost transmission for an imaginary acutely infecting virus (see text).

Under the conditions of this model a virus may also exhibit A > B, in which case an antiviral response is (effectively) never activated and the virus will be transmitted satisfactorily as long as its concentration remains above B. This latter strategy might only be evolutionarily stable for viruses of limited pathogenicity since host mortality usually limits sustainable viral spread. Also, subpopulations within a given host might differ in the shape of their respective curves and values for A, B, C, D, and E.

c. *Immunologic Cat and Mouse*.   Yet another adaptation used to evade host antiviral defenses that has arisen in many viruses with RNA genomes is rapid and continual production of antigenic variation. High mutation rates in some viruses (HIV, FMDV) as well as reassortment of genomic segments of others (influenza virus) give rise to a quasi-species (also referred to as a mutant or variant swarm) population structure within an infected host and/or host population (Holland *et al.*, 1982; Both *et al.*, 1983; Mateu *et al.*, 1989; Bachrach, 1968; Nowak *et al.*, 1990, 1995; Wolinsky *et al.*, 1996). The result is genomic variation that is manifest as phenotypic diversity in the form of multiple antigenic types present in a single host at one time (see Fig. 2). This heterogeneity may cause high overall mortality rates for the virus, but at the same time it provides assurance that at least a small

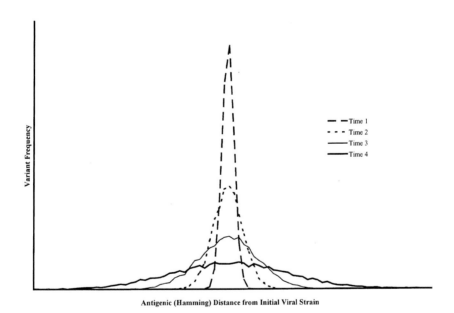

Fig. 2.   Simulated quasi-species neutral diversification. Antigenic divergence (measured accurately as Hamming distance) between virus strains increases with time.

proportion of variants will avoid host neutralization due to the inability of the specific antiviral response to act both effectively and simultaneously on every antigenic type that exists or arises. Domingo and Holland (1997) reviewed the fitness and evolution of RNA viruses in relation to their mutability quite thoroughly.

This adaptation has perhaps no duplicate in species predator–prey evolutionary ecology. It may, however, be thought of in relation to negative frequency-dependent selection, a condition that does exist in many species predator–prey interactions. In this case, rarer phenotypes (e.g., antigenic variants) are favored as long as they remain relatively uncommon. Once a variant rises in frequency, the probability that it will be recognized by a predator species (or host immune system) increases accordingly, and thus its growth rate will be reduced, usually to the advantage of (immunologically) less conspicuous mutants. It is in this way that the virus evades contact with antiviral components and persists within a host as a heterogeneous population. It appears to be a unique strategy, however, to avoid predation in that the adaptation is indeed the process of mutability (the virus damages itself rather than the host tissue!).

In fact, viruses such as coliphage Qβ FMDV, VSV, and HIV have such high mutation rates that they are near the point of error threshold; that is to say, slight increases in their mutation rates would result in failure of sufficiently accurate (stable) genomic perpetuation and subsequent extinction of the viral strains (Nowak and Schuster, 1989; Eigen and Biebricher, 1988; Holland *et al.*, 1990). The biological nonviability of mutation rates beyond the point of error threshold has even been demonstrated experimentally for VSV and poliovirus (Holland *et al.*, 1990).

5. Intrahost Population Fitness

Concepts such as quasi-species and population fitness evoke notions of group selection, a somewhat controversial topic among evolutionists. Group selection favors those characters that promote the survival of populations (or taxonomic levels higher than that of the individual) but either decrease or do not affect individual fitness. The actual contributions of group versus individual selection to the evolutionary process are difficult to discern, especially with respect to situations in which the two appear to act in opposite directions. However, when considering viruses and other microparasites, this idea may be helpful if not necessary to explain the evolutionary maintenance of certain traits.

Quasi-species actually provide very good examples of group selection in action. Under these circumstances, selection on the individual is weak when compared to that acting on the entire spectrum of mutants since the representation of a given genotype is determined by a combination of: (1) its own relative replication rate and (2) the frequency with which it arises from the mutation of other parental forms. In a mathematical sense, selection is thereby both acting on and shaping the largest eigenvalue of an immense mutation transition matrix that represents the mean replication rate of the quasi-species (after Nowak, 1992).

The influence of group selection on quasi-species evolution has been examined experimentally by de la Torre and Holland (1990). In their study, small numbers of highly fit VSV mutants were introduced into large quasi-species populations of lower relative fitness. It was found that these mutants only rose to dominance when they were planted above threshold quantities; otherwise, they were displaced by the original population. Such evolutionary dynamics may appear to contradict the classic neo-Darwinian doctrines regarding competition and survival. However, it is a reality of viral existence that the reproductive success of the individual oftentimes is determined by the characteristics of the population in which it exists.

The previous discussion focused on adaptive strategies used by individual viruses, or strains as the case may be, to compete with other such units within a particular host. It should be evident that many such traits may, both singly and in combination, exclusively improve the fitness of the virus(es) in possession of those traits. Viral phenotypes that affect individual host cells or cell products on an intracellular level usually contribute to the fitness of the bearer. These possibilities for fitness enhancement include most strategies of individual replication improvement such as receptor specificity, use of transcription factors, and enhancers. Such possibilities also include many techniques of localized host antiviral response impairment, such as disruption of antigen processing or presentation, intracellular IFN signaling, and diminished antiviral recognition.

There are, however, other types of individual viral characteristics that appear to confer survival advantage to the viral population as a whole, typically by having more widespread effects on the host organism. As an example, many viral gene products that interfere with host antiviral response stimulation act as decoys for immune product binding or sequester such products. These tend to have extracellular site-of-action effects.

As elegantly demonstrated by Bonhoeffer and Nowak (1994), the consequence of these intracellular/extracellular trait relationships is maintenance of such populational phenotypes when they are associated with increased transmission probabilities (and thus long-term evolutionary success). But this maintenance can only exist under conditions in which mutation rates are below a threshold that is determined to a large extent by relative transmission advantages (i.e., interhost selection) of the new mutant versus its ancestral population. The authors show that at the intrahost level mutants defective for the "unselfish" character are selectively neutral and thus will drift to fixation at a rate approximately equal to the rate of production of such mutants. These authors point out that such strategies are more likely to *arise* under unusual stochastic circumstances such as those of transmission founder effects or intrahost genetic drift yet to *persist* in viral types with low mutation rates (such as DNA viruses).

When considering competition between viruses within an infected host, it is crucial not to forget that the one and only sought-after evolutionary reward is transmission to a new susceptible host. In addition, host replication rates (generation times) are usually much longer than those for the individual viruses infecting them. Because of this, infecting viruses are capable of rapid comparative rates of adaptation and evolution. Transmission probability is commonly assumed to be positively (although by no means linearly) correlated with within-host virus concentration (as well as infected host life-span, which will be discussed below). As such, with respect to intrahost viral competition it has been demonstrated that individual viral fitness is a composite of both reproductive rate and ability to withstand (or evade) host antiviral defenses. Accordingly, viruses increase their fitness by increasing the ratio between reproductive rate and antiviral avoidance (i.e., r/a from Eq. (2)) relative to other viruses against which they compete (Bonhoeffer and Nowak, 1994).

6. The Spatiotemporal Transmission Window

Although rarely if ever discussed, transmission probability also involves a substantial spatiotemporal component. In other words, for a functional virus to be transmitted to a new host (i.e., *from* the old host), it must be present in the right place at the right time. Clearly, there is a strong stochastic element at work here; however, differential selection of specific tissue types and alterations in timing of progeny release may have significant ramifications in terms of the fitness of viral strains. It is in this way that viruses may enhance their evolutionary success without specifically changing their population sizes or growth rates and thereby not causing excess host damage.

Long-distance intrahost movement in plant viruses is an essential component of reproduction. For example, viruses often need to be shed on multiple plant surfaces that may be spatially diffuse in order to facilitate vector-mediated transmission. Intrahost transport is perhaps most "extreme" in systemic plant viruses since distances can be very large (e.g., as in coniferous trees), and this subject has been investigated (for reviews, see Carrington et al., 1996; Deom et al., 1992; Maule, 1991; Mushegian and Koonin, 1993). Long-distance movement in plant viruses usually takes place in the sieve elements and cells of the phloem. Plant viruses are known to utilize movement proteins (MPs) that facilitate both cell-to-cell and long-distance transport. MPs have been identified in many plant viruses, but the biochemical and molecular mechanisms involved unfortunately are largely unknown. MPs are frequently capsid proteins, as has been demonstrated in tobacco mosaic virus (Dawson et al., 1988; Saito et al., 1990), red clover necrotic mosaic dianthovirus (Xiong et al., 1993; Vaewhongs and Lommel, 1995), cucumber mosaic virus (CMV) (Taliansky and Garcia-Arenal, 1995), and tobacco etch potyvirus (Dolja et al., 1994, 1995), among others. Replication proteins such as those involved in genome duplication have also been identified as necessary for long-distance movement in brome mosaic virus (Traynor *et al.*, 1991) and CMV (Gal-On *et al.*, 1994).

Much less research has focused on intrahost replication of virus that is temporally determined. It is conceivable that a virus type might maximize its intrahost replication only within a very specific time-frame in order to maximize the probability of transmission. For example, this may be pertinent with respect to solitary animal species that congregate only at very specific times of the year for breeding. Under such circumstances, natural selection might favor a reproductive strategy in which a resident virus population exists at low, persistent concentrations during most of the year but "turns on" high-level replication in order to be shed during a time of year when transmission probability is consistently (and predictably) highest. There are numerous avian and mammalian hormones, for instance, whose release is dependent on photoperiod that could realistically activate or suppress viral replication.

Furthermore, high replication might even be coordinated with the onset of virus-induced disease symptoms in infected individuals that might be associated with or promote host–host contact and thus viral transmission. These examples are of course merely conjectural but may warrant further investigation since such temporal coordination of replication is not only biologically realistic but also evolutionarily intriguing.

Of relevance here is polyomavirus infection in mice. McCance and Mims (1979) report that virus that infects female mice at birth is only detectable at high levels late in gestation when it is transmitted to newborns and persists at low levels otherwise. In addition, virus replication was also activated in nonpregnant females injected with sex hormones related to pregnancy, implicating them as the stimulators of viral reproduction. This appears to be a viral replication strategy that is timed to coincide with a high-probability transmission window.

E. Summary: Importance of the Host as an Agent of Selection

In summary, the host cellular, tissue, and organismal environments are vitally important selective realms that contribute profoundly to the adaptation and diversity of viruses. Intrahost selective pressure may be compared with that within a species ecosystem. In an overall sense, resources require optimal, not maximal, consumption. On an individual level, negative interspecific interactions (predation) should be minimized and reproductive rates high. Just as with cellular species, viral traits that may arise or be selected for in one particular environmental scenario (time and place) do not guarantee indefinite evolutionary stability. Selective pressure within a particular host may similarly be contradictory to selective pressure imposed by requirements of interhost long-term survival. For instance, as with cellular species, where ecological continuance will not occur under conditions of resource overexploitation, virus populations will not persist evolutionarily when host resources are consumed beyond the point of sustainability. The effect is a balance between adaptations that allows sufficient harvest of host tissue without jeopardizing prospective abundance of the host as a source of reproductive substrate.

IV. INTERHOST EVOLUTION

While the amount of evolutionary and ecological research focusing on the intrahost selective environment may be somewhat sparse, this is certainly not the case when considering viral interhost evolutionary dynamics. Chiefly because of its close association with infectious disease epidemiology as well as an increasing sophistication of ecological theory in the 1970s, interhost population biology of viruses and other microparasites has been the focus of substantial investigation throughout the past three decades (Burnet and White, 1972; May and Anderson, 1983; Anderson and May, 1991; Nowak, 1991; Ewald, 1994a). It is difficult to make generalizations about the reproductive problems faced by viruses whether inside or between hosts, but something that we can perhaps expect about interhost (especially horizontal) transmission is relative unpredictability in the external environment, especially with respect to the host population. It is important to emphasize here that intra- and interhost selective pressures differ considerably in character. The intrahost environment is more or less a closed system somewhat analogous to a natural selective "microcosm," except that the intrahost conditions are comparatively less heterogeneous temporally and spatially than the interhost "macrocosm."

Yet, while these two selective environments (intra- and interhost) may differ substantially in their properties, the viral evolutionary responses that either generates carry implications for fitness in both venues. For example, this may be no more apparent than in an examination of virulence. Virus consumption of host resources (an attribute determined primarily by intrahost adaptations) frequently results in pathogenesis and may lead to host death. This may cause decreased transmission possibilities and thus affect interhost fitness. Likewise, adaptations that arise as a result of interhost selection may influence a virus's intrahost fitness. The challenge is to maximize fitness in both domains through a balance of adaptations whose effects may not be as favorable in their evolutionarily "unintended" selective environments.

A. Interhost Fitness

Interhost fitness is measured by the relative ability of viruses within a single host to successfully occupy new hosts. Examination and comprehension of this is best accomplished mathematically using simplified models of population change. It is first useful to partition the total host population into the number of individuals who are susceptible to viral infection (X), those who are infected and infectious (Y), and those recovered from infection that are permanently immune to subsequent infection (Z). We can show how these respective subpopulations change using simple differential equations (after Anderson and May, 1991; Garnett and Antia, 1994; Bulmer, 1994) that resemble Eqs. (2) and (3):

$$\frac{dX}{dt} = \mu N - (\mu_X + \beta Y)\, X \tag{4}$$

$$\frac{dY}{dt} = \beta XY - (\mu_Y + \sigma + \alpha)\, Y \tag{5}$$

$$\frac{dZ}{dt} = \sigma_Y - (\mu_Z)\, Z \tag{6}$$

where N is the total host population size (i.e., $X + Y + Z$), β is the coefficient of virus transmission (i.e., probability of new host occupation upon contact between a susceptible and infectious host), μ is the per-capita virus-independent host birth and mortality rate (this is partitioned into susceptible host birth/mortality rate μ_X, infected host birth/mortality rate μ_Y, and recovered host birth/mortality rate μ_Z, so that $\mu = \mu_X + \mu_Y + \mu_Z$), σ is the rate of host recovery from infection (i.e., 1/average duration of infection), and parameter α is the per-capita virus-induced mortality rate (virulence). Here we are assuming population equilibrium and therefore equality between birth and mortality rates. The overall dynamic can be visualized in Figure 3.

Individual interhost fitness for viruses can be measured as intrinsic fitness using Fisher's "net reproductive rate" (R_0), which has been adapted by Anderson and May (1982) for microparasites as

$$R_0 = \frac{\beta XY}{\alpha + \mu + \sigma} \tag{7}$$

If $R_0 < 1$, a virus cannot persist in a totally susceptible host population; alternatively, if $R_0 > 1$, all such host individuals will eventually become infected. Likewise, these relationships give rise to a threshold susceptible host population

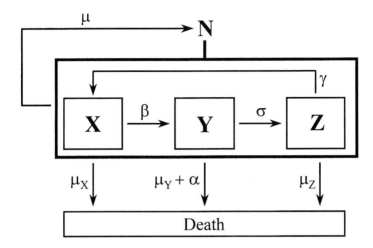

Fig. 3.   Graphic illustration of virus movement through a susceptible host population. Here host population equilibrium is assumed such that birth rate equals death rate and $\mu_X + \mu_Y + \mu_Z = \mu$. Other variables are as described in the text from Eqs. (4) to (6).

size required for survival of a virus given the specific nature of its virulence, transmission, and immune clearance characteristics as

$$X > \frac{\alpha + \mu + \sigma}{\beta} \tag{8}$$

An infecting virus population with $R_0 > 1$ will thus grow in size while Eq. (8) holds true, and if the susceptible host population falls below this threshold the viral population size will decrease.

Interhost R_0 instantaneously measures the absolute fitness of an infecting viral population, that is, regardless of variation in generation time length. However, by itself R_0 is not valuable as an estimate of relative fitness, as it provides no information about a strain's reproductive output in comparison to that of strains infecting other members of the host population. Therefore, it is helpful to use a term that defines the instantaneous reproductive rate of a virus population within a host population; in other words, the instantaneous rate of spread of a viral infection through a host population, represented by Eq. (5). For an infecting virus strain we now can examine its relative fitness using one value by finding the difference between R_0 and dY/dt since they are both measured instantaneously (as with Φ described above).

Although a highly simplified model that admittedly lacks certain biological realities, Eqs. (4)–(6) effectively demonstrate common interhost selective pressures and represent a good starting point for discussing interhost viral evolutionary ecology. Merely for convenience, we describe here a continuous-time model. It is, however, certainly possible (and perhaps desirable) to employ discrete-time models based on virus, host, and/or vector generation times, days, seasons, or other temporal intervals relevant to the specific ecology of the infection cycle.

Although perhaps not necessary for illustrative purposes, it would also be possible to increase this model's accuracy (and its subsequent complexity) by the incorporation of additional variables representing host and viral population growth and other factors. For instance, a rate parameter for individual loss of immunity may be included (i.e., γ as in Fig. 3) that serves to change the dynamic by enlarging the presence of susceptible individuals (and consequent transmission opportunities). Also, intraspecific competition (density dependence) may be included in growth of both virus and host populations; that is to say, host population sizes are regulated by extrinsic variables that are independent of viral infections yet dependent on the relative presence of other conspecific individuals. Therefore, population growth will in theory slow asymptotically as population size reaches the species carrying capacity. Other factors relevant to the dynamics of viral infection include population structure and spatiotemporal distribution of the potential hosts, environmental heterogeneity, and genetic makeup of the host population (for relevant and insightful reviews, see Anderson, 1991, and Frank, 1996, 1997).

Equation (7), with the help of Figure 3, outlines the prime adaptive pathways available for viral interhost fitness enhancement. That is, favorable adaptations can be ones that increase host–host transmission (β), decrease virulence (α), or decrease host recovery (σ). It also illustrates a complicated interrelationship of fitness characters. Fitness can be improved via effects on a number of attributes, and the magnitude of the improvement is directly related to the status of other selective conditions and adaptations (as well as ecological conditions). The effects of such adaptive changes also depend critically on the type and degree of interdependence of these factors.

B. Viral Virulence

1. Virulence and Viral Intrahost Density

The relationship between viral virulence and transmission is one that has been investigated extensively. At first glance, it might seem unusual that all viruses are not avirulent since relative fitness appears to increase as α approaches zero. It may also seem intuitive that harm to the host should obviously be minimized since viruses are obligate parasites requiring living tissue for propagation. However, the reality of virus–host biology is that transmission is generally positively and host recovery negatively related to virus density, and as virus density increases so too does pathogenicity/mortality.

Pathogenic effects have numerous etiologies, most of which ostensibly fall under the dominion of intrahost characters (cytolytic effects, viral protein toxicity, tissue tropism, immune responses, etc.), but these may have numerous implications for interhost fitness as well. In almost all models of parasite ecology it is assumed that, while transmission coefficients increase as intrahost parasite (virus) density increases, this relationship is rarely linear. However, host mortality (or morbidity) will occur near some threshold virus density, at which point the de facto transmission of intrahost viruses is zero. The effect is a balancing of pathogenic effects and overall transmission that is closely related to conditions in the host population.

2. Virulence and Transmission Opportunities (Host Density)

For many years an accepted presumption was that high virulence indicated the evolutionary youth of a parasite–host relationship, because it was assumed that not enough time had elapsed for such parasites to acquire abilities of benign host occupation. An abundance of theoretical (and, to a lesser degree, empirical) work has since shown how virulence can be selected for in virus populations. A particularly important model of virulence is related to host population size and the corresponding prospects for viral transmission.

It has been observed that highly virulent (>99% fatality) myxomavirus evolved to stable, lower (albeit still very high) levels of virulence (50–70% fatality) following introduction into a completely susceptible rabbit population in Australia (Fenner and Ratcliffe, 1965; Anderson and May, 1982). Extensive virus-induced diminishment of the host population resulted in evolutionarily stable emergence of a less virulent myxomavirus strain (see Fig. 4). In this instance, myxomavirus strains of high virulence killed their hosts too soon after inoculation and as a result were transmitted to new hosts infrequently since transmission occurs via a mosquito vector that only feeds on live rabbits. Strains of very low virulence also exhibit an interhost fitness disadvantage because they are usually cleared by host antiviral defenses before their populations reached sizes necessary for mounting of effective transmission stages (B in Fig. 1). From this example the influence of host population structure can be seen. That is, levels of virulence are often determined by (and proportional to) the frequency with which interhost transmission opportunities arise, so that low virulence can typically be selected for when host–host contact is infrequent and vice versa (Bull, 1994). Myxomavirus virulence is discussed in more detail below.

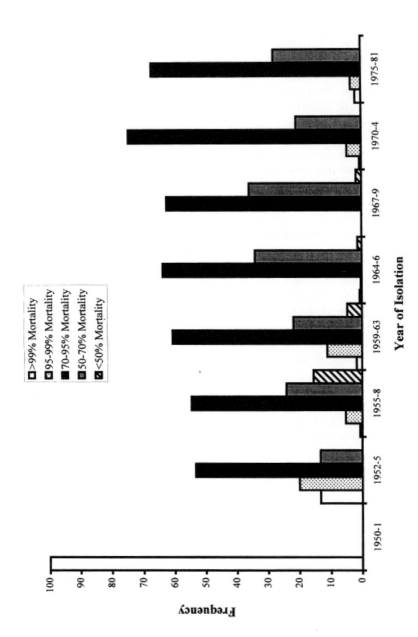

Fig. 4. Decreasing virulence with increasing time after introduction of myxomavirus into Australian rabbit populations (data taken from Fenner, 1983).

3. Virulence and Transmission Pathway

Levels of viral virulence are also often associated with transmission pathway (Fine, 1975; Herre, 1993; Lipsitch *et al.*, 1996). It is well accepted that exclusive vertical transmission (from parents to progeny through the germ cell line) of viruses will strongly select against parasite-induced host injury. This is intuitive, since a reduction in host fertility of any kind will mean a reduction in virus fertility. This can be seen in many plant viruses that replicate to very high levels with little pathogenic effects (Cooper, 1995). In addition, this opens the door for mutualistic relationships between virus and host.

Exceptions to this rule have been hypothesized to exist in cases where strictly endogenous viruses (EVs) confer on the host resistance to similar, more pathogenic exogenous viruses or those in which sex ratio distortion is used to counter the negative effects on host fitness (Hurst, 1993; Weiss, 1993; Lipsitch *et al.*, 1996; Löwer *et al.*, 1996). More frequently disputed, however, is the existence in viruses and other parasites that rely on both horizontal and vertical transmission of a virulence continuum that is established by relative dependence on both types of transmission (Ewald, 1987, 1994a). Evidence does exist in support of such a continuum in nematode parasites of fig wasps (Herre, 1993) and mosquito microsporidian parasites (Ewald and Schubert, 1989). But this has never been demonstrated in viruses, despite the existence of numerous viral groups that rely on both modes of transmission (e.g., roseplant, citrus, HIV, HTLV). Furthermore, Lipsitch *et al.* (1996) have constructed mathematical models that demonstrate how increased opportunities for horizontal transmission can actually select for decreased virulence at virus–host equilibrium by increasing viral prevalence and shifting a greater fraction of actual transmission events to the vertical route.

4. Virulence as an Adaptation

The preceding discussion of virulence neglects to mention virulence as an adaptation that has actually arisen or persisted due to specific selective pressure. For example, it has been hypothesized that virulence represents a way to increase vector-mediated transmission by disabling a normally mobile host, thereby allowing easier vector feeding access (Ewald, 1983, 1994a).

Arboviruses such as dengue virus (togaviridae) and yellow fever virus (flaviviridae) rely on mosquito vectors for transmission and cause fever, muscular aches, and/or encephalitis, which typically result in host inactivity. The host's virally induced sedentary state may facilitate mosquito vector feeding and therefore interhost transmission. In this case, viral virulence may actually enlarge the transmission coefficient (β) and as such increase viral fitness. So far, however, large-scale empirical investigation of such adaptive virulence has not been undertaken.

Additionally, it is likely that many specific symptoms of viral infection that have negative effects on host fitness but which increase transmission are viral adaptations maintained by natural selection. For instance, such symptoms as coughing and sneezing caused by viruses infecting respiratory regions (adenoviridae, rhinoviruses, coronaviridae, paramyxoviridae, orthomyxoviridae), diarrhea caused by enteric viruses (rotaviruses, caliciviridae, astroviruses), and lesions caused by many sexually transmitted viruses, as well as some poxviruses, all have clear adaptive significance when considering interhost transmission routes since they shed infectious particles into a suitable spatial (and often temporal) configuration for new host invasion.

5. Virulence as a Byproduct of "Abnormal" Infection

In many cases, virulence may be the result of viral reproduction within atypical hosts or host tissue. In such cases, the virus is not utilizing cell types to which its replication characters have naturally adapted. Usually, the result is an evolutionary dead-end because nonproductive infection, premature host death, or improper viral shedding all preclude sufficient interhost transmission and persistence.

Many human hemorrhagic fever viruses — including some hantaviruses, Machupo virus, Sin Nombré virus, and Ebola virus, to name just a few — are thought to use nonhuman mammalian or vertebrate hosts for their natural maintenance, often as avirulent vertically transmitted occupants (Childs and Peters, 1993; Peters and LeDuc, 1995; Peters, 1997). When these viruses occasionally infect humans (and presumably other mammal species), the evolutionary response in the virus, if any, must be minuscule since viral progeny produced in these nonnatural hosts are rarely transmitted effectively within the new host population.

It is interesting and also important to note that the myxoma viral strains introduced into Australia were actually of South American origin and that the rabbit pests (*Oryctolagus cuniculus*) on which they were used were of European origin. The virus actually exhibits very low (almost "silent") pathogenicity in its natural hosts (New World *Sylvilagus* species) (Regnery and Miller, 1972); yet, as mentioned, it was extremely lethal to the virgin rabbit populations in Australia. This is an excellent example of increased viral pathogenicity in relation to reproduction in taxa to which a virus has not evolved.

Similarly, during a normal host infection cycle, many viruses may be exposed to other tissue types that support replication or at least viral entry. Such encounters might result in irregular, reduced, or no viral progeny production and as such could have catastrophic overall fitness effects for the virus. This is often the case for a number of viruses — such as poliovirus, HIV, and reoviruses — which can, upon proper exposure, infect cells of the central nervous system and exhibit a high degree of virulence yet acquire no epidemiological persistence.

It must be stated that incidences of "rogue" replication such as these are important evolutionary mechanisms that rarely but significantly may lead to the genesis of new stable virus–host relationships (and will be discussed in detail below). In a general sense, however, this type of virulence may only be relevant to the host population, whose existence may be affected by losses from perhaps frequent zoonotic encounters (see Morse, 1997). The viral strains that replicate under these circumstances are almost always doomed, yet this circumstance may occasionally (but essentially) lead to a successful host taxon jump or even adaptive radiation.

6. (Unrefined) Natural Illustrations

Interpretation of our interhost relative fitness discussion and model may be "nurtured" using real examples of virus–host relationships that exhibit divergent life-history strategies. We have selected three well-researched virus–host model systems, each of which appears to be evolutionarily well established. Measles virus (MeV) infection of human populations has been examined from many biological perspectives, ranging from the molecular to the epidemiological, throughout this century. Human T-cell leukemia virus type 1 (HTLV-1) is a persistent virus that infects immune cells and may or may not cause disease in the host. Polydnavirus (PDV) is a mutualistic virus of parasitoid wasps that is actually required for host reproduction.

a. **Measles Virus.**   MeV represents an infectious agent that is very transmissible horizontally, exists stably in human populations of suitable size, often causes a decrease in host fitness, and is usually fully cleared from the host with lifelong immunity. MeV is very effective at occupation of new hosts because it is transmitted as an aerosol and thus does not require direct host–host contact, is not inactivated upon drying, and merely requires susceptible host inhalation for establishment of a new infection. Because of this, MeV has a very high transmission rate (β) in fully susceptible populations, which subsequently results in potentially explosive reproduction (R_0). This is exemplified by the 1951 MeV outbreak in southwest Greenland (Christensen et al., 1953). In this instance, the extraordinary transmissibility of MeV is demonstrated in the fact that one infected host gave rise to 250 new infections (i.e., $R_0 = 250$) following exposure to many individuals during a large social congregation. In the end, 98.5% of the local population of 4320 people were infected.

MeV-induced mortality, used here as a measure of host fitness decrease, during "virgin" host epidemics (i.e., epidemics in host populations that lack previous exposure to the virus) can be as high as 0.5 per infected individual (Black et al., 1971; Cliff and Haggett, 1985). This is thought to be positively related to size of inoculum (Aaby, 1988), a breakdown in nursing care during epidemics (Neel et *al.*, 1970), malnutrition (Aaby, 1988; Scrimshaw *et al.*, 1966; Chen *et al.*, 1980), and early or late age at infection (Carrada Bravo and Velazquez Diaz, 1980; Aaby, 1988; Panum, 1940).

In modern-day immunologically experienced populations, MeV mortality per infected individual rarely exceeds 10%, and it is typically far below 1% in developed nations (Babbitt *et al.*, 1963; Black, 1991). However, infants are at an elevated risk for MeV-induced mortality, which significantly affects host population growth since these individuals will obviously not reproduce before death. Therefore, MeV clearly has a negative effect on host fitness, which subsequently decreases future transmission opportunities. In addition, MeV results in lifelong immunity that additionally decreases transmission potential.

The interaction among MeV's transmission dynamics, virulence, and immune response causes the virus to be maintained with widely fluctuating ("boom and bust") population sizes through time. At the heart of this dynamic is the influence of herd immunity, host population sizes, host population structure, dispersal mechanisms, and seasonality, as well as many other extrinsic variables on MeV population change. It is estimated that host population sizes of less than 250,000 are incapable of permanently sustaining MeV transmission and existence (Bartlett, 1957). MeV's potentially high virulence and generation of lifelong immunity both work to decrease its transmission potential and evolutionary fitness, yet this is counterbalanced by its extreme infectivity, which ensures that under the proper conditions MeV will thrive. For reviews on measles epidemiology, see Garenne (1994) and Black (1991).

b. *Human T-Cell Leukemia Virus Type 1.*   HTLV-1 is a persistent virus that infects CD4[+] lymphocytes and is capable of transforming infected cells after approximately 10–40 years of host occupation, thereby giving rise to adult T-cell leukemia/lymphoma (ATL). Another HTLV-1 disease, known as tropical spastic paraparesis (TSP) (also called HTLV-1-associated myelopathy), results from progressive demyelination of long motor neurons, causing paralysis, and is seen mainly in 20- to 50-year-old female hosts. HTLV-1 infection may also induce immunosuppression, resulting in an increase in opportunistic infections (Nakada *et al.*, 1984; Newton *et al.*, 1992; O'Doherty *et al.*, 1984). Most HTLV-1 infections are asymptomatic, but approximately 1–4% of infected hosts will develop disease symptoms (White and Fenner, 1994).

The human immune system cannot clear HTLV-1 from the body, so that the duration of infectiousness is lifelong. The number of unsusceptible hosts in a population is therefore always effectively zero. HTLV-1 differs from viruses such as MeV in that occupation of a host is long term and interhost transmission is both inefficient and infrequent. HTLV is transmitted between hosts almost exclusively via infected CD4[+] lymphocytes (Yamamoto *et al.*, 1984; Yamade *et al.*, 1993; Seto *et al.*, 1991; Ruscetti *et al.*, 1983; de Rossi *et al.*, 1985). Replication is very slow, and this is most likely due to tight gene regulation by viral proteins. Host–host transmission takes place through sexual intercourse (Nakano *et al.*, 1984; Tajima *et al.*, 1982), via blood transfusion (Okochi *et al.*, 1984; Sato and Okochi, 1986), and from mother to fetus or infant by passage of infected lymphocytes through the placenta or in breast milk (Kinoshita *et al.*, 1984; Komuro *et al.*, 1983; Tajima *et al.*, 1982; Yamanouchi *et al.*, 1985; Nakano *et al.*, 1984).

Rates of seroconversion following infected-male-to-female sexual transmission have been estimated at 60.8% and for infected-female-to-male at 0.4% (Murphy *et al.*, 1989). Seroconversion following contaminated blood transfusion (a relatively infrequently used pathway) has been reported at 30–60% (Manns and Blattner, 1991). Although the primary route of HTLV transmission appears to be from mother to offspring, this method does not appear to be highly efficient. For instance, Hirata and colleagues (1992) investigated the offspring of HTLV-1-infected mothers and found that 16% were infected at birth. They also observed among breast-fed infants that 27% of those breast-fed for more than 3 months and 5% of those breast-fed for less than 3 months were infected and 13% of bottle-fed infants were positive for HTLV-1.

HTLV-1 is a slowly replicating virus that is maintained exclusively in human populations. The lack of an effective host antiviral response as well as the virus's low replication rate (i.e., resource consumption) allows an intrahost population to successfully persist for many years. A very low probability (per unit time) of interhost transmission as well as a low probability of host damage (death) accompanies this persistence. Thus, HTLV-1 is an example of a parasite that employs an alternative life history to that of a more opportunistic replicator like MeV. HTLV-1 manages to keep $R_0 > 1$ by exhibiting a very low transmission probability while remaining relatively unobtrusive within its host for a very long time. This long period of infectiousness will probably, but not necessarily, result in transmission to a new host(s). There is thus heterogeneity in an HTLV-1 population's overall transmission but not so much that it is lost from a host population, because this variance is counteracted by HTLV-1's host availability (susceptibility) and its protracted "sit and quietly wait" intrahost replication tactics.

c. *Polydnavirus.* PDVs have segmented genomes containing up to 28 covalently closed dsDNA circlets (Fleming and Krell, 1993; Krell *et al.*, 1982; Blissard *et al.*, 1986a,b; Stoltz, 1993; Stoltz *et al.*, 1984). They are found only in endoparasitic wasp species of the families Brachonidae and Ichneumonidae. PDV genomes are integrated throughout the chromosomes of both male and female wasps (Fleming and Krell, 1993) and are only transmitted vertically in Mendelian fashion (Fleming and Krell, 1993; Stoltz, 1990). The viruses replicate exclusively within cells of the female's ovarian calyx during a specific stage in development and are subsequently released into the oviduct (Fleming and Summers, 1986, 1991; Fleming and Krell, 1993; Webb and Summers, 1992). During oviposition within a susceptible larval lepidopteran host, the eggs are coated with viral particles and thus transferred into the host body cavity (Vinson, 1977a; Norton and Vinson, 1977; Vinson and Scott, 1974, 1975). Once inside the lepidopteran, PDV genes are expressed in various tissues (Fleming and Krell, 1993; Blissard *et al.*, 1986a,b, 1989; Strand *et al.*, 1992; Theilman and Summers, 1986, 1988; Harwood and Beckage, 1994; Harwood *et al.*, 1994), but no further viral replication takes place (Stoltz, 1990; Blissard *et al.*, 1989; Strand *et al.*, 1992).

The PDV-derived proteins assist development of wasp larva(e) by impairing lepidopteran immune attack against the wasp eggs, thus serving a protective function. In addition, viral products actually alter the developmental rate of the lepidopteran host so as to allow for optimal tissue consumption by the young wasps. It has been shown that PDVs alone (or in combination with wasp venom) provide these functions that are actually essential to reproduction of their parasitoid hosts (Asgari *et al.*, 1997; Shelby and Webb, 1997; Vinson, 1977b; Davies *et al.*, 1987; Davies and Vinson, 1988; Vinson and Stoltz, 1986; Edson *et al.*, 1980; Vinson *et al.*, 1979; Guzo and Stoltz, 1985; Kitano, 1982, 1986; Stoltz *et al.*, 1988; Tanaka, 1987; Tanaka and Vinson, 1991). PDV replication within the wasp is highly localized both in space and time and essentially causes no significant adverse organismal effects (Fleming and Summers, 1986; Norton and Vinson, 1983; Theilmann and Summers, 1986). Not surprisingly, there is no antiviral response from the wasp.

The PDV life history incorporates a reproductive strategy that differs dramatically from viruses like MeV and HTLV-1, since wasp host fitness is actually enhanced by PDV survival and reproduction. This might be equivalent to assigning a negative value to α in Eqs. (5), (7), and (8). PDV evolutionary persistence is thus maintained by small but predictable transmission rates between hosts. Under some life-historical/ecological circumstances this strategy might not be viable, but since infection is lifelong, there is no virus-induced host fitness loss (but rather gain), and these conditions lack great variance, and the virus–host relationship is stable.

7. A Shifting Pathogen–Mutualist Continuum?

It is clear that viruses need not exhibit pathogenicity to exist. It is unfortunate, however, that our knowledge about viral taxa appears to increase with the pathogenicity of those taxa! Nevertheless, what is of great interest to the evolutionary ecologist is reconstruction of the "virulence history" of virus lineages to determine what conditions either have led or may lead to an increase or decrease in the negative effects on host fitness caused by viral infection.

For example, was occupation of ancestral wasp species by some PDV predecessors actually pathogenic? And, along those lines, perhaps modern PDVs descended from a once benign commensal parasite that gained a replicative advantage over its competitors by assisting with reproduction of its host. One may also speculate that MeV ancestors were at one time harmless human occupants and that some change in their replication that resulted in host damage conferred upon the virus a reproductive advantage. This fitness advantage may have resulted in cladogenesis, which gave rise to the MeV we see today.

Likewise, there is a growing body of evidence suggesting that HIV-1 has existed in the human population for many years prior to the AIDS epidemic of the late twentieth century (Williams *et al.*, 1960, 1983; Corbit *et al.*, 1990; Zhu *et al.*, 1998). It has been suggested that its previously silent residence was altered in response to changes in human social conditions (e.g., population density, urbanization, transportation, sexual promiscuity, and intravenous drug use), which allowed for greater transmission opportunities and thus permitted, even promoted, changes in virulence (Ewald, 1994a,b; Lipsitch and Nowak, 1995; Levin and Bull, 1994; Zhu *et al.*, 1998). Such a change in life cycle may have resulted in a modern adaptive radiation that is today manifest as the HIV/AIDS epidemic.

Piecing together the viral evolutionary past to resolve such patterns might often be a formidable task, to say the least. Yet the existence and presumed stability of multiple levels of virulence in extant virus–host relationships as well as the observable short-term changes in such levels are strong evidence that variation in pathogenicity is a common evolutionary maneuver.

C. Transmission Enhancement via Alteration in Patterns of Exposure

Viruses also employ many techniques to optimize transmission between hosts that do not per se involve changes in intrahost viral density (and thus host fitness). These techniques may include adaptations that enhance the spatial and perhaps temporal presence of the virus that result in an increased probability of new host infection. For example, an increase in the amount of time during which a virus can remain infectious while between hosts will subsequently increase opportunities for new host contact. Also, increases in the area over which a virus is shed may produce a similar effect.

1. Temporal Presence

Many viruses of terrestrial organisms whose transmission routes rely on direct exposure to the environment are surprisingly sensitive to degradation from ultraviolet radiation, desiccation, and temperature (Cooper, 1995). There are, however, some viruses that rely on contact spread that actually remain infectious through long periods of environmental exposure. Canine parvovirus is shed in feces and is extremely stable in the external environment, including prolonged exposure at 60°C and pH 3–9, which may actually contribute to its propensity for host-range shifts (Parrish, 1993). In fact, canine parvovirus was disseminated worldwide less than 2 years after its emergence. Also, smallpox is known to remain infectious within dried skin crusts for many years (Tyler and Fields, 1990).

Many insect viruses utilize multiparticle occlusion within proteinaceous crystalline structures to enhance environmental stability (Blissard and Rohrmann, 1990; Adams and McClintock, 1991). These include the nuclear polyhidrosis viruses (NPVs) and granulosis viruses (GVs), both of which are mostly lepidopteran-associated baculoviruses and also cytoplasmic polyhidrosis viruses (CPVs) and entomopox viruses, both of which infect a broader range of host taxa. Each such occlusion body is derived from either a nuclear or cytoplasmic locale and may include only a few or as many as tens of thousands of viral particles each (Cooper, 1995). Occluded viruses may remain infectious in the external environment for long periods of time and under extremely harsh conditions. This is most likely related to their reliance upon environmental exposure on plant surfaces and in soil prior to their subsequently being ingested by a potential host insect. Occluded NPV particles have been observed to remain infectious after 21 years of laboratory storage and following exposure to pH 2–9 (Cooper, 1995).

2. Spatial Presence

Viruses may also improve their interhost fitness without host damage by influencing their spatial presence. This may involve dissemination over large areas or shedding into very small but appropriate locales for new host occupation. The employment of vectors to effectively transmit small quantities of virus represents a very specific mode of interhost transmission, while the use of vehicles such as wind broadcasting of many aerosol-transmitted viruses illustrates a more general or even random approach.

Respiratory viruses of vertebrates are acquired by inhalation and thus transmitted through the air. Such viruses that have adapted to transmission during circumstances of close host proximity are usually inactivated fairly easily when desiccated or exposed to light and ionizing radiation (Chessin, 1972; Killick, 1990; Cooper, 1995). However, they may remain infectious for long periods under the proper conditions (high humidity, little direct sunlight, cool temperatures) and likewise be transported over great distances. Infectious foot-and-mouth-disease virus (FMDV) has been known to be distributed via air currents over 150 km across the English Channel (Donaldson et al., 1982). It is perhaps even conceivable that sensitivity to ionizing solar radiation may contribute to the seasonality of many human respiratory viral infections.

In addition to the use of weather conditions for improving their spatial presence, many viruses are also shed within body fluids or fecal material, which may aid their existence in the appropriate habitat for transmission. For instance, most human enteroviruses are ingested through the oropharynx and shed within feces to the external environment. Sexually transmitted viruses both enter and exit the host via the urogenital tract and are contained within bodily fluids that are associated with or even employ these portals.

D. Transmission via Vectors

The use of vectors to transmit viruses between hosts was an extremely important adaptation for diversification of many viruses. Vectors represent a tremendous dispersal opportunity that was destined to be exploited evolutionarily. They offer a method of viral transport between potential hosts that is indirect but does not necessarily involve exposure of the virus to an unpredictable environment. Vectors actively seek out the same or similar viral hosts for their own existence and therefore offer an accurate system of delivery. Vectors often take the form of flying insects but may also include fungi, mites, nematodes, or even other viruses.

1. Vector Utilization and Host Population Structure

An important aspect of vector-mediated transmission for the virus is its alleviation on host density dependence. That is to say, the reproductive fitness of a virus that relies on some (mobile) vector is influenced not only by the number of available hosts but also the number of vectors such that R_0 now becomes

$$R_0 = \beta^2 \left(\frac{\hat{N}}{N} \right) \left(\frac{f_v f_h}{\sigma_v \, \sigma_h} \right) \tag{9}$$

where $\hat{N}$ represents overall vector population density; f_v and f_h represent proportions of infected vectors and hosts that survive into a stage of infectiousness, respectively; $1/\sigma_v$ and $1/\sigma_h$ represent the duration of vector and host infectiousness, respectively (i.e., the inverse of recovery rates). Probability of transmission (β) is squared because it is applied to both transmission from an infected host to a vector and from that vector to a new host (the equivalence of these values is unrealistic but satisfactory for our purposes).

From Eq. (9) it can be seen how vector-mediated transmission can actually be partially independent of host density (Ross, 1911; Bailey, 1975; Anderson, 1982). This is especially true of mobile vectors in which viruses are transmitted as an incidental effect of feeding. Under such circumstances, survival of the vector depends on host visitation (feeding) at a relatively fixed rate regardless of how many hosts are present. Vectors will therefore engage in a constant number of host contacts that is largely independent of host density.

The above depiction gives rise to a vector/host density threshold required for stability of this type of system that is dependent on the overall magnitude of transmission as:

$$\frac{\hat{N}}{N} = \frac{1}{\beta^2}\left(\frac{\sigma_v\,\sigma_h}{f_v f_h}\right) \tag{10}$$

(after Begon *et al.*, 1990). Because of this, such a system of transmission may be used to a virus's advantage only under certain environmental conditions. For instance, when host populations are small or very dispersed, direct host-to-host contact may either be too infrequent or the probability of identical habitat usage by hosts may be too small to support necessary levels of direct host–virus contact transmission. A vector in these situations would be valuable since it often will possess search abilities and ensure proper virus dispersal regardless of host spatial distribution.

2. Vector Abundance and Activity

An additional situation where vector-mediated transmission may arise and persist is when vectors are abundant or make frequent host-to-host contact. In such cases, although the hosts may have either direct frequent contact, common space utilization, or both, certain viruses may increase their transmission efficiency by employing the exclusive or partial use of vectors for additional host occupation and access to portals of entry.

Such may be the case in many well-studied tropical arboviruses, such as the dengue viruses and yellow fever virus (YFV). These viruses are transmitted via the mosquito *Aedes aegypti* in both urban and sylvan areas of Africa and South America, where the vector flourishes because of climate (warm temperatures, large amounts of standing water, etc.). The co-occurrence of the existent host population structures and environmental conditions may alone be conducive to direct transmission of viruses in general, but the large presence of the mosquito must contribute substantially to its utilization in virus transmission. Not surprisingly, YFV transmission is known to increase with increased fecundity, longevity, and density of *Ae. aegypti* (Monath, 1989). In fact, indices exist that use mosquito presence and activity to measure risk of yellow fever epidemic (Service, 1974; Bang *et al.*, 1979, 1981; Monath, 1989).

Specific cases are known of human-induced environmental changes that have fostered mosquito population growth, which afterward resulted in increased viral disease prevalence (e.g., Adames *et al.*, 1979). In addition, numerous examples exist of viral epidemics following unusual weather events that are attributed to changes in vector populations, such as the *Culex pipiens*-facilitated St. Louis encephalitis outbreak of 1933. Holland (1996) proposed that dengue (a human–mosquito–human-transmitted virus) outbreaks in recent human history are likely driven by changes in human population size and mobility, and inadequate vector control. Unfortunately, detailed studies examining the precise relationship between absolute transmission rates and vector populations are limited, but there is growing concern regarding the impact of global climate change on insect populations and resultant increases in viral infections (see Lovejoy, 1993; McMichael and Beers, 1994; Jetten and Focks, 1997). It is conceivable, even likely, that future increased opportunities for vector-mediated transmission may bring about evolutionary adaptations that result in vector usage by viruses that previously did not rely on this inoculation route.

3. Sessile Hosts and Vector Utilization

A very important system of vector-mediated virus transmission occurs in sessile organisms, of which plants are particularly well researched. In this case, viruses cannot rely on host mobility to assist in spatial dissemination, and direct contact between plants is often (but not always) too infrequent to support sustained host–host transmission. Thus, many plant viruses rely on vertical transmission, intrahost persistence, large intrahost populations, environmental stability, vector-mediated transmission, or some combination thereof to sustain their existence within plant populations.

Plant virus transmission via aphids (Order: Homoptera, family: Aphididae) is especially noteworthy. According to one investigation, of 288 aphid species examined, 227 were found to transmit plant viruses (Eastop, 1983). In fact, the green pea aphid *Myzus persicae* is known to solely transmit over 34 different potyviruses (Kennedy *et al.*, 1962). Most aphid-borne viruses are carried by mechanical, noncirculative modes, meaning they are retained externally on the vector's feeding apparatuses or body surfaces, and as such are acquired and lost with great ease (especially during molting). For this type of mechanical vectoring, vector-mediated viral transmission is usually complete within a very short time (seconds or minutes), depending on environmental conditions. Therefore, viruses transported by this mechanism will exist in populations where great distances do not separate individual hosts such that viral transport times through adverse environmental conditions are minimal.

4. Propagative Vectors

Arthropod vectors may remain infectious for longer periods when the virus is retained inside the insect whether circulating in the hemolymph (in which case the arthropod serves as a mechanical vector) or actually replicating within cells of the vector (in which case the arthropod serves as a propagative vector). Such internal carriage techniques result in longer periods of vector (and hence virus) infectious-ness (thereby increasing σ_v in Eqs. (9) and (10)) and are more common in the arboviruses of animals than in the vector-transmitted viruses of plants. This includes the upper extreme of vector infectivity, transovarial transmission, in which vertical (and even venereal) infection of progeny and other conspecific arthropods results in their ability to transmit the virus. It is very likely that this association has arisen as an adaptation for viruses to overwinter in temperate regions (Reeves, 1974; McLintock, 1978; Rosen, 1987).

Viruses that employ complex transmission procedures such as transovarial transmission are exposed to yet another dimension of natural selection: intravector selection. Selection is certainly taking place between viral units for uptake into, survival within, as well as transmission from feeding vectors. There is perhaps no finer demonstration of viral interhost adaptive efficiency than the adeno-associ-ated viruses (genus Dependovirus) and the satellite viruses of plants that actually rely on replication of another virus for facilitation of their own replication.

E. Infectivity Optimization

Another potential "virulence-free" way to improve viral fitness via host–host transmission is to increase β by maximizing the infectivity of individual viral progeny. An increase in the capability of virus progeny to occupy susceptible cells upon contact would result in a high probability of infection per contact. This will subsequently decrease the number of descendent virions needed and therefore decrease virus-induced host mortality (α) in cases where virulence and virus replication are linked. Some viruses, like reovirus and adenovirus, can have particle-to-plaque-forming-unit ratios approaching 1, meaning that nearly all viral progeny are able to originate a new infection given the proper host cell exposure. However, the paucity of data on this topic prevents more exhaustive historical discussion about virion infectivity as an evolutionary adaptation.

F. The Interhost Evolutionary Objective

The venue of interhost competition is illustrated in Figure 5. This demonstrates the main evolutionary challenge for virus populations within an infected host as follows. The overall virus objective is occupation of a new susceptible host that will give rise to further such occupation. The obstacle is transport, both accurate

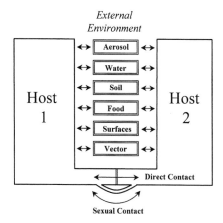

Fig. 5. Pathways of horizontal virus transmission.

and abundant, sometimes through abiotic or non-host media. The solution is adaptative mechanisms that: (1) do not significantly diminish intrahost fitness and (2) do not result in parasitism so potent that host resources are devastated. The adaptations that arise will have effects on the evolution of hosts, vectors, and other parasites, which will in turn feedback to influence the virus's selective environment. Direction of selection for interhost viral adaptation is governed fundamentally by the ease with which viruses pass between hosts, which in turn is determined by susceptible host density, host behavior, antiviral abilities, host population structure, environmental makeup, spatiotemporal heterogeneity, vector characteristics, and virus–host genetics, as well as myriad other factors.

V. THE ADAPTIVE LANDSCAPE AND RESPONSE TO SELECTION

The fate of any viral phenotype, whether derived from a nonadaptive source or from selective pressure of either an intra- or interhost nature, depends on its net evolutionary effect. Wright (1932, 1982) viewed evolution of units (individuals, populations, species) as movement through a multidimensional adaptive landscape composed of peaks of high fitness and valleys (and plateaus) of lower fitness. An individual's (or population's) location on the plane of this hyperspace is determined by its genomic makeup (collection of genes or genotypes). Its degree of elevation (i.e., fitness) is determined by the effect of their interactions with each other as well as the environment. Populations will thus move by way of natural selection up fitness peaks that are "local" in terms of recombinational or mutational distance (probability) and cannot (via natural selection alone) move through valleys to get to potentially higher peaks. Valleys can, however, be crossed and higher isolated peaks reached with the help of genetic drift (causing random allele fixations), genetic exchange, gene flow, high population variability (broad fitness peaks or ridges), or a changing adaptive topography.

A. Phenotypic Net Evolutionary Effect

For any taxonomic unit (individual population, species, etc.) the adaptive land-scape may be viewed as composed of the relationship of fitness to the collection of possible phenotypes that influence reproduction (this is discussed in greater depth below). As illustrated in Figure 6, each individual character or phenotype can be viewed as having fitness effects that are typically greatest in one dimension only (i.e., for one or a few replicative functions). The net evolutionary effect, however, is a composite of a particular trait's influence on fitness in an aggregate adaptive landscape that takes into account the combined relative survival and

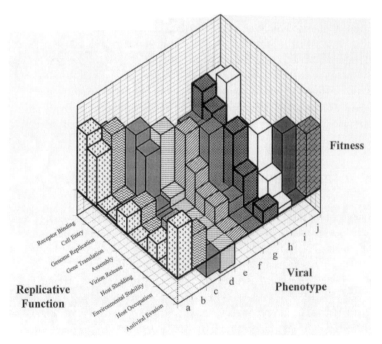

Fig. 6.   Differential phenotype fitness values with respect to functions of viral reproduction. Every phenotype is expected to have one (or a few) functions for which its fitness is highest.

reproduction probabilities of the characters. Movement through these landscapes may be facilitated by the potential for high genetic drift of some viruses especially due to random sampling of genotypes during transmission (Domingo *et al.*, 1985; Gebauer *et al.*, 1988). Also, high mutation rates of many RNA viruses may likewise hasten ascension of adaptive peaks (Eigen and Biebricher, 1988). However, landscapes are not static; they are perpetually shifting through evolutionary time, and this will promote movement to (and subsequent movement up) optimal peaks. Transformation of the topography is driven by abiotic (environmental) variation, viral evolutionary change, and by host evolution, including responses to these changes.

B. Habitat and Ecological Niche

The evolutionary ecologist focuses on the variables that cause changes in the adaptive landscape (the selective environment) and the evolutionary outcome of existence in these landscapes. While viruses presumably exist in every cellular organism, individual virus types may only occupy a very narrow range of conditions. Rarely, if ever, addressed in the field of virology is Hutchinson's (1957) concept of ecological niche. Since the same fundamental laws of evolutionary change apply across all forms of life, the ecological niche is an essential component of any discussion about virus natural history. Indeed, the seemingly unbounded viral diversity was and is patterned by the existence of manifold selective backgrounds in which viruses are able to survive and reproduce.

To begin with, a working definition of ecological niche is required. All organisms are capable of existence within a specific range of conditions, both abiotic and biotic. For example, an organism can survive within a certain temperature range. Survival success usually follows a specific probability distribution (e.g., normal) with increasing temperature. Another environmental condition, say pH, may confer a different survival distribution. Therefore, when considered concurrently, an organism's area of potential survival with respect to these two variables may be represented as a distinct two-dimensional space, the dimensions being temperature and pH. The number of conditions is increased to generate a multidimensional (*n*-dimensional) space or hypervolume in which the organism's population is capable of continued existence. This represents the *fundamental* niche: the set of conditions under which an organism is capable of survival and reproduction in the absence of competition or predation. If the dimensions are expanded to include not only abiotic but also interactive biotic factors, the result is an organism's *realized* niche: the subset of the fundamental niche that can be occupied in the presence of competition and predation.

Ecological niche should not be confused with habitat, which refers to the actual location of or set of conditions that give rise to a niche. Ecological niche is therefore a species characteristic and habitat is a place, population, organism, or tissue that can (and usually does) contain more than one niche. For instance, a vertebrate body contains numerous habitats (tissue types, etc.) that jointly or separately contain distinct niches (sets of conditions that allow for replication of certain viruses). Furthermore, as habitats are often composed of numerous niches, it is the "obligation" of the organism (i.e., the virus) to, by way of adaptive characters, evolve strategies for utilization of those niches and in the process perhaps give rise to new taxa. In this way, viral and host diversification is guaranteed given the existence of niches.

C. Niche Width

Virus types, like all organisms, display varying degrees of restriction with respect to the conditions under which they will reproduce. Obviously, there is no virus that can successfully replicate in all cell types of all tissues of all host species. Complete identification of all the niche components of even a simple organism is nearly impossible, but determination of a few of the primary ones is often sufficient to uniquely characterize a taxon. Accurate description of a generalized "virus niche," distinguishing the viruses as a unique taxonomic group, separate from all other parasites (and life-forms), has not been attempted. What can be proclaimed, however, is that all viruses require a functional intracellular host environment that includes: (1) protein translation machinery, (2) energy resources (ATP), (3) amino acids, and (4) nucleotides. Many viruses also require additional host cellular resources: DNA replication machinery, RNA splicing machinery, transport mechanisms, assembly mechanisms, membrane structure, membrane functions, and other distinctive translational mechanisms. Using this as a starting point, one can construct an exclusive portrait of specific viruses in terms of their replication optima. Relative to other organisms, viruses may seem highly specialized with respect to the niches they occupy (viruses frequently replicate in one cell type of one host species only), but there exists great variation in degrees of viral specificity across families as will be demonstrated.

D. Tissue Tropism and Host Range

Two very crucial attributes defining viral ecology and subsequently driving evolution are tissue tropism and host range (or specificity). The tissue tropism of a virus includes all the tissues that can naturally be infected by that virus regardless of the degree of pathogenicity or productivity. Host range includes all of the taxa in which a virus can replicate, also regardless of pathogenicity or reproductive potential. There are, of course, viral generalists who rely on more than one tissue type of more than one host organism and viral specialists that can only infect one or a few tissue types in one host species or subspecies. The process of adaptive radiation (discussed below) may occur when a virus type happens to successfully exploit a new niche by way of expanded host range or novel tissue occupation and is therefore a primary source of viral diversity (and even disease emergence).

The proximate determinants of tissue tropism include ability of the virion to bind to, enter, and release its contents within cells; ability of the viral genes to be translated and replicated; ability of the viral macromolecules (nucleic acids, structural and nonstructural proteins) to be transported and assembled; and the ability of the virus to remain functional (infectious) in the extracellular environment. Virus tissue tropism is therefore strongly influenced by the types of cellular membrane receptors (carbohydrates, glycoproteins, transport proteins, etc.), intracellular functional proteins (replicases, transcription factors, splicing and translation machinery), organelle function, host genetic makeup, and components that cause viral replication dysfunction (proteases, nucleases, modification enzymes) or virus neutralization. Since host range is primarily determined by the permissivity of the cell types occurring therein, viruses obviously can only occupy species that house tissue types for which the virus exhibits some degree of tropism. Other determinants of natural host range include antiviral capabilities as well as more "inconstant" host characteristics that may influence sustainable transmission such as population size, spatiotemporal population structure, habitat usage, diet, and behavior.

E. Peripheral Host Tissue and Taxa

Viruses are known to temporarily infect host species and host tissues that fall outside of the normal reproductive maintenance cycles to which those viruses have evolved. Some viruses are in this way able to replicate in a variety of cell types. For instance, hantaviruses naturally persist in rodent populations, where they are maintained as avirulent infections of the lung and kidneys (Lee *et al.*, 1981). Under certain conditions, humans who come into close contact with aerosolized hantaviruses can be infected, and this is often manifest as a hemorrhagic fever with occasional kidney failure due to viral tropism for and damage to renal tissue.

Ebola viruses are an excellent illustration of the ability of a virus to replicate in both diverse species and tissue. The natural history of Ebola viruses is largely mysterious, but the four known subtypes are thought to be maintained in African nonhuman primates or other mammal populations and can sporadically be transmitted to humans and many other species of primates, mice, and guinea pigs. The extraordinary feature of Ebola virus replication is its ability to use many host tissue types. For instance, Ebola virus particles or antigens have been found in such diverse tissues as liver, spleen, lymph nodes, lung, gastrointestinal tract, and even the sweat glands of the skin (Murphy *et al.*, 1990; Peters, 1997).

It cannot be said that viral groups capable of such aberrant replication are ecological "generalists" since this activity is rarely evolutionarily stable; hantavirus and Ebola virus transmission between humans is not epidemiologically sustainable. However, as is discussed below, the ability of viruses to explore novel habitats in an evolutionary search of fertile or empty niche space is a major contributor to their diversification.

F. Evolution of Tissue Tropism and Host Range

An important question arises. What are the relative contributions of constraints on tissue tropism and host specificity to sustainable viral reproduction? In other words, is a given virus more likely to occupy a new tissue type within its normal host species or a homologous tissue type within a different host species/taxon? There are certainly examples of both types of diversification. One might predict, however, that viruses naturally occupying tissue 1 within host species 1 would have a higher probability of replicating in tissue 2 within this species since the likelihood of exposure to that tissue is very high relative to appropriate exposure to the homologous tissue in host species 2 (per Fahrenholz's rule; see Brooks and McLennan, 1993). The relevant evidence instead appears to favor diversification driven by tissue type, not only for viruses but also for many parasites. In other words, a virus is often more likely to diversify into similar tissue types in new host species than to infect new tissue types in the same host. Along these lines, the probability of successful host jumping can be viewed as inversely related to phylogenetic distance between hosts, just as the probability of new tissue infection might be inversely related to factors such as molecular and developmental similarity of the tissues.

Figure 7 shows a hypothetical phylogeny and the probability of viral transmission from one taxon (e.g., host species) to another as decreasing with increasing taxonomic/evolutionary divergence. Also shown is an arbitrary measure of tissue permissivity in relation to phylogenetic distance. The shapes of the two curves were deliberately drawn as seen to illustrate a point. Tissue permissivity for a virus is assumed to be directly related to overall evolutionary divergence of the cellular and molecular characteristics of tissues within a host clade. Therefore (assuming a constant rate of molecular evolutionary change), as host taxa diverge, their

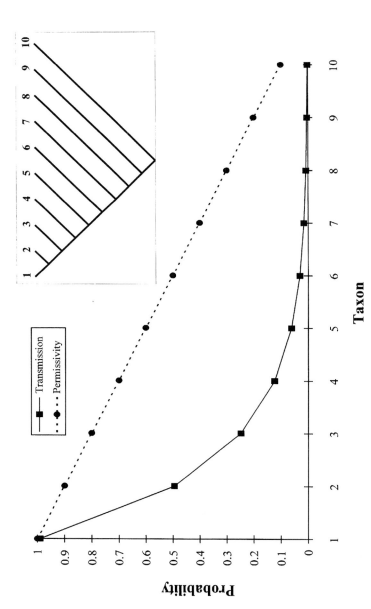

Fig. 7. An illustrative model of the probability of viral replication in two different taxa. It is assumed here that a virus type has evolved to replicate stably in species 1 (see text). "Transmission" refers to the probability that this virus will be transmitted from species 1 to stably replicate in another species. "Permissivity" refers to the probability that the homologous tissue of another species allows replication of this virus.

homologous tissues continually and regularly accumulate morphological differences that are "recognizable" to the virus. The exponentially decreasing probability of stable interspecies transmission by the virus reflects both the effects of changes in cell permissivity and the probability of necessary contact between disparate taxa. Interspecific contact between potential viral hosts is assumed to decrease at a rate greater than that of morphological change since as taxa diverge they often will become separated spatially (and perhaps temporally) by geographic movement, habitat usage changes, niche segregation, etc. This model obviously will not be applicable to all virus–host relationships, and both transmission and permissivity curves may take on many shapes depending on the specific virus–host biology represented therein.

1. Constraints on Tissue Tropism

There are numerous examples of virus types that have evolved to replicate within new host species but in cell types that are part of homologous tissue. For example, using molecular phylogenetic techniques, Taber and Pease (1990) demonstrated how the host species usage of paramyxoviruses evolves faster than does tissue tropism. By mapping tissue tropisms and host species onto a phylogeny derived from the F glycoprotein DNA sequence, they were able to show how viruses that primarily infect respiratory tissues (parainfluenza viruses, Sendai virus) are evolutionarily clustered despite the fact that they occur in diverse vertebrate species.

Papillomaviruses often cluster phylogenetically into monophyletic groups whose viral taxa exploit one tissue type (e.g., genital versus epidermal epithelial) yet encompass more than one host species (Chan et al., 1995, 1997). In addition, no single papillomavirus type is known to infect more than one host species. Despite being evolutionarily related (Webster et al., 1995), most influenza A virus strains — which naturally occur in horses, humans, whales, mink, seals, pigs, and a variety of birds — primarily infect either the respiratory or gastrointestinal tracts of all these taxa. Many (but not all) polyomavirus types persistently infect the kidney epithelium of humans, macaques, baboons, bovines, and mice. The alphaherpesviruses are found in many vertebrate species, including primates, canines, bovines, cervids, catfish, felines, rodents, and birds (Roizman, 1993). These viruses will remain latent only in the sensory ganglia during most of these infections.

2. Constraints on Host Species

This is not to say, however, that change in a virus's cellular tropism does not occur or is even infrequent. To the contrary, many viruses not only utilize more than one intrahost tissue type during the normal course of infection but the process of tissue tropism expansion is a commonly observed phenomenon in viruses. For instance, through *in vivo* and *in vitro* studies HIV has been found to replicate within practically any human tissue, including most hematopoietic cells, fibroblasts, bowel, retina, cervix, testes, lung, brain, kidney, gastrointestinal epithelium, skin, and heart (Armstrong and Horne, 1984; Cohen *et al.*, 1989; Ho *et al.*, 1984, 1985; Levy *et al.*, 1985; Nelson *et al.*, 1988; Popovic and Gartner, 1987; Tschachler *et al.*, 1987; Levy, 1993).

MeV exhibits both obligate and nonessential multitissue utilization during its course of its infection. MeV replicates in tissues throughout the body, including monocytes, macrophages, and T cells, as well as in respiratory tissue, and epithelial tissues of the oropharynx, conjunctiva, skin, and the alimentary canal (Grist, 1950; Hall *et al.*, 1971; Katz *et al.*, 1965; Kempe and Fulginiti, 1965; Frazer and Martin, 1978). It is also capable of replication in the bladder, central nervous system, spleen, and local lymph nodes.

Bowen and colleagues (1997) constructed and compared arenavirus molecular phylogenies with those of their rodent hosts and detected multiple instances of virus–host cospeciation. In addition, they determined that pathogenicity had arisen independently many times during arenavirus evolution.

3. Diversity in Host Species Tissue Type

There also exist "generalist" viruses that have evolved to rely on more than one tissue type of more than one host species. The most obvious example is viruses that depend on propagative vector-mediated transmission such as arboviruses of vertebrates (e.g., many alphaviruses, flaviviruses, bunyaviruses) and some rhabdoviruses of plants. Such viruses must replicate within disparate tissue types of phylogenetically very distant organisms (i.e., arthropods and vertebrates or arthropods and plants) in order to achieve successful interhost transmission.

For instance, within hematophagous mosquito vectors, arboviruses such as the equine encephalitides (Chamberlain *et al.*, 1954; Chamberlain and Sudia, 1955), yellow fever virus (Bates and Roca-Garcia, 1946), Semliki Forest virus (Davies and Yoshpe-Purer, 1954), vesicular stomatitis (Mussgay and Suarez, 1962), and the dengue viruses (Kuberski, 1979) to name a few, replicate within the mesenteronal epithelial cells of their mosquito vectors. Afterwards they are transferred via the hemolymph to the salivary glands for further replication (Hardy, 1988). During hemocoel translocation, mosquito-borne arboviruses may also replicate within alimentary tissue, reproductive organs, ganglia, muscle, pericardium, and Malpighian tubules (Hardy, 1988). In addition to their tissue versatility, many arboviruses are also capable of using more than one vector species. For instance, YFV has been shown to be transmitted via *Aedes aegypti* (Beaty *et al.*, 1980; Aitken *et al.*, 1979), *Aedes dorsalis/taylori* (Monath, 1979), *Aedes mascarensis*

(Beaty *et al.*, 1980), *Haemagogus equinis* (Dutary and LeDuc, 1981), and *Amblyomma variegatum* (Saluzzo *et al.*, 1980).

The distant phylogenetic relationship between arthropods and vertebrates makes obligate replication in both taxa a fascinating evolutionary feat, and yet even some plant viruses rely on replication within arthropod vectors. These include rhabdoviruses such as lettuce necrotic yellows virus (Stubbs and Grogan, 1963), sowthistle yellow vein virus (Behncken, 1973), and strawberry crinkle virus (Richardson *et al.*, 1972), which all use aphid vectors.

Some virus types appear to rely on more than one nonvector host species for normal maintenance. An excellent and well-investigated example of this is influenza A, which is naturally perpetuated through populations of many wild birds (especially aquatic species), wherein it is rarely pathogenic despite the fact that all the known subtypes are represented (Hinshaw *et al.*, 1980; Alexander, 1986). Strains of influenza A (especially H1N1) have been known to move between swine and humans (Webster *et al.*, 1992; Hinshaw *et al.*, 1978; Schnurrenburger *et al.*, 1970) and are all but conclusively determined as being derived from avian reservoirs (Webster *et al.*, 1992, 1995). Influenza A is also found in many other vertebrate species, including horses and sea mammals, and, based on serological and phylogenetic analysis, it is thought that interspecies transmission is frequent in these cases as well (Gorman *et al.*, 1991; Webster *et al.*, 1992, 1995). Thus, influenza A is capable of existence in more than one species and is quite frequently transmitted between species. This characteristic is thought to be related to the mutability of its genome through reassortment and rapid mutation accumulation (Scholtissek *et al.*, 1977, 1978, 1993; Scholtissek, 1995; Yamnikova *et al.*, 1993; Reinacher *et al.*, 1983; Domingo and Holland, 1994) and thus may not be as common in DNA viruses or viruses with nonsegmented genomes.

G. Stimuli for Evolutionary Expansion of Reproductive Substrate (Emergence)

The successful transition or extension of a virus into usage of new cell types, tissues, organs, or even host taxa can be brought about by four main virus–host interactive stimuli. First, increased contact between host and virus can result in infection outside the normal site of replication. Second, a change may occur in specific characteristics of previously unsusceptible individuals that results in their transformation into infectible hosts. Third, speciation (cladogenesis) of a host and the ensuing cladogenesis of viruses that infected the ancestral taxon (a.k.a. phylogenetic tracking, parallel cladogenesis). These three causes may or may not involve adaptive or other evolutionary change in the virus. Lastly, alteration in an existing viral phenotype or acquisition of a new phenotype via mutation or genetic exchange could allow occupation of new viral niches and habitats (host species, tissue type, etc.) and possibly bring about adaptive radiation.

New phenotypes that allow ecological expansion have been termed "key innovations" since they may subsequently give rise to multiple new phyletic types by way of allowing entry into new niches or adaptive zones (Miller, 1949; Van Valen, 1971; Simpson, 1953; Cracraft, 1982; Mishler and Churchill, 1984; Brooks *et al.*, 1985; Levinton, 1988; Rosensweig and McCord, 1991; Baum and Larson, 1991; Erwin, 1992; Heard and Hauser, 1995; Hunter, 1998). Given the enormous degrees of variation found in viruses, their population sizes, and their ubiquity, this last mechanism probably represents a primary method of adaptive response in viruses.

1. Key Innovations and Adaptive Radiation

Evolution through acquisition of new phenotypes possibly representing key innovations has been observed in many viruses. For example, feline panleukopenia virus (FPV) is incapable of sustained replication outside its normal host range. Canine parvovirus (CPV) is strongly suspected as being derived from FPV since it is genetically very similar and is capable of replication in both canines and felines. CPV did not exist before the late 1970s and spread worldwide in less than two years (Parrish *et al.*, 1988). CPV and FPV differ in only three amino acids of their VP2 capsid proteins: position 93 is asparagine in CPV and lysine in FPV; position 103 is alanine in CPV and valine in FPV; and position 323 is asparagine in CPV and aspartic acid in FPV (Parrish, 1997). Using site-directed mutagenesis, Parrish (1997) was able to demonstrate that CPV viruses that were otherwise normal but contained lysine at position 93 were able to replicate in feline cells but lost their ability to replicate in canine cells. Furthermore, otherwise normal FPV that contained asparagine at positions 93 and 323 were able to replicate in both canine and feline cells. It is highly probable that these latter two mutations represent a key ecological innovation that recently facilitated parvoviral occupation of a new host niche and therefore enabled rapid (and extensive) spread of parvovirus into the world population of canines.

Porcine respiratory coronavirus (PRCV) emerged in 1983–1984 and became enzootic in almost all European countries within 2 years (Pensaert *et al.*, 1986; Brown and Cartwright, 1986; Jestin *et al.*, 1987; Have, 1990). Rasschaert *et al.* (1990) have determined that this virus differs from the more established porcine transmissible gastroenteritis virus (TGEV) by only four genomic deletions.

In a third example, cell tropism in HIV has been found to be changed by few mutations in the V3 region of the gp120 envelope glycoprotein, which can give rise to strains capable of replicating in macrophages (Hwang *et al.*, 1991; Shioda *et al.*, 1991, 1992). In addition, during the normal course of HIV infection, different tissue-tropic strains may arise that are distinguishable from their parental viruses not only by the tissue from which they are retrieved but also by their phenotypes and genotypes. For instance, HIV isolates collected from blood are very capable of syncytia formation and are also T-cell tropic. Those collected from bowel tissue are only slightly capable of syncytia formation and only slightly T-cell tropic. However, HIV collected from brain tissue seem incapable of syncytia formation and are not T-cell tropic (Barnett *et al.*, 1991; Cheng-Meyer *et al.*, 1989). These examples illustrate heritable viral variation in (differential) tissue tropism and host range arising in very short time periods.

2. *"Nonadaptive" Radiation*

Emerging properties such as multiple tissue tropisms, large host ranges, and propagative vector-mediated transmission provide illustrations of how the great adaptability of viruses enables them to withstand ecological pressures and to exploit opportunity. From these examples, it is clear that viruses will evolve to occupy new habitats (hosts, tissues, etc.) by way of heritable characters that allow them to invade niches contained therein.

Adaptive radiation has been defined as the diversification into new ecological niches of taxa derived from a common ancestor (Futuyma, 1986). It should be emphasized, however, that diversification (especially for viruses and other parasites) need not be driven by adaptive viral mechanisms and may merely be the result of host mechanisms such as diversification and phylogenetic tracking, geographic isolation, and asymmetrical cladogenesis. Care must therefore be used when interpreting viral evolutionary change as adaptive, that is, whether a viral group has diversified due to the acquisition of expanded reproductive abilities driven by ecological obstacles or openings presented to it, or merely because of the probabilistic conditions of its existence. It is necessary to authenticate the presence of synapomorphic adaptive characters that provide some specific, selected function to conclusively demonstrate that viral adaptation is the source of a phylogenetic radiation.

VI. INFLUENCE OF OTHER INTERACTIONS

An important driving force in the evolution of viruses is their interactions with other, usually coinfecting viruses. This includes both interactions between "conspecific" viruses derived from the same or different inocula (as discussed in section III.A) as well as interaction between taxonomically or phylogenetically distinct viruses. Virus–virus interaction may take the form of competition for resources (i.e., hosts and host tissues), replication interference (e.g., as in defective interfering particles), pathogenic modification, or of "evolutionary assistance" as in intra- or intergenomic genetic exchange for instance. Potential virus–non-host species interactions may include predation (perhaps rarely) and coinfection between a virus and another nonviral parasite.

A. Competition: Displacement versus Coexistence

1. The Red Queen's Hypothesis

The "Red Queen's Hypothesis" (Van Valen, 1973) predicts that, when two or more populations exist in direct competition for some resource, the processes of adaptation and counter-adaptation must be continually occurring in order for both populations to survive. The name is derived from Lewis Carroll's *Through the Looking Glass*, in which the Red Queen states, "Now here, you see, it takes all the running you can do, to keep in the same place." Clarke and colleagues (1994) examined fitness changes in distinct competing clonal populations of vesicular stomatitis virus (VSV) cultured *in vitro*. To accomplish this, they used quantitative relative fitness assays to measure fitness changes in VSV populations that initially were competitively equal. When sequentially passaged together, one of the VSV populations was inevitably totally displaced by the other. "Winner" and "loser" strains were then mixed with the ancestral wild-type strains to examine absolute fitness changes. In all cases, the derived virus populations were able to outcompete and displace their ancestral populations. In this important experiment, evolutionary change continually proceeded in such a way that overall survivability of present generations surpassed that of previous ones and was lower than those of the future. In other words, populations are competing in a "zero sum game," where they must adaptively "run" just to stay at the same relative fitness level (Van Valen, 1973). This effect has only been experimentally examined in VSV; therefore, whether or not it occurs in the evolution of other viruses (especially DNA viruses) cannot been addressed here.

2. Competitive Exclusion

The barley yellow dwarf virus (BYDV) system provides another interesting example of virus–virus competition. These viruses are aphid transmitted and infect a wide range of grass species worldwide. Historical records of BYDV isolated from oats in New York State between 1957 and 1976 indicate that, of the two main viral subtypes, BYDV–MAV and BYDV–PAV, the percentage of diseased plants infected with BYDV–MAV have decreased from 90 to 0% while the prevalence of BYDV–PAV increased during the same period from 3 to 98% (Rochow, 1979) and apparently still shows signs of superiority (Miller *et al.*, 1991).

Power (1996) has examined direct and indirect viral interactions that may be contributing to the competitive dominance of BYDV–PAV. She maintains that BYDV–PAV may be winning an evolutionary battle because of its interactions with aphid vector species that are favoring higher transmission rates. For instance, BYDV–PAV can potentially use more than one aphid species for transmission (*Sitobion avenae* and *Rhopalosiphum padi*), whereas BYDV–MAV only uses *S. avenae*. In addition, *R. padi* was found to be a more efficient transmitter of virus than *S. avenae* (Gray *et al.*, 1991; Power *et al.*, 1991). Morphological alteration of vectors as a result of viral infection may also be enhancing the spread of BYDV–PAV. Aphids are polymorphic species, having both winged and nonwinged adults. Gildow (1980, 1983) has shown that *R. padi* and *S. avenae* adults are winged with significantly higher frequency when feeding on BYDV-infected plants. Furthermore, *S. avenae* adults were found (in separate experiments) to be more frequently winged when feeding on BYDV–PAV-infected plants than when feeding on BYDV–MAV-infected plants (Gildow, 1980). Power *et al.* (1991) point out that the competitive benefits of such viral-induced changes may include increased vector dispersal, which may therefore provide BYDV–PAV with a transmission advantage over BYDV–MAV.

3. Niche Differentiation and Character Displacement

The VSV experiment and BYDV-group observations illustrate two possible outcomes of competitive viral interactions: coexistence with ongoing adaptive change and competitive exclusion. Niche differentiation is another possible outcome of competitive interactions that allows stable coexistence between competitors but has so far not been examined in viral systems. Competition between two viral taxa for an obligate but finite resource (niche component) can and does eventually result in exclusion of one by the other, as demonstrated above in Clarke and colleagues' (1994) research. However, intertaxonomic competition in the presence of more than one fundamental niche may result in alteration (including expansion or contraction) of the niches used by the individual competing populations. Niches can be differentiated by resource partitioning, in which coexisting populations evolve the ability to utilize different resources, thereby avoiding direct competition. For viruses, this type of partitioning might involve replication in separate cell or tissue types of a particular host.

Niche differentiation may also occur by utilization of similar host resources but in different temporal or spatial contexts. This may include either infection of the same host population but at different times of the year or perhaps infection of different populations of a single host species that are geographically or otherwise environmentally separate.

Character displacement may be an observable condition indicative of niche differentiation. This involves the evolution of measurable phenotypic differences between taxa existing in sympatry that do not exist in populations that are not involved in competition. This may be the result of (ultimately) and give rise to (proximately) viral usage of different resources (cell types, vectors, etc.). It is easy to imagine sympatric disparities in a viral surface protein, for instance. Examples of sympatric disparity patterns are, however, rare in those species of macroorganisms traditionally investigated ecologically and not surprisingly have never been sought in microorganisms.

B. Noncompetitive Interactions: Coinfection and Altered Pathogenicity

Competition may not be the only virus–virus interaction that exists that has a direct (and often directional) impact on individual viral fitness. Coinfection of a single host with two or more distinct virus types may also result in a change in the type or degree of pathogenic effects caused by one or the other's replication. An increase in virulence caused by coinfection may decrease the length of host infectiousness, thereby diminishing transmission rates of one or all resident virus types. This effect may be interpreted mathematically by inclusion of additional terms in Eq. (7) and Figure 3, which define the rate of coinfection of two or more viruses as well as the effects of such coinfection on host fitness (virulence) and immunity.

Examples of multiple viral infections producing symptoms that are not found during single infections with any of the individual viral types involved have been documented. Contag and colleagues found that lactate dehydrogenase-elevating virus induced paralytic poliomyelitis in mouse strains that expressed endogenous murine leukemia virus but not in other mouse strains (Contag *et al.*, 1989; Contag and Plagemann, 1989). Pedersen *et al.* (1990) discovered that coinfection with feline leukemia virus (FeLV) and feline immunodeficiency virus (FIV) amplified FIV-induced immunodeficiency. In 1995, Atencio and colleagues discovered that 100% of newborn mice coinfected with Moloney virus, a retrovirus that causes murine leukemia (MoMLV) and A2 strain polyomavirus (PyV), exhibited stunted growth (runting). This effect was not observed in mice singly infected with MoMLV and was observed with much lower frequency in mice singly infected with PyV.

Although poorly understood, altered or enhanced pathogenicity caused by mixed infections may be the result of direct virus–virus interactions such as potentiation of replication of one virus when the gene products of another virus are present in the same host tissue. It may also be caused by indirect effects such as presence of the host's immune-related molecules (cytokines, antibodies, etc.) that are induced by infection of one virus and that might potentiate replication of another. However, regardless of the mechanism, viruses whose overall fitness is weakened by the intrahost presence of another viral type must adapt to avoid, overcome, or tolerate the coinfection-induced variation in pathogenicity and co-infecting replication rates if they are to persist.

C. Indirect Interactions

Competitive virus–virus interactions need not be direct in the sense of two or more viruses coinfecting a host simultaneously. Infection of a host or host population with one virus type can conceivably influence infection of that same host with a different virus either at some time in the future or simultaneously by inducing crossreactive adaptive immune responses against other viruses or by altering cells such that they are rendered uninfectible.

For instance, as reviewed by Weiss (1993) and Best *et al.* (1997), proteins derived from endogenous retrovirus genes may act as replication inhibitors of similar exogenous agents by blocking host cell receptors. This has been observed in the case of the *Fv-4* gene of mice, which originated from an endogenous retrovirus. *Fv-4* is an envelope protein that appears to prohibit infection by potentially pathogenic ecotropic murine leukemia viruses by binding to host cell surface receptors, thereby preventing fusion and entry of other, closely related virus types (Kozak *et al.*, 1984; Weiss, 1993). This type of effect is also thought to occur in the chicken *ev-3* gene, also of retroviral origin, which blocks the cell receptors used by Rous-associated virus-O (Crittenden *et al.*, 1974).

In the case of protective endogenous retroviruses, it is easy to see the fitness advantage conferred upon both the host and the virus itself and the resultant ecological barrier to which potentially infecting exogenous viral agents are exposed. This generates selective pressure reinforcing the role of endogenous retroviruses as host guardians that exclude phylogenetically similar exogenous potential competitors while at the same time favoring divergence in the reproductive capabilities among those potentially pathogenic viruses under exclusion. This may also provide a competitive exclusion explanation for the observation of a lack of exogenous viruses that are phylogenetically close to endogenous viruses of humans as well as an apparent lack of endogenous versions of the lentiviridae and spumaviridae (Löwer *et al.*, 1996).

Specific host immune response to infection with one virus type may also be crossreactive to other distinct viruses, thus preventing their occupation of host individuals or populations. Such forms of indirect competition may also bring about, and perhaps even drive, viral evolutionary change. The most famous example of this occurrence is likely the exclusionary relationship between cowpox and smallpox infections. In the late eighteenth century, Edward Jenner demonstrated how humans exposed to relatively avirulent cowpox were resistant to infection with its more deadly relative, smallpox virus (Jenner, reprinted in 1959). Immunological effects such as this are usually the result of similarity in antigenic makeup between the viruses, which causes crossreactivity of specific antiviral defense mechanisms against related types.

D. (Non-Host) Viral Predation?

A common interspecies or "intertaxonomic" interaction that has profound effects on evolutionary change is predation. Despite the ubiquity and quantity of viruses on our planet, there have been very few attempts to explore the possible existence of grazing organisms that rely on viral biomass for nourishment.

In a 1993 publication, González and Suttle looked at phagotrophic consumption by marine nanoflagellates and determined that viruses and virus-like particles made up a significant portion of these microorganisms' total nutritional intake. The authors further note that these nanoflagellates engage in selective grazing of viruses in that they actually ingested different viral particles at different rates, perhaps implying an ecological and evolutionary stability of this feeding practice (González and Suttle, 1993). While this is an interesting phenomenon, there exists no additional evidence to suggest that host-unrelated predation has any influence on the course or pace of viral evolutionary change as it does in numerous other organisms.

E. Genetic Exchange as Viral Interaction

Virus–virus interactions need neither be exclusively competitive or negative. For instance, the coinfection of a cell by two or more viruses can result in interviral genetic recombination or reassortment. Discussions of ecological interaction rarely include treatment of taxa as sources of genetic (indeed evolutionary) diversity. Yet, genetic exchange between individual viruses is clearly an interaction that has obvious measurable effects on virus evolution to such an extent that virus survival and diversification is often propelled by, if not dependent on, such genetic exchange.

1. Recombination and Reassortment

It is important here to make a distinction between these two processes, something that in the literature is occasionally neglected. Reassortment only occurs when two or more viruses with segmented (or diploid, such as lentiviruses) genomes coinfect a host cell. The genome segments are randomly packaged into progeny virions, and as a result the progeny may be composed of genomic segments from more than one parent. The outcome of this reassortment is a genomically diverse assemblage of offspring. As cleverly pointed out by Cann (1995), two coinfecting influenza viruses each with eight original genomic segments may give rise to $2^8 = 256$ potential progeny genomes.

Recombination is a process whereby genetic material is exchanged between individual nucleic acid strands of the viral genome, which results in the formation of descendant nucleic acid strands composed of material from two or more parental strands. Recombination may be caused by polymerase copy choice or by strand breakage and re-ligation and can occur between either homologous or nonhomologous regions of the genomic strands. Detailed molecular descriptions of these processes are beyond the scope of this chapter, but see Nagy and Simon (1997) and Scholtissek (1995) for reviews on these subjects.

2. Jumping through Sequence Space

Many important evolutionary leaps have undoubtedly been influenced as well as initiated by close viral interactions. Interviral recombination and reassortment generate diversity in infecting viral populations and thus, in addition to mutation, represent mechanisms that produce the phenotypic raw material upon which selection acts. The evolutionary role of genetic exchange differs from mutation, however, in that exchange allows for more rapid and extreme evolutionary modification. In other words, the sequence space of an organism can be thought of as every possible combination of nucleotides in its genome (e.g., 4^v possible genome sequences may exist, where v is the genome length). Most of these potential genome compositions are not biologically viable and are thus never observed in nature. Likewise, other gene combinations may allow viral replication but only under specific circumstances determined by environmental conditions that may or may not exist at a given time. Still other genetic progeny may exhibit very efficient and successful replication (comparatively), therefore possessing high fitness and existing (symbolically) near the top of a fitness or adaptive peak.

However, existence atop a fitness peak may not be as optimal as it sounds. For instance, other taller peaks may simultaneously be present in a particular landscape or may emerge in the future that may not be accessible by way of the normal mutational mechanisms no matter how high the mutation rates. This is because the probabilistic attributes of the mutational process preclude quick and prodigious alteration of the genome and will usually produce small stepwise changes that correspond to movement of short distances through sequence space. Therefore, once a genome has "summitted" an adaptive peak, any other small motion would constitute a decrease in fitness — an improbable occurrence save for the effects of random drift. General movement through adaptive landscapes is discussed more thoroughly and eloquently by Wright (1931, 1932, 1982), Kauffman and Levin (1987), and Coyne et al. (1997). The specific adaptive landscape of viruses is also discussed by Chao (1994), Scott et al. (1994), Eigen and Biebricher (1988), and Domingo and Holland (1997). Niche expansion or shift in the form of changes in host range, tissue type preferences, vector usage, pathogenicity, or combinations of these and myriad other biological traits may represent movement of viral populations between fitness peaks or hills throughout their adaptive space.

Genetic exchange between viruses is very evident both *in vitro* and *in situ*. Of greatest evolutionary significance is the generation of recombinants that exhibit phenotypes that differ from both (all) individual viral parents. The ensuing changes may yield expansion of replication capabilities, including changes in cell type preference, interspecies host–host transmission, or even changes in vector utilization. Mixed infections in the laboratory between parent viruses whose replication might otherwise be restricted to infection of only one or another cell type will frequently result in production of viral recombinants capable of replication within both cell types. As an example, using mixed infections of lepidopteran cell lines, Kondo and Maeda (1991) developed recombinants between various NPVs that subsequently exhibited expanded host ranges. The outcome was that coinfection of *Bombyx mori* NPV (BmNPV) with *Autographa californica* NPV (AcMNPV) resulted in recombinants that could replicate in previously nonpermissive cell types. Another example is provided by the efforts of Spector *et al.* (1990), who found that experimental coinfection of HIV and a murine amphotropic retrovirus resulted in HIV pseudo-types capable of infecting both CD4$^-$ human cells and mouse cells. These retroviral recombinants thus exhibited expansion in potential host range as well as tissue tropism.

3. Genetic Exchange in Nature

The occurrence of interviral genetic exchange in nature is supported by many lines of direct and indirect evidence and is thought to be a major contributor to host range expansion, tissue tropism changes, periodic epidemic and epizootic infection, "emerging" infections, and changes in viral pathogenicity. HIV and other primate lentiviruses have been investigated quite extensively in this respect. As reviewed by Katz and Skalka (1990), Temin (1991), Sharp *et al.* (1995), Robertson *et al.* (1995a,b), Burke (1997), and Dornburg (1997), recombination (especially involving regions of the envelope gene) within and between primate immunodeficiency viruses has been an important evolutionary mechanism contributing to successful host-range expansion and widespread diversification of the group.

Retroviruses are known to recombine ("have sex") not only within populations but also with host endogenous retroviruses (Weiss *et al.*, 1973; Coffin, 1982; Risser *et al.*, 1983; Stoye and Coffin, 1985; Stoye *et al.*, 1988; Temin, 1991), with some herpesviruses (Isfort *et al.*, 1994), and with other exogenous retroviruses (Sharp *et al.*, 1995; Robertson *et al.*, 1995a,b). As illustrated in Figure 8, lentivirus phylogenies constructed using single genomic regions often exhibit consistent branching patterns, but incongruencies can be strongly apparent when trees for identical taxa are built using different regions of the genome (Li *et al.*, 1988; Gao

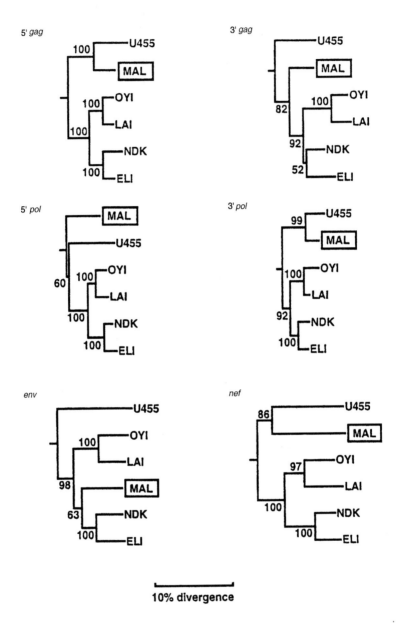

10% divergence

Fig. 8. Phylogenetic relationships of isolates U455 (HIV-1 subtype A), OYI (subtype B), LAI (subtype B), NDK (subtype D), ELI (subtype D) and MAL as indicated by *gag*, *pol*, *env*, and *nef* DNA sequences. Numbers at nodes indicate bootstrap percentages. Reproduced with permission from Robertson *et al.* (1995a).

et al., 1992; Robertson *et al.*, 1995a,b; Jin *et al.*, 1994; Louwagie *et al.*, 1995; McCutchan *et al.*, 1996; Sabino *et al.*, 1994; Salminen *et al.*, 1997; Dornburg, 1997). Such phylogenetic disparities indicate that many lentivirus genomes are actually "mosaics" representing the products of hybridization events between distinct viral types. These interactions then gave rise to the evolutionarily successful ("new") viruses we observe today, perhaps exhibiting aggregate or even novel phenotypes.

The frequency of recombination in lentivirus envelope genes has a strong parallel in the phage world. Tail fiber proteins are the functional analogues of lentivirus envelope glycoproteins in that they are both involved in host cell attachment, specificity, and genome entry. As discussed by Kutter *et al.* (1996) and Haggård-Ljungquist *et al.* (1992), the makeup of tail fiber genes of many phages (including Mu, P1, P2, and T4) are clearly derived from genetic exchange between viral types. This should not be surprising, as viral proteins that control cell host range can be thought of as "passkeys" that allow entry into reproductive niches.

4. Genetic Exchange as Adaptation

It is highly likely that recombination in the lentiviruses, phages, and perhaps other viruses is not only a significant cause of host range/tissue expansion and emergence but is also maintained and promoted to some extent by selective pressure related to this very attribute. That is to say, niche exploration, perhaps in the form of interspecies transmission (reproductive substrate variation), itself may be imposing directional selective pressures that facilitate and preserve the characteristic of recombination.

Newly occupied adaptive peaks may be quickly ascended by RNA and retroviruses since these viruses have high mutation rates and exist as quasi-species that can evolve very rapidly due to the presence of enormous genetic variance. Under such circumstances, voyages through sequence space either pay off quickly (as when an adaptive peak is discovered) resulting in viral diversification or cladogenesis, or fail quickly, resulting in viral extinction. In these cases, the interaction (coinfection with recombination) of different viral types may be highly favorable. There may also be similar selective mechanisms operating on the integrase proteins of many viruses including retroviruses.

The equine encephalomyelitis complex alphaviruses represent an excellent example of viral use of recombination for evolutionary diversification. The evolution of these viruses is reviewed and discussed extensively by Weaver *et al.* (1994, 1997), Strauss (1993), and Scott *et al.* (1994). Phylogenetic analysis has revealed how recombination in these viruses has resulted in the acquisition and exchange of genes (and traits) between different viral types, which has given rise to many evolutionarily successful taxonomic groups which persist today. Strauss (1993) has shown how western equine encephalitis virus (WEEV) is actually a recombinant between eastern equine encephalitis virus (EEEV) and Sindbis virus. The recombination is manifest biologically not only in the pathogenicity of WEEV, which is similar to that of EEEV, but also in the virus's vector specificity (*Culex* mosquitoes), which differs from that of EEEV (*Culiseta* mosquitoes).

Genetic reassortment has often been implicated as a primary driving force in the evolution of influenza viruses, allowing both for host species jumps and for the evasion of specific host antiviral responses (reviewed in Webster *et al.*, 1992; Scholtissek, 1995).

F. Mutation and Random Genetic Drift

Mutation is not only a frequently inadequate method of generating requisite evolutionary diversity, it also has the potential, when coupled with random genetic drift, to cause significant decreases in fitness. Following a drastic reduction in population size, the probability of fixation of deleterious mutations (which occur with much higher frequency than advantageous ones) is enhanced by the improbability of subsequent reversions (Haigh, 1978; Chao, 1991) and a lack of restorative mechanisms (Muller, 1964). This is termed "Muller's ratchet" since mutations in subsequent (colonizing) populations are accumulated in an irreversible fashion similar to the advance of a ratchet (Muller, 1964). The end-result is a decrease in fitness caused by an increasing mutational load, and this has been observed in numerous viral systems *in vitro*. Significant fitness decreases following plaque-to-plaque transfers (bottlenecks) have been demonstrated for VSV (Duarte *et al.*, 1992, 1993), RNA phage Φ6 (Chao, 1990; Chao *et al.*, 1992), FMDV (Escarmis *et al.*, 1996), as well as in the DNA-based cellular microorganism, *Salmonella typhimurium* (Andersson and Hughes, 1996).

Viral bottlenecks occurring in nature have been investigated to a limited extent (see Domingo *et al.*, 1985; Gebauer *et al.*, 1988), but their frequency is most likely common during certain host–host transmission events and intrahost colonization of secondary replication sites (Domingo and Holland, 1994; Domingo *et al.*, 1996). In viruses susceptible to the effects of Muller's ratchet due to transmission dynamics or other aspects of their host colonization biology, genetic exchanges via recombination and reassortment may represent a "buffer" that prevents population fitness loss due to an otherwise extensive accumulation of deleterious mutations. This mechanistic "relief valve" has been suggested as a selective determinant supporting the evolution and maintenance of genetic exchange in RNA viruses by Chao (1991, 1992) and Chao and colleagues (1992).

G. Genetic Exchange: The Upshot

Genetic exchange in viruses, whether through recombination or reassortment, translates into the production of genetic and phenotypic diversity, which is of undeniable significance in viral survival as well as viral evolution. Although largely ignored as an ecological interaction, genetic exchange in viruses is a process that involves extremely close association (coinfection of a single cell) between individual viruses whose phylogenetic relationship may be somewhat arbitrary. This interaction has the potential to influence the evolutionary path of the existing viruses and even the power to create new viral taxa. The degree and direction of the resulting evolutionary change is intimately dependent on the virus types involved, the conditions of the interaction, both host and virus population characteristics, other coevolutionary interactions, as well as many other probabilistic components.

VII. CONCLUSION: VIRAL DIVERSITY DATA AND VIRAL DIVERSIFICATION

Unfortunately, this review represents an inadequate introduction to the evolutionary ecology of viruses, which remains largely unexplored as a discipline. The primary basis of this shortcoming is the representational scope of scientific data that exists for viral taxa. As is apparent by our recurrent use of specific virus types in illustration (e.g., HIV), there is a clear overrepresentation of information on viruses that (1) infect humans or economically important biota and (2) are pathogenic. The skewed distribution of data type may be an inaccurate representation of the "viral universe" that exists on our planet. Therefore, while the existent virology literature may be immense and thus allow elucidation of amazing phenomena, our knowledge about chemistry, biology, and evolution will certainly be enhanced by investigation of new and unusual viruses. This limitation also precludes the types of large-scale generalizations about virus biology and evolution that can occasionally be made regarding other organisms.

What hopefully has emerged from our discussion as presented in this chapter is the condition and capacity of the selective environment in determining the evolutionary path of viruses. This selective environment is partitioned into many compartments, each of which has a role in shaping the overall adaptive landscape. In addition, we hopefully have demonstrated the extreme range of virus adaptability that allows these beings to parasitize genomes in essentially any of their possible manifestations and to subsequently exhibit overwhelming diversity. The

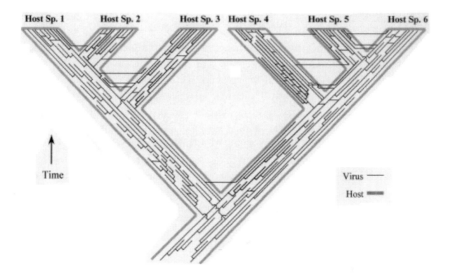

Fig. 9. Hypothetical host–virus phylogeny illustrating various patterns of cladogenesis. As discussed in the text, virus diversification takes place within host taxa not only via adaptive and nonadaptive radiation, but may also occur through virus–host cospeciation, host jumping, and genetic exchange.

mechanisms that are employed in diversification (adaptation, genetic exchange, mutation, niche segregation, cospeciation) contribute fundamentally to the broad range of their variation. We have attempted to illustrate this in Figure 9, which demonstrates the evolutionary history of imaginary virus taxa in relation to their hosts.

The existence of viruses will also induce evolutionary responses in their hosts, which in turn feeds back as changes in the virus adaptive landscape. Changes in this landscape rarely appear to be so profound as to end all genome parasitization. The study of viral evolutionary ecology will assist researchers in many disciplines as they endeavor to explain and predict changes in viral reproduction, including the emergence of viruses as pathogens in the more "conspicuous" species such as *Homo sapiens*.

ACKNOWLEDGMENTS

The authors would like to thank Steve Frank, Christon Hurst, Esteban Domingo, Donald Burke, Mark Lipsitch, and Walter Fitch for their valuable and insightful comments.

REFERENCES

Aaby, P. (1988). Malnutrition and overcrowding/intensive exposure in severe measles infection: Review of community studies. *Rev. Infect. Dis.* **10**, 478–491.

Ackrill, A. M., and Blair, G. E. (1988). Regulation of major histocompatibility class I gene expression at the level of transcription in highly oncogenic adenovirus transformed rat cells. *Oncogene* **3**, 483–487.

Ackrill, A. M., Foster, G. R., Laxton, C. D., Flavell, D. M., Stark, G. R., and Kerr, I. M. (1991). Inhibition of the cellular response to interferons by products of the adenovirus type 5 E1A oncogene. *Nucleic Acids Res.* **19**, 4387–4393.

Adames, A. J., Peralta, P. H., Saenz, R., Johnson, C. M., and Galindo, P. (1979). Brote de encefalomyletis equina Venezolana (VEE) durante la formacion de Lag Bayano, en Panama, 1977. *Rev. Med. Panama* **4**, 246–257.

Adams, J. R., and McClintock, J. T. (1991). Nuclear polyhidrosis viruses of insects. *In* "Atlas of Invertebrate Viruses" (J. R. Adams and J. R. Bonami, eds.), pp. 87–204. CRC Press, Boca Raton, FL.

Ahmed, R., Salmi, A., Butler, L. D., Chiller, J. M., and Oldstone, M. B. (1984). Selection of genetic variants of lymphocytic choriomeningitis virus in spleens of persistently infected mice: Role in suppression of cytotoxic T lymphocyte response and viral persistence. *J. Exp. Med.* **160**, 521–540.

Ahmed, R., Butler, L. D., and Bhatti, L. (1988). T4$^+$ T helper cell function in vivo: Differential requirement for induction of antiviral cytotoxic T-cell and antibody responses. *J. Virol.* **62**, 2102–2106.

Ahmed, R., Morrison, L. A., and Knipe, D. M. (1997). Viral persistence. *In* "Viral Pathogenesis" (N. Nathanson, R. Ahmed, F. Gonzalez-Scarano, D. E. Griffin, K. V. Holmes, F. A. Murphy, and H. L. Robinson, eds.), pp. 181–205. Lippincott-Raven, Philadelphia.

Aitken, T. H., Tesh, R. B., Beaty, B. J., and Rosen, L. (1979). Transovarial transmission of yellow fever virus by mosquitoes (*Aedes aegypti*). *Am. J. Trop. Med. Hyg.* **28**, 119–121.

Albrecht, J. C., and Fleckenstein, B. (1992). New member of the multigene family of complement control proteins in herpesvirus saimiri. *J. Virol.* **66**, 3937–3940.

Albrecht, J. C., Nicholas, J., Cameron, K. R., Newman, C., Fleckenstein, B., and Honess, R. W. (1992). Herpesvirus saimiri has a gene specifying a homologue of the cellular membrane glycoprotein CD59. *Virology* **190**, 527–530.

Alcami, A., and Smith, G. L. (1992). A soluble receptor for interleukin-1 beta encoded by vaccinia virus: A novel mechanism of virus modulation of the host response to infection. *Cell* **71**, 153–167.

Alexander, D. J. (1986). Avian diseases: Historical aspects. *In* "Proceedings of the 2nd International Symposium on Avian Influenza," pp. 4–13. U.S. Animal Health Association, Richmond, VA.

Anderson, M., McMichael, A., and Peterson, P. A. (1987). Reduced allorecognition of adenovirus-2 infected cells. *J. Immunol.* **138**, 3960–3966.

Anderson, R. M. (1982). Transmission dynamics and control of infectious disease agents. *In* "Population Biology of Infectious Diseases" (R. M. Anderson and R. M. May, eds.), pp. 149–176. Springer, Berlin.

Anderson, R. M. (1991). Populations and infectious diseases: Ecology or epidemiology? *J. Anim. Ecol.* **60**, 1–50.

Anderson, R. M., and May, R. M. (1982). Coevolution of hosts and parasites. *Parasitology* **85**, 411–426.

Anderson, R. M., and May, R. M. (1991). "Infectious Diseases of Humans: Dynamics and Control." Oxford University Press, New York.

Andersson, D. I., and Hughes, D. (1996). Muller's ratchet decreases fitness of a DNA-based microbe. *Proc. Natl. Acad. Sci. U.S.A.* **93**, 906–907.

Armstrong, J. A., and Horne, R. (1984). Follicular dendritic cells and virus-like particles in AIDS-related lymphadenopathy. *Lancet* **ii**, 370.

Asgari, S., Schmidt, O., and Theopold, U. (1997). A polydnavirus-encoded protein of an endoparasitoid wasp is an immune suppressor. *J. Gen. Virol.* **78**, 3061–3070.

Atencio, I. A., Belli, B., Hobbs, M., Cheng, S. F., Villarreal, L. P., and Fan, H. (1995). A model for mixed virus disease: Co-infection with Moloney murine leukemia virus potentiates runting induced by polyomavirus (A2 strain) in Balb-c and NIH Swiss mice. *Virology* **212**, 356–366.

Babbitt, F. L., Galbraith, N. S., McDonald, J. C., Shaw, A., and Zuckerman, A. J. (1963). Deaths from measles in England and Wales in 1961. *Mon. Bull. Minist. Pub. Health Lab. Serv.* **22**, 167–175.

Bachrach, H. L. (1968). Foot-and-mouth disease. *Annu. Rev. Microbiol.* **22**, 201–244.

Bagasra, O., Seshamma, T., Oakes, J. W., and Pomerantz, R. J. (1993). High percentages of CD4-positive lymphocytes harbor the HIV-1 provirus in the blood of certain infected individuals. *AIDS* **7**, 1419–1425.

Bailey, N. J. T. (1975). "The Mathematical Theory of Infectious Diseases." Macmillan, New York.

Bang, Y. H., Bown, D. N., Onwubiko, A. O., and Lambrecht, F. L. (1979). Prevalence of potential vectors of yellow fever in the vicinity of Enugu, Nigeria. *Cah. ORSTOM Ser. Entomol. Med. Parasitol.* **17**, 139.

Bang, Y. H., Bown, D. N., and Onwubiko, A. O. (1981). Prevalence of larvae of potential yellow fever vectors in domestic water containers in south-east Nigeria. *Bull. World Health Organiz.* **59**, 107–114.

Bangham, C. R. M. (1995). The influence of immunity on virus evolution. *In* "Molecular Basis of Virus Evolution" (A. J. Gibbs, C. H. Calisher, and F. Garcia–Arenal, eds.), pp. 150-164. Cambridge University Press, Cambridge.

Barnett, S. W., Barboza, A., Wilcox, C. M., Forsmark, C. E., and Levy, J. A. (1991). Characterization of human immunodeficiency virus type 1 strains recovered from the bowel of infected individuals. *Virology* **182**, 802–809.

Bartlett, M. S. (1957). Measles periodicity and community size. *J. R. Stat. Soc. Ser. A* **120**, 48–70.

Bates, M., and Roca-Garcia, M. (1946). The development of the virus of yellow fever in *Haemagogus* mosquitoes. *Am. J. Trop. Med.* **26**, 585.

Baum, D. A., and Larson, A. (1991). Adaptation reviewed: A phylogenetic methodology for studying character macroevolution. *Syst. Zool.* **40**, 1–18.

Beaty, B. J., Tesh, R. B., and Aitken, T. H. (1980). Transovarial transmission of yellow fever virus in *Stegomyia* mosquitoes. *Am. J. Trop. Med. Hyg.* **29**, 125–132.

Begon, M., Harper, J. L., and Townsend, C. R. (1990). "Ecology: Individuals, Populations, and Communities." Blackwell Scientific, Boston.

Behncken, G. M. (1973). Evidence of multiplication of sowthistle yellow vein virus in an inefficient aphid vector, *Macrosiphum euphorbiae*. *Virology* **53**, 405–412.

Best, S., Le Tissier, P. R., and Stoye, J. P. (1997). Endogenous retroviruses and the evolution of resistance to retroviral infection. *Trends Microbiol.* **5**, 313–318.

Bhat, R. A., Metz, B., and Thimmappaya, B. (1983). Organization of the noncontiguous promoter components of adenovirus VAI RNA gene is strikingly similar to that of eucaryotic tRNA genes. *Mol. Cell. Biol.* **3**, 1996–2005.

Black, F. L. (1991). Epidemiology of *Paramyxoviridae*. *In* "The Paramyxoviruses" (D. W. Kingsbury, ed.), pp. 509–536. Plenum, New York.

Black, F. L., Hierholzer, W., Woodall, J. P., and Pinhiero, F. (1971). Intensified reactions to measles vaccine in unexposed populations of american Indians. *J. Infect. Dis.* **124**, 306–317.

Blissard, G. W., and Rohrmann, G. F. (1990). Baculovirus diversity and molecular biology. *Annu. Rev. Entomol.* **35**, 127–155.

Blissard, G. W., Fleming, J. G. W., Vinson, S. B., and Summers, M. D. (1986a). *Campoletis sonorensis* virus: Expression in *Heliothis virescens* and identification of expressed sequences. *J. Insect Physiol.* **32**, 351–359.

Blissard, G. W., Vinson, S. B., and Summers, M. D. (1986b). Identification, mapping, and in vitro translation of *Campoletis sonorensis* virus mRNAs from parasitized *Heliothis virescens* larvae. *J. Virol.* **57**, 318–327.

Blissard, G. W., Theilmann, D. A., and Summers, M. D. (1989). Segment W of *Campoletis sonorensis* virus: Expression, gene products, and organization. *Virology* **169**, 78–89.

Bonhoeffer, S., and Nowak, M. A. (1994). Intra-host versus inter-host selection: Viral strategies of immune function impairment. *Proc. Natl. Acad. Sci. U.S.A.* **91**, 8062–8066.

Both, G. W., Sleigh, M. J., Cox, N. J., and Kendal, A. P. (1983). Antigenic drift in influenza virus H3 hemagglutinin from 1968 to 1980: Multiple evolutionary pathways and sequential amino acid changes at key antigenic sites. *J. Virol.* **48**, 52–60.

Bowen, M. D., Peters, C. J., and Nichol, S. T. (1997). Phylogenetic analysis of the Arenaviridae: Patterns of virus evolution and evidence for cospeciation between arenaviruses and their rodent hosts. *Mol. Phylogen. Evol.* **8**, 301–316.

Brooks, D. R., and McLennan, D. A. (1993). "Parascript: Parasites and the Language of Evolution." Smithsonian Institution Press, Washington DC.

Brooks, D. R., O'Grady, R. T., and Glen, D. R. (1985). The phylogeny of the Cercomeria Brooks, 1982 (Platyhelminthes). *Proc. Helminthol. Soc. Wash.* **52**, 1–20.

Brown, I., and Cartwright, S. (1986). New porcine coronavirus? [letter]. *Vet. Rec.* **119**, 282–283.

Brown, J. P., Twardzik, D. R., Marquardt, H., and Todaro, G. J. (1985). Vaccinia virus encodes a polypeptide homologous to epidermal growth factor and transforming growth factor. *Nature* **313**, 491–492.

Buchmeier, M. J., Welsh, R. M., Dutko, F. J., and Oldstone, M. B. (1980). The virology and immunobiology of lymphocytic choriomeningitis virus infection. *Adv. Immunol.* **30**, 275–331.

Bull, J. J. (1994). Perspective: Virulence. *In* "Evolution," Vol. 48, pp. 1423–1437.

Bulmer, M. (1994). "Theoretical Evolutionary Ecology." Sinauer, Sunderland, MA.

Burke, D. S. (1997). Recombination in HIV: An important viral evolutionary strategy. *Emerg. Infect. Dis.* **3**, 253–259.

Burkhardt, A. L., Bolen, J. B., Kieff, E., and Longnecker, R. (1992). An Epstein–Barr virus transformation-associated membrane protein interacts with src family tyrosine kinases. *J. Virol.* **66**, 5161–5167.

Burnet, F. M., and White, D. O. (1972). "Natural History of Infectious Disease." Cambridge University Press, London.

Byrne, J. A., and Oldstone, M. B. (1984). Biology of cloned cytotoxic T lymphocytes specific for lymphocytic choriomeningitis virus: Clearance of virus in vivo. *J. Virol.* **51**, 682–686.

Cann, A. J. (1995). "Principles of Molecular Virology." Academic Press, San Diego.

Carrada Bravo, T., and Velazquez Diaz, G. (1980). El impacto del sarampion en México. *Salud. Pub. Mex.* **22**, 359–405.

Carrington, J. C., Kasschau, K. D., Mahajan, S. K., and Schaad, M. C. (1996). Cell-to-cell and long-distance transport of viruses in plants. *Plant Cell Physiol.* **8**, 1669–1681.

Chamberlain, R. W., and Sudia, W. D. (1955). The effects of temperature upon the extrinsic incubation of equine encephalitis in mosquitoes. *Am. J. Hyg.* **62**, 295.

Chamberlain, R. W., Corristan, E. C., and Sikes, R. K. (1954). Studies on the North American arthropod-borne encephalitides, V: The extrinsic incubation of eastern and western equine encephalitis in mosquitoes. *Am. J. Hyg.* **60**, 269.

Chan, S. Y., Delius, H., Halpern, A. L., and Bernard, H. U. (1995). Analysis of genomic sequences of 95 papillomavirus types: Uniting typing, phylogeny, and taxonomy. *J. Virol.* **69**, 3074–3083.

Chan, S. Y., Bernard, H. U., Ratterree, M., Birkebak, T. A., Faras, A. J., and Ostrow, R. S. (1997). Genomic diversity and evolution of papillomaviruses in rhesus monkeys. *J. Virol.* **71**, 4938–4943.

Chang, H. W., Watson, J. C., and Jacobs, B. L. (1992). The E3L gene of vaccinia virus encodes an inhibitor of the interferon-induced, double-stranded RNA-dependent protein kinase. *Proc. Natl. Acad. Sci. U.S.A.* **89**, 4825–4829.

Chang, W., Upton, C., Hu, S. L., Purchio, A. F., and McFadden, G. (1987). The genome of Shope fibroma virus, a tumorigenic poxvirus, contains a growth factor gene with sequence similarity to those encoding epidermal growth factor and transforming growth factor alpha. *Mol. Cell. Biol.* **7**, 535–540.

Chao, L. (1990). Fitness of RNA virus decreased by Muller's ratchet [see comments]. *Nature* **348**, 454–455.

Chao, L. (1991). Fitness of RNA decreased by Muller's ratchet. *Nature* **348**, 454–455.

Chao, L. (1992). Evolution of sex in RNA viruses. *Trends Ecol. Evol.* **7**, 147–151.

Chao, L. (1994). Evolution of genetic exchange in RNA viruses. *In* "The Evolutionary Biology of Viruses" (S. S. Morse, ed.), pp. 233–250. Raven Press, New York.

Chao, L., Tran, T., and Matthews, C. (1992). Muller's ratchet and the advantage of sex in the RNA virus f6. *Evolution* **46**, 289–299.

Chen, L. C., Chowdhury, A., and Huffman, S. L. (1980). Anthropometric assessment of energy–protein malnutrition and subsequent risk of mortality among preschool aged children. *Am. J. Clin. Nutr.* **33**, 1836–1845.

Cheng-Mayer, C., Weiss, C., Seto, D., and Levy, J. A. (1989). Isolates of human immunodeficiency virus type 1 from the brain may constitute a special group of the AIDS virus. *Proc. Natl. Acad. Sci. U.S.A.* **86**, 8575–8579.

Chessin, M. (1972). Effect of radiation on viruses. *In* "Principles and Techniques in Plant Virology" (C. I. Kado and H. O. Agrawal, eds.), pp. 531–545. Van Nostrand Reinhold, New York.

Childs, J. C., and Peters, C. J. (1993). Ecology and epidemiology of arenaviruses and their hosts. *In* "The Arenaviridae" (M. S. Salvato, ed.), pp. 331–373. Plenum, New York.

Choi, Y., Kappler, J. W., and Marrack, P. (1991). A superantigen encoded in the open reading frame of the 3′ long terminal repeat of mouse mammary tumour virus. *Nature* **350**, 203–207.

Christensen, P. E., Schmidt, H., Bang, H. O., Andersen, V., Jordal, B., and Jensen, O. (1953). An epidemic of measles in southern Greenland, 1951. Measles in virgin soil, II: The epidemic proper. *Acta Med. Scand.* **144**, 430–449.

Clarke, D. K., Duarte, E. A., Elena, S. F., Moya, A., Domingo, E., and Holland, J. (1994). The red queen reigns in the kingdom of RNA viruses. *Proc. Natl. Acad. Sci. U.S.A.* **91**, 4821–4824.

Cliff, A. D., and Haggett, P. (1985). "The Spread of Measles in Fiji and the Pacific." Department of Human Geography, Australian National University, Canberra, Australia.

Coffin, J. M. (1982). Endogenous viruses. *In* "RNA Tumor Viruses" (R. A. Weiss, N. Teich, H. Varmus, and J. Coffin, eds.), Vol. 1, pp. 1109–1203. Cold Spring Harbor Laboratory, Cold Spring Harbor, NY.

Cohen, A. H., Sun, N. C., Shapshak, P., and Imagawa, D. T. (1989). Demonstration of human immunodeficiency virus in renal epithelium in HIV-associated nephropathy. *Mod. Pathol.* **2**, 125–128.

Contag, C. H., and Plagemann, P. G. (1989). Age-dependent poliomyelitis of mice: Expression of endogenous retrovirus correlates with cytocidal replication of lactate dehydrogenase-elevating virus in motor neurons. *J. Virol.* **63**, 4362–4369.

Contag, C. H., Harty, J. T., and Plagemann, P. G. (1989). Dual virus etiology of age-dependent poliomyelitis of mice. A potential model for human motor neuron diseases. *Microb. Pathog.* **6**, 391–401.

Cooper, J. I. (1995). "Viruses and the Environment." Chapman and Hall, London.

Corbitt, G., Bailey, A. S., and Williams, G. (1990). HIV infection in Manchester, 1959 [letter] [see comments]. *Lancet* **336**, 51.

Coyne, J. A., Barton, N. H., and Turelli, M. (1997). Perspective: A critique of Sewall Wright's shifting balance theory of evolution. *Evolution* **51**, 643–671.

Cracraft, J. (1982). Geographic differentiation, cladistics, and vicariance biogeography: Reconstructing the tempo and mode of evolution. *Amer. Zool.* **22**, 411–424.

Crittenden, L. B., Smith, E. J., Weiss, R. A., and Sarma, P. S. (1974). Host gene control of endogenous avian leukosis virus production. *Virology* **57**, 128–138.

Davies, A. M., and Yoshpe-Purer, Y. (1954). The transmission of Semliki Forest virus by *Aedes aegypti*. *J. Trop. Med. Hyg.* **57**, 273.

Davies, D. H., and Vinson, S. B. (1988). Interference with function of plasmatocytes of *Heliothis virescens* in vivo by calyx fluid of the parasitoid *Campoletis sonorensis*. *Cell Tissue Res.* **251**, 467–475.

Davies, D. H., Strand, M. R., and Vinson, S. B. (1987). Changes in differential heamocyte count and in vitro behavior of plasmatocytes from host *Heliothis virescens* caused by *Campoletis sonorensis* polydnavirus. *J. Insect Physiol.* **33**, 143–153.

Davies, M. V., Chang, H. W., Jacobs, B. L., and Kaufman, R. J. (1993). The E3L and K3L vaccinia virus gene products stimulate translation through inhibition of the double-stranded RNA-dependent protein kinase by different mechanisms. *J. Virol.* **67**, 1688–1692.

Dawson, W. O., Bubrick, P., and Grantham, G. L. (1988). Modification of the tobacco mosaic virus coat protein gene affecting replication, movement, and symptomatology. *Phytopathology* **78**, 783–789.

de la Torre, J. C., and Holland, J. J. (1990). RNA virus quasispecies populations can suppress vastly superior mutant progeny. *J. Virol.* **64**, 6278–6281.

del Val, M., Hengel, H., Hacker, H., Hartlaub, U., Ruppert, T., Lucin, P., and Koszinowski, U. H. (1992). Cytomegalovirus prevents antigen presentation by blocking the transport of peptide-loaded major histocompatibility complex class I molecules into the medial-Golgi compartment. *J. Exp. Med.* **176**, 729–738.

de Rossi, A., Aldovini, A., Franchini, G., Mann, D., Gallo, R. C., and Wong-Staal, F. (1985). Clonal selection of T lymphocytes infected by cell-free human T-cell leukemia/lymphoma virus type I: Parameters of virus integration and expression. *Virology* **143**, 640–645.

Deom, C. M., Lapidot, M., and Beachy, R. N. (1992). Plant virus movement proteins. *Cell* **69**, 221–224.

Dolja, V. V., Haldeman, R., Robertson, N. L., Dougherty, W. G., and Carrington, J. C. (1994). Distinct functions of capsid protein in assembly and movement of tobacco etch potyvirus in plants. *EMBO J.* **13**, 1482–1491.

Dolja, V. V., Haldeman-Cahill, R., Montgomery, A. E., Vandenbosch, K. A., and Carrington, J. C. (1995). Capsid protein determinants involved in cell-to-cell and long distance movement of tobacco etch potyvirus. *Virology* **206**, 1007–1016.

Domingo, E., and Holland, J. J. (1994). Mutation rates and rapid evolution of RNA viruses. *In* "The Evolutionary Biology of Viruses" (S. S. Morse, ed.), pp. 161-184. Raven Press, New York.

Domingo, E., and Holland, J. J. (1997). RNA virus mutations and fitness for survival. *Annu. Rev. Microbiol.* **51**, 151–178.

Domingo, E., Martinez-Salas, E., Sobrino, F., de la Torre, J. C., Portela, A., Ortin, J., Lopez-Galindez, C., Perez-Brena, P., Villanueva, N., Najera, R., *et al.* (1985). The quasispecies (extremely heterogeneous) nature of viral RNA genome populations: Biological relevance — a review. *Gene* **40**, 1–8.

Domingo, E., Escarmis, C., Sevilla, N., Moya, A., Elena, S. F., Quer, J., Novella, I. S., and Holland, J. J. (1996). Basic concepts in RNA virus evolution. *FASEB J.* **10**, 859–864.

Donaldson, A. I., Gloster, J., Harvey, L. D. J., and Deans, D. H. (1982). Use of prediction models to forecast and analyse airborne spread during the foot-and-mouth disease outbreaks in Brittany, Jersey and the Isle of Wight in 1981. *Vet. Record* **110**, 53–57.

Dornburg, R. (1997). From the natural evolution to the genetic manipulation of the host-range of retroviruses. *Biol. Chem.* **378**, 457–468.

Dover, B. A., Davies, D. H., Strand, M. R., Gray, R. S., Keeley, L. L., and Vinson, S. B. (1987). Ecdysteroid-titre reduction and developmental arrest of last-instar *Heliothis virescens* larvae by calyx fluid from the parasitoid *Campoletis sonorensis. J. Insect Physiol.* **33**, 333–338.

Duarte, E., Clarke, D., Moya, A., Domingo, E., and Holland, J. (1992). Rapid fitness losses in mammalian RNA virus clones due to Muller's ratchet. *Proc. Natl. Acad. Sci. U.S.A.* **89**, 6015–6019.

Duarte, E. A., Clarke, D. K., Moya, A., Elena, S. F., Domingo, E., and Holland, J. (1993). Many-trillionfold amplification of single RNA virus particles fails to overcome the Muller's ratchet effect. *J. Virol.* **67**, 3620–3623.

Dutary, B. E., and Leduc, J. W. (1981). Transovarial transmission of yellow fever virus by a sylvatic vector, *Haemagogus equinis* [letter]. *Trans. R. Soc. Trop. Med. Hyg.* **75**, 128.

Eastop, V. F. (1983). The biology of the principal aphid virus vectors. *In* "Plant Virus Epidemiology" (R. T. Plumb and J. M. Thresh, eds.), pp. 115–132. Blackwell, Oxford.

Edson, K. M., Vinson, S. B., Stoltz, D. B., and Summers, M. D. (1980). Virus in a parasitoid wasp: Suppression of the cellular immune response in the parasitoid's host. *Science* **211**, 582–583.

Eigen, M., and Biebricher, C. K. (1988). Sequence space and quasispecies distribution. *In* "RNA Genetics," Vol. 3, pp. 211–245.

Erwin, D. H. (1992). A preliminary classification of evolutionary radiations. *Historical Biol.* **6**, 133–147.

Escarmis, C., Davila, M., Charpentier, N., Bracho, A., Moya, A., and Domingo, E. (1996). Genetic lesions associated with Muller's ratchet in an RNA virus. *J. Mol. Biol.* **264**, 255–267.

Essani, K., Chalasani, S., Eversole, R., Beuving, L., and Birmingham, L. (1994). Multiple anti-cytokine activities secreted from tanapox virus-infected cells. *Microb. Pathog.* **17**, 347–353.

Ewald, P. W. (1983). Host-parasite relations, vectors, and the evolution of disease severity. *Annu. Rev. Ecol. Syst.* **14**, 465–485.

Ewald, P. W. (1987). Transmission modes and evolution of the parasitism-mutualism continuum. *Ann. N. Y. Acad. Sci.* **503**, 295–306.

Ewald, P. W. (1994a). "Evolution of Infectious Disease." Oxford University Press, New York.

Ewald, P. W. (1994b). Evolution of mutation rate and virulence among human retroviruses. *Philos. Trans. R. Soc. Lond. B. Biol. Sci.* **346**, 333–341 [discussion 341–343].

Ewald, P. W., and Schubert, J. (1989). Vertical and vector-borne transmission of insect endosymbionts, and the evolution of benignness. *In* "CRC Handbook of Insect Endosymbiosis: Morphology, Physiology, Genetics and Evolution" (W. Schwemmler, ed.), pp. 21–35. CRC Press, Boca Raton, FL.

Fenner, F. (1983). The Florey lecture, 1983. Biological control, as exemplified by smallpox eradication and myxomatosis. *Proc. R. Soc. Lond. B. Biol. Sci.* **218**, 259–285.

Fenner, F., and Ratcliffe, F. N. (1965). "Myxomatosis." Cambridge University Press, Cambridge.

Fields, B. N., Knipe, D. M., Howley, P. M., *et al.* (1996a). "Virology," Vol. 1. Lippincott-Raven, Philadelphia.

Fields, B. N., Knipe, D. M., Howley, P. M., *et al.* (1996b). "Virology," Vol. 2. Lippincott-Raven, Philadelphia.

Fine, P. E. F. (1975). Vectors and vertical transmission: An epidemiological perspective. *Ann. N. Y. Acad. Sci.* **266**, 173–194.

Fisher, R. A. (1930). "The Genetical Theory of Natural Selection." Clarendon, Oxford.

Fleming, J. G. W., and Krell, P. J. (1993). Polydnavirus genome organization. *In* "Parasites and Pathogens of Insects" (N. E. Beckage, S. M. Thompson, and B. A. Federici, eds.), pp. 189–225. Academic Press, San Diego.

Fleming, J. G. W., and Summers, M. D. (1986). *Campoletic sonorensis* endoparasitic wasps contain forms of *C. sonorensis* virus DNA suggestive of integrated and extrachromosomal polydnavirus DNAs. *J. Virol.* **57**, 552–562.

Fleming, J. G. W., and Summers, M. D. (1991). Polydnavirus DNA is integrated in the DNA of its parasitoid wasp host. *Proc. Natl. Acad. Sci. U.S.A.* **88**, 9770–9774.

Foster, G. R., Ackrill, A. M., Goldin, R. D., Kerr, I. M., Thomas, H. C., and Stark, G. R. (1991). Expression of the terminal protein region of hepatitis B virus inhibits cellular responses to interferons alpha and gamma and double-stranded RNA *Proc. Natl. Acad. Sci. U.S.A.* **88**, 2888–2892 [published erratum appears in *Proc. Natl. Acad. Sci. U.S.A.*, 1995, **92**, 3632].

Frank, S. A. (1996). Models of parasite virulence. *Q. Rev. Biol.* **71**, 37–78.

Frank, S. A. (1997). Spatial processes in host–parasite genetics. *In* "Metapopulation Biology," pp. 325–352. Academic Press, San Diego.

Frazer, K. B., and Martin, S. J. (1978). "Measles Virus and Its Biology." Academic Press, London.

Fries, L. F., Friedman, H. M., Cohen, G. H., Eisenberg, R. J., Hammer, C. H., and Frank, M. M. (1986). Glycoprotein C of herpes simplex virus 1 is an inhibitor of the complement cascade. *J. Immunol.* **137**, 1636–1641.

Futuyma, D. J. (1986). "Evolutionary Biology." Sinauer Associates, Sunderland, MA.

Gal-On, A., Kaplan, I., Roossinck, M. J., and Palukaitis, P. (1994). The kinetics of infection of zucchini squash by cucumber mosaic virus indicate a function for RNA 1 in virus movement. *Virology* **205**, 280–289.

Gao, F., Yue, L., White, A. T., Pappas, P. G., Barchue, J., Hanson, A. P., Greene, B. M., Sharp, P. M., Shaw, G. M., and Hahn, B. H. (1992). Human infection by genetically diverse SIVSM-related HIV-2 in West Africa. *Nature* **358**, 495–499.

Garenne, M., Glasser, J., and Levins, R. (1994). Disease, population and virulence. Thoughts about measles mortality. *Ann. N. Y. Acad. Sci.* **740**, 297–302.

Garnett, G. P., and Antia, R. (1994). Population biology of virus–host interactions. *In* "The Evolutionary Biology of Viruses" (S. S. Morse, ed.), pp. 51–73. Raven Press, New York.

Ge, R., Kralli, A., Weinmann, R., and Ricciardi, R. P. (1992). Down-regulation of the major histocompatibility complex class I enhancer in adenovirus type 12-transformed cells is accompanied by an increase in factor binding. *J. Virol.* **66**, 6969–6978.

Gebauer, F., de la Torre, J. C., Gomes, I., Mateu, M. G., Barahona, H., Tiraboschi, B., Bergmann, I., de Mello, P. A., and Domingo, E. (1988). Rapid selection of genetic and antigenic variants of foot-and-mouth disease virus during persistence in cattle. *J. Virol.* **62**, 2041–2049.

Germain, M., Saluzzo, J. F., Cornet, J. P., Herve, J. P., Sureau, P., Camicas, J. L., Robin, Y., Salaun, J. J., and Heme, G. (1979). Isolation of the yellow fever virus from an egg-cluster and the larvae of the tick *Amblyomma variegatum. C. R. Séances Acad. Sci. D* **289**, 635–637.

Gildow, F. E. (1980). Increased production of alatae by aphids reared on oats infected with barley yellow dwarf virus. *Ann. Entomol. Soc. Am.* **73**, 343–347.

Gildow, F. E. (1983). Influence of barley yellow dwarf virus infected oats and barley on morphology of aphid vectors. *Phytopathology* **73**, 1196–1199.

Giske, J., Aksnes, D. L., and Forland, B. (1993). Variable generation times and Darwinian fitness measures. *Evolut. Ecol.* **7**, 233–239.

Golovkina, T. V., Chervonsky, A., Dudley, J. P., and Ross, S. R. (1992). Transgenic mouse mammary tumor virus superantigen expression prevents viral infection. *Cell* **69**, 637–645.

González, J. M., and Suttle, C. A. (1993). Grazing by marine nanoflagellates on viruses and virus-sized particles ingestion and digestion. *Marine Ecol. Prog. Ser.* **94**, 1–10.

Gooding, L. R., Elmore, L. W., Tollefson, A. E., Brady, H. A., and Wold, W. S. (1988). A 14,700-MW protein from the E3 region of adenovirus inhibits cytolysis by tumor necrosis factor. *Cell* **53**, 341–346.

Gooding, L. R., Aquino, L., Duerksen-Hughes, P. J., Day, D., Horton, T. M., Yei, S. P., and Wold, W. S. (1991). The E1B 19,000-molecular-weight protein of group C adenoviruses prevents tumor necrosis factor cytolysis of human cells but not of mouse cells. *J. Virol.* **65**, 3083–3094.

Gorman, O. T., Bean, W. J., Kawaoka, Y., Donatelli, I., Guo, Y., and Webster, R. G. (1991). Evolution of influenza A virus nucleoprotein genes: Implications for the origin of J1Ni human and classical swine viruses. *J. Virol.* **65**, 3704–3714.

Gray, S. M., Power, A. G., Smith, D. M., Seaman, A. J., and Altman, N. S. (1991). Aphid transmission of barley yellow dwarf virus: Acquisition access periods and virus concentration requirements. *Phytopathology* **81**, 539–545.

Griffin, D. E. (1997). Virus-induced immune suppression. *In* "Viral Pathogenesis" (N. Nathanson, ed.), pp. 207–233. Lippincott-Raven, Philadelphia.

Grist, N. R. (1950). The pathogenesis of measles: A review of the literature and discussion of the problem. *Glasg. Med. J.* **31**, 431.

Gutch, M. J., and Reich, N. C. (1991). Repression of the interferon signal transduction pathway by the adenovirus E1A oncogene. *Proc. Natl. Acad. Sci. U.S.A.* **88**, 7913–7917.

Guzo, D., and Stoltz, D. B. (1985). Obligatory multiparasitism in the tussock moth, *Orgyia leucostigma*. *Parasitology* **90**, 1–10.

Haggård-Ljungquist, E., Halling, C., and Calendar, R. (1992). DNA sequences of the tail fiber genes of bacteriophage P2: Evidence for horizontal transfer of tail fiber genes among unrelated bacteriophages. *J. Bacteriol.* **174**, 1462–1477.

Haibach, F., Hata, J., Mitra, M., Dhar, M., Harmata, M., Sun, P., and Smith, D. (1991). Purification and characterization of a *Coffea canephora* alpha-D-galactosidase isozyme. *Biochem. Biophys. Res. Commun.* **181**, 1564–1571.

Haigh, J. (1978). The accumulation of deleterious genes in a population: Muller's ratchet. *Theor. Popul. Biol.* **14**, 251–267.

Hall, W. C., Kovatch, R. M., Herman, P. H., and Fox, J. C. (1971). Pathology of measles in Rhesus monkeys. *Vet. Pathol.* **8**, 307–319.

Hanski, I., and Simberloff, D. (1997). The metapopulation approach, its history, conceptual domain, and application to conservation. *In* "Metapopulation Biology: Ecology, Genetics, and Evolution" (I. Hanski and M. E. Gilpin, eds.). Academic Press, San Diego.

Hardy, J. L. (1988). Susceptibility and resistance of vector mosquitoes. *In* "The Arboviruses: Epidemiology and Ecology" (T. P. Monath, ed.), Vol. 1, pp. 87–126. CRC Press, Boca Raton, FL.

Harris, S. L., Frank, I., Yee, A., Cohen, G. H., Eisenberg, R. J., and Friedman, H. M. (1990). Glycoprotein C of herpes simplex virus type 1 prevents complement-mediated cell lysis and virus neutralization. *J. Infect. Dis.* **162**, 331–337.

Harwood, S. H., and Beckage, N. E. (1994). Purification and characterization of an early-expressed polydnavirus-induced protein from the hemolymph of *Manduca sexta* larvae parasitized by *Cotesia congregata*. *Insect Biochem. Mol. Biol.* **24**, 685–698.

Harwood, S. H., Grosovsky, A. J., Cowles, E. A., Davis, J. W., and Beckage, N. E. (1994). An abundantly expressed hemolymph glycoprotein isolated from newly parasitized *Manduca sexta* larvae is a polydnavirus gene product. *Virology* **205**, 381–392.

Have, P. (1990). Infection with a new porcine respiratory coronavirus in Denmark: Serologic differentiation from transmissible gastroenteritis virus using monoclonal antibodies. *Adv. Exp. Med. Biol.* **276**, 435–439.

Heard, S. B., and Hauser, D. L. (1995). Key evolutionary innovations and their ecological mechanisms. *Historical Biol.* **10**, 151–173.

Held, W., Shakhov, A. N., Izui, S., Waanders, G. A., Scarpellino, L., MacDonald, H. R., and Acha-Orbea, H. (1993). Superantigen-reactive CD4+ T cells are required to stimulate B cells after infection with mouse mammary tumor virus. *J. Exp. Med.* **177**, 359–366.

Herre, E. A. (1993). Population structure and the evolution of virulence in nematode parasites of fig wasps. *Science (Washington, DC)* **259**, 1442–1445.

Hinshaw, V. S., Bean, W. J., Webster, R. G., and Easterday, B. C. (1978). The prevalence of influenza viruses in swine and the antigenic and genetic relatedness of influenza viruses from man and swine. *Virology* **84**, 51–62.

Hinshaw, V. S., Webster, R. G., and Turner, B. (1980). The perpetuation of orthomyxoviruses and paramyxoviruses in Canadian waterfowl. *Can. J. Microbiol.* **26**, 622–629.

Hirata, M., Hayashi, J., Noguchi, A., Nakashima, K., Kajiyama, W., Kashiwagi, S., and Sawada, T. (1992). The effects of breastfeeding and presence of antibody to p40tax protein of human T cell lymphotropic virus type-I on mother to child transmission. *Int. J. Epidemiol.* **21**, 989–994.

Ho, D. D., Schooley, R. T., Rota, T. R., Kaplan, J. C., Flynn, T., Salahuddin, S. Z., Gonda, M. A., and Hirsch, M. S. (1984). HTLV-III in the semen and blood of a healthy homosexual man. *Science* **226**, 451–453.

Ho, D. D., Rota, T. R., Schooley, R. T., Kaplan, J. C., Allan, J. D., Groopman, J. E., Resnick, L., Felsenstein, D., Andrews, C. A., and Hirsch, M. S. (1985). Isolation of HTLV-III from cerebrospinal fluid and neural tissues of patients with neurologic syndromes related to the acquired immunodeficiency syndrome. *New Engl. J. Med.* **313**, 1493–1497.

Holland, J., Spindler, K., Horodyski, F., Grabau, E., Nichol, S., and VandePol, S. (1982). Rapid evolution of RNA genomes. *Science* **215**, 1577–1585.

Holland, J. J. (1996). Evolving virus plagues [comment]. *Proc. Natl. Acad. Sci. U.S.A.* **93**, 545–546.

Holland, J. J., MacLaren, L. C., and Sylverton, J. T. (1959). The mammalian cell-virus relationship, IV: Infection of naturally insusceptible cells with enterovirus nucleic acid. *J. Exp. Med.* **110**, 65–80.

Holland, J. J., Domingo, E., de la Torre, J. C., and Steinhauer, D. A. (1990). Mutation frequencies at defined single codon sites in vesicular stomatitis virus and poliovirus can be increased only slightly by chemical mutagenesis. *J. Virol.* **64**, 3960–3962.

Hsia, K., and Spector, S. A. (1991). Human immunodeficiency virus DNA is present in a high percentage of CD4+ lymphocytes of seropositive individuals. *J. Infect. Dis.* **164**, 470–475.

Hunter, J. P. (1998). Key innovations and the ecology of macroevolution. *Trends Ecol. Evol.* **13**, 31–36.

Hurst, L. D. (1993). The incidences, mechanisms and evolution of cytoplasmic sex ratio distorters in animals. *Biol. Rev. Camb. Philos. Soc.* **68**, 121–194.

Hutchinson, G. E. (1957). Concluding remarks. *Cold Spring Harbor Symp. Quant. Biol.* **22**, 415–427.

Hwang, S. S., Boyle, T. J., Lyerly, H. K., and Cullen, B. R. (1991). Identification of the envelope V3 loop as the primary determinant of cell tropism in HIV-1. *Science* **253**, 71–74.

Imani, F., and Jacobs, B. L. (1988). Inhibitory activity for the interferon-induced protein kinase is associated with the reovirus serotype 1 sigma 3 protein. *Proc. Natl. Acad. Sci. U.S.A.* **85**, 7887–7891.

Isaacs, S. N., Kotwal, G. J., and Moss, B. (1992). Vaccinia virus complement-control protein prevents antibody-dependent complement-enhanced neutralization of infectivity and contributes to virulence. *Proc. Natl. Acad. Sci. U.S.A.* **89**, 628–632.

Isfort, R. J., Witter, R., and Kung, H. J. (1994). Retrovirus insertion into herpesviruses. *Trends Microbiol.* **2**, 174–177.

Jamieson, B. D., Somasundaram, T., and Ahmed, R. (1991). Abrogation of tolerance to a chronic viral infection. *J. Immunol.* **147**, 3521–3529.

Jenner, E. (1959 [reprinted]). An inquiry into the causes and effects of the variolae vaccinae, a disease discovered in some of the western countries of England, particularly Gloucestershire, and known by the name of the cowpox. *In* "Classics of Medicine and Surgery" (L. N. B. Camac, ed.). Dover, New York.

Jestin, A., Le Forban, Y., Vannier, P., Madec, F., and Gourreau, J.-M. (1987). Un nouveau coronavirus porcin. Études sero-epidemiologiques retrospectives dans les elevages de Bretagne. *Rec. Med. Vet.* **163**, 567–571.

Jetten, T. H., and Focks, D. A. (1997). Potential changes in the distribution of dengue transmission under climate warming. *Am. J. Trop. Med. Hyg.* **57**, 285–297.

Jin, M. J., Hui, H., Robertson, D. L., Muller, M. C., Barre-Sinoussi, F., Hirsch, V. M., Allan, J. S., Shaw, G. M., Sharp, P. M., and Hahn, B. H. (1994). Mosaic genome structure of simian immunodeficiency virus from West African green monkeys. *EMBO J.* **13**, 2935–2947.

Kalvakolanu, D. V., Bandyopadhyay, S. K., Harter, M. L., and Sen, G. C. (1991). Inhibition of interferon-inducible gene expression by adenovirus E1A proteins: Block in transcriptional complex formation. *Proc. Natl. Acad. Sci. U.S.A.* **88**, 7459–7463.

Katz, R. A., and Skalka, A. M. (1990). Generation of diversity in retroviruses. *Annu. Rev. Genet.* **24**, 409–445.

Katz, S. L., and Enders, J. F. (1965). Measles virus. *In* "Viral and Rickettsial Infections of Man" (F. L. J. Horsfall and I. Tamm, eds.). Lippincott, Philadelphia.

Katze, M. G. (1992). The war against the interferon-induced dsRNA-activated protein kinase: Can viruses win? *J. Interferon Res.* **12**, 241–248.

Kauffman, S. A., and Levin, S. (1987). Towards a general theory of adaptive walks on rugged landscapes. *J. Theor. Biol.* **128**, 11–45.

Kempe, C. H., and Fulginiti, V. A. (1965). The pathogenesis of measles virus infection. *Arch. Ges. Virusforsch* **16**, 103–128.

Kennedy, J. S., Day, M. F., and Eastop, V. F. (1962). "A Conspectus of Aphids as Vectors of Plant Viruses." Commonwealth Institute of Entomology, London.

Kilbourne, E. D. (1994). Host determination of viral evolution: A variable tautology. *In* "The Evolutionary Biology of Viruses" (S. S. Morse, ed.), pp. 253–271. Raven Press, New York.

Killick, H. J. (1990). Influence of droplet size, solar ultraviolet light and protectants, and other factors on the efficacy of baculovirus sprays against *Panolis flammea* (Schiff.) (Lepidoptera: Noctuidae). *Crop Prot.* **9**, 21–28.

Kinoshita, K., Hino, S., Amagaski, T., Ikeda, S., Yamada, Y., Suzuyama, J., Momita, S., Toriya, K., Kamihira, S., and Ichimaru, M. (1984). Demonstration of adult T-cell leukemia virus antigen in milk from three sero-positive mothers. *Gann* **75**, 103–105.

Kitano, H. (1982). Effect of venom of the gregarious parasitoid *Apanteles glomeratus* on its hemocytic encapsulation by the host, *Pieres*. *J. Invertebr. Pathol.* **40**, 61–67.

Kitano, H. (1986). The role of *Apanteles glomeratus* venom in the defensive response of its host, *Pieres. J. Invertebr. Pathol.* **32**, 369–375.

Komuro, A., Hayami, M., Fujii, H., Miyahara, S., and Hirayama, M. (1983). Vertical transmission of adult T-cell leukaemia virus [letter]. *Lancet* **1**, 240.

Kondo, A., and Maeda, S. (1991). Host range expansion by recombination of the baculoviruses *Bombyx mori* nuclear polyhidrosis virus and *Autographa californica* nuclear polyhidrosis virus. *J. Virol.* **65**, 3625–3632.

Korman, A. J., Bourgarel, P., Meo, T., and Rieckhof, G. E. (1992). The mouse mammary tumour virus long terminal repeat encodes a type II transmembrane glycoprotein. *EMBO J.* **11**, 1901–1905.

Kotwal, G. J., Isaacs, S. N., McKenzie, R., Frank, M. M., and Moss, B. (1990). Inhibition of the complement cascade by the major secretory protein of vaccinia virus. *Science* **250**, 827–830.

Kozak, C. A., Gromet, N. J., Ikeda, H., and Buckler, C. E. (1984). A unique sequence related to the ecotropic murine leukemia virus is associated with the Fv-4 resistance gene. *Proc. Natl. Acad. Sci. U.S.A.* **81**, 834–837.

Krell, P. J., Summers, M. D., and Vinson, S. B. (1982). Virus with a multipartite superhelical DNA genome from the ichneumonid parasitoid *Campoletis sonorensis. J. Virol.* **43**, 859–870.

Kuberski, T. (1979). Fluorescent antibody studies in the development of dengue-2 virus in *Aedes albopictus* (Diptera: Culicidae). *J. Med. Entomol.* **16**, 343–349.

Kusters, J. G., Niesters, H. G., Lenstra, J. A., Horzinek, M. C., and van der Zeijst, B. A. (1989). Phylogeny of antigenic variants of avian coronavirus IBV. *Virology* **169**, 217–221.

Kutter, E., Gachechiladze, K., Poglazov, A., Marusich, E., Shneider, M., Aronsson, P., Napuli, A., Porter, D., and Mesyanzhinov, V. (1996). Evolution of T4-related phages. *Virus Genes* **11**, 285–297.

Lee, H. W., Lee, P. W., and Baek, L. H. (1981). Intraspecific transmission of Hantaan virus, etiologic agent of Korean hemorrhagic fever, in the rodent *Apodemus agrarius. Am. J. Trop. Med. Hyg.* **30**, 1106–1112.

Lee, H. W., Lee, P. W., and Baek, L. H. (1981). Intraspecific transmission of Hantaan virus, etiologic agent of Korean hemorrhagic fever, in the rodent *Apodemus agrarius. Am. J. Trop. Med. Hyg.* **30**, 1106–1112.

Levin, B. R., and Bull, J. J. (1994). Short-sighted evolution and the virulence of pathogenic microorganisms. *Trends Microbiol.* **2**, 76–81.

Levinton, J. S. (1988). "Genetics, Paleontology, and Macroevolution." Cambridge University Press, Cambridge.

Levy, J. A. (1993). Pathogenesis of human immunodeficiency virus infection. *Microbiol. Rev.* **57**, 183–289.

Levy, J. A., Shimabukuro, J., Hollander, H., Mills, J., and Kaminsky, L. (1985). Isolation of AIDS-associated retroviruses from cerebrospinal fluid and brain of patients with neurological symptoms. *Lancet* **2**, 586–588.

Lewontin, R. (1970). The units of selection. *Annu. Rev. Ecol. Syst.* **1**, 1–17.

Li, W.-H., Tanimura, M., and Sharp, P. M. (1988). Rates and dates of divergence between AIDS virus nucleotide sequences. *Mol. Biol. Evol.* **5**, 313–330.

Lipsitch, M., and Nowak, M. A. (1995). The evolution of virulence in sexually transmitted HIV/AIDS. *J. Theor. Biol.* **174**, 427–440.

Lipsitch, M., Siller, S., and Nowak, M. A. (1996). The evolution of virulence in pathogens with vertical and horizontal transmission. *Evolution* **50**, 1729–1741.

Lloyd, R. M., and Shatkin, A. J. (1992). Translational stimulation by reovirus polypeptide sigma, 3: Substitution for VAI RNA and inhibition of phosphorylation of the alpha subunit of eukaryotic initiation factor 2. *J. Virol.* **66**, 6878–6884.

Lomas, D. A., Finch, J. T., Seyama, K., Nukiwa, T., and Carrell, R. W. (1993). Alpha 1-antitrypsin Siiyama (Ser53→Phe). Further evidence for intracellular loop-sheet polymerization. *J. Biol. Chem.* **268**, 15333–15335.

Lotka, A. J. (1932). The growth of mixed populations: Two species competing for a common food supply. *J. Wash. Acad. Sci.* **22**, 461–469.

Louwagie, J., Janssens, W., Mascola, J., Heyndrickx, L., Hegerich, P., van der Groen, G., McCutchan, F. E., and Burke, D. S. (1995). Genetic diversity of the envelope glycoprotein from human immunodeficiency virus type 1 isolates of African origin. *J. Virol.* **69**, 263–271.

Lovejoy, T. E. (1993). Global change and epidemiology: Nasty synergies. *In* "Emerging Viruses" (S. S. Morse, ed.), pp. 261–268. Oxford University Press, New York.

Löwer, R., Löwer, J., and Kurth, R. (1996). The viruses in all of us: Characteristics and biological significance of human endogenous retrovirus sequences. *Proc. Natl. Acad. Sci. U.S.A.* **93**, 5177–5184.

Macen, J. L., Upton, C., Nation, N., and McFadden, G. (1993). SERP1, a serine proteinase inhibitor encoded by myxoma virus, is a secreted glycoprotein that interferes with inflammation. *Virology* **195**, 348–363.

Manns, A., and Blattner, W. A. (1991). The epidemiology of the human T-cell lymphotrophic virus type I and type II: Etiologic role in human disease. *Transfusion* **31**, 67–75.

Mateu, M. G., Martinez, M. A., Rocha, E., Andreu, D., Parejo, J., Giralt, E., Sobrino, F., and Domingo, E. (1989). Implications of a quasispecies genome structure: Effect of frequent, naturally occurring amino acid substitutions on the antigenicity of foot-and-mouth disease virus. *Proc. Natl. Acad. Sci. U.S.A.* **86**, 5883–5887.

Matloubian, M., Concepcion, R. J., and Ahmed, R. (1994). CD4+ T cells are required to sustain CD8+ cytotoxic T-cell responses during chronic viral infection. *J. Virol.* **68**, 8056–8063.

Matthews, M. B., and Shenk, T. (1991). Adenovirus virus-associated RNA and translation control. *J. Virol.* **65**, 5657–5662.

Maule, A. J. (1991). Virus movement in infected plants. *Crit. Rev. Plant Sci.* **9**, 457–473.

May, R. M. (1995). The co-evolutionary dynamics of viruses and their hosts. *In* "Molecular Basis of Virus Evolution" (A. J. Gibbs, C. H. Calisher, and F. Garcia-Arenal, eds.), pp. 192–212. Cambridge University Press, Cambridge.

May, R. M., and Anderson, R. M. (1983). Parasite–host coevolution. *In* "Coevolution" (D. J. Futuyma and M. Slatkin, eds.). Sinauer Associates, Sunderland, MA.

McCance, D. J., and Mims, C. A. (1979). Reactivation of polyoma virus in kidneys of persistently infected mice during pregnancy. *Infect. Immunol.* **25**, 998–1002.

McCutchan, F. E., Artenstein, A. W., Sanders-Buell, E., Salminen, M. O., Carr, J. K., Mascola, J. R., Yu, X. F., Nelson, K. E., Khamboonruang, C., Schmitt, D., Kieny, M. P., McNeil, J. G., and Burke, D. S. (1996). Diversity of the envelope glycoprotein among human immunodeficiency virus type 1 isolates of clade E from Asia and Africa. *J. Virol.* **70**, 3331–3338.

McLaren, L. C., Holland, J. J., and Syverton, J. T. (1959). The mammalian cell–virus relationship, I: Attachment of poliovirus to cultivated cells of primate and non-primate origin. *J. Exp. Med.* **109**, 475.

McLean, A. R., Rosado, M. M., Agenes, F., Vasconcellos, R., and Freitas, A. A. (1997). Resource competition as a mechanism for B cell homeostasis. *Proc. Natl. Acad. Sci. U.S.A.* **94**, 5792–5797.

McLintock, J. (1978). Mosquito–virus relationships of American encephalitides. *Annu. Rev. Entomol.* **23**, 17–37.

McMichael, A. J., and Beers, M. Y. (1994). Climate change and human population health: Global and South Australian perspectives. *Trans. R. Soc. So. Aust.* **118**, 91–98.

McNearney, T. A., Odell, C., Holers, V. M., Spear, P. G., and Atkinson, J. P. (1987). Herpes simplex virus glycoproteins gC-1 and gC-2 bind to the third component of complement and provide protection against complement-mediated neutralization of viral infectivity. *J. Exp. Med.* **166**, 1525–1535.

Miller, A. H. (1949). Some ecologic and morphologic considerations in the evolution of higher taxonomic categories. *In* "Ornithologie als Biologische Wissenschaft" (E. Mayr and E. Schuz, eds.), pp. 84–88. Carl Winter.

Miller, C. L., Longnecker, R., and Kieff, E. (1993). Epstein–Barr virus latent membrane protein 2A blocks calcium mobilization in B lymphocytes. *J. Virol.* **67**, 3087–3094.

Miller, N. R., Bergstrom, G. C., and Gray, S. M. (1991). Identity, prevalence, and distribution of viral diseases in winter wheat in New York in 1988 and 1989. *Plant Disease* **75**, 1105–1108.

Mishler, B. D., and Churchill, S. P. (1984). A cladistic approach to the phylogeny of the "Bryophytes." *Brittonia* **36**, 406–424.

Mold, C., Bradt, B. M., Nemerow, G. R., and Cooper, N. R. (1988). Epstein–Barr virus regulates activation and processing of the third component of complement. *J. Exp. Med.* **168**, 949–969.

Monath, T. P. (1989). Yellow fever. *In* "The Arboviruses: Epidemiology and Ecology" (T. P. Monath, ed.), Vol. 5, pp. 139–231. CRC Press, Boca Raton, FL.

Moore, K. W., Vieira, P., Fiorentino, D. F., Trounstine, M. L., Khan, T. A., and Mosmann, T. R. (1990). Homology of cytokine synthesis inhibitory factor (IL-10) to the Epstein–Barr virus gene BCRFI [published erratum appears in *Science 1990* Oct. 26, **250**, 494]. *Science* **248**, 1230–1234.

Moore, A. H. (1949). Some ecologic and morphologic considerations in the evolution of higher taxonomic categories. *In* "Ornithologie als Biologische Wissenschaft" (E. Mayr and E. Schuz, eds.), pp. 84–88. Carl Winter.

Morse, S. S. (1997). The public health threat of emerging viral disease. *J. Nutr.* **127**, 951S–957S.

Moskophidis, D., Cobbold, S. P., Waldmann, H., and Lehmann-Grube, F. (1987). Mechanism of recovery from acute virus infection: Treatment of lymphocytic choriomeningitis virus-infected mice with monoclonal antibodies reveals that Lyt-2$^+$ T lymphocytes mediate clearance of virus and regulate the antiviral antibody response. *J. Virol.* **61**, 1867–1874.

Moya, A. (1997). RNA viruses as model systems for testing the population genetics view of evolution. In press.

Muller, H. (1986). Replication of infectious bursal disease virus in lymphoid cells. *Arch. Virol.* **87**, 191–203.

Muller, H. J. (1964). The relation of recombination to mutational advance. *Mutat. Res.* **1**, 2–9.

Murphy, E. L., Figueroa, J. P., Gibbs, W. N., Brathwaite, A., Holding-Cobham, M., Waters, D., Cranston, B., Hanchard, B., and Blattner, W. A. (1989). Sexual transmission of human T-lymphotropic virus type I (HTLV-1). *Ann. Intern. Med.* **111**, 555–60.

Murphy, F. A., Kiley, M. P., and Fisher-Hoch, S. P. (1990). Filoviridae: Marbug and Ebola viruses. *In* "Fields Virology" (B. N. Fields, D. M. Knipe, R. M. Chanock, M. S. Hirsch, J. L. Melnick, T. P. Monath, and B. Roizman, eds.), Vol. 1, pp. 933–944. Raven Press, New York.

Murphy, F. A., Fauquet, C. M., Bishop, D. H. L., Ghabrial, S. A., Jarvis, A. W., Martelli, M. A., and Summers, M. D. (1995). "Virus Taxonomy: Classification and Nomenclature of Viruses." Springer-Verlag, New York.

Mushegian, A. R., and Koonin, E. V. (1993). Cell-to-cell movement of plant viruses. Insights from amino acid sequence comparisons of movement proteins and from analogies with cellular transport systems. *Arch. Virol.* **133**, 239–57.

Mussgay, M., and Suarez, O. (1962). Multiplication of vesicular stomatitis virus in *Aedes aegypti* (L.) mosquitoes. *Virology* **17**, 202–204.

Nagy, P. D., and Simon, A. E. (1997). New insights into the mechanisms of RNA recombination. *Virology* **235**, 1–9.

Nakada, K., Kohakura, M., Komoda, H., and Hinuma, Y. (1984). High incidence of HTLV antibody in carriers of *Strongyloides stercoralis* [letter]. *Lancet* **1**, 633.

Nakano, S., Ando, Y., Ichijo, M., Moriyama, I., Saito, S., Sugamura, K., and Hinuma, Y. (1984). Search for possible routes of vertical and horizontal transmission of adult T-cell leukemia virus. *Gann* **75**, 1044–1045.

Neel, J. V., Centerwall, W. R., Chagnon, N. A., and Casey, H. L. (1970). Notes on the effect of measles and measles vaccine in a virgin-soil population of South American Indians. *Am. J. Epidemiol.* **91**, 418–429.

Nelson, J. A., Wiley, C. A., Reynolds-Kohler, C., Reese, C. E., Margaretten, W., and Levy, J. A. (1988). Human immunodeficiency virus detected in bowel epithelium from patients with gastrointestinal symptoms. *Lancet* **1**, 259–262.

Neote, K., DiGregorio, D., Mak, J. Y., Horuk, R., and Schall, T. J. (1993). Molecular cloning, functional expression, and signaling characteristics of a C—C chemokine receptor. *Cell* **72**, 415–425.

Newton, R. C., Limpuangthip, P., Greenberg, S., Gam, A., and Neva, F. A. (1992). Strongyloides stercoralis hyperinfection in a carrier of HTLV-1 virus with evidence of selective immunosuppression [see comments]. *Am. J. Med.* **92**, 202–208.

Norton, W. N., and Vinson, S. B. (1977). Encapsulation of a parasitoid egg within its habitual host: An ultrastructural investigation. *J. Invertebr. Pathol.* **30**, 55–67.

Norton, W. N., and Vinson, S. B. (1983). Correlating the initiation of virus replication with a specific pupal developmental phase of an ichneumonid parasitoid. *Cell Tissue Res.* **231**, 387–398.

Nowak, M. (1991). The evolution of viruses. Competition between horizontal and vertical transmission of mobile genes. *J. Theor. Biol.* **150**, 339–347.

Nowak, M. A. (1992). What is a Quasispecies? *TREE* **7**, 118–121.

Nowak, M., and Schuster, P. (1989). Error thresholds of replication in finite populations mutation frequencies and the onset of Muller's ratchet. *J. Theor. Biol.* **137**, 375–395.

Nowak, M. A., May, R. M., and Anderson, R. M. (1990). The evolutionary dynamics of HIV-1 quasispecies and the development of immunodeficiency disease. *AIDS* **4**, 1095–1103.

Nowak, M. A., May, R. M., Phillips, R. E., Rowland-Jones, S., Lalloo, D. G., McAdam, S., Klenerman, P., Koppe, B., Sigmund, K., Bangham, C. R., *et al.* (1995). Antigenic oscillations and shifting immunodominance in HIV-1 infections [see comments]. *Nature* **375**, 606–611.

O'Doherty, M. J., Van de Pette, J. E., Nunan, T. O., and Croft, D. N. (1984). Recurrent *Strongyloides stercoralis* infection in a patient with T-cell lymphoma-leukaemia [letter]. *Lancet* **1**, 858.

Okochi, K., Sato, H., and Hinuma, Y. (1984). A retrospective study on transmission of adult T cell leukemia virus by blood transfusion: Seroconversion in recipients. *Vox Sang.* **46**, 245–253.

Pääbo, S., Bhat, B. M., Wold, W. S., and Peterson, P. A. (1987). A short sequence in the COOH-terminus makes an adenovirus membrane glycoprotein a resident of the endoplasmic reticulum. *Cell* **50**, 311–317.

Panum, P. L. (1940). "Observations Made During the Epidemic of Measles on the Faroe Islands in the Year 1946." American Publishing Association, New York.

Parrish, C. R. (1993). Canine parvovirus 2: A probable example of interspecies transfer. *In* "Emerging Viruses" (S. S. Morse, ed.), pp. 194–202. Oxford University Press, New York.

Parrish, C. R. (1997). How canine parvovirus suddenly shifted host range. *ASM News* **63**, 307–311.

Parrish, C. R., Have, P., Foreyt, W. J., Evermann, J. F., Senda, M., and Carmichael, L. E. (1988). The global spread and replacement of canine parvovirus strains. *J. Gen. Virol.* **69**, 1111–1116.

Patterson, B. K., Till, M., Otto, P., Goolsby, C., Furtado, M. R., McBride, L. J., and Wolinsky, S. M. (1993). Detection of HIV-1 DNA and messenger RNA in individual cells by PCR-driven in situ hybridization and flow cytometry. *Science* **260**, 976–979.

Peart, A. F. W., and Nagler, F. P. (1954). Measles in the Canadian arctic, 1952. *Can. J. Pub. Health* **45**, 146–157.

Pedersen, N. C., Torten, M., Rideout, B., Sparger, E., Tonachini, T., Luciw, P. A., Ackley, C., Levy, N., and Yamamoto, J. (1990). Feline leukemia virus infection as a potentiating cofactor for the primary and secondary stages of experimentally induced feline immunodeficiency virus infection. *J. Virol.* **64**, 598–606.

Pensaert, M., Callebaut, P., and Vergote, J. (1986). Isolation of a porcine respiratory, non-enteric coronavirus related to transmissible gastroenteritis. *Vet. Quart.* **8**, 257–261.

Peters, C. J. (1997). Viral hemorrhagic fevers. *In* "Viral Pathogenesis" (N. Nathanson, R. Ahmed, F. Gonzalez-Scarano, D. E. Griffin, K. V. Holmes, F. A. Murphy, and H. L. Robinson, eds.), pp. 779–799. Lippincott-Raven Publishers, Philadelphia.

Peters, C. J., and LeDuc, J. W. (1995). Viral hemorrhagic fevers: Persistent problems, persistence in resevoirs. *In* "Immunobiology and Pathogenesis of Persistent Virus Infections" (B. W. J. Mahy and R. W. Compans, eds.). Harwood Academic Publishers, Chur, Switzerland.

Petursson, G., Nathanson, N., Georgsson, G., Panitch, H., and Palsson, P. A. (1976). Pathogenesis of visna, I: Sequential virologic, serologic, and pathologic studies. *Lab. Invest.* **35**, 402–412.

Pickup, D. J., *et al.* (1993). Soluble tumor necrosis factor receptors of two types are encoded by cowpox virus. *J. Cell Biochem.* **17B**, 81.

Pircher, H., Burki, K., Lang, R., Hengartner, H., and Zinkernagel, R. M. (1989). Tolerance induction in double specific T-cell receptor transgenic mice varies with antigen. *Nature* **342**, 559–561.

Popovic, M., and Gartner, S. (1987). Isolation of HIV-1 from monocytes but not T lymphocytes [letter]. *Lancet* **2**, 916.

Power, A. G. (1996). Competition between viruses in a complex plant–pathogen system. *Ecology (Washington, DC)* **77**, 1004–1010.

Power, A. G., Seaman, A. J., and Gray, S. M. (1991). Aphid transmission of barley yellow dwarf virus: Inoculation access periods and epidemiological implications. *Phytopathology* **81**, 575–548.

Rasschaert, D., Duarte, M., and Laude, H. (1990). Porcine respiratory coronavirus differs from transmissible gastroenteritis virus by a few genomic deletions. *J. Gen. Virol.* **71**, 2599–2607.

Rawle, F. C., Tollefson, A. E., Wold, W. S., and Gooding, L. R. (1989). Mouse anti-adenovirus cytotoxic T lymphocytes. Inhibition of lysis by E3 gp19K but not E3 14.7K. *J. Immunol.* **143**, 2031–2037.

Ray, C. A., Black, R. A., Kronheim, S. R., Greenstreet, T. A., Sleath, P. R., Salvesen, G. S., and Pickup, D. J. (1992). Viral inhibition of inflammation: Cowpox virus encodes an inhibitor of the interleukin-1 beta converting enzyme. *Cell* **69**, 597–604.

Reeves, W. C. (1974). Overwintering of arboviruses. *Prog. Med. Virol.* **17**, 193–220.

Regnery, D. C., and Miller, J. H. (1972). A myxoma virus epizootic in a brush rabbit population. *J. Wildl. Dis.* **8**, 327–331.

Reinacher, M., Bonin, J., Narayan, O., and Scholtissek, C. (1983). Pathogenesis of neurovirulent influenza A virus infection in mice. Route of entry of virus into brain determines infection of different populations of cells. *Lab. Invest.* **49**, 686–692.

Richardson, J., Frazier, N. W., and Sylvester, E. S. (1972). Rhabdoviruslike particles associated with strawberry crinkle virus. *Phytopathology* **62**, 491–492.

Risser, R., Horowitz, J. M., and McCubrey, J. (1983). Endogenous mouse leukemia viruses. *Annu. Rev. Genet.* **17**, 85–121.

Robertson, D. L., Hahn, B. H., and Sharp, P. M. (1995a). Recombination in AIDS viruses. *J. Mol. Evol.* **40**, 249–259.

Robertson, D. L., Sharp, P. M., McCutchan, F. E., and Hahn, B. H. (1995b). Recombination in HIV-1 [letter]. *Nature* **374**, 124–126.

Rochow, W. F. (1979). Field variants of barley yellow dwarf virus: Detection and fluctuation during twenty years. *Phytopathology* **69**, 655–660.

Roizman, B. (1993). The family herpesviridae: A brief introduction. *In* "The Human Herpesviruses" (B. Roizman, R. J. Whitley, and C. Lopez, eds.), pp. 1–10. Raven Press, New York.

Rosen, L. (1987). Overwintering mechanisms of mosquito-borne arboviruses in temperate climates. *Am. J. Trop. Med. Hyg.* **37**, 69S–76S.

Rosensweig, M. L., and McCord, R. D. (1991). Incumbent replacement: Evidence for long term evolutionary progress. *Paleobiology* **17**, 202–213.

Ross, R. (1911). "The Prevention of Malaria." Murray, London.

Roy, S., Katze, M. G., Parkin, N. T., Edery, I., Hovanessian, A. G., and Sonenberg, N. (1990). Control of the interferon-induced 68-kilodalton protein kinase by the HIV-1 tat gene product. *Science* **247**, 1216–1219.

Ruscetti, F. W., Robert-Guroff, M., Ceccherini-Nelli, L., Minowada, J., Popovic, M., and Gallo, R. C. (1983). Persistent in vitro infection by human T-cell leukemia–lymphoma virus (HTLV) of normal human T-lymphocytes from blood relatives of patients with HTLV-associated mature T-cell neoplasms. *Int. J. Cancer* **31**, 171–180.

Sabino, E. C., Shpaer, E. G., Morgado, M. G., Korber, B. T., Diaz, R. S., Bongertz, V., Cavalcante, S., Galvao-Castro, B., Mullins, J. I., and Mayer, A. (1994). Identification of human immunodeficiency virus type 1 envelope genes recombinant between subtypes B and F in two epidemiologically linked individuals from Brazil. *J. Virol.* **68**, 6340–6346.

Saif, Y. M. (1991). Immunosuppression induced by infectious bursal disease virus. *Vet. Immunol. Immunopathol.* **30**, 45–50.

Saito, T., Yamanaka, K., and Okada, Y. (1990). Long-distance movement and viral assembly of tobacco mosaic virus mutants. *Virology* **176**, 329–336.

Saknimit, M., Inatsuki, I., Sugiyama, Y., and Yagami, K. (1988). Virucidal efficacy of physico-chemical treatments against coronaviruses and parvoviruses of laboratory animals. *Jikken Dobutsu* **37**, 341–345.

Salminen, M. O., Carr, J. K., Robertson, D. L., Hegerich, P., Gotte, D., Koch, C., Sanders-Buell, E., Gao, F., Sharp, P. M., Hahn, B. H., Burke, D. S., and McCutchan, F. E. (1997). Evolution and probable transmission of intersubtype recombinant human immunodeficiency virus type 1 in a Zambian couple. *J. Virol.* **71**, 2647–2655.

Saluzzo, J. F., Herve, J. P., Salaun, J. J., Germain, M., Cornet, J. P., Camicas, J. L., Heme, G., and Robin, J. (1980). Caracteristiques des souches du virus de la fièvre jaune isolées des oeufs et des larves d'une tique *Amblyomma variegatum*, recoltée sur le betail a Bangui (Centrafrique). *Ann. Virol. (Inst. Pasteur)* **131**, 155.

Sato, H., and Okochi, K. (1986). Transmission of human T-cell leukemia virus (HTLV-1) by blood transfusion: Demonstration of proviral DNA in recipients' blood lymphocytes. *Int. J. Cancer* **37**, 395–400.

Schnittman, S. M., Greenhouse, J. J., Psallidopoulos, M. C., Baseler, M., Salzman, N. P., Fauci, A. S., and Lane, H. C. (1990). Increasing viral burden in CD4+ T cells from patients with human immunodeficiency virus (HIV) infection reflects rapidly progressive immunosuppression and clinical disease. *Ann. Intern. Med.* **113**, 438–443.

Schnurrenberger, P. R., Woods, G. T., and Martin, R. J. (1970). Serologic evidence of human infection with swine influenza virus. *Am. Rev. Respir. Dis.* **102**, 356–361.

Scholtissek, C. (1995). Molecular evolution of influenza viruses. *Virus Genes* **11**, 209–215.

Scholtissek, C., Rott, R., Orlich, M., Harms, E., and Rohde, W. (1977). Correlation of pathogenicity and gene constellation of an influenza A virus (fowl plague), I: Exchange of a single gene. *Virology* **81**, 74–80.

Scholtissek, C., Koennecke, I., and Rott, R. (1978). Host range recombinants of fowl plague (influenza A) virus. *Virology* **91**, 79–85.

Scholtissek, C., Ludwig, S., and Fitch, W. M. (1993). Analysis of influenza A virus nucleoproteins for the assessment of molecular genetic mechanisms leading to new phylogenetic virus lineages. *Arch. Virol.* **131**, 237–250.

Schrier, P. I., Bernards, R., Vaessen, R. T., Houweling, A., and van der Eb, A. J. (1983). Expression of class I major histocompatibility antigens switched off by highly oncogenic adenovirus 12 in transformed rat cells. *Nature* **305**, 771–775.

Scott, T. W., Weaver, S. C., and Mallampalli, V. L. (1994). Evolution of mosquito-borne viruses. *In* "The Evolutionary Biology of Viruses" (S. S. Morse, ed.), pp. 293–324. Raven Press, New York.

Scrimshaw, N. S., Salomon, J. B., Bruch, H. A., and Gordon, J. E. (1966). Studies of diarrheal disease in Central America, 8: Measles, diarrhea, and nutritional deficiency in rural Guatemala. *Am. J. Trop. Med. Hyg.* **15**, 625–631.

Sen, G. C., and Ransohoff, R. M. (1993). Interferon-induced antiviral actions and their regulation. *Adv. Virus Res.* **42**, 57–102.

Service, M. W. (1974). Survey of the relative prevalence of potential yellow fever vectors in north-west Nigeria. *Bull. World Health Organiz.* **50**, 487–494.

Seto, A., Isono, T., and Ogawa, K. (1991). Infection of inbred rabbits with cell-free HTLV-1. *Leuk. Res.* **15**, 105–110.

Sharp, P. M., Robertson, D. L., and Hahn, B. H. (1995). Cross-species transmission and recombination of 'AIDS' viruses. *Philos. Trans. R. Soc. London B. Biol. Sci.* **349**, 41–47.

Shelby, K. S., and Webb, B. A. (1997). Polydnavirus infection inhibits translation of specific growth-associated host proteins. *Insect Biochem. Mol. Biol.* **27**, 263–270.

Shioda, T., Levy, J. A., and Cheng-Mayer, C. (1991). Macrophage and T cell-line tropisms of HIV-1 are determined by specific regions of the envelope gp120 gene. *Nature* **349**, 167–169.

Shioda, T., Levy, J. A., and Cheng-Mayer, C. (1992). Small amino acid changes in the V3 hypervariable region of gp120 can affect the T-cell-line and macrophage tropism of human immunodeficiency virus type 1. *Proc. Natl. Acad. Sci. U.S.A.* **89**, 9434–9438.

Simpson, G. G. (1953). "The Major Features of Evolution." Columbia University Press, New York.

Smith, C. A., Davis, T., Wignall, J. M., Din, W. S., Farrah, T., Upton, C., McFadden, G., and Goodwin, R. G. (1991). T2 open reading frame from the Shope fibroma virus encodes a soluble form of the TNF receptor. *Biochem. Biophys. Res. Commun.* **176**, 335–342.

Spector, D. H., Wade, E., Wright, D. A., Koval, V., Clark, C., Jaquish, D., and Spector, S. A. (1990). Human immunodeficiency virus pseudotypes with expanded cellular and species tropism. *J. Virol.* **64**, 2298–2308.

Spriggs, M. K., Hruby, D. E., Maliszewski, C. R., Pickup, D. J., Sims, J. E., Buller, R. M., and VanSlyke, J. (1992). Vaccinia and cowpox viruses encode a novel secreted interleukin-1-binding protein. *Cell* **71**, 145–152.

Stoltz, D. B. (1990). Evidence for chromosomal transmission of polydnavirus DNA. *J. Gen. Virol.* **71**, 1051–1056.

Stoltz, D. B. (1993). The polydnavirus life cycle. *In* "Parasites and Pathogens of Insects" (N. E. Beckage, S. M. Thompson, and B. A. Federici, eds.), pp. 167–187. Academic Press, San Diego.

Stoltz, D. B., Krell, P., Summers, M. D., and Vinson, S. B. (1984). Polydnaviridae—a proposed family of insect viruses with segmented, double-stranded, circular DNA genomes. *Intervirology* **21**, 1–4.

Stoltz, D. B., Guzo, D., Belland, E. R., Lucarotti, C. J., and MacKinnon, E. A. (1988). Venom promotes uncoating in vitro and persistence in vivo of DNA from a braconid polydnavirus. *J. Gen.
Virol.* **69**, 903–907.

Stoye, J. P., and Coffin, J. M. (1985). Endogenous retroviruses. *In* "RNA Tumor Viruses" (R. A. Weiss,
N. Teich, H. Varmus, and J. Coffin, eds.), Vol. 2, pp. 357–404. Cold Spring Harbor Lab, Cold
Spring Harbor, NY.

Stoye, J. P., Fenner, S., Greenoak, G. E., Moran, C., and Coffin, J. M. (1988). Role of endogenous
retroviruses as mutagens: The hairless mutation of mice. *Cell* **54**, 383–391.

Strand, M. R., McKenzie, D. I., Grassl, V., Dover, B. A., and Aiken, J. M. (1992). Persistence and
expression of *Microplitis demolitor* polydnavirus in *Pseudoplusia includens*. *J. Gen. Virol.* **73**,
1627–1635.

Strauss, J. H. (1993). Recombination in the evolution of RNA viruses. *In* "Emerging Viruses" (S. S.
Morse, ed.), pp. 241–251. Oxford University Press, New York.

Stubbs, L. L., and Grogan, R. G. (1963). Necrotic yellows: A newly recognized virus disease of lettuce.
Aust. J. Agric. Res. **14**, 439–459.

Taber, S. W., and Pease, C. M. (1990). Paramyxovirus phylogeny: Tissue tropism evolves slower than
host specificity. *Evolution* **44**, 435–438.

Tajima, K., Tominaga, S., Suchi, T., Kawagoe, T., Komoda, H., Hinuma, Y., Oda, T., and Fujita, K.
(1982). Epidemiological analysis of the distribution of antibody to adult T-cell leukemia-virus-
associated antigen: Possible horizontal transmission of adult T-cell leukemia virus. *Gann* **73**,
893–901.

Taliansky, M. E., and Garcia-Arenal, F. (1995). Role of cucumovirus capsid protein in long-distance
movement within the infected plant. *J. Virol.* **69**, 916–922.

Tanaka, T. (1987). Effect of venom of the endoparasitoid, *Apanteles kariyai*, on the cellular defence
reaction of the host, *Pseudaletia separata* Walker. *J. Insect Physiol.* **33**, 413–420.

Tanaka, T., and Vinson, S. B. (1991). Interaction of venoms with the calyx fluids of three parasitoids,
Cardiochiles nigriceps, *Microplitis croicepes* (Hymenoptera: Braconidae), and *Campoletis
sonorensis* (Hymenoptera: Ichneumonidae) in effecting a delay in the pupation of *Heliothis
virescens* (Lepidoptera: Noctuidae). *Ann. Entomol. Soc. Am.* **84**, 87–92.

Tate, H., Kodama, H., and Izawa, H. (1990). Immunosuppressive effect of infectious pancreatic
necrosis virus on rainbow trout (*Oncorhynchus mykiss*). *Nippon Juigaku Zasshi* **52**, 931–937.

Temin, H. M. (1991). Sex and recombination in retroviruses. *Trends Genet.* **7**, 71–74.

Theilmann, D. A., and Summers, M. D. (1986). Molecular analysis of Campoletis sonorensis virus
DNA in the lepidopteran host *Heliothis virescens*. *J. Gen. Virol.* **67**, 1961–1969.

Theilmann, D. A., and Summers, M. D. (1988). Identification and comparison of *Campoletis sonorensis* virus transcripts expressed from four genomic segments in the insect hosts *Campoletis
sonorensis* and *Heliothis virescens*. *Virology* **167**, 329–341.

Traynor, P., Young, B. M., and Ahlquist, P. (1991). Deletion analysis of brome mosaic virus 2a protein:
Effects on RNA replication and systemic spread. *J. Virol.* **65**, 2807–2815.

Tschachler, E., Groh, V., Popovic, M., Mann, D. L., Konrad, K., Safai, B., Eron, L., diMarzo Veronese,
F., Wolff, K., and Stingl, G. (1987). Epidermal Langerhans cells—a target for HTLV-III/LAV
infection. *J. Invest. Dermatol.* **88**, 233–237.

Tyler, K. L., and Fields, B. N. (1990). Pathogenesis of viral infections. *In* "Fields Virology" (B. N.
Fields, D. M. Knipe, R. M. Chanock, M. S. Hirsch, J. L. Melnick, T. P. Monath, and B. Roizman,
eds.), Vol. 2, pp. 191–239. Raven Press, New York.

Upton, C., Macen, J. L., Wishart, D. S., and McFadden, G. (1990). Myxoma virus and malignant rabbit
fibroma virus encode a serpin-like protein important for virus virulence. *Virology* **179**, 618–631.

Upton, C., Macen, J. L., Schreiber, M., and McFadden, G. (1991). Myxoma virus expresses a secreted protein with homology to the tumor necrosis factor receptor gene family that contributes to viral virulence. *Virology* **184**, 370–82.

Upton, C., Mossman, K., and McFadden, G. (1992). Encoding of a homologue of the IFN-gamma receptor by myxoma virus. *Science* **258**, 1369–1372.

Vaewhongs, A. A., and Lommel, S. A. (1995). Virion formation is required for the long-distance movement of red clover necrotic mosaic virus in movement protein transgenic plants. *Virology* **212**, 607–613.

Van Valen, L. M. (1971). Adaptive zones and the orders of mammals. *Evolution* **25**, 420–428.

Van Valen, L. (1973). A new evolutionary law. *Evol. Theory* **1**, 1–30.

Vinson, S. B. (1977a). Insect host responses against parasitoids and the parasitoid's resistance: With emphasis on the Lepidoptera–Hymenoptera association. *Comp. Pathobiol.* **3**, 103–125.

Vinson, S. B. (1977b). *Microplitis croceipes*: Inhibitions of the *Heliothis zea* defense reaction to *Cardiochiles nigriceps*. *Exp. Parasitol.* **41**, 112–117.

Vinson, S. B., and Scott, J. R. (1974). Parasitoid egg shell changes in a suitable and unsuitable host. *J. Ultrastruct. Res.* **47**, 1–15.

Vinson, S. B., and Scott, J. R. (1975). Particles containing DNA associated with the oocyte of an insect parasitoid. *J. Invertebr. Pathol.* **25**, 375–378.

Vinson, S. B., and Stoltz, D. B. (1986). Cross-protection experiments with two parasitoid (Hymenoptera: Ichneumonidae) viruses. *Ann. Entomol. Soc. Am.* **79**, 216–218.

Vinson, S. B., Edson, K. M., and Stoltz, D. B. (1979). Effect of virus associated with the reproductive system of the parasitoid wasp, *Campoletis sonorensis*, on host weight gain. *J. Invertebr. Pathol.* **34**, 133–137.

Volterra, V. (1926). Variations and fluctuations of the numbers of individuals in animal species living together. *In* "Animal Ecology" (Chapman, ed.). McGraw-Hill, New York.

Wassom, D. L. (1993). Immunoecological succession in host–parasite communities. *J. Parasitol.* **79**, 483–487.

Weaver, S. C., Hagenbaugh, A., Bellew, L. A., Gousset, L., Mallampalli, V., Holland, J. J., and Scott, T. W. (1994). Evolution of alphaviruses in the eastern equine encephalomyelitis complex. *J. Virol.* **68**, 158–169.

Weaver, S. C., Kang, W., Shirako, Y., Rumenapf, T., Strauss, E. G., and Strauss, J. H. (1997). Recombinational history and molecular evolution of western equine encephalomyelitis complex alphaviruses. *J. Virol.* **71**, 613–623.

Webb, B. A., and Summers, M. D. (1992). Stimulation of polydnavirus replication by 20-hydroxyecdysone. *Experientia* **48**, 1018–1022.

Webster, R. G., Bean, W. J., Gorman, O. T., Chambers, T. M., and Kawaoka, Y. (1992). Evolution and ecology of influenza A viruses. *Microbiol. Rev.* **56**, 152–179.

Webster, R. G., Bean, W. J., and Gorman, O. T. (1995). Evolution of influenza viruses: Rapid evolution and stasis. *In* "Molecular Basis of Virus Evolution" (A. J. Gibbs, C. H. Calisher, and F. Garcia-Arenal, eds.), pp. 531–543. Cambridge University Press, Cambridge.

Weiss, R. A. (1993). Cellular receptors and viral glycoproteins involved in retrovirus entry. *In* "The Retroviridae" (J. A. Levy, ed.), Vol. 2, pp. 1–108. Plenum, New York.

Weiss, R. A., Mason, W. S., and Vogt, P. K. (1973). Genetic recombinants and heterozygotes derived from endogenous and exogenous avian RNA tumor viruses. *Virology* **52**, 535–552.

White, D. O., and Fenner, F. J. (1994). "Medical Virology." Academic Press, San Diego.

White, E., Sabbatini, P., Debbas, M., Wold, W. S., Kusher, D. I., and Gooding, L. R. (1992). The 19-kilodalton adenovirus E1B transforming protein inhibits programmed cell death and prevents cytolysis by tumor necrosis factor alpha. *Mol. Cell. Biol.* **12**, 2570–2580.

Williams, G., Stretton, T. B., and Leonard, J. C. (1960). Cytomegalic inclusion disease and *pneumocystis cairnii* infection in an adult. *Lancet* **2**, 951–955.

Williams, G., Stretton, T. B., and Leonard, J. C. (1983). AIDS in 1959? [letter]. *Lancet* **2**, 1136.

Williams, G. C. (1966). "Adaptation and Natural Selection." Princeton University Press, Princeton, NJ.

Winkler, C., Modi, W., Smith, M. W., Nelson, G. W., Wu, X. Y., Carrington, M., Dean, M., Honjo, T., Tashiro, K., Yabe, D., Buchbinder, S., Vittinghoff, E., Goedert, J. J., Obrien, T. R., Jacobson, L. P., Detels, R., Donfield, S., Willoughby, A., Gomperts, E., Vlahov, D., Phair, J., and Obrien, S. J. (1998). Genetic restriction of AIDS pathogenesis by an SDF-1 chemokine gene variant. *Science* **279**, 389–393.

Wolinsky, S. M., Korber, B. T., Neumann, A. U., Daniels, M., Kunstman, K. J., Whetsell, A. J., Furtado, M. R., Cao, Y., Ho, D. D., Safrit, J. T., *et al.* (1996). Adaptive evolution of human immunodeficiency virus-type 1 during the natural course of infection [see comments]. *Science* **272**, 537–542.

Wright, S. (1931). Evolution in Mendelian populations. *Genetics* **16**, 97–159.

Wright, S. (1932). The roles of mutation, inbreeding, crossbreeding, and selection in evolution. *Proc. XI Int. Congr. Genetics* **1**, 356–366.

Wright, S. (1982). Character change, speciation and the higher taxa. *Evolution* **36**, 427–443.

Xiong, Z., Kim, K. H., Giesman-Cookmeyer, D., and Lommel, S. A. (1993). The roles of the red clover necrotic mosaic virus capsid and cell-to-cell movement proteins in systemic infection. *Virology* **192**, 27–32.

Yamade, I., Isono, T., Ishiguro, T., and Yoshida, Y. (1993). Comparative study of human and rabbit cell infection with cell-free HTLV-1. *J. Med. Virol.* **39**, 75–79.

Yamamoto, N., Hayami, M., Komuro, A., Schneider, J., Hunsmann, G., Okada, M., and Hinuma, Y. (1984). Experimental infection of cynomolgus monkeys with a human retrovirus, adult T-cell leukemia virus. *Med. Microbiol. Immunol. (Berlin)* **173**, 57–64.

Yamanouchi, K., Kinoshita, K., Moriuchi, R., Katamine, S., Amagasaki, T., Ikeda, S., Ichimaru, M., Miyamoto, T., and Hino, S. (1985). Oral transmission of human T-cell leukemia virus type-I into a common marmoset (*Callithrix jacchus*) as an experimental model for milk-borne transmission. *Jpn. J. Cancer Res.* **76**, 481–487.

Yamnikova, S. S., Mandler, J., Bekh-Ochir, Z. H., Dachtzeren, P., Ludwig, S., Lvov, D. K., and Scholtissek, C. (1993). A reassortant H1N1 influenza A virus caused fatal epizootics among camels in Mongolia. *Virology* **197**, 558–563.

Zhou, R., Daar, I., Ferris, D. K., White, G., Paules, R. S., and Vande Woude, G. (1992). pp39mos is associated with p34cdc2 kinase in c-mosxe-transformed Nih 3t3 cells. *Mol. Cell. Biol.* **12**, 3583–3589.

Zhu, T. F., Korber, B. T., Nahmias, A. J., Hooper, E., Sharp, P. M., and Ho, D. D. (1998). An African HIV-1 sequence from 1959 and implications for the origin of the epidemic. *Nature* **391**, 594–597.

Zinkernagel, R. M., and Welsh, R. M. (1976). H-2 compatibility requirement for virus-specific T cell-mediated effector functions in vivo, I: Specificity of T cells conferring antiviral protection against lymphocytic choriomeningitis virus is associated with H-2K and H-2D. *J. Immunol.* **117**, 1495–502.

II

Viruses of Other Microorganisms

5

Ecology of Bacteriophages in Nature

JOHN H. PAUL and CHRISTINA A. KELLOGG

Department of Marine Science
University of South Florida
St. Petersburg, Florida 33701

I. INTRODUCTION

The role of bacteriophages (viruses that infect bacteria) in the environment has been the subject of intense investigation over the past several years. The development of techniques to study natural viral populations *in situ* has progressed tremendously. Various aspects of bacteriophage ecology in nature — including abundance, role in microbial mortality and water column trophodynamics, viral decay rates, repair mechanisms, and lysogeny — are now becoming or are nearly understood. However, most of these studies have been performed in aquatic

VIRAL ECOLOGY

environments. Thus, this review will mainly be limited to a discussion of aquatic environments. For reviews of the earlier literature, the reader is referred to Moebus (1987), Goyal et al. (1987), Fuhrman and Suttle (1993), Ackermann and DuBow (1987), and Proctor (1997).

II. DISTRIBUTION

Not long after the discovery of phages by d'Herelle in 1917, viruses from aquatic environments were isolated, but usually on enteric bacterial hosts (Zobell, 1946). For the first 50 years, the study of bacteriophages in nature was sporadic. Studies could be divided into those just describing the presence of one or several phage isolates from the environment (Spencer, 1955; Wiebe and Liston, 1968; Sklarow et al., 1973) and exhaustive surveys where large numbers of bacteria were screened for phage susceptibility (Hidaka, 1977; Baross et al., 1978). In all of these studies, phages were seldom isolated from unenriched seawater, but required enrichment by nutrient stimulation of the endogenous hosts (Moebus, 1980, 1992; Ackermann and Nguyen, 1983) or addition of mixtures of host cultures (Spencer, 1955). In fact, filter-feeding invertebrates (mollusks and shell-fish) were observed to concentrate vibriophages and provide an environment for their replication (Baross et al., 1978). Because of the need for concentration prior to isolation, bacteriophages were viewed as quantitatively insignificant in aquatic environments.

The problems of quantitating bacterial viruses in nature were aptly put by Goyal et al. (1987): "to date, all studies of heterogenous populations of phages in natural systems suffer from the inability to determine the total number of phages present. Plaque assay is no good until every host bacterium is used and conditions are optimized in each case." Electron microscopy as a counting tool was not viewed as practical because of the need to first efficiently concentrate viruses, which was usually accomplished by nutrient enrichment and selection. An elegant method involving ultracentrifugation, which is the basis of modern phage concentration techniques, was developed by Sharp (1960). It involved centrifugation of water samples onto collodion film on a glass coverslip. Adaptations of this basic technique were employed by Ewert and Paynter (1980) for sewage samples. Clarified sewage was spun on agar blocks and then collodion coated. The collodion was floated off the blocks and the resulting "pseudoreplicas" stained with uranyl acetate, and 10^6 to 10^7 viruses/ml were counted by transmission electron microscopy (TEM). Hidaka (1977) utilized ultracentrifugation to concentrate viral particles into a suspension, followed by phosphotungstic acid staining.

Modern methods for aquatic viral enumeration employ direct counting of concentrated samples either by TEM or epifluorescence microscopy. True direct counting of viruses in aquatic environments was reported as early as 1978 by Johnson and Sieburth, and TEM counts in excess of 10^4/ml were reported for bacteriophage particles in Yaquina Bay, Oregon (Torrella and Morita, 1979). Sieburth *et al.* (1988) reported viral direct counts of 5.8×10^9/liter in Narragansett Bay, Rhode Island, using epifluorescence microscopy of DAPI-stained preparations. The quantitative significance of viruses was not appreciated, however, until the paper of Bergh *et al.* (1989) and the associated press releases appeared. Using ultracentrifugation of natural waters onto TEM grids followed by electron microscopy, similar to the technology of Sharp (1960), Bergh and colleagues found viral concentrations from 6×10^4/ml for the Barents Sea, $\sim 6 \times 10^6$/ml for fjords, and $>10^8$/ml for eutrophic Lake Plussee, Germany. Other papers have reported the use of a variety of methods for concentrating viruses, including vortex flow filtration (Paul *et al.*, 1991), ultrafiltration (Suttle *et al.*, 1991; Wommack *et al.*, 1995), and no concentration (Demuth *et al.*, 1993). Enumeration has been by TEM or epifluorescence microscopy using DAPI (Hara *et al.*, 1991), Yo-Pro (Hennes and Suttle, 1995; Xenopoulos and Bird, 1997), and Sybr Green (Noble and Fuhrman, 1998). There seems to be good agreement among all these methods, although they have not all been systematically compared with identical water samples. In general, epifluorescent cell counts are greater than TEM counts. Hara *et al.* (1991) found 1- to 1.6-fold greater, Hennes and Suttle (1995) 1.2- to 7.1-fold, Xenopoulos and Bird (1997) 1.6-fold, and Noble and Fuhrman (1998) 1.28-fold greater. However, it is not clear if more is better. The recent concern that DAPI staining of bacteria may result in enumeration of non-nucleoid-containing and therefore dead cells (Zweifel and Hägstrom, 1995) may also apply in theory to epifluorescent viral counts. It is not known if all viral-sized particles detected by fluorescence microscopy are in fact intact viruses or some other DNA-containing particles (Jiang and Paul, 1995).

Despite the diversity of methodologies used to enumerate viruses, consistent trends in viral abundance have been found. Hypereutrophic estuarine or lake environments have viral direct counts (VDCs) in excess of 10^8/ml [e.g., Lake Plussee (Bergh *et al.*, 1989; Demuth *et al.*, 1993) and Chesapeake Bay (Proctor and Fuhrman, 1990; Wommack *et al.*, 1992)]. One sample taken over a cyanobacterial mat in the Gulf of Mexico and counted with Yo-Pro exceeded 10^9/ml, the highest reported VDC for a natural community (Hennes and Suttle, 1995). Other reports for estuarine environments are in the range of 10^7/ml (Jiang and Paul, 1994; Heldal and Bratbak, 1991; Hara *et al.*, 1991; Wommack *et al.*, 1992). Coastal oceanic environments have been reported to contain between 10^6 and 10^7 virus-like particles (VLPs)/ml (Suttle *et al.*, 1990; Paul *et al.*, 1991; Cochlan *et al.*, 1993; Boehme *et al.*, 1993; Jiang and Paul, 1996; Weinbauer and Peduzzi, 1995; Weinbauer *et al.*, 1993; Steward *et al.*, 1996). Oligotrophic oceanic environments and deep-sea samples (depth 1500–5000 m) usually contain 10^4 to 10^5 VLPs/ml (Paul *et al.*, 1997a; Bergh *et al.*, 1989; Boehme *et al.*, 1993; Hara *et al.*, 1996). When seasonal studies have been performed, there has been a strong indication of seasonal variation in VDC, often with a tenfold decrease in winter compared to summer months (Jiang and Paul, 1994; Cochran and Paul, submitted; Weinbauer *et al.*, 1993). Variation in bacterial direct counts (BDCs) is usually only two- to threefold, often not reflecting any obvious seasonal trend, and indicating an uncoupling of viral and bacterial populations.

Several studies have systematically sized VLPs, and these have found the majority of viral head sizes to be between 30 and 60 nm (Bergh *et al.*, 1989; Cochlan *et al.*, 1993; Wommack *et al.*, 1992). Very few studies have attempted to systematically classify observed phages by morphology. One such study performed in the hypertrophic Lake Plussee indicated that the majority of the viruses were Siphoviridae, and that nearly all VLPs observed had tails (Demuth *et al.*, 1993). Other reports generally do not indicate an abundance of tailed phages. In the Chesapeake Bay, approximately 50% of the VLPs observed were tailed phages (Wommack *et al.*, 1992). In our field observations of estuarine and marine samples, most VLPs lack tails (Jiang and Paul, unpublished observations). However, the loss of tails may be an artefact of sample concentration (ultracentrifugation or ultrafiltration). In the study by Demuth *et al.* (1993), the environment sampled had such a high phage concentration that TEM grids were just floated on the sample and the phages adsorbed to the grid. Alternatively, this procedure may select for the most adsorptive phages, which may be those possessing tails.

The presence of tailed phages suggests that these are DNA-containing bacterio-phages (Ackermann and DuBow, 1987). However, there has as yet to be developed a method to rapidly enumerate RNA-containing viruses. Because phages are seemingly efficiently enumerated with DNA-specific fluorochromes such as DAPI, it appears that DNA phages comprise the overwhelming majority of VLPs observed in aquatic environments.

Because phages are in the <0.45 μm size range, they are operationally in the "dissolved" fraction of seawater. Therefore, it seemed important to determine if viruses were a significant component of the dissolved DNA. Beebee (1991) used an ultracentrifugation technique to characterize dissolved DNA, found that 80% was high molecular weight (in the high-speed pellet), and assumed that it was viral DNA. Work in our lab (Jiang and Paul, 1995) indicated that purified calf thymus DNA when ultracentrifuged also appeared in the pellet (~60%) and therefore would be viral DNA by this definition. Mauryama *et al.* (1993) found most of the filterable DNA to be DNase insensitive ("coated DNA") and also to be of high MW (>300,000) and concluded that this was viral DNA. Work in our laboratory has focused on the origin of dissolved DNA, using a combination of ultrafiltration, ultracentrifugation, DNase digestion, ethanol precipitation, and rRNA probing. We have concluded that ~50% of the <0.2 μm DNA was free DNA, and that 50% was "coated" or bound in a DNase-insensitive form, 17 to 30% of which was viral DNA (Jiang and Paul, 1995; Paul et al. 1996). Thus, as a percentage of total dissolved (<0.2 μm) DNA, viral DNA (vDNA) ranged from 0.8 to 21.8%, averaging 7.7% (Jiang and Paul, 1995). Weinbauer et al. (1993), using differing techniques for DNA, estimated that 17.1% of the dissolved DNA was vDNA, again indicating that viruses are a relatively small component of the dissolved DNA pool.

III. ECOSYSTEM DYNAMICS

A major question posed to microbial ecologists concerning the tremendous number of viruses in aquatic environments is: So what? Are viruses causing significant bacterial mortality, short-circuiting the Microbial Loop, and causing DOM (dissolved organic matter) release? Calculations of bacterial mortality, phage decay rates, and phage production rates varied widely in initial attempts to quantify these parameters. Initial estimates of 100% of the daily bacterial mortality being viral-mediated or 24%/hr bacterial mortality (Proctor and Fuhrman, 1990; Heldal and Bratbak, 1991) seem to have been replaced by more reasonable numbers. Weinbauer et al. (1993) estimated viral-induced bacteria mortality to be 3.5 to 24% for mesotrophic environments, and 7 to 64.3% for eutrophic areas of the North Adriatic Sea. Similarly, Steward et al. (1996) estimated viral mortality in the Chukchi Sea to range from 10% at the northern oligotrophic regions to 23% at the more eutrophic southern areas. In sinking particles, Proctor and Fuhrman (1991) estimated that 2 to 37% of the particle-associated bacteria were killed by viral lysis. Jiang and Paul (1994) estimated 3 to 53% of the bacterial lysis (average 25%) was caused by viruses based on analysis of 4-hr incubations during a 48-hr diel experiment. To put this in perspective with flagellate grazing (bacterivory), Steward et al. (1996) reported that viral-induced lysis was approximately equal to the significance of grazing, but the relative importance of each often varied with depth in the water column. Fuhrman and Noble (1995) estimated that viral lysis and bacterivory contributed approximately equally to bacterial mortality, each at about 25%. Collectively, these papers indicate that bacterivory and viral-induced mortality are of equal importance as agents of bacterial mortality (~25% each), but that the relative importance of either may vary in space and time and with environmental characteristics. For example, Weinbauer and Hofle (1998) found varying importance of bacterivory and viral lysis in stratified Lake Plussee. In the epilimnion (well-oxygenated upper water column), bacterivory dominated, contributing to 90% of the bacterial mortality, with viral-induced mortality being a mere 9.2%. In the metalimnion, viral lysis and grazing contributed 53.6 and 46.4% to the bacterial mortality, respectively, while in the anoxic hypolimnion viral lysis caused 87.5% of mortality. The latter observation was most likely caused by inhibition of flagellates by anoxic conditions.

Similar results were found in a series of salt ponds (salinities ranging from 37 to 372‰), where the importance of bacterivory was inversely related to salinity (Guixa-Boixareau et al., 1996). When salinity was >250‰, bacterivory was essentially nonexistent, yet the frequency of virally infected cells (FVICs) remained relatively constant across the salinity gradient. At extremely high salinities, square Archaea were abundant, as well as a lemon-shaped phage. Thus, it seems that bacterivory is often limited by conditions that define optimal eukaryotic protozoal growth; in the above examples, salinity and oxygen tension defined these limitations. In extreme environments not conducive to eukaryotic aerobic growth, viral lysis is the favored (and perhaps only) mechanism for bacterial mortality. Therefore, extreme environments have altered food webs, with little link of bacteria to higher trophic levels and a lot of viral "short-circuiting" of carbon flow. This can serve to perpetuate a primarily prokaryotic food web in these environments.

A. Phage Production and Decay

Calculations of bacterial mortality require either a measure of phage production or decay. Initial studies on decay rates used cyanide to inhibit bacterial activity and phage production. This approach yielded some very high decay rates (0.26–1.1/hr; Heldal and Bratbak, 1991). The problem with this approach is that it impairs bacterial (and other microbial) processes that contribute to viral decay. Bratbak et al. (1990) estimated viral decay rates from deceases in viral concentrations observed while following a spring diatom bloom. These decay rates are net, because no corrections for viral predation were made, and ranged from 0.05 to 0.5 day^{-1} (averaging 0.28 day^{-1}). Bratbak et al. (1992) estimated a higher decay rate of 0.3 hr^{-1} or 7.3 day^{-1}. Using cultivated marine phages, Noble and Fuhrman (1997) found inactivation rates of 0.96 to 1.73 day^{-1} for phages isolated from Santa Monica in Californian coastal waters. They investigated the role of microbes, heat-labile organic compounds, and sunlight on viral inactivation rates and found sunlight to be the greatest factor contributing to inactivation.

B. Reactivation of Phages Damaged by UV Radiation

Inactivation due to UV radiation in sunlight (particularly UVB, 320–290 nm) is believed to be a major factor in the decay of natural bacteriophage populations. The radiation causes DNA damage, the most serious forms of which are cyclobutane pyrimidine dimers and pyrimidine–pyrimidone (6–4) photoproducts (Friedberg et al., 1995). Over four decades of intensive study on E. coli and many of its phages have revealed three major repair mechanisms that allow irradiated viruses to be reactivated.

The earliest known of these mechanisms was the host-encoded enzyme photolyase. The action of this enzyme, which catalyzes a single-step reversal of pyrimidine dimers, was first observed in T phages (Dulbecco, 1949). The enzyme binds to DNA at the lesion, two chromophores absorb visible light (>300 nm), and then the dimer is "unwound" like a knot, restoring the original DNA (Friedberg et al., 1995). This process is referred to as photoreactivation.

The second host-mediated repair system is a more complex operation, requiring multiple enzymes to remedy the damage (Friedberg et al., 1995; Hanawalt et al., 1979; Lindahl, 1982). This system, known as nucleotide excision repair or host cell reactivation, is not specific for photoproducts alone, but responds to many types of DNA damage. A damage-specific endonuclease nicks the DNA on both sides of the lesion, creating an oligonucleotide (usually about 12 nucleotides long), which is excised. DNA polymerase fills in the gap and DNA ligase seals the repair in place (Friedberg et al., 1995). This repair is not dependent on "light and is sometimes called "dark repair," to distinguish it from photoreactivation.

The final method of repair is novel in that it is not a cellular process, but is virally encoded. Coliphage T4 encodes an endonuclease that is specific for the excision of pyrimidine dimers (Friedberg, 1972; Minton et al., 1975; Yasuda and Sekiguchi, 1976). The action of this enzyme explains the observation that T4 is approximately twice as resistant to laboratory-induced DNA damage than either T2 or T6 (Luria, 1947).

These three repair mechanisms are now being studied in marine phage–host systems to determine the extent of UV DNA damage and repair in the ocean. Photoreactivation has been observed in culture experiments with marine phage–host systems (Kellogg and Paul, 1997; Weinbauer et al., 1997; Wilhelm et al., 1998), as well as in the total viral community (Weinbauer et al., 1997). It appears that a larger fraction of the viral population (41 to 52%) undergoes repair in coastal waters than that in offshore oligotrophic environments (21 to 26%). This may be due to less penetration of UVB in turbid near-shore waters, resulting in less severe, and thus repairable, damage (Weinbauer et al., 1997). Wilhelm et al. (1998) correlated DNA damage (number of cyclobutane pyrimidine dimers) to infectivity in a cultured marine isolate, and then used that data to extrapolate to the proportion of the total marine viral community that is infective (>50%). This high level of infectivity in spite of a large degree of solar UV damage is attributed to repair by photolyase. In neither study (Weinbauer *et al.*, 1997; Wilhelm *et al.*, 1998) was dark repair (nucleotide excision repair) quantified, but it appeared to play little role, if any. It was suggested that this may be because photolyase is either more efficient, as well as less energetically demanding (one enzyme required rather than several), or that the amount of DNA damage was not great enough to induce the excision repair system (Weinbauer *et al.*, 1997).

We have observed both photoreactivation and excision repair in a marine phage–host system in culture. In the absence of photoreactivating light, nucleotide excision repair can reactivate UVC-damaged phages two to four orders of magnitude above the survival seen in a host where that system was repressed with caffeine (Kellogg and Paul, 1997).

Although not a bacteriophage, a freshwater virus that infects the green algae *Chlorella* has recently been discovered to encode a homologue of the T4 endonuclease (Furuta *et al.*, 1997; Lu *et al.*, 1995). The genes are 41% similar and execute the same function-specific excision of pyrimidine dimers. This is the only virus other than T4 ever found to encode its own repair enzyme.

C. Contribution of Viruses to Food Webs

In terms of the contribution of viruses to food webs, Bratbak *et al.* (1994) said that viruses did not add any new processes or connections to the food web; rather, the viruses just affected the rate of production of particulate matter and the rate of production of dissolved matter. In contrast to this view of viruses in water column trophodynamics, Gonzalez and Suttle (1993) showed that native flagellate assemblages could feed on viral particles when viral concentrations were in the range of 10^7 to 10^8/ml. In fact, viruses could represent an additional 0.2 to 9% of the carbon, 0.3 to 14% of the nitrogen, and 0.6 to 28% of the phosphorus beyond the amounts that flagellates obtain from ingestion of bacteria. The phosphorus values reflect the relatively high relative contribution that DNA makes to the total organic carbon content of a virus particle.

To summarize, in normal nonextreme aquatic environments, viral-induced mortality is roughly equal to flagellate grazing (bacterivory), and both account for 25% of the bacterial mortality, on the average. In extreme environments unfavorable to flagellate growth, viral lysis can account for nearly all of the bacterial mortality. Thus, viruses can strongly influence the carbon flow in pelagic food webs, and may themselves be a source of carbon for higher tropic levels, particularly in environments where they are abundant, such as estuaries and eutrophic lakes.

IV. MOLECULAR ECOLOGY

A. Genetic Diversity

As the international scientific community focuses on issues in microbial diversity, interest in viral diversity has also emerged. Phage morphology has been used for many years to separate and identify viral isolates (Adams, 1952; Bradley, 1967; Delbruck, 1946; Grimont *et al.*, 1978; Lindstrom and Kaijalainen, 1991; Werquin *et al.*, 1988; Wilson *et al.*, 1993) and remains a significant criterion in phage characterization studies. However, with the advent of DNA technology, diversity on the genetic level has opened a new, more detailed, level of comparison. It allows us to address the question of genetic diversity within and among the established morphological groups.

Some early studies that incorporated DNA methods when comparing phages used the broad technique of DNA–DNA homology to look at diversity. Because DNA from morphologically similar phages infective for the same host cross-hybridized, it was suggested that, in first approximation, genetic diversity mirrored morphological diversity (Grimont and Grimont, 1981; Jarvis, 1984; Werquin *et al.*, 1988).

Thirty-three bacteriophages of *Rhizobium meliloti*, bacteria that form root nodules in legumes, were characterized by their morphology, host range, serology, restriction endonuclease digestion patterns, DNA–DNA homologies, and DNA molecular weights (Werquin *et al.*, 1988). The 33 phages were divided into five morphotypes based on transmission electron microscopy, and a each type of phage was used for further study. Each of the five phage types had a unique restriction digest pattern, corroborating the above-mentioned conclusion of unrelatedness derived from the lack of DNA–DNA homology. However, a similar study in Finland looked at seven bacteriophages of *Rhizobium galegae* (Lindstrom and Kaijalainen, 1991), with very different results. The seven phages segregated into three morphotypes: myovirus (Φ1R, Φ10W, Φ3R, Φ30W), siphovirus with elongated head (Φ1261M, Φ1261V), and siphovirus with hexagonal head (Φgor3V). Four restriction patterns were observed: one for each of the siphovirus morphologies, and two within the myovirus group, dividing it into Φ1R/Φ10W and Φ3R/Φ30W. Using Φ1R genome as a probe, dot blots of the viral DNAs of each restriction group were assayed. All groups hybridized at some level, but even more interesting, Φ1R was most closely related to Φ1261 (a different morphology) than to Φ3R (same morphology but different restriction pattern). It is important to note that these different genotypes were all capable of cross-infecting each other's host, which suggests the similarity may stem from horizontal phage gene transfer (see IV.D.2, Gene Transfer: Phage Genes).

Shifting our focus from diversity between morphological groups, several studies have focused on the diversity of *Streptococcus thermophilus* bacteriophages, all the known isolates of which are siphoviruses (Brussow and Bruttin, 1995; Brussow et al., 1994a,b; Le Marrec et al., 1997). One-hundred and eleven phages infective for S. thermophilus have been characterized by restriction profiles, host ranges, DNA homology, and some by additional criteria (Brussow et al., 1994a,b; Le Marrec et al., 1997). Out of these more than 100 phages, there were 76 unique restriction profiles, showing considerable heterogeneity among the genomes. However, a 6-kb fragment cloned from ΦS1 (Brussow et al., 1994a) hybridized weakly with all known S. thermophilus phages. A 2.2-kb EcoRI fragment from another phage was found to be conserved across the range of four lytic groups, with those groups established on the basis of serology and host range by Brussow et al. (1994b), a temperate phage (rare in S. thermophilus) (Brussow and Bruttin, 1995), and 57% (17 of the 30) of Le Marrec et al.'s (1997) phages. Sequencing of this 2.2-kb fragment from several isolates revealed differences at only 3 of 2207 positions; a similarity of 99.86%. Further investigations into the similarity among these bacteriophages exposed more diversity, as Brussow and Bruttin (1995) used nine labeled restriction fragments (which add up to the entire linear genome) of the temperate phage ΦSfi21 to probe their four lytic groups (divided by serology and host range). The resulting hybridization patterns were extremely varied; some groups hybridized to consecutive fragments, others to a collection of fragments from different parts of the genome. These data are indicative of common modular genomic "cassettes" that have been shuffled (see IV.D.2, Gene Transfer: Phage Genes). Le Marrec et al. (1997) found that S. thermophilus phages could not be grouped effectively by host range and serology, but could reliably be divided into two large groups based on their genome packaging mechanism. Of their 30 phages, 19 used a cohesive ends (cos) mechanism and 11 used a pac-like mechanism. Le Marrec and coworkers labeled DNA from one cos-phage and one pac-phage and found each hybridized to all phages, again indicating genetic relatedness among all S. thermophilus phages. However, they found no similarity between the areas that encode for the packaging mechanisms, and the packaging region probes hybridized more strongly to their own type of phage (cos to cos-phage, pac to pac-phage). In summary, it appears that all S. thermophilus phages are related, stemming from two original ancestors (one with a cos-mechanism, one with a pac-mechanism). Subsequent exchange of genetic modules between these groups could then have given rise to the great diversity of current phages. Also, considering that these phages were all isolated from dairy product processing plants, their environment (many large rapidly growing cultures of bacteria) provides plenty of opportunities for both horizontal gene transfer by coinfection of hosts, and for mutations to occur through successive generations of phage propagation.

An understanding of genetic diversity in phages in larger environments — soil, freshwater, the ocean — is just beginning. An early suggestion of the magnitude of bacteriophage diversity in the marine environment came from a study in which 733 bacterial strains and 258 phages from the Atlantic Ocean were isolated

(Moebus and Nattkemper, 1981). Many bacteria isolated from water between Europe and the Azores (eastern Atlantic) were sensitive to phages isolated west of the Azores, but fewer "western" bacteria were sensitive to "eastern" phages. The sensitivity of hosts to viruses implied that there were two groups of phage–host systems in this ocean basin, with the dividing line occurring near the Azores. The authors interpreted this to mean that the microbial populations were undergoing genetic changes as they were transported by currents across the Atlantic. Their findings hint at the diversity of both marine bacteria and phages.

The first study to look at diversity of specific marine isolates dealt with phages that infect the marine cyanobacterium *Synechococcus* (Wilson *et al.*, 1993). Five phages were divided into three morphologies; two different sizes of Myoviridae, and one type of Siphoviridae. Three restriction endonuclease digestion patterns or "profiles" were observed, one for each morphology. Labeled DNA of one virus was used to probe a Southern transfer of the restriction digests, and under low stringency, hybridization was observed for all five phages. To extrapolate from this small data set, it appears that on a more limited basis genetic diversity mirrors morphology, but that some degree of similarity is common to many or all cyano-phages. For more discussion of cyanophages and eukaryotic algal viruses, see Chapter 6.

Our lab has been conducting research focused on the genetic diversity of a widespread group of marine *Vibro parahaemoliticus* phages (Kellogg *et al.*, 1995). Like the study of *S. thermophilus* phages discussed before, the viruses have one uniform morphology, but in our case it is that of the Myoviridae, the contractile-tailed phages. Our original investigation involved isolating 73 vibriophages on a common host, *V. parahaemoliticus* st. 16. A 1.5-kb *Eco*RI restriction enzyme digestion fragment was cloned from the type phage (Φ16) and used as a hybridi-zation probe to examine dot blots of DNA from the other phages, which had been isolated from Hawaii, the Gulf of Mexico, the Florida Keys, and Tampa Bay, Florida. All isolates were shown to hybridize to the Φ16 fragment, indicating some level of genetic relatedness among this group of phages. Restriction endonuclease profiles were generated with *Eco*RI, and Southern transfers of the gel were probed with the 1.5-kb fragment, highlighting different banding patterns. Phages were segregated into six groups based on these hybridization patterns. The 34 Hawaiian isolates formed one group, but no geographic pattern was obvious in the groupings of the 39 Florida isolates (Kellogg *et al.*, 1995).

We have recently concluded a year-long study to assess the seasonal component of genetic diversity in marine phages. That is, how much diversity would we see in one place over time, compared to the diversity seen over large geographic areas at single time points? We sampled a single site bimonthly for a year, screening 165 phages by PCR amplifying a 500-bp fragment, digesting it with the 4-bp cutter HhaI, and grouping the isolates into operational taxonomic units (OTUs) based on their restriction patterns. Seven OTUs were detected during the seasonal study, with one consistently dominant (71% of the total isolates), and the others more rare (16 to 1%) (Kellogg *et al.*, 1997). We then turned to our 73 geographic isolates (which had been grouped by *Eco*R1 restriction of the entire phage genome), amplified and sorted them using HhaI for comparison: the 34 Hawaiian phage formed two new OTUs, the 39 Florida isolates appeared to be similar to the dominant seasonal OTU (Kellogg and Paul, unpublished). DNA sequencing has been done on one isolate from each of the six *Eco*RI groups (Kellogg *et al.*, 1995), and at least two isolates of each seasonal OTU. The nucleotide sequence similarities ranged from 83 to 100% of the 500-bp fragment analyzed. A prelimi- nary phylogram, constructed by parsimony, shows some of the seasonal isolates are no more related to each other than they are to other geographic isolates (Fig. 1). This suggests that diversity on a temporal scale can be as high as diversity on a spatial scale.

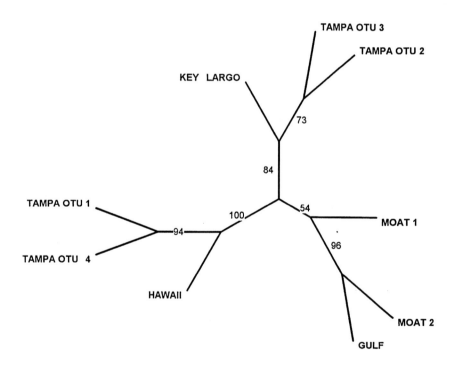

Fig. 1.   Phylogenetic distance tree generated by parsimony analysis (PHYLIP package; Felsen- stein, 1989) of 483-bp DNA sequences of phages infecting *Vibrio parahaemoliticus* st. 16. Viruses are identified by the location of their isolation; phages from the seasonal study in Tampa Bay are identified as separate operational taxonomic units (OTUs). Numbers at the branch points represent the bootstrap values for 100 replicate trees.

B. Lysogeny

Lysogeny occurs when a phage enters into a stable symbiosis with its host bacterium (Ackermann and DuBow, 1987). The bacterium is called a lysogen, which literally means "capable of generating phages that can lyse." A phage capable of lysogenizing a host is called a temperate phage. A key feature of lysogenic interactions is integration of the phage genome as a prophage into one of the host's replicons (chromosome, plasmid, or other temperate phage genome).

The lysogenic state is a highly evolved state, involving coevolution of the host and the temperate phage (Levin and Lenski, 1983). The factors favoring selection of lysogeny over lytic existence include: (1) lytic extinction of many virulent phage–host systems, (2) advantages to the host, including immunity to superinfection, conversion by phage genes, including restriction modification systems, antibiotic resistance and others properties, (3) general increase in cellular fitness (Edlin *et al.*, 1975), and (4) ensuring survival of the virus at low host cell densities (Levin and Lenski, 1983).

Surveys of bacterial isolates and phages indicate that lysogeny is widespread. Ackermann and DuBow (1987) indicated that, of 1200 strains of diverse bacteria, an average of 47% contained inducible prophage. Nearly 100% of naturally occurring *Pseudomonas* strains were lysogenic for some temperate viruses (Levin and Lenski, 1983). Freifelder (1987) has indicated that over 90% of all known bacterial viruses were temperate in nature. Additionally, polylysogeny, or the process by which a lysogen contains two or more different viral prophages, is also common. Up to 68% of the strains of *Bacillus megaterium* were polylysogenic in nature. In contrast to these studies, Moebus (1983) showed that, out of 300 marine phage strains examined, only 29 were found to be temperate. Therefore, it would seem that lysogenic interactions would be common in nature; that is, in natural microbial communities.

On infection with a temperate phage, a lysogenic decision is made whether to enter a virulent or lysogenic interaction. Factors influencing this molecular decision include multiplicity of infection (MOI, or the attack ratio of infectious phage particles per host cell; high MOIs favor lysogeny), nutrient deprivation, and other factors which seem particularly environmentally relevant (Levin and Lenski, 1983).

However, few studies have approached the issue of the significance of lysogeny in the environment by actual experimentation, although many papers have expressed opinions on this subject. Borsheim *et al.* (1990) thought that handling of samples resulted in prophage induction from marine bacterial populations. Heldal and Bratbak (1991) suggested that "light and nutrients" were among environmental factors that could induce lysogenic bacteria to liberate viruses. Bratbak *et al.* (1994) indicated that "even modest manipulation of the bacterial communities can cause prophage induction."

We don't know what factors in fact cause prophage induction in natural populations. Ackermann and DuBow (1987) pointed out that most inducing agents act on DNA replication, and usually at the replicating fork. Although UV light (UVC) and mitomycin C are the most widely used agents in laboratory studies, these two may not induce the same prophages. Other agents include radiation, chemical carcinogens, mutagens, base analogues, antibiotics, pressure, and temperature, to name a few. In λ, activation of the recA protein is involved in induction.

How natural and anthropogenic factors cause induction of prophage remains unknown. However, before these factors can be considered, the abundance of lysogens in aquatic systems needs to be known. Jiang and Paul (1994) examined 62 marine and estuarine bacterial isolates for prophage induction by mitomycin C, and found that 43% had inducible prophage or bacteriocins. In an extension of this work to include 110 bacterial isolates (Jiang and Paul, 1998a,b), the same approximate percentage (~40%) was lysogenized. Although this is in agreement with the original observation of the distribution of lysogeny among terrestrial isolates (Ackermann and DuBow, 1987), it may be argued that the isolation process selects for lysogens, that is, perhaps only actively growing cells are selected in the isolation procedure, and such cells might be preferentially lysogenized. To answer this question using natural populations, Wilcox and Fuhrman (1994) attempted to determine the relative significance of lysogenic compared to lytic phage production in the waters off Southern California. The microbial population of the water was removed by either 0.2 μm filtration (removes bacteria but leave viral particles) or 0.02 μm filtration (removes bacteria and viral particles). The water was then re-inoculated with 0.6 μm filtered water, which should have contained both bacteria as well as viral particles, and exposed to sunlight in pulses. Because no phage production occurred in sunlight nor when 0.02 μm filtering was used, Wilcox and Fuhrman (1994) concluded that lytic rather than lysogenic phage production was occurring. However, more recent studies by Jiang and Paul (1996) indicated that sunlight did not cause prophage induction in natural populations, most likely because of the low amount of <302 nm UV in sunlight and thereby the relatively small amount of damage caused. Such a level of damage may not induce the SOS response (Weinbauer et al., 1997). Additionally, prefiltration by 0.6 μm (sensu Wilcox and Fuhrman, 1994) may have removed the larger more actively growing bacteria, which would be more susceptible to induction by mitomycin C compared to dormant or starved cells.

Weinbauer and Suttle (1996) studied lysogeny in natural populations from a pier in Texas and in offshore Gulf of Mexico waters by prophage induction using mitomycin C. Using a burst size and the number of viruses produced in the presence of mitomycin C compared to a control, they estimated that 1.5% of the bacterial population was lysogenized. If all the lysogens were induced, this would represent 3% of bacterial mortality. However, seven of nine samplings were taken during winter months, and the two from June were taken at offshore stations, a sampling strategy that might result in no detection of lysogeny (see below). Tapper and Hicks (1998) investigated the proportion of lysogens in natural populations of Lake Superior and found 0.1 to 7.4% of the bacteria to be inducible by mitomycin C. Jiang and Paul (1996) investigated the distribution of lysogens in estuarine, near-shore, and offshore environments of the Gulf of Mexico. Eight of ten eutrophic estuarine and near-shore environments showed prophage induction by mutagens, whereas only 3 of 11 offshore environments contained inducible prophage. Whether this means that lysogeny is uncommon in offshore environments or that offshore populations were too nutrient depleted (i.e., there is a requirement for active growth for induction) is not known. A variety of inducing treatments were employed, including mitomycin C, UV radiation, sunlight, temperature, pressure, polynuclear aromatic hydrocarbons (PAHs), Bunker C fuel oil, and trichloroethylene (Table I). PAHs were most efficient in prophage induction, eliciting a positive response for 73% of samples. Sunlight and pressure did not cause prophage induction in any experiment. A temperature increase to just 30°C for 30 min caused induction of lysogens (369% increase in VDC compared to controls) from coastal waters of the Atlantic near Cape Hatteras. Temperature has long been known to induce λ because of the thermal instability of the cI repressor/DNA complex (Murooka and Mitani, 1985). It may be that many natural repressors of prophage promoters are unstable at elevated temperatures. This may explain in part the tremendous number of phages observed in estuaries in summer.

The fraction of the ambient microbial community containing inducible prophage has been calculated by two methods: the "mortality method," whereby the net change in bacterial abundance after induction is divided by the total population [Eq. (1)] and the "burst size" method [Eq. (2)]:

$$\frac{B_C - B_I}{B_C} \times 100 \tag{1}$$

$$\frac{(V_1 - V_C)/B_Z}{B_C} \times 100 \tag{2}$$

where B_C and B_I are bacterial direct counts in the control and induced samples, respectively, V_I and V_C are the viral direct counts in induced and control samples, and B_Z is the average burst size.

The mortality method yields larger values for percentage lysogenic bacteria, probably because of toxicity and lysis of non-induced cells, whereas the burst size method suffers from uncertainties regarding average burst size. Using the mortality method, the average percentage of the population estimated to be lysogenized

TABLE I

Efficiency of Treatments for Induction of Natural Populations of Lysogens in the
Marine Environment

Inducing agent	No. of samples tested	No. of samples induced	Efficiency (%)	Average viral increase for positive response (%)
Mitomycin C	33	16	48.5	424
UV radiation (254 nm)	12	6	50	202
Sunlight	9	0	0	N/A
Temperature	6	2	33	256
Elevated hydrostatic pressure	3	0	0	N/A
PAH[a] chemicals	11	8	73	215
Bunker C fuel oil	16	1	6.3	343
Trichloroethylene	6	1	17	206
PCB[b] mixture	4	3	75	286
Arochlor 1248	4	3	75	552
Pesticide mix	6	3	50	570

[a]Polynuclear aromatic hydrocarbons.
[b]Polychlorinated biphenyls.

ranged from 10.5 to 78.3% (average 35%), whereas using the burst size method
the range was from 1.5 to 38% (average 8.8%).

A second study of lysogeny in natural populations (Cochran et al., submitted)
investigated polychlorinated biphenyls (PCBs), pesticides, and Bunker C fuel oil
as inducing agents for natural populations of bacteria that had been collected from
a variety of near- and offshore environments. Treatment of samples from the
majority (78%) of environments resulted in prophage induction by one or more of
the inducing agents tested. Arochlor 1248, a PCB, was the most efficient agent
tested. Using the burst size method, the percentage of the bacterial population
found to have been lysogenized ranged from 1.32 to over 100%, with an average
of 16.8%.

A 13-month biweekly seasonal study of prophage induction in natural bacterial populations was performed in Tampa Bay (Cochran and Paul, submitted). Prophage induction only occurred during warmer months (February to October), when water temperatures were >19°C. No induction was observed in November, December, and January (Fig. 2). The percentage of the population lysogenized as estimated by the burst size method ranged from 0 to 37%, averaging 6.9%. To determine the potential impact an induction event might have on the phage population, it was estimated that, if 30% of the population was lysogenized, and ⅓ of these induced by some event, then this would produce 1.2×10^7 phage/ml, equivalent to 100% of the indigenous phage population. If only 8% of the bacteria were lysogenized and half of these induced, this would produce 5×10^6 phage/ml, or nearly half the ambient viral population. Thus, by our calculations, if induction of only a portion of the ambient bacterial population occurs, this could contribute significantly to the water column viral population, and potentially increase bacterial mortality. This research also emphasizes a problem with detection of prophage by mitomycin C induction. It may be that the proportion of the bacterial population lysogenized does not change with the seasons, but rather our ability to detect prophage induction, which requires actively dividing bacteria. In the winter months, the bacterial population may be at a lower metabolic state than in the summer months. Thus, the true proportion of lysogens may be underestimated in the winter. In the summer months, if a proportion of the lysogens are already induced by elevated temperatures, salinity, or environmental mutagens, then a relatively small portion of the remaining lysogenized population could be induced by mitomycin C. This hypothesis is based on observations from the seasonal study, whereby the first spring samples gave the greatest number of inducible phages and the largest proportion of the bacteria as lysogens (Cochran and Paul, 1998).

C. Pseudolysogeny

Pseudolysogeny is a condition in which a bacterial strain has a chronic phage infection and the viral DNA is not integrated into any of the host's replicons (Ackermann and DuBow, 1987). This condition is sometimes equated to the "carrier state," although Ackermann and DuBow (1987) have equated it to cells containing a plasmid-like prophage. In pseudolysogeny, the prophages are by definition not inducible, because no integration of the prophages has occurred, induction being the result of DNA repair mechanisms activating the excision of prophages. Pseudolysogeny is thought to occur when either the phage–host system is a mixture of sensitive and resistant cells, or a mixture of prophage-containing

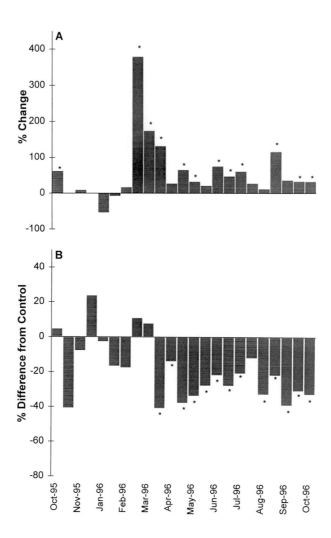

Fig. 2. Results of a seasonal study of lysogeny in Tampa Bay. Samples were taken biweekly and aliquots treated with mitomycin C for 24 hr. Treated and control samples were assayed for viral direct counts by TEM and bacterial direct counts by epifluorescence microscopy. In panel A, the percentage change in viral direct counts between the mitomycin C and control treatments are given. In panel B, the percentage difference in bacterial counts compared to control is given. A significant increase in viral direct counts (as indicated by asterisks) is considered a positive response for prophage induction. Therefore, no lysogens could be detected during the winter months in Tampa Bay.

cells and virulent phages. It may also be that pseudolysogeny encompasses several distinct biological phenomenon of phage–host systems, which results in constant phage production without a host population crash and in the absence of prophage integration.

Moebus (1997a–c) described the lysogeny of marine bacterium H24. Apparently, at low nutrient concentrations true lysogeny was favored. Growth at high nutrient concentrations in culture favored phage mutation, which resulted in pseudolysogeny. These findings are in contrast to those of Ripp and Miller (1997), who describe pseudolysogeny in *Pseudomonas aeruginosa* phage UT1. In the latter phage–host system, pseudolysogeny was an unstable state that only occurred under extreme starvation conditions. The phage genome resided inside the cell, without enough energy to make "the lysogenic decision," such that neither lysogeny nor a virulent condition occurred. Unlike classic psuedolysogeny whereby viral particles are copiously produced, this state produces few or no virions.

We have described four phage–host systems from Hawaii (Jiang and Paul, 1998a,b), three of which are capable of establishing lysogenic interactions between the host and the phage. One system, the HSIC host and ΦHSIC phage, were described in detail and appeared to be a typical lysogenic interaction. However, the HSIC/ΦHSIC system could also establish a pseudolysogenic state, with constant production of high numbers of phages (10^8 to 10^9/ml) without induction by mitomycin C (McLaughlin and Paul, unpublished).

The above observations of pseudolysogeny were for isolates in culture, and it is not known if this process occurs in natural populations. It may be that aquatic microbial populations are comprised of nonlysogenized bacteria, lysogens, and pseudolysogens (Fig. 3). Phages and their host may be able to shift from lytic, lysogenic, and pseudolysogenic lifestyles depending on nutrient levels, temperature, and levels of toxic agents in their environment. It may also be that our current method for detection of lysogens (mitomycin C induction) may underestimate the fraction of lysogens present if the cells are not rapidly growing, or in a dormant or resting state. The hypothesis that lysogeny is not important (Wilcox and Fuhrman, 1994; Weinbauer and Suttle, 1996) was based on the lack of prophage induction by sunlight or by observations of field populations in winter or from offshore oligotrophic environments. Even if the fraction of the bacteria that is truly lysogenized is small, their induction can have a dramatic impact on the total number of VLPs present. Also, the processes of "natural" induction of ambient populations as well as the proportion of bacteria in a state of pseudolysogeny are not known. Although Moebus (1997c) and Ripp and Miller (1997) have a different understanding of the role of nutrient deprivation in pseudolysogeny, both believe that pseudolysogeny and lysogeny are responsible for the tremendous number of VLPs observed in aquatic environments.

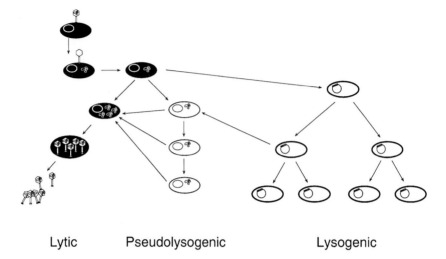

Fig. 3. Relationship between lytic, pseudolysogenic, and lysogenic lifestyles of bacteriophage. After phage adsorption, the phage nucleic acid is injected into the host (represented by a twisted line; the host chromosome appears as a circle). In the lysogenic mode, the viral genome becomes incorporated into the host genome (represented as a straight bar in the host genome), which is passed on to progeny. Induction can occur, and the prophage is excised from the host genome. In the pseudolysogenic lifestyle, the phage genome is maintained as a plasmid and can result in cell lysis, propagation of the plasmid in daughter cells, or segregation and loss of the plasmid. In the lytic lifestyle, the phages are produced at the expense of the host, and the host does not benefit. Under varying conditions of nutrients and other factors, it is hypothesized that members of the phage–host microbial community can interconvert between these three lifestyles. This results in a high level of immunity to superinfection on the part of the indigenous microbial population to the phages present, while resulting in a steady state of high numbers of viruses and hosts.

D. Gene Transfer

1. Host Genes

Transduction, which means "carrying over," is a naturally occurring mechanism of phage-mediated gene transfer. A piece of host DNA, either chromosomal or plasmid, is picked up by the bacteriophage prior to lysis and transferred to another cell on subsequent infection (Zinder and Lederberg, 1952). This phenomenon was first observed by Lederberg *et al.* (1951) during an experiment to test for recombination in *Salmonella typhimurium*.

The overall process is more specifically divided into generalized and specialized transduction. Generalized transduction can be mediated by either virulent or temperate phages. It is due to a packaging error, in which the bacteriophage mistakenly incorporates a piece of host DNA. This is termed "general transduction" because any bacterial gene or fragment can be transduced. Specialized transduction is only mediated by temperate phages. Here the transducing particle is created by faulty separation of the prophage from the host chromosome, resulting in the packaging of a chimera that includes both bacterial and viral genes. It is termed "specialized" because these prophages always insert at a particular location on the host chromosome, and therefore only the specific bacterial genes on either side of that locus will be transduced. Transduction of *Pseudomonas aeruginosa* chromosomal markers has been observed with a number of different bacteriophages (Krishnapillai, 1971; Morgan, 1979). Transduction of plasmids has been recorded in strains of *Escherichia coli* (Gyles and Palchaudhuri, 1977; Hashimoto and Fujisawa, 1988; Mise and Nakaya, 1977), *Salmonella typhimurium* (Mise and Nakaya, 1977; Monroe and Kredich, 1988; Watanabe *et al.*, 1972), *Bacillus thuringiensis* (Lecadet *et al.*, 1992), and *Lactobacillus acidophilus* ADH (Luchansky *et al.*, 1989).

While transduction, as described above, has mainly been used as a laboratory tool to map bacterial chromosomes or characterize strains, there is increasing interest in the role it plays in gene transfer in the natural environment. Some attempt has been made to estimate the frequency of transduction in soil (Germida and Khachatourians, 1988; Zeph *et al.*, 1988) by seeding the environment with a specialized transduction system of *E. coli*. Reported frequencies of observed transduction events have varied by three orders of magnitude; 10^{-3} transductants/recipient (Zeph *et al.*, 1988) versus 10^{-6} (Germida and Khachatourians, 1988).

Prior to the now numerous reports (Bergh *et al.*, 1989; Borsheim *et al.*, 1990; Bratbak *et al.*, 1990; Proctor and Fuhrman, 1990; Suttle *et al.*, 1990) that have found viral concentrations in the marine water column of 10^4 to 10^8 particles per ml of seawater, it was postulated that the "vast dilution potential" of an environmental body of water would make phage–cell contact too infrequent for transduction to occur at levels of significance (Morrison *et al.*, 1978). It was supposed that transduction would only be important in areas where phages were concentrated, such as sewer effluents or the tissues of filter-feeding mollusks (Morrison *et al.*, 1978).

Evidence that this was not necessarily true exists in a group of studies on transduction in aquatic (freshwater) environments such as lakes and reservoirs (Morrison *et al.*, 1978; Ogunseitan *et al.*, 1990; Saye *et al.*, 1987, 1990). The *Pseudomonas aeruginosa* model system developed in freshwater microcosms has demonstrated phage-mediated transfer of both chromosomal and plasmid genes, occurring at significant frequencies (Table II) (Morrison *et al.*, 1978; Saye *et al.*, 1987), optimally at a pH of 7.0 (Amin and Day, 1988). Later studies (Ripp and Miller, 1995; Ripp *et al.*, 1994) have determined that transduction in freshwater systems by a different *P. aeruginosa* phage increased in frequency in the presence of suspended particulates. These particulates are thought to promote aggregation

TABLE II

Transduction Frequencies in Aquatic and Marine Environments

Environment	Transduction frequency	Multiplicity of infection (MOI)	Recipient	Phage	Type of transduced DNA	Reference
Aquatic	2.16×10^{-7}/pfu	Unknown	*P. aeruginosa*	F116	Chromosomal	Amin and Day (1988)
	4.60×10^{-7}/pfu	0.02	*P. aeruginosa*	F116L	Chromosomal	Saye *et al.* (1990)
	4.70×10^{-7}/pfu	0.02	*P. aeruginosa*	DS1	Chromosomal	Saye *et al.* (1990)
	3.30×10^{-7}/pfu	0.02	*P. aeruginosa*	F116L	Plasmid	Saye *et al.* (1990)
	8.20×10^{-7}/pfu	0.02	*P. aeruginosa*	DS1	Plasmid	Saye *et al.* (1990)
Marine	1.33×10^{-7}/pfu	0.01	Strain HSIC	T-ΦHSIC	Plasmid	Jiang (1996)
	1.57×10^{-8}/pfu	Unknown	Natural marine community	T-ΦD1B	Plasmid	Jiang (1996)

of hosts and phages, allowing increased contact and infection, and therefore facilitating transduction.

Four reports of transduction systems isolated from the marine environment are known, the older three all involving virulent phages of *Vibrio* species. The first described an unusual system in which specific transduction of tryptophan synthesis genes was observed (Keynan *et al.*, 1974). The mechanism of this specialized transduction in the apparent absence of lysogeny is unknown. Levisohn *et al.* (1987) and Ichige *et al.* (1989) reported additional marine transduction systems, each capable of generalized transduction. These transduction systems were primarily used as descriptive tools in the laboratory rather than in environmental studies.

The fourth study was conducted by our lab (Jiang, 1996) to examine the potential for gene transfer occurring in the marine environment. The nonconjugative plasmid pQSR50 (Meyer *et al.*, 1982) was introduced to two lysogenic bacterial hosts, HSIC and D1B, isolated from waters off Oahu, Hawaii. These hosts, designated HOPE-1 and HOPE-2, respectively, were then used as donors in the experiments performed, to generate transducing lysates of their phages that we designated T-ΦHSIC and T-ΦD1B. Both wild-type hosts and natural marine bacterial communities were used as recipients. Transduction of the plasmid was confirmed by hybridization of the transductants with a gene probe specific to a fragment of the neomycin phosphotransferase (*nptII*) gene of pQSR50. As with plasmid transformation of natural populations (Williams *et al.*, 1997), there was evidence of some genetic rearrangement of the plasmid after transduction. Extrapolation of these results to the whole of Tampa Bay suggested that 1.3×10^{14} transduction events per year could occur there (Jiang and Paul, 1998a,b). As can be seen in Table II, the frequencies of transduction for the two marine systems are very similar to those seen in the aquatic *P. aeruginosa* system.

2. Phage Genes

Efforts to sequence and better characterize the coliphage family of bacteriophages have led to a hypothesis of frequent horizontal gene transfer between otherwise unrelated viruses. Haggard-Ljungquist *et al.* (1992) sequenced two tail fiber genes of phage P2 and discovered during computational alignments of the genes with other coliphages that parts of the P2 H gene contained elements related to λ, Mu, P1, K3, and T2. Further comparisons of tail fibers from other coliphages revealed that they appear to consist of scrambled cassettes, portions of which were held in common by unrelated bacteriophages. Tail fibers are used by phages to attach to their bacterial host, and therefore the specificity of the fibers determines the host range of a given phage. It would be possible for a phage to extend its host range by acquiring a piece of a tail fiber gene from an unrelated virus. Haggard-Ljungquist *et al.* (1992) feel this is the most logical explanation for their findings. More evidence of phages exchanging modular bits of genetic information is compiled in a review paper by Casjens *et al.* (1992). In addition to the tail fiber example, lytic enzymes, integrases, and head-assembly genes are discussed. Studies of lysozymes and integrases of coliphages found that, while the genes were fairly diverse, they were more like each other than like nonphage enzymes, but the enzymes of a particular group (lambdoids) were not always more closely related to one another than they were to another phage group.

Ackermann *et al.* (1995) examined 150 protein sequences from 35 tailed phages looking for additional examples of "gene blocks" that had been exchanged among phage genomes. Sequences that contained more than 30% identity were considered to be significantly related. The commonalities identified included: (a) λ, T4, P1, and P2 tail fibers, (b) P1 and Mu invertases, (c) λ, P1, P2, P22 and Mu lysozymes, muramidases, and integrases, (d) T4 and Φ3T thymidylate synthases, and (e) T7 and SP6 RNA polymerases. From these results, it is clear that some gene exchange has occurred in all four of the major categories of phage proteins: constitutional (tail fibers), lysis-related (lysozymes), replication-related (polymerases), and lysogeny-related (integrases).

V. NON-INDIGENOUS PHAGE

A. Coliphages as Fecal Indicators

Coliphages are viruses that are infective for *E. coli*. These viruses have recently been considered indicators of fecal contamination, in addition to the standard fecal indicator, fecal coliform bacteria. However, there have been problems with the fecal coliform standard, particularly in tropical environments, because fecal coliforms can be present in the absence of any fecal contamination (Hazen, 1988). This may be due to the fact that in tropical environments the bacteria may occur naturally in soil and multiply in streams (Whitman *et al.*, 1995). Additionally, fecal coliforms do not accurately model the transport and fate of pathogenic viruses and protozoa (Alhajjar *et al.*, 1988). Therefore, coliphages have been investigated as indicators of fecal pollution (Gerba, 1987).

Coliphages are thought to be good indicators of fecal pollution because they model the fate and transport of at least the enteroviruses (Havelaar *et al.*, 1984, 1990, 1993). An examination of a variety of pristine and sewage-impacted freshwater and tropical marine environments indicated that coliphages were better sewage indicators than fecal bacteria (Hernandez-Delgado *et al.*, 1991). Fecal bacterial indicators were higher in the water that collected on the leaves of bromeliads than in sewage-contaminated waters, whereas coliphages were only isolated from waters used for recreational purposes and known to be contaminated with domestic sewage. Coliphages have also been used as indicators of fecal pollution in the coastal waters of Oahu, Hawaii (Paul *et al.*, 1997a). Coliphages were always present near an offshore sewage outfall but were only abundant in near-shore waters during rainfall events, indicating stormwater input. Coliphages were not detected at the beaches of Waikiki, the major tourist attraction of Oahu (Paul *et al.*, 1997a).

Coliphages are usually differentiated as somatic (those infecting non-pilis-containing *E. coli* cells) and male-specific (those infecting cells with the F pilis). The latter are usually a less numerous subset of total present coliphages (Dhillon *et al.*, 1970; Ackermann and Nguyen, 1983). Whether the somatic or the male-specific coliphages are better indicators of fecal pollution in waters is a matter of debate. Havelaar *et al.* (1990) examined the distribution of fecal indicator bacteria and coliphages (both somatic and F-specific) in fecal samples and a variety of wastewaters. Somatic coliphages were abundant in all fecal samples, whereas F-specific coliphages were seldom found in feces from humans or cattle, but abundant in pig and bird fecal matter. F-specific coliphages were found in a variety of wastewaters and were deemed a better indicator of wastewater pollution than fecal pollution. The origin of the F-specific phage in wastewater was not known (Havelaar *et al.*, 1990). Cornax *et al.* (1991) found *E. coli* C somatic phages to be a better indicator of fecal pollution than either fecal indicator bacteria or F-specific coliphage, with the F-specific coliphages considered too rare to be useful as indicators. Morinigo *et al.* (1992) also found somatic coliphages to be a better indicator of fecal pollution than F-specific RNA phages. However, the F-specific RNA phages were deemed a good model for enterovirus abundance in a wastewater treatment plant, despite their not being a good an indicator of fecal pollution.

Differentiation between human and animal waste is needed because the risk of infection from enteric viruses is much greater for contact with human fecal material than with animal fecal material. Despite their general lower abundance than somatic phages, F-specific RNA coliphages have been proposed as indicators of fecal contamination and have been used to differentiate sources of fecal pollution. There are four serological groups of F-specific RNA phages, including group I (MS2-like), group II (GA-like), group III (Qβ-like), and group IV (SP, FI, and ID2; Havelaar *et al.* 1990). Osawa *et al.* (1981) found that groups II and III tend to be found in human feces, groups I and II are generally found in pigs, and feces from all other mammals contain group I-like coliphage. These results are in conflict with Havelaar *et al.* (1990), who found that the F-specific RNA coliphages from human samples are only of serotype IV.

Hsu *et al.* (1995) developed a series of oligonucleotide probes to differentiate the classes of F-specific coliphage in an attempt to differentiate human fecal contamination from animal fecal contamination. The assumption made was that group I RNA coliphages were from nonhuman animals, groups II and III were from human origin, group II could also be from pigs, and group IV was from birds, in contrast to Havelaar's findings. Hsu *et al.* (1995) examined 203 phage isolates that had been classified by serotyping and found that 96.6% were correctly classified by genotyping (specific oligonucleotide hybridization). Although their paper describes successful oligonucleotide detection from plaque lifts, in our hands we have only had success with detection of Northern blots of relatively high titers of the phage.

Another approach to F-specific RNA coliphage detection has been taken by Rose and coworkers (1997), who have designed PCR primers to the replicase gene. There is apparently enough homology among the groups at this locus to enable one primer set to detect all the groups. A comparison of plaque titer and PCR techniques for 68 environmental samples from Hawaii indicated that 44 samples were in agreement between the two methods while 14 samples differed.

Clearly, coliphages are good indicators of fecal pollution, particularly in tropical and subtropical waters where the fecal bacterial indicators are inaccurate. Despite the differences in the literature, the general consensus seems to be that total coliphages or somatic coliphages may be the best indicators of fecal contamination, whereas the F-specific RNA coliphages may be more representative of domestic wastewater contamination. The latter group also may be useful in differentiation of animal and human waste input.

B. Viral Tracers

Viral tracers (i.e., non-indigenous bacteriophages) have been used to measure the velocity and direction of groundwater flow to gain knowledge of complex groundwater environments (Gerba, 1984). However, the focus of recent studies has shifted to understanding transport of the microorganisms themselves, because of the importance of microbial transport in aquifer restoration, pathogen contamination of potable groundwater supplies, and contamination of surface waters by waste disposal practices (Harvey, 1997). When trying to ascertain pathogen movement, phages are superior to chemical tracers such as fluorescein dye because they more correctly model microbial movement (Gerba, 1984). Additionally, phages can be added at a high abundance (10^{14} per experiment) for an effective dynamic range of 14 orders of magnitude, which neither chemical tracers nor fluorescent beads can approach. Fletcher and Myers (1974) used coliphage T4 in carbonate rock from the Ozark region of Missouri and found viral movement of 1600 m within about 16 hr. T4 has also been used to identify a leaky sewer (Skilton and Wheeler, 1988). Phages of *Serratia marcescens*, *Enterobacter cloacae*, and the coliphage MS2 were used to detect a leak in a sewage pipe that ran parallel to a beach that formed a part of the English coast (Martin, 1988). Viral tracers have also been used to locate leaks in a dam and to measure river flow rates (Martin, 1988).

Frederick and Lloyd (1996) used phages of *Serratia marcescens* to determine retention time in a waste stabilization pond system in the Cayman Islands. The ponds had been engineered for a retention time of 11.5 days, but tracer studies indicated that the retention time was less than 2 days, with 30% of the inoculum transiting in less than 6 hours.

Paul *et al.* (1995, 1997b) used viral tracer technology to study wastewater movement in the Florida Keys. Septic tanks or waste disposal injection wells were seeded with ΦHSIC, *Salmonella* phage PRD1, and coliphage MS2. Viral tracers appeared in surface waters within surprisingly short periods of time (8–10.5 hr). Thus, viral tracers can show the ineffectiveness of sewage treatment processes, or the inappropriateness of certain treatment strategies for particular environments. In the Florida Keys, the porous karst geology is not suitable for waste disposal practices such as septic tanks and shallow injection wells, although >25,000 and 700 of each, respectively, can be found there.

VI. SUMMARY

Viruses are now known to be the most numerically abundant form of life in the surface waters of this planet. Direct counting techniques including epifluorescence microscopy and TEM have brought this fact to light, even though epifluorescent cell counts are usually greater than TEM counts. The potential for other forms of "coated" DNA aside from viruses may account for these discrepancies. In many environments, there are strong seasonal trends in viral counts (usually highest in summer and lowest in winter), which often seem uncoupled from bacterial abundance.

Viruses are believed, on average, to contribute to approximately 25% of total bacterial mortality, roughly equivalent to that caused by flagellate grazing. However, this percentage can shift under extreme conditions (high salinity or anoxia) to favor viral-induced mortality. Phage decay rates often average between 0.2 and 1.0 day^{-1}, with sunlight being the greatest factor contributing to phage mortality. UV damage by sunlight may be the greatest factor in phage decay, while photoreactivation plays a major role in phage repair.

The genetic diversity of phages is becoming known. Often differences in phage morphology are reflected in differences in DNA hybridization or restriction patterns for phages sharing a common host. There is also conservation of blocks of viral DNA ("modular cassettes") in morphologically variable phage isolated on the same host. In fact, evidence for gene transfer of viral genes is based on blocks of common or similar genes in widely differing phages. Transduction of host genes has been observed in isolates in culture, in isolates added to microcosms, and in natural populations. Although the frequency seems low, when extended across an entire estuary, transduction could result in 10^{14} gene transfer events annually.

Lysogeny is a common occurrence among marine bacteria, and estimates in natural populations based on mitomycin C induction indicate that, on average, ≤10% of the population contains inducible prophage. The ability to detect prophage induction in natural populations has been shown to have a seasonal pattern, with little induction capability in the winter. A variety of environmentally significant pollutants can cause prophage induction, and it may be that during summer months a proportion of the population undergoes induction by indigenous (natural and anthropogenic) factors. Induction of only a portion of the indigenous bacterial population could result in a doubling of the phage population.

Non-indigenous phages play important roles in the study of aquatic microbial ecology. Coliphage, the viruses infective for *E. coli*, are useful indicators of fecal and wastewater contamination. Differentiation between human- and animal-derived wastes is possible by oligonucleotide probing of F-specific RNA coliphages. Lastly, non-indigenous viral tracers are useful to determine groundwater and surface water flow, as well as to detect malfunctioning wastewater disposal systems.

Future directions in viral research might include basic questions about the diversity of ambient viral populations: How many different viral "species" are present when we encounter 10^7 or 10^8/ml? How many of these are infective for heterotrophic bacteria or phytoplankton, including cyanobacteria? How infective are they? What is the relative importance of lysogeny and pseudolysogeny (chronic infection) on maintaining phage populations? In one sense, entire aquatic ecosystems behave like a pseudolysogen in continuous culture, with a constant production of viral particles and bacterial cells while maintaining relatively steady-state concentrations of both, without "wipeout" of the bacterial host(s). This is in contrast to viral/algal interactions in the environment, with blooms that result in sudden termination and population crash. It may be that various bacterial species are forming blooms followed by viral-mediated bloom termination, but they go unnoticed because of our inability to recognize such species-specific blooms. A different bacterial species then replaces the one that was virally terminated. These and other interesting questions will hopefully be answered as we continue to unravel the role of bacteriophages in nature.

REFERENCES

Ackermann, H.-W., and DuBow, M. S. (1987). "Viruses of Prokaryotes," Vol. 1: "General Properties of Bacteriophages." CRC Press, Boca Raton, FL.

Ackermann, H.-W, and Nguyen, T.-M. (1983). Sewage coliphages studied by electron microscopy. *Appl. Environ. Microbiol.* **45**, 1049–1059.

Ackermann, H.-W., Elzanowski, A., Fobo, G., and Stewart, G. (1995). Relationships of tailed phages: A survey of protein sequence identity. *Arch. Virol.* **140**, 1871–1884.

Adams, M. H. (1952). Classification of bacterial viruses: Characteristics of the T5 species and of the T2, C16 species. *J. Bacteriol.* **64**, 387–396.

Alhajjar, B. J., Stramer, S. L., and Harkin, J. M. (1988). Transport modeling of biological tracers from septic systems. *Wat. Res.* **22**, 907–915.

Amin, M. K., and Day, M. J. (1988). Influence of pH value on viability and transduction frequency of *Pseudomonas aeruginosa* phage F116. *Lett. Appl. Microbiol.* **6**, 93–96.

Baross, J., Liston, J., and Morita, R. Y. (1978). Incidence of *Vibrio parahaemolyticus* bacteriophages and other *Vibrio* bacteriophages in marine samples. *Appl. Environ. Microbiol.* **36**, 492–499.

Beebee, T. J. (1991). Analysis, purification and quantification of extracellular DNA from aquatic environments. *Freshwat. Biol.* **25**, 525–532.

Bergh, O., Borsheim, K. Y., Bratbak, G., and Heldal, M. (1989). High abundances of viruses found in aquatic environments. *Nature* **340**, 467–468.

Boehme, J., Frischer, M. E., Jiang, S. C., Kellogg, C. A., Pichard, S., Rose, J. B., Steinway, C., and Paul, J. H. (1993). Viruses, bacterioplankton, and phytoplankton in the southeastern Gulf of Mexico: Distribution and contribution to oceanic DNA pools. *Mar. Ecol. Prog. Ser.* **97**, 1–10.

Borsheim, K. Y., Bratbak, G., and Heldal, M. (1990). Enumeration and biomass estimation of planktonic bacteria and viruses by transmission electron microscopy. *Appl. Environ. Microbiol.* **56**(2), 352–356.

Bradley, D. E. (1967). Ultrastructure of bacteriophages and bacteriocins. *Bacteriol. Rev.* **31**, 230–314.

Bratbak, G., Heldal, M., Norland, S., and Thingstad, F. (1990). Viruses as partners in spring bloom microbial trophodynamics. *Appl. Environ. Microbiol.* **56**, 1400–1405.

Bratbak, G., Heldal, M., Thingstad, T. F., Riemann, B., and Haslund, P. H. (1992). Incorporation of viruses into the budget of microbial C-transfer: A first approach. *Mar. Ecol. Prog. Ser.* **83**, 273–280.

Bratbak, G., Thingstad, F., and Heldal, M. (1994). Viruses and the microbial loop. *Microb. Ecol.* **28**, 209–221.

Brussow, H., and Bruttin, A. (1995). Characterization of a temperate *Streptococcus thermophilus* bacteriophage and its genetic relationship with lytic phages. *Virology* **212**, 632–640.

Brussow, H., Fremont, M., Bruttin, A., Sidoti, J., Constable, A., and Fryder, V. (1994a). Detection and classification of *Streptococcus thermophilus* bacteriophages isolated from industrial milk fermentation. *Appl. Environ. Microbiol.* **60**, 4537–4543.

Brussow, H., Probst, A., Fremont, M., and Sidoti, J. (1994b). Distinct *Streptococcus thermophilus* bacteriophages share an extremely conserved DNA fragment. *Virology* **200**, 854–857.

Casjens, S., Hatfull, G., and Hendrix, R. (1992). Evolution of dsDNA tailed-bacteriophage genomes. *Sem. Virol.* **3**, 383–397.

Cochlan, W. P., Wikner, J., Steward, G. F., Smith, D. C., and Azam, F. (1993). Spatial distribution of viruses, bacteria and chlorophyll *a* in neritic, oceanic and estuarine environments. *Mar. Ecol. Prog. Ser.* **92**, 77–87.

Cochran, P. K., and Paul, J. H. (1998). Seasonal abundance of lysogenic bacteria in a subtropical estuary. *Appl. Environ. Microbiol.* **64**, 2308–2312.

Cochran, P. K., Kellogg, C. A., and Paul, J. H. (1998). Prophage induction of indigenous marine lysogenic bacteria by environmental pollutants. *Mar. Ecol. Prog. Ser.* **164**, 125–133.

Cornax, R., Morinigo, M. A., Balebona, M. C., Castro, D., and Borrego, J. J. (1991). Significance of several bacteriophage groups as indicators of sewage pollution in marine waters. *Wat. Res.* **25**, 673–678.

Delbruck, M. (1946). Bacterial viruses or bacteriophages. *Biol. Rev.* **21**, 30–40.

Demuth, J., Neve, H., and Witzel, K. (1993). Direct electron microscopic study on the morphological diversity of bacteriophage populations in Lake Plubsee. *Appl. Environ. Microbiol.* **59**, 3378–3384.

d'Herelle, F. (1917). Sur un microbe invisible antagoniste des bacilles dysenteriques. *C. R. Acad. Sci. Paris.* **165**, 373–375.

Dhillon, T. S., Chan, Y. S., Sun, S. M., and Chau, W. S. (1970). Distribution of coliphages in Hong Kong sewage. *Appl. Microbiol.* **20**, 187–191.

Dulbecco, R. (1949). Reactivation of ultraviolet inactivated bacteriophage by visible light. *Nature* **163**, 949–950.

Edlin, G., Lin, L., and Kudrna, R. (1975). λ lysogens of *E. coli* reproduce more rapidly than non-lysogens. *Nature* **255**, 735.

Ewert, D. L., and Paynter, M. J. B. (1980). Enumeration of bacteriophages and host bacteria in sewage and the activated-sludge treatment process. *Appl. Environ. Microbiol.* **39**, 576–583.

Felsenstein, J. (1989). PHYLIP — Phylogeny inference package (version 3.2). *Cladistics* **5**, 164–166.

Fletcher, M. V., and Myers, R. L. (1974). Groundwater tracing in karst terrain using bacteriophage T4. *Abst. 74th Ann. Meet. Amer. Soc. Microbiol.*, Chicago, G194.

Frederick, G. L., and Lloyd, B. J. (1996). An evaluation of retentiaon time and short-circuiting in waste stabilisation ponds using *Serratia marcescens* bacteriophage as a tracer. *Wat. Sci. Tech.* **33**, 49–56.

Freifelder, D. (1987). "Molecular Biology." Jones and Bartlett, Boston.

Friedberg, E. C. (1972). Studies on the substrate specificity of the T4 excision repair endonuclease. *Mutat. Res.* **15**, 113–123.

Friedberg, E. C., Walker, G. C., and Siede, W. (1995). "DNA Repair and Mutagenesis." ASM Press, Washington, DC.

Fuhrman, J. A., and Noble, R. (1995). Viruses and protists cause similar bacterial mortality in coastal seawater. *Limnol. Oceanogr.* **40**, 1236–1242.

Fuhrman, J. A., and Suttle, C. A. (1993). Viruses in marine planktonic systems. *Oceanography* **6**, 57–63.

Furuta, M., Schrader, J. O., Schrader, H. S., Kokjohn, T. A., Nyaga, S., McCullough, A. K., Lloyd, R. S., Burbank, D. E., Landstein, D., Lane, L., and Van Etten, J. L. (1997). *Chlorella* virus PBCV-1 encodes a homolog of the bacteriophage T4 UV damage repair gene *denV*. *Appl. Environ. Microbiol.* **63**(4), 1551–1556.

Gerba, C. P. (1984). Microorganisms as groundwater tracers. *In* "Groundwater Pollution Microbiology" (G. Bitton and C. P. Gerba, eds.), pp. 225–234. Wiley-Interscience, New York.

Gerba, C. P. (1987). Distribution of coliphages in the environment: General considerations. *In* "Phage Ecology" (S. M. Goyal, C. P. Gerba, and G. Bitton, eds), pp. 87–123. Wiley-Interscience, New York.

Germida, J. J., and Khachatourians, G. G. (1988). Transduction of *Escherichia coli* in soil. *Can. J. Microbiol.* **34**, 190–193.

Gonzalez, J. M., and Suttle, C. A. (1993). Grazing by marine nanoflagellates on viruses and virus-sized particles: Ingestion and digestion. *Mar. Ecol. Prog. Ser.* **94**, 1–10.

Goyal, S. M., Gerba, C. P., and Bitton, G. (1987). "Phage Ecology." Wiley-Interscience, New York.

Grimont, F., and Grimont, P. A. D. (1981). DNA relatedness among bacteriophages of the morphological group C3. *Curr. Microbiol.* **6**, 65–69.

Grimont, F., Grimont, P. A. D., and du Pasquier, P. (1978). Morphological study of five bacteriophages of yellow-pigmented enterobacteria. *Curr. Microbiol.* **1**, 37–40.

Guixa-Boixareu, N, Calderon-Paz, J. I., Heldal, M., Bratbak, G., and Pedros-Alio, C. (1996). Viral lysis and bacterivory as prokaryotic loss factors along a salinity gradient. *Microb. Ecol. Aquat. Microb. Ecol.* **11**, 215–227.

Gyles, C. L., and Palchaudhuri, S. (1977). Naturally occurring plasmid-carrying genes for enterotoxin production and drug resistance. *Science* **113**, 198–199.

Haggard-Ljungquist, E., Halling, C., and Calendar, R. (1992). DNA sequences of the tail fiber genes of bacteriophage P2: Evidence for horizontal transfer of tail fiber genes among unrelated bacteriophages. *J. Bacteriol.* **174**, 1462–1477.

Hanawalt, P. C., Cooper, P. K., Ganesan, A. K., and Smith, C. A. (1979). DNA repair in bacteria and mammalian cells. *Annu. Rev. Biochem.* **48**, 783–836.

Hara, S., Terauchi, K., and Koike, I. (1991). Abundance of viruses in marine waters: Assessment by epifluorescence and transmission electron microscopy. *Appl. Environ. Microbiol.* **57**, 2731–2734.

Hara, S., Koike, I., Terauchi, K., Kamiya, H., and Tanoue, E. (1996). Abundance of viruses in deep oceanic waters. *Mar. Ecol. Prog. Ser.* **145**, 269–277.

Harvey, R. W. (1997). Microorganisms as tracers in groundwater injection and recovery experiments: A review. *FEMS Microbiol. Rev.* **20**, 461–472.

Hashimoto, C., and Fujisawa, H. (1988). Packaging and transduction of non-T3 DNA by bacteriophage T3. *Virology* **166**, 432–439.

Havelaar, A. H., Pot-Hogeboom, W. M., and Pot, R. (1984). F-specific RNA bacteriophages in sewage: Methodology and occurrence. *Wat. Sci. Technol.* **17**, 645–655.

Havelaar, A. H., Pot-Hogeboom, W. M., Furuse, K., Pot, R., and Hormann, M. P. (1990). F-specific RNA bacteriophages and sensitive host strains in faeces and wastewater of human and animal origin. *J. Appl. Bacteriol.* **69**, 30–37.

Havelaar, A. H., van Olphen, M., and Drost, Y. C. (1993). F-specific RNA bacteriophages are adequate model organisms for enteric viruses in fresh water. *Appl. Environ. Microbiol.* **59**, 2956–2962.

Hazen, T. C. (1988). Fecal coliforms as indicators in tropical waters: A review. *Toxic Assess.* **3**, 461–477.

Heldal, M., and Bratbak, G. (1991). Production and decay of viruses in aquatic environments. *Mar. Ecol. Prog. Ser.* **72**, 205–212.

Hennes, K. P., and Suttle, C. A. (1995). Direct counts of viruses in natural waters and laboratory cultures but epifluorescence microscopy. *Limnol. Oceanogr.* **40**, 1050–1055.

Hernandez-Delgado, E. A., Sierra, M. L., and Toranzos, G. A. (1991). Coliphages as alternate indicators of fecal contamination in tropical waters. *Environ. Toxicol. Wat. Qual.* **6**, 131–143.

Hidaka, T. (1977). Detection and isolation of marine bacteriophage systems in the southwestern part of the Pacific Ocean. *Mem. Fac. Fish, Kagoshima Univ.* **26**, 55–62.

Hsu, F.-C., Shieh, Y.-S. C., van Duin, J., Beekwilder, M. J., and Sobsey, M. (1995). Genotyping male-specific RNA coliphages by hybridization with oligonucleotide probes. *Appl. Environ. Microbiol.* **61**, 3960–3966.

Ichige, A., Matsutani, S., Oishi, K., and Mizushima, S. (1989). Establishment of gene transfer systems and construction of the genetic map of a marine *Vibrio* strain. *J. Bacteriol.* **171**, 1825–1834.

Jarvis, A. W. (1984). Differentiation of lactic streptococcal phages into phage species by DNA–DNA homology. *Appl. Environ. Microbiol.* **47**(5), 343–349.

Jiang, C. (1996). PhD Dissertation. University of South Florida, St. Petersburg.

Jiang, S. C., and Paul, J. H. (1994). Seasonal and diel abundance of viruses and occurrence of lysogeny/bacteriocinogeny in the marine environment. *Mar. Ecol. Prog. Ser.* **104**, 163–172.

Jiang, S. C., and Paul, J. H. (1995). Viral contribution to dissolved DNA in the marine environment as determined by differential centrifugation and kingdom probing. *Appl. Environ. Microbiol.* **61**, 317–325.

Jiang, S. C., and Paul, J. H. (1996). Occurrence of lysogenic bacteria in marine microbial communities as determined by prophage induction. *Mar. Ecol. Prog. Ser.* **142**, 27–38.

Jiang, S. C., and Paul, J. H. (1998a). Significance of lysogeny in the marine environment: Studies with isolates and a model for viral production. *Aquat. Microb. Ecol.*

Jiang, S. C., and Paul, J. H. (1998b). Gene transfer by transduction in the marine environment. *Appl. Environ. Microbiol.* **64**, 2780–2787.

Johnson, P. W., and Sieburth, J. McN. (1978). Morphology of non-cultured bacterioplankton from estuarine, shelf, and open ocean water. *Abst. Ann. Meet. Am. Soc. Microbiol.* **95**, 178.

Kellogg, C. A., and Paul, J. H. (1997). *ASLO, Aquatic Sciences Meeting*, Santa Fe, NM.

Kellogg, C. A., Rose, J. B., Jiang, S. C., Thurmond, J. M., and Paul, J. H. (1995). Genetic diversity of related vibriophages isolated from marine environments around Florida and Hawaii, USA. *Mar. Ecol. Prog. Ser.* **120**, 89–98.

Kellogg, C. A., Cochran, P. K., and Paul, J. H. (1997). *Abst. Ann. Meet. Am. Soc. Microbiol.*, Miami Beach, N052.

Keynan, A., Nealson, K., Sideropoulos, H., and Hastings, J. W. (1974). Marine transducing bacteriophage attacking a luminous bacterium. *J. Virol.* **14**, 333–340.

Krishnapillai, V. (1971). A novel transducing phage. *Mol. Gen. Genet.* **114**, 134–143.

Lecadet, M. M., Chaufaux, J., Ribier, J., and Lereclus, D. (1992). Construction of novel *Bacillus thuringiensis* strains with different insecticidal activities by transduction and transformation. *Appl. Environ. Microbiol.* **58**(3), 840–849.

Lederberg, J., Lederberg, E. M., Zinder, N. D., and Lively, E. R. (1951). Recombination analysis of bacterial heredity. *Cold Spring Harbor Symp. Quant. Biol.* **16**, 413–441.

Le Marrec, C., Van Sinderen, D., Walsh, L., Stanley, E., Vlegels, E., Moineau, S., Heinze, P., Fitzgerald, G., and Fayard, B. (1997). Two groups of bacteriophages infecting *Streptococcus thermophilus* can be distinguished on the basis of mode of packaging and genetic determinants for major structural proteins. *Appl. Environ. Microbiol.* **63**, 3246–3253.

Levin, B. R., and Lenski, R. E. (1983). Coevolution in bacteria and their viruses and plasmids. *In* "Coevolution" (D. J. Futuyma and M. Slaktin, eds.), pp. 99–127. Sinauer Associates, Sunderland, MA.

Levisohn, R., Moreland, J., and Nealson, K. (1987). Isolation and characterization of a generalized transducing phage for the marine luminous bacterium *Vibrio fischeri*. *J. Gen. Microbiol.* **133**, 1577–1582.

Lindahl, T. (1982). DNA repair enzymes. *Annu. Rev. Biochem.* **51**, 61–87.

Lindstrom, K., and Kaijalainen, S. (1991). Genetic relatedness of bacteriophage infecting *Rhizobium galegae* strains. *FEMS Microbiol. Lett.* **82**, 241–246.

Lu, Z., Li, Y., Zhang, Y., Kutish, G. F., Rock, D. L., and Van Etten, J. L. (1995). Analysis of 45 kb of DNA located at the left end of the chlorella virus PBCV-1 genome. *Virology* **206**, 339–352.

Luchansky, J. B., Kleeman, E. G., Raya, R. R., and Klaenhammer, T. R. (1989). Genetic transfer systems for delivery of plasmid deoxyribonucleic acid to *Lactobacillus acidophilus* ADH: Conjugation, electroporation, and transduction. *J. Dairy Sci.* **72**, 1408–1417.

Luria, S. E. (1947). Reactivation of irradiated bacteriophage by transfer of self-reproducing units. *Proc. Natl. Acad. Sci. U.S.A.* **33**, 253–264.

Martin, C. (1988). The application of bacteriophage tracer techniques in southwest water. *J. IWEM* **2**, 638–642.

Mauryama, A., Oda, M., and Higashihara, T. (1993). Abundance of virus-sized non-DNase-digestible DNA (coated DNA) in eutrophic seawater. *Appl. Environ. Microbiol.* **59**, 712–717.

McLaughlin, M. R., and Paul, J. H. Unpublished observations.

Meyer, R., Laux, R., Boch, G., Hinds, M., Bayly, R., and Shapiro, J. A. (1982). Broad-host-range IncP-4 plasmid R1162: Effects of deletions and insertions on plasmid maintenance and host range. *J. Bacteriol.* **152**, 140–150.

Minton, K., Durphy, M., Taylor, R., and Friedberg, E. C. (1975). The ultraviolet endonuclease of bacteriophage T4. Further characterization. *J. Biol. Chem.* **250**, 2823–2829.

Mise, K., and Nakaya, R. (1977). Transduction of R plasmids by bacteriophages P1 and P22. *Mol. Gen. Genet.* **157**, 131–138.

Moebus, K. (1980). A method for the detection of bacteriophages from ocean water. *Helgoland. Meeres.* **34**, 1–4.

Moebus, K. (1983). Lytic and inhibition responses to bacteriophages among marine bacteria, with special reference to the origin of phage–host systems. *Helgoland. Meeres.* **36**, 375–391.

Moebus, K. (1987). Ecology of marine bacteriophages. *In* "Phage Ecology" (S. M. Goyal, C. P. Gerba, and G. Bitton, eds), pp. 137–156. Wiley-Interscience, New York.

Moebus, K. (1992). Laboratory investigations on the survival of marine bacteriophages in raw and treated seawater. *Helgoland. Meeres.* **46**, 252–273.

Moebus, K. (1997a). Investigations of the marine lysogenic bacterium H24, I: General description of the phage–host system. *Mar. Ecol. Prog. Ser.* **148**, 217–228.

Moebus, K. (1997b). Investigations of the marine lysogenic bacterium H24, II: Development of pseudolysogeny in nutrient-rich broth culture. *Mar. Ecol. Prog. Ser.* **148**, 229–240.

Moebus, K. (1997c). Investigations of the marine lysogenic bacterium H24, III: Growth of bacteria and production of phage under nutrient-limited conditions. *Mar. Ecol. Prog. Ser.* **148**, 241–250.

Moebus, K., and Nattkemper, H. (1981). Bacteriophage sensitivity patterns among bacteria isolated from marine waters. *Helgoland. Meeres.* **34**, 375–385.

Monroe, R. S., and Kredich, N. M. (1988). Isolation of *Salmonella typhimurium cys* genes by transduction with a library of recombinant plasmids packaged in bacteriophage P22HT capsids. *J. Bacteriol.* **170**, 42–47.

Morgan, A. F. (1979). Transduction of *Pseudomonas aeruginosa* with a mutant of bacteriophage E79. *J. Bacteriol.* **139**(1), 137–140.

Morinigo, M. A., Wheeler, D., Berry, C., Jones, C., Munoz, M. A., Cornax, R., and Borrego, J. J. (1992). Evaluation of different bacteriophage groups as faecal indicators in contaminated natural waters in southern England. *Wat. Res.* **26**, 267–271.

Morrison, W. D., Miller, R. V., and Sayler, G. S. (1978). Frequency of F116-mediated transduction *Pseudomonas aeruginosa* in a freshwater environment. *Appl. Environ. Microbiol.* **36**, 724–730.

Murooka, Y., and Mitani, I. (1985). Efficient expression of a promoter-controlled gene: Tandem promoters of lambda p_R and p_L functional gene in bacteria. *J. Biotechnol.* **2**, 303–316

Noble, R. T., and Fuhrman, J. A. (1997). Virus decay and its causes in coastal waters. *Appl. Environ. Microbiol.* **63**, 77–83.

Noble, R. T., and Fuhrman, J. A. (1998). SYBR Green I counts of viruses and bacteria. *Aquat. Microb. Ecol.* **14**, 113–118.

Ogunseitan, O. A., Sayler, G. S., and Miller, R. V. (1990). Dynamic interactions of *Pseudomonas aeruginosa* and bacteriophage in lake water. *Microb. Ecol.* **19**, 171–185.

Osawa, S., Furuse, K., and Watanabe, I. (1981). Distribution of ribonucleic acid coliphages in animals. *Appl. Environ. Microbiol.* **41**, 164–168.

Paul, J. H., Jiang, S. C., and Rose, J. B. (1991). Concentration of viruses and dissolved DNA from aquatic environments by vortex flow filtration. *Appl. Environ. Microbiol.* **57**, 2197–2204.

Paul, J. H., Rose, J. B., Brown, J., Shinn, E. A., Miller, S., and Farrah, S. R. (1995). Viral tracer studies indicate contamination of marine waters by sewage disposal practices in Key Largo, Florida. *Appl. Environ. Microbiol.* **61**, 2230–2234.

Paul, J. H., Kellogg, C. A., and Jiang, S. C. (1996). Viruses and DNA in marine environments. *In* "Microbial Diversity in Time and Space (R. R. Colwell, U. Simidu, and K. Ohwada, eds.), pp. 115–124. Plenum Press, New York.

Paul, J. H., Rose, J. B., Jiang, S. C., London, P., Xhou, X., and Kellogg, C. A. (1997a). Coliphage and indigenous phage in Mamala Bay, Oahu, Hawaii. *Appl. Environ. Microbiol.* **63**, 133–138.

Paul, J. H., Rose, J. B., Jiang, S. C., Xhou, X., Cochran, P., Kellogg, C., Kang, J. B., Griffin, D., Farrah, S., and Lukasik, J. (1997b). Evidence for groundwater and surface marine water contamination by waste disposal wells in the Florida Keys. *Wat. Res.* **31**, 1448–1454.

Proctor, L. M. (1997). Advances in the study of marine viruses. *Microscop. Res. Tech.* **37**, 136–161.

Proctor, L. M., and Fuhrman, J. A. (1990). Viral mortality of marine bacteria and cyanobacteria. *Nature* **343**, 60–62.

Proctor, L. M., and Fuhrman, J. A. (1991). Roles of viral infection in organic particle flux. *Mar. Ecol. Prog. Ser.* **69**, 133–142.

Ripp, S., and Miller, R. V. (1995). Effects of suspended particulates on the frequency of transduction among *Pseudomonas aeruginosa* in a freshwater environment. *Appl. Environ. Microbiol.* **61**, 1214–1219.

Ripp, S., and Miller, R. V. (1997). The role of pseudolysogeny in bacteriophage–host interactions in a natural freshwater environment. *Microbiology* **143**, 2065–2070.

Ripp, S., Ogunseitan, O. A., and Miller, R. V. (1994). Transduction of a freshwater microbial community by a new *Pseudomonas aeruginosa* generalized transducing phage, UT1. *Mol. Ecol.* **3**, 121–126.

Rose, J. B., Zhou, X., Griffin, D. W., and Paul, J. H. (1997). Comparison of PCR and plaque assay for detection and enumeration of coliphage in polluted marine waters. *Appl. Environ. Microbiol.* **63**, 4564–4566.

Saye, D. J., Ogunseitan, O., Sayler, G. S., and Miller, R. V. (1987). Potential for transduction of plasmids in a natural freshwater environment: Effect of plasmid donor concentration and a natural microbial community on transduction in *Pseudomonas aeruginosa*. *Appl. Environ. Microbiol.* **53**, 987–995.

Saye, D. J., Ogunseitan, O., Sayler, G. S., and Miller, R. V. (1990). Transduction of linked chromosomal genes between *Pseudomonas aeruginosa* strains during incubation in situ in a freshwater habitat. *Appl. Environ. Microbiol.* **56**, 140–145.

Sharp, D. G. (1960). Sedimentation counting of particles via electron microscopy. *In* "Proceedings of the Fourth International Congress on Electron Microscopy, Berlin, 1958," pp. 542–548. Springer-Verlag, Berlin.

Sieburth, J. McN., Johnson, P. W., and Hargraves, P. E. (1988). Ultrastructure and ecology of *Aureococcus anophagefferens* gen. et sp. nov. (Chrysophyceae): The dominant picoplankter during a bloom in Narragansett Bay, Rhode Island, summer 1985. *J. Phycol.* **24**, 416–425.

Skilton, H., and Wheeler, D. (1988). Bacteriophage tracer experiments in groundwater. *J. Appl. Bacteriol.* **66**, 549–557.

Sklarow, S. S., Colwell, R. R., Chapman, G. B., and Zane, S. F. (1973). Characteristics of a new *Vibrio parahaemolyticus* bacteriophage isolated from Atlantic coast sediment. *Can. J. Microbiol.* **19**, 1519–1520.

Spencer, R. (1955). A marine bacteriophage. *Nature* **175**, 690.

Steward, G. F., Smith, D. C., and Azam, F. (1996). Abundance and production of bacteria and viruses in the Bering Seas and Chukchi Seas. *Mar. Ecol. Prog. Ser.* **131**, 287–300.

Suttle, C. A., Chan, A. M., and Cottrell, M. T. (1990). Infection of phytoplankton by viruses and reduction of primary productivity. *Nature* **347**, 467–469

Suttle, C. A., Chan, A. M., and Cottrell, M. T. (1991). Use of ultrafiltration to isolate viruses from seawater which are pathogens of marine phytoplankton. *Appl. Environ. Microbiol.* **57**, 721–726.

Tapper, M. A., and Hicks, R. E. (1998). Morphology and abundance of free and temperate viruses in Lake Superior. *Limnol. Oceanogr.* **43**, 95–103.

Torrella, F., and Morita, R. Y. (1979). Evidence by electron micrographs for a high incidence of bacteriophage particles in the water of Yaquina Bay. *Appl. Environ. Microbiol.* **37**, 774–778.

Watanabe, T., Ogata, Y., Chan, R. K., and Botstein, D. (1972). Specialized transduction of tetracycline resistance by phage P22 in *Salmonella typhimurium. Virology* **50**, 874–882.

Weinbauer, M. G., and Hofle, M. G. (1998). Significance of viral lysis and flagellate grazing as controlling factors of bacterioplankton production in an eutrophic lake. *Appl. Environ. Microbiol.* **64**, 431–438.

Weinbauer, M. G., and Peduzzi, P. (1995). Significance of viruses versus heterotrophic nanoflagellates for controlling bacterial abundance in the northern Adriatic Sea. *J. Plankton Res.* **17**, 1851–1856.

Weinbauer, M. G., and Suttle, C. A. (1996). Potential significance of lysogeny to bacteriophage production and bacterial mortality in coastal waters of the Gulf of Mexico. *Appl. Environ. Microbiol.* **62**, 4374–4380.

Weinbauer, M. G., Fuks, D., Puskaric, S., and Peduzzi, P. (1993). Distribution of viruses and dissolved DNA along a coastal trophic gradient in the northern Adriatic Sea. *Appl. Environ. Microbiol.* **59**, 4074–4082.

Weinbauer, M. G., Wilhelm, S. W., Suttle, C. A., and Garza, D. R. (1997). Photoreactivation compensates for UV damage and restores infectivity to natural marine virus communities. *Appl. Environ. Microbiol.* **63**(6), 2200–2205.

Werquin, M., Ackermann, H.-W., and Levesque, R. C. (1988). A study of 33 bacteriophages of *Rhizobium meliloti. Appl. Environ. Microbiol.* **54**(1), 188–196.

Whitman, R. L., Gochee, A. L., Dustman, W. As, and Kennedy, K. J. (1995). Use of coliform bacteria in assessing human sewage contamination. *Nat. Areas J.* **15**, 22–24.

Wiebe, W. J., and Liston, J. (1968). Isolation and characterization of a marine bacteriophage. *Mar. Biol.* **1**, 244–249.

Wilcox, R. M., and Fuhrman, J. A. (1994). Bacterial viruses in coastal seawater: Lytic rather than lysogenic production. *Mar. Ecol. Prog. Ser.* **114**, 35–45.

Wilhelm, S. W., Weinbauer, M. G., Suttle, C. A., Pledger, R. J., and Mitchell, D. L. (1998). Measurements of DNA damage and photoreactivation imply that most viruses in marine surface waters are infective. *Aquat. Microb. Ecol.* **14**, 215–222.

Williams, H. G., Benstead, J., Frischer, M. E., and Paul, J. H. (1997). Alteration in plasmid DNA following natural transformation to populations of marine bacteria. *Mol. Mar. Biol. Biotechnol.* **6**, 238–247.

Wilson, W. H., Joint, I. R., Carr, N. G., and Mann, N. H. (1993). Isolation and molecular characterization of five marine cyanophages propagated on *Synechococcus* sp. strain WH7803. *Appl. Environ. Microbiol.* **59**(11), 3736–3743.

Wommack, K. E., Hill, T. T., Kessel, M., Tussek-Cohen, E., and Colwell, R. R. (1992). Distribution of viruses in the Chesapeake Bay. *Appl. Environ. Microbiol.* **58**, 2965–2970.

Wommack, K. E., Hill, T. T., and Colwell, R. R. (1995). A simple method for the concentration of viruses from natural water samples. *J. Microbiol. Meth.* **22**, 57–67.

Xenopoulos, M. A., and Bird, D. F. (1997). Virus a la sauce Yo-Pro-enhanced staining for counting viruses by epifluorescence microscopy. *Limnol. Oceanogr.* **42**, 1648–1650.

Yasuda, S., and Sekiguchi, M. (1976). Further purification and characterization of T4 endonuclease V. *Biochim. Biophys. Acta* **442**, 197–207.

Zeph, L. R., Onaga, M. A., and Stotzky, G. (1988). Transduction of *Escherichia coli* by bacteriophage P1 in soil. *Appl. Environ. Microbiol.* **54**, 1731–1737.

Zinder, N. D., and Lederberg, J. (1952). Genetic exchange in *Bacillus subtilis* by bacteriophage SPP1. *J. Bacteriol.* **64**, 679–699.

Zobell, C. (1946). Marine microbiology. *In Chron. Bot.*, pp. 82–83. Waltham, MA.

Zweifel, U. L., and Hägstrom, A. (1995). Total counts of marine bacteria include a large fraction of non-nucleoid-containing bacteria (ghosts). *Appl. Environ. Microbiol.* **61**, 2180–2185.

6

Ecological, Evolutionary, and Geochemical Consequences of Viral Infection of Cyanobacteria and Eukaryotic Algae

CURTIS A. SUTTLE

Departments of Earth and Ocean Sciences (Oceanography),
Botany, and Microbiology and Immunology
The University of British Columbia
Vancouver, British Columbia, Canada V6T 1Z4

247

I. INTRODUCTION

More than 35 years ago, Safferman and Morris (1963) reported the isolation of a virus (cyanophage) that infected a freshwater filamentous "blue-green alga." This discovery stimulated research that led to isolation of many viruses that infect freshwater cyanobacteria. Although the potential for viruses to control cyanobacterial blooms was recognized (Safferman and Morris, 1964; Shilo, 1971), much of this work focused on the biology rather than the ecology of cyanophages (reviewed in Brown, 1972; Padan and Shilo, 1973; Safferman, 1973; Stewart and Daft, 1977; Sherman and Brown, 1978; Martin and Benson, 1988). At about the same time, evidence for viral infection of eukaryotic algae was beginning to emerge with reports by Zavarzina (1961, 1964) of lysis of *Chlorella pyrenoidosa* cultures. However, it was not until a decade later that observations of viruslike particles (VLPs) in eukaryotic algae began to appear in the literature (e.g., Lee, 1971; Chapman and Lang, 1973; Lemke, 1976; Dodds, 1979, 1983), and shortly thereafter a virus (CCV) was isolated that infected the macroalga *Chara corallina* (Gibbs *et al.*, 1975; Skotnicki *et al.*, 1976). This was followed a few years later by isolation of a virus (MPV) that infected the marine photosynthetic flagellate *Micromonas pusilla* (Mayer, 1978; Mayer and Taylor, 1979). Despite the ecological implications of viruses infecting major primary producers in aquatic environments, interest in viruses that infect eukaryotic algae was slow to gather. In fact, after the work by Waters and Chan (1982), there was little further interest in MPV, and the virus was lost from culture (F. J. R. Taylor, personal communication). Furthermore, by the 1980s ecological interest in cyanophages also began to wane as cyanobacterial blooms were brought under control by regulations reducing nutrient inputs to lakes. The decline in interest was exacerbated by the lack of appreciation by many algal and aquatic ecologists of the ecological importance of microbes in general and of viruses in particular. Although there was little interest in the ecological aspects, there were major advances in our understanding of the biology of a group of viruses that infect *Chlorella*-like algae that are symbionts of *Hydra viridis* and *Paramecium bursaria*. These viruses were isolated in the early 1980s and possess a number of unusual features (reviewed in Van Etten *et al.*, 1991; Reisser, 1993; Van Etten, 1995; Van Etten and Meints, 1999) that led to the creation of a new family of viruses, Phycodnaviridae (Van Etten and Ghabrial, 1991). As well, work has progressed on a group of widely distributed viruses that infect filamentous brown algae belonging to order Ectocarpales (Oliveira and Bisalputra, 1978; Müller, 1991; Henry and Meints, 1992; Müller and Frenzer, 1993; Müller, 1996; Van Etten and Meints, 1999).

Much more effort has been expended on trying to understand the ecological role of viruses that infect algae, especially phytoplankton. A reason for this change in emphasis was a group of observations that there are typically more than a million bacteria per milliliter in aquatic systems (e.g., Hobbie *et al.*, 1977). This has led to increasing awareness of the ecological significance of microbial processes in aquatic environments (Pomeroy, 1974; Conover, 1982; Azam *et al.*, 1983). The result has been a much larger constituency of scientists interested in the role of microbes in aquatic systems. Indeed, prokaryotes constitute by far the largest biomass in the nutrient-depleted regions that make up most of the world's oceans (Fuhrman *et al.*, 1989; Cho and Azam, 1990; Li *et al.*, 1992; Caron *et al.*, 1995; Whitman *et al.*, 1998). Complementing observations of abundant microbial life were numerous studies showing that heterotrophic bacteria consume a large portion of the carbon fixed by primary producers (reviewed in Cole *et al.*, 1988; Ducklow and Carlson, 1992). As well, prokaryotes were shown to be responsible for much of the primary productivity in large areas of the ocean, and in many lakes (e.g., Joint and Pomroy, 1983; Glover *et al.*, 1986; Stockner and Antia, 1986). By the late 1980s, the importance of microbial processes had been firmly established. Consequently, there was great interest in reports that, on average, there are about 10 million viruses per milliliter of lake and ocean water (Bergh *et al.*, 1989; Proctor and Fuhrman, 1990). Furthermore, observations that viruses infecting phytoplankton could be readily isolated from seawater (Suttle *et al.*, 1990) suggested that viruses are important in the ecology of algal communities.

This contribution explores our emerging understanding of the ecology of viruses that infect algae. For the purpose of this chapter, algae are treated as a functional group rather than as a taxonomic unit. There could be some debate on whether cyanophages should be included in a chapter on algal viruses, especially given that another chapter in this book focuses on bacteriophages. Clearly, blue-green algae are bacteria, and their evolutionary history is more accurately reflected by the term *cyanobacteria* rather than *cyanophyta*. However, the ecological role of cyanobacteria is more closely tied to that of eukaryotic algae than it is to heterotrophic bacteria; hence, it is appropriate that cyanophages and viruses that infect eukaryotic algae be considered together. The scope of this chapter will be on the ecology, and to some extent, evolutionary relationships and geochemical consequences of viruses that infect prokaryotic and eukaryotic algae. The biology, biochemistry, and molecular biology of algal viruses will not be considered in detail, and other sources should be consulted (e.g., Martin and Benson, 1988; Van Etten *et al.*, 1991; Reisser, 1993; Van Etten and Meints, 1999). As well, the primary focus of the chapter is on viruses that infect phytoplankton, as this is the group in which our ecological knowledge is advancing most quickly. The landscape of algal virus ecology is rapidly changing as new viruses are isolated on a regular basis and new methods for quantifying the effect of viruses in the natural environment are developed.

II. TAXONOMY, MORPHOLOGY, AND EVOLUTION OF ALGAL VIRUSES

A. Cyanophages

The morphology and taxonomy of cyanophages is relatively well studied and has been extensively reviewed elsewhere (Safferman *et al.*, 1983; Suttle, 1999). All known cyanophages have tails, contain double-stranded (ds) DNA, and belong to three families of viruses that also infect heterotrophic bacteria (Murphy *et al.*, 1995). The families are the Myoviridae, whose members possess a contractile tail, Siphoviridae (also referred to as Styloviridae), which have a long noncontractile tail, and the Podoviridae, which have short noncontractile tails (Fig. 1a,b,c, respectively). Representatives from each of these families infect both unicellular and filamentous cyanobacteria. In the scheme of Safferman *et al.* (1983), it was argued that cyanophages should be placed in their own genera (*Cyanomyovirus*, *Cyanopodovirus*, and *Cyanostylovirus*), because the host range of cyanophages extends across genera, whereas coliphages are typically genus specific. However, as the systematics of cyanobacteria become more refined (e.g., Turner, 1997), it is

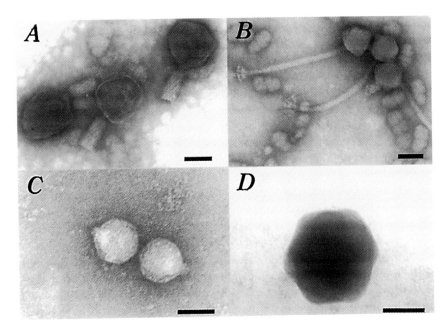

Fig. 1. Typical morphologies of algal viruses. (**A–C**) Cyanophages belonging to the families Myoviridae, Siphoviridae, and Podoviridae, respectively. (**D**) *Micromonas pusilla* virus, which is representative of the morphology of the Phycodnaviridae. Scale bar = 50 nm.

becoming increasingly apparent that the broad host range attributed to cyano-phages is likely the result of errors in cyanobacterial taxonomy (Suttle, 2000). In its most recent overviews of viral taxonomy, the International Committee on Taxonomy of Viruses (ICTV) did not recognize specific cyanophage genera (Murphy *et al.*, 1995; ICTV, 1999; Pringle, 1999).

The relationship of cyanophages to other bacteriophages raises interesting evolutionary questions (Suttle, 1999), particularly because each of the three families of viruses to which cyanophages belong are thought to form a monophyletic group, the Caudovirales (Maniloff and Ackermann, 1998). Consequently, the origin of cyanophages probably predates the divergence of cyanobacteria from other bacteria. Cyanobacteria are an ancient branch of evolution (Woese, 1987) that is thought to have been well established by about 3.5 billion years ago (Schopf and Packer, 1987), implying that the origin of cyanophages is even older. An interesting consequence of cyanophages being evolutionarily ancient is that their origin likely predates that of eukaryotes. This suggests that cyanophages were among the earliest predators of cyanobacteria, and may have regulated cyanobacterial populations, as well as being a major force in cyanobacterial evolution.

B. Viruses Infecting Eukaryotic Algae

In contrast to cyanophages, much remains to be learned about the taxonomy of viruses that infect eukaryotic algae. Based on present understanding, at least two unrelated groups of viruses infect eukaryotic algae. The first of these to be described infects the charophyte *Chara corallina* and is a rod-shaped (ca. 530 nm long), single-stranded (ss) RNA virus (Gibbs *et al.*, 1975; Skotnicki *et al.*, 1976). This is the only clearly documented example of an RNA virus that infects algae. It is morphologically similar to tobacco mosaic virus but has been reported to be more closely related to furoviruses (cited in Van Etten, 1991). The system is also significant because the family Charaphyceae has been postulated to be the evolutionary pathway to higher plants. Furthermore, tobamoviruses and furoviruses are only known to infect higher plants, although they are transmitted by other organisms.

More is known about a group of large dsDNA viruses that infect a variety of microalgae, and possibly filamentous brown algae. The first representative of these viruses was isolated by Mayer and Taylor (1979). It was a large (ca. 130 nm) polyhedral virus that caused the lysis of a marine photosynthetic picoflagellate, *Micromonas pusilla* (Prasinophyceae). Several years later, a morphologically similar but larger (185 nm) virus (HVCV-1) was discovered that infected a *Chlorella*-like symbiont isolated from *Hydra viridis* (Meints *et al.*, 1981). This led to the discovery of a genetically diverse but morphologically indistinguishable suite of viruses that infects symbiotic *Chlorella*-like algae, isolated from *H. viridis* and *Paramecium bursaria*. These viruses shared features with poxviruses, iridoviruses, and African swine fever virus (reviewed in Van Etten *et al.*, 1991; Van Etten and Meints, 1999), but nonetheless were clearly distinct from each of these families. This led to their formal recognition as a new family, Phycodnaviridae, with the *Chlorella* viruses being its sole members (Van Etten and Ghabrial, 1991). At the time, no other eukaryotic algal viruses were sufficiently well characterized to determine their taxonomic position.

At about the time that family Phycodnaviridae was being established, evidence was emerging that viruses that infect phytoplankton could be readily isolated from seawater (Suttle *et al.*, 1990). Shortly thereafter, viruses (MpV) infecting *Micromonas pusilla* were re-isolated (Cottrell and Suttle, 1991). The viruses were large polyhedra (ca. 115 nm), similar to the original MPV, and abundant in nature, genetically diverse, and morphologically indistinguishable from each other (Fig. 1d). They also strongly resembled viruses that infected *Chlorella*-like symbionts, suggesting that the two groups of viruses were related. This hypothesis was addressed by comparing the DNA polymerase (*pol*) gene sequences from the two groups of viruses (reviewed in Short and Suttle, 1999). This seemed like a good approach, as complete *pol* sequences were available for the *Chlorella* viruses PBCV-1 and NY-2A (Grabherr *et al.*, 1992). Moreover, as many dsDNA viruses contain *pol* (Braithwaite and Ito, 1993), there was a high probability of finding one in MpV. As *pol* contains conserved and variable regions, it is relatively easy to find and clone the gene, and infer genetic distances by aligning the sequences with the large database of *pol* sequences. A comparison of inferred amino-acid sequences showed that there were regions conserved within the *pol* genes of MpV-SP1, PBCV-1, and NY-2A that were absent in other *pol* sequences (Zhang and Suttle, 1994). This strongly supported the inclusion of MpV within family Phycodnaviridae. Further analysis demonstrated that the Phycodnaviridae were monophyletic and contained the *Micromonas* and *Chlorella* viruses (Chen and Suttle, 1996). As well, a group of morphologically similar viruses that infects marine *Chrysochromulina* (Prymnesiophyceae) (Suttle and Chan, 1995) was found to fall within family Phycodnaviridae. One of the surprising outcomes of the analyses was that herpesviruses were the closest relatives to the Phycodnaviridae (Chen and Suttle, 1996). Most recently, sequence analysis of a *pol* gene demonstrated that a virus (FsV), which infects the filamentous brown alga *Feldmannia*, likely belongs within family Phycodnaviridae (Lee *et al.*, 1998a).

The discovery that FsV is probably a member of family Phycodnaviridae suggests that this may also be true of several other morphologically similar viruses that infect genera within order Ectocarpales. All of the viruses are large icosahedra (Müller, 1996) that contain DNA genomes ranging in size from 158 to 320 kbp

(Lanka *et al.*, 1993; Ivey *et al.*, 1996; Kapp *et al.*, 1997). Although the genome of FsV was originally reported to be linear (Henry and Meints, 1992), it was later shown to be circular (Ivey *et al.*, 1996), as is the case for EsV-1 that infects *Ectocarpus siliculosus* (Lanka *et al.*, 1993). In addition, the genomes of FsV (Lee *et al.*, 1998b) and EsV-1 (Müller, 1991) appear to be associated with the DNA of their hosts. A difference that has been reported among these viruses is that the genome of FsV consists of two size classes and is dsDNA throughout (Ivey *et al.*, 1996), whereas the genome of EsV-1 is of a single size and contains dsDNA interrupted by single-stranded regions (Klein *et al.*, 1994).

A striking observation of potential evolutionary significance is that viruses that infect *Ectocarpus siliculosus* only infect zoospores (Müller, 1991). Traditionally, order Ectocarpales is considered to be the most primitive within the brown algae (e.g., Bold and Wynne, 1978; Lee, 1980), and its zoospores are morphologically and ultrastructurally very similar to unicellular algae within family Chrysophyceae. This raises the possibility that the viruses represent an evolutionarily ancient association that stems from the time that ancestors of the Ectocarpales were unicellular algae. Resolution of whether all viruses infecting Ectocarpales fall within family Phycodnaviridae or belong to more than one taxonomic group awaits further study.

Family Phycodnaviridae has now been subdivided into the genera *Chlorovirus*, *Prasinovirus*, *Prymnesiovirus*, and *Phaeovirus* based on the class of algae that the viruses infect (ICTV, 1998). It seems likely that there will be further additions to family Phycodnaviridae. Viruses that infect the eukaryotic phytoplankton *Heterosigma akashiwo* (Nagasaki *et al.*, 1994a,b; Nagasaki and Yamaguchi, 1997), *Phaeocystis* (Jacobsen *et al.*, 1996), *Emiliania huxleyi* (Bratbak *et al.*, 1993, 1996; Brussaard *et al.*, 1996), and *Aureococcus anophagefferens* are also large polyhedral viruses (diameters ca. 200, 130–160, 140–180, and 140–160 nm, respectively), similar to others of family Phycodnaviridae.

There is some uncertainty whether there is one or two morphological types of viruses that infect the eukaryotic photosynthetic picoplankter *Aureococcus anophagefferens* (Pelagophyceae). Large polyhedral Phycodnaviridae-like VLPs have been seen in field populations of *A. anophagefferens* (Sieburth *et al.*, 1988; Nixon *et al.*, 1994). In contrast, Milligan and Cosper (1994) claimed to have isolated a pathogen that caused lysis of *A. anophagefferens* that was phagelike. However, the published electron micrograph showed a cell that was smaller than *A. anophagefferens* with an attached phagelike particle (Milligan and Cosper, 1994), suggesting that the micrograph illustrated a phage attached to a contaminating bacterium. Subsequently, Garry *et al.* (1998) went on to purify and describe this virus (AaV-1) in more detail, confirming its phagelike appearance. There have been occasional reports of tailed VLPs in eukaryotic algae (e.g., Pienaar, 1976; Dodds and Cole, 1980; Gromov and Mamkaeva, 1981), although none of the particles were typical of phage. An investigation of viral infection in *A. anophagefferens* (Gastrich *et al.*, 1998) using virus purified by Garry *et al.* (1998) found large (140 to 160 nm in diameter) tail-less intracellular VLPs that were morphologically typical of the Phycodnaviridae. These particles were similar to those described in field populations of *A. anophagefferens* (Sieburth *et al.*, 1988; Nixon *et al.*, 1994). Clearly, Milligan and Cosper (1994) isolated a lytic agent that infects *A. anophagefferens*, but it is unlikely to be phagelike. When pathogens are isolated by screening natural virus communities against potential algal hosts (Suttle *et al.*, 1991; Milligan and Cosper, 1994), invariably phage that infect contaminant bacteria in the cultures are also isolated. It can be extremely difficult to separate the bacteriophage from other viruses, as they will frequently copurify. Another difficulty is that lytic bacteria can also be copurified. Suttle *et al.* (1990, 1991) described the isolation of a putative virus that causes lysis of the pennate diatom *Navicula* sp. Although the pathogen was 0.2-µm filterable and species specific, it was bacterial (Chan *et al.*, 1997). Hence, to conclusively demonstrate that a lytic agent is viral, it must be propagated on an axenic host. If this is not possible, morphologically similar viral particles must be shown to occur within infected cells.

It has been clearly established that at least two unrelated types of viruses infect eukaryotic alga. These are the ssRNA viruses that infect Charaphytes (Gibbs *et al.*, 1975), and the large dsDNA viruses belonging to family Phycodnaviridae that infect photosynthetic protists and some brown algae (ICTV, 1998). Other large polyhedral viruses that infect photosynthetic protists and brown algae are of unknown taxonomy, but may be related to the Phycodnaviridae. In addition, tailed viruses that do not closely resemble currently recognized viral taxa of eukaryotes infect unicellular (Gromov and Mamkaeva, 1981) and filamentous (Dodds and Cole, 1980) green algae and possibly pelagophytes (Milligan and Cosper, 1994; Garry *et al.*, 1998). Viruses that infect eukaryotic algae represent an enormous and largely unexplored diversity of unknown and poorly studied viruses.

III. ECOLOGICAL IMPLICATIONS OF VIRAL REPLICATION

One of the most important considerations in understanding the effects of viruses on organisms and aquatic ecosystems is viral replication. A number of aspects of viral life cycles and replication have direct implications on the role of viruses in aquatic communities. One of the most important of these considerations is whether a virus can form latent or lysogenic infections.

Viruses can interact with their algal hosts either through a lytic cycle or via a lysogenic or latent cycle. When infections are lytic, the virus immediately begins to replicate following introduction of the viral nucleic acid into the host cell. Viral nucleic acids and proteins are made, the viruses are "assembled" and released through lysis of the host cell. An important implication from an ecological perspective of this type of infection is that it results in cell death, thereby imposing strong selective pressure on organisms, especially unicellular ones! As well, lysis results in release of cellular material into the surrounding water, which supplies nutrients to other organisms. This can be an important source of nutrients for other organisms, and places viruses in a key position in nutrient cycling in aquatic communities (see §VIII).

A. Lysogeny and Latency

One replication strategy for viruses is to form latent or lysogenic infections. Lysogenic infections refer to infections where bacteriophage DNA forms a stable association with the DNA of its host. The viral DNA is archived and replicated along with that of its host. Not only does this protect the virus from environmental conditions that might destroy the capsid, or damage the nucleic acids within the capsid, it confers immunity to the host cell from lytic infections by similar viruses. Moreover, the virus can carry host DNA from one cell to another and serve as a vector of genetic information. Viruses in the lysogenic state can persist indefinitely until an environmental or other signal triggers reversion to the lytic pathway. The signal indicating that it is time to leave can be damage to the host DNA, or an indication that the host is doing well (e.g., rapid host cell growth rate). Rapid growth implies a high probability of the virus finding other host cells.

In freshwater filamentous cyanobacteria, the phenomenon of lysogenic infections has been known for more than 25 years. Most of the lysogens have been found in the *LPP* (genera *Lyngbya*, *Plectonema*, and *Phormidium*) group of cyanobacteria (e.g., Cannon *et al.*, 1971; Padan *et al.*, 1972; Khudyakov and Gromov, 1973; Rimon and Oppenheim, 1974). The earlier literature has been extensively reviewed (Sherman and Brown, 1978; Martin and Benson, 1988). Recently, the first example was found of a lysogenic cyanophage that infects a marine filamentous cyanobacterium (Ohki and Fujita, 1996). As is the case with many other lysogens, exposure to mitomycin C or ultraviolet light could induce the lytic cycle.

In contrast to temperate cyanophages of filamentous cyanobacteria, lysogens in unicellular cyanobacteria were unknown until recently. Preliminary evidence suggested lysogeny in *Anacystis nidulans* (Bisen *et al.*, 1985); however, it was almost a decade later that lysogeny in a unicellular cyanobacterium was clearly documented (Sode *et al.*, 1994). This virus infects an isolate of marine *Synechococcus* and can be induced to enter the lytic cycle by exposure to ultraviolet light or Cu^{2+} (Sode *et al.*, 1994, 1997). Although observations of lysogeny in *Synechococcus* are confined to a single strain, there is no reason to assume that this phenomenon is rare. Wilson *et al.* (1996) provided evidence that a virus established a lysogenic association with marine *Synechococcus* strain DC2 (WH7803) when grown under phosphate limitation.

Systematic efforts to find lysogens in natural cyanobacterial communities have been lacking, although their presence could be of great environmental significance. Unicellular cyanobacteria (including prochlorophytes) are major primary producers in the nutrient deplete waters that make up much of the world's oceans (Li *et al.*, 1983, 1992; Joint and Pomroy, 1983; Liu *et al.*, 1997). If a substantial proportion of cyanobacterial communities is lysogenized, then environmental triggers such as nutrient availability or ultraviolet light could induce the lytic cycle. This could lead to synchronous and large-scale mortality, as well as release substantial nutrients into the surrounding water. These nutrients could stimulate heterotrophic bacterial production by increasing the availability of dissolved organic compounds (Middelboe *et al.*, 1996). In addition, lysogens in these communities could be important agents of genetic transfer. Lysogeny occurs in marine microbial communities (Jiang and Paul, 1994; Weinbauer and Suttle, 1996; Wilson and Mann, 1997), and at times lysogens can be relatively abundant (Jiang and Paul, 1996). Yet, our understanding of the ecological significance of lysogeny, particularly with respect to cyanobacteria, remains rudimentary.

The significance of latency in eukaryotic algae is unknown. Evidence for the occurrence of latent infections in eukaryotic algae is circumstantial, with the exception of the Ectocarpales (Müller, 1991; Lee *et al.*, 1998b). A number of studies have noted an increase in phytoplankton cells containing viruslike particles toward the end of blooms (Bratbak *et al.*, 1993; Nagasaki *et al.*, 1994a; Brussaard *et al.*, 1996). Such observations are consistent with latent infections that have become lytic as the result of an environmental trigger (Wilson and Mann, 1997). However, it is not possible to distinguish rapid propagation of lytic viruses at high host densities (Wiggins and Alexander, 1985; Suttle and Chan, 1994) from induction of viruses archived within the host. Latency can be very difficult to demonstrate, even in culture. Unlike lysogeny, where the viral DNA is integrated into the host's genome, latent infections need not be associated with the DNA of the host. An observation that is often taken as evidence for latency or lysogeny is the production of virus in the presence of an abundant host population. The explanation is that some lysogens are being induced to enter the lytic cycle, while the majority of cells carry prophage. Although it seems counterintuitive, such observations are also consistent with a lytic infection. For example, when MpV is added to cultures of *M. pusilla*, lysis occurs. This is generally followed by the regrowth of a resistant subset of the population (Waters and Chan, 1982). If a small aliquot of this culture is transferred to a new medium, continued growth occurs along with viral production (Cottrell and Suttle, unpublished). This might be interpreted as evidence of resistance conferred by a lysogenic association; however, kinetic data indicate undetectable adsorption of the viruses to host cells. Therefore, although the viruses are still produced by lytic infection, the efficiency of infection is extremely low. Many collisions must occur between the viruses and their host before successful infection occurs. Consequently, in order to demonstrate latency or lysogeny, it is necessary to show that the viral nucleic acid is being carried within the host. Although good evidence of latent infections in unicellular algae is lacking, this should not be taken as evidence that latent infections are unlikely. This area is poorly studied, but it is worthy of investigation. If latent infections in eukaryotic phytoplankton can be demonstrated, it raises the potential of termination mechanisms for algal blooms and lateral gene transfer.

B. Burst Size and Lytic Cycle

A number of factors related to the lytic cycle can have important influences on the ecological effects of viral replication, particularly the number of viruses produced per lytic event (burst size) and the length of the lytic cycle. Burst size is particularly important because the number of viruses produced represents the potential for other cells to be infected. In addition, estimates on the impact of viruses on mortality often require accurate estimates of burst size. Despite their importance, there is little data on the effect of physiological status on the length of the lytic cycle or burst size in algae. In marine heterotrophic bacteria, burst size can range from a few viruses to >500 viruses produced per lysed cell (Borsheim, 1993). Burst size and length of the lytic cycle are typically sensitive to the physiological status of the host cells, and can be affected by factors such as growth rate, salinity, and temperature (e.g., Zachary 1976, 1978; Proctor et al., 1993).

Limited data on the effect of light on viral replication in cyanobacteria and eukaryotic algae have been collected. Cyanophage replication requires ATP, and therefore the production of cyanophages is reduced in darkness (Padan et al., 1970; Sherman and Haselkorn, 1971; Adolph and Haselkorn, 1972). There is variation among cyanobacteria in the effect of darkness on viral replication. In some filamentous cyanobacteria such as *Plectonema boryanum*, the length of the lytic cycle is increased and burst size is decreased if infected cells are placed in darkness (Padan et al., 1970; Sherman and Haselkorn, 1971). Nonetheless, viral replication is able to proceed. The effect on viral replication in other cyanobacteria can be more severe. In *Nostoc muscorum* (Adolph and Haselkorn, 1972) and unicellular cyanobacteria (MacKenzie and Haselkorn, 1972a; Allen and Hutchinson, 1976), viral replication is essentially aborted. In unicellular eukaryotic algae, viral replication is also affected by the availability of light. Viral replication in

Micromonas pusilla is severely curtailed or prevented entirely in darkness (Waters and Chan, 1982; Cottrell and Suttle, unpublished data). Bratbak *et al.* (1998a) demonstrated that viral replication in *Phaeocystis pouchetti* was able to proceed in darkness, although burst size was reduced from about 370 to 100 viruses produced per infected cell.

Even less data are available examining the effect of nutrient limitation on viral replication in algae. Wilson *et al.* (1996) has shown that in marine *Synechococcus* viral replication is affected by phosphorus availability. They found that burst size was reduced by about 80%, and cell lysis was delayed by about 18 hr relative to the nutrient-replete controls. The effects were much less pronounced in nitrate-depleted cultures, in which a 25% reduction in final viral titer was observed relative to the nutrient-replete controls. Viral replication in nitrogen- and phosphate-depleted cultures of *P. pouchetti* was also affected relative to cells grown in nutrient-replete medium (Bratbak *et al.*, 1998a). Burst size was reduced from about 130 to 70, and viral lysis appeared to be somewhat slower. Environmental factors clearly affect the viral life cycle in prokaryotic and eukaryotic algae, and much more work needs to be done in this area.

IV. DIVERSITY

Viral diversity can be considered to be the number of morphologically distinct groups of viruses that infect cyanobacteria and eukaryotic algae. More relevant from an ecological perspective is the diversity of viruses that infect a given host or group of hosts. One approach for examining diversity that has direct ecological implications is to examine the host range of viral isolates. Alternatively, phenotypic (e.g., major capsid proteins or morphological criteria) or genetic (e.g., restriction fragment length polymorphism, hybridization or nucleic acid sequence analysis) approaches can be used.

A. Host Range and Host Cell Resistance

1. Cyanophages

Host range is an important aspect of diversity that has ecological repercussions. Viruses that have broad host ranges are less likely to be dependent on a specific host for replication, and their effect on community structure may be less than viruses with very narrow host ranges. In general, viruses have relatively narrow host ranges and only infect members of a single species, and are often restricted to strains within a species.

There are numerous reports of freshwater cyanophages that infect members of different genera. As stressed by Safferman *et al.* (1983), this is in contrast to other bacteriophages, which typically show a narrow host range. In many cases, observations of broad host range appear to have resulted from taxonomic problems within cyanobacteria, rather than providing direct evidence of a wide host range for cyanophages (Johnson and Potts, 1985; Suttle, 1999). Progress on the molecular taxonomy of cyanobacteria (e.g., Wilmotte, 1994; Turner, in press) should allow cyanophage host range to be better defined in the future. One study provides evidence that some cyanophages may infect both filamentous and unicellular cyanobacteria. Moisa *et al.* (1981) isolated nine cyanophages from the Black Sea that caused lysis of both filamentous and unicellular cyanobacteria. Unfortunately, the authors provided few details on the host strains used or phage purification procedures. Two other studies have examined the host range of marine cyanophages that infect *Synechococcus* (Suttle and Chan, 1993; Waterbury and Valois, 1993). Most marine cyanophages appear to infect *Synechococcus* strains within the taxonomic group Marine-cluster A. Cyanobacteria within this cluster contain phycoerythrin, have an obligate salt requirement, and are dominant members of marine cyanobacterial communities (Waterbury and Rippka, 1989). Isolates within this cluster include strains DC2 (= WH7803), SYN48 (= WH6501), and WH8012. Some cyanophages that infect Marine-cluster A also infect *Synechococcus* isolates outside of this taxonomic group. Two myoviruses isolated by Waterbury and Valois (1993) were able to infect isolates from Marine-cluster A as well as an isolate from Marine-cluster B (WH8101). Cells from Cluster B do not possess phycoerythrin. Similarly, a myovirus that infected several phycoerythrin- and phycocyanin-dominant strains of cyanobacteria was isolated from the Gulf of Mexico by Suttle and Chan (1993). The results of these studies indicate that the host range of cyanophages can be extremely complex. Some cyanophage isolates appear to have very restricted host ranges and only infect a single strain of *Synechococcus*, while other isolates appear to have broad host ranges that overlap with those of other cyanophages.

Superimposed over the story of host range is that of host cell resistance. Wide variations occur in the extent to which specific strains of cyanobacteria are susceptible to infection by cyanophages. In coastal waters, titers of viruses infecting some strains of *Synechococcus* reach 10^4 to 10^6 ml^{-1}, whereas titers infecting other strains range from undetectable to a few per milliliters (Suttle and Chan 1993, 1994; Waterbury and Valois, 1993). Isolates that have the least resistance to infection belong to Marine-cluster A, which is comprised of phycoerythrin-rich strains that occur in open ocean waters. In contrast, phycocyanin-dominant isolates that are exclusively found in coastal waters are resistant to infection by most cyanophages in natural waters. This makes evolutionary sense. In near-shore waters the titer of infectious cyanophages can exceed 10^5 ml^{-1}, resulting in very high contact rates between infectious cyanophages and *Synechococcus* spp. (Suttle and Chan, 1994). Estimates from the coastal waters of Texas indicated that 83% of *Synechococcus* were contacted by cyanophages each day. Consequently, there would be strong selection to be resistant to infection. This is consistent with estimates that only about 0.5% of *Synechococcus* were infected by viruses on a daily basis, despite the high contact rates. In contrast, in offshore waters only about 5% of *Synechococcus* were contacted by infectious cyanophages on a daily basis, and mortality factors other than viral infection would dominate. Therefore, there would be less selection for resistance. This idea is supported by calculations showing that most contacts resulted in infection. Waterbury and Valois (1993) studied the abundance of infectious cyanophages in the coastal waters of the northeastern United States and also concluded that most *Synechococcus* from these waters were resistant to infection by the most abundant cyanophages.

2. Viruses Infecting Eukaryotic Algae

Few host-range studies have been done on viruses that infect eukaryotic algae. Mayer and Taylor (1979) reported that MPV infected four isolates of *M. pusilla*, including representatives from the Atlantic and Pacific oceans, but did not infect 35 other algal species tested. Their results are consistent with MPV having less host specificity than cyanophages. However, there is no indication that the MPV isolate was clonal. Furthermore, Sahlsten (1998) has recently shown that the host range of MpV isolates is complex, and that lysis by individual isolates of MpV was restricted to specific host strains. This indicates a continuum of MpV isolates that differ in their ability to replicate among different strains of *M. pusilla*. Conversely, a continuum of host strains exists that differs in ability to support replication of any given MpV strain. These ranges do not overlap perfectly, and replication of all MpV isolates does not occur in all *M. pusilla* strains.

A complex pattern of host ranges is also supported by other studies. An isolate of HaV that infects *Heterosigma akashiwo* was shown not to lyse three other species of Raphidophytes or representatives from 15 other classes of algae (Nagasaki and Yamaguchi, 1997). Moreover, the virus did not lyse three of five isolates of *H. akashiwo*, demonstrating that the virus was strain specific. This work was extended (Nagasaki and Yamaguchi, 1998a) to include 14 isolates of HaV and 13 additional strains of *H. akashiwo*. In a similar pattern to isolates of *Synechococcus*, some strains of *H. akashiwo* were not lysed by any of the viruses, other strains were lysed by every virus, and some strains were only sensitive to a few of the viruses. Studies on viruses infecting *Phaeocystis pouchetii* are less extensive, but the results show a similar pattern. Viruses infected two strains of *P. pouchetii* isolated from northern European waters but did not infect other species within the genus (Jacobsen *et al.*, 1996). In contrast, some viruses infecting Ectocarpales have an intergeneric host range; *Ectocarpus siliculosus* virus is also able to infect *Kuckuckia kylinii* (Müller and Schmid, 1996). Although there is some taxonomic confusion within order Ectocarpales, *Ectocarpus* and *Kuckuckia* appear to be valid genera (Stache-Crain *et al.*, 1997).

B. Genetic Diversity

1. Restriction Fragment Polymorphism

The limited amount of work that has been done suggests that there is an enormous amount of genetic variation in viruses that infect prokaryotic and eukaryotic algae. For example, restriction digests of DNA from five freshwater cyanophages that infect an isolate of *Nostoc* indicated that each isolate was genetically distinct (Muradov *et al.*, 1990). In another study, Wilson *et al.* (1993) used restriction fragment length polymorphism (RFLP) analysis to examine five cyanophages that infect marine *Synechococcus* strain DC2 (synonym WH7803). The viruses fell into three groups. Interestingly, viruses that were isolated from the Sargasso Sea and the English Channel had similar restriction patterns. A similar study (cited in Suttle, 1999) examined three cyanophages that were isolated from a similar environment using *Synechococcus* strain DC2 as the host. These cyano-phages belonged to the same family, differed in host range, and were genetically distinct based on RFLP analysis. The data from these studies demonstrate that cyanophages from distant locations can be genetically similar, while viruses from the same environment can be quite different. One of the next steps in this area of research should be to quantify this diversity by sequence analysis.

The genetic diversity that is observed among cyanophages is also found in viruses that infect *Micromonas pusilla* and exsymbiotic *Chlorella*-like algae. Viruses infecting *M. pusilla* reach abundances of 10^4 to 10^5 ml^{-1} in seawater (Waters and Chan, 1982; Cottrell and Suttle, 1995a), leading to the obvious question of whether viruses isolated from the same water sample were genetically more similar than those isolated from widely separated areas. In order to answer this question, viruses that infected a single strain of *M. pusilla* were isolated from the coastal waters of Texas, California, New York, British Columbia, and the central Gulf of Mexico and subjected to RFLP analysis (Cottrell and Suttle, 1991). Every virus isolate, including those from the same water sample, had a different restriction-banding pattern, indicating high genetic diversity. *Chlorella* viruses are also geographically widespread, reach high titers in natural waters, and are genetically diverse (Van Etten *et al.*, 1985a; Zhang *et al.*, 1988). They have been broken down into eight groups based on resistance to digestion by specific nucleases (Van Etten *et al.*, 1991).

2. Hybridization, Sequence Analysis, and Denaturing Gradient Gel Electrophoresis

RFLP analysis demonstrates that the viruses that infect cyanobacteria and eukaryotic algae are genetically diverse, but it does not allow the degree of diversity or genetic distance among different viruses to be easily quantified. Two approaches that can be used to quantify genetic similarity are hybridization of total genomic DNA and sequence analysis of specific genes. Hybridization of total genomic DNA has the advantage that assumptions of gene transfer are of less concern, but it is restricted to comparisons among relatively closely related groups of viruses. It is also an enormous amount of work.

Hybridization of total genomic DNA was used to quantify genetic differences among eight isolates of MpV obtained from the five locations mentioned above (Cottrell and Suttle, 1995b). The hypothesis was that viruses from the same water sample would be more closely related than viruses from geographically distant locations. Surprisingly, the study showed that the genetic variability among viruses in the same water sample was as great as the variability among viruses from different oceans. Therefore, factors other than location appear to control viral genotype.

Hybridization does not allow comparisons to be made among more distantly related viruses. For example, hybridization of MpV DNA with that of *Chlorella* virus PbCV-1 was not significantly different from the degree of hybridization of DNA from MpV and phage lambda (Cottrell and Suttle, 1995b). Comparisons among more distantly related viruses were accomplished using the polymerase chain reaction (PCR) to amplify a region of the viral DNA-polymerase (*pol*) genes and analyzing the sequences (Chen and Suttle, 1996; see §II.B). The analysis confirmed those from the hybridization study (Chen and Suttle, 1996) and revealed the same pattern of genetic diversity within MpV isolates. As well, *pol* sequences from viruses that infect three genera of eukaryotic microalgae formed distinct clades. This demonstrated that, although genetic diversity is high within individual groups of viruses, it is not as large as the differences among viruses that infect different species.

The identification of algal-virus-specific *pol* gene sequences also allows rapid determination of genetic diversity in natural samples without the need to isolate the viruses. PCR has been used to amplify alga-viruslike DNA *pol* sequences from natural virus communities concentrated from the Gulf of Mexico (Chen and Suttle, 1995). RFLP analysis of cloned fragments from a Gulf of Mexico sample yielded five different restriction-banding patterns. Sequence analysis of these fragments showed that four clustered within the MpV group, while one was from a virus that was within family Phycodnaviridae but outside the previously established groups (Chen *et al.*, 1996). These results demonstrate that the PCR-based method is useful for examining genetic diversity in natural marine viral communities.

Information on diversity, when combined with data on specific virus isolates, can lead to an understanding of the effects of these pathogens (Short and Suttle, 1999). As more *pol* sequences are collected, specific natural viral populations can be identified and related to specific host assemblages. Such data can also lead to inferences on the role of viruses in regulating algal community structure and primary production. For example, knowledge of the community structure of algal viruses can be coupled with data on their abundance and turnover rates. This information can be used to infer the effect of viruses on the mortality of specific subsets of the phytoplankton community.

An even more powerful approach for examining diversity in natural viral communities is to couple PCR amplification of *pol* sequences with denaturing gradient gel electrophoresis (DGGE) (Short and Suttle, 1999). DGGE has been used to analyze the diversity of complex microbial communities (e.g., Muyzer *et al.*, 1993; Ferris *et al.*, 1996), and can be used to resolve DNA fragments of the same size that differ in as little as a single base pair. Consequently, this method can be used to fingerprint natural viral communities. As well, DGGE eliminates the need to clone PCR products, and offers a rapid means of detecting predominant populations.

The utility of DGGE was examined in a comparison of algal virus isolates and natural virus communities from the Gulf of Mexico and from the coastal waters of British Columbia by Short and Suttle (1999). Algal virus PCR products from viral isolates that could not be resolved by standard agarose gel electrophoresis were clearly separated by DGGE. Similarly, PCR products from natural samples that could not be resolved by agarose gel electrophoresis revealed several bands using DGGE. Bands from the Gulf of Mexico sample were not present in samples from British Columbia, implying that virus communities were genetically distinct. Moreover, samples collected from the same location but at different times showed similar banding patterns, and had the same dominant band. This result is interesting because previous results have suggested that the genetic diversity of algal viruses within the same sample is as great as between oceans (Cottrell and Suttle, 1991). The results from DGGE suggest that this may not be true if the entire community is considered.

Molecular biological techniques provide powerful approaches for quantifying diversity in virus communities. Ultimately, these tools should provide the resources required to understand the factors that regulate the relative distribution and abundance of different viral genotypes.

V. DISTRIBUTION, ABUNDANCE, AND POPULATION
DYNAMICS OF NATURAL ALGAL VIRUS COMMUNITIES

A. Cyanophages in Freshwaters

A number of studies have examined the distribution of cyanophages in freshwaters (reviewed in Suttle, 1999). These investigations have demonstrated that viruses infecting filamentous cyanobacteria in the *LPP* group occurred throughout the world (e.g., Safferman and Morris, 1967; Singh and Singh, 1967; Daft *et al.*, 1970), although titers were generally less than a few hundred per milliliter. Although high titers of cyanophages were not observed, there is good evidence that the abundance of the viruses varied seasonally. For example, in ponds used for waste stabilization and fish rearing, cyanophage titers varied more than tenfold (Safferman and Morris, 1967; Padan and Shilo, 1969; Safferman, 1973). Seasonal changes in cyanophage titer were likely related to environmental changes that affect host cell abundance, although this was not been conclusively demonstrated.

Viruses infecting the *Anabaena–Nostoc* group (e.g., Adolph and Haselkorn, 1971; Mendzhul *et al.*, 1973; Hu *et al.*, 1981; Moisa *et al.*, 1981) and the *Anacystis–Synechococcus* group (Safferman *et al.*, 1972; Gromov, 1983; Kim and Choi, 1994) are also found in lakes, reservoirs, and sewage settling ponds. However, they are less frequently isolated, and appear to be less abundant than *LPP* cyanophages. A group of cyanophages that is widely distributed in the former Soviet Union and that is reported to infect *Nostoc* and *Plectonema* is an exception (Muradov *et al.*, 1990), and can reach titers as high as 10^4 ml^{-1}.

The viruses that have been isolated that infect freshwater isolates of *Synechococcus* infect phycocyanin-rich strains. Yet, in marine systems, phycoerythrin-rich strains are most susceptible to viral infection (Suttle and Chan, 1993; Waterbury and Valois, 1993). Although phycoerythrin-rich *Synechococcus* can be dominant in many freshwater systems (e.g., Caron *et al.*, 1985; Stockner and Shortreed, 1991), few attempts have been made to isolate viruses infecting these cyanobacteria. Nonetheless, viruses infecting phycoerythrin-rich cyanobacteria exist in lake communities (Suttle, 1999), and more effort should be spent to document their distribution and abundance.

B. Cyanophages in Marine Surface Waters

Cyanophages in the marine environment were first reported by Moisa *et al.* (1981) following isolation from the Black Sea of viruses that infected filamentous and unicellular cyanobacteria. This report went largely unnoticed. Not until almost a decade later was it shown that viruses infected a significant proportion of the cyanobacteria in seawater (Proctor and Fuhrman, 1990). This was followed by observations that infectious cyanophages could be readily isolated form seawater (Suttle et al., 1990) and that they were frequently present in high titers (Suttle and Chan, 1993, 1994; Waterbury and Valois, 1993). In the Gulf of Mexico, viruses infecting *Synechococcus* strains DC2 and SYN48 routinely occur at concentrations of 10^5 ml^{-1} (Suttle and Chan, 1993, 1994), and have been reported to be as high as 10^6 ml^{-1} (Suttle et al., 1996). Infectious cyanophages can represent up to about 10% of the entire viral community. In contrast to viruses that infect *Synechococcus* strains DC2 and SYN48, the abundance of cyanophages infecting non-phycoerythrin-containing *Synechococcus* are frequently undetectable. When detectable, however, the pattern of seasonal change is similar to that of viruses that infect strains DC2 and SYN48, but the maximum abundance is much less (Suttle and Chan, 1994).

The abundance of cyanophages follows distinct seasonal patterns and is highest in summer and fall when water temperatures are warmest and *Synechococcus* is most abundant (Waterbury and Valois, 1993; Suttle and Chan, 1994). These changes are probably tied to the abundance of *Synechococcus*, and not directly to temperature. A regression analysis on seasonal data showed that temperature only explained 53 to 70% of the variance in the abundance of cyanophages (Suttle and Chan, 1993). Moreover, the abundances of cyanophages and *Synechococcus* along a seaward transect from the coast of Texas decreased more than tenfold, whereas the temperature increased by about 1°C (Suttle and Chan, 1994). This change was associated with a threshold in the abundance of *Synechococcus* of ~10^3 cells ml^{-1} beyond which infectious cyanophages increased from ~10^2 ml^{-1} to 10^5 ml^{-1} (Suttle and Chan 1994). These data are consistent with the idea that, when host cell abundance exceeds a critical level, host–virus contacts increase greatly and infectious viruses propagate rapidly (Wiggins and Alexander, 1985). These results are supported by those of a seasonal study along the same transect in which Principle Component Analysis demonstrated that cyanophage abundance was most closely related to that of *Synechococcus* followed by temperature (Rodda, 1996).

Viruses also exist in marine environments that infect marine filamentous cyanobacteria (Moisa et al., 1981), although there are no reports on their abundance or distribution.

C. Cyanophages in Offshore Sediments

Recent evidence indicates that infectious cyanophages can occur at considerable depth in marine sediments. As indicated above, cyanophages are abundant in the western Gulf of Mexico. In May, August, and December 1995, at a station 45 km offshore, cyanophages infecting *Synechococcus* strain DC2 ranged from 1.4×10^5 ml^{-1} at the sea surface to 1.0×10^4 ml^{-1} at 46.6 m, just above the sediment surface (Suttle, 1999). In the sediment, the titers decreased exponentially with depth from 9.4×10^4 ml^{-1} at the water–sediment interface to 3.0×10^2 ml^{-1} at 30 cm below the sediment surface (Suttle, 1999). Possible sources of the cyanophages are production in the sediment, diffusion or transport in the pore water, or sedimentation on particles and subsequent burial. The concentration of cyanophages in the water immediately overlying the sediment surface was about an order of magnitude less than at the water–sediment interface. This suggests that the viruses probably attached to particles, either directly or via infected cells, and sank to the bottom, where they were buried. Absolute dating indicates that sedimentation rates in this region of the Gulf of Mexico are ~0.33 cm per year (Snedden, 1985; Berryhill, 1986); therefore, viruses at 30 cm were approximately 100 years old. These estimates are supported by little evidence for physical disturbance or reworking of the sediments (Snedden, 1985), and calculated rates of diffusion that would require >300 years for a virus to diffuse 30 cm. The persistence of cyanophages is likely similar to that of other viruses; therefore, sediments are probably an important reservoir of infectious viruses that infect algae as well as other organisms. These results imply that resuspension of sediments by dredging or natural processes will reintroduce infectious viruses into the overlying water column.

D. Viruses Infecting Eukaryotic Algae

The most comprehensive studies on the distribution, abundance or seasonal dynamics of viruses that infect eukaryotic algae have examined pathogens of *Micromonas pusilla*. Waters and Chan (1982) briefly reported that the concentration of lytic agents for *M. pusilla* ranged from 10 to 10^4 ml^{-1} in 10 samples from the coastal waters of British Columbia. Subsequently, *Micromonas pusilla* viruses (MpV) were shown to be geographically widespread and at times abundant in Gulf of Mexico waters (Cottrell and Suttle, 1991). The highest titer of MpV (4600 ml^{-1}) was recorded in near-shore waters during spring, while the lowest were recorded offshore along a transect between Texas and Florida (undetectable [<0.02 ml^{-1}] to 0.1 ml^{-1}). A more comprehensive look at the seasonal dynamics of MpV was obtained during a study in inshore waters of the Gulf of Mexico (Cottrell and Suttle, 1995a). The abundance of viruses was measured on a weekly basis for several months, and ranged from ~10^5 ml^{-1} in January to ~10^3 ml^{-1} at the end of April. The abundance of viruses was tied to water temperature and decreased exponentially in February and March, as the water temperature rose from 15 to 22°C. As was the case for cyanophages, viral abundance was probably tied to that of the host, rather than directly to temperature. Although *M. pusilla* can be abundant and widely distributed in the plankton, many strains grow poorly at warm temperatures. Consequently, host abundance would be expected to decrease as temperature increased. Rodda (1996) followed the abundance of *M. pusilla* and MpV in the coastal waters of Texas over an annual cycle. She found that *M. pusilla* was only detectable during November through March, when water temperatures were <25°C. Infectious MpV followed a similar pattern and was detectable shortly after the appearance of *M. pusilla*, and disappeared when the host reached detection limits. The occurrences of MpV and *M. pusilla* were strongly correlated with each other and negatively correlated with temperature. Most recently, the seasonal (Sahlsten, 1998) and vertical (Sahlsten and Karlson, 1998) distributions of MpV have been examined in the coastal waters of the Skagerrak–Kattegat, between Sweden and Norway. It was found that the seasonal abundance of viruses infecting an MpV isolate from the Oslofjord was at least tenfold higher than viruses infecting an isolate from the Gulf of Maine. MpV was found in all the samples examined, with the highest abundances of viruses (about 2500 ml^{-1}) occurring in the spring, although seasonal trends were not clearly discernible.

The only other published studies on the seasonal distribution of viruses infecting marine eukaryotic phytoplankton reported on changes in the abundances of viruses (CbV) that infect *Chrysochromulina brevifilum* (Suttle and Chan, 1995). The highest titers of these viruses also occurred in the spring, but only reached maximum abundances of 10 ml^{-1}. In many samples the virus was below the detection limit of 0.002 ml^{-1}.

In freshwater systems, viruses that infect symbiotic *Chlorella*-like algae that were isolated from *Paramecium* and *Hydra* are widely distributed and can be isolated from North America (Van Etten *et al.*, 1985a), China (Zhang *et al.*, 1988), and Europe (Reisser, 1993). Few studies have quantitatively reported the distribution and abundance of *Chlorella* viruses. In a survey of a variety of lakes and ponds in North America (Van Etten *et al.*, 1985a), infectious viruses were detected in 37% of 35 samples, with the highest titer measured in a sample collected from a drainage ditch (3200 ml^{-1}). The highest titers reported for *Chlorella* viruses in North America were 10^4 ml^{-1} in river water from North Carolina (Van Etten *et al.*, 1985b).

VI. THE FATE OF VIRUSES IN THE ENVIRONMENT

A. Effects of Ultraviolet Radiation

Studies on bacteriophages have demonstrated that a number of factors destroy viral particles and infectivity in aquatic environments. These factors include solar radiation, attachment to particles and host cells, consumption by protozoan grazers, and digestion by bacterial enzymes. In near-surface waters, destruction by solar radiation is the most significant of these processes (Suttle and Chen, 1992; Suttle *et al.*, 1993; Wommack *et al.*, 1996; Noble and Fuhrman, 1997). Sunlight damages viral infectivity by causing pyrimidine dimers to be formed in their DNA. Cyanophages and eukaryotic algal viruses are also very sensitive to solar radiation (Suttle and Chan, 1994; Cottrell and Suttle, 1995a), implying high rates of viral turnover when the levels of UVB (290 to 320 nm) are high. Decay rates of infectivity of natural cyanophage communities in full sunlight range from <0.1 to ~0.4 hr^{-1} at the surface and decrease strongly with depth (Garza and Suttle, 1998), along with the levels of UVB. At two stations on the continental shelf of the Gulf of Mexico, sunlight was responsible for removal rates of cyanophage infectivity of 0.53 and 0.75 per day, when integrated over the surface mixed layer (28.2 and 10.4 m, respectively) and averaged over 24 hr. High rates of decay in sunlight for viruses infecting *Micromonas pusilla* have also been reported (Cottrell and Suttle, 1995a). One of the most important parameters dictating the relative importance of UV to the survival of viruses is the depth of the mixed layer (the layer of the ocean where the water is homogeneously mixed due to wind and convection). As the mixed depth increases, the average exposure of the viral community to UVB will decrease (Murray and Jackson, 1993). Deeper mixing also shifts spectral quality to the longer wavelengths required for DNA repair by photoreactivation (Wilhelm *et al.*, 1998b; see §VII.B.3), which can reverse the UVB-induced damage. The significance of sunlight relative to other causes of viral decay depends on a number of factors, including the amount of damaging radiation at the surface, the optical clarity of the water, and the depth of the mixed layer.

1. Selection for UV-Resistant Viral Communities

A recent study has shown that cyanophage infectivity is more resistant to damaging solar radiation in the summer, than in winter and spring (Garza and Suttle, 1998). Decay rates of the infectivity of cyanophage isolates and natural communities were compared when exposed to full sunlight. Decay rates of isolates and natural communities were generally similar in spring and winter, but during summer the decay rates of isolates were much higher than the natural communities

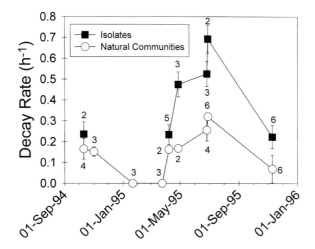

Fig 2.   Seasonal changes in decay rates of infectivity of natural cyanophage communities. Cyanophage isolates were maintained at *in situ* water temperatures and exposed to full sunlight at 27°50′ N
latitude. The data demonstrate that the resistance of natural viral communities to damaging radiation
increases in summer when compared to that of control isolates. Error bars represent standard deviations
of the number of replicates indicated.

(Fig. 2). As the titer of viruses was always determined using the same host under
identical conditions, changes in relative sensitivity to damaging radiation between
isolates and natural communities was the result of changes in the natural viral
community. Infectivity of the natural cyanophage community was most resistant
to UV damage when solar radiation was greatest.

The presence of high concentrations of infectious algal viruses in surface waters
is in part attributable to viral resistance to damaging radiation during periods of
high UVB flux. In addition, repair of damaged DNA by host cell- and viral-encoded mechanisms is undoubtedly important (Friedberg *et al.*, 1995).

2. Diel Patterns of Viral Replication and Lysis

Algal viruses may also persist in surface waters by having strong diel cycles of
infection and lysis. Viruses that infect algae have a disadvantage over those that
infect heterotrophic bacteria, because algal viruses infect organisms that are
dependent on sunlight for survival, and sunlight destroys viral infectivity. One way
that some cyanophages and other algal viruses may minimize the damaging effects
of solar radiation is to have a diel cycle of infection and replication. Studies of
cyanophage production and decay in the western Gulf of Mexico indicated that
viral production during the day did not balance the destruction of viral infectivity

by solar radiation (Garza and Suttle, 1998). It was hypothesized that most of the viral production occurred at night. This was supported by results from a parallel diel study (Suttle *et al.*, unpublished) in which a sharp decline in the abundance of *Synechococcus* was seen at the onset of darkness.

A proposed pattern for viral replication of cyanophages infecting *Synechococcus* and possibly for viruses infecting other phytoplankton is depicted in Figure 3. The cycle begins at the onset of darkness with cell lysis and the release of viral progeny. Viral production at dusk gives the viruses until daylight to contact a suitable host, without exposure to damaging radiation from sunlight. Viral replication begins following injection of the viral DNA into a host cell, and when there is sufficient light for photosynthetic energy production (e.g., Sherman and Haselkorn, 1971; Adolph and Haselkorn, 1972). The repair mechanisms of the host protect the viral DNA from damage by solar radiation. In *Synechococcus*, photosynthesis is unaffected by viral infection until the point of cell lysis (MacKenzie and Haselkorn, 1972a; Suttle and Chan, 1993). In contrast, photosynthesis is suppressed shortly after infection in filamentous cyanobacteria (Padan *et al.*,

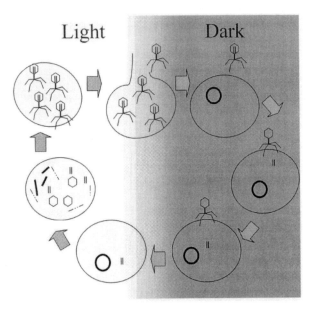

Fig. 3.   A hypothesis for the strategy of replication in nature of viruses that infect *Synechococcus* spp. Viral progeny are released near dusk and have the night to encounter a suitable host. This protects the free viral particles from damage from solar radiation. The viral DNA is injected into the host cell, but replication of the virus is arrested until there is sufficient light for photosynthesis. Viral replication proceeds during daylight, during which time damage to the viral DNA from solar radiation is repaired by the host cell- or viral-encoded repair mechanisms. Viral replication is complete at the end of the light period, and the new progeny are released.

1970) and in other unicellular algae (Waters and Chan, 1982; Van Etten *et al.*, 1983). In the absence of photosynthesis, unicellular cyanobacteria likely do not contain adequate energy reserves to allow completion of the lytic cycle. The relatively long lytic cycle of cyanophages in unicellular cyanobacteria (MacKenzie and Haselkorn, 1972a; Safferman *et al.*, 1972; Suttle and Chan, 1993) suggests that viral replication will continue throughout the day and be complete near dusk. At this point, the viral progeny are released and the cycle of infection begins again.

B. Light-Independent Decay of Viral Infectivity

In contrast to solar radiation, light-independent factors that destroy infectivity vary less with depth in the surface mixed layer. The virucidal properties of seawater can be eliminated or reduced by heating or filtering (e.g., Lycke *et al.*, 1965; Matossian and Garabedian, 1967; Mitchell and Jannasch, 1969). Heat-labile particulate material is responsible for much of the light-independent decay. Bacteria can be associated with some of this decay (e.g., Moebus, 1992); however, much of the virucidal activity is not directly attributable to the predominant bacteria, and appears to be associated with heat-labile colloidal material in the >1.0 μm size fraction (Suttle and Chen, 1992). Colloidal heat-labile virucidal material has also been observed in the <0.2 μm size fraction (Noble and Fuhrman, 1997). Flagellate grazing and attachment of viral particles to host cells also removes viruses from seawater, although these processes generally account for <1% of measured removal rates in natural waters (Suttle and Chen, 1992; González and Suttle, 1993). Although grazing by flagellates has little effect on the removal rate of viruses, viruses can be a significant source of nutrients to flagellate grazers. In fact, some flagellates in natural communities were only observed to consume viral size particles (Suttle and Chen, 1992; González and Suttle, 1993).

Decay rates of algal virus communities are also temperature dependent. In the dark, decay rates for natural communities of viruses infecting *Micromonas pusilla* varied from ~0.058 to 0.085 per day between 4 and 25°C (Cottrell and Suttle, 1995a). The decay rates of a natural cyanophage community were similar and ranged from 0.034 per day at 4°C to 0.126 per day at 25°C (Garza and Suttle, 1998). These corresponded to turnover times of 29 and 8 days, respectively. Decay rates were similar at 4 and 14°C, but increased sharply at higher temperature. These rates are significant, but they are still much less than those measured in sunlight. Consequently, even in warm waters solar radiation is responsible for much of the decay of viral infectivity.

Other algal viruses may be much less stable than cyanophages and MpV (Nagasaki and Yamaguchi, 1998b). Viruses that infect *Heterosigma akashiwo* were found to decay quite rapidly, even when stored at 5°C. At temperatures between 15 and 25°C, all detectable infectivity was lost within 18 days.

VII. THE EFFECT OF VIRUSES ON THE MORTALITY OF ALGAL POPULATIONS

One of the most important aspects of algal virus ecology that remains to be adequately explored is the effect of viruses on the mortality of algal populations and communities. Viral infection has important implications at a number of levels. It may reduce primary productivity and decrease the amount of carbon and other nutrients that are available for transfer to higher trophic levels. Conversely, viral lysis may liberate nutrients that are tied up in bacterial or phytoplankton biomass, and stimulate primary productivity. As taxon-specific agents of mortality, viruses can also play a primary role in controlling community structure, and preventing or terminating algal blooms. This strong selective pressure has probably been a major factor in the evolution of phytoplankton and may even have been responsible for extinction events.

A. Primary Productivity

Some of the first evidence that viruses are important mortality agents of phytoplankton was the observation that, when concentrated viral size material from seawater was added back to natural plankton communities, rates of photosynthesis were decreased by up to 78% (Suttle et al., 1990; Suttle, 1992). The decreases occurred quickly and were proportional to declines in chlorophyll fluorescence. Microautoradiography demonstrated that the observed decreases in photosynthetic rates were associated primarily with cells in the 3 to 10 μm and >3 μm size fractions. In contrast, cells in the <3 μm size fraction (primarily *Synechococcus*) were less affected. This was probably because the lytic cycle in *Synechococcus* is typically longer than 8 hours, and photosynthetic rates in *Synechococcus* are largely unaffected until the onset of cell lysis (MacKenzie and Haselkorn, 1972a; Safferman et al., 1972; Suttle and Chan, 1993). In these experiments, because the incubations were 8 hours or less, the rates of photosynthesis in virus-infected *Synechococcus* would not have decreased.

Decreased photosynthetic rates in response to elevated concentrations of the natural viral community do not conclusively demonstrate that viruses were directly responsible for the effect; however, a number of observations are consistent with this interpretation. First, only a subset of the phytoplankton were affected. Above a certain point, further increases in virus concentration did not result in further decreases in photosynthesis. As well, microautoradiography indicated that many of the phytoplankton continued to photosynthesize at elevated virus concentrations, suggesting that the effect was not a toxic one. Second, the bioactive component was heat labile and greater than about 30,000 daltons (ca. 2 nm). With the exception of very large proteins or protein complexes, there are few candidates other than viruses in the 2-to-200-nm size range that could inhibit primary production and be both heat labile and taxon specific. Third, in hundreds of experiments to screen natural communities for viruses that infect phytoplankton (e.g., Suttle *et al.*, 1991), decreases in chlorophyll fluorescence invariably have been associated with the presence of a pathogen.

Further evidence that viral lysis of phytoplankton is quantitatively important was provided in a paper by Agustí *et al.* (1998). They determined the dissolved esterase activity in seawater samples collected in the northwestern Mediterranean to estimate the release of esterases from phytoplankton as the result of cell lysis. The release rates were substantial, yielding estimated phytoplankton lysis rates up to 1.9 per day, corresponding to as much as 50% of the estimated growth rate of the phytoplankton.

Quantifying the effect of viruses on primary productivity in aquatic systems is a difficult task. Viruses may reduce primary productivity by removing phytoplankton or they may stimulate productivity by liberating nutrients that can be used to fuel growth. One approach that has been used to deduce the effect of viruses on primary production is to extrapolate from the relationship between photosynthetic rate and viral abundance, to an inferred photosynthetic rate in the absence of viruses. The *y*-intercept of a linear regression through the initial slope of the relationship between photosynthetic rate and virus concentration yields the photosynthetic rate in the absence of the viruses. In four experiments, it was estimated that primary productivity in eukaryotic phytoplankton would have been an average of 2.1% (range, 1.2 to 3.7%) higher in the absence of the viral size fraction (Suttle, 1994).

There is also evidence that viruses can increase rates of primary productivity in natural communities of phytoplankton. In a study in the western Gulf of Mexico (Suttle *et al.*, 1996), a monospecific bloom of *Synechococcus* (10^5 cells ml^{-1}) was found to be associated with high concentrations of cyanophages (up to 10^6 ml^{-1}). Observations by transmission electron microscopy indicated that ~1.1% (SD = 0.32) of the cyanobacteria-contained viral particles, similar to values reported in the western Atlantic and Caribbean (Proctor and Fuhrman, 1990), and implying viral-induced mortality. In an effort to quantify the effect of viruses on cyanobacterial mortality, triplicate natural samples were diluted with either virus-free seawater (30,000 daltons ultrafiltered), or seawater containing viruses (0.2-μm pore-size membrane filtered). The final range of dilutions was 10, 30, 60, 80, and 100% natural seawater. The abundances of *Synechococcus* and cyanophages, as well as the frequency of dividing cyanobacteria (FDC) were followed over 24 hr in simulated *in situ* shipboard incubations. The abundances of cyanobacteria remained relatively constant over the first 10 hr (dark period), and approximately doubled the following morning, irrespective of treatment. At the onset of the next dark period, there was a strong treatment effect. Cyanobacterial growth rates as indicated by FDC were much higher in samples diluted with seawater containing viruses than in samples to which viral-free seawater was added. This was followed by a decrease in cyanobacterial numbers at the onset of darkness in the samples in which virus concentrations were greatly reduced. These results indicate that the growth rates of *Synechococcus* were enhanced by the presence of viruses, presumably because of nutrient recycling.

B. Mortality of Specific Populations

As viruses are obligate pathogens that typically cause cell lysis when they reproduce in microorganisms, their presence indicates that they are responsible for mortality in phytoplankton. However, the tools available to access the impact of viruses on natural phytoplankton populations are limited. An elegant approach is to infer mortality from the frequency of infected cells, which is the proportion of cells that contain viruslike particles. Alternatively, potential rates of infection can be inferred from contact rates, which are the calculated collision frequencies between viruses and host cells. A final approach is to measure the rates at which specific subsets of viral populations are removed, and assume that these viral decay rates are balanced by viral production stemming from host cell mortality. If viral production can be measured directly, then it is not necessary to assume that rates of destruction and production are balanced.

1. Frequency of Infected Cells

This approach is based on the premise that, if the proportion of cells that contain visible viral particles is known, then the proportion of the population that is infected by viruses, and the number of cells that will die as the result of viral lysis, can be estimated. The greatest strength of the method is that incubations or culturing are unnecessary; however, there are a number of assumptions required that pose difficulties. A critical assumption is that the portion of the lytic cycle during which viral particles are visible is known. This is a required assumption because the proportion of cells containing viral particles is multiplied by the proportion of the lytic cycle during which the particles are visible in order to calculate the proportion of the population that is infected. This approach was first used to estimate viral-mediated mortality in heterotrophic bacteria and cyanobacteria in samples from the western Atlantic and Caribbean (Proctor and Fuhrman, 1990). On average, 0.8 to 2.8% of cyanobacteria contained phage particles. This translated to 8 to 28% of *Synechococcus* cells being infected assuming that viral particles are visible within the host cells for the last 10% of the latent period. Waterbury and Valois (1993) argued that these were probably overestimates, as viruses are only visible within infected cyanobacteria for about half the latent period (Padan and Shilo, 1973; Sherman *et al.*, 1976). Consequently, only about 2 to 6% of *Synechococcus* cells would have been infected using the more conservative assumption.

A major obstacle to using the proportion of visibly infected cells to estimate viral-mediated mortality in cyanobacteria is that few data are available on the proportion of the lytic cycle during which viruses are visible within infected cells. In *Plectonema boryanum* (UTEX 594), viruses are visible approximately halfway through the lytic cycle (Padan *et al.*, 1970), while in freshwater *Synechococcus* (PCC 6911) viruses appear about 75% through the cycle (MacKenzie and Haselkorn, 1972b). If data for *Synechococcus* strain PCC 6911 can be applied to natural marine communities, then 0.8 to 2.8% of *Synechococcus* cells containing VLPs would translate into 3 to 12% of the population being infected. It is also difficult to translate from the proportion of cells that are infected to a daily mortality rate. However, in a steady-state system in which 50% of the cells live to reproduce and 50% die, then something that removes 50% of the cells will account for 100% of the mortality (Proctor and Fuhrman, 1990). Therefore, if 3 to 12% of *Synechococcus* cells are infected, and the lytic cycle is less than the doubling time of the cells, then 6 to 24% of the total mortality may be the result of viral infection. This assumes that all infected cells are killed by viruses.

Despite the difficulties with estimating viral-induced mortality from the percent of infected cells, the approach has considerable promise. Reliable estimates of the impact of viruses on mortality in cyanobacteria and eukaryotic algae will depend on acquiring more data on the proportion of the latent period during which viral particles are visible. The data for cyanobacteria are very limited and those for eukaryotic algae nonexistent. Such studies will need to take into account the effect of host physiology on the viral replication cycle. There will be additional challenges in using this approach with eukaryotic algae, because it will be much more difficult to identify infected cells and quantify their relative abundance in natural populations. Nonetheless, it will likely be possible to make such estimates in specific situations such as during phytoplankton blooms.

2. Contact Rates

Transport theory can be used to calculate contact rates between viruses and host cells (Murray and Jackson, 1992) and is particularly useful because it provides an estimate of the maximum potential rate for infection (Suttle, 1994; Suttle and Chan, 1994). Contact rates are calculated as $[(S_h \cdot 2\pi \cdot d \cdot D_v) V \cdot P]$, where S_h is the Sherwood number (dimensionless), which for *Synechococcus* is about 1.01 and for *Micromonas pusilla* about 1.53 (Murray and Jackson, 1992); d is cell diameter (cm); D_v is the diffusivity of viruses (cm^2 day^{-1}); V is the concentration of infectious viruses (cm^{-3}); and P is the concentration of host cells (cm^{-3}). The potential infection rate is directly proportional to the product of the viral and host cell abundances. As the infection rate cannot exceed the rate at which viruses encounter or adsorb to host cells, these rates provide maximum estimates of the viral infection rate or the potential for viral infection.

Contact rates have only occasionally been used to estimate the potential effect of viruses on the mortality of phytoplankton. Waterbury and Valois (1993) argued that cyanophages could only be responsible for the daily mortality of a few percent of *Synechococcus* in Woods Hole Harbor and waters farther offshore. The highest abundances of cyanophages and *Synechococcus* were 1.14×10^4 and 3.8×10^5 ml^{-1}, respectively, which yielded contact rates of only 3.8% of the cells per day (Suttle, 1994). In the western Gulf of Mexico cyanophage abundances, and hence encounter rates, can be much higher, resulting in most *Synechococcus* being contacted by infectious cyanophage on a daily basis (Suttle and Chan, 1994). These calculations assume that the estimated abundances of infectious cyano-phages are accurate, and that all cyanophages have the potential to infect all *Synechococcus*. Although the estimated abundances of cyanophages are probably reasonably accurate, it is likely that the proportion of the *Synechococcus* popula-tion that can be infected by the most abundant cyanophages varies.

Calculated contact rates also provide insights into host cell resistance in natural populations, if independent estimates are available for viral-mediated mortality. Studies on cyanophages indicate that fewer collisions result in infection when contact rates between infectious viruses and potential host cells are high. For example, in the western Gulf of Mexico there is evidence of a threshold in *Synechococcus* abundance of about 10^3 ml^{-1} beyond which the concentration of infectious cyanophages rapidly increases (Suttle and Chan, 1994). Near the shore, where the abundances of cyanophages and *Synechococcus* were high, approxi-mately 83% of *Synechococcus* cells were contacted by infectious cyanophages daily. Yet, only 0.2 to 1.0% of cells needed to be contacted to balance the calculated removal rates resulting from solar radiation. Consequently, few of the contacts resulted in infection, and either the efficiency of infection was very low or most cells were resistant. It should be pointed out that, although the efficiency of infection may be low, it does not necessarily follow that most cells are resistant. Most host cells may still be susceptible to lytic infection, but the probability of successful infection is low for a single collision. In contrast, farther offshore only a few percent of *Synechococcus* cells were contacted by infectious cyanophages each day. Almost all of these contacts had to result in infection to produce enough viruses to balance the estimated decay rates of cyanophage infectivity. Therefore, most of the offshore cyanobacterial community appeared to have little resistance to infection. A similar conclusion was drawn for natural populations of MpV and *M. pusilla*, which also persist in coastal waters of Texas (Cottrell and Suttle, 1995a).

3. Rates of Viral Decay and Production

The effect of viruses on the mortality of phytoplankton populations can also be inferred from either the rates of decay of viral particles (Heldal and Bratbak, 1991) or infectivity (Suttle and Chen, 1992). This approach assumes that removal rates of virus particles or infectivity are balanced by production rates. In addition, the number of viruses produced per cell that is lysed (burst size) must be known. The mortality rate attributable to viral lysis is simply the decay rate of the viruses divided by the burst size. This approach was originally used to estimate phage-induced mortality rates in marine bacteria, but it can easily be adapted for phytoplankton.

There is a caveat to the approach of using decay rates of viral infectivity to infer the effect of viruses on host cell mortality. Exposure to sunlight can result in infectivity loss rates of the viral particles that greatly exceed their physical destruction (Suttle *et al.*, 1993; Wommack *et al.*, 1996). This is because the loss of infectivity is primarily the result of damage to the viral DNA (Wilhelm *et al.*, 1998a), which does not destroy viral particles. A problem arises if infectivity is destroyed more rapidly than viral particles, because infectious viruses would have to be produced more rapidly than particles to balance production and destruction rates. However, if the production of infectious viruses is solely the result of the production of new viral particles, then the particle production rate will exceed the removal rate, and viral particles will accumulate in the environment. This means that the difference between particle destruction rates and infectivity cannot be balanced by production of new virus particles. Consequently, infectivity must be restored to sunlight-damaged viruses.

One mechanism whereby infectivity can be restored to viruses that have been exposed to sunlight is photoreactivation. Photoreactivation is a host-mediated light-dependent repair process, originally described by Dulbecco (1949) as the restoration of infectivity to UV-irradiated viruses on exposure to visible light, in the presence of host bacteria. The process is catalyzed by the enzyme photolyase

(Friedberg *et al.*, 1995). It was shown more recently that photoreactivation can restore much of the lost infectivity in natural virus communities that have been exposed to sunlight (Weinbauer *et al.*, 1997; Wilhelm *et al.*, 1998b). Therefore, when measuring decay rates of infectivity it is essential that titers of infectious units be incubated in the light. Fortunately, photoreactivation should not be a problem for assays done with cyanophages or algal viruses, as titers are always determined under photoreactivating conditions (i.e., in the light). Nonetheless, there is undoubtedly a great deal of photoreactivation occurring in algal virus communities. Evidence of this has been seen in filamentous and unicellular cyanobacteria, which have very efficient light-dependent mechanisms that can repair UV-damaged cyanophage DNA (e.g., Wu *et al.*, 1967; Amla, 1979; Levine and Thiel, 1987).

The first attempts to estimate mortality in natural cyanophage communities were based on decay rates of bacteriophage infectivity and suggested that cyano-bacterial mortality was very high in surface waters (Suttle *et al.*, 1993). Sub-sequently, destruction rates of marine cyanophage isolates were used to infer viral-imposed mortality on populations of *Synechococcus* along a series of stations in the western Gulf of Mexico (Suttle and Chan, 1994). The estimates were based on decay rates of infectivity as a function of attenuation coefficients for damaging radiation in seawater. For optically transparent waters off the coast of Texas, the decay rate of cyanophage infectivity averaged over 24 hours and integrated over the surface mixed layer (7.5 m) was estimated to be 2 per day. This decay rate, along with a burst-size estimate of 250 (Suttle and Chan, 1993), and data on the abundances of *Synechococcus* and infectious cyanophages, were used to estimate that viruses caused the lysis of about 6.5% of the *Synechococcus* population each day. The estimated decay rates of the viruses decreased as the stations approached shore, because of the lower transparency of the water. Consequently, the percent of the *Synechococcus* cells estimated to be infected by viruses per day decreased to 5.1, 3.0, and 0.2%, respectively.

These results have been corroborated by studies at two stations off the coast of Texas in which the *in situ* decay rates of infectious natural cyanophage communi-ties were measured (Garza and Suttle, 1998). These rates, averaged over 24 hours and integrated over the surface mixed layer (28.2 and 10.4 m, respectively), were estimated to be 0.53 and 0.75 per day. Observations by transmission electron microscopy indicated that the average burst size was about 81 (SD = 17.3) viruses produced per cell lysed (Suttle *et al.*, 1996). These estimates of decay rates and burst size were used to calculate that approximately 1 and 8% of the *Synechococ-cus* populations were lost daily at each of the respective stations as the result of viral lysis. These are minimum estimates, as they only include viruses that lyse *Synechococcus* strain DC2, and the assumption is made that infectious cyano-phages are produced by all *Synechococcus* cells. In reality, subsets of the population may produce most of the viruses. These estimates are similar to those obtained by Suttle and Chan (1994), although the data from the more recent study indicate that the decay rate and burst size were originally overestimated by about threefold. However, the net effect of these overestimates was essentially neutral because they have opposite effects on calculated mortality rates (Suttle, 1999). The decay rate experiments led to conclusions that were very similar to those based on the proportion of visibly infected *Synechococcus* and indicated that cyanophages lyse several percent of the cells in natural *Synechococcus* communities on a daily basis.

The percentage of total mortality in natural populations of *Synechococcus* that can be attributed to viruses depends on the growth rate of the cells. The relative importance of viruses on mortality can be approximated by a simple model. If, on average, 50% of the cells survive to divide again, and 50% die, then something that removes 5% of the cells would account for 10% of the mortality (Proctor and Fuhrman, 1990). For a *Synechococcus* population growing at a typical rate of 0.5 to 1.0 per day, in which viruses cause the lysis of 5% of the population per day, viruses would be responsible for approximately 10 to 20% of the total mortality.

A somewhat different approach has been used to infer the effect of MpV on *Micromonas pusilla* populations (Cottrell and Suttle, 1995a). The abundance of viruses infecting *M. pusilla* can exceed 10^5 ml^{-1}, suggesting that they might be important agents of mortality. Decay rates of infectivity of natural communities of MpV in the dark ranged from 0.6 to 0.9 per day and in full sunlight from 6.9 to 7.1 per day. As the rate of change in the abundance of MpV in the environment (0.035 per day) was much less than the calculated turnover rate (ca. 0.82 per day), it implied that the production and removal of MpV were approximately in balance. As *M. pusilla* is difficult to identify by light microscopy, the abundance of *M. pusilla* required to support the inferred viral production rates was calculated based on measured adsorption kinetics of viruses to host cells. These results indicated that MpV was likely responsible for the daily mortality of 2.0 to 10.0% of the *M. pusilla* population. This could represent 8 to 40% of the total mortality if the host population was growing at 0.5 per day.

Results from studies on cyanophages and MpV suggest that these viruses exist in a relatively stable equilibrium with the hosts they infect, and are a significant source of mortality. In contrast, some viruses probably exist in a much more dynamic relationship with the phytoplankton they infect, and are a large but variable source of host cell mortality.

C. Community Structure and Bloom Termination

As has been previously argued, one of the most important effects of viruses on phytoplankton is their potential to influence community structure and population dynamics (e.g., Cottrell and Suttle, 1991; Suttle and Chan, 1993; Zingone, 1995). Viral replication depends on contact with a suitable host. Consequently, the probability of successful encounters increases as host and viral abundance increases. When host density is high, viruses can propagate rapidly through a population. When host density is low, viral infections will spread more slowly, or may not spread at all. In this way, viruses can exert a powerful influence on the community structure. This may also be the reason that phytoplankton blooms are the exception rather than the rule. Yet, evidence that viruses regulate phytoplankton community structure remains largely circumstantial.

The best evidence that viruses may be involved in bloom termination is from studies on the coccolithophorid *Emiliania huxleyi*. Viruslike particles (VLPs) in *E. huxleyi* have been known for some time (Manton and Leadbeater, 1974); however, observations on the collapse of blooms in mesocosms provided the first evidence that viruses may be involved in bloom termination. The collapse of the blooms was coincident with the appearance of large polyhedral VLPs (ca. 180 nm in diameter) within *E. huxleyi* and in the surrounding water, and a shift in the composition of the phytoplankton community (Bratbak *et al.*, 1993). Interestingly, the collapse of the blooms and the appearance of the VLPs occurred in mesocosms in which nutrients were nonlimiting, or in which nitrogen was in short supply, but did not occur when phosphate concentrations were low. Further evidence that viruses may be involved in termination of *E. huxleyi* blooms was obtained from field observations. Studies of a fjord in Norway and on a cruise in the North Sea have documented the appearance in the water column of Phycodnaviridae-like VLPs associated with the collapse of *E. huxleyi* blooms (Bratbak *et al.*, 1995, 1996). Other data have also been collected from the North Sea during the decay of a bloom (Brussaard *et al.*, 1996), in which up to 50% of *E. huxleyi* cells contained VLPs that fell into two size classes. It has also been reported that a lytic agent for *E. huxleyi* was brought into culture, but could not be maintained (Bratbak *et al.*, 1996). Although the evidence is circumstantial, it is quite convincing that viruses are involved in disintegration of *E. huxleyi* blooms.

There is also considerable evidence that viral infection plays a role in blooms of the raphidophyte *Heterosigma akashiwo*. This alga forms toxic blooms that are responsible for large fish kills in temperate and subarctic areas. VLPs were first reported in *H. akashiwo* at the final stages of a bloom in Hiroshima Bay, Japan (Nagasaki *et al.*, 1994a,b). Although <1% of the cells near the end of the bloom contained VLPs, when the cells were brought into the laboratory and cultured, the proportion of cells containing VLPs increased to ~11% (Nagasaki *et al.*, 1994a), implying significant viral mortality. More recently, viruses infecting this alga have been cultured (Nagasaki and Yamaguchi, 1997), providing further evidence for viral lysis of *H. akashiwo* populations. Nonetheless, in some studies there has been no evidence of viral infection during the demise of *H. akashiwo* blooms (Nagasaki *et al.*, 1996), suggesting that other mechanisms such as cyst formation may also be involved in the termination of blooms.

It is also likely that viruses affect blooms of other phytoplankton. For example, viruses have been isolated that infect *Chrysochromulina brevifilum* and *C. strobilus* (Suttle and Chan, 1995). Other species within this genus have been responsible for massive mortality in fisheries (Dundas *et al.*, 1989). Although the abundance of viruses infecting *C. brevifilum* and *C. strobilus* is relatively low in the western Gulf of Mexico (Suttle and Chan, 1995), Brussaard *et al.* (1996) has reported that *Chrysochromulina* at two stations in the North Sea were heavily infected with VLPs. *Phaeocystis pouchetii* and *Aureococcus anophagefferens* also form extensive blooms, which may be regulated in part by viral infection. Evidence of viral infection in blooms of *A. anophagefferens* have been reported (Sieburth *et al.*, 1988; Nixon *et al.*, 1994) and a virus infecting this alga subsequently isolated (Milligan and Cosper, 1994). More recently, a lytic virus has also been isolated that infects *P. pouchetii* (Jacobsen *et al.*, 1996). Observations on a number of bloom-forming phytoplankton suggest that viruses are important in bloom termination.

Given that viruses in the natural environment infect a variety of bloom-forming phytoplankton, it is puzzling that blooms occur. One mechanism that has been proposed is that bloom-forming phytoplankton release dissolved organic carbon (DOC) that fuels the growth of bacteria, which intercept the viruses by nonspecific binding before they can adsorb to host cells (Murray, 1995). This is an intuitively attractive model that extends the suggestion by Murray and Jackson (1992) that adsorption of viruses to nonhost cells is a major loss factor for viral particles in natural waters. The data suggest, however, that adsorption of viruses to nonhost cells is relatively rare. For example, when several strains of fluorescently tagged marine viruses were added to natural communities of bacteria containing $>10^6$ bacteria ml^{-1}, adsorption only occurred to the host cells (Hennes *et al.*, 1995). Moreover, many viral capsids are extremely resistant to digestion by bacterial enzymes. Murray (1995) has also suggested that increased bacterial growth resulting from DOC exudation would lead to higher abundances of heterotrophic flagellates that will consume viruses. Grazing by heterotrophic flagellates on viruses is documented; however, their impact on viral turnover is typically relatively small (Suttle and Chen, 1992; González and Suttle, 1993).

The interactions between bloom-forming organisms and the viruses that infect them are undoubtedly complex. For example, during a bloom of *Synechococcus* in the Gulf of Mexico during which cell abundances were in excess of 10^5 ml^{-1}, the abundance of infectious cyanophages was greater than 10^6 ml^{-1} (Suttle *et al.*, 1996). Yet, the bloom was able to persist, and the impact of viruses on the mortality of *Synechococcus* appeared to be low. Another example of a complex interaction between a bloom of *Synechococcus* and cyanophages has been described for a mesocosm experiment in a Norwegian fjord (Wilson *et al.*, 1998). In this experiment, a *Synechococcus* bloom developed in a phosphate-limited enclosure. Following the addition of phosphate to this enclosure, there was a marked increase in the abundance of infectious cyanophages, followed by a collapse of the bloom. A possible explanation for this observation was that addition of phosphate caused the induction of lysogenic phage and subsequent termination of the bloom.

The potential for strong interactions also exists among bloom formation, light attenuation, and viral destruction rate. In this scenario, the abundance of viruses infecting the bloom-forming organism is initially low because of low host cell biomass. This results in low virus production, and high light penetration, which leads to high viral removal rates. As host cell biomass increases, penetration of UVB would decrease, resulting in lower infectivity decay rates. As well, the increase in the abundance of host cells allows for efficient viral propagation. Eventually, a critical point could be reached where viral production rates will greatly exceed their removal rates, resulting in collapse of the bloom. The demise of the bloom would result in higher rates of viral decay, because of increased light penetration, and lower rates of viral production. This could ultimately lead to conditions where the host cell population can reestablish itself.

Possibly, one of the most important roles of algal viruses is to maintain community diversity by preventing bloom formation. Perhaps the reason that blooms are the exception rather than the rule is because of increased viral replication at elevated host abundance. It has been postulated that population control as a result of species-specific viral attack is the reason that most widely distributed phytoplankton do not form blooms (Cottrell and Suttle, 1995a).

D. Evolution and Extinction of Host Populations

Viruses also have the potential to affect phytoplankton community structure over much longer time scales. The dogma in the virological literature is that viruses do not eliminate their hosts. As obligate pathogens, contemporary viruses must have a stable coexistence with the organisms they infect, or they would not be present. Consequently, most virus–host relationships are highly coevolved. Over evolutionary time, however, there is ample opportunity for mutations to occur that allow for the expansion of host range. Evidence of this is seen when viruses cross the species barrier into human populations. Such expansions in host range have undoubtedly occurred often during geologic time; hence, most extant species will likely be able to accommodate these incursions. In theory, however, there is nothing preventing an expansion of host range resulting in an extremely virulent pathogen to which the alternate host has little or no resistance. This could drive the host to extinction by imposing a much higher mortality rate than the species can absorb.

Evidence of large changes in the dominant Foraminifers and Coccolithophorids are relatively common in the fossil record. Relatively rare are massive extinctions such as occurred at the end of the Cretaceous. Much more common are the synchronous worldwide disappearances of individual taxa that were dominant members of the plankton (e.g., Gartner, 1972; Thierstein et al., 1977). These observations, along with the discovery of lytic algal viruses (Mayer and Taylor, 1979) led Emiliani (1982) to propose that viruses were the causative agents of these extinctions. He termed this process "extinctive evolution" and argued strongly (Emiliani, 1993a,b, 1995) that infection by viruses was the only available explanation for background extinctions. Extinction as the result of viral infection is also consistent with the observation that the species that disappeared were the most widely distributed and abundant ones. An abundant and ubiquitous host population would lead to high host–virus contact rates and rapid viral propagation. Elimination of taxa as the result of viral infection would open up niches that could be exploited by competing species, and potentially lead to rapid evolution.

VIII. THE ROLE OF ALGAL VIRUSES IN GLOBAL GEOCHEMICAL CYCLES

It is becoming increasingly clear that viruses are major players in geochemical cycles in oceans and lakes (Wilhelm and Suttle, 1999). Viruses destroy a significant proportion of the biomass in the world's oceans on a daily basis (e.g., Fuhrman and Suttle, 1993; Bratbak et al., 1994; Suttle, 1994), and as a consequence are major recyclers of organic material on a global scale. Therefore, it is essential to incorporate viral-mediated processes into models of nutrient fluxes and geochemical cycles.

There are a number of ways in which viral lysis of phytoplankton will affect the bulk flow of nutrients and energy in aquatic systems. One of the most significant effects is to convert living particulate organic material into the products of cell lysis (virus particles, particulate and colloidal cellular debris, and dissolved organic and inorganic nutrients). This uncouples the transfer of primary production to higher trophic levels and reduces the resources available to higher-level consumers. In turn, the dissolved and particulate material that is released into the surrounding water fuels production of bacteria and phytoplankton. Clear evidence of the bioavailability of the products of cell lysis was seen in the efficient transfer of nutrients from lysed bacterial (Middelboe et al., 1996) and algal (Gobler et al., 1997; Bratbak et al., 1998b) cultures into new bacterial biomass. The released nutrients can also fuel new phytoplankton production. In laboratory studies, viral lysis of the bloom-forming alga *Aureococcus anophagefferens* alleviated nutrient limitation in a diatom culture (Gobler *et al.*, 1997). Further evidence comes from a field study in which the growth rate of *Synechococcus* during a bloom was reduced when the concentration of viruses was reduced (Suttle *et al.*, 1996). Another consequence of viral-mediated nutrient recycling is that the concentration of carbon will be reduced, relative to that of other nutrients, because of bacterial respiration (Wilhelm and Suttle, 1999).

Increasing concern with the buildup of CO_2 in the atmosphere and its potential impact on global warming has placed emphasis on understanding the global carbon cycle. One of the largest active reservoirs of organic carbon on earth is dissolved organic matter in the ocean, with an average age of about 6000 years in the deep ocean (Williams and Druffel, 1987). Although the exact composition of the bulk of this material remains largely unknown, there is increasing evidence that much of the material is composed of polysaccharides (Benner *et al.*, 1992; McCarthy *et al.*, 1996). Although much of the material that is released by viral lysis will be available to microbial food webs, other components are much more recalcitrant and are not directly available for use by other organisms. This recalcitrant material would include complex polysaccharides associated with cell wall material, as well as lipids and other materials. These can be very resistant to microbial breakdown, and have the potential to accumulate in the environment. I suggest that viral lysis of phytoplankton and bacteria is a primary source of recalcitrant dissolved organic material in the deep oceans.

Infection of phytoplankton by viruses may also affect the flux of particulate organic matter to deeper waters (Wilhelm and Suttle, 1999). If cellular material is released into the surface waters through viral lysis, it can be rendered biologically available by photochemical degradation (Mopper *et al.*, 1991). In fact, solar radiation is possibly the major factor making nonlabile dissolved organic matter biologically available in surface waters. In contrast, phytoplankton not lysed will sediment to deeper water, exporting organic carbon from the surface waters and greatly slowing the remobilization of the organic material.

Viral lysis of phytoplankton may play another role in global warming by triggering the release of greenhouse gasses. Dimethylsulfoniopropionate (DMSP) released by marine phytoplankton is thought to the major source of dimethylsulfide (DMS), a major greenhouse gas (Keller *et al.*, 1989). Bratbak *et al.* (1995) attempted to follow in mesocosms and a fjord the release of DMS and DMSP during the collapse of blooms of *Emiliania huxleyi* attributed to viral lysis. They were unable to demonstrate evidence of DMSP release associated with the demise of the blooms even though *E. huxleyi* is reported to be a major DMSP producer (Keller *et al.*, 1989). There are a number of possibilities why DMSP production was not documented, including the physiological state of the cells, the presence of bacteria in the samples, and the dynamics of DMSP production and conversion to DMS. In contrast, a recent study by Hill *et al.* (1998) clearly demonstrated mass balance in the release of DMSP by cultures of *Micromonas pusilla* infected by viruses. Hill *et al.* (1998) concluded that viral lysis of phytoplankton may be a major source of DMSP production in the environment. Their conclusions are also supported by observations showing that lysis of cultures of DMSP-containing *Phaeocystis pouchetii* resulted in a buildup of DMS in the media (Malin *et al.*, 1990).

IX. SUMMARY

It is increasingly apparent that infection of algae by viruses has major ecological and environmental ramifications, ranging from autoecological interactions between viruses and their hosts to effects on carbon and geochemical cycles that have global consequences. For many years most reports of viral infection in algae were anecdotal reports of viruslike particles in electron micrographs. More intensive investigations in recent years have demonstrated that algal viruses are a ubiquitous component of marine and freshwater ecosystems. This has heightened interest in the effects of viruses on algal populations in nature, and has led to isolation of previously unknown viruses. Evidence continues to accumulate that viral infection is common in algal populations, and ultimately it may be shown that viruses are at least as important as other factors in regulating the abundance and distribution of populations in nature. They may also be extremely important as vectors of genetic information among algal populations in nature. As microalgae are major primary producers in aquatic ecosystems, viral lysis also has major implications for our understanding of nutrient and energy cycling on a global scale. Viruses may have significant effects on pathways of cycling of organic and inorganic nutrients, and may even be involved in the release of greenhouse gases. The next decade promises to be extremely exciting as the central role of algal viruses in aquatic ecosystems continues to emerge.

ACKNOWLEDGMENTS

I am grateful to the many individuals who have worked in my laboratory and shared their ideas about viruses and their roles in the ecosystem. In particular, I would like to thank Feng Chen, Sean Brigden, André Comeau, Matthew Cottrell, Hudson DeYoe, Randy Garza, Kilian Hennes, Janice Lawrence, Alice Ortmann, Kristen Rodda, Steven Short, Vera Tai, Steven Wilhelm, Markus Weinbauer, and Yanping Zhang. Special thanks are reserved for Amy Chan, who has formed the core of the research group, and for Jed Fuhrman and Lita Proctor, who introduced me to the world of viruses in the sea. The early encouragement of Jim Van Etten was also instrumental in encouraging our examination of algal viruses. The support of the U.S. Office of Naval Research, the U.S. National Science Foundation, and the Natural Science and Engineering Research Council of Canada is gratefully acknowledged.

REFERENCES

Adolph, K. W., and Haselkorn, R. (1971). Isolation and characterization of a virus infecting the blue-green alga *Nostoc muscorum*. *Virology* **46**, 200–208.

Adolph, K. W., and Haselkorn, R. (1972). Photosynthesis and the development of the blue-green algal virus N-1. *Virology* **47**, 370–374.

Agustí, S., Satta, M. P., Mura, M. P., and Benavent, E. (1998). Dissolved esterase activity as a tracer of phytoplankton lysis: Evidence of high phytoplankton lysis rates in the northwestern Mediterranean. *Limnol. Oceanogr.* **43**, 1836–1849.

Allen, M. M., and Hutchinson, F. (1976). Effect of some environmental factors on cyanophage AS-1 development in *Anacystis nidulans*. *Arch Microbiol.* **110**, 55–60.

Amla, D. V. (1979). Photoreactivation of ultraviolet irradiated blue-green alga *Anacystis nidulans* and cyanophage AS-1. *Arch Virol.* **59**, 173–179.

Azam, F., Fenchel, T., Field, J. G., Gray, J. S., Meyer-Reil, L. A., and Thingstad, F. (1983). The ecological role of water-column microbes in the sea. *Mar. Ecol. Prog. Ser.* **10**, 257–263.

Benner, R., Pakulski, J. D., McCarthy, M., Hedges, J. I., and Hatcher, P. G. (1992). Bulk chemical characteristics of dissolved organic matter in the ocean. *Science* **255**, 1561–1564.

Bergh, O., Borsheim, K. Y., Bratbak, G., and Heldal, M. (1989). High abundance of viruses found in aquatic environments. *Nature* **340**, 467–468.

Berryhill Jr., H. L. (1986). "Late Quaternary Facies and Structure, Northern Gulf of Mexico: Interpretations from Seismic Data." American Association of Petroleum Geologists, Tulsa, OK.

Bisen, P. S., Audholia, S., and Bhatnagar, A. K. (1985). Mutation to resistance for virus AS-1 in the cyanobacterium *Anacystis nidulans*. *Microbiol. Lett.* **29**, 7–13.

Bold, H. C., and Wynne, M. J. (1978). "Introduction to the Algae." Prentice-Hall, Englewood Cliffs, NJ.

Borsheim, K. Y. (1993). Native marine bacteriophages. *FEMS Microbiol. Ecol.* **102**, 141–159.

Braithwaite, D. K., and Ito, J. (1993). Compilation, alignment, and phylogenetic relationships of DNA polymerases. *Nucleic Acids Res.* **21**, 787–802.

Bratbak, G., Egge, J. K., and Heldal, M. (1993). Viral mortality of the marine alga *Emiliania huxleyi* (Haptophyceae) and termination of algal blooms. *Mar. Ecol. Prog. Ser.* **93**, 39–48.

Bratbak, G., Thingstad, T. F., and Heldal, M. (1994). Viruses and the microbial loop. *Microb. Ecol.* **28**, 209–221.

Bratbak G., Levasseur, M., Michaud, S., Cantin, G., Fernández, E., Heimdal, B. R., and Heldal, M. (1995). Viral activity in relation to *Emiliania huxleyi* blooms: A mechanism of DMSP release? *Mar. Ecol. Prog. Ser.* **128**, 133–142.

Bratbak, G., Wilson, W., and Heldal, M. (1996). Viral control of *Emiliania huxleyi* blooms? *J. Mar. Syst.* **9**, 75–81.

Bratbak, G., Jacobsen, A., Heldal, M., Nagasaki, K., and Thingstad, F. (1998a). Virus production in *Phaeocystis pouchetii* and its relation to host cell growth and nutrition. *Aquat. Microb. Ecol.* **16**, 1–9.

Bratbak, G., Jacobsen, A., and Heldal, M. (1998b). Viral lysis of *Phaeocystis pouchetii* and bacterial secondary production. *Aquat. Microb. Ecol.* **16**, 11–16.

Brown Jr., R. M. (1972). Algal Viruses. *Adv. Virus Res.* **17**, 243–277.

Brussaard, C. P. D., Kempers, R. S., Kop, A. J., Riegman, R., and Heldal, M. (1996). Virus-like particles in a summer bloom of *Emiliania huxleyi* in the North Sea. *Aquat. Microb. Ecol.* **10**, 105–113.

Cannon, R. E., Shane, M. S., and Bush, V. N. (1971). Lysogeny of a blue-green alga *Plectonema boryanum. Virology* **45**, 149–153.

Caron, D. A., Pick, F. R., and Lean, D. R. S. (1985). Chroococcoid cyanobacteria in Lake Ontario: Vertical and seasonal distributions during 1982. *J. Phycol.* **21**, 171–175.

Caron, D. A., Dam, H. G., Kremer, P., Lessard, E. J., Madin, L. P., Malone, T. C., Napp, J. M., Peele, E. R., Roman, M. R., and Youngbluth, M. J. (1995). The contribution of microorganisms to particulate carbon and nitrogen in surface waters of the Sargasso Sea near Bermuda. *Deep Sea Res.* **42**, 943–972.

Chan, A. M., Kacsmarska, I., and Suttle, C. A. (1997). Isolation and characterization of a species-specific bacterial pathogen which lyses the marine diatom *Navicula pulchripora. Abst. Amer. Soc. Limnol. Oceanogr., Santa Fe, NM.* p. 121.

Chapman, R. L., and Lang, N. J. (1973). Virus-like particles and nuclear inclusions in the red alga *Porphoridium purpureum* (Bory) Drew et Ross. *J. Phycol.* **9**, 117–122.

Chen, F., and Suttle, C. A. (1995). Amplification of DNA polymerase gene fragments from viruses infecting microalgae. *Appl. Environ. Microbiol.* **61**, 1274–1278.

Chen, F., and Suttle, C. A. (1996). Evolutionary relationships among large double-stranded DNA viruses that infect microalgae and other organisms as inferred from DNA polymerase genes. *Virology* **219**, 170–178.

Chen, F., Suttle, C. A., and Short, S. M. (1996). Genetic diversity in marine algal virus communities as revealed by sequence analysis of DNA polymerase genes. *Appl. Environ. Microbiol.* **62**, 2869–2874.

Cho, B. C., and Azam, F. (1990). Biogeochemical significance of bacterial biomass in the ocean's euphotic zone. *Mar. Ecol. Prog. Ser.* **63**, 253–259.

Cole, J. J., Findlay, S., and Pace, M. L. (1988). Bacterial production in fresh and saltwater ecosystems: A cross system overview. *Mar. Ecol. Prog. Ser.* **43**, 1–10.

Conover, R. J. (1982). Interrelations between microzooplankton and other plankton organisms. *Annu. Inst. Oceanogr. Paris* 58(s), 31–46.

Cottrell, M. T., and Suttle, C. A. (1991). Widespread occurrence and clonal variation in viruses which cause lysis of a cosmopolitan, eukaryotic marine phytoplankter, *Micromonas pusilla. Mar. Ecol. Prog. Ser.* **78**, 1–9.

Cottrell, M. T., and Suttle, C. A. (1995a). Dynamics of a lytic virus infecting the photosynthetic marine picoflagellate, *Micromonas pusilla. Limnol. Oceanogr.* **40**, 730–739.

Cottrell, M. T., and Suttle, C. A. (1995b). Genetic diversity of algal viruses which lyse the photosynthetic picoflagellate *Micromonas pusilla* (Prasinophyceae). *Appl. Environ. Microbiol.* **61**, 3088–3091.

Daft, M. J., Begg, J., and Stewart, W. D. P. (1970). A virus of blue-green algae from freshwater habitats in Scotland. *New Phytologist* **69**, 1029–1038.

Dodds, J. A. (1979). Viruses of marine algae. *Experientia* **35**, 440–442.

Dodds, J. A. (1983). New viruses of eukaryotic algae and protozoa. In "A Critical Appraisal of Viral Taxonomy" (R. E. F. Mathews, ed.), pp. 177–188. CRC Press, Boca Raton, FL.

Dodds, J. A., and Cole, A. (1980). Microscopy and biology of *Uronema gigas*, a filamentous eucaryotic green alga, and its associated tailed virus-like particle. *Virology* **100**, 156–165.

Ducklow, H. W., and Carlson, C. A. (1992). Oceanic bacterial production. In "Advances in Microbial Ecology" (K. C. Marshall, ed.), Vol. 12, pp. 113–181. Plenum, New York.

Dulbecco, R. (1949). Reactivation of ultra-violet-inactivated bacteriophage by visible light. *Nature* **163**, 949–950.

Dundas, I. O., Johannessen, M., Berge, G., and Heimdal, B. (1989). Toxic algal bloom in Scandinavian waters, May–June 1988. *Oceanography* **2**, 9–14.

Emiliani, C. (1982). Extinctive evolution: Extinctive and competitive evolution combine into a unified model of evolution. *J. Theor. Biol.* **97**, 13–33.

Emiliani, C. (1993a). Viral extinctions in deep-sea species. *Nature* **366**, 217–218.

Emiliani, C. (1993b). Extinction and viruses. *Biosystems* **31**, 155–159.

Emiliani, C. (1995). Evolution: A composite model. *Evolut. Theory* **10**, 299–303.

Ferris, M. J., Muyzer, G., and Ward, D. M. (1996). Denaturing gradient gel electrophoresis profiles of 16S rRNA-defined populations inhabiting a hot spring microbial mat community. *Appl. Environ. Microbiol.* **62**, 340–346.

Fuhrman, J. A., and Suttle, C. A. (1993). Viruses in marine planktonic systems. *Oceanography* **6**, 51–63.

Fuhrman, J. A., Sleeter, T. D., Carlson, C. A., and Proctor, L. M. (1989). Dominance of bacterial biomass in the Sargasso Sea and its ecological implications. *Mar. Ecol. Prog. Ser.* **57**, 207–217.

Fuller, N. J., Wilson, W. H., Joint, I. R., and Mann, N. H. (1998). Occurrence of a sequence in marine cyanophages similar to that of T4 g20 and its application to PCR-based detection and quantification techniques. *Appl. Environ. Microbiol.* **64**, 2051–2060.

Friedberg, E. C., Walker, G. C., and Siede, W. (1995). "DNA Repair and Mutagenesis." ASM Press, Washington, DC.

Garry, R. T., Hearing, P., and Cosper, E. M. (1998). Characterization of a lytic virus infectious to the bloom-forming microalga *Aureococcus anophagefferens* (Pelagophyceae). *J. Phycol.* **34**, 616–621.

Gartner, S. (1972). Late Pleistocene calcareous nanofossils in the Caribbean and their interoceanic correlation. *Paleogeogr. Paleoclimatol. Paleoecol.* **12**, 169–191.

Garza, D. R., and Suttle, C. A. (1998). The effect of cyanophages on the mortality of *Synechococcus* spp. and seasonal changes in the resistance of natural viral communities to UV radiation. *Microb. Ecol.* **36**, 281–292.

Gastrich, M. D., Anderson, O. R., Benmayor, S. S., and Cosper, E. M. (1998). Ultrastructural analysis of viral infection in the brown-tide Alga, *Aureococcus anophagefferens* (Pelagophyceae). *Phycologia* **37**, 300–306.

Gibbs, A., Skotnicki, A. H., Gardiner, J. E., Walker, E. S., and Hollings, M. (1975). A tobamovirus of a green alga. *Virology* **64**, 571–574.

Glover, H. E., Campbell, L., and Prezelin, B. B. (1986). Contribution of *Synechococcus* spp. to size-fractionated primary productivity in three water masses of the northwest Atlantic Ocean. *Mar. Biol.* **91**, 193–203.

Gobler, C. J., Hutchins, D. A., Fisher, N. S., Cosper, E. M., and Sanudo-Wilhelmy, S. (1997). Release and bioavailability of C, N, P, Se, and Fe following viral lysis of a marine Chrysophyte. *Limnol. Oceanogr.* **42**, 1492–1504.

González, J. M., and Suttle, C. A. (1993). Grazing by marine nanoflagellates on viruses and virus-sized particles: Ingestion and digestion. *Mar. Ecol. Prog. Ser.* **94**, 1–10.

Grabherr, R., Strasser, P., and Van Etten, J. L. (1992). The DNA polymerase gene from *Chlorella* viruses PBCV-1 and NY-2A contains an intron with nuclear splicing sequences. *Virology* **188**, 721–731.

Gromov, B. V. (1983). Cyanophages. *Ann. Microbiol.* **134**, 43–59.

Gromov, B. V., and Mamkaeva, K. A. (1981). A virus infection in the synchronized population of the *Chlorococcum minutum* zoospores. *Arch. Hydrobiol. Suppl.* **60**, 252–259.

Heldal, M., and Bratbak, G. (1991). Production and decay of viruses in aquatic environments. *Mar. Ecol. Prog. Ser.* **72**, 205–212.

Hennes, K. P., Chan, A. M., and Suttle, C. A. (1995). Fluorescently labeled virus probes show that natural virus populations can control the structure of marine microbial communities. *Appl. Environ. Microbiol.* **61**, 3623–3627.

Henry, E. C., and Meints, R. H. (1992). A persistent virus infection in *Feldmannia* (Phaeophyceae). *J. Phycol.* **28**, 517–526.

Hill, R. W., White, B. A., Cottrell, M. T., and Dacey, J. W. H. (1998). Virus-mediated total release of dimethylsulfoniopropionate from marine phytoplankton: A potential climate process. *Aquat. Microb. Ecol.* **14**, 1–6.

Hobbie, J. E., Daley, R. J., and Jasper, S. (1977). Use of nucleopore filters for counting bacteria by fluorescence microscopy. *Appl. Environ. Microbiol.* **33**, 1225–1228.

Hu, N.-T., Thiel, T., Gidding Jr., T. H., and Wolk, C. P. (1981). New *Anabaena* and *Nostoc* cyanophages from sewage settling ponds. *Virology* **114**, 236–246.

ICTV (1999). "ICTVindex: The dsDNA Viruses" [Web Page], Available at http://www.ncbi.nlm. nih.gov/ICTV/viruslist/dsdna_viruses.pdf. Accessed 19 October February 1999.

Ivey, R. G., Henry, E. C., Lee, A. M., Klepper, L., Krueger, S. K., and Meints, R. H. (1996). A *Feldmannia* algal virus has two genome size-classes. *Virology* **220**, 267–273.

Jacobsen, A., Bratbak, G., and Heldal, M. (1996). Isolation and characterization of a virus infecting *Phaeocystis pouchetii* (Prymnesiophyceae). *J. Phycol.* **32**, 923–927.

Jiang, S. C., and Paul, J. H. (1994). Seasonal and diel abundance of viruses and occurrence of lysogeny/bacteriocinogeny in the marine environment. *Mar. Ecol. Prog. Ser.* **104**, 163–172.

Jiang, S. C., and Paul, J. H. (1996). Occurrence of lysogenic bacteria in marine microbial communities as determined by prophage induction. *Mar. Ecol. Prog. Ser.* **142**, 27–48.

Johnson, D. W., and Potts, M. (1985). Host range of LPP cyanophages. *Int. J. Syst. Bacteriol.* **35**, 76–78.

Joint, I. R., and Pomroy, A. J. (1983). Production of picoplankton and small nanoplankton in the Celtic Sea. *Mar. Biol.* **77**, 19–27.

Kapp, M., Knippers, R., and Müller, D. G. (1997). New members of a group of DNA viruses infecting brown algae. *Phycol. Res.* **45**, 85–90.

Keller, M. D., Bellows, W. K., and Guilland, R. R. L. (1989). Dimethyl production in marine phytoplankton. *In* "Biogenic Sulfur in the Environment" (E. S. Saltzman and W. J. Cooper, eds.), pp. 167–182. American Chemical Society, Washington, DC.

Khudyakov, I. Y., and Gromov, B. V. (1973). The temperate cyanophage A-4 (L) of the blue-green alga *Anabaena variabilis*. *Mikrobiologiya* 904–907.

Kim, M., and Choi, Y. K. (1994). A new *Synechococcus* cyanophage from a reservoir in Korea. *Virology* **204**, 338–342.

Klein, M., Lanka, S. T. J., Müller, D. G., and Knippers, R. (1994). Single-stranded regions in the genome structure of the *Ectocarpus siliculosus* virus. *Virology* **202**, 1076–1078.

Lanka, S. T. J., Klein, M., Ramsperger, U., Müller, D. G., and Knippers, R. (1993). Genome structure of a virus infecting the marine brown alga *Ectocarpus siliculosus*. *Virology* **193**, 802–811.

Lee, A. M., Ivey, R. G., and Meints, R. H. (1998a). The DNA polymerase gene of a brown algal virus: Structure and phylogeny. *J. Phycol.* **34**, 608–615.

Lee, A. M., Ivey, R. G., and Meints, R. H. (1998b). Repetitive DNA insertion in a protein kinase ORF of a latent FSV (*Feldmannia* sp. Virus) genome. *Virology* **248**, 35–45.

Lee, R. E. (1971). Systemic viral material in the cells of the freshwater red alga *Sirodotia tenuissima* (Holden) Skuja. *J. Cell Sci.* **8**, 623–631.

Lee, R. E. (1980). "Phycology." Cambridge University Press, New York.

Lemke, P. A. (1976). Viruses of eucaryotic microorganisms. *Annu. Rev. Microbiol.* **30**, 105–145.

Levine, E., and Thiel, T. (1987). UV-inducible DNA repair in the cyanobacteria *Anabaena* spp. *J. Bacteriol.* **169**, 3988–3993.

Li, W. K. W., Subba Rao, D. V., Harrison, W. G., Smith, J. C., Cullen, J. J., Irwin, B., and Platt, T. (1983). Autotrophic picoplankton in the tropical ocean. *Science* **219**, 292–295.

Li, W. K. W., Dickie, P. M., Irwin, B. D., and Wood, A. M. (1992). Biomass of bacteria, cyanobacteria, prochlorophytes and photosynthetic eukaryotes in the Sargasso Sea. *Deep Sea Res.* **39**, 501–519.

Liu, H. B., Nolla, H. A., and Campbell, L. (1997). Prochlorococcus growth rate and contribution to primary production in the equatorial and subtropical North Pacific Ocean. *Aquat. Microb. Ecol.* **12**, 39–47.

Lycke, E., Magnusson, S., and Lund, E. (1965). Studies on the nature of the virus inactivating capacity of sea water. *Arch. Gesamte. Viruschung* **17**, 409–413.

MacKenzie, J. J., and Haselkorn, R. (1972a). Photosynthesis and the development of blue-green algal virus SM-1. *Virology* **49**, 517–521.

MacKenzie, J. J., and Haselkorn, R. (1972b). An electron microscope study of infection by the blue-green algal virus SM-1. *Virology* **49**, 505–516.

Malin, G., Wilson, W. H., Bratbak, G., Liss, P. S., and Mann, N. H. (1998). Elevated production of dimethylsulfide resulting from viral infection of cultures of *Phaeocystis pouchetii*. *Limnol. Oceanogr.* **43**, 1389–1393.

Maniloff, J., and Ackermann, H. W. (1998). Taxonomy of bacterial viruses: Establish of the order Caudovirales. *Arch. Virol.* **143**, 2051–2063.

Manton, I., and Leadbeater, B. S. C. (1974). Fine-structural observations on six species of *Chrysochromulina* from wild Danish marine nanoplankton, including a description of *C. campanulifera* sp. nov. and a preliminary summary of the nanoplankton as a whole. *Det Kongelige Danske Videnskabernes Selskab Biologiske Skrifter* **20**, 1–26.

Martin, E. L., and Benson, R. (1988). Phages of cyanobacteria. *In* "The Bacteriophages" (R. Calendar, ed.), pp. 607–645. Plenum, New York.

Matossian, A. M., and Garabedian, G. A. (1967). Virucidal action of sea water. *Am. J. Epidemiol.* **85**, 1–8.

Mayer, J. A. (1978). Isolation and ultrastructural study of a lytic virus in the small phytoflagellate *Micromonas pusilla* (Prasinophyceae). PhD Dissertation, University of British Columbia.

Mayer, J. A., and Taylor, F. J. R. (1979). A virus which lyses the marine nanoflagellate *Micromonas pusilla*. *Nature* **281**, 299–301.

McCarthy, M., Hedges, J., and Benner, R. (1996). Major biochemical composition of dissolved high-molecular-weight organic matter in seawater. *Mar. Chem.* **55**, 281–297.

Meints, R. H., Van Etten, J. L., Kuczmarski, D., Lee, K., and Ang, B. (1981). Viral infection of the symbiotic chlorella-like alga present in *Hydra viridis*. *Virology* **113**, 698–703.

Mendzhul, M. I., Lysenko, T. G., Bobrovnik, S. A., and Spivak, M. Ya. (1973). Detection of A-1 virus of blue-green alga *Anabaena variabilitis* in the Kremenchug artificial reservoir. *Mikrobiol. Zh.* **35**, 747–751.

Middelboe, N., Jorgensen, N. O. G., and Kroer, N. (1996). Effects of viruses on nutrient turnover and growth efficiency of non-infected marine bacterioplankton. *Appl. Environ. Microbiol.* **62**, 1991–1997.

Milligan, K. L. D., and Cosper, E. M. (1994). Isolation of virus capable of lysing the brown tide microalga, *Aureococcus anophagefferens*. *Science* **266**, 805–807.

Mitchell, R., and Jannasch, H. W. (1969). Processes controlling virus inactivation in seawater. *Environ Sci Technol.* **3**, 941–943.

Moebus, K. (1992). Laboratory investigations on the survival of marine bacteriophages in raw and treated seawater. *Helgol. Meeres.* **46**, 251–273.

Moisa, I., Sotropa, E., and Velehorschi, V. (1981). Investigations on the presence of cyanophages in fresh and sea waters of Romania. *Rev. Roum. Med.–Virol.* **32**, 127–132.

Mopper, K., Zhou, X., Kieber, R. J., Kiever, D. J., Sikorski, R. J., and Jones, R. D. (1991). Photochemical degradation of dissolved organic carbon and its impact on the oceanic carbon cycle. *Nature* **353**, 60–62.

Müller, D. G. (1991). Mendelian segregation of a virus genome during host meiosis in the marine brown alga *Ectocarpus siliculosus*. *J. Plant Physiol.* **137**, 739–743.

Müller, D. G. (1996). Host–virus interactions in marine brown algae. *Hydrobiologia* **326/327**, 21–28.

Müller, D. G., and Frenzer, K. (1993). Virus infections in three marine brown algae: *Feldmannia irregularis, F. simplex,* and *Ectocarpus siliculosus. Hydrobiologia* **260/261**, 37–44.

Müller, D. G., and Schmid, C. E. (1996). Intergeneric Infection and persistence of *Ectocarpus* virus DNA in *Kuckuckia* (Phaeophyceae, Ectocarpales). *Bot. Mar.* **39**, 401–405.

Muradov, M. M., Cherkasova, G. V., Akhmedova, D. U., Kamilova, F. D., Mukhamedov, R. S., Abdukarimov, A. A., and Khalmuradov, A. G. (1990). Comparative study of NP-IT cyanophages, which lysogenize nitrogen-fixing bacteria of the genera *Nostoc* and *Plectonema. Microbiology* (Translation of *Mikrobiologiya*) **59**, 558–563.

Murphy, F. A., Fauquet, C. M., Bishop, D. H. L., Ghabrial, S. A., Jarvis, A. W., Martelli, G. P., Mayo, M. A., and Summers, M. D. (1995). The classification and nomenclature of viruses. The sixth report of the International Committee on Taxonomy of Viruses. *Arch. Virol.* **1** (Suppl.), Springer, Vienna.

Murray, A. G. (1995). Phytoplankton exudation: Exploitation of the microbial loop as a defence against algal viruses. *J. Plankton Res.* **17**, 1079–1094.

Murray, A. G., and Jackson, G. A. (1992). Viral dynamics: A model of the effects of size, shape, motion and abundance of single-celled planktonic organisms and other particles. *Mar. Ecol. Prog. Ser.* **89**, 103–116.

Murray, A. G., and Jackson, G. A. (1993). Viral dynamics, II: A model of the interaction of ultraviolet light and mixing processes on virus survival in seawater. *Mar. Ecol. Prog. Ser.* **102**, 105–114.

Muyzer, G., Wall, E. C. D., and Uitterlinden, A. G. (1993). Profiling of complex microbial populations by denaturing gradient gel electrophoresis analysis of polymerase chain reaction-amplified genes coding for 16S rRNA. *Appl. Environ. Microbiol.* **59**, 695–700.

Nagasaki, K., and Yamaguchi, M. (1997). Isolation of a virus infectious to the harmful bloom causing microalga *Heterosigma akashiwo* (Raphidophyceae). *Aquat. Microb. Ecol.* **13**, 135–140.

Nagasaki, K., and Yamaguchi, M. (1998a). Intra-species host specificity of HaV (*Heterosigma akashiwo* virus) clones. *Aquat. Microb. Ecol.* **14**, 109–112.

Nagasaki, K., and Yamaguchi, M. (1998b). Effect of temperature on the algicidal activity and the stability of HaV (*Heterosigma akashiwo* virus). *Aquat. Microb. Ecol.* **15**, 211–216.

Nagasaki, K., Ando, M., Itakura, S., Imai, I., and Ishida, Y. (1994a). Viral mortality in the final stage of *Heterosigma akashiwo* (Raphidophyceae) red tide. *J. Plankton Res.* **16**, 1595–1599.

Nagasaki, K., Ando, M., Imai, I., Itakura, S., and Ishida, Y. (1994b). Virus-like particles in *Heterosigma akashiwo* (Raphidophyceae): A possible red tide disintegration mechanism. *Mar. Biol.* **119**, 307–312.

Nagasaki, K., Itakura, S., Imai, I., Nakagiri, S., and Yamaguchi, M. (1996). The disintegration process of a *Heterosigma akashiwo* (Raphidophyceae) red tide in northern Hiroshima Bay, Japan, during the summer of 1994. *In* "Harmful and Toxic Algal Blooms" (T. Yasumoto, Y. Oshima, and Y. Fukuyo, eds.), pp. 251–254. UNESCO, Paris.

Nixon, S. W., Granger, S. L., Taylor, D. I., Johnson, P. W., and Buckley, B. A. (1994). Subtidal volume fluxes, nutrient inputs and the brown tide — an alternate hypothesis. *Estuar. Coast Shelf Sci.* **39**, 303–312.

Noble, R. T., and Fuhrman, J. A. (1997). Virus decay and its causes in coastal waters. *Appl. Environ. Microbiol.* **63**, 77–83.

Ohki, K., and Fujita, Y. (1996). Occurrence of a temperate cyanophage lysogenizing the marine cyanophyte *Phormidium persicinum. J. Phycol.* **32**, 365–370.

Oliveira, L., and Bisalputra, T. (1978). A virus infection in the brown alga *Sorocarpus uvaeformis* (Lyngbye) Pringsheim (Phaeophyta, Ectocarpales). *Ann. Bot.* **42**, 439–445.

Padan, E., and Shilo, M. (1969). Distribution of cyanophages in natural habitats. *Vehr. Int. Verein. Limnol.* **17**, 747–751.

Padan, E., and Shilo, M. (1973). Cyanophages–viruses attacking blue-green algae. *Bacteriol. Rev.* **37**, 343–370.

Padan, E., Ginzburg, D., and Shilo, M. (1970). The reproductive cycle of cyanophage LPP1-G in *Plectonema boryanum* and its dependence on photosynthetic and respiratory systems. *Virology* **40**, 514–521.

Padan, E., Shilo, M., and Oppenheim, A. B. (1972). Lysogeny of the blue-green alga *Plectonema boryanum* by LPP2-SPI cyanophage. *Virology* **47**, 525–526.

Pienaar, R. N. (1976). Virus-like particles in three species of phytoplankton from San Juan Island, Washington. *Phycologia* **15**, 185–190.

Pomeroy, L. R. (1974). The ocean's food web, a changing paradigm. *Bioscience* **24**, 499–504.

Pringle, C. R. (1999). Virus taxonomy, 1999: The universal system of virus taxonomy, updated to include the new proposals ratified by the International Committee on Taxonomy of Viruses during 1998. *Arch. Virol.* **144**, 421–429.

Proctor, L. M., and Fuhrman, J. A. (1990). Viral mortality of marine bacteria and cyanobacteria. *Nature* **343**, 60–62.

Proctor, L. M., Okubo, A., and Fuhrman, J. A. (1993). Calibrating estimates of phage-induced mortality in marine bacteria: Ultrastructural studies of marine bacteriophage development from one-step growth experiments. *Microb. Ecol.* **25**, 161–182.

Reisser, W. (1993). Viruses and virus-like particles of freshwater and marine eukaryotic algae: A review. *Arch. Protistenkd.* **143**, 257–265.

Rimon, A., and Oppenheim, A. B. (1974). Isolation and genetic mapping of temperature-sensitive mutants of cyanophage LPP2-SP1. *Virology* **62**, 567–569.

Rodda, K. M. (1996). Temporal and spatial dynamics of *Synechococcus* spp. and *Micromonas pusilla* host–viral systems. MSc Thesis, University of Texas at Austin.

Safferman, R. S. (1973). Phycoviruses. *In* "The Biology of Blue-Green Algae" (N. G. Carr and B. A. Whitton, eds.), pp. 214–237. University of California Press, Berkeley.

Safferman, R. S., and Morris, M. E. (1963). Algal virus: Isolation. *Science* **140**, 679–680.

Safferman, R. S., and Morris, M. E. (1964). Control of algae with viruses. *J. Am. Water Works Assoc.* **56**, 1217–1224.

Safferman, R. S., and Morris, M. E. (1967). Observations on the occurrence, distribution, and seasonal incidence of blue-green algal viruses. *Appl. Microbiol.* **15**, 1219–1222.

Safferman, R. S., Diener, T. O., Desjardins, P. R., and Morris, M. E. (1972). Isolation and characterization of AS-1, a phycovirus infecting the blue-green algae, *Anacystis nidulans* and *Synechococcus cedrorum. Virology* **47**, 105–113.

Safferman, R. S., Cannon, R. E., Desjardins, P. R., Gromov, B. V., Haselkorn, R., Sherman, L. A., and Shilo, M. (1983). Classification and nomenclature of viruses of cyanobacteria. *Intervirology* **19**, 61–66.

Sahlsten, E. (1998). Seasonal abundance in Skagerrak–Kattegat coastal waters and host specificity of viruses infecting the marine photosynthetic flagellate *Micromonas pusilla. Aquat. Microb. Ecol.* **16**, 103–108.

Sahlsten, E., and Karlson, B. (1998). Vertical distribution of virus-like particles (VLP) and viruses infecting *Micromonas pusilla* during late summer in the southeastern Skagerrak, North Atlantic. *J. Plankton Res.* **20**, 2207–2212.

Schopf, J. W., and Packer, B. M. (1987). Early Archaen (3.3 billion to 3.5 billion years old) microfossils from Warrawoona group, Australia. *Science* **137**, 70–73.

Sherman, L. A., and Brown Jr., R. M. (1978). Cyanophages and viruses of eukaryotic algae. *In* "Comprehensive Virology" (H. Fraenkel-Conrat and R. R. Wagner, eds.), Vol. 12, pp. 145–234. Plenum, New York.

Sherman, L. A., and Haselkorn, R. (1971). Growth of the blue-green algae virus LPP-1 under conditions which impair photosynthesis. *Virology* **45**, 739–746.

Sherman, L. A., Connelly, M., and Sherman, D. M. (1976). Infection of *Synechococcus cedrorum* by the cyanophage AS-1M, I: Ultrastructure of infection and phage assembly. *Virology* **71**, 1–16.

Shilo, M. (1971). Biological agents which cause lysis of blue-green algae. *Vehr. Int. Verein. Limnol.* **19**, 206–213.

Short, S. M., and Suttle, C. A. (1999). Use of the polymerase chain reaction and denaturing gradient gel electrophoresis to study diversity in natural virus communities. *Hydrobiologia* **401**, 19–33.

Sieburth, J. McN., Johnson, P. W., and Hargraves, P. E. (1988). Ultrastructure and ecology of *Aureococcus anophagofferens* Gen. Set. Sp. Nov. (Chrysophyceae): The dominant picoplankton during a bloom in Narragansett Bay, Rhode Island, Summer 1985. *J. Phycol.* **24**, 416–425.

Singh, R. N., and Singh, P. K. (1967). Isolation of cyanophages from India. *Nature* **216**, 1020–1021.

Skotnicki, A., Gibbs, A., and Wrigley, N. G. (1976). Further studies on *Chara corrallina* virus. *Virology* **75**, 457–468.

Snedden, J. W. (1985). Origin and sedimentary characteristics of discrete sand beds in modern sediments of the Central Texas continental shelf. PhD Dissertation, Louisiana State University, Baton Rouge.

Sode, K., Oozeki, M., Asakawa, K., Burgess, J. G., and Matsunaga, T. (1994). Isolation of a marine cyanophage infecting the marine unicellular cyanobacterium, *Synechococcus* sp. NKBG 042902. *J. Mar. Biotech.* **1**, 189–192.

Sode, K., Oonari, R., and Oozeki, M. (1997). Induction of a temperate marine cyanophage by heavy metal. *J. Mar. Biotech.* **5**, 178–180.

Stache-Crain, B., Müller, D. G., and Goff, L. J. (1997). Molecular systematics of *Ectocarpus* and *Kuckuckia* (Ectocarpales, Phaeophyceae) inferred from phylogenetic analysis of nuclear- and plastid-encoded DNA sequences. *J. Phycol.* **33**, 152–168.

Stewart, W. D. P., and Daft, M. J. (1977). Microbial pathogens of cyanophycean blooms. *In* "Advances in Aquatic Microbiology" (M. R. Droop and H. W. Jannasch, eds.), Vol. 1, pp. 177–218. Academic Press, New York.

Stockner, J. G., and Antia, N. J. (1986). Algal picoplankton from marine and freshwater ecosystems: Multidisciplinary perspective. *Can. J. Fish Aquat. Sci.* **43**, 2472–2503.

Stockner, J. G., and Shortreed, K. R. S. (1991). Autotrophic picoplankton: Community composition, abundance and distribution across a gradient of oligotrophic British Columbia and Yukon Territory lakes. *Int. Rev. Ges. Hydrobiol.* **76**, 581–601.

Suttle, C. A. (1992). Inhibition of photosynthesis in phytoplankton by the submicron size fraction concentrated from seawater. *Mar. Ecol. Prog. Ser.* **87**, 105–112.

Suttle, C. A. (1994). The significance of viruses to mortality in aquatic microbial communities. *Microb. Ecol.* **28**, 237–243.

Suttle, C. A. (2000). Cyanophages and their role in the ecology of cyanobacteria. *In* "The Ecology of Cyanobacteria: Their Diversity in Time and Space" (B. A. Whitton and M. Potts, eds.). Kluwer Academic, Boston. In press.

Suttle, C. A., and Chan, A. M. (1993). Marine cyanophages infecting oceanic and coastal strains of *Synechococcus*: Abundance, morphology, cross-infectivity and growth characteristics. *Mar. Ecol. Prog. Ser.* **92**, 99–109.

Suttle, C. A., and Chan, A. M. (1994). Dynamics and distribution of cyanophages and their effect on marine *Synechococcus* spp. *Appl. Environ. Microbiol.* **60**, 3167–3174.

Suttle, C. A., and Chan, A. M. (1995). Viruses infecting the marine Prymnesiophyte *Chrysochromulina* spp.: Isolation, preliminary characterization and natural abundance. *Mar. Ecol. Prog. Ser.* **118**, 275–282.

Suttle, C. A., and Chen, F. (1992). Mechanisms and rates of decay of marine viruses in seawater. *Appl. Environ. Microbiol.* **58**, 3721–3729.

Suttle, C. A., Chan, A. M., and Cottrell, M. T. (1990). Infection of phytoplankton by viruses and reduction of primary productivity. *Nature* **347**, 467–469.

Suttle, C. A., Chan, A. M., and Cottrell, M. T. (1991). Use of ultrafiltration to isolate viruses from seawater which are pathogens of marine phytoplankton. *Appl. Environ. Microbiol.* **57**, 721–726.

Suttle, C. A., Chan, A. M., Chen, F., and Garza, D. R. (1993). Cyanophages and sunlight: A paradox. *In* "Trends in Microbial Ecology" (R. Guerrero and C. Pedrós-Alió, eds.), pp. 303–307. Spanish Society of Microbiology, Barcelona.

Suttle, C. A., Chan, A. M., Rodda, K. M., Short, S. M., Weinbauer, M. G., Garza, D. R., and Wilhelm, S. W. (1996). The effect of cyanophages on *Synechococcus* spp. during a bloom in the western Gulf of Mexico. *Eos* **76** (Suppl.), OS207–OS208.

Thierstein, H. R., Geitzenauer, J. R., Molfino, B., and Shackelton, N. S. (1977). Global synchroneity of Late Quaternary coccolith datum levels: Validation by oxygen isotopes. *Geology* **5**, 400–404.

Turner, S. (1997). Molecular systematics of oxygenic photosynthetic bacteria. *Plant Sys. Evol.* (Suppl.) **11**, 13–52.

Van Etten, J. L. (1995). Giant *Chlorella* viruses. *Mol. Cells* **5**, 99–106.

Van Etten, J. L., and Ghabrial, S. A. (1991). Phycodnaviridae. *In* "Classification and Nomenclature of Viruses: Archives of Virology Supplement" (R. I. B. Francki, C. M. Fauguet, D. L. Knudson, and F. Brown, eds.), Vol. 2, pp. 137–139. Springer-Verlag, Vienna.

Van Etten, J. L., and Meints, R. H. (1999). Giant algal viruses. *Annu. Rev. Microbiol.* **53**, 447–494.

Van Etten, J. L., Burbank, D. E., Xia, Y., and Meints, R. H. (1983). Growth cycle of a virus, PBCV-1, that infects *Chlorella*-like algae. *Virology* **126**, 117–125.

Van Etten, J. L., Van Etten, C. H., Johnson, J. K., and Burbank, D. E. (1985a). A survey of viruses from fresh water that infect a eukaryotic Chlorella-like green alga. *Appl. Environ. Microbiol.* **49**, 1326–1328.

Van Etten, J. L., Burbank, D. E., Schuster, A. M., and Meints, R. H. (1985b). Lytic viruses infecting a *Chlorella*-like alga. *Virology* **140**, 135–143.

Van Etten, J. L., Lane, L. C., and Meints, R. H. (1991). Viruses and viruslike particles of eukaryotic algae. *Microbiol. Rev.* **55**, 586–620.

Waterbury, J. B., and Rippka, R. (1989). Subsection I. Order Chroococcales Wettstein 1924, emend. Rippka *et al.*, 1979. *In* "Bergey's Manual of Systematic Bacteriology" (J. T. Staley, M. P. Bryant, N. Pfennig, and J. B. Holt, eds.), pp. 1728–1746. Williams and Wilkins, Baltimore.

Waterbury, J. B., and Valois, F. W. (1993). Resistance to co-occurring phages enables marine *Synechococcus* communities to coexist with cyanophages abundant in seawater. *Appl. Environ. Microbiol.* **59**, 3393–3399.

Waters, R. E., and Chan, A. T. (1982). *Micromonas pusilla* virus: The virus growth cycle and associated physiological events within the host cells, host range mutation. *J. Gen. Virol.* **63**, 199–206.

Weinbauer, M. G., and Suttle, C. A. (1996). Potential significance of lysogeny to bacteriophage production and bacterial mortality in coastal waters of the Gulf of Mexico. *Appl. Environ. Microbiol.* **62**, 4374–4380.

Weinbauer, M. G., Wilhelm, S. W., Suttle, C. A., and Garza, D. R. (1997). Photoreactivation compensates for UV damage and restores infectivity to natural marine viral communities. *Appl. Environ. Microbiol.* **63**, 2200–2205.

Whitman, W. B., Coleman, D. C., and Wiebe, W. J. (1998). Prokaryotes: The unseen majority. *Proc. Natl. Acad. Sci. U.S.A.* **95**, 6578–6583.

Wiggins, B. A., and Alexander, M. (1985). Minimum bacterial density for bacteriophage replication: Implications for significance of bacteriophages in natural systems. *Appl. Environ. Microbiol.* **49**, 19–23.

Wilhelm, S. W., and Suttle, C. A. (1999). Viruses and nutrient cycles in the sea. *Bioscience* **49**, 781–788.

Wilhelm, S. W., Weinbauer, M. G., Suttle, C. A., Pledger, R. J., and Mitchell, D. L. (1998a). Measurements of DNA damage and photoreactivation imply that most viruses in marine surface waters are infective. *Aquat. Microb. Ecol.* **14**, 215-222.

Wilhelm, S. W., Weinbauer, M. G., Suttle, C. A., and Jeffrey, W. H. (1998b). The role of sunlight in the removal and repair of viruses in the sea. *Limnol. Oceanogr.* **43**, 586–592.

Williams, P. M., and Druffel, E. R. M. (1987). Radiocarbon in dissolved organic matter in the central North Pacific Ocean. *Nature* **330**, 246–248.

Wilmotte, A. (1994). Molecular evolution and taxonomy of the cyanobacteria. *In* "The Molecular Biology of Cyanobacteria" (D. A. Bryant, ed.), pp. 1–25. Kluwer Academic, Boston.

Wilson, W. H., and Mann, N. H. (1997). Lysogenic and lytic viral production in marine microbial communities. *Aquat. Microb. Ecol.* **13**, 95–100.

Wilson, W. H., Joint, I. R., Carr, N. G., and Mann, N. H. (1993). Isolation and molecular characterization of five marine cyanophages propagated on *Synechococcus* sp. strain WH7803. *Appl. Environ. Microbiol.* **59**, 3736–3743.

Wilson, W. H., Carr, N. G., and Mann, N. H. (1996). The effect of phosphate status on the kinetics of cyanophage infection in the oceanic cyanobacterium *Synechococcus* sp. WH7803. *J. Phycol.* **32**, 506–516.

Wilson, W. H., Turner, S., and Mann, N. H. (1998). Population dynamics of phytoplankton and viruses in a phosphate-limited mesocosm and their effect on DMSP and DMS production. *Estuar. Coast. Shelf Sci.* **46**, 45–59.

Woese, C. R. (1987). Bacterial evolution. *Microbiol. Rev.* **51**, 221–271.

Wommack, K. E., Hill, R. T., Muller, T. A., and Colwell, R. R. (1996). Effects of sunlight on bacteriophage viability and structure. *Appl. Environ. Microbiol.* **62**, 1336–1341.

Wu, J. H., Lewin, R. A., and Werbin, H. (1967). Photoreactivation of UV-irradiated blue-green alga virus LPP-1. *Virology* **31**, 657–664.

Zachary, A. (1976). Physiology and ecology of bacteriophages of the marine bacterium *Beneckea natriegens*: Salinity. *Appl. Environ. Microbiol.* **31**, 415–422.

Zachary, A. (1978). An ecological study of bacteriophages of *Vibrio natriegens*. *Can. J. Microbiol.* **24**, 321–324.

Zavarzina, N. B. (1961). A lytic agent in cultures of *Chlorella pyrenoidosa* Pringh. *Dokl. Akad. Nauk SSSR Ser. Biol.* **137**, 435–437.

Zavarzina, N. B. (1964). Lysis of *Chlorella* cultures in the absence of bacteria. *Mikrobiologiya* **33**, 561–564.

Zhang, Y., and Suttle, C. A. (1994). Design and use of PCR primers for B-family DNA polymerase genes to detect and identify viruses and microbes. *Abst. Amer. Soc. Limnol. Oceanogr., Miami, FL.*

Zhang, Y., Burbank, D. E., and Van Etten, J. L. (1988). *Chlorella* viruses isolated in China. *Appl. Environ. Microbiol.* **54**, 2170–2173.

Zingone, A. (1995). The role of viruses in the dynamics of phytoplankton blooms. *Giorn. Bot. Ital.* **129**, 415–423.

7

Viruses of Fungi and Protozoans: Is Everyone Sick?

JEREMY A. BRUENN

Department of Biological Sciences
State University of New York at Buffalo
Buffalo, New York 14260

Hauptman–Woodward Medical Research Institute
Buffalo, New York 14203

I. INTRODUCTION

A. The *Totiviruses*

There are two extreme strategies for survival pursued by pathogens: kill the host and spread rapidly, and coexist with the host, sometimes providing a selective advantage. Most viruses have lifestyles that lie between these two extremes. Among the RNA viruses, the double-stranded RNA (dsRNA) viruses of protozoans and fungi represent the extreme of coexistence. This group of viruses, with few exceptions, has no infectious phase. The viruses exist as permanent persistent

VIRAL ECOLOGY

infections, passing from cell to cell only by mating and cell division. In general, they seem to have no deleterious effects on their hosts, and in some cases they provide the host cells with a selective advantage, providing replication functions and encapsidation for satellite viruses that encode cellular toxins lethal to cells without the virus. In this respect, they have adopted a strategy similar to that of certain plasmids and DNA viruses of prokaryotes that encode toxins, restriction enzymes, or antibiotic resistance. This is the only example among RNA viruses of such a mutually beneficial symbiosis.

The *Totiviruses* comprise a subgroup of the fungal and protozoan viruses that is even more remarkable. With a single essential dsRNA, the *Totiviruses* are present in phyla separated by a billion years of evolution and are recognizably related to each other and to no other class of viruses (except possibly the partitiviruses of fungi). They have been discovered in at least nine genera of fungi (*Saccharomyces, Ustilago, Helminthosporium, Gaeumannomyces, Mycogone, Yarrowia, Aspergillus, Thielaviopsis,* and probably *Agaricus*) and at least four genera of protozoans (*Leishmania, Eimeria, Giardiavirus,* and *Trichomonas*) (Buck, 1986; Francki *et al.,* 1991; Ghabrial, 1994; Roditi *et al.,* 1994; Wang and Wang, 1991). The partitiviruses are also dsRNA viruses that exist solely as persistent infections. They have been discovered in an additional nine genera of fungi (Francki *et al.,* 1991). There are limited sequence data available for the partitiviruses, so their relationship to the *Totiviruses* remains unclear. In almost all genera of protozoans and fungi in which a systematic search has been made, there are representatives of these related dsRNA viruses. Only in *Candida* (which lacks any known sexual cycle) and *Neurospora* have serious attempts at finding resident dsRNA viruses failed.

The prototypical *Totivirus* is the *Saccharomyces cerevisiae* virus L1 (or LA), abbreviated ScVL1 (or ScVLA). This is the most extensively studied fungal virus, and in many respects its genomic organization and life cycle are typical of the fungal and protozoan dsRNA viruses.

B. The ScV System

1. Gene Expression

Most laboratory *Saccharomyces cerevisiae* (yeast) strains contain virus particles in the cytoplasm (the *Saccharomyces cerevisiae* virus, or ScV), in which segmented double-stranded RNAs (dsRNAs) are separately encapsidated, and there is a single essential dsRNA. The large essential dsRNA of one viral family, L1, is 4580 base pairs (Diamond *et al.,* 1989; Icho and Wickner, 1989) and encodes on the plus strand in reading frames that overlap by 130 bases a major capsid protein, Cap or P1 (Hopper *et al.,* 1977), and an RNA-dependent RNA polymerase, or Pol

(Diamond *et al.*, 1989; Fujimura and Wickner, 1988a; Pietras *et al.*, 1988), which are used by all dsRNAs of this family. The Pol protein is translated as a Cap–Pol fusion product by a frameshift event (Diamond *et al.*, 1989; Dinman *et al.*, 1991). The L1 Cap–Pol frameshift is similar to frameshifting in the retroviruses, both in site specificity and in RNA secondary and tertiary structure requirements (Dinman *et al.*, 1991; Tzeng *et al.*, 1992).

As with *Coronavirus* frameshifting (Brierley *et al.*, 1989) and retroviral frameshifts (Chamorro *et al.*, 1992), ScVL1 translational frameshifting requires a slippery site (GGGUUUA) and a 3′ pseudoknot (Dinman *et al.*, 1991; Tzeng *et al.*, 1992). The yeast system has been very useful in understanding translational frameshifting (Dinman, 1995). It was in this system that it was first shown that +1 translational frameshifting in eukaryotes is due to rare tRNAs (Belcourt and Farabaugh, 1990; Farabaugh *et al.*, 1993), and a number of nuclear genes affecting frameshifting efficiency have been identified in yeast (and in no other eukaryote) by two groups (Dinman and Wickner, 1994; Lee *et al.*, 1995). Some of these are now known to affect mRNA stability (Cui *et al.*, 1996). In the ScV system, our group first showed that ribosomes pause at a slippery site due to the presence of a downstream pseudoknot (Tu *et al.*, 1992).

The structure of the viral dsRNAs is outlined in Figure 1. A second dsRNA virus, ScVLa, is also present in most strains and has a single essential dsRNA of 4615 bp with a similar genomic organization and mode of expression (Park *et al.*, 1996b). The L1 family of RNAs are packaged exclusively in particles with the L1 Cap protein, due to the highly specific recognition of these RNAs by the L1 Cap–Pol protein at a viral binding site (VBS) located in each of the RNAs (Fujimura *et al.*, 1990; Shen and Bruenn, 1993; Yao and Bruenn, 1995; Yao *et al.*, 1995, 1997).

Replication and transcription in this system are of general interest, since the Pol protein shares conserved domains with the RNA-dependent RNA polymerases (RDRPs) of the plus-strand RNA viruses (Bruenn, 1991; Diamond *et al.*, 1989; Icho and Wickner, 1989; Pietras *et al.*, 1988). L1 is also known as L-A and La as L-BC. *In vivo* expression of the genome is apparently by translation of the entire plus strand (Haylock and Bevan, 1981), although there is one report of a 5′, but not a 3′, subgenomic fragment (Bostian *et al.*, 1983). There are no precedents for subgenomic mRNAs from dsRNA viral segments. Translation of the L1 plus strand synthesized by the nuclear DNA-dependent RNA polymerase (RNAP) from a cDNA expression vector is capable of producing the requisite viral gene products in appropriate proportions (Wickner *et al.*, 1991).

The dsRNAs of ScV replicate conservatively. As in reovirus, transcription within viral particles is followed by extrusion of the newly synthesized plus strand, which then interacts with cellular and viral proteins. The plus strand is packaged in viral particles after recognition by the Cap–Pol fusion protein, and replicated

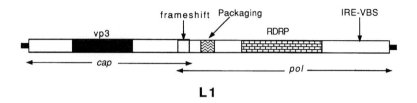

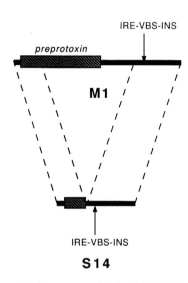

Fig. 1. Genome structure of L1, the one essential viral dsRNA. M1 = the toxin encoding dsRNA; S14 = a defective interfering dsRNA derived from M1 by internal deletion (dashed lines). Known coding regions are in italics. Arrows indicate *cis*-acting sites. Functional domains are shown in L1. VBS = viral binding site; IRE = internal replication enhancer; INS = interference sequence; vp3 refers to the region of L1 and La cap similar to vp3 of poliovirus.

by synthesis of minus strand, resulting in duplex formation (Bruenn, 1986; Fujimura *et al.*, 1986; Nemeroff and Bruenn, 1986; Williams and Leibowitz, 1987). Although some viral particles may contain more than one copy of a viral dsRNA when transcription is not followed by extrusion of the new plus strand (Esteban and Wickner, 1986, 1988), new viral particles are only formed as described. That is, if a viral dsRNA is small enough and its transcript is not extruded from a particle after transcription, ensuing replication of the transcript will result in a particle with two molecules of the same dsRNA. This process may continue until multiple copies of very small dsRNAs are present in some particles (Esteban and Wickner, 1988).

2. Satellite Viruses

In k1 killer strains there is a second dsRNA, M1, of about 1.9 kb, which encodes a secreted toxin that kills sensitive cells (Bostian *et al.*, 1980, 1984; Lolle and Bussey, 1986; Skipper *et al.*, 1984). The M1 preprotoxin functions as an immunity protein (Boone *et al.*, 1986; Hanes *et al.*, 1986). This provides a convenient phenotype for determining the effect of viral and host mutations on the viral life cycle. Suppressive sensitive mutants (Somers, 1973) of ScV contain L1 and smaller, or S, dsRNAs (Sweeny *et al.*, 1976; Vodkin, 1977). These are defective interfering versions of M1 that are derived from M1 by internal deletion, some-times followed by tandem duplication (Bruenn and Kane, 1978; Bruenn and Brennan, 1980; Fried and Fink, 1978; Huan *et al.*, 1991; Kane *et al.*, 1979; Lee *et al.*, 1986; Ridley and Wickner, 1983; Thiele *et al.*, 1984). DI segments of rota-viruses have a similar structure (Scott *et al.*, 1989). A second killer toxin, k2, is encoded by M2, a dsRNA that can also be packaged by L1 Cap (Dignard *et al.*, 1991; El-Sherbeini *et al.*, 1984). The sequences necessary for interference (INSs) are located within an 89-bp region that includes two sequences homologous to the IRE-VBS in L1 (Huan *et al.*, 1991).

There are chromosomal genes (MAK) required for M1 and S maintenance, but usually not for L1 maintenance (Wickner, 1986, 1989). At least some of these are also necessary for an L1 DI mutant (Esteban and Wickner, 1988). The viral binding site (VBS), internal replication enhancer (IRE) (Esteban *et al.*, 1989), and interference sequence (INS) (Huan *et al.*, 1991; Shen and Bruenn, 1993) are probably synonymous, and consist of a 19–31-bp sequence with defined primary and secondary structure (Shen and Bruenn, 1993; Yao and Bruenn, 1995; Yao *et al.*, 1995, 1997). The *cis*-acting sites of the viral dsRNAs are also outlined in Figure 1.

C. Host–Viral Interaction

ScV replication must be under direct or indirect host control, since the virus neither kills the cells nor fails to infect all progeny of infected cells. There are no DNA copies of the dsRNA genome present in host strains (Hastie *et al.*, 1978; Wickner and Leibowitz, 1977). ScV replication may (Sclafani and Fangman, 1984; Zakian *et al.*, 1981) or may not (Newman *et al.*, 1981) be confined to the G_1 phase of the cell cycle. Nothing is now known of the mechanism(s) of host regulation of ScV replication. However, there are many host genes known to affect maintenance of ScV: the MAK, SPE, SKI, KRB, MKT, and DET genes (Toh-e *et al.*, 1978; Toh-e and Wickner, 1979, 1980; Wickner, 1986, 1987; Wickner and Leibowitz, 1977). Most of the MAK genes are necessary for ScV-M1 but not for

ScV-L1. The host functions, but not the viral functions, of some of these are known: MAK7 encodes ribosomal protein L4 (Ohtake and Wickner, 1995a); MAK8 encodes ribosomal protein L3 (Wickner *et al.*, 1982), SPE2 encodes a product required for polyamine synthesis (Cohn *et al.*, 1978); and MAK1 encodes DNA topoisomerase I (Thrash *et al.*, 1984). Several of the MAK gene products (MAK3, MAK27, and MAK10) are necessary for maintenance of some ScVL1 viruses (El-Sherbeini and Bostian, 1987; Field *et al.*, 1982; Sommer and Wickner, 1982), as well as for maintenance of ScVM1. MAK3 codes for an *N*-acetyltransferase responsible for blocking the N-terminus of P1 (Cap), which is apparently necessary to produce functional P1 (Tercero *et al.*, 1993; Tercero and Wickner, 1992). Some 18 of the 30 known MAK genes affect the supply of free 60S ribosomal subunits, and hence the efficiency of translation of the uncapped un-polyadenylated L1 mRNA (Ohtake and Wickner, 1995b). ScV Cap de-caps cellular mRNAs, decoying the host degradation of uncapped mRNAs and thereby permitting translation of the uncapped viral mRNAs (Ohtake and Wickner, 1995b). Most of the interactions of ScV with host proteins appear to reflect the unusual requirements for translation of the uncapped unpolyadenylated viral RNAs, so it is not clear whether specific host functions are necessary for persistence. This was dramatized by the success in establishing brome mosaic virus as a persistent infection in yeast (Janda and Ahlquist, 1993; Quadt *et al.*, 1995). However, a host factor does appear to be accessory to ScV replication (Fujimura and Wickner, 1988b).

II. RELATIONSHIPS AMONG dsRNA VIRUSES OF LOWER EUKARYOTES

A. Genome Expression

The *Totiviruses*, by definition, and *Giardiaviruses* encode all viral proteins on one strand of a single dsRNA. However, they all make a capsid polypeptide without Pol domains and must consequently have a mechanism by which to produce two polypeptides from one mRNA. ScV does this by −1 translational frameshifting, as described above. So does the *Giardia lamblia* virus GlV (Wang *et al.*, 1993). However, *Leishmania* virus LrV1 and *Trichomonas* virus TvV use +1 translational frameshifting (Stuart *et al.*, 1992; Su and Tai, 1996; Tai and Ip, 1995). *Ustilago* virus UmVP1H1 appears to use proteolytic processing of a Cap–Pol precursor (Park *et al.*, unpublished), while *Helminthosporium* virus Hv190SV appears to use a "stop-and-go" translation (S. Ghabrial, personal communication) similar to that by which the chestnut blight fungus dsRNA virus produces its RDRP (Shapira *et al.*, 1991). Gene expression strategies in CyV and AbVL1 (see below) are not yet clear. In short, this family of closely related viruses uses a variety of strategies for gene expression — the end justifies the means.

B. The Polymerase Domain in the dsRNA Viruses of Lower Eukaryotes

In contrast to the viral RDRPs as a whole (Bruenn, 1991), the *Totivirus* RDRPs are easily aligned with statistical significance (Bruenn, 1993). Eight conserved motifs were identified in five *Totivirus* RDRPs (Bruenn, 1993). Five more sequences not included in this collection but shown here (Table I; see also Fig. 4) are also 19–22% identical overall with this set, and show highly significant alignments as evaluated by BLAST. AbV1L1 is a viral segment from a dsRNA mycovirus in *Agaricus bisporus* that may or may not be a *Totivirus* (Van der Lende *et al.*, 1996). There is no definitive evidence differentiating the possible presence of a number of *Totiviruses* and satellite RNAs (as in yeast or *Ustilago*) from the proposed multisegmented virus in *Agaricus bisporus* (Van der Lende *et al.*, 1996). The sequence of the L1 segment (used here) is incomplete, and the open reading frame may extend at least 300 bases further. This would give at least 500 amino acids prior to the first conserved motif (LIGRR, motif 1), so that the encoded protein could be a fusion protein combining Cap with Pol and providing for cleavage of the precursor to generate a separate Cap, as appears to be the case with UmVP1H1 (Park *et al.*, unpublished). CyV is an incomplete sequence derived (by RT-PCR) from a single-segmented dsRNA virus that was thought to be a plant virus (Coffin and Coutts, 1995), but it is probably a fungal virus contaminant (R. H. A. Coutts, personal communication). Another protozoan dsRNA virus that may be a *Totivirus* has been isolated from *Eimeria nieschulzi* and a partial sequence obtained from the region around motif 5 (Roditi *et al.*, 1994). The tertiary structure of the polio RDRP suggests that motifs 3–7 are in the active site of the enzyme (S. Schultz, personal communication).

Figure 2 plots percentage identity across the entire RDRP alignment. Conserved motifs show up as peaks in identity, of which eight are numbered consecutively. The sequences of the conserved motifs and their positions in the proteins are shown in Table I. Although polio polymerase cannot be adequately aligned with the *Totivirus* RDRPs, it is possible to pick out all the conserved motifs of the *Totiviruses* in polio (with the possible exception of motifs 1, 2, and 8). Although it is clear that motifs 4, 5, and 6 are properly identified in polio, identification of motifs 7 and 8 from the structural determination will require careful consideration of all known polymerase structures (S. Schultz, personal communication).

TABLE I

Conserved Motifs in RDRPs of *Totiviruses* Compared to Polio[a]

RDRP	1	2	3	4	5	6	7	8
LRV1	LLGRG 59	WAAN.GS.HS 44	KLEH..GK..TRLLL 57	DYDDFNSQHT 46	TLMSGHRATSFINSVLNRA.YI 11	HVGDDILM 33	EFLRV 9	YLAR
TvV	LLGRG 58	WSKS.GS.HY 41	KLEH..GK..ERFIY 50	DYTDFNSQHT 43	TLPSGHRATTFINPVLNWC.YT 11	CAGDDVIL 31	EFLRK 9	YPCR
EnV					TLMSGHRGTSFINSVLNAA.YV			
Hv190	LQGRY 64	WCVN.GSQNA 46	KLEN.GK..DRAIF 56	DYDNFNSQHS 50	TLMSGHRATTFTNSVLNAA.YI 14	HAGDDVYL 35	EFLRL 9	YLCR
ScVL1	LMNRG 57	WVPG.GSVHS 46	KYEW..GK..QRAIY 52	DYDDFNSQHS 52	TLLSGWRLTTFMNTVLNWA.YM 15	HNGDDVMI 33	EFLRV 13	YLSR
ScVLa	LENGV 58	IMPG.GSVHS 46	KYEW..GK..VRALY 51	DFDDFNSQHS 52	TLFSGWRLTTFENTALNYC.YL 13	HNGDDVFA 33	EFLRV 11	YLTR
UmVH1	LYGRG 66	WLVS.GSSAG 55	KLNETGGK..ARAIY 55	DYPDFNSMHT 63	GLYSGDRDTTLINTLLNIA.YA 20	CHGDDIIT 34	EYLRI 10	CLAR
UmVH2	PFNRV 59	RMPT.GSTVS	KYEW..GK..QRAIY					
CyV		WLSS.GSAAG 58	KYEV..GK..PRALY 58	DYADFNVQIR 54	GMFSGTKATDLLNTLLNKA.YF 24	HQGDDVWI 35	EYLRI 10	YLQR
AbV1L	LIGRR 79	ADPSAGELLT 44	KHEV..GKNASRSLW 63	DYANFNEQHS 54	GLLSGWRCTAYINNLINIAQYE 21	TGGDDGCA 35	EFFRL 10	SVIR
	*	*	* **	** ** **	* ** *** * *** *	***	*** *	*
GlV	LLGKV 65	WGTT.GSGYI 37	KPEL..TK..VRAVI 55	DQSNFDRQPD 59	GLPSGWKWTALLGALINT.QLL 16	VQGDDIAL 33	EFLRR 9	YPAR
Polio	VFEGV 64	YGTD.GLEAL 49	KDEL.RSK..TKVEQ 64	DYTGYDASLS 42	GMPSGCSGTSIFNSMINNL.II 18	AYGDDVIA 38	TFLKR 9	FLIH

[a] The amino-acid sequences of the conserved motifs identified in the multiple sequence alignment of Figure 2 are shown in the standard single-letter amino-acid code. Periods indicate where insertions have been made to maximize alignment. Some sequences (e.g., EnV and UmVH2) are only partially determined in this region. Spacing between motifs (in amino acids) is indicated for each RDRP. Asterisks indicate residues identical in all *Totivirus* RDRPs. The numbers 1–8 indicate the conserved motifs. This alignment includes a region of about 400 amino acids (Fig. 2).

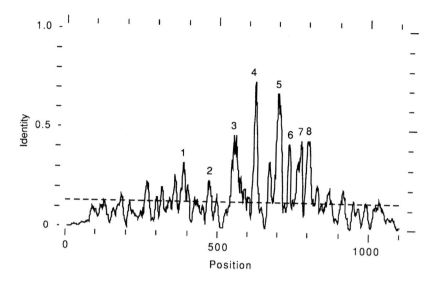

Fig. 2. PLOTSIMILARITY diagram for the *Totivirus* RDRPs aligned by PILEUP (Genetics Computer Group, 1991). Motifs 1–8 are identified as peaks in similarity. This is modified from a previous, similar figure (Bruenn, 1993) by inclusion of three more sequences.

A number of studies have been done on mutations within viral RDRPs that generally confirm that the highly conserved motifs (such as 5 and 6 below) are required for enzyme function (Mills *et al.*, 1988; Ribas and Wickner, 1992). Alanine substitutions have demonstrated the importance of motifs 3, 5, and 6 for replication or transcription in ScV (Ribas *et al.*, 1994; Ribas and Wickner, 1992). Recently, the structure of most of the polio RDRP has been determined by X-ray diffraction, and its structure is quite similar to that of the DNA-dependent RNA polymerases, the DNA polymerases, and the reverse transcriptases (Kohlstaedt *et al.*, 1992; Schultz *et al.*, 1996). The five conserved motifs (3–7) appear in the "palm" of the polymerase (Hansen *et al.*, 1997).

GlV, the *Giardia lamblia* virus, is distinctly different from the *Totiviruses*, although it is still close enough for accurate alignment. This is interesting, since it alone among the known protozoan dsRNA viruses is infectious to its host cells. Substitution of GlV motifs for ScVL1 motifs in the ScV RDRP demonstrates that motifs 3–7 are part of the polymerase function and that motifs 1, 2, and 8 have other functions (Routhier and Bruenn, 1998). As expected from previous results (Fujimura *et al.*, 1990, 1992; Ribas *et al.*, 1994; Yao *et al.*, 1995), packaging is independent of motifs 1–8: all of these constructs gave normal packaging of VBS containing transcripts in a two-plasmid assay (Routhier and Bruenn, 1998). Replication and/or transcription of ScV RNAs strongly requires the region of Pol represented in all the RNA-dependent RNA polymerases (motifs 3–7) (Routhier and Bruenn, 1998).

The mapping of a cryptic RNA-binding domain between conserved motifs 2 and 4, which includes conserved motif 3 (Ribas *et al.*, 1994), is probably indicative of a universal function of the polymerase, perhaps a conformational change during RNA translocation. One mutant tested by Wickner is a change of conserved motif 3 from KYEWGKQRAIY to AAEWGAQAAIY. This change obliterates support of ScVM1 (Ribas *et al.*, 1994), consistent with GlV substitutions in this region (Routhier and Bruenn, 1998). This supports extension of the essential RDRP domain at least to motif 3.

A small region of the UmVP6H1 segment has been sequenced (C. M. Park and J. A. Bruenn, unpublished manuscript), and its comparison to UmVP1H1 shows only two conservative changes in 162 amino acids (total sequenced region of Pol in UmVP6H1). Over the sequenced region of UmVP1H2, UmVP1H2 is on average 34% identical to ScVLa and only 21.5% identical to UmVP1H1. In these same regions, ScVL1 and ScVLa are only 30.1% identical, so that UmVP1H2 and ScVLa are closer to each other than either is to a second virus present in the same cells. This is clear evidence of a common origin of UmVP1H2 and ScVLa prior to evolutionary separation of *Ustilago* and *Saccharomyces*. Similarly, the protozoan viruses EnV and LrV1, rather than the other fungal *Totiviruses*, are the closest relatives of fungal *Totivirus* Hv190SV. This relationship is confirmed by a similar analysis of the capsid polypeptide sequences; the capsid polypeptide of Hv190SV is clearly related to that of LrV1 (Fig. 3), but not to any of the other sequenced capsid polypeptides among the *Totiviruses* (not shown).

A phylogenetic analysis of motifs 2–8, combined with analysis of the sequenced regions of UmVP1H2 and EnV, gives a tree that summarizes these relationships (Fig. 4). Clearly, viruses present in the same cells may be much more closely related to viruses in cells evolutionarily separated by billions of years than they are to each other (e.g., UmVP1H2 to ScVLa and EnV to Hv190SV). Since these are noninfectious viruses, this can only be interpreted to mean that they arose very early in evolution and that their sequences have been well preserved for a very long time.

III. SATELLITE VIRUSES: KILLER TOXINS

Some of the dsRNA viruses of lower eukaryotes, like some plasmids and bacteriophages in the prokaryotes, confer a selective advantage on their hosts. In several genera of fungi, all of which are yeasts or have a yeastlike (nonfilamen-

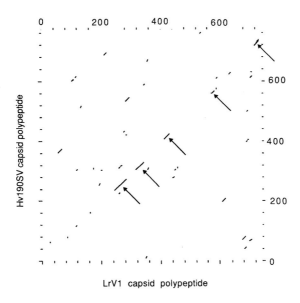

Fig. 3.   Comparison of the Hv190SV capsid polypeptide sequence with that of Lrv1 by the GCG program COMPARE and plotted by the program DOTPLOT, using the default matrix, a window of 20 amino acids, and a stringency of 16. Arrows indicate regions of significant similarity.

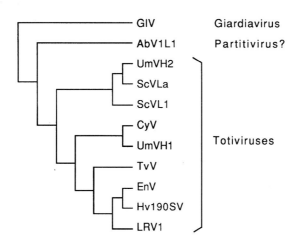

Fig. 4.   Phylogenetic tree of protozoan and fungal viruses with known RDRP sequences. Derived from motifs 2–8 and sequenced regions of UmVH2 and EnV by the protein parsimony algorithm of the PHYLIP programs (Felsenstein, 1989).

tous) mode of growth, satellite dsRNA viruses may encode secreted protein toxins that kill sensitive cells. In some cases (e.g., the *S. cerevisiae* k1 and k2 killer systems), the satellite virus itself provides the host cell with immunity to the toxin. This is the result of a function of the preprotoxin in the case of k1 (Boone *et al.*, 1986). In some cases — for instance, with the *U. maydis* killer toxins — nuclear mutations are required to provide the host with resistance to the toxin (Finkler *et al.*, 1992; Ginzberg and Koltin, 1994; Koltin and Kandel, 1978; Koltin *et al.*, 1978). There are at least five genera of fungi in which such toxins have been well documented (Table II). Most of these are encoded by satellite dsRNA viruses, but some are encoded by DNA plasmids (*Kluyveromyces*) or by the nuclear genome (*Williopsis*). Most are processed by homologues of the Golgi enzymes (kex2p and kex1p) responsible for processing of serum proteins in mammals (Bloomquist *et al.*, 1991; Brenner and Fuller, 1992; Fuller *et al.*, 1988, 1989; Van de Ven *et al.*, 1991; Zhu *et al.*, 1992).

The origin of the toxin genes encoded by dsRNA satellite viruses is probably the nuclear genome of the host. Two lines of evidence justify this conclusion. First, the dsRNA genome of the k1 killer toxin gene (M1) in *S. cerevisiae* includes an internal polyA sequence of variable length (Hannig *et al.*, 1986). The origin of this sequence is not viral: none of the dsRNA viruses have polyadenylated RNAs, although the viral polymerase commonly adds one nontemplated A or G residue at the 3′ end of its transcripts, both plus and minus strands (Brennan *et al.*, 1981; Diamond *et al.*, 1989). The location of this sequence in M1 is highly significant: it comes immediately after the coding sequence for the preprotoxin (Hannig *et al.*, 1986) and before the 3′ region necessary for packaging of viral plus strands (Fujimura *et al.*, 1990; Huan *et al.*, 1991; Shen and Bruenn, 1993; Yao *et al.*, 1995, 1997). The most likely explanation for this structure is that the viral replicase switched from its normal template to an erroneously packaged cellular mRNA after copying the several hundred 3′ bases of L1. Subsequent selection may have resulted in duplication of the 3′ packaging signal in M1 (Shen and Bruenn, 1993) and the addition of the 5′ terminal GAAAAA RNA sequence to the M1 plus strand, as well as loss of all L1 sequences not essential for packaging. The other two dsRNA-encoded killer toxins in *S. cerevisiae* (k2 and k28) also have internal polyA sequences in their genomic dsRNAs (Dignard *et al.*, 1991; Hannig *et al.*, 1984; Schmitt, 1995; Schmitt and Tipper, 1990; Skipper, 1983), so that it is possible that all arose from one initial miscopying event.

Many of the killer toxins may have arisen from a common ancestor. Most of the killer toxins have two polypeptides (α and β) that may be separate (*U. maydis* KP6), noncovalently associated in solution (*Pichia farinosa* KK1), or linked by intermolecular disulfides (*S. cerevisiae* k1). The exceptional toxin, which is a single polypeptide (*U. maydis* KP4), appears from its structure to be a tandem duplication of one sequence that has subsequently diverged (Gu *et al.*, 1995), and

TABLE II

Known Killer Toxins: Sequence, Structure, Function, and Coding Genomes

Organism	Type	Peptide	Size (aa)	Genome	Mechanism of action	Processing	3D structure	References
Saccharomyces cerevisiae	k1	α	103	dsRNA	Channel	kex2p, kex1p	no	Bostian et al., 1984; Martinac et al., 1990; Skipper et al., 1984
S. cerevisiae	k2	β	83	dsRNA	?	kex2p, kex1p	no	Dignard et al., 1991
S. cerevisiae	k28	α	178	dsRNA	?	kex2p, kex1p	no	Schmitt, 1995
		β	159	dsRNA				
		β	100	dsRNA				
Ustilago maydis	KP1	β	117	dsRNA	?	kex2p, kex1p	no	Park et al., 1996a
U. maydis	KP4	β	105	dsRNA	Ca^{2+} inh.	—	α–β sandwich	Gu et al., 1995; Park et al., 1994
U. maydis	KP6	α	79	dsRNA	Channel	kex2p, kex1p	4-strand β/2α	Li et al., 1997; Tao et al., 1990
Williopsis mrakii	WmKT	β	81	dsRNA	β-glucan inh.	kex2p	like KP6α? 2×4 strand β sheet	Antuch et al., 1996; Kimura et al., 1993
(*Hansenula mraki*)	(HMK)		88	DNA				
Pichia farinosa	KK1	α	63	DNA	?	kex2p, kex1p	like KP4	Kashiwagi et al., 1997; Suzuki and Nikkuni, 1994
Pichia inositovora	NRRL18709	β	77	DNA	?	?	no	Hayman and Bolen, 1991
Pichia kluyveri		?	DNA plas.	?	Channel	?	no	Kagan, 1983
		?	?	?				
Kluyveromyces lactis		α	865	DNA plas.	?	kex2p, kex1p	no	Stark et al., 1990; Stark and Boyd, 1986
		β	252	DNA plas.	?	kex2p, kex1p	no	Stark et al., 1990; Stark and Boyd, 1986
		γ	229	DNA plas.	?	kex2p, kex1p	no	Stark et al., 1990; Stark and Boyd, 1986

Three-dimensional structures determined by NMR or X-ray crystallography (3D structure) are abbreviated α for alpha helix and β for beta sheet. "No" indicates no structural determination has been done. In the "mechanism of action" column, "inh." indicates inhibition. The "processing" column lists the enzymes (other than signal peptidase) known or suspected to be involved in processing of the preprotoxin.

it could easily have arisen from two polypeptides processed by kex2p by a single deletion of the genomic sequence encoding the intervening sequence and the cleavage sites in the prepropolypeptide. This model is supported by the tertiary sequence identity between KP4 and the KK1 toxin, in which α and β together have the structure of the single polypeptide in KP4 (Kashiwagi et al., 1997). Models for KP6 β assume that it arose by a tandem duplication of the ancestral form of the KP6 α sequence followed by divergence as well (Li et al., 1997). Further structural analysis of killer toxins may help derive an evolutionary tree for them.

The second line of evidence indicating that the dsRNA killer toxins arose from nuclear genes is the relationship between the U. maydis KP4 killer toxin and the Pichia farinosa KK1 toxin. These have no evident primary sequence similarity, but their tertiary structures are essentially identical (Gu et al., 1995; Kashiwagi et al., 1997; Park et al., 1994; Suzuki and Nikkuni, 1994), even though KP4 is a single polypeptide and KK1 is two polypeptides. The KP4 toxin is encoded by a satellite dsRNA virus in U. maydis and the KK1 toxin by the nuclear genome of P. farinosa. The obvious explanation for this data is that the two toxins retain the same function, in which a few key residues inhibit the Ca^{2+} channel when held in appropriate position by the protein scaffolding, so that only the tertiary structure of the toxin is conserved by evolution. Since one of these toxins is encoded by a nuclear gene, it may be this gene that is the prototype for all the killer toxins.

The existence of dsRNA segments encoding killer toxins may partially explain the prevalence of endogenous dsRNA viruses in the fungi, since these segments clearly provide a selective advantage to the cells harboring their parental viruses. However, the majority of isolates of fungi in which killer toxins have been discovered lack the dsRNAs encoding the toxins but still have the parental viruses.

IV. OTHER VIRAL SYMBIONTS IN FUNGI

In addition to the *Totiviruses* and *Partitiviruses*, fungi are replete with a number of other viruslike symbionts. The hypovirulence-associated dsRNA virus of chestnut blight fungus is an endosymbiont like the *Totiviruses* but has no capsid polypeptide, instead being enclosed in a membrane (Koonin et al., 1991; Shapira et al., 1991). There are two single-stranded RNA elements in *S. cerevisiae* encoding their own RDRPs but without any capsid or membrane (Garcia-Cuellar et al., 1995; Matsumoto et al., 1990; Matsumoto and Wickner, 1991; Widner et al., 1991). None of these elements is related to the *Totiviruses* as judged by genome structure, expression, or RDRP sequence, but they are related to other viral families (Koonin et al., 1991). Similarly, there are numerous retrotransposons in the fungi, the best characterized of which is the Ty element. Ty encodes a single capsid polypeptide and a reverse transcriptase (Clare and Farabaugh, 1985); its single-stranded RNA genome is encapsidated within *S. cerevisiae* cells; and its reverse transcriptase, which is related to those of the retroviruses, is synthesized by a +1 translational frameshift followed by proteolytic processing (Belcourt and Farabaugh, 1990; Farabaugh, 1996; Farabaugh *et al.*, 1993).

This collection of viruses and degenerate viruses testifies to the ubiquity of viruses present very early in evolution, some of which appear to have been trapped in the fungi and protozoans, where the only sure strategy for survival is to become permanent residents of host cells.

In summary, most of the fungi and protozoans may have dsRNA virus symbionts, and this symbiosis predates differentiation of the single-celled organisms, so that it is a symbiosis of very ancient origin.

REFERENCES

Antuch, W., Guntert, P. and Wuthrich, K. (1996). Ancestral βγ-crystallin precursor structure in a yeast killer toxin. *Nature Struct. Biol.* **3**, 662–665.

Belcourt, M. F., and Farabaugh, P. J. (1990). Ribosomal frameshifting in the yeast retrotransposon Ty: tRNAs induce slippage on a 7 nucleotide minimal site. *Cell* **62**, 339–352.

Bloomquist, B. T., Eipper, B. A., and Mains, R. E. (1991). Prohormone-converting enzymes: Regulation and evaluation of function using antisense RNA. *Mol. Endocrinol.* **5**, 2014–2024.

Boone, C., Bussey, H., Greene, D., Thomas, D. Y., and Vernet, T. (1986). Yeast killer toxin: Site-directed mutations implicate the precursor protein as the immunity component. *Cell* **46**, 105–113.

Bostian, K. A., Hopper, J. E., Rogers, D. T., and Tipper, D. J. (1980). Translational analysis of the killer-associated virus-like particle dsRNA genome of *S. cerevisiae*: M dsRNA encodes toxin. *Cell* **19**, 403–414.

Bostian, K. A., Burn, V. E., Jayachandran, S., and Tipper, D. J. (1983). Yeast killer dsRNA plasmids are transcribed in vivo to produce full and partial-length plus-stranded RNAs: Models for their synthesis and for the functional sequence of M1-dsRNA. *Nucleic Acids Res.* **11**, 1077–1097.

Bostian, K. A., Elliott, Q., Bussey, H., Burn, V., Smith, A., and Tipper, D. J. (1984). Sequence of the preprotoxin dsRNA gene of type 1 killer yeast: Multiple processing events produce a two-component toxin. *Cell* **36**, 741–751.

Brennan, V. E., Field, L., Cizdziel, P., and Bruenn, J. A. (1981). Sequences at the 3' ends of yeast viral dsRNAs: Proposed transcriptase and replicase initiation sites. *Nucleic Acids Res.* **25**, 4007–4021.

Brenner, C., and Fuller, R. S. (1992). Structural and enzymatic characterization of a purified prohormone-processing enzyme: Secreted, soluble Kex2 protease. *Proc. Natl. Acad. Sci. U.S.A.* **89**, 922–926.

Brierley, I., Digard, P., and Inglis, S. C. (1989). Characterization of an efficient coronavirus ribosomal frameshifting signal: Requirement for an RNA pseudoknot. *Cell* **57**, 537–547.

Bruenn, J. A. (1986). The killer systems of *Saccharomyces cerevisiae* and other yeasts. *In* "Fungal Virology" (K. W. Buck, ed.), pp. 85–108. CRC Press, Boca Raton, FL.

Bruenn, J. A. (1991). Relationships among the positive-strand and double-strand RNA viruses as viewed through their RNA-dependent RNA polymerases. *Nucleic Acids Res.* **19**, 217–226.

Bruenn, J. A. (1993). A closely related group of RNA-dependent RNA polymerases from double-stranded RNA viruses. *Nucleic Acids Res.* **21**, 5667–5669.

Bruenn, J. A., and Brennan, V. E. (1980). Yeast viral double-stranded RNAs have heterogeneous 3′ termini. *Cell* **19**, 923–933.

Bruenn, J., and Kane, W. (1978). Relatedness of the double-stranded RNAs present in yeast virus-like particles. *J. Virol.* **26**, 762–772.

Bruenn, J. A., Diamond, M. E., and Dowhanick, J. J. (1989). Similarity between the picornavirus VP3 capsid polypeptide and the *Saccharomyces cerevisiae* virus capsid polypeptide. *Nucleic Acids Res.* **17**, 7487–7493.

Buck, K. W. (1986). Fungal virology: An overview. In "Fungal Virology" (K. W. Buck, ed.), pp. 1–84. CRC Press, Boca Raton, FL.

Chamorro, M., Parkin, N., and Varmus, H. E. (1992). An RNA pseudoknot and an optimal heptameric shift site are required for highly efficient ribosomal frameshifting on a retroviral messenger RNA. *Proc. Natl. Acad. Sci. U.S.A.* **89**, 713–717.

Clare, J., and Farabaugh, P. J. (1985). Nucleotide sequence of a yeast Ty element: Evidence for an unusual mechanism of gene expression. *Proc. Natl. Acad. Sci. U.S.A.* **82**, 2829–2833.

Coffin, R. S., and Coutts, R. H. A. (1995). Relationships among *Trialeurodes vaporariorum*-transmitted yellowing viruses from Europe and North America. *J. Phytopathol.* **143**, 375–380.

Cohn, M. S., Tabor, C. W., Tabor, H., and Wickner, R. B. (1978). Spermidine or spermine requirement for killer double-stranded RNA plasmid replication in yeast. *J. Biol. Chem.* **253**, 5225–5227.

Cui, Y., Dinman, J. D., and Peltz, S. W. (1996). Mof4-1 is an allele of the UPF1/IFS2 gene which affects both mRNA turnover and –1 ribosomal frameshifting efficiency. *EMBO J.* **15**, 5726–5736.

Diamond, M. E., Dowhanick, J. J., Nemeroff, M. E., Pietras, D. F., Tu, C.-L., and Bruenn, J. A. (1989). Overlapping genes in a yeast dsRNA virus. *J. Virol.* **63**, 3983–3990.

Dignard, D., Whiteway, M., Germain, D., Tessier, D., and Thomas, D. Y. (1991). Expression in yeast of a cDNA copy of the K2 killer toxin gene. *Mol. Gen. Genet.* **227**, 127–136.

Dinman, J. D. (1995). Ribosomal frameshifting in yeast viruses. *Yeast* **11**, 1115–1127.

Dinman, J. D., and Wickner, R. B. (1994). Translational maintenance of frame: Mutants of *Saccharomyces cerevisiae* with altered –1 ribosomal frameshifting efficiencies. *Genetics* **136**, 75–86.

Dinman, J. D., Icho, T., and Wickner, R. B. (1991). A –1 ribosomal frameshift in a double-stranded RNA virus of yeast forms a *gag–pol* fusion protein. *Proc. Natl. Acad. Sci. U.S.A.* **88**, 174–178.

El-Sherbeini, M., and Bostian, K. A. (1987). Viruses in fungi: Infection of yeast with the k1 and k2 dsRNA killer viruses. *Proc. Natl. Acad. Sci. U.S.A.* **84**, 4293–4297.

El-Sherbeini, M., Tipper, D. J., Mitchell, D. J., and Bostian, K. A. (1984). Virus-like particle capsid proteins encoded by different L-dsRNAs of *S. cerevisiae*: Their roles in maintenance of M-dsRNA killer plasmids. *Mol. Cell. Biol.* **4**, 2818–2827.

Esteban, R., and Wickner, R. B. (1986). Three different M1 RNA-containing viruslike particle types in *Saccharomyces cerevisiae*: In vitro M1 double-stranded RNA synthesis. *Mol. Cell. Biol.* **6**, 1552–1561.

Esteban, R., and Wickner, R. B. (1988). A deletion mutant of L–A double-stranded RNA replicates like M1 double-stranded RNA. *J. Virol.* **62**, 1278–1285.

Esteban, R., Fujimura, T., and Wickner, R. B. (1989). Internal and terminal *cis*-acting sites are necessary for in vitro replication of the L-A double-stranded RNA virus of yeast. *EMBO J.* **8**, 947–954.

Farabaugh, P. J. (1996). Programmed translational frameshifting. *Microbiol. Rev.* **60**, 103–134.

Farabaugh, P. J., Zhao, H., and Vimaladithan, A. (1993). A novel programmed frameshift expresses the *POL3* gene of retrotransposon Ty3 of yeast: Frameshifting without tRNA slippage. *Cell* **74**, 93–103.

Felsenstein, J. (1989). Cladistics: PHYLIP — Phylogeny Inference Package (Version 3.2). *Cladistics* **5**, 164–166.

Field, L. J., Bobek, L., Brennan, V., Reilly, J. D., and Bruenn, J. (1982). There are at least two yeast viral dsRNAs of the same size: An explanation for viral exclusion. *Cell* **31**, 193–120.

Finkler, A., Peery, T., Tao, J., Bruenn, J. A., and Koltin, Y. (1992). Immunity and resistance to the KP6 toxin of *Ustilago maydis*. *Mol. Gen. Genet.* **233**, 393–403.

Francki, R. I. B., Fauquet, C. M., Knudson, D. L., and Brown, F. (1991). "Classification and Nomenclature of Viruses: Fifth Report of the International Committee on Taxonomy of Viruses." Springer-Verlag, Vienna.

Fried, H. M., and Fink, G. R. (1978). Electron microscopic heteroduplex analysis of "killer" double-stranded RNA species from yeast. *Proc. Natl. Acad. Sci. U.S.A.* **75**, 4224–4228.

Fujimura, T., and Wickner, R. B. (1988a). Gene overlap results in a viral protein having an RNA binding domain and a major coat protein domain. *Cell* **55**, 663–671.

Fujimura, T., and Wickner, R. B. (1988b). Replicase of L–A virus-like particles of *Saccharomyces cerevisiae*: In vitro conversion of exogenous L-A and M1 single-stranded RNAs to double-stranded form. *J. Biol. Chem.* **263**, 454–460.

Fujimura, T., Esteban, R., and Wickner, R. B. (1986). In vitro L–A double-stranded RNA synthesis in virus-like particles from *Saccharomyces cerevisiae*. *Proc. Natl. Acad. Sci. U.S.A.* **83**, 4433–4437.

Fujimura, T., Esteban, R., Esteban, L. M., and Wickner, R. B. (1990). Portable encapsidation signal of the L–A double-stranded RNA virus of *S. cerevisiae*. *Cell* **62**, 819–828.

Fujimura, T., Ribas, J. C., Makhov, A. M., and Wickner, R. B. (1992). Pol of *gag–pol* fusion protein required for encapsidation of viral RNA of yeast L–A virus. *Nature* **359**, 746–749.

Fuller, R. S., Sterne, R. E., and Thorner, J. (1988). Enzymes required for yeast prohormone processing. *Annu. Rev. Physiol.* **50**, 345–362.

Fuller, R. S., Brake, A., and Thorner, J. (1989). Yeast prohormone processing enzyme (KEX2 gene product) is a Ca^{2+}-dependent serine protease. *Proc. Natl. Acad. Sci. U.S.A.* **86**, 1434–1438.

Garcia-Cuellar, M. P., Esteban, L. M., Fujimura, T., Rodriguez-Cousino, N., and Esteban, R. (1995). Yeast viral 20 S RNA is associated with its cognate RNA-dependent RNA polymerase. *J. Biol. Chem.* **270**, 20084–20089.

Ghabrial, S. A. (1994). New developments in fungal virology. *Adv. Virus Res.* **43**, 303–388.

Genetics Computer Group (1991). "Program Manual for the GCG Package." Version 7. GCG, Madison, WI.

Ginzberg, I., and Koltin, Y. (1994). Interaction of the *Ustilago maydis* KP6 killer toxin with sensitive cells. *Microbiology* **140**, 643–649.

Gu, F., Khimani, A., Rane, S., Flurkey, W. H., Bozarth, R. F., and Smith, T. J. (1995). Structure and function of a virally encoded fungal toxin from *Ustilago maydis*: A fungal and mammalian Ca^{2+} channel inhibitor. *Structure* **3**, 805–814.

Hanes, S. D., Burn, V. E., Sturley, S. L., Tipper, D. J., and Bostian, K. A. (1986). Expression of a cDNA derived from the yeast killer preprotoxin gene: Implications for processing and immunity. *Proc. Natl. Acad. Sci U.S.A.* **83**, 1675–1679.

Hannig, E. M., Thiele, D. J., and Leibowitz, M. J. (1984). *Saccharomyces cerevisiae* killer virus transcripts contain template-coded polyadenylate tracts. *Mol. Cell. Biol.* **4**, 101–109.

Hannig, E. M., Williams, T. L., and Leibowitz, M. J. (1986). The internal polyadenylate tract of yeast killer virus M1 double-stranded RNA is variable in length. *Virology* **152**, 149–158.

Hansen, J. L., Long, A. M., and Schultz, S. C. (1997). Structure of the RNA-dependent RNA polymerase of poliovirus. *Structure* **5**, 1109–1122.

Hastie, N. D., Brennan, V., and Bruenn, J. (1978). No homology between double-stranded RNA and nuclear DNA of yeast. *J. Virol.* **28**, 1002–1005.

Haylock, R. W., and Bevan, E. A. (1981). Characterization of the L dsRNA encoded mRNA of yeast. *Curr. Genet.* **4**, 181–186.

Hayman, G. T., and Bolen, P. L. (1991). Linear DNA plasmids of *Pichia inositovora* are associated with a novel killer toxin activity. *Curr. Genet.* **19**, 389–393.

Hopper, J. E., Bostian, K. A., Rowe, L. B., and Tipper, D. J. (1977). Translation of the L-species dsRNA genome of the killer-associated virus-like particles of *Saccharomyces cerevisiae*. *J. Biol. Chem.* **252**, 9010–9017.

Huan, B.-F., Shen, Y., and Bruenn, J. A. (1991). *In vivo* mapping of a sequence required for interference with the yeast killer virus. *Proc. Natl. Acad. Sci. U.S.A.* **88**, 1271–1275.

Icho, T., and Wickner, R. B. (1989). The double-stranded RNA genome of yeast virus L–A encodes its own putative RNA polymerase by fusing two open reading frames. *J. Biol. Chem.* **264**, 6716–6723.

Janda, M., and Ahlquist, P. (1993). RNA-dependent replication, transcription, and persistence of brome mosaic virus RNA replicons in *S. cerevisiae*. *Cell* **72**, 961–970.

Kagan, B. (1983). Mode of action of yeast killer toxins: Channel formation in lipid bilayer membranes. *Nature* **302**, 709–711.

Kane, W. F., Pietras, D., and Bruenn, J. (1979). Evolution of defective interfering double-stranded RNAs in yeast. *J. Virol.* **32**, 692–696.

Kashiwagi, T., Kunishima, N., Suzuki, C., Tsuchiya, F., Nikkuni, S., Arata, Y., and Morikawa, K. (1997). The novel acidophilic structure of the killer toxin from halotolerant yeast demonstrates remarkable folding similarity with a fungal killer toxin. *Structure* **5**, 81–94.

Kimura, T., Kitamoto, N., Matsuoka, K., Nakamura, K., Iimura, Y., and Kito, Y. (1993). Isolation and nucleotide sequences of the genes encoding killer toxins from *Hansenula mrakii* and *H. saturnus*. *Gene* **137**, 265–270.

Kohlstaedt, L. A., Wang, J., Friedman, J. M., Rice, P. A., and Steitz, T. A. (1992). Crystal structure at 3.5 Å resolution of HIV-1 reverse transcriptase complexed with an inhibitor. *Science* **256**, 1783–1790.

Koltin, Y., and Kandel, J. (1978). Killer phenomenon in *Ustilago maydis*: The organization of the viral genome. *Genetics* **88**, 267–276.

Koltin, Y., Mayer, I., and Steinlauf, R. (1978). Killer phenomenon in *Ustilago maydis*, mapping of viral functions. *Mol. Gen. Genet.* **166**, 181–186.

Koonin, E. V., Choi, G. H., Nuss, D. L., Shapira, R., and Carrington, J. C. (1991). Evidence for common ancestry of a chestnut blight hypovirulence-associated double-stranded RNA and a group of positive-strand RNA plant viruses. *Proc. Natl. Acad. Sci. U.S.A.* **88**, 10647–10651.

Lee, M., Pietras, D. F., Nemeroff, M., Corstanje, B., Field, L., and Bruenn, J. (1986). Conserved regions in defective interfering viral double-stranded RNAs from a yeast virus. *J. Virol.* **58**, 402–407.

Lee, S., Umen, J., and Varmus, H. (1995). A genetic screen identifies cellular factors involved in retroviral −1 frameshifting. *Proc. Natl. Acad. Sci. U.S.A.* **92**, 6587–6591.

Li, N., Erman, M., Pangborn, W., Duax, W. L., Bruenn, J. A., Park, C.-M., and Ghosh, D. (1999). Structure of *Ustilago maydis* killer toxin KP6 α-subunit: A multimeric assembly with a cental pore. *J. Biol. Chem.* **274**, 20425–20431.

Lolle, S. J., and Bussey, H. (1986). In vivo evidence for posttranslational translocation and signal cleavage of the killer preprotoxin of *Saccharomyces cerevisiae*. *Mol. Cell. Biol.* **6**, 4274–4280.

Martinac, B., Zhu, H., Kubalski, A., Zhou, X. L., Culbertson, M., Bussey, H., and Kung, C. (1990). Yeast k1 killer toxin forms ion channels in sensitive yeast spheroplasts and in artificial liposomes. *Proc. Natl. Acad. Sci. U.S.A.* **87**, 6228–6232.

Matsumoto, Y., and Wickner, R. B. (1991). Yeast 20-S RNA replicon-replication intermediates and encoded putative RNA polymerase. *J. Biol. Chem.* **266**, 12779–12783.

Matsumoto, Y., Fishel, R., and Wickner, R. B. (1990). Circular single-stranded RNA replicon in *Saccharomyces cerevisiae*. *Proc. Natl. Acad. Sci. U.S.A.* **87**, 7628–7632.

Mills, D. R., Priano, C., DiMauro, P., and Binderow, B. D. (1988). Q-beta replicase: Mapping the functional domains of an RNA-dependent RNA polymerase. *J. Mol. Biol.* **205**, 751–764.

Nemeroff, M., and Bruenn, J. (1986). Conservative replication and transcription of yeast viral double-stranded RNAs in vitro. *J. Virol.* **57**, 754–758.

Newman, A. M., Elliott, S. G., McLaughlin, C. S., Sutherland, P. A., and Warner, R. C. (1981). Replication of double-stranded RNA of the virus-like particles in *Saccharomyces cerevisiae*. *J. Virol.* **38**, 263–271.

Ohtake, Y., and Wickner, R. B. (1995a). KRB1, a suppressor of mak7-1 (a mutant RPL4A), is RPL4B, a second ribosomal protein L4 gene, on a fragment of *Saccharomyces* chromosome XII. *Genetics* **140**, 129–137.

Ohtake, Y., and Wickner, R. B. (1995b). Yeast virus propagation depends critically on free 60S ribosomal subunit concentration. *Mol. Cell. Biol.* **15**, 2772–2781.

Park, C.-M., Bruenn, J. A., Ganesa, C., Flurkey, W. F., Bozarth, R. F., and Koltin, Y. (1994). Structure and heterologous expression of the *Ustilago maydis* viral toxin KP4. *Mol. Microbiol.* **11**, 155–164.

Park, C.-M., Banerjee, N., Koltin, Y., and Bruenn, J. A. (1996a). The *Ustilago maydis* virally encoded KP1 killer toxin. *Mol. Microbiol.* **20**, 957–963.

Park, C.-M., Lopinski, J., Masuda, J., Tzeng, T.-H., and Bruenn, J. A. (1996b). A second double-stranded RNA virus from yeast. *Virology* **216**, 451–454.

Park, C.-M., Routhier, E., and Bruenn, J. A. (Unpublished). Processing of a polyprotein by a *Ustilago maydis* double-stranded RNA virus. Genbank accession no. U01059.

Pietras, D. F., Diamond, M. E., and Bruenn, J. A. (1988). Identification of a putative RNA-dependent RNA polymerase encoded by a yeast double-stranded RNA virus. *Nucleic Acids Res.* **16**, 6226.

Quadt, R., Ishikawa, M., Janda, M., and Ahlquist, P. (1995). Formation of brome mosaic virus RNA-dependent RNA polymerase in yeast requires coexpression of viral proteins and viral RNA. *Proc. Natl. Acad. Sci. U.S.A.* **92**, 4892–4896.

Ribas, J. C., and Wickner, R. B. (1992). RNA-dependent RNA polymerase consensus sequence of the L-A double-stranded RNA virus: Definition of essential domains. *Proc. Natl. Acad. Sci. U.S.A.* **89**, 2185–2189.

Ribas, J. C., Fujimura, T., and Wickner, R. B. (1994). A cryptic RNA-binding domain in the pol region of the L–A double-stranded RNA virus *gag–pol* fusion protein. *J. Virol.* **68**, 6014–6020.

Ridley, S. P., and Wickner, R. B. (1983). Defective interference in the killer system of *Saccharomyces cerevisiae*. *J. Virol.* **45**, 800–812.

Roditi, I., Wyle, R. T., Smith, N., and Braun, R. (1994). Virus-like particles in *Eimeria nieschulzi* are associated with multiple RNA segments. *Mol. Biochem. Parasitol.* **63**, 275–282.

Routhier, E., and Bruenn, J. A. (1998). Functions of conserved motifs in the RNA-dependent RNA polymerase of a yeast double-stranded RNA virus. Submitted for publication. *J. Virol.* **72**, 4427–4429.

Schmitt, M. J. (1995). Cloning and expression of a cDNA copy of the viral k-28 killer toxin gene in yeast. *Mol. Gen. Genet.* **246**, 236–246.

Schmitt, M. J., and Tipper, D. J. (1990). K28, a unique double-stranded RNA killer virus of *Saccharomyces cerevisiae*. *Mol. Cell. Biol.* **10**, 4807–4815.

Sclafani, R. A., and Fangman, W. L. (1984). Conservative replication of yeast double-stranded RNA by displacement of progeny single strands. *Mol. Cell. Biol.* **4**, 1618–1626.

Scott, G. E., Tarlow, O., and McCrae, M. A. (1989). Detailed structural analysis of a genome rearrangement in bovine rotavirus. *Virus Res.* **14**, 119–128.

Shapira, R., Choi, G. H., and Nuss, D. L. (1991). Virus-like genetic organization and expression strategy for a double-stranded RNA genetic element associated with biological control of chestnut blight. *EMBO J.* **10**, 731–739.

Shen, Y., and Bruenn, J. A. (1993). RNA structural requirements for RNA binding in a double-stranded RNA virus. *Virology* **195**, 481–491.

Skipper, N. (1983). Synthesis of a double-stranded cDNA transcript of the kiler toxin-coding region of the yeast M1 double-stranded RNA. *Biochem. Biophys. Res. Comm.* **114**, 518–525.

Skipper, N., Thomas, D. Y., and Lau, P. C. (1984). Cloning and sequencing of the preprotoxin-coding region of the yeast M1 double-stranded RNA. *EMBO J.* **3**, 107–111.

Somers, J. M. (1973). Isolation of suppressive sensitive mutants from killer and neutral strains of *Saccharomyces cerevisiae. Genetics* **74**, 571–579.

Sommer, S. S., and Wickner, R. B. (1982). Yeast L dsRNA consists of at least three distinct RNAs: Evidence that the non-Mendelian genes [HOK], [NEX], and [EXL] are on one of these dsRNAs. *Cell* **31**, 429–441.

Stark, M. J., and Boyd, A. (1986). The killer toxin of *Kluyveromyces lactis*: Characterization of the toxin subunits and identification of the genes which encode them. *EMBO J.* **5**, 1995–2002.

Stark, M. J., Boyd, J. R., Mileham, A. J., and Romanos, M. A. (1990). The plasmid-encoded killer system of *Kluyveromyces lactis*: A review. *Yeast* **6**, 1–29.

Stuart, K. D., Weeks, R., Guilbride, L., and Myler, P. J. (1992). Molecular organization of *Leishmania* RNA virus 1. *Proc. Natl. Acad. Sci. U.S.A.* **89**, 8596–8600.

Su, H.-M., and Tai, J.-H. (1996). Genomic organization and sequence conservation in type I *Trichomonas vaginalis* viruses. *Virology* **222**, 470–473.

Suzuki, C., and Nikkuni, S. (1994). The primary and subunit structure of a novel type killer toxin produced by a halotolerant yeast, *Pichia farinosa. J. Biol. Chem.* **269**, 3041–3046.

Sweeny, T. K., Tate, A., and Fink, G. R. (1976). A study of the transmission and structure of dsRNAs associated with the killer phenomenon in *Saccharomyces cerevisiae. Genetics* **84**, 27–42.

Tai, J. H., and Ip, C. F. (1995). The cDNA sequence of *Trichomonas vaginalis* virus-T1 double-stranded RNA. *Virology* **206**, 773–776.

Tao, J., Ginsberg, I., Banerjee, N., Koltin, Y., Held, W., and Bruenn, J. A. (1990). The *Ustilago maydis* KP6 killer toxin: Structure, expression in *Saccharomyces cerevisiae* and relationship to other cellular toxins. *Mol. Cell. Biol.* **10**, 1373–1381.

Tercero, J. C., and Wickner, R. B. (1992). *Mak3* encodes an *N*-acetyltransferase whose modification of the L-A *gag* N-terminus is necessary for virus particle assembly. *J. Biol. Chem.* **267**, 20277–20281.

Tercero, J. C., Dinman, J. D., and Wickner, R. B. (1993). Yeast MAK3 *N*-acetyltransferase recognizes the N-terminal four amino acids of the major coat protein (gag) of the L–A double-stranded RNA virus. *J. Bacteriol.* **175**, 3192–3194.

Thiele, D. J., Hannig, E. M., and Leibowitz, M. J. (1984). Multiple L double-stranded RNA species of *Saccharomyces cerevisiae*: Evidence for separate encapsidation. *Mol. Cell. Biol.* **4**, 92–100.

Thrash, C., Voelkel, K. A., DiNardo, S., and Sternglanz, R. (1984). Identification of *Saccharomyces cerevisiae* DNA topoisomerase I mutants. *J. Biol. Chem.* **259**, 1375–1377.

Toh-e, A., and Wickner, R. B. (1979). A mutant killer plasmid whose replication is dependent upon a chromosomal "superkiller" mutation. *Genetics* **91**, 673–682.

Toh-e, A., and Wickner, R. B. (1980). Superkiller mutations suppress chromosomal mutations affecting dsRNA killer plasmid replication in *S. cerevisiae. Proc. Natl. Acad. Sci. U.S.A.* **77**, 527–530.

Toh-e, A., Guerry, P., and Wickner, R. B. (1978). Chromosomal superkiller mutants of *Saccharomyces cerevisiae*. *J. Bacteriol.* **136**, 1002–1007.

Tu, C.-L., Tzeng, T.-H., and Bruenn, J. A. (1992). Ribosomal movement impeded at a pseudoknot required for frameshifting. *Proc. Natl. Acad. Sci. U.S.A.* **89**, 8636–8640.

Tzeng, T.-H., Tu, C.-L., and Bruenn, J. A. (1992). Ribosomal frameshifting requires a pseudoknot in the yeast double-stranded RNA virus. *J. Virol.* **66**, 999–1006.

Van der Lende, T. R., Duitman, E. H., Gunnewijk, M. G. W., Yu, L., and Wessels, J. G. H. (1996). Functional analysis of dsRNAs (L1, L3, L5, and M2) associated with isometric 34 nm virions of *Agaricus bisporus* (white button mushroom). *Virology* **217**, 88–96.

Van de Ven, W. J., Creemers, J. W., and Roebroek, A. J. (1991). Furin: The prototype mammalian subtilisin-like proprotein-processing enzyme. Endoproteolytic cleavage at paired basic residues of proproteins of the eukaryotic secretory pathway. *Enzyme* **45**, 257–270.

Vodkin, M. (1977). Induction of yeast killer factor mutations. *J. Bacteriol.* **132**, 346–348.

Wang, A. L., and Wang, C. C. (1991). Viruses of protozoa. *Annu. Rev. Microbiol.* **45**, 251–263.

Wang, A. L., Yang, H.-M., Shen, K. A., and Wang, C. C. (1993). Giardiavirus double-stranded RNA genome encodes a capsid polypeptide and a *gag–pol*-like fusion protein by a translational frameshift. *Proc. Natl. Acad. Sci. U.S.A.* **90**, 8595–8599.

Wickner, R. B. (1986). Double-stranded RNA replication in yeast: The killer system. *Annu. Rev. Biochem.* **55**, 3732–3795.

Wickner, R. B. (1987). MKT1, a nonessential *Saccharomyces cerevisiae* gene with a temperature-dependent effect on replication of M2 double-stranded RNA. *J Bacteriol.* **169**, 4941–4945.

Wickner, R. B. (1989). Yeast virology. *FASEB J.* **3**, 2257–2265.

Wickner, R. B., and Leibowitz, M. J. (1977). Dominant chromosomal mutant bypassing chromosomal genes needed for killer RNA plasmid replication in yeast. *Genetics* **87**, 453–469.

Wickner, R. B., Ridley, S. P., Fried, H. M., and Ball, S. G. (1982). Ribosomal protein L3 is involved in replication or maintenance of the killer double-stranded RNA genome of *Saccharomyces cerevisiae*. *Proc. Natl. Acad. Sci. U.S.A.* **79**, 4706–4708.

Wickner, R. B., Icho, T., Fujimura, T., and Widner, W. R. (1991). Expression of yeast L–A double-stranded RNA virus proteins produces derepressed replication: A *ski⁻* phenocopy. *J. Virol.* **65**, 155–161.

Widner, W. R., Matsumoto, Y., and Wickner, R. B. (1991). Is 20s RNA naked? *Mol. Cell. Biol.* **11**, 2905–2908.

Williams, T. L., and Leibowitz, M. J. (1987). Conservative mechanism of the *in vitro* transcription of killer virus of yeast. *Virology* **158**, 231–234.

Yao, W., and Bruenn, J. A. (1995). Interference with replication of two double-stranded RNA viruses by production of N-terminal fragments of capsid polypeptides. *Virology* **214**, 215–221.

Yao, W., Muqtadir, K., and Bruenn, J. A. (1995). Packaging in a yeast double-stranded RNA virus. *J. Virol.* **69**, 1917–1919.

Yao, W.-S., Adelman, K., and Bruenn, J. A. (1997). In vitro selection of packaging sites in a double-stranded RNA virus. *J. Virol.* **71**, 2157–2162.

Zakian, V. A., Wagner, D. W., and Fangman, W. L. (1981). Yeast L double-stranded RNA is synthesized during the G_1 phase but not the S phase of the cell cycle. *Mol. Cell. Biol.* **1**, 673–679.

Zhu, Y.-S., Zhang, X.-Y., Cartwright, C. P., and Tipper, D. J. (1992). Kex2-dependent processing of yeast K1 killer preprotoxin includes cleavage at ProArg-44. *Mol. Microbiol.* **6**, 511–520.

III

Viruses of Macroscopic Plants

8

Ecology of Plant Viruses, with Special Reference to Whitefly-Transmitted Geminiviruses (WTGs)

ANJU CHATTERJI and CLAUDE M. FAUQUET

Donald Danforth Plant Science Center
St. Louis, Missouri 63105

Plant virus diseases result from a complex interplay of host, virus, and vector over time under favorable environmental conditions that allow efficient spread and survival of both the virus and its vector. This chapter relates to the ecology of plant

321

viruses with emphasis on geminiviruses transmitted by the whitefly *Bemisia tabaci*. The ability of the virus to spread and reach epidemic proportions depends on several ecological factors that directly or indirectly influence the population dynamics of the aerial vectors, its feeding behavior, or its movement over long distances, and certainly the availability of susceptible cultivated or wild hosts. In the light of these ecological factors, management strategies to prevent virus infections are discussed. Finally, the role of man by manipulating the agroclimatic environment and indiscriminate movement of plant material is highlighted, which, in some instances, has led to the emergence of new recombinant viruses.

I. INTRODUCTION

Viruses are infectious nucleoproteins that must have a living cell to multiply in and establish themselves. All viruses are obligate parasites containing either an RNA or DNA genome enveloped in a protective protein coat. Approximately one-fourth of all the known viruses cause diseases in plants. A plant virus may infect a variety of plants, thus having a broad host range, or may be restricted to a single species of plant.

Plant viruses occur in different sizes and shapes but can be broadly categorized as rods (rigid or flexuous), or spherical (isometric or polyhedral), or bacilliform, that is, bullet shaped. Some elongated viruses are rigid rods about 15×300 nm, but most are long flexible threads 10–13 nm wide and 480–2000 nm in length. Most spherical viruses are actually icosahedral, about 13–60 nm in diameter. The virus particles consist of a definite number of protein subunits that package the nucleic acid genome and are spirally arranged in elongated viruses and packed along the sides in polyhedral viruses. Some viruses also have an outer lipoprotein envelope surrounding the coat protein.

Each plant virus particle consists of at least one molecule of a nucleic acid and a protein shell. The nucleic acid of most plant viruses consists of RNA, but some are also known to have DNA, which may be single or double stranded. The nucleic acid makes up 5–40% of the virus mass, and the protein makes up the remaining 60–95%. The genome of many plant viruses are split into two or more nucleic acid components, each of which may be individually encapsidated. The coat protein of the virus not only provides protection for the nucleic acid but also plays a role in vector transmission of the virus, the kind of symptoms it causes, and its movement in the plant. Most of the proteins encoded by plant viruses, therefore, have special functions either in vector transmission, movement in the plant, in cleavage of the viral polyproteins, or to produce inclusion bodies in the cells.

A. Classification of Plant Viruses

Viruses are classified into different genera using a set of established criteria based on their morphological, physical, and biochemical properties. The International Committee on Taxonomy of Viruses (ICTV) has set aside the following criteria for classification of different viruses.

Genome type: DNA versus RNA, number of strands, linear, circular, or superhelical

Morphology: size, shape, presence, or absence of an envelope, capsid size, and structure

Physicochemical properties: mass, buoyant densities, sedimentation velocity, and stability

Protein: content, number, size, and function of structural and nonstructural proteins, amino acid sequences, and glycosylation

Lipids and carbohydrates: content and nature

Genome organization and replication: gene number, characterization of transcription and translation, posttranslational control, site of viral assembly, type of release

Antigenic and biological properties: host range, transmission, distribution, etc.

Detailed descriptions on different virus genera can be found in van Regenmortel *et al.*, 1999; Matthews, 1991). Figure 1 provides the categorization of different plant viruses based on the type of nucleic acid and the structure of virus particles, and Table I lists some of the important plant virus families and the diseases they cause.

Of the different viruses that affect plants, one of the emerging and economically very important are the viruses belonging to the family Geminiviridae. Even though one of the earliest known viruses reported in the literature was a geminivirus (quoted in Osaki *et al.*, 1985), it was not until the late 1970s that the etiology of the first geminivirus was revealed (Galvez and Castano, 1976; Matyis *et al.*, 1975), and then again it took almost another decade before the molecular details of these viruses were unraveled (Goodman, 1981; Harrison, 1985). In recent years, geminiviruses have proved to be the single most important group of plant viruses with respect to the potential to spread rapidly throughout fields and cause tremendous economic losses in cereals, legumes, and several vegetable crops (Brown and Nelson, 1984, 1986, 1988; Brown and Bird, 1992, Brown, 1994; Munniyappa, 1980; Duffus, 1987). Most geminiviruses are transmitted by whiteflies, while others are transmitted by leafhoppers. All geminiviruses consist of twinned isometric (geminate) particles (Fig. 1) containing single-stranded circular DNA genomes (Stanley, 1985; Lazarowitz, 1992). They replicate via double-stranded DNA intermediates in the nuclei of infected cells. Viruses in the family Gemini-

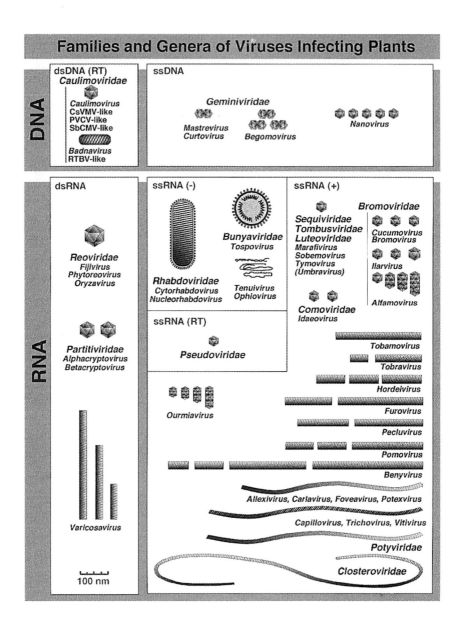

Fig. 1. Families and genera of viruses infecting plants. Relative variation in size and morphology between the genera can be distinguished. Reprinted with permission from van Regenmortel *et al.* (2000).

TABLE I

The Order of Presentation of Plant Viruses

Order Family	Genus	Type species	Host

The ssDNA Viruses

Geminiviridae	*Mastrevirus*	Maize streak virus	Plants
	Curtovirus	*Beet curly top virus*	Plants
	Begomovirus	*Bean golden mosaic virus – Puerto Rico*	Plants
	Nanovirus	Subterranean clover stunt virus	Plants

The DNA and RNA Reverse-Transcribing Viruses

Caulimoviridae	*Caulimovirus*	*Cauliflower mosaic virus*	Plants
	"Petunia vein-clearing-like viruses"	*Petunia vein-clearing virus*	Plants
	"Soybean chlorotic mottle-like viruses"	*Soybean chlorotic mottle virus*	Plants
	"Cassava vein mosaic virus-like viruses"	*Cassava vein mosaic virus*	Plants
	Badnavirus	*Commelina yellow mottle virus*	Plants
	"Rice tungro bacilliform-like viruses"	*Rice tungro bacilliform virus*	Plants
Pseudoviridae	*Pseudovirus*	*Saccharomyces cerevisiae Ty1 virus*	Fungi[c]
	Hemivirus	*Drosophila melanogaster copia virus*	Invertebrate[b]
Metaviridae	*Metavirus*	*Saccharomyces cerevisiae Ty3 virus*	Fungi[c]

The RNA Viruses
The dsRNA Viruses

Reoviridae	*Fijivirus*	*Fiji disease virus*	Plants[d]
	Phytoreovirus	*Rice dwarf virus*	Plants*
	Oryzavirus	*Rice ragged stunt virus*	Plants*
Partitiviridae	*Alphacryptovirus*	*White clover cryptic virus 1*	Plants
	Betacryptovirus	*White clover cryptic virus 2*	Plants
	Varicosavirus	*Lettuce big-vein virus*	Plants

The Negative-Sense ssRNA Viruses
Mononegavirales

Rhabdoviridae	*Cytorhabdovirus*	*Lettuce necrotic yellows virus*	Plants*
	Nucleorhabdovirus	*Potato yellow dwarf virus*	Plants*
Bunyaviridae	*Tospovirus*	*Tomato spotted wilt virus*	Plants*
	Tenuivirus	*Rice stripe virus*	Plants*
	Ophiovirus	*Citrus psorosis virus*	Plants

continued

Order Family	Genus	Type species	Host

The Positive-Sense ssRNA Viruses

Sequiviridae	*Sequivirus*	*Parsnip yellowfleck virus*	Plants
	Waïkavirus	*Rice tungro spherical virus*	Plants
Comoviridae	*Comovirus*	*Cowpea mosaic virus*	Plants
	Fabavirus	*Broad bean wilt virus 1*	Plants
	Nepovirus	*Tobacco ringspot virus*	Plants
Potyviridae	*Potyvirus*	*Potato virus Y*	Plants
	Ipomovirus	*Sweet potato mild mottle virus*	Plants
	Macluravirus	*Maclura mosaic virus*	Plants
	Rymovirus	*Ryegrass mosaic virus*	Plants
	Tritimovirus	*Wheat streak mosaic virus*	Plants
	Bymovirus	*Barley yellow mosaic virus*	Plants
	Sobemovirus	*Southern bean mosaic virus*	Plants
	Marafivirus	*Maize rayado fino virus*	Plants
Luteoviridae	*Luteovirus*	*Barley yellow dwarf virus – PAV*	Plants
	Polerovirus	*Potato leafroll virus*	Plants
	Enamovirus	*Pea enation mosaic virus-1*	Plants
	Umbravirus	*Carrot mottle virus*	Plants
Tombusviridae	*Aureusvirus*	*Pothos latent virus*	Plants
	Avenavirus	*Oat chlorotic stunt virus*	Plants
	Carmovirus	*Carnation mottle virus*	Plants
	Dianthovirus	*Carnation ringspot virus*	Plants
	Machlomovirus	*Maize chlorotic mottle virus*	Plants
	Necrovirus	*Tobacco necrosis virus*	Plants
	Panicovirus	*Panicum mosaic virus*	Plants
	Tombusvirus	*Tomato bushy stunt virus*	Plants
	Tobamovirus	*Tobacco mosaic virus*	Plants
	Tobravirus	*Tobacco rattle virus*	Plants
	Hordeivirus	*Barley stripe mosaic virus*	Plants
	Furovirus	*Soil-borne wheat mosaic virus*	Plants
	Pomovirus	*Potato mop-top virus*	Plants
	Pecluvirus	*Peanut clump virus*	Plants
	Benyvirus	*Beet necrotic yellow vein virus*	Plants
Bromoviridae	*Alfamovirus*	*Alfalfa mosaic virus*	Plants
	Bromovirus	*Brome mosaic virus*	Plants
	Cucumovirus	*Cucumber mosaic virus*	Plants
	Ilarvirus	*Tobacco streak virus*	Plants
	Oleavirus	*Olive latent virus 2*	Plants
	Ourmiavirus	*Ourmia melon virus*	Plants
	Idaeovirus	*Raspberry bushy dwarf virus*	Plants

continued

Order Family	Genus	Type species	Host
Closteroviridae	Closterovirus	Beet yellows virus	Plants
	Crinivirus	Lettuce infectious yellows virus	Plants
	Capillovirus	Apple stem grooving virus	Plants
	Trichovirus	Apple chlorotic leaf spot virus	Plants
	Vitivirus	Grapevine virus A	Plants
	Tymovirus	Turnip yellow mosaic virus	Plants
	Carlavirus	Carnation latent virus	Plants
	Potexvirus	Potato virus X	Plants
	Allexivirus	Shallot virus X	Plants
	Foveavirus	Apple stem pitting virus	Plants

Subviral Agents: Viroids

Pospiviroidae	Pospiviroid	Potato spindle tuber viroid	Plants
	Hostuviroid	Hop stunt viroid	Plants
	Cocadviroid	Coconut cadang-cadang viroid	Plants
	Apscaviroid	Apple scar skin viroid	Plants
	Coleviroid	Coleus blumei viroid 1	Plants
Avsunviroidae	Avsunviroid	Avocado sunblotch viroid	Plants
	Pelamoviroid	Peach latent mosaic virus	Plants

Subviral Agents: Satellites

Satellites			Plants

[a]Hosts also include plants (genus *Pseudovirus*).
[b]Hosts also include fungi (genus *Hemivirus*).
[c]Hosts also include fungi, plants, and invertebrates (genus *Metavirus*).
[d]Plant arthropod-borne viruses are listed according to their plant hosts, and are identified by an asterisk.

viridae (Briddon and Markham, 1995; Rybicki, 1999) are classified into three genera: *Begomovirus*, *Mastrevirus* and *Curtovirus*, based on their genome organization, host range, and vector species. Begomoviruses (e.g., the *Bean golden mosaic virus* and the *Tomato golden mosaic viruses*) have bipartite genomes, are transmitted by whiteflies, and infect dicotyledonous plants. Mastreviruses like the *Maize streak virus* have monopartite genomes, are transmitted by leafhoppers, and infect monocotyledonous plants. Curtoviruses, typified by *Beet curly top virus* have monopartite genomes, are transmitted by leafhoppers, and infect mostly dicotyledonous plants. The genome organization of different members of the Geminiviridae are illustrated in Figure 2. The genome size of the geminiviruses

Geminiviridae

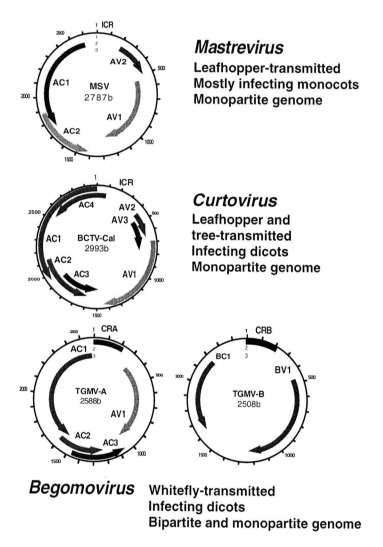

Mastrevirus
Leafhopper-transmitted
Mostly infecting monocots
Monopartite genome

Curtovirus
Leafhopper and
tree-transmitted
Infecting dicots
Monopartite genome

Begomovirus Whitefly-transmitted
Infecting dicots
Bipartite and monopartite genome

Fig. 2. Genome organization of different genera belonging to the family Geminiviridae. The open reading frames are highlighted as solid arrows. The major viral sense genes are the coat protein (AV1) and the precoat protein (AV2), while the complementary sense genes are represented as the replication-associated protein (AC1 or Rep), the transcriptional activator protein (AC2 or TrAP) and the replication enhancer (AC3 or Ren). In case of bipartite geminiviruses, the DNA-B encodes the movement protein (BC1) and the nuclear shuttle protein (BV1) responsible for systemic spread of the virus in the plant.

varies from 2.7 to 3.6 kb. In the case of monopartite viruses, the essential functions for virus replication and movement are located on a single genomic component. Bipartite geminiviruses divide these functions onto two DNA molecules, termed as DNA-A and DNA-B. DNA-A encodes for the replication-associated protein, the replication enhancer, and the coat protein, while DNA-B encodes the two movement proteins responsible for cell-to-cell and long-distance spread of the virus in the plant.

B. Symptomatology of Virus-Infected Plants

Plant viruses cause a wide range of diseases, inducing a variety of symptoms. Developmental abnormalities such as stunting, uneven growth, and leaf curling are some of the striking symptoms of plant virus infections. The most easily recognizable symptoms, however, are mosaics and mottles of green and yellow and ringspots that can appear on leaves as well as the fruits. Some viruses cause swellings in the stem (e.g., the cocoa swollen shoot disease) or, in other cases, certain tumorous outgrowths, known as enations, may appear from the upper or lower surface of the leaf. They may be small ridges of tissues or large irregular leaflike tissues associated with the veins. In some cases, a virus may infect a plant without causing any visible symptoms (latent infections). Aside from the symptoms produced on their cultivated crop hosts, most viruses also induce a hypersensitive response in certain plant species manifested by necrotic lesions that limit the spread of the virus at the site of inoculation. Such plants are called the "local lesion hosts" and are used in quantitation of the virus concentration in infectivity assays.

Plants infected by geminiviruses often exhibit symptoms ranging from bright mottles to golden yellow mosaics and sometimes the leaves are puckered and curled (Fig. 3). When plants are infected at an early stage, they become stunted and bushy and plant yield is very poor. When infected at older stages of plant growth, even though reduction in yield does occur, crop losses are not so staggering. Geminiviruses spread rapidly throughout the field because their whitefly vectors are voracious feeders and retain the virus for days. In recent years, both the prevalence and distribution of whitefly-transmitted geminiviruses have increased tremendously, and, depending on the crop, stage of infection, and the whitefly population, the yield losses can range from 20 to 100% (Brown, 1994). Moreover, these viruses can occur singly or in mixtures, the latter of which makes their detection and accurate diagnosis very difficult.

The visible symptoms induced by virus infection of plants are often a reflection of the histological changes occurring within the plant. These changes could be manifested as hypoplasia (reduction in cell size), hyperplasia (enlargement of

Fig. 3. Symptomatology of some of the common geminivirus diseases: (**A**) African cassava mosaic virus, (**B**) cowpea golden mosaic virus, (**C**) pepper golden mosaic virus, (**D**) tomato yellow leaf curl virus, and (**E**) A purified preparation of geminivirus particles. Note the twinned particle morphology characteristic of these viruses.

cells), or necrosis. Leaves with mosaic symptoms frequently show hypoplasia in their yellowed areas, while the leaf lamina layer is thinner in the dark green areas. Also, in a typical mosaic pattern, the mesophyll cells are less differentiated, with fewer chloroplasts and few or no intercellular spaces. In contrast, the hyperplastic cells are larger than normal, with no intercellular spaces. The vascular tissues associated with the mosaic undergo abnormal cell division, causing proliferation of the phloem cells.

The cytological effects of virus infection have been associated with appearance of crystalline arrays or plates in the nuclei (potyviruses), hypertrophy of the nucleolus, deposition of electron-dense particles in the nucleus, and accumulation of fibrillar rings in the nucleus (geminiviruses; Kim and Flores, 1979). The chloroplasts become aggregated, cup shaped, and accumulate starch grains, and sometimes show large vesicles or fragmentation of chloroplast structure. Besides these characteristics, the cell walls get thickened because of deposition of callose, and in some cases the mitochondria may become affected as well.

Most of the cytological changes induced by viruses are not due to depletion of nutrients diverted toward synthesis of the virus itself but rather are due to disruption of normal metabolic processes in the host cells. Virus infections generally cause a decrease in photosynthesis because of a reduction in the amount of chlorophyll and leaf area per plant. In addition, soluble nitrogen levels drop during rapid virus synthesis and in mosaic-infected tissues carbohydrate levels go down drastically.

C. Transmission of Plant Viruses

Plant viruses are transmitted to healthy plants through a variety of ways, including mechanical transmission via sap, seed, pollen, vegetative propagation; a number of insects like aphids, leafhoppers, and the whiteflies; mites, thrips, fungi, and nematodes. Because plant cells have a thick wall, viruses cannot easily penetrate the cells directly, and most often enter the plants through wounds by mechanical means or by vector transmission. In other cases, they are transmitted through seeds. Transmission through seeds is an effective means of introducing a virus in the field at an early stage, resulting in a randomized distribution of infection throughout the field. Viruses may persist in seeds for a very long time, enabling commercial distribution and spread of the virus over long distances. For seed transmission to occur efficiently, the virus may reside within the tissues of the embryo; otherwise, the virus simply may be present on the surface of the seed, as is known to occur for *Tobacco mosaic virus* (TMV). Vegetative propagation of plant materials is a very effective method for perpetuating and spreading viruses. A plant infected systemically with a virus may remain so throughout its life, and therefore all vegetative parts of the plant subsequently used for propagation may be infected. Transmission through grafts and dodder (*Cuscuta* spp.) are other ways by which viruses can be transmitted and spread in nature.

For the purpose of this chapter, we shall henceforth limit ourselves to the discussion of insect transmission only. Our reason for this is that in the field vectors are the most common means of spread of virus diseases and thereby constitute an important component of viral ecology. Of the different known insect vectors, aphids are by far the largest group and transmit a great majority of all stylet-borne viruses, followed by whiteflies, leafhoppers, and thrips. All of these vectors have piercing and sucking mouthparts and carry viruses on their stylets. These vectors can acquire or inoculate the viruses in a matter of seconds to minutes while feeding on infected or healthy plants.

D. Virus–Vector Relationships

Depending on how long a virus can persist in association with the vector, the viruses are categorized as nonpersistent, semipersistent, or persistent. Nonpersistent viruses are acquired within seconds, retained by the vector for only few hours, and, because they are mostly associated with the mouthparts of the vector, transmitted within minutes of being acquired. Transmission of nonpersistent viruses occurs mostly because the vectors tend to probe different plants in quick succession and do not settle on any single plant for prolonged, continuous feeding. Most of the aphid-borne viruses belonging to the genera *Potyvirus, Cucumovirus, Caulimovirus*, and *Carlavirus* are transmitted in a nonpersistent manner. In the case of semipersistent viruses, the virus may be acquired over several minutes to hours, and get accumulated in the insect's gut before it is released again through the stylets. Such viruses typically persist in the vector for 1–4 days. Typical examples of semipersistent viruses are the *Beet yellows virus* and the *Citrus tristeza virus* transmitted by the aphids, and the *Maize chlorotic dwarf virus* transmitted by the leafhoppers. Persistent viruses, for example, the luteoviruses, are accumulated internally in the hemocoel, or body cavity, of the vector before they are inoculated back in plants thorough the insect's mouthparts. Some of these circulative viruses may even multiply within their vectors such that their vector is by definition a biological vector and are then referred to as circulative-propagative viruses. These biological vectors can retain and transmit the virus for a long period of time. Most of the geminiviruses are believed to be circulative but nonpropagative viruses, but the *Tomato yellow leaf curl virus* (TYLCV) has been shown to be retained through the molt or eggs (Ghanim *et al.*, 1998), making it a circulative-propagative virus. A very small percentage of viruses are transmitted by nematodes and fungi. Nematode vectors transmit the viruses by feeding on the roots of the infected plants and then moving on to healthy plants. Both juveniles as well as adult nematodes can acquire and transmit viruses, but the virus is not carried through the insect's eggs.

All geminiviruses belonging to the Begomovirus group are transmitted by *Bemisia tabaci*, also known as the cotton, tobacco, or sweet potato whitefly (Duffus, 1987; Gerling, 1990). *B. tabaci* is believed to have originated either in the orient or in Pakistan and subsequently spread to other parts of the world (Russell, 1957; Mound, 1963; Cock, 1986). This whitefly species is a major pest of numerous crop species because of its polyphagous nature, and at least 506 plant species in 74 families of dicots and monocots have been reported as hosts of this species (Butler and Henneberry, 1985; Cock, 1986). This gives the viral vector ample flexibility in being able to adapt itself to different hosts under unfavorable conditions. These plants then serve as carriers or reservoirs for the whitefly, allowing its survival for the next season. The viruses carried by the whiteflies are persistent and circulative, which means the whitefly can efficiently transmit the virus during a period of from 5 to 20 days (Duffus, 1987).

E. Replication and Movement of the Virus

On entering a host cell, the viruses uncoat, replicate their genomes, and spread to adjacent cells through plasmodesmata. On reaching the vascular tissues, they use the phloem to infect the entire plant. Once the virus enters the phloem, it moves rapidly toward the growing regions and other food-utilizing parts of the plant (Samuel, 1934). All systemic virus infections, as is the case with the geminiviruses, depend on some viral gene products to regulate movement of the virus in the plant. For some viruses, functional coat protein may be required for long-distance spread in the plant, while other virus types, like the geminiviruses, encode specific movement proteins to assist cell-to-cell and long-distance movement of the virus in the plant.

II. VIRUS–VECTOR–PLANT ECOSYSTEMS: ECOLOGY OF WTGs

Since the last decade, attention has focused on virus survival systems as ecological systems (Thresh, 1980; Harrison, 1981; Gibbs, 1982). In case of soil-inhabiting vectors like nematodes or fungi, a well-developed ability of the viral vector to survive at a site seems to compensate for a vector having limited ability to spread to new sites. In contrast, some viruses transmitted by aerial vectors spread readily to new sites but perennate (survive under adverse conditions) inefficiently at existing ones. Possession of effective means both for spreading and perennating is a feature of some of the most consistently prevalent viruses (Harrison, 1983). Because viruses in the same taxonomic group tend to have similar survival systems, and different types of plant communities support different kinds of survival systems, it is to be expected that specific groups of viruses will be best adapted to specific types of plant communities. Mechanically transmitted viruses like the potexviruses and the tobamoviruses occur in high concentrations and are very stable in their hosts — features that allow them to perpetuate under adverse conditions. Other plant viruses, like the ilarviruses, which are mainly found in woody species, or pollen- and fungi-transmitted viruses that persist in the resting spores of their vectors, are favored by monoculture. Viruses like the tobra-, nepo-, gemini-, and luteoviruses can survive in a variety of wild plants because they have wide host ranges and long persistence in their vectors — features that make them fit to survive in communities that contain many plant species (Harrison, 1983).

A. Ecological Factors Affecting Virus Survival and Spread

Plant virus diseases are not simply an outcome of interplay among the host, pathogen, and the environment. Instead, the interactions observed are complex and involve a multitude of factors. Figure 4 provides a comprehensive picture of all the major factors involved in the development, spread, and survival of a plant viral disease. Evidently, the key components involved include the crops and the conditions in which they grow, the availability of vectors that spread infection, the sources of infection, the types of soil and water, and the climate. Whether or not a crop will suffer from a virus disease depends on its susceptibility, on the presence of sources of infection, and certainly on the availability and behavior of vectors. For viruses that are transmitted by airborne vectors like the aphids, leafhoppers, or whiteflies, several subfactors — like weather conditions, wind speed, and direction and the presence of barrier crops — may play a role in the eventual spread and survival of a virus. Crop plants themselves as well as adjacent weeds may act as reservoirs for spread of infection. The growing conditions that plants are subjected to can also greatly influence crop susceptibility and the sensitivity of crops to infection, as well as affect vector behavior. The opportunity for disease spread within crops and epidemic development also depends on the type of planting pattern. A homogenous crop may yield uniform produce, but if the genotype is susceptible the effects of virus infection can be drastic (Bos, 1983). Introduction of new plant genotypes or an increase in the cropping area can also lead to either a sudden outbreak of new diseases or allow a tremendous increase in the spread of an existing one.

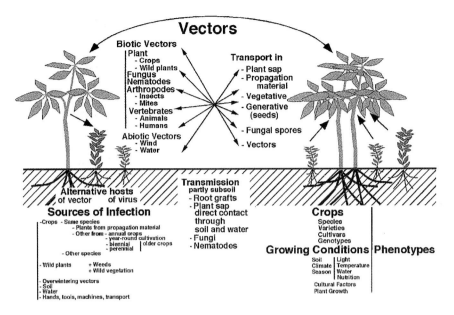

Fig. 4. Interrelationship between the various ecological factors that affect virus survival and spread in the field. Adapted from Bos (1981).

Among the biological factors that affect the survival and spread of a virus in the field and determine whether the virus can spread and survive under adverse conditions are different properties of the virus, including its physical characteristics, like the stability in the plant or soil, the concentrations that the virus can reach in plants, and their rate of movement and distribution within host plants. Viruses that can go systemic rapidly or can move into plant seeds have a far greater chance for survival than those viruses that spread slowly. Furthermore, a virus that can mutate, recombine, or otherwise quickly adapt itself to changes in the environment will be selected naturally for better survival and dispersal. When combined with the potential to infect a diverse host range, such viruses have a better opportunity to maintain themselves and spread efficiently. The pattern of spread within the crop and the rate and extent of spread may depend on available sources of viral inoculum, the size or concentration of inoculum load, the availability of vector species, and the persistence of the virus in the vector. In the case of WTGs, the spread of viruses is by winged adults. Therefore, factors that affect the survival and behavior of adult whiteflies can have an effect on virus spread. Several examples of geminivirus diseases have provided evidence for a positive correlation between *B. tabaci* population size and the spread of persistent and semipersistent viruses. Examples of these diseases are the *Horsegram yellow mosaic virus* in India (Munniyappa, 1983); *African cassava mosaic virus* (ACMV) in the Ivory Coast (Fargette *et al.*, 1985, 1988; Fauquet *et al.*, 1986, 1987; Fauquet and Fargette, 1990; Fishpool and Burban, 1994), and *Tomato yellow leaf curl virus* (Cohen *et al.*, 1988) in Israel.

1. Temperature and Relative Humidity

The population of adult whiteflies can vary significantly between different seasons, primarily in association with changes in temperature, rainfall, and relative humidity (Fig. 5). Higher temperatures and increased relative humidity favor whitefly population buildup, while lower temperatures and heavy rainfall can lead to a decline in the whitefly population (Fargette *et al.*, 1992, 1993, 1994). The development time of *B. tabaci* varies greatly between different host plant species,

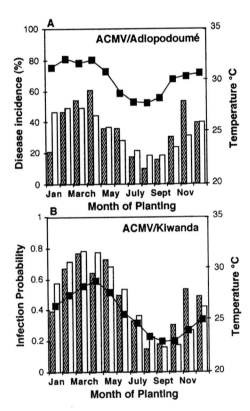

Fig. 5. Correlation between ACMV incidence and mean monthly temperatures in the Ivory Coast (**A**) and Tanzania (**B**). The observed disease incidence is represented by stippled histograms, and the calculated disease incidence values are shown by the open histograms.

and the rate of development is positively correlated to temperature, with its maximum at 28°C. The longevity of *B. tabaci* adults is 10–15 days in the field during summer and 30–60 days during winter.

2. Wind Speed and Direction

Whiteflies are not uniformly distributed within the cassava fields. These are mostly concentrated at the margins of the crops, especially on the upwind borders (Fargette *et al.*, 1985, 1986a,b). The number of whiteflies are relatively small within fields irrespective of the density of the whitefly population or the size of the field. In the Ivory Coast, a higher incidence of ACMV was measured at the upwind edges of the cassava fields that correspond to the vector distribution gradient (Fig. 6) (Fargette *et al.*, 1985). The mean number of whiteflies correlated to an increase in the incidence of disease until the crops were about 3–5 months old. Subsequent increases in the incidence of viral disease to a maximum at crop maturity were not related to whitefly numbers, because with age the nutritional

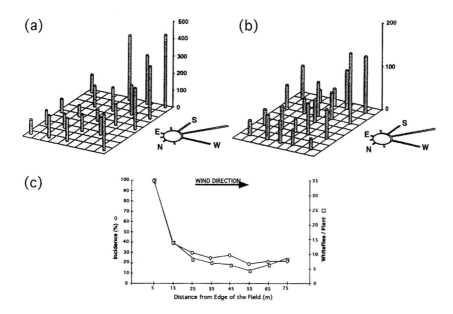

Fig. 6.   Distribution of adult whiteflies caught in yellow traps in a cassava field during the first two and the third month of cassava growth. The two directional symbols on the right side of (**A**) and (**B**) indicate the relative frequency of winds from each direction. ACMV incidence (open diamonds) and number of whiteflies per plant (solid squares) at different distances in meters from the edge of cassava fields. Reprinted with permission from Fauquet and Fargette, 1990.

quality of the plants decreases and the older plants are no longer suitable for whitefly feeding.

3. Movement of the Vector

Spread of infection in time and space is determined by the movement of adult whiteflies and is positively correlated with the size of the vector population (Fig. 7). There are two main ways by which *B. tabaci* move: a short active movement over distances measured in meters (Fauquet *et al.*, 1986) and long-distance passive movement controlled mainly by the wind (Melamed-Madjar *et al.*, 1982; Gerling and Horowitz, 1983; Berlinger and Dahan, 1986; Fargette *et al.*, 1986a,b; Fauquet *et al.*, 1986; Youngman *et al.*, 1986). The major flight hours for the vector are during the morning hours, but sometimes a short peak is observed in the afternoon (Fauquet *et al.*, 1986). Short-range migration takes place regularly between culti-vated and weed hosts, and this constant movement greatly contributes to persist-ence of the flies in cropping systems and to their ability as vectors. The adult whiteflies are reported to be relatively shortsighted, and, in contrast to aphids, most of them fly close to the ground (Gerling and Horowitz, 1983; Byrne *et al.*, 1995). Movement between crops and weed hosts is accomplished by those white-fly populations that tumble along very close to the ground in the manner where direction of movement is determined by the wind. The average flight speed of an adult whitefly within the canopy is estimated to be about 0.2 m/sec (Fig. 8) (Yao *et al.*, 1987).

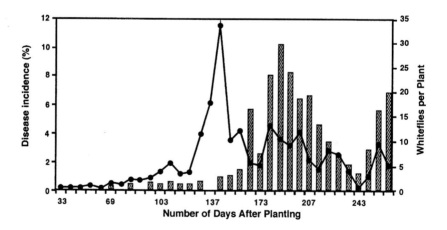

Fig. 7. Relationship between the incidence of ACMV and adult whiteflies per cassava plant at different stages of cassava growth in the Ivory Coast. Reprinted with permission from Fargette *et al.* (1994).

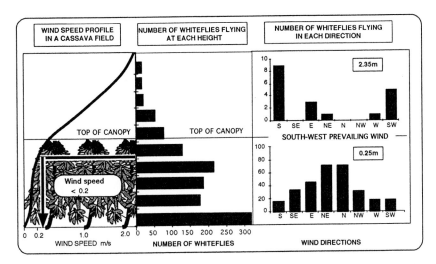

Fig. 8.   The effect of wind speed, direction, and the presence of cassava canopy on the number of adult whiteflies as estimated by the number of vector whiteflies trapped. Reprinted with permission from Fauquet and Fargette (1990).

4. Feeding Behavior of the Vector

Vector distribution, virus concentration, and the susceptibility of plants to virus inoculation are all related to leaf age. Up to 95% of all the adult whiteflies in cassava fields are found to be concentrated on the lower surface of the youngest leaves of the shoot apices (Fauquet and Fargette, 1990). It is important to note that *B. tabaci* adults prefer to feed on young leaves, and this preferential feeding practice of whiteflies has increased their chances to acquire and transmit these viruses. ELISA results have demonstrated that the highest titer may be found in the young leaves of cassava plants in comparison with older leaves. Virus particles could not be detected serologically in older leaves even though the leaves were symptomatic (Fig. 6) (Fargette *et al.*, 1987). Young cassava leaves are also the ones most susceptible to infection. A significantly higher rate of transmission was achieved by whiteflies that had previously fed on young leaves of *Datura stramonium* infected with TYLCV than by those flies fed on mature leaves. Czosnek *et al.* (1988) have shown that the highest concentration of TYLCV viral genomic DNA is located in the shoot apex. The relatively high efficiency of whiteflies to acquire virus from young leaves is probably due to their better feeding behavior on younger leaves, which results from the relatively large amounts of soluble nitrogenous compounds available in the younger leaves (Mound, 1983).

5. Host Genotype

Aside from the above-mentioned physical and biological factors that determine vector buildup in a field, host plant differences most often determine if the vector will colonize a particular field. In Uganda, mean whitefly numbers were significantly greater on the susceptible plant varieties than on ACMV-resistant varieties (Fauquet *et al.*, 1987; Otim-Nape *et al.*, 1994, 1995). Up to 1.7% of adult whiteflies were shown to carry the virus when collected in heavily infected cassava fields (Fargette *et al.*, 1990).

6. Impact of Host Reservoirs on Virus Infection and Spread

Wild plants and weeds greatly assist in virus survival through the adverse periods. Several crops are short-lived and absent from the field during winters or dry summers, or during crop rotations. At these times, wild alternative hosts may be essential for virus survival. Viruses may also perennate in annual weeds if they can pass through the seeds of such weed hosts. It is likely that viruses that infect embryos remain infective in seeds for as long as the seeds remain viable, and that this infection may not reduce seed viability and vitality (Bos, 1981). When a virus is transmitted by weed seeds, it has a tremendous potential to survive in the soil even for extended periods of time. Virus-infected seedlings of weeds act as major reservoirs of infection for efficient short-distance spread by vectors to the crop and other plants. Virus infection in the wild hosts is often symptomless. Aside from harboring crop viruses and other pathogens, wild plants act as important reservoirs and sources of insects, mites, and nematodes. Certain wild plant species may be indispensable to a vector as its alternate host, acting as an essential intermediary in the ecology of the virus. Arthropod vectors often have diverse food plants, including several wild ones, and may probe many different species on which they happen to alight. Wild plants and weeds thus may often also serve as reservoirs of both virus and vector.

The prevalence and distribution of sources of infection also affect the spread of virus in the field. For ACMV in Uganda, high population densities of adult whiteflies were positively correlated to increased incidence of the disease, suggesting that differences in rates of spread were because of differences in vector population (Otim-Nape *et al.*, 1995). In similar studies, Fauquet *et al.* (1988) have found large differences in the rate of spread of ACMV between sites in the lowland coastal area of the Ivory Coast and attributed these to differences in the potency, prevalence, and distribution of nearby sources of infection rather than to differences in the whitefly population. Correlations were observed between high incidence of the disease and the occurrence of infected cassava plants further up, suggesting spread over considerable distances (Fauquet *et al.*, 1988).

B. The Factoring in of Weed Plants

While weeds undoubtedly play an important role in maintaining whitefly populations in agroecosystems, cultivated hosts are often equally important. This complex of continuously available hosts and the movement of *B. tabaci* among them is a vital link in the dissemination of several viruses. Populations of *B. tabaci* are maintained, albeit at relatively low levels, on a series of cultivated and weed hosts through the winter and spring months. They subsequently migrate into cotton during the summer, where their populations build up exponentially (Byrne *et al.*, 1995). In every situation where whiteflies are a serious problem, wild and cultivated hosts grow in proximity to one another, so that whiteflies have little difficulty in finding new sites when existing conditions on any individual plants become less hospitable (Gerling, 1984). Host preference has also been observed among those whiteflies that prefer cucumbers to tomatoes and in turn prefer tomatoes to either corn or eggplants. As a host, cucumber is preferred to eggplants, bean, tobacco, tomato, squash, pepper, and watermelon (Al-Hitty and Sharif, 1987). The hosts that are favored by the adults are probably also the best hosts for breeding.

III. CONTROL OF GEMINIVIRUS DISEASES

Understanding the factors involved in the ecology and epidemiology of certain virus diseases can help to design effective ways on managing them. In most of the whitefly-transmitted geminivirus diseases, different cultural practices — like roughing, manipulation of planting dates, and removal of reservoir plants and weeds within and around the fields — may limit the spread of virus infection. Crops also can be protected from damage by viruses through (a) avoiding or removing sources of infection, (b) preventing or reducing virus spread, and (c) improving crop resistance.

A. Roughing of Wild and Cultivated Hosts

Wild plants are widely recognized as important direct sources of viruses as well as virus vectors. Their removal eliminates both active sources of infection and subsequent virus spread in the seeds, as well as preventing vectors from having breeding sites. Studies showing the effect of reservoirs in initiation and subsequent establishment of virus infection in the field have been conducted in roughed (weeded) and unroughed fields. Fargette *et al.* (1990) reported that a larger source of infection resulted in an increase in the aerial spread of African cassava mosaic disease spread over greater distances as compared to smaller plots. For those fields where the diseased plants were periodically removed, the spread of infection was checked as compared to that occurring in unroughed fields, signifying the need to adopt good cultural practices as a means to contain virus spread (Robertson, 1987).

WTGs are not seed transmitted, but for other viruses that are, seed transmission is another important source of infection as it introduces the disease at a very early stage of plant growth, allowing infection to spread to neighboring plants while they are still young. In such cases, it may be prudent to start with a virus-free stock as an effective means to control virus diseases. This same prevention concept holds true for many vegetatively propagated plants, where the main source of infection is the plant itself, so that development, propagation, and maintenance of virus-free stock is essential. There are several schemes now in operation that help us obtain virus-free stocks for a variety of agricultural and horticultural plants, among which is that it is now possible to obtain rapid initial multiplication of virus-free material. Maintaining general hygiene in the field to ensure that all tools, knives, clothing, and hands are cleaned can prevent the spread of a large number of those viruses that are mechanically transmitted.

B. Manipulation of Planting and Harvesting Dates

In areas where the same crops or a group of related crops are grown repeatedly and there are chances that these may act as potential volunteers for introducing infection to the next crop, it may be possible to limit virus infection by introducing a break in the cycle where no susceptible plants are grown. For those viruses that are transmitted by airborne vectors, a change in the sowing cycle may influence the time and amount of disease incidence. The best time for planting will depend on the time of maximum vector influx. If the vector influx is a late migration, early sowing may allow the plants to be past their vulnerable stages of infection by the time the vectors invade the field. Altering the spacing between plants — or plant density — can be a useful strategy to reduce the movement of a vector population within and around the crop canopy (Narasimhan and Arjunan, 1976; Fargette *et al.*, 1990; Fargette and Fauquet, 1988).

C. Avoidance of Vectors

Coupled with a reliable disease forecasting system, judicious use of systemic insecticides can lower the population of vectors within the crop. Good control of the spread of tomato leaf curl and the yellow vein mosaic of okra has been achieved through control of whitefly vector by use of insecticides (Shastri and

Singh, 1973, 1974; Singh *et al.*, 1973, 1979). For whiteflies, aside from chemical control, biological control can be achieved through the use of their natural predators, and parasitoids can help reduce the whitefly population to a great extent (Gerling, 1990).

D. Plant Resistance

One of the most effective means of controlling virus infections is to grow either resistant cultivars, that is, crop varieties that are resistant to pathogens, or use crop varieties that do not allow vector populations to build up. For most WTG diseases, a combination of phytosanitation (treatment of the above-ground parts of the crop plants) and use of resistant varieties is encouraged. Improved varieties of cassava raised at the International Institute for Tropical Agriculture (IITA) show some degree of resistance to ACMV and are now widely adopted in Nigeria, Uganda, and other African countries (Mahungu *et al.*, 1994). Virus-free plant stocks have been obtained by rigorous selection and through the use of meristem tip culture. These stocks have been shown to be free of cassava mosaic disease in India, and a substantial increase in yield has been achieved by their use (Nair, 1990; Nair and Thankappan, 1990). Similarly, tomato cultivars with improved tolerance to tomato leaf curl virus have been selected (Moustaffa and Nakhla, 1990), and some of these cultivars have been released for commercial use (Pilowsky *et al.*, 1989). Crop resistance to insect vectors is likely to alter the population size, activity, and probing and feeding behavior of the vectors, thereby influencing the pattern of virus spread. Whitefly resistance has been used to minimize virus transmission in various diseases, such as those cause by cowpea golden mosaic virus, bean golden mosaic virus, and cotton leaf curl virus (Vetten and Allen, 1983). Considering the potential for devastation due to whitefly-transmitted geminiviruses, and the ineffectiveness of alternative controls, resistance breeding is a very useful strategy to control geminiviruses whenever available.

IV. ROLE OF MAN IN VIRUS ECOLOGY

There are several examples of agricultural attempts by humans that inadvertently have resulted in the appearance and spread of new diseases either by introduction of new germplasm or by introduction of new vector species. For instance, *Cotton leaf curl virus* has attained epidemic proportions in Pakistan by the introduction of new host genotypes that turned out to be extremely susceptible to the local virus, resulting in a severe outbreak of the viral disease and wiping out most of the cotton crop (Fauquet, unpublished results). Other examples of outbreaks include those of *Turnip mosaic virus* and *Cauliflower mosaic virus* in Britain caused by the introduction of new, highly vulnerable genotypes of Brussels sprouts (Tomlinson and Ward, 1982), of *Lettuce yellow vein virus* in lettuce in California (Zink and Duffus, 1975), and of *Rice yellow mottle virus* in Africa as a result of introducing high-yielding rice cultivars (Raymundo and Buddenhagen, 1976; Bakker, 1974). In addition, introduction of new crops over large areas can have dramatic influence over vector population and spread of diseases, as observed for whitefly-transmitted golden yellow mosaics of legumes in Brazil as a result of planting soybean, which is a preferred host for the whiteflies, over extensive areas (Costa, 1975). The spread of the sweet potato whitefly *Bemisia tabaci* (B biotype) to Europe and the Americas has been associated with human transport of plant materials (e.g., soybean, okra, and eggplant) into the new world (Cock, 1986). A new population of *B. tabaci*, termed biotype B to distinguish it from the biotype A indigenous to North America, is now dispersed worldwide (Bedford *et al.*, 1992; Brown *et al.*, 1992, 1994), primarily because of its extraordinary ability to adapt to an extremely broad host range of plant species (Bedford *et al.*, 1994).

Furthermore, the role of man in the dissemination of virus diseases is exemplified by free trade and trafficking of plant material. These activities resulted in the appearance and spread of *East African cassava mosaic virus* in coastal and inland countries of East and Southern Africa because of considerable movement of cuttings along the roads, railways, and other transportation routes (Thresh *et al.*, 1998). Such risks are obvious when considering vegetatively propagated plant material, or seedborne viruses that are introduced in new areas because of exchange or import of plant material. Similarly, *Tomato yellow leaf curl virus* from Israel was apparently imported into the Dominican Republic a few years ago along with the tomato seedlings, and it has been reported in other Caribbean countries (e.g., Cuba) and in Florida (Polston and Anderson, 1997; Polston *et al.*, 1998).

V. EMERGENCE OF NEW RECOMBINANT VIRUSES

Natural selection of the most fit variant is a basic concept of evolution; thus, every change in the environment or replicative niche of the virus may imply different fitness requirements. This becomes especially important in many plant viruses that have a broad host range or that can use several different vector species for transmission. Those viruses that replicate both in plants and in the insect vectors have dramatically different selection pressures (Roossnick, 1997). Natural transmission of some plant viruses involves a vector, which variously may be an insect, fungus or a nematode, and considerations of natural fitness must include the constraints of the vector. The uptake of virus by the vector always involves some degree of specificity, and the viral proteins must provide an appropriate fit in order for transmission to occur (Gray, 1996).

Considering that viruses are extremely adaptable and capable of rapid change, it is not surprising that most of the recently emerging viruses are recombinants (Padidam et al., 1999, submitted). Since geminiviruses appear to exhibit a considerable degree of sequence variation, both within and between populations, and appear to have a rapidly increasing host range, DNA recombination is likely to be responsible for some of the variation seen in the geminiviruses, allowing related viruses to exchange genes or parts thereof in mixed infections, potentially generating ever-more-fit variants. Evidence has been found for both homologous and nonhomologous recombination. This evidence includes the release of infectious viral DNA from monomer-containing recombinant plasmids, deletion of foreign sequences, reversion of deletion mutants to generate a wild-type genome size, and production of defective subgenomic DNA molecules. Nonhomologous recombination may result in deletions, insertions, and repetitions of viral sequences (Bisaro, 1994).

Whenever the hosts of two or more whitefly-transmitted geminiviruses are present together in the same area (Brown and Bird, 1992), intergenetic viral recombination provides a mechanism for the production of new forms, and this process may already have played a key role in the genesis of the forms that exist today (Zhou et al., 1997). A novel type of recombinant virus, UgV (Uganda variant), has been associated with a recent epidemic on cassava in Uganda (Harrison et al., 1997; Deng et al., 1997; Fauquet et al., 1998a). This recombinant virus typically has even more severe effects than UgV alone and is an example of interspecific recombination between ACMV and *East African cassava mosaic virus* (EACMV), having led to emergence of a new geminivirus pathogen (Fig. 9). Another example of recombination-related emergence of viral variants represents Pakistani isolates of *Cotton leaf curl virus* and *Okra yellow vein mosaic virus* (OYVMV) (Zhou et al., 1998). The cotton leaf curl epidemic in Pakistan is caused by several distinct viral variants, with recombination events involving OYVMV and other unspecified geminiviruses probably having been involved in their evolution (Fauquet et al., 1998b; Zhou et al., 1998). Infections by *Potato yellow mosaic virus* in Trinidad and Tobago (Umaharan et al., 1998) and *Tomato yellow leaf curl virus* in Italy and Spain not only caused severe damage but spread

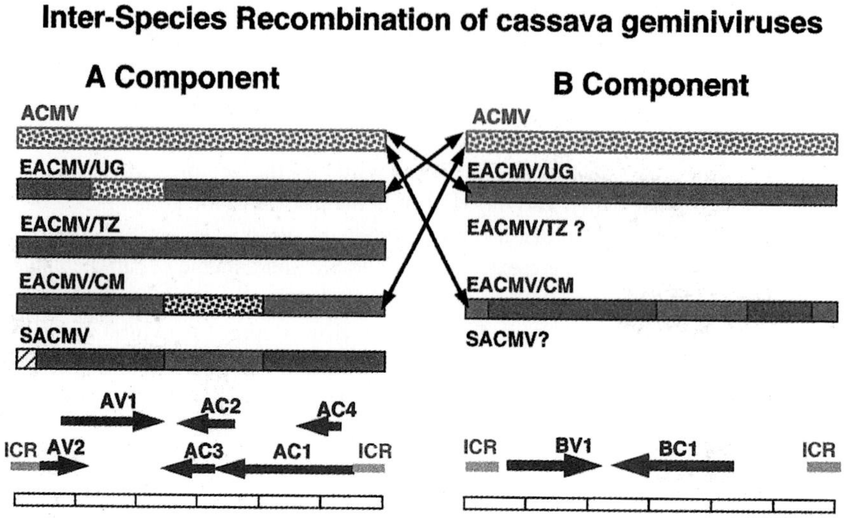

Inter-Species Recombination of cassava geminiviruses

A Component   B Component

Fig. 9.   Interspecies recombination of cassava geminiviruses involving different DNA components.

throughout those respective countries (Noris *et al.*, 1994). Like the other examples, these viruses are recombinants that presumably evolved in response to changes in the ecosystem.

VI. CONCLUSIONS

   It is obvious that the best approach to understand and solve virus problems is ecological. While man is responsible for inducing several virus epidemics by interfering with crop ecosystems, he has continued to develop various cultural control measures to minimize the spread of viruses. Finally, it is in our interest to realize that virus ecosystems are very dynamic and that we will need to continually improve the phytosanitary methods used to contain viral diseases. Increased awareness with respect to the role of man in spread of new diseases, and the ability of the vector to adapt itself to harsh environments, and, not in the least, the extraordinary ability of the viruses to mutate, recombine, and transcomplement each other in a race to survive offer a very real challenge that to some extent can be managed by advanced techniques of detection and diagnosis and prediction models for disease advent and spread, but will also require an integrated approach based on all the factors considered here.

REFERENCES

Al-Hitty, A., and Sharrif, H. L. (1987). Studies on the host preference of *Bemisia tabaci* (Genn.) on some crops and effects of using traps on the spread of tomato yellow leaf curl virus to tomato in plastic house. *Arab J. Plant Prot.* **5**, 19–23.

Bakker, W. (1974). "Characterization and Ecological Aspects of Rice Yellow Mottle Virus in Kenya." Agricultural Research Report No. 829. Center For Agricultural Publishing and Documentation, Wageningen, Netherlands.

Bedford, I. D., Briddon, R. W., Markham, P. G., Brown, J. K., and Rosell, R. C. (1992). A new species of *Bemisia tabaci* (Genn.) as a future pest of European agriculture. *Proc. Plant Health Euro. Single Market (BCPC Monograph)* **54**, 381–386.

Bedford, I., Briddon, R. W., Brown, J. K., Rosell, R. C., and Markham, P. G. (1994). Geminivirus transmission and biological characterization of *Bemisia tabaci* (Gennadius) biotypes from different geographic regions. *Ann. Appl. Biol.* **125**, 311–325.

Berlinger, M. J., and Dahan, R. (1986). Flight patterns of *Bemisia tabaci*, a vector of plant viruses. *Abstr. Int. Conf. Trop. Entomol.*, Nairobi, Kenya, p. 23.

Bisaro, D. M. (1994). Recombination in the geminiviruses: Mechanisms for maintaining the genome size and generating genomic diversity. *In* "Homologous Recombination in Plants" (J. Paszkowski, ed.), pp. 39–60. Kluwer, Amsterdam.

Bos, L. (1981). Wild plants in the ecology of virus diseases. *In* "Plant diseases and vectors" (K. F. Harris and K. Maramorosch, eds.), pp. 1–33. Academic Press, New York.

Bos, L. (1983). Plant virus ecology: The role of man and the involvement of governments and international organizations. *In* "Plant Virus Epidemiology" (R. T. Plumb and J. M. Thresh, eds.), pp. 7–23. Blackwell Scientific, Oxford.

Briddon, R. W., and Markham, P. G. (1995). Geminiviridae. *In* "The Classification and Nomenclature of Viruses: Sixth Report of the International Committee on Taxonomy of Viruses" (F. A. Murphy, C. M. Fauquet., M. A. Mayo, A. W. Javies, S. A. Ghabriel., M. D. Summers., G. P. Martelli., and D. H. L. Bishop, eds.), pp. 158–165. Springer-Verlag, Vienna.

Brown, J. K. (1994). Current status of *Bemisia tabaci* as a plant pest and virus vector in agro-ecosystems worldwide. *FAO Plant Prot. Bull.* **42**, 3–32.

Brown, J. K., and Bird, J. (1992). Whitefly-transmitted geminiviruses and associated disorders in the Americas and the Caribbean basin. *Plant Dis.* **76**, 220–225.

Brown, J. K., and Nelson, M. R. (1984). Geminate particles associated with cotton leaf crumple disease in Arizona. *Phytopathology* **74**, 987–990.

Brown, J. K., and Nelson, M. R. (1986). Whitefly-borne viruses of melons and lettuce in Arizona. *Phytopathology* **76**, 236–239.

Brown, J. K., and Nelson, M. R. (1988). Transmission, host range and virus vector relationships of Chine-del tomato virus, a whitefly-transmitted geminivirus from Sinaloa, Mexico. *Plant Dis.* **72**, 866–869.

Butler Jr., G. D., and Henneberry, T. J. (1985). *Bemisia tabaci* (Genn.), a pest of cotton in southwestern United States. *U.S. Dept. Agric. Tech. Bull.*, No. 1707.

Byrne, D. N., Bellows Jr., T. S., and Parrella, M. P. (1995). Whiteflies in agricultural systems. *In* "Whiteflies: Their Bionomics, Pest Status and Management" (D. Gerling, ed.), pp. 227–262. Intercept, London.

Cock, M. J. W. (1986). *Bemisia tabaci*: "A Literature Survey on the Cotton Whitefly with an Annotated Bibliography." UKYAO/International Institute of Biological Control, Ascot.

Cohen, S., and Berlinger, M. J. (1986). Transmission and cultural control of whitefly-borne viruses. *Agric. Ecosyst. Environ.* **17**, 89–97.

Cohen, S., Duffus, J. E., and Berlinger, M. J. (1988). "Epidemiological Studies of Whitefly-Transmitted Viruses in California and Israel." Final Report, BARD Project No. 1-589-83, ARO, Bet Dagan, Israel.

Costa, A. S. (1975). Increase in the population density of *Bemisia tabaci*, a threat of widespread virus infection of legume crop in Brazil. *In* "Tropical Diseases of Legumes" (J. Maramorosch, ed.), pp. 27–49. Academic Press, New York.

Czosnek, H., Ber, R., Navot, N., Zamir, D., Antignus, Y., and Cohen, S. (1988). Detection of tomato yellow leaf curl virus in lysates and of plants and insects by hybridization with a viral DNA probe. *Plant Dis.* **72**, 949–951.

Deng, D., Otim-Nape, J. P., Legg, J. P., Sngare, S., Ogwal, S., Beachy, R. N., and Fauquet, C. (1997). Presence of a new virus closely related to East African cassava mosaic geminivirus associated with cassava mosaic virus outbreak in Uganda. *Afr. J. Root Tuber Crops* **2**, 23–27.

Duffus, J. E. (1987). Whitefly transmission of plant viruses. *In* "Current Topics in Vector Research" (K. F. Harris, ed.), Vol. 4, pp. 73–91. Springer-Verlag, New York.

Fargette, D., and Fauquet, C. (1988). A preliminary study on the influence of intercropping maize and cassava on the spread of African cassava mosaic virus by whiteflies. *Aspects Appl. Biol.* **17**, 195–202.

Fargette, D., Fauquet, C., and Thouvenel, J. C. (1985). Field studies on the spread of African cassava mosaic. *Ann. Appl. Biol.* **106**, 285–294.

Fargette, D., Fauquet, C., Leccoustre, R., and Thouvenel, J. C. (1986a). Primary and secondary spread of African cassava mosaic virus. *Proc. Workshop Epidemiol. Plant Virus Dis.*, Orlando, Vol. 7, pp. 19–21.

Fargette, D., Fauquet, C., Noirot, M., Raffailac, J. P., and Thouvenel, J. C. (1986b). Temporal pattern of African cassava mosaic virus spread. *Proc. Workshop Epidemiol. Plant Virus Dis.*, Orlando, Vol. 7, pp. 25–27.

Fargette, D., Thouvenel, J. C., and Fauquet, C. (1987). Virus content of leaves of cassava infected by African cassava mosaic virus. *Ann. Appl. Biol.* **110**, 65–73.

Fargette, D., Fauquet, C., and Thouvenel, J. C. (1988). Yield losses induced by African cassava mosaic virus in relation to the mode and date of infection. *Trop. Pest Mgmt.* **34**, 89–91.

Fargette, D., Fauquet C., Grenier, E, and Thresh, J. M. (1990). The spread of African cassava mosaic virus into and within cassava fields. *J. Phytopathol.* **130**, 289–302.

Fargette, D., Fauquet, C., Fishpool, L. D. C., and Thresh, J. M. (1992). Ecology of African cassava mosaic virus in relation to the climatic and agricultural environment. *In* "Viruses, Vectors and the Environment" (R. Martelli, ed.), pp. 107–108. *Proc. 5th Int. Plant Virus Epidemiol. Symp.*, Valenzano, Bari, Italy. 27–31 July 1992.

Fargette, D., Munniyappa, V., Fauquet, C. N., Guessan, P., and Thouvenel, J. C. (1993). Comparative epidemiology of three whitefly-transmitted geminiviruses. *Biochimie* **75**, 547–554.

Fargette, D., Jeger, M., Fauquet, C., and Fishpool, L. D. C. (1994). Analysis of temporal disease progress of African cassava mosaic virus. *Phytopathology* **84**, 91–98.

Fauquet, C., and Fargette, D. (1990). African cassava mosaic virus: Etiology, epidemiology and control. *Plant Dis.* **74**, 404–411.

Fauquet, C., Fargette, D., Helden, M., Vanhelder, I., and Thouvenel, J. C. (1986). Field dispersal of *Bemisia tabaci* vector of African cassava mosaic virus. *Proc. Workshop Epidemiol. Afr. Cassava Mosaic Virus*, Orlando, Vol. 7, pp. 10–12.

Fauquet, C., Fargette, D., and Thouvenel, J. C. (1987). The resistance of Cassava to African cassava mosaic. *Proc. Int. Sem. Afr. Cassava Mosaic Dis. and Its Control*, Yamoussoukro, Ivory Coast, 4–8 May 1987, pp. 183–188.

Fauquet, C., Fargette, D., and Thouvenel, J. C. (1988). Some aspects of the epidemiology of African cassava mosaic geminivirus in Ivory Coast. *Trop. Pest Mgmt.* **34**, 92–96.

Fauquet, C. M., Pita, J., Deng, D., Torez-Jerez, I., Otim-Nape, W. G., Ogwal, S., Sangare, A., Beachy, R. N., and Brown, J. K. (1998a). The East African cassava mosaic virus epidemic in Uganda. *Abstr. Int. Workshop Bemisia Geminiviruses*, Puerto Rico, 7–12 June 1998, L-4.

Fauquet, C. M., Padidam, M., and Sawyer, S. (1998b). Taxonomy and evolution of geminiviruses: Interspecies recombination is a major force for geminivirus evolution. *Abstr. Int. Workshop Bemisia Geminiviruses*, Puerto Rico, 7–12 June 1998, L-87.

Fishpool, L. D. C., and Burban, C. (1994). *Bemisia tabaci*: The whitefly vector of African cassava mosaic virus. *Trop. Sci.* **34**, 55–72.

Galvez, G. E., and Castano, M. (1976). Purification of the whitefly-transmitted bean golden mosaic virus. *Turrialba* **26**, 205–207.

Gerling, D. (1984). The overwintering mode of *Bemisia tabaci* and its parasitoids in Israel. *Phytoparasitica* **12**, 109–119.

Gerling, D. (1990). "Whiteflies: Their Bionomics, Pest Status and Management." Intercept, London.

Gerling, D., and Horowitz, R. (1983). Flight of adult *Bemisia tabaci* as determined in yellow trap catches. *Phytoparasitica* **12**, 64.

Ghanim M., Morin, S., Zeidan, M., and Czosnek, H. (1998). Evidence for transovarial transmission of tomato yellow leaf curl virus by its vector, the whitefly, *B. tabaci. Virology* **240**, 295–303.

Gibbs, A. J. (1982). Virus ecology: Struggle for the genes. *In* "Physiological Plant Ecology," Vol. 3: "Responses to the Chemical and Biological Environment" (O. L. Lange, P. S. Nobel, C. B. Osmond and H. Zeigler, eds.). Springer-Verlag, Berlin.

Goodman, R. (1981). Geminiviruses. *J. Gen. Virol.* **54**, 9–21.

Gray, S. M. (1996). Plant virus proteins involved in natural vector transmission. *Trends Microbiol.* **4**, 259–264.

Harrison, B. D. (1981). Plant virus ecology: Ingredients, interactions and environmental influences. *Ann. Appl. Biol.* **99**, 195–205.

Harrison, B. D. (1983). Epidemiology of plant virus diseases. *In* "Plant Virus Epidemiology" (R. T. Plumb and J. M. Thresh, eds.), pp. 1–6, Blackwell Scientific, Oxford.

Harrison, B. D. (1985). Advances in geminivirus research. *Annu. Rev. Phytopathol.* **23**, 55–82.

Harrison, B. D., Zhou, X., Otim-Nape, G. W., Liu, Y., and Robinson, D. J. (1997). Role of a novel type of a double infection in the geminivirus-induced epidemic of a severe cassava mosaic in Uganda. *Ann. Appl. Biol.* **131**, 437–448.

Kim, K. S., and Flores, E. M. (1979). Nuclear changes associated with euphorbia mosaic virus transmitted by the whitefly. *Phytopathology* **69**, 980.

Lazarowitz, S. G. (1992). Geminiviruses: Genomes, structure and gene function. *Crit. Rev. Plant Sci.* **11**, 227–234.

Mahungu, N. M., Dixon, A. G. O., and Kumbira, J. M. (1994). Breeding cassava for multiple pest resistance in Africa. *Afr. Crop Sci. J.* **2**, 539–552.

Matthews, R. E. F. (1991). "Plant Virology," 3rd ed. Academic Press, San Diego.

Matyis, J. C., Silva, D. M., Oleeira, A. S., and Costa, A. S. (1975). Purificaco e morphologia do virus do mosaico dorado do tomateiro. *Summa Phytopathol.* **1**, 267–274.

Melamed-Madjar, V., Cohen, S., Chen, M., Tam, S., and Rosilio, D. (1979). A method for monitoring *Bemisia tabaci* and timing spray applications against the pest in cotton fields in Israel. *Phytoparasitica* **10**, 85–91.

Mound, L. A. (1963). Host correlated variation in *Bemisia tabaci* (Gennadius) (Homoptera: Aleyorididae). *Proc. Roy. Entomol. Soc. London A* **38**, 171–180.

Mound, L. A. (1983). Biology and identity of whitefly vectors of plant pathogens. *In* "Plant Virus Epidemiology" (R. T. Plumb and J. M. Thresh, eds.), pp. 305–313. Blackwell Scientific, Oxford.

Moustafa, S. E., and Nakhala, M. K. (1990). An attempt to develop a new tomato variety resistant to tomato yellow leaf curl virus (TYLCV). *Assuit J. Agric. Sci.* **21**, 167–184.

Munniyappa, V. (1980). Whiteflies. *In* "Vectors of Plant Pathogens" (K. F. Harris and K. Maramorosch, eds.), pp. 39–85. Academic Press, New York.

Munniyappa, V. (1983). Epidemiology of yellow mosaic disease of horsegram (*Macrotyloma uniflorum*) in South India. *In* "Plant Virus Epidemiology" (R. T. Plumb and J. M. Thresh, eds.), pp. 331–335. Blackwell, Oxford.

Nair, N. G. (1990). Performance of virus-free cassava (*Manihot esculenta* Crantz) developed through meristem tip culture. *J. Root Crops* **16**, 123–131.

Nair, N. G., and Thankappan, M. (1990). Spread of Indian cassava mosaic disease under different agroclimatic conditions. *J. Root Crops*, Special Volume: ISRC National Symposium, pp. 216–219.

Narasimhan, V., and Arjunan, G. (1976). The effect of plant density and cultivation method on the incidence of mosaic disease of cassava. *Indian J. Mycol. Plant Pathol.* **6**, 189–190.

Noris, E., Hidalgo, E., Accotto, G. P., and Moriones, E. (1994). High similarity among the tomato yellow leaf curl virus isolates from the west mediterranean basin: The nucleotide sequence of an infectious clone from Spain. *Arch. Virol.* **135**, 165–170.

Osaki, T., Yamada, M., and Inouye, T. (1985). Whitefly-transmitted viruses from three plant species. *Ann. Phytopathol. Soc. Japan* **51**, 82–83.

Otim-Nape, G. W., Shaw, M. W., and Thresh, J. M. (1994). The effects of African cassava mosaic geminivirus on the growth and yield of cassava in Uganda. *Trop. Sci.* **34**, 43–54.

Otim-Nape, G. W., Thresh, J. M., and Fargette, D. (1995). *Bemisia tabaci* and cassava mosaic disease in Africa. *In* "Bemisia, 1995: Taxonomy, Biology, Damage, Control and Management" (D. Gerling and R. T. Meyer, eds.), pp. 319–350. Intercept, London.

Padidam, M., Sawyer, S., and Fauquet, C. M. (1999). Possible emergence of new geminiviruses by frequent recombination. *Virology*. Submitted for publication.

Pilowsky, M., Cohen, S., Ben-Joseph, R., Shlomo, A., Chen, L., Nahon, S., and Krikun, J. (1989). TY-20: A tomato cultivar tolerant to tomato leaf curl virus. *Hassadeh* **69**, 1212–1215.

Polston, J. E., and Anderson, P. K. (1997). The emergence of whitefly-transmitted geminiviruses in tomato in the Western Hemisphere. *Plant Dis.* **77**, 1181–1184.

Polston, J. E., Bois, D., Ano, J., Poliakoff, F., and Urbino, C. (1998). Occurrence of a strain of potato yellow mosaic geminivirus infecting tomato in the eastern Caribbean. *Plant Dis.* **82**, 126.

Raymundo, S. A., and Buddenhagen, I. W. (1976). A rice virus disease in west Africa. *International Rice Commission Newsletter*, p. 25.

Robertson, L. A. D. (1987). The role of *Bemisia tabaci* Gennadius in the epidemiology of ACMV in East Africa: Biology, population dynamics and interactions with cassava varieties. *Proc. Int. Sem. Afr. Cassava Mosaic Dis. and Its Control*, Yamoussoukro, Ivory Coast, 4–8 May 1987, pp. 57–63.

Roossnick, M. J. (1997). Mechanisms of plant virus evolution. *Annu. Rev. Phytopathol.* **35**, 191–200.

Russell, L. M. (1957). Synonyms of *Bemisia tabaci* (Genn.) (Homoptera: Aleyrodidae). *Bull. Brooklyn Entomol. Soc.* **52**, 122–123.

Rybicki, E. (1999). Geminiviridae. *In* "Virus Taxonomy: Seventh Report on International Committee on Taxonomy of Viruses" (M. H. V. van Regenmortel, C. M. Fauquet, D. H. L. Bishop, E. Carstens, M. K. Estes, S. Lemon, D. McGeoch, R. B. Wickner, M. A. Mayo, C. R. Pringle, and J. Maniloff, eds.). Academic Press, New York.

Samuel, G. (1934). The movement of tobacco mosaic virus within the plant. *Ann. Appl. Biol.* **22**, 90–111.

Sharaf, N. (1982). Factors limiting the abundance of the tobacco whitefly (*Bemisia tabaci*) Gennadius (Homoptera: Aleyrodidae) on tomatoes during the spring season in the Jordan Valley. *Dirasat* **9**, 97–103.

Shastri, K. M., and Singh, S. J. (1973). Restriction of yellow vein mosaic virus spread of okra through the control of vector whitefly, *Indian J. Mycology Plant Path.* **3**, 76–81.

Shastri, K. M., and Singh, S. J. (1974). Control of the spread of tomato leaf curl virus by controlling the whitefly *Bemisia tabaci. Indian J. Hort.* **31**, 178–181.

Singh, S. J., Shastri, K. S. M., and Shastri, K. S. (1973). Effect of oil sprays on the control of tomato leaf curl virus in the field. *Indian J. Agric. Sci.* **43**, 669–672.

Singh, S. J., Shastri, K. S., and Shastri, K. S. M. (1979). Efficacy of different insecticides and oil in the control of the leaf curl virus diseases of chilli. *J. Plant Dis. Prot.* **86**, 253–256.

Stanley, J. (1985). The molecular biology of geminiviruses. *Adv. Virus Res.* **30**, 139–177.

Thresh, J. M. (1980). An ecological approach to the epidemiology of plant virus diseases. *In* "Comparative Epidemiology" (J. Palti and J. Kranz, eds.), pp. 57–70. Centre for Agricultural Publishing and Documentation, Wageningen, Netherlands.

Thresh, J. M., Otim-Nape, G. W., Thankappan, M., and Munniyappa, V. (1998). The mosaic diseases of cassava in Africa and India caused by whitefly-borne geminiviruses. *Rev. Plant Path.* **77**, 937–945.

Tomlinson, J. A., and Ward, C. M. (1982). Selection for immunity in Swede (*Brassica napus*) to infection by turnip mosaic virus. *Ann. Appl. Biol.* **101**, 43–50.

Umaharan, P., Padidam, M., Phelps, R. H., Beachy, R. N., and Fauquet, C. M. (1998). Distribution and diversity of geminiviruses in Trinidad and Tobago. *Phytopathology* **88**, 1262–1268.

van Regenmortel, M. H. V., Fauquet, C. M., Bishop, D. H. L., Carstens, E., Estes, M. K., Lemon, S., McGeoch, D., Wickner, R. B., Mayo, M. A., Pringle, C. R., and Maniloff, J. (1999). *In* "Virus Taxonomy: Seventh Report on International Committee on Taxonomy of Viruses" (M. H. V. van Regenmortel, C. M. Fauquet, D. H. L. Bishop, E. Carstens, M. K. Estes, S. Lemon, D. McGeoch, R. B. Wickner, M. A. Mayo, C. R. Pringle, and J. Maniloff, eds.). Academic Press, New York.

Vetten, H. J., and Allen, F. J. (1983). Effects of environment and host on vector biology and incidence of two whitefly-spread diseases of legumes in Nigeria. *Ann. Appl. Biol.* **102**, 219–227.

Yao, N. R., Fargette, D., and Fauquet, C. (1987). Microclimate of a cassava canopy. *Proc. Int. Sem. Afr. Cassava Mosaic Dis. and Its Control*, Yamoussoukro, Ivory Coast, 4–8 May 1987, pp. 92–104.

Youngman, R. R., Toscano, N. C., Jones, V. P., Kido, K., and Natwick, E. T. (1986). Correlations of seasonal trap counts of *Bemisia tabaci* (Homoptera: Aleyrodidae) in Southeast California. *J. Econ. Entomol.* **79**, 67–70.

Zink, F. W., and Duffus, J. E. (1975). Reaction of downy mildew resistant lettuce cultivars to infection by turnip mosaic virus. *Phytopathology* **65**, 243–245.

Zhou, X., Liu, Y., Calvert, L., Munoz, C., Otim-Nape, G. W., Robinson, D. J., and Harrison, B. D. (1997). Evidence that DNA-A of a geminivirus associated with severe cassava mosaic disease in Uganda has arisen by interspecific recombination. *J. Gen. Virol.* **78**, 2101–2111.

Zhou, X., Liu, Y., Robinson, D. J., and Harrison, B. D. (1998). Four DNA-A variants among Pakistani isolates of cotton leaf curl virus and their affinities to DNA-A of geminivirus isolates from okra. *J. Gen. Virol.* **79**, 915–923.

9

Viroid Diseases of Plants

R. A. OWENS

Molecular Plant Pathology Laboratory
United States Department of Agriculture
Beltsville Agricultural Research Center
Beltsville, Maryland 20705

I. INTRODUCTION

Viroids are the smallest known agents of infectious disease — small, single-stranded, highly structured RNA molecules containing 246–399 nucleotides and lacking both a protein capsid and detectable messenger RNA activity. Whereas viruses have been described by Diener (1987a) as "obligate parasites of the cell's

VIRAL ECOLOGY

translational system" that supply some or most of the genetic information required for their replication, viroids can be regarded as "obligate parasites of the cell's transcriptional machinery." Thus far, viroids are only known to infect plants.

While scientific investigation of diseases now known to be caused by plant viruses did not begin until the late nineteenth century, the earliest written records of such diseases date to the eighth century CE. Potato spindle tuber disease, the first viroid disease to be studied by plant pathologists, was discovered less than 100 years ago. In 1923, its infectious nature and ability to spread under field conditions led Schultz and Folsom (1923) to group potato spindle tuber with several other "degeneration diseases" of potatoes. Nearly 50 years were to elapse between this first published description of potato spindle tuber disease and the demonstration by Diener (1971a) of the fundamental differences between the physical structure of its causal agent, potato spindle tuber viroid (or PSTVd), and those of conventional plant viruses.

II. VIROID STRUCTURE

Viroids possess rather unusual properties for single-stranded RNAs (e.g., a pronounced resistance to digestion by ribonuclease and a highly cooperative thermal denaturation profile), leading to an early realization that they might have an unusual higher-order structure. Efforts to understand how viroids replicate and cause disease without the assistance of viroid-encoded polypeptides has prompted detailed analysis of their structure (reviewed by Riesner, 1990). Known viroids are single-stranded circular RNA molecules containing 246–399 unmodified nucleotides and lacking any unusual $2',5'$-phosphodiester bonds or $2'$-phosphate moieties. A combination of electron microscopy, optical melting, and other physical-chemical studies and theoretical calculations indicates that all except avocado sunblotch, peach latent mosaic, and chrysanthemum chlorotic mottle viroids assume a highly base-paired rodlike conformation *in vitro*. Figure 1 compares the "native" structure of potato spindle tuber viroid (PSTVd, the best characterized of all viroids) with that of peach latent mosaic viroid (PLMVd).

To date, the nucleotide sequences of more than 20 different "species" of viroids have been determined (see Table I). Several other plant diseases are believed to be caused by viroid infection, but characterization of the causal agents remains incomplete (see Table II). As first pointed out by Keese and Symons (1985), pairwise comparison of the nucleotide sequences of PSTVd and related viroids suggests that their native structures are organized into five "domains" whose boundaries are defined by sharp changes in sequence similarity. The "conserved central domain" is the most highly conserved of these domains and is believed to

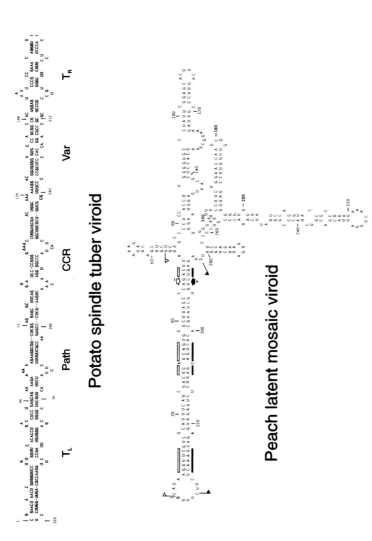

Fig. 1.   Secondary structures of potato spindle tuber (PSTVd) and peach latent mosaic (PLMVd) viroids. (**Above**) A rodlike secondary structure for PSTVd is supported by a variety of physical studies as well as chemical and enzymatic mapping data (Gross *et al.*, 1978; Riesner, 1990). Boundaries of the terminal-left (T$_L$), pathogenicity (Path), central conserved (CCR), variable (Var), and terminal-right (T$_R$) domains are indicated by vertical lines. (**Below**) The lowest free energy structure for PLMVd containing two branching points from which five and three stem-loops emerge. Cleavage domains of the (+) and (−)strands are delimited by filled and open flags. Arrows = locations of the respective cleavage sites; bars = 13 conserved residues present in all hammerhead ribozymes. Reprinted with permission from Hernandez and Flores (1992).

TABLE I

Viroids of Known Nucleotide Sequence (1998)

Family[a]	Genus[a]	Name	Abbreviation	Variants	Nucleotides[b]
Avsunviroidae	Avsunviroid	Avocado sunblotch	ASBVd	19	246–251
		Peach latent mosaic	PLMVd	1	336–337
	Pelamoviroid	Chrysanthemum chlorotic mottle	CChMVd	2	398, 399
Pospiviroidae	Pospiviroid	Chrysanthemum stunt	CSVd	5	354–356
		Citrus exocortis[c]	CEVd	5	368–375 (463)
		Columnea latent	CLVd	2	370–373
		Hop latent	HLVd	1	256
		Iresine	IrVd	1	370
		Mexican papita	MPVd	9	359–360
		Potato spindle tuber	PSTVd	29	358–361 (341)
		Tomato apical stunt	TASVd	3	360–363
		Tomato planta macho	TPMVd	10	359–360
	Cocadiviroid	Citrus viroid IV	CVdIV	1	284
		Coconut cadang-cadang	CCCVd	7	246–247 (287–301)
		Coconut tinangaja	CTiVd	2	254
	Hostuviroid	Hop stunt[d]	HSVd	32	294–303
	Apscaviroid	Apple dimple fruit	ADFVd	1	306
		Apple scar skin[e]	ASSVd	4	329–334
		Australian grapevine	AGVd	1	369
		Citrus bent leaf	CBLVd	2	315, 318
		Citrus viroid III	CVd III	4	291–297
		Grapevine yellow speckle[f]	GYSVd1	35	363–368
		Pear blister canker	PBCVd	1	315
	Coleviroid	Coleus blumei viroid 1	CbVd1	4	248–251
		Coleus blumei 2	CbVd2	1	301
		Coleus blumei 3	CbVd3	3	361–364

[a]Classification scheme essentially that of Koltunow and Rezaian (1989a) as modified by Flores et al. (1998).

[b]Sizes of of naturally occurring variants containing sequence deletions or duplications are shown in parentheses.

[c]Includes Indian bunchy top viroid; see Mishra et al. (1991).

[d]Includes cucumber pale fruit and citrus cachexia viroids.

[e]Includes pear rusty skin and dapple apple viroids; see Zhu et al. (1995).

[f]Includes grapevine viroid 1B; see Koltunow and Rezaian (1989b).

TABLE II

Diseases of Known or Suspected Viroid Etiology

Name	Reported host(s)	Reference
Apple fruit crinkle	*Malus pumila*	Ito *et al.* (1993)
Burdock stunt	*Arctium tomentosum, A. lappa*	Tien and Chen (1987)
Nicotiana glutinosa stunt	*Nicotiana glutinosa*	Bhattiprolu (1991)
Pigeon pea mosaic mottle	*Cajunus cajan*	Flores (1995)
Tomato bunchy top	*Lycopersicon esculentum* (South Africa only?)	Diener (1987b)

contain the site where multimeric viroid RNAs are cleaved and ligated to form circular progeny. The "pathogenicity domain" contains one or more structural elements that modulate symptom expression, and the relatively small "variable domain" exhibits the greatest sequence variability between otherwise closely related viroids. The two "terminal domains" appear to play an important role in viroid replication and evolution.

Several viroids, including *Coleus blumei* viroid 2 (CbVd 2; Spieker, 1996) appear to be "mosaic" molecules formed by rearrangement of domains between two or more viroids infecting the same cell. RNA rearrangement/recombination can also occur within individual domains, leading, in the case of coconut cadang-cadang viroid (CCCVd), to duplication of the left terminal domain plus part of the variable domain as disease progresses (reviewed by Hanold and Randles, 1991). Pending identification of additional group members, the presence of domains in avocado sunblotch viroid (ASBVd) and other self-cleaving viroids remains uncertain.

III. VIROID TAXONOMY

Marked differences in the structural and functional properties of their genomes have led to the assignment of viroids and conventional plant viruses to separate taxa (see ICTV, 1995). To highlight these differences, viroid acronyms now contain the suffix "Vd" rather than the "V" used for conventional virus. Groups of one or more independently replicating sequence variants showing <90% similarity in pairwise comparisons are assumed to represent individual viroid species.

Based on a combination of overall sequence similarity and the presence or absence of certain structural features within the conserved central domain, several different viroid classification schemes have been proposed. All separate ASBVd, chrysanthemum chlorotic mottle viroid (CChMVd), and PLMVd (i.e., the only viroids known to undergo spontaneous self-cleavage) from three or more additional groupings whose type members include apple scar skin viroid (ASSVd) and CbVd. Koltunow and Rezaian (1989a) consider the remaining viroids to form one large grouping whose type member is PSTVd. The consensus phylogenetic tree of Elena *et al.* (1991), in contrast, distributes these same species among three smaller groupings: the PSTVd-like viroids, the CCCVd-like viroids, and HSVd-like viroids. Inclusion of several viroid-like satellite RNAs and the viroid-like domain of hepatitis delta virus in this analysis has provided evidence for an evolutionary link between viroids and viral satellite RNAs (see also Bussière *et al.*, 1995). Carnation small viroid-like RNA (CarSV RNA), a 275-nucleotide circular molecule with self-cleaving hammerhead structures in its strands of both polarities, is a novel retroviroid-like element that shares certain features with both viroids and a small RNA transcript from the newt (Daròs and Flores, 1995; Flores *et al.*, 1997).

IV. PRINCIPAL AND ALTERNATE HOSTS

All but three of the more than 20 viroid species listed in Table I (i.e., *Columnea* latent viroid [CLVd], CbVd, and hop latent viroid [HLVd]) were discovered as a result of their ability to cause disease in cultivated plants. Subsequent transmission studies with additional hosts have shown, however, that many viroid infections are *latent* (symptomless). As illustrated in Table III, individual viroids vary greatly in their ability to infect different plant species. Viroids like PSTVd and HSVd are able to replicate in a wide variety of plants, while the host ranges of CCCVd and coconut tinangaja viroid (CTiVd) are restricted to certain palm species. With the possible exception of PSTVd, where sequence changes in the lower portion of the central conserved region appear to control the ability to replicate in tobacco (Wassenegger *et al.*, 1996), the structural features controlling these host range differences are completely unknown.

All viroids are mechanically transmissible, and most are naturally transmitted from plant to plant by man and his tools. PSTVd, for example, was originally isolated from the cultivated potato and has subsequently been shown to replicate in approximately 160 additional (primarily solanaceous) species. In contrast, only two members of the Lauraceae other than avocado are known to support ASBVd replication. HSVd has a particularly wide host range that includes several herbaceous species as well as many woody perennials (Shikata, 1990; Flores, 1995). Principal hosts of viroids include such vegetatively propagated crops as chrysan-

TABLE III

Viroid Host Ranges

Viroid	Principal host(s)	Alternate hosts	Experimental hosts
CSVd	Chrysanthemum (*Dendranthema morifolium*)	*Ageratum, Petunia, Senecio*	Chrysanthemum, tomato
CEVd	Citrus	Grapevine, tomato, and several other vegetable species	Etrog citron (*Citrus medica*), tomato, *Gynura*
CVd-IV	Citrus	—	Etrog citron
CCCVd	Coconut (*Cocos nucifera*)	Several members of the Palmae	Coconut
CLVd	*Columnea erytrophae* and *Nemanthus wettsteinii* (Gesneriaceae)	*Brunfelsia undulata* (Solanaceae)	Tomato, cucumber
HLVd	Hop (*Humulus lupulus*)	—	Hop
HSVd	Hop, citrus, grapevine	Cucumber, pear, apricot, raspberry, banana, *Hibiscus*, and croton	Cucumber, tomato
IrVd	*Iresine herbstii* (Solanaceae)	—	Tomato
PSTVd	Potato (*Solanum tuberosum*)	Several wild *Solanum* species, avocado	Tomato
TASVd	Tomato (*Lycopersicon esculentum*)	Jerusalem cherry (*Solanum pseudocapsicum*)	Tomato
TPMVd	Several wild members of the Solanaceae	Tomato and other cultivated solanaceous species	Tomato
ADFVd	Apple (*Malus* spp.)		Apple
ASSVd	Apple	Pear (*Pyrus* spp.)	Apple
AGVd	Grapevine (*Vitis vinifera*)	—	Grapevine
CBLVd	Citrus	—	Etrog citron
CVd-III	Citrus	—	Etrog citron
GYSVd	Grapevine	—	Grapevine
PBCVd	Pear	—	
CbVd	Several *Coleus* species and cultivars	—	*Coleus* spp.
ASBVd	Avocado (*Persea americana*)	Several other members of the Lauraceae	Avocado
CChMoVd	Chrysanthemum (*Dendranthema morifolium*)	—	Chrysanthemum
PLMVd	Peach (*Prunus persica*), nectarine	Cherry, plum, apricot	Peach cultivar GF-305

themum and potato and crops such as citrus, grapes, and hops that are subjected to repeated grafting or dressing operations. The role of man rather than an insect, fungus, or nematode as the principal vector for viroid movement gives a somewhat different connotation to the terms "principal host" and "alternative host" than when they are applied to conventional plant or animal viruses.

Identification of the *natural host* of PSTVd or any other viroid is not a simple matter. Genomes of plant pathogens (i.e., fungus, bacteria, virus, or viroid) are generally assumed to coevolve with those of their hosts in such a way as to minimize the effects of infection on plant growth and development (see Matthews, 1991, chap. 15). Most viroid diseases appear to be of only recent origin, but viroids themselves are likely to have existed for much longer periods in wild host species, causing little or no disease and occasionally escaping into cultivated crops (Diener, 1987a). Although man has practiced agriculture for at least 10,000 years, the modern emphasis on plant breeding and the large-scale monoculture of genetically uniform plants would be expected to greatly increase the risk of widespread outbreaks of viroid disease.

In the late nineteenth century, plant breeders made extensive collections of solanaceous plants in Mexico in an effort to identify genes specifying resistance against *Phytophthora infestans*, the late blight fungus responsible for the Irish potato famine. Martinez-Soriano *et al.* (1996) have described the isolation and characterization of nine sequence variants of tomato planta macho (TPMVd) viroid from *Solanum cardiophyllum* growing wild in the Mexican state of Aguas-calientes. The lack of disease symptoms observed in infected plants is consistent with (but does not prove) the coevolution of TPMVd and *S. cardiophyllum*. Previous studies by Galindo *et al.* (1989) had provided circumstantial evidence for the transfer of TPMVd from several native endemically infected solanaceous species to cultivated tomato. Such a transfer could occur via cultivation-associated wounding; alternatively, an aphid vector (i.e., *Myzus persicae*) was shown to transmit TPMVd from uncultivated *Physalis* aff. *foetens* to tomato.

The center of origin for the cultivated potato (*Solanum tuberosum* L.) is the Andean region of South America. Surveys of wild species growing either in the Andes or in the northeastern United States, where the spindle tuber disease was first reported, have failed to identify similar reservoir species for PSTVd, and the natural host of PSTVd remains unknown. Because TPMVd and PSTVd share ~90% sequence similarity, Martinez-Soriano *et al.* (1996) have suggested that PSTVd may have evolved following chance transfer of a TPMVd sequence variant to cultivated potatoes from a wild host species imported for use as germplasm. Both the naturally occurring intermediate strain and various mutant derivatives of PSTVd have been shown to accumulate mutations *in vivo* (Wassenegger *et al.*, 1994, 1996; Qu *et al.*, 1993), but direct evidence for such speciation events may be difficult to obtain.

Viroid diseases may also arise by transfer between cultivated crop species. Studies conducted in the People's Republic of China have shown that pears provide a latent reservoir for ASSVd, the causal agent of scar skin disease in apples. Likewise, while there is no obvious correlation between disease status and the presence of HSVd in grapes, this viroid is known to cause severe disease in hops. In both instances, the two crops are often grown in close proximity. Finally, archeological evidence from Israel and other parts of the Near East suggests that the original source of certain viroids affecting citrus may have been *Vitis vinifera* L., the cultivated grape (Bar-Joseph and Yang, 1997).

Grapevines are known to harbor low levels of citrus exocortis viroid (CEVd) and HSVd, often as symptomless carriers (Semancik, 1991). Viticulture originated in the Near East, and grapes were widely cultivated in the Mediterranean region long before the introduction of the Etrog citron around 300 CE. A mosaic from the floor of an ancient synagogue at Maon in the northern Negev (Israel) dating to the early sixth century CE depicts malformed Etrog fruits similar to those caused by viroid infection. Etrog citron is native to the foothills of the Himalayas, and early introductions and propagation were from seeds. As none of the presently known citrus viroids is seed-borne in Etrog, this gap of 200–300 years may reflect the time required for infection of viroid-free plants by viroids already endemic in the area. As pointed out by Bar-Joseph and Yang (1997), certain Jewish teachings prohibit close cultivation of nonrelated plant species such as fruit trees and vegetables in vineyards as well as the heterologous grafting of different plant species. In part, these proscriptions may represent very early phytosanitary regulations aimed at preventing the spread of plant disease.

V. TRANSMISSION STRATEGIES

A. Laboratory Methods

Neither the epidemiological properties nor the symptoms of viroid diseases differ greatly from those of conventional virus diseases (Diener, 1987a). Both types of agent are readily transmissible using crude sap extracts, and, unlike defective viruses lacking a coat protein, viroids usually spread rapidly throughout the infected plant. Commonly used techniques for the experimental transmission of viroids include (1) standard leaf abrasion methods developed for use with conventional viruses, (2) various "razor slashing" methods in which phloem tissue in the stem or petiole is inoculated via cuts made with a razor blade previously dipped into inoculum, and (3) high-pressure injection of nucleic acid inoculum into folded apical leaves (for CCCVd). PSTVd and HSVd have also been experimentally transmitted by "agroinoculation," a technique in which a modified Ti plasmid from *Agrobacterium tumefaciens* is used to introduce a full-length viroid cDNA into the potential host cell. In some cases, such inoculations are able to overcome marked host resistance to mechanical inoculation (Salazar *et al.*, 1986).

B. Natural Transmission Routes

Most viroids are naturally transmitted from plant to plant by man (or his tools). At least three different viroids — PSTVd (Fernow *et al.*, 1969; Singh, 1970), ASBVd (Wallace and Drake, 1962), and CbVd (Singh and Boucher, 1991) — can be vertically transmitted through true seed and/or pollen. In most cases, however, seed transmission is believed to play only a minor role in the natural spread of viroid disease. Under field conditions, viroid transmission has also been documented following direct contact of healthy plants with infected foliage, contaminated tractors, or other farm machinery (Merriam and Bonde, 1954) as well as various grafting, pruning, and harvesting procedures (Garnsey and Jones, 1967; Yaguchi and Takahashi, 1984).

Many attempts have been made to identify insect vectors for viroid diseases, but only a few positive results have been reported. Querci *et al.* (1997) have recently shown that *M. persicae* is able to acquire and transmit PSTVd from potato plants that are doubly infected with PSTVd and potato leafroll virus (PLRV). The amount of PSTVd associated with purified PLRV virions was low, but its resistance to micrococcal nuclease digestion indicated that the viroid RNA was encapsidated within the virus particles. Epidemiological surveys carried out at three locations in China revealed a strong correlation between PSTVd infection and the presence of PLRV, suggesting that the virus can facilitate PSTVd spread under field conditions.

Galindo *et al.* (1989) have reported rates of TPMVd transmission as high as 97% when the aphid vector *Myzus persicae* was allowed to acquire the viroid from *Physalis* aff. *foetens* before transfer to cultivated tomato. All instars were able to transmit the viroid, and transmission was persistent following a 24-hr acquisition period. Unfortunately, neither the donor nor the recipient plants in these experiments appear to have been examined for the presence of latent virus(es). *Physalis* spp. have been reported to harbor at least two luteoviruses (i.e., *Physalis* mild chlorosis and *Physalis* vein blotch viruses), and, as described above for PSTVd and PLRV (also a luteovirus), aphid transmission of TPMVd may also require transencapsidation by a coinfecting virus.

Available epidemiological data suggest that CCCVd may spread by a variety of means (Hanold and Randles, 1991). Symptoms of CCCVd infection develop quite slowly (i.e., over a period of several years), and infected palms cannot be reliably identified on the basis of a single observation. Rates of disease spread vary greatly, some sites showing little expansion, whereas the borders of active epidemics may advance at rates of 0.5 kilometers per year. Diseased palms are not clustered, and repeated trials have failed to identify an insect vector. CCCVd is seed-transmitted at a low rate (approximately 1 in 300 nuts) and can also be detected in purified pollen. The possibility that CCCVd may be transmitted by mechanical damage from cultural practices has not been adequately tested.

Like TPMVd and the Mexican papita viroids in wild solanaceous plants, RNA molecules related to CCCVd have also been identified in oil palms, coconut palms, and several other monocotyledonous species growing in other parts of the Southwest Pacific. Affected oil palms show symptoms similar to those in naturally CCCVd-infected oil palms in the Philippines, but the coconut palms do not show typical symptoms of cadang-cadang disease. cDNA cloning and sequence analysis will be required to determine how closely these molecules are related to known isolates of CCCVd.

VI. SURVIVAL STRATEGIES

A. Survival in the Host

The contagious nature of viroid diseases and their ready mechanical transmissibility would appear incompatible with the existence of a free RNA pathogen. Indeed, results from early subcellular fractionation studies (e.g., Diener, 1971b) as well as more recent *in situ* hybridization and confocal laser scanning microscopy of nuclei isolated from PSTVd-infected tomato leaves (Harders *et al.*, 1989) have shown that the great majority of PSTVd-related RNA is localized in the nucleoli of infected cells, where it is associated with several cellular proteins. The interaction of PSTVd with nuclear proteins has been studied by reconstitution *in vitro* as well as characterization of complexes isolated *in vivo* under nondenaturing conditions (Wolff *et al.*, 1985; Klaff *et al.*, 1989). While in such complexes PSTVd is relatively resistant to nuclease attack (Diener and Raymer, 1969), and even after deproteinization, its quasi-double-stranded structure renders the viroid much less susceptible to inactivation by ribonuclease than a typical single-stranded viral RNA. Interestingly, ASBVd appears to be associated with chloroplast thylakoid membranes rather than the nucleoli of infected cells, thereby raising the possibility that a host-encoded enzyme other than RNA polymerase II is responsible for its replication (Bonfiglioli *et al.*, 1994a,b; Lima *et al.*, 1994).

Mezitt and Lucas (1996) discuss the role of cell-to-cell trafficking of cellular proteins and nucleic acids via the plasmodesmata in regulating plant growth and development. Formation of viroid–protein complexes as described above is believed to be essential for their ability to replicate, to move from cell to cell and for long distances within their hosts, and to induce disease. Unfortunately, very little information is currently available about the molecular mechanisms involved.

B. Persistence in the Environment

As discussed by Matthews (1991), several factors contribute to the persistence of plant viruses in nature. Tobacco mosaic virus has no significant invertebrate vectors but survives because of four characteristics, that is, the resistance of its virions to inactivation, the high titers generally reached in its hosts, its wide host range, and the ease with which the virus is mechanically transmitted. The persistence of animal viruses often depends on their ability to evade immune surveillance by generating virus variants. Because plants lack an immune system, their viruses are not subject to immune surveillance. The ability to coevolve with their hosts (and thereby temporarily escape the inhibitory effects of various host resistance genes) is nevertheless crucial to the long-term survival of plant viruses and viroids.

Among the most important factors responsible for viroid persistence *in vivo* are (1) the large number of perennial hosts and (2) the vegetative propagation of many herbaceous hosts. With the exception of the tomato planta macho and (possibly) coconut cadang-cadang diseases, wild plant reservoirs and insect transmission do not appear to play an important role in disease persistence. Many viroid infections are essentially symptomless, and "mild" strains of several viroids have been described; for example, PSTVd (Fernow, 1967) and CEVd (Visvader and Symons, 1985). Vegetative propagation from plants with acceptable horticultural performance tends to eliminate severe strains while allowing mild strains or certain viroid mixtures to persist/accumulate. So-called "old clone" citrus trees often contain complex mixtures of viroids (Duran-Vila *et al.*, 1988), and these infections may result in a higher tolerance to root diseases (Rossetti *et al.*, 1980; Ashkenazi and Oren, 1988). Viroid infection of citrus can also lead to dwarfing, an increasingly desirable phenotype that can result in considerable savings in management costs via reduced labor costs and more efficient use of irrigation, fertilizers, and herbicides (Hutton *et al.*, 1999).

VII. REPLICATION STRATEGIES

Viroids were described in the introduction as "obligate parasites of the cell's transcriptional machinery." This phrase neatly summarizes several fundamental differences between the replication strategies employed by viroids and conventional RNA viruses. Unlike viruses with negative- or double-stranded RNA genomes, unencapsidated viroid RNAs do not carry their own replicase enzyme with them. Theoretically, either viroids themselves (arbitrarily designated as the (+)strand RNA) or their complementary (–)strand RNAs could be translated into small polypeptides able to convert a host enzyme into a viroid-specific RNA replicase. Sequence analysis has revealed, however, that (1) both the (+) and (–) strands of many viroids lack an AUG initiation codon, and (2) neither the size nor the amino-acid composition of potential open reading frames are conserved. Viroids thus appear to be completely dependent on host-encoded proteins for replication.

A. Cellular Level

At the cellular level, the first obstacle faced by an incoming viroid is how to move through the cytoplasm of the potential host cell without degradation and enter the nucleus to initiate replication. Active transport of proteins and RNA between the cytoplasm and nucleus is a major process in eukaryotic cells (reviewed by Görlich and Mattaj, 1996), but the nature of the interaction(s) responsible for viroid movement into the nucleus is completely unknown. A variety of evidence, including *in situ* hybridization using strand-specific probes followed by confocal laser scanning microscopy (Harders *et al.*, 1989), indicates that PSTVd and related viroids accumulate in the nucleolus. As mentioned above, ASBVd (and possibly other ribozyme-containing viroids) accumulate in the chloroplast.

Northern blotting experiments with RNAs extracted from infected leaf tissue (e.g., Branch *et al.*, 1981; Owens and Diener, 1982) have shown that PSTVd replication involves the synthesis of multimeric linear (–)PSTVd RNAs via an RNA-directed rolling circle mechanism (see Fig. 2). The absence of monomeric circular (–)PSTVd RNA in infected tissues indicates that the replicative pathway is asymmetric rather than symmetric, like the pathway used by ASBVd and several viral satellite RNAs (reviewed by Branch and Robertson, 1984). Low levels of α-amanitin inhibit the synthesis of both (+) and (–)PSTVd RNAs, thereby implicating DNA-dependent RNA polymerase II as the enzyme responsible for replication of PSTVd and related viroids (Spiesmacher *et al.*, 1985; Schindler and Mühlbach, 1992). For viroids like ASBVd, which appear to replicate in the chloroplast, replication is presumably catalyzed by the prokaryote-like organellar DNA-dependent RNA polymerase. Rohde *et al.* (1982) have shown that *E. coli*

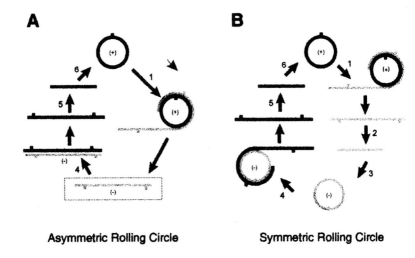

Asymmetric Rolling Circle   **Symmetric Rolling Circle**

Fig. 2. Rolling circle mechanisms for viroid and viral satellite RNA replication. Replication of viroids and other subviral RNAs may involve as many as six separate steps; that is, synthesis of multimeric (–) and (+)RNAs (steps 1 and 4), cleavage of these multimeric RNAs (steps 2 and 5), and circularization of the resulting linear (–) and (+)RNA monomers (steps 3 and 6). PSTVd and related viroids use an asymmetric mechanism (**A**) in which multimeric linear (–)RNAs are believed to serve as templates for synthesis of the (+) viroid progeny. In the symmetric rolling circle mechanism used by ASBVd and certain nepoviral satellite RNAs (**B**), the corresponding multimeric (–)RNAs are cleaved and circularized before acting as template for synthesis of (+)RNA progeny.

RNA polymerase is able to synthesize longer-than-unit-length RNAs complementary to PSTVd when provided with monomeric circular (+)PSTVd RNA as template.

The final step in viroid replication involves cleavage/ligation of the newly synthesized multimeric (+)strand RNAs to produce mature circular progeny. For ASBVd, PLMVd, and CChMVd, both multimeric (+) and (–)strand RNAs have been shown to undergo spontaneous cleavage via formation of "hammerhead" ribozyme structures *in vitro* (Hutchins *et al.*, 1986; Hernandez and Flores, 1992; Navarro and Flores, 1997). Most of these cleavage reactions are not readily reversible; thus, a host-encoded RNA ligase activity is presumably required for progeny circularization. Lafontaine *et al.* (1995) have shown, however, that both monomeric (+) and (–)PLMVd RNAs are capable of undergoing nonenzymatic ligation *in vitro*. Longer-than-unit-length (+)PSTVd RNAs do not undergo spontaneous self-cleavage *in vitro*, and accurate processing to form mature circular progeny in cell-free nuclear extracts requires rearrangement of the central conserved region to form a complex multihelix structure (Baumstark *et al.*, 1997). Recently, Liu and Symons (1998) have shown that RNA transcripts prepared from the conserved central domain of a related viroid (CCCVd) are able to undergo specific cleavage after denaturation *in vitro*.

B. Organismal Level

In addition to supporting viroid replication in the initially infected cells, systemic hosts also allow the resulting progeny to move from cell to cell as well as long distances through the vascular system. Plant viruses move systemically by modifying preexisting host pathways for macromolecular movement, and the underlying molecular mechanisms are currently the subject of intensive investigation (reviewed by Carrington *et al.*, 1996; see also Chapter 8, this volume). Figure 3 schematically illustrates several key points along these transport pathways.

Cell-to-cell movement of most plant viruses does *not* require synthesis of coat protein, but only the presence of a viral-encoded movement protein that allows the viral RNA (or DNA) to pass through the plasmodesmata connecting adjacent cells. Long-distance virus transport from the inoculated leaf occurs via the phloem, where transport of photoassimilate occurs by bulk flow from source to sink. Movement into the shoot and root apices and young leaves via the phloem is much more rapid than cell-to-cell movement, that is, rates are measured in centimeters per hour rather than cells per day

Very little information is currently available about viroid movement in infected plants. Microinjection experiments using infectious RNA transcripts synthesized *in vitro* and labeled with the nucleotide-specific fluorescent dye TOTO-1 iodide have shown that PSTVd can move rapidly from cell to cell via the plasmodesmata in tobacco mesophyll (Ding *et al.*, 1997). PSTVd RNA was also observed to accumulate in the nuclei of both the injected cell and neighboring cells. Using dot-blot hybridization to monitor PSTVd distribution in infected tomato seedlings, Palukaitis (1987) found that the movement pattern of PSTVd was indistinguishable from that of most plant viruses at the whole plant level. Hammond (1994) used an *Agrobacterium*-mediated inoculation strategy to examine the biological effects of apparently lethal mutations located in either the left or right terminal loops of PSTVd (see Figure 1). The presence of mutations in the right terminal loop resulted in infections that were primarily restricted to tomato gall and root tissues. Progeny were only occasionally detectable in newly emerging leaves, and, unlike wild-type PSTVd, which was present in all stem tissues, they were localized in the vascular cambium and outer phloem tissues. Many of the steps illustrated in Figure 3 are presumably dependent on formation of specific RNA–protein complexes. As mentioned above (see §VI.A), the nature (and existence) of such complexes remains to be established.

A

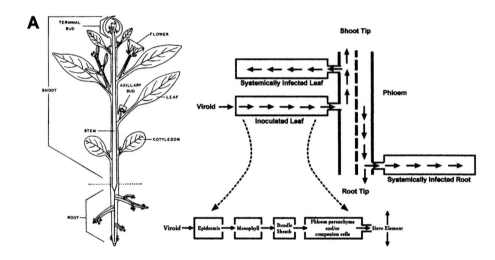

B

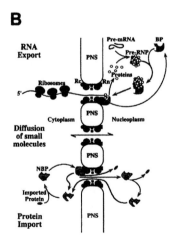

Nucleocytoplasmic transport

C

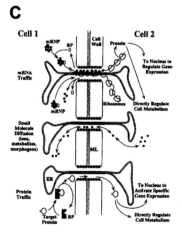

Plasmodesmal transport

C. Host Population

All viroids are mechanically transmissible, and some (e.g., PSTVd) can also be transmitted via true seed or pollen. With the exception of HSVd, natural infections appear to be limited to one or a very few cultivated hosts. Many of these hosts are vegetatively propagated, and most are woody perennials. Only in the case of TPMVd (and possibly CCCVd) does the outbreak of disease appear to depend on viroid transmission from a wild plant reservoir.

As mentioned above, viroid infections are persistent. The absence of an immune system leaves the infected plant without a means to inactivate/remove infectious viroid progeny; thus, once a host becomes infected by a viroid, it usually remains so. Many viroid infections are not accompanied by obvious disease symptoms, and vegetative propagation of these apparently healthy plants leads to a buildup of viroids in the population. Indeed, woody perennials such as citrus or grapevine may be infected with more than one viroid. Viroids share with plant viruses the phenomenon of "cross-protection," where infection with a mild strain of a viroid protects the host from the effects of superinfection with a severe strain of the same viroid (Fernow *et al.*, 1969; Niblett *et al.*, 1978; Semancik *et al.*, 1995). Cross-protection is observed only between closely related viroids, and superinfection by members of different viroid groups (see Table I) is not affected.

Despite many efforts (e.g., Manzer *et al.*, 1964; Singh, 1985), no useful sources of resistance to PSTVd have been identified. This fact has been interpreted by Diener (1979, 1987a) as evidence that the cultivated potato is a comparatively new host for PSTVd. Presumably, had the host and pathogen coevolved, gene-for-gene vertical resistance should be identifiable. Very little effort has been made to identify viroid resistance in other crop species; instead, efforts have centered on the development of sensitive and rapid diagnostic tests to detect infected plants, so that they can be removed and destroyed.

Fig. 3 (opposite). Pathways of viroid movement. (A) Cell-to-cell and long-distance movement in a systemically infected host plant. (**Left**) Diagram of the structure of a typical dicot. Within the vascular system, transport of water and minerals from root to shoot occurs in the xylem; transport of photoassimilate from source (mature leaves) to sink (roots and terminal portions of the stem) occurs in the phloem. Reprinted with permission from Cronquist (1961). (**Right**) On leaving an infected epidermal cell in the inoculated leaf, viroids must pass through (and infect) several different cell types in order to reach the vascular system. In systemically infected tissue, the pattern of viroid movement is thought to be the reverse of that observed in the inoculated leaf; that is, phloem sieve elements → companion cells → → epidermis. For certain viroids (e.g., citrus viroids), successful inoculation requires direct introduction into the phloem or cambial tissues (Garnsey and Jones, 1967). Note that long-distance movement in the phloem may be either basipetal (toward the root) or acropetal (toward the shoot apex). (**B,C**) Functional similarities between nucleocytoplasmic and plasmodesmatal transport of proteins and RNA. Nucleocytoplasmic transport through the nuclear pore complex (**B**) involves binding to receptor proteins located in the nucleoplasmic ring (Rn). Although pathways for RNA export and protein import are shown, either pathway can also function in reverse. Rc = cytoplasmic ring; PNS = perinuclear space; BP = binding protein; NBP = nuclear localization signal binding protein. A similar pattern of interaction between binding proteins and a transport complex embedded in the appressed endoplasmic reticulum (ER) are thought to be responsible for transport of RNAs and proteins from cell to cell via the plasmodesmata (**C**). Reprinted with permission from Lucas *et al.* (1993).

Four types of diagnostic tests have been widely used for viroid detection, namely bioassay (on suitable indicator hosts, polyacrylamide gel electrophoresis, dot-blot and other nucleic acid hybridization assays, and polymerase chain reaction-based assays. Each type of test has certain advantages and disadvantages, and no single assay can provide rapid, specific, and reliable detection for all viroid/host combinations. Dot-blot and tissue printing assays (e.g., Owens and Diener, 1981; Podleckis *et al.*, 1993) have been widely adopted for routine screening of potato, chrysanthemum, and various fruit tree species for viroid infection. The absence of a coat protein means that serological assays such as ELISA (*E*nzyme-*L*inked *I*mmuno*S*orbent *A*ssay) cannot be used to detect viroids.

VIII. EFFECTS OF VIROID INFECTION ON THE HOST

A. Direct Effects (Pathogenicity, Symptom Expression)

Studies describing the macroscopic symptoms associated with viroid infection as well as the underlying physiological and ultrastructural alterations have been reviewed in detail elsewhere (e.g., Diener, 1979, 1987a; Semancik and Conejero-Tomas, 1987). These symptoms — stunting, leaf distortion and epinasty, chlorosis and vein clearing, as well as necrotic reactions sometimes leading to death of the host — are very similar to those associated with many plant virus infections (see Fig. 4). Unlike many RNA viruses, however, viroids do not serve as mRNAs; thus, the abnormalities occurring as a result of infection must be due to *direct interaction of the viroid* (or its complementary RNA) *with host cell constituents.* As more and more information has become available about viroid structure, the possible nature of such interactions has become the subject of intense speculation. Unfortunately, this speculation has often provided more heat than light, and we are still far from understanding the molecular processes responsible for symptom production.

Viroid replication and symptom production have similar (and rather high) temperature optima: 28–32°C. Once disease symptoms begin to appear, viroid-infected plants usually remain symptomatic. One exception to this generalization is the so-called "symptomless carrier" state of ASBVd-infected avocado (Wallace and Drake, 1962; Desjardins, 1987). Only a low frequency of seed transmission is observed for ASBVd-infected trees whose leaves, stems, or fruit showed the characteristic blotching symptoms. In contrast, virtually 100% seed transmission was reported for fruit collected from either symptomless trees or symptomless branches on symptomatic trees. The resulting seedlings (and their succeeding generations) remained symptomless, but healthy plants grafted with buds excised from such plants developed typical disease symptoms. ASBVd titers in asymptomatic tissues are lower than those in symptomatic tissue, but the physiological mechanisms responsible for the establishment and maintenance of this symptomless carrier state are presently unknown.

Stunting and epinasty (a downward curling of a leaf caused by more vigorous growth of its upper surface) are often regarded as "typical" symptoms of viroid infection. As discussed by Semancik and Conejero-Tomas (1987), their appearance reflects viroid-induced disturbances in gibberellin, auxin, and ethylene metabolism. The reduced root formation often observed in viroid-infected plants is also likely to reflect these changes in hormone concentrations/ratios and may contribute to the overall reduction in plant size. Viroid-induced leaf epinasty and stunting resembles the effects of ethylene exposure on healthy plants, and CEVd infection has been shown to lead to increased ethylene production by *Gynura* tissues (Duran-Vila, 1982). Foliar applications of either ethephon (2-chloroethyl-phosphonic acid), which liberates ethylene, or $AgNO_3$ also result in the appearance of viroid-like symptoms in *Gynura*. The ethylene signal transduction pathway is known to play a crucial role in many plant stress responses and developmental processes (Ecker, 1995).

At the cellular level, viroid infection has been associated with abnormal development of the cell wall, alterations in the plasma membrane leading to formation of so-called "paramural bodies" or "plasmalemmasomes" and various chloroplast abnormalities. Formation of chlorotic lesions is a major feature of both ASBVd infection of avocado and CSVd infection of chrysanthemum. As ASBVd has been shown to accumulate in the chloroplast (Bonfiglioli *et al.*, 1994a,b; Lima *et al.*, 1994), it is not surprising that these organelles contain distorted thylakoid membranes and electron-dense deposits. As pointed out by Diener (1987a), chloroplast abnormalities may be a general feature of viroid infection, however. Tomatoes infected by HSVd or the closely related cucumber pale fruit viroid (CPFVd) show no visible signs of viroid infection, but chloroplast disintegration has also been observed in such plants (Kojima *et al.*, 1983). Likewise, the development of thickened or distorted cell walls cannot be directly correlated with plasma membrane alterations. With the exception of a possible moderate degree of hypertrophy observed with nuclei isolated from PSTVd-infected tomato leaf tissue (Takahashi and Diener, 1975), viroid infection appears to have little or no effect on nuclear structure.

In their discussion of viroid pathogenesis, Semancik and Conejero-Tomas (1987) emphasize two points: first, the spatial separation between sites of replication and pathogenesis for PSTVd and related viroids; and second, the absence of a unique viroid-specified cytoplasmic lesion. The effect of viroid infection on several cell membrane systems was also noted. Because membrane surfaces provide recognition sites for many regulatory processes, they speculate that either

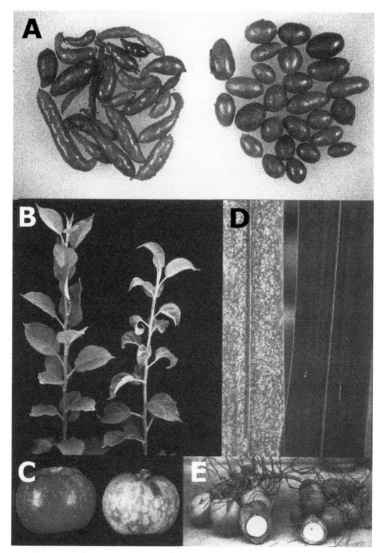

Fig. 4(A–E).   Symptoms associated with viroid infection. (**A**) Elongation of Kennebec microtubers caused by PSTVd infection. Tubers on the right were collected from uninfected control plants. Note also the prominent bud scales or "eyebrows" on the infected tubers. (**B,C**) Leaf epinasty and fruit dappling in apple caused by ASSVd infection. Shoots were taken from growth chamber-grown, graft-inoculated "Stark's Earliest" approximately 6 weeks post-inoculation; they were collected from 5-year-old orchard-grown "Lord Lambourne" apple trees. In each case, samples on the left were collected from uninfected trees. (**D,E**) Non-necrotic chlorotic leaf spotting and nut abnormalities associated with CCCVd infection of coconut palm. Samples on the left were collected from infected palms. Note the rounding, equatorial scarifications, and reduced husks in the nuts from diseased palms.

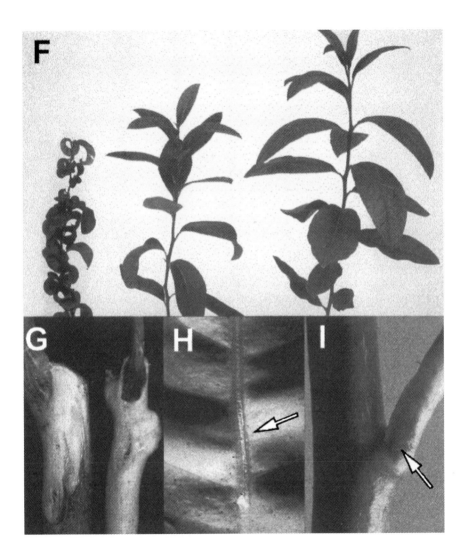

Fig. 4(F–I). Characteristic symptoms of viroid infection in citrus. (**F**) Leaf epinasty in "Etrog" citron induced by CEVd (left) or citrus viroid III (middle). The plant on the right is an uninfected control. (**G**) Cachexia (or xyloporosis) at the graft union between rootstock and scion caused by citrus viroid IIb, a sequence variant of HSVd. Plant on the left is uninfected. (**H,I**) Necrosis of leaf midveins (**H**) and petioles (**I**) caused by citrus viroid III. Panels **A–C** reprinted through the courtesy of E. V. Podleckis. Panels **F–I** reprinted through the courtesy of S. M. Garnsey. Panels **D** and **E** Reprinted with permission from Hanold and Randles (1991).

"the viroid itself or a viroid-induced 'signal' may be capable of interacting with these reactive interfaces." Preliminary data from two laboratories suggest that at least one such interaction may involve cellular protein kinase(s).

Many viruses induce the synthesis of doubled-stranded (ds) RNA at some point during their replication cycle, and the presence of this dsRNA appears to trigger many of the host responses to virus infection, probably through activation of such dsRNA-dependent enzymes as an interferon-inducible protein kinase (PKR) and $2',5'$-oligoadenylate synthetase (reviewed by Jacobs and Langland, 1996). Interestingly, adenovirus and several other animal viruses also synthesize certain small and highly structured RNAs (e.g., VAI RNA) that are able to block the activation and/or activity of these enzymes. Hiddinga *et al.* (1988) identified a host-encoded 68-kDa protein that appeared to be preferentially phosphorylated in leaf extracts prepared from PSTVd-infected tobacco. This protein appeared to be immunologically related to the similar-sized interferon-inducible PKR in mammalian cells. Diener *et al.* (1993) have subsequently presented evidence for differential activation of human PKR by PSTVd strains of varying pathogenicity. If confirmed, these results would provide the first convincing explanation for how infection by one viroid can result in severe symptoms while infection by a closely related species is asymptomatic. Several more recent reviews discuss the role of protein kinases and other molecules in host-pathogen recognition and subsequent defense responses (e.g., Bent, 1996; Hammond-Kosack and Jones, 1996; Dangl *et al.*, 1996; Baker *et al.*, 1997).

Several viroid diseases have had major impact on the production of specific crops, and the review by Garnsey and Randles (1987) provides an excellent overview of the effects of viroid infection at the crop or population level. Since it was first recognized in the 1930s, coconut cadang-cadang disease has killed more than 30 million palms within relatively small areas in the Philippines and Guam. Losses continue at the rate of several hundred thousand trees per year, and estimates of the resulting loss in copra production range between U.S. $80 and $100 per tree. Damage would be far greater if the disease were to spread to other production areas. Shortly after the end of World War II, the U.S. chrysanthemum industry was nearly destroyed by the rapid spread of chrysanthemum stunt viroid. In addition to stunting, important symptoms of this disease include chlorotic blotching of the foliage, premature flowering and poor flower quality, and reduced root formation by cuttings. Commercial chrysanthemum production depends almost exclusively on vegetative propagation.

The ready transmission of PSTVd by vegetative propagation, foliar contact, and true seed or pollen poses a potentially serious threat to germplasm collections and breeding programs as well as potato production. Exact losses are hard to document because effects of infection vary with PSTVd strain, host cultivar, and environment. In addition to tuber elongation/deformation, overall yield can also be reduced. Currently, the most significant effects of PSTVd infection are twofold: the costs associated with certification testing and the significant delays in movement of seed stock and plants for both commercial and experimental purposes associated with mandatory quarantine testing. PSTVd and other viroid diseases can be contained by rigorous testing and certification protocols, but there is yet no proven strategy for their permanent control. Quarantine bioassays for viroids affecting fruit trees can take as long as 6–18 months to complete.

B. Indirect Effects (Fitness, Competitiveness)

Many examples of synergistic interactions involving two (or more) plant viruses or an individual virus and its associated satellite RNA have been described, and the molecular mechanisms underlying these interactions are currently under investigation (e.g., Sriskinda et al., 1996). Synergistic interactions between viroids and viruses have also been reported; for example, Singh (1982) has described a severe necrotic disease in Kennebec potatoes caused by dual infection with PSTVd and potato virus Y. Neither agent alone was able to produce the severe disease symptoms, and the combination was effective only when the initial infecting agent was PSTVd.

Other indirect effects of viroid infection are best illustrated by several phenomena associated with CEVd. Infection of tomato by either CEVd or PSTVd induces a phenomenon known as "systemic acquired resistance" that involves an array of non pathogen-specific host responses (Camacho Henriquez and Sänger, 1982; Vera et al., 1994; Domingo et al., 1994). These include stimulation of several phenylpropanoid metabolic pathways, modification of cell wall properties, and increased synthesis and/or accumulation of antimicrobial proteins and phytoalexins (reviewed by Ryals et al., 1996). It is not surprising, therefore, that CEVd infection has been reported to increase the resistance of citrus to at least one fungal disease, *Phytophthora* root rot (Rossetti et al., 1980).

Finally, viroid infection may also have other beneficial effects on citrus production. Labor costs associated with harvesting, spraying, and other grove care operations represent major production expenses, and such costs can be reduced by utilizing trees with smaller, more compact canopies. Even though mechanical pruning may reduce total fruit yield per unit area, the resulting decrease in production and harvesting costs can actually increase net financial return. The dwarfing associated with propagation of viroid-infected trees on susceptible rootstocks provides another means to reduce production costs while maintaining fruit quality and yield per unit volume of tree canopy.

Deliberate release of a known pathogen into the environment is not without risk (see Garnsey and Randles, 1987); nevertheless, field trials using naturally derived citrus "dwarfing agents" have been carried out in Australia (Hutton *et al.*, 1999), Israel (Ashkenazi and Oren, 1988), and the United States (Semancik *et al.*, 1997). In deciding whether or not to use viroid-induced dwarfing, growers must carefully consider the choice of rootstock. Not all rootstock–scion combinations are susceptible to viroid-induced dwarfing, and specific production areas may require the use of certain rootstocks that are resistant/tolerant to blight, tristeza virus, or *Phytophthora* infection.

ACKNOWLEDGMENTS

I thank E. V. Podleckis (USDA/APHIS, Riverdale, MD), S. M. Garnsey (USDA/ARS, Orlando, FL), and L. F. Salazar (International Potato Center, Lima, Peru) for their ideas and suggestions offered during critical review of this manuscript.

REFERENCES

Ashkenazi, S., and Oren, Y. (1988). The use of citrus exocortis virus for tree size control in Israel: Practical aspects. *Soc. Citriculture* **2**, 917–919.

Baker, B., Zambryski, P., Staskawicz, B., and Dinesh-Kumar, S. P. (1997). Signalling in plant–microbe interactions. *Science* **276**, 726–733.

Bar-Joseph, M., and Yang, G. (1997). Attempts to team citrus viroids for horticultural purposes. *In* "Proceedings of the Workshop on Plant Viroids and Viroid-Like RNAs from Plants, Animals, and Fungi" (R. Flores and H. L. Sänger, eds.), p. 47. Instituto Juan March, Madrid.

Baumstark, T., Schröder, A. R. W., and Riesner, D. (1997). Viroid processing: Switch from cleavage to ligation is driven by a change from a tetraloop to a loop E conformation. *EMBO J.* **16**, 599–610.

Bent, A. F. (1996). Plant disease resistance genes: Function meets structure. *Plant Cell Physiol.* **8**, 1757–1771.

Bhattiprolu, S. L. (1991). Studies on a newly recognized disease of *Nicotiana glutinosa* of viroid etiology. *Plant Dis.* **75**, 1068–1071.

Bonfiglioli, R. G., McFadden, G. I., and Symons, R. H. (1994a). *In situ* hybridization localizes avocado sunblotch viroid on chloroplast thylakoid membranes and coconut cadang cadang viroid in the nucleus. *Plant J.* **6**, 99–103.

Bonfiglioli, R. G, Webb, D. R., and Symons, R. H. (1994b). Tissue and intracellular distribution of coconut cadang cadang viroid and citrus exocortis viroid determined by *in situ* hybridization and confocal laser scanning and transmission electron microscopy. *Plant J.* **9**, 457–465.

Branch, A. D., and Robertson, H. D. (1984). A replication cycle for viroids and other small infectious RNAs. *Science* **223**, 450–455.

Branch, A. D., Robertson, H. D., and Dickson, E. (1981). Longer-than-unit-length viroid minus strands are present in RNA from infected plants. *Proc. Natl. Acad. Sci. U.S.A.* **78**, 6381–6385.

Bussière, F., Lafontaine, D., Côté, F., Beaudry, D., and Perrault, J.-P. (1995). Evidence for a model ancestral viroid. *Nucleic Acids Symp. Ser.* **33**, 143–144.

Camacho Henriquez, A., and Sänger, H. L. (1982). Analysis of acid-extractable tomato leaf proteins after infection with a viroid, two viruses and a fungus and partial purification of the "pathogenesis-related" protein p14. *Arch. Virol.* **74**, 181–196.

Carrington, J. C., Kasschau, K. D., Mahajan, S. K., and Schaad, M. C. (1996). Cell-to-cell and long-distance transport of viruses in plants. *Plant Cell Physiol.* **8**, 1669–1681.

Cronquist, A. (1961). "Introductory Botany." Harper and Row, New York.

Dangl, J. L., Dietrich, R. A., and Richberg, M. H. (1996). Death don't have no mercy: Cell death programs in plant–microbe interactions. *Plant Cell Physiol.* **8**, 1793–1807.

Darós, J. A., and Flores, R. (1995). Identification of a retroviroid-like element from plants. *Proc. Natl. Acad. Sci. U.S.A.* **92**, 6856–6860.

Desjardins, P. R. (1987). Avocado sunblotch. *In* "The Viroids" (T. O. Diener, ed.), pp. 299–313. Plenum, New York.

Diener, T. O. (1971a). Potato spindle tuber virus, IV: A replicating low-molecular-weight RNA. *Virology* **45**, 411–428.

Diener, T. O. (1971b). Potato spindle tuber virus: A plant virus with properties of a free nucleic acid, III: Subcellular location of PSTV-RNA and the question of whether virions exist in extracts or in situ. *Virology* **43**, 75–89.

Diener, T. O. (1979) "Viroids and Viroid Diseases." Wiley-Interscience, New York.

Diener, T. O. (1987a). Biological properties. *In* "The Viroids" (T. O. Diener, ed.), pp. 9–35. Plenum, New York.

Diener, T. O. (1987b). Potato spindle tuber. *In* "The Viroids" (T. O. Diener, ed.), pp. 329–331. Plenum, New York.

Diener, T. O., and Raymer, W. B. (1969). Potato spindle tuber virus: A plant virus with properties of a free nucleic acid, II: Characterization and partial purification. *Virology* **37**, 351–366.

Diener, T. O., Hammond, R. W., Black, T., and Katze, M. G. (1993). Mechanism of viroid pathogenesis: Differential activation of the interferon-induced, double-stranded RNA-activated. M_r 68,000 protein kinase by viroid strains of varying pathogenicity. *Biochimie* **75**, 533–538.

Ding, B., Kwon M.-O., Hammond, R., and Owens, R. (1997). Cell-to-cell movement of potato spindle tuber viroid. *Plant J.* **12**, 931–936.

Domingo, C., Conejero, V., and Vera, P. (1994). Genes encoding acidic and basic class III β-1,3-glucanases are expressed in tomato plants upon viroid infection. *Plant Mol. Biol.* **24**, 725–732.

Duran-Vila, N. (1982). "Studies of Viroid Pathogenesis: Responses of CEV-Containing Tissues and Cells." PhD dissertation, University of California, Riverside.

Duran-Vila, N., Roistacher, C. N., Rivera-Bustamante, R., and Semancik, J. S. (1988). A definition of citrus viroid groups and their relationship to the exocortis disease. *J. Gen. Virol.* **69**, 3069–3080.

Ecker, J. R. (1995). The ethylene signal transduction pathway in plants. *Science* **268**, 667–675.

Elena, S. F., Dopazo, J., Flores, R., Diener, T. O., and Moya, A. (1991). Phylogeny of viroids, viroidlike satellite RNAs, and the viroidlike domain of hepatitis δ virus RNA. *Proc. Natl. Acad. Sci. U.S.A.* **88**, 5631–5634.

Fernow, K. H. (1967). Tomato as a test plant for detecting mild strains of potato spindle tuber virus. *Phytopathology* **57**, 1347–1352.

Fernow, K. H., Peterson, L. C., and Plaisted, R. L. (1969). Spindle tuber virus in seeds and pollen of infected potato plants. *Am. Potato J.* **46**, 424–429.

Flores, R. (1995). Subviral agents: Viroids. *Arch. Virol.* (Suppl. 10), pp. 495–497.

Flores, R., Di Serio, F., and Hernandez, C. (1997). Viroids: The noncoding genomes. *Sem. Virol.* **8**, 65–73.

Flores, R., Randles, J. W., Bar-Joseph, M., and Diener, T. O. (1998). A proposed scheme for viroid classification and nomenclature. *Arch. Virol.* **143**, 623–629.

Galindo, J., Lopez, C., and Aguilar, T. (1989). Discovery of the transmitting agent of tomato planta macho viroid. *Revista Mexicana de Fitopatologia* **7**, 61–65.

Garnsey, S. M., and Jones, S. M. (1967). Mechanical transmission of exocortis virus with contaminated budding tools. *Plant Dis. Reporter* **51**, 410–413.

Garnsey, S. M., and Randles, J. W. (1987). Biological interactions and agricultural implications of viroids. *In* "Viroids and Viroid-Like Pathogens" (J. S. Semancik, ed.), pp. 127–160. CRC Press, Boca Raton, FL.

Görlich, D., and Mattaj, I. W. (1996). Nucleocytoplasmic transport. *Science* **271**, 1513–1518.

Gross, H. J., Domdey, H., Lossow, C. Jank, P., Raba, M., Alberty, H., and Sänger, H. L. (1978). Nucleotide sequence and secondary structure of potato spindle tuber viroid. *Nature (London)* **273**, 203–208.

Hammond, R. W. (1994). Agrobacterium-mediated inoculation of PSTVd cDNAs onto tomato reveals the biological effect of apparently lethal mutations. *Virology* **201**, 36–45.

Hammond-Kosack, K. E., and Jones, J. D. G. (1996). Resistance gene-dependent plant defense responses. *Plant Cell Physiol.* **8**, 1773–1791.

Hanold, D., and Randles, J. W. (1991). Coconut cadang-cadang disease and its viroid agent. *Plant Dis.* **75**, 330–335.

Harders, J., Lukács, N., Robert-Nicoud, M., Jovin, T. M., and Riesner, D. (1989). Imaging of viroids in nuclei from tomato leaf tissue by in situ hybridization and confocal laser scanning microscopy. *EMBO J.* **8**, 3941–3949.

Hernandez, C., and Flores, R. (1992). Plus and minus RNAs of peach latent mosaic viroid self-cleave in vitro via hammerhead structures. *Proc. Natl. Acad. Sci. U.S.A.* **89**, 3711–3715.

Hiddinga, H. J., Grum, C. J., Hu, J., and Roth, D. A. (1988). Viroid-induced phosphorylation of a host protein related to a dsRNA-dependent protein kinase. *Science* **241**, 451–453.

Hutchins, C. J., Rathjen, P. D., Forster, A. C., and Symons, R. H. (1986). Self-cleavage of plus and minus RNA transcripts of avocado sunblotch viroid. *Nucleic Acids Res.* **14**, 3627–3640.

Hutton, R. J., Broadbent, P., and Berington, K. B. (1999). Viroid dwarfing for high-density citrus plantings. *Hort. Rev.* **24**, 277–317.

ICTV (International Committee on Taxonomy of Viruses) (1995). "Virus Taxonomy: Sixth Report of the International Committee on Taxonomy of Viruses" (F. A. Murphy, C. M. Fauquet, D. H. L. Bishop, S. A. Ghabrial, A. W. Jarvis, G. P. Martelli, M. A. Mayo, and M. D. Summers, eds.). Springer-Verlag, Vienna.

Ito, T., Kanematsu, S., Koganezawa, H., Tsuchizaki, T., and Yoshida, K. (1993). Detection of a viroid associated with apple fruit crinkle disease. *Ann. Phytopathol. Soc. Japan* **59**, 520–527.

Jacobs, B. L., and Langland, J. O. (1996). When two strands are better than one: The mediators and modulators of the cellular responses to double-stranded RNA. *Virology* **219**, 339–349.

Keese, P., and Symons, R. H. (1985). Domains in viroids: Evidence of intermolecular RNA rearrangements and their contribution to viroid evolution. *Proc. Natl. Acad. Sci. U.S.A.* **82**, 4582–4586.

Klaff, P., Gruner, R., Hecker, R., Sättler, A., Theissen, G., and Riesner, D. (1989). Reconstituted and cellular viroid-protein complexes. *J. Gen. Virol.* **70**, 2257–2270.

Kojima, M., Murai, M., and Shikata, E. (1983). Cytopathic changes in viroid-infected leaf tissues. *J. Fac. Agric. Hokkaido Univ.* **61**, 219–223.

Koltunow, A. M., and Rezaian, M. A. (1989a). A scheme for viroid classification. *Intervirology* **30**, 194–201.

Koltunow, A. M., and Rezaian, M. A. (1989b). Grapevine viroid Ib, a new member of the apple scar skin viroid group contains the left terminal region of tomato planta macho viroid. *Virology* **170**, 575–578.

Lafontaine, D., Beaudry, D., Marquis, P., and Perraullt, J.-P. (1995). Intra- and intermolecular nonenzymatic ligations occur within transcripts derived from the peach latent mosaic viroid. *Virology* **212**, 705–709.

Lima, M. I., Fonseca, M. E. N., Flores, R., and Kitajima, E. W. (1994). Detection of avocado sunblotch viroid in chloroplasts of avocado leaves by in situ hybridization. *Arch. Virol.* **138**, 385–390.

Liu, Y.-H., and Symons, R. H. (1998) Specific RNA self-cleavage in coconut cadang cadang viroid: Potential for a role in rolling circle replication. *RNA* **4**, 418–429.

Lucas, W. F., Ding, B., and Van der Schoot, C. (1993). Tansley Review No. 58: Plasmodesmata and the supracellular nature of plants. *New Phytol.* **125**, 435–476.

Manzer, F. M., Akeley, R. V., and Merriam, D. (1964). Resistance in *Solanum tuberosum* to mechanical inoculation with the potato spindle tuber virus. *Am. Potato J.* **41**, 411–416.

Martinez-Soriano, J. P., Galindo-Alonzo, J., Maroon, C. J. M., Yucel, I., Smith, D. R., and Diener, T. O. (1996). Mexican papita viroid: Putative ancestor of crop viroids. *Proc. Natl. Acad. Sci. U.S.A.* **93**, 9397–9401.

Matthews, R. E. F. (1991). "Plant Virology," 3rd ed. Academic Press, New York.

Merriam, D., and Bonde, R. (1954). Dissemination of spindle tuber by contaminated tractor wheels and by foliage contact with diseased plants. *Phytopathology* **44**, 111.

Mezitt, L. A., and Lucas, W. J. (1996). Plasmodesmal cell-to-cell transport of proteins and nucleic acids. *Plant Mol. Biol.* **32**, 251–273.

Mishra, M. D., Hammond, R. W., Owens, R. A., Smith, D. R., and Diener, T. O. (1991). Indian bunchy top disease of tomato plants is caused by a distinct strain of citrus exocortis viroid. *J. Gen. Virol.* **72**, 1781–1785.

Navarro, B., and Flores, R. (1997). Chrysanthemum chlorotic mottle viroid: Unusual structural properties of a subgroup of self-cleaving viroids with hammerhead ribozymes. *Proc. Natl. Acad. Sci. U.S.A.* **94**, 11262–11267.

Niblett, C. L., Dickson, E., Fernow, K. H., Horst, R. K., and Zaitlin, M. (1978). Cross-protection among four viroids. *Virology* **91**, 198–203.

Owens, R. A., and Diener, T. O. (1981). Sensitive and rapid diagnosis of potato spindle tuber viroid disease by nucleic acid hybridization. *Science* **213**, 60–672.

Owens, R. A., and Diener, T. O. (1982). RNA intermediates in potato spindle tuber viroid replication. *Proc. Natl. Acad. Sci. U.S.A.* **79**, 113–117.

Palukaitis, P. (1987). Potato spindle tuber viroid: Investigation of the long-distance, intra-plant transport route. *Virology* **158**, 239–241.

Podleckis, E. V., Hammond, R. W., Hurtt, S. S., and Hadidi, A. (1993). Chemiluminescent detection of potato and pome fruit virids by digoxygenin-labeled dot-blot and tissue blot hybridization. *J. Virol. Methods* **43**, 147–158.

Qu, F., Heinrich, C., Loss, P., Steger, G., Tien, P., and Riesner, D. (1993). Multiple pathways of reversion in viroids for conservation of structural elements. *EMBO J.* **12**, 2129–2139.

Querci, M., Owens, R. A., Bartolini, I., Lazarte, V., and Salazar, L. F. (1997). Evidence for heterologous encapsidation of potato spindle tuber viroid in particles of potato leafroll virus. *J. Gen. Virol.* **78**, 1207–1211.

Riesner, D. (1990). Structure of viroids and their replicative intermediates. Are thermodynamic domains also functional domains. *Sem. Virol.* **1**, 83–99.

Rohde, W., Rackwitz, H.-R., Boege, F., and Sänger, H. L. (1982). Viroid RNA is accepted as a template for in vitro transcription by DNA-dependent DNA polymerase I and RNA polymerase from *Escherichia coli. Biosci. Rep.* **2**, 929–939.

Rossetti, V., Pompeu Jr., J., Rodriguez, O., Vechiato, M. H., Da Veiga, M. L., Oliveira, D. A., and Sobrinho, J. T. (1980). Reaction of exocortis-infected and healthy trees to experimental Phytophthora inoculations. *Proc. Int. Organiz. Citrus Virologists* **8**, 209–214.

Ryals, J. A., Neuenschwander, U. H., Willits, M. G., Molina, A., Steiner, H.-Y., and Hunt, M. D. (1996). Systemic acquired resistance. *Plant Cell Physiol.* **8**, 1651–1688.

Salazar, L. F., Hammond, R. W., Diener, T. O., and Owens, R. A. (1986). Analysis of viroid replication following Agrobacterium-mediated inoculation of non-host species with potato spindle tuber viroid cDNA. *J. Gen. Virol.* **69**, 879–889.

Schindler, I. M., and Mühlbach, H. P. (1992). Involvement of nuclear DNA-dependent RNA polymerases in potato spindle tuber viroid replication: A reevaluation. *Plant Sci.* **84**, 221–229.

Schultz, E. S., and Folsom, D. (1923). Transmission, variation, and control of certain degeneration diseases of Irish potatoes. *J. Agric. Res.* **25**, 43–118.

Semancik, J. S. (1991). Progress and perspectives in grapevine viroid research 1985–1990. *In* "Proceedings of the 10th Meeting of the International Council for the Study of Viruses and Virus Diseases of the Grapevine (ICVG)" (I. C. Rumbos, R. Bovey, D. Gonsalves, W. B. Hewitt, and G. P. Martelli, eds.), pp. 260–269. Volos, Greece.

Semancik, J. S., and Conejero-Tomas, V. (1987). Viroid pathogenesis and expression of biological activity. *In* "Viroids and Viroid-Like Pathogens" (J. S. Semancik, ed.), pp. 71–126. CRC Press, Boca Raton, FL.

Semancik, J. S., Gumpf, D. J., and Bash, J. A. (1995). Interactions amoung the group II citrus viroids: A potential for protection from the cachexia disease. *Proc. Int. Organiz. Citrus Virologists* **12**, 189–195.

Semancik, J. S., Rakowski, A. G., Bash, J. A., and Gumpf, D. J. (1997). Application of selected viroids for dwarfing and enhancement of production of "Valencia" orange. *J. Hort. Sci.* **72**, 563–570.

Shikata, E. (1990). New viroids from Japan. *Sem. Virol.* **1**, 107–115.

Singh, R. P. (1970). Seed transmission of potato spindle tuber virus in tomato and potato. *Am. Potato J.* **47**, 225–227.

Singh, R. P. (1982). A unique interaction of spindle tuber viroid and virus Y in potatoes. *Phytopathology* **72**, 962.

Singh, R. P. (1985). Clones of *Solanum berthaultii* resistant to potato spindle tuber viroid. *Phytopathology* **75**, 1432–1434.

Singh, R. P., and Boucher, A. (1991). High incidence of transmission and occurrence of a viroid in commercial seeds of *Coleus* in Canada. *Plant Dis.* **75**, 184–187.

Spieker, R. L. (1996). *In vitro*-generated "inverse" chimeric *Coleus blumei* viroids evolve *in vivo* into infectious RNA replicons. *J. Gen. Virol.* **77**, 2839–2846.

Spiesmacher, E., Mühlbach, H.-P., Tabler, M., and Sänger, H. L. (1985). Synthesis of (+) and (–)RNA molecules of potato spindle tuber viroid (PSTV) in isolated nuclei and its impairment by transcription inhibitors. *Biosci. Rep.* **3**, 251–265.

Sriskinda, V. S., Pruss, G., Ge, X., and Vance, V. B. (1996). An eight-nucleotide sequence in the potato virus X 3′ untranslated region is required for both host protein binding and viral multiplication. *J. Virol.* **70**, 5266–5271.

Takahashi, T., and Diener, T. O. (1975). Potato spindle tuber viroid, XIV: Replication in nuclei isolated from infected leaves. *Virology* **64**, 106–114.

Tien, P., and Chen, W. (1987). Burdock stunt. *In* "The Viroids" (T. O. Diener, ed.), pp. 333–339. Plenum, New York.

Vera, P., Tornero, P., and Conejero, V. (1994). Cloning and expression analysis of a viroid-induced peroxidase from tomato plants. *Mol. Plant-Microbe Interact.* **6**, 790–794.

Visvader, J. E., and Symons, R. H. (1985). Eleven new sequence variants of citrus exocortis viroid and the correlation of sequence with pathogenicity. *Nucleic Acids Res.* **13**, 2907–2920.

Wallace, J. M., and Drake, R. J. (1962). A high rate of seed transmission of avocado sunblotch virus from symptomless trees and the origin of such trees. *Phytopathology* **52**, 237–241.

Wassenegger, M., Heimes, S., and Sänger, H. L. (1994). An infectious viroid RNA replicon evolved from an in vitro-generated non-infectious viroid deletion mutant via a complementary deletion *in vivo*. *EMBO J.* **13**, 6172–6177.

Wassenegger, M., Spieker, R. L., Thalmeir, S., Gast, F.-U., Riedel, L., and Sänger, H. L. (1996). A single nucleotide substitution converts potato spindle tuber viroid (PSTVd) from a noninfectious to an infectious RNA for *Nicotiana tabacum*. *Virology* **226**, 191–197.

Wolff, P., Gilz, R., Schumacher, J., and Riesner, D. (1985). Complexes of viroids with histones and other proteins. *Nucleic Acids Res.* **13**, 355–367.

Yaguchi, S., and Takahashi, T. (1984). Survival of hop stunt viroid in the hop garden. *Phytopathol. Z.* **109**, 32–44.

Zhu, S. F., Hadidi, A., Yang, X., Hammond, R. W., and Hansen, A. J. (1995). Nucleotide sequence and secondary structure of pome fruit viroids from dapple apple diseased apples, pear rusty skin diseased pears and apple scar skin symptomless pears. *Acta Horticulturae* **386**, 554–559.

IV

Viruses of Macroscopic Animals

10

Ecology of Insect Viruses

LORNE D. ROTHMAN* and JUDITH H. MYERS†

*SAS Institute (Canada) Inc.
Toronto, Ontario M5J 2T3, Canada

†Centre for Biodiversity Research
Departments of Zoology and Plant Science
University of British Columbia
Vancouver, British Columbia V6T 1Z4, Canada

I. INTRODUCTION

Insects, mostly in the orders Diptera (flies), Hymenoptera (e.g., sawflies, wasps, bees), Coleoptera (beetles), and particularly Lepidoptera (moths and butterflies)

VIRAL ECOLOGY

are hosts to a variety of viruses. Research on insect viruses has focused on the more virulent pathogens of pests of forests and agriculture. However, the effects of viruses are extremely varied. Some cause spectacular epizootics and extensive mortality, while others are more benign and cause less obvious characteristics of disease. Recent advances in molecular biology have greatly improved our ability to identify and describe insect viruses and to study their behavior within the host. But to predict viral epizootics, or to successfully use virus to control pest species, we require a greater understanding of viral ecology. We do not provide an exhaustive review of insect host–virus interactions. Rather, we focus on some of the fundamental aspects of transmission, and effects of viruses on individuals and on host populations.

II. TYPES OF VIRUSES

We outline in Table I the major categories of insect viruses. Insect viruses can be categorized as being occluded or nonoccluded, of being DNA or RNA viruses, and of replicating in the nucleus or cytoplasm of cells. Except for the Baculoviridae and the Nudaurelia β virus group, families including entomoviruses also have forms that infect vertebrates or plants. The baculoviruses are specific to insects and have been the focus of the most study (Cory et al., 1997). Two major groups of baculoviruses are the nuclear polyhedral viruses (NPVs) and granulosis viruses (GVs), DNA viruses that replicate in the nuclei of cells. The virions of baculoviruses are encapsulated in protein occlusion bodies (OBs), which can be observed with a light microscope: 1–4 μm for NPVs and 0.1–0.5 μm for GVs. The protein matrix of the OBs protects the virions in the environment following the death of infected individuals.

III. TRANSMISSION

Two general pathways of virus transmission are horizontal, among individuals in the same generation, and between generations through environmental contamination, and vertical, directly from parents to offspring (Andreadis, 1987; Kukan, 1996).

TABLE I

Families and Genera of Viruses Associated with Insects Either as Pathogens or as Viruses that Replicate within the Insect and Are Transmitted to Vertebrate (V) or Plant (P) Hosts[a]

Family	Genus infecting insects[b]	Nucleic acid	Occlusion bodies	Replicates in	Hosts of other genera of the family (common example)
Baculoviridae	*Nucleopolyhidrovirus* (NPV)	dsDNA	Yes	Nucleus	None
	Granulovirus (GV)				
Unassigned	*Oryctes virus* (OrV)	dsDNA	No	Nucleus	None
Poxviridae	*Entomopox* (EPV)	dsDNA	Yes	Cytoplasm	V (smallpox)
Reoviridae	*Cytoplasmic polyhidrosis* (CPV)	dsRNA	Yes	Cytoplasm	V (blue tongue), P
Iridoviridae	*Iridovirus* (IV)	dsDNA	No	Cytoplasm	V (*Ranavirus*)
Parvoviridae	*Densovirus* (DNV)	ssDNA	No	Nucleus	V (*Parvovirus*)
Picornaviridae	Unassigned (PV)	ssRNA	No	Cytoplasm	V (hepatitis A)
Nodaviridae	*Alphanodavirus*	ssRNA	No	Cytoplasm	Mammalian cells
	Betanodavirus				
Tetraviridae	*Betatetravirus*	ssRNA	?	Cytoplasm and nucleus?	None
	Omegatetravirus				
Rhabdoviridae	*Sigmavirus*	ssRNA	No	Cytoplasm	V (rabies), P (many)
Polydnaviridae	*Ichnovirus*	dsDNA	No	Nucleus (host ovary)	None
	Bracovirus				
Birnaviridae	*Entomobirnavirus*	dsRNA	No	Cytoplasm	V
Replication in and transmission by insects to vertebrate and plant hosts					
Togaviridae	*Alphavirus*	ssRNA	No	Intracellular	V (encephalitis, rubella)
Flaviviridae	*Flavivirus*	ssRNA	No	Cytoplasm	V (yellow fever, hepatitis C)
Bunyaviridae	*Bunavirus*	smRNA	No	Cytoplasm	V, P (tomato spotted wilt)

[a]Based on ICTV (1995).

[b]Abbreviations of genera are in parentheses.

[c]May sometimes be incorporated in occlusion bodies of baculoviruses (Moore, 1991).

A. Horizontal Transmission

Virulent pathogens such as nuclear polyhidrosis viruses (NPVs) and granulosis viruses (GVs) of Lepidoptera are primarily transmitted both within and between host generations through the release of OBs into the environment following death of infected individuals (Myers, 1993). Low levels of infectious virus can also be released in feces and regurgitate from the midgut in the late stages of infections (Vasconcelos, 1996). Less lethal pathogens are transmitted primarily in feces and regurgitate. This mode of transmission is important in *Oryctes* virus, cytoplasmic polyhidrosis viruses (CPVs), and entomopox viruses (EPVs) of some Lepidoptera, NPVs of sawflies, and small RNA viruses infecting bees (Sikorowski *et al.*, 1973; Granados, 1981; Kelly, 1981; Payne, 1981, 1982; Zelazny and Alfiler, 1991; Myers and Rothman, 1995a). *Oryctes* virus is excreted by infected adult coconut palm rhinoceros beetles, *Oryctes rhinoceros* at breeding sites (coconut trunks) and while mating (Zelazny 1973, 1976; Zelazny and Alfiler, 1991; Zelazny *et al.*, 1992; Kelly, 1981). The small RNA viruses, like the picornaviruses and sacbrood virus, cause chronic and acute honey bee paralysis. These viruses are likely secreted from salivary and hypopharyngeal glands into the liquid added by bees to the pollen as they collect it (Bailey, 1973; Kelly, 1981). In this way they are transmitted to other individuals. Although horizontal transmission usually occurs by ingestion of contaminated food such as foliage or egg chorion on hatching (discussed in what follows), iridescent viruses (IVs) of *Aedes taeniarhynchus* and *Tipula oleracea* (Diptera) (Linley and Nielson, 1968b; Carter, 1973a,b), and an NPV of *Heliothis armigera* (Dhandapani *et al.*, 1993a) can be transmitted by cannibalism.

An example of host mortality following horizontal transmission of an NPV is illustrated in Figure 1. In this experiment, young tent caterpillars were fed leaves contaminated with NPV in the laboratory and then introduced to host colonies at two densities in the field. The first peak in mortality involves the lab-infected individuals that released NPV occlusion bodies when they died. Susceptible individuals ingested contaminated foliage and died about 2 weeks later, causing the second peak in mortality. Bi- and trimodal peaks of host mortality have been recorded for *Pieris brassicae* following experimental introduction of GV (Hochburg, 1991a) and for field populations of gypsy moth (Woods and Elkinton, 1987).

1. Persistence

Persistence of active virus in the environment or in the host population is important to the dynamics of within- and between-generation infection. Viruses in occlusion bodies are likely to persist longer than nonoccluded viruses. However,

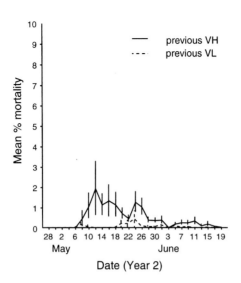

Fig. 1. Mean percentage mortality (+S.E.) of *Malacosoma californicum pluviale* (Lepidoptera) larvae from NPV observed at 2-day intervals in the year of introduction of NPV and host colonies at high (VH) and low (VL) larval and viral densities. Adapted with permission from Rothman (1997).

a nonoccluded small RNA virus of aphids was found to persist and be transmitted through plants (Gildow and D'Arcy, 1988). This interaction is more likely for sucking than chewing insects, but it suggests that more than plant surfaces may be involved in the persistence of insect viruses.

Foliage is an important substrate for short-term persistence of viruses. For example, Pringle and Lewis (1997) found that NPV of the celery looper *Anagrapha falcifera* was still 80% active after 9 days when sprayed onto silk of sweet corn. However, virus on foliage is generally deactivated in a matter of hours to days through exposure to ultraviolet sunlight (David *et al.*, 1968; Jaques, 1972; Ignoffo *et al.*, 1977; see Benz [1987] and Cory *et al.* [1997] for reviews). Roland and Kaupp (1995) and Rothman and Roland (1998) found reduced infection by NPV of the forest tent caterpillar *Malacosoma disstria* in forest-edge habitats compared to that in forest-interior habitats, which they attributed to faster deactivation by sunlight.

Many of the most lethal insect viruses produce infective stages that can persist outside the host for one or many seasons in the soil (see Benz, 1987; Entwistle and Evans, 1985; Cory *et al.*, 1997), on tree bark, or on host eggs. In agricultural and pasture habitats, where hosts and host food plants are in close proximity, soil may be an important source of inoculum for host infections for NPVs, GVs, and EPVs (Hurpin and Robert, 1972; Jaques, 1975; Crawford and Kalmakoff, 1977).

In forest systems, the mechanisms by which virus is transferred from soil to caterpillars are unclear, but wind, rain splash, and contamination of hosts, predators, and parasitoids are possible (Entwistle and Evans, 1985; Olofsson, 1988).

Host food plants are also important substrates for transgeneration virus transmission, particularly in forest systems. Woods *et al.* (1989) released neonate larvae of the gypsy moth *Lymantria dispar* (Lepidoptera) onto sterilized, untreated, and NPV-treated bark surfaces in areas where NPV epizootics occurred during the previous year. Larvae were collected after walking on the bark for 15–90 minutes and reared in the laboratory on an artificial diet. Percentage mortality was related to the degree of NPV contamination (NPV treated > untreated > sterilized bark). Infection was greater in colonies adjacent to treated areas than in those on control trees. Figure 2 illustrates an example based on horizontal transmission that is likely to have occurred by way of host tree surfaces (e.g., tree bark or possibly remains of colony material) in the western tent caterpillar. Virus-free colonies of tent caterpillars were obtained by removing viral contamination of egg masses with bleach. These were deployed on trees that supported infected caterpillars the previous year. Resultant infection of these caterpillars was presumed to have been caused by environmental contamination of NPV on bark and remaining tent material.

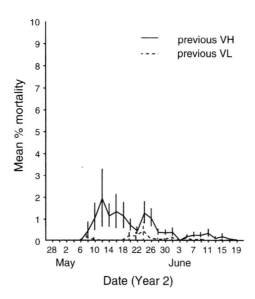

Fig. 2.   Mean percentage mortality (+S.E.) of *M. c. pluviale* larvae from NPV observed at 2-day intervals in the year following introduction of NPV and host colonies at high (VH) and low (VL) larval and viral densities. Adapted with permission from Rothman (1997).

Virus can also overwinter on host egg or egg mass surfaces following environmental contamination according to work by Murray and Elkinton (1989), who found that egg masses of gypsy moth acquired most of the NPV inoculum from their environment during or within 3 days of oviposition in contaminated habitats. They suggest that NPV may be acquired by incorporation of virus from the substrate into the egg mass as the moth rubs her abdomen on the substrate during oviposition. Eggs of other Lepidoptera can acquire virus via the genitalia of moths that are contaminated by the environment during an epizootic (Tatchell, 1981). Larvae ingest the contaminated egg chorion and egg mass hairs on hatching (Doane, 1975). Ingestion of externally contaminated egg mass material should not to be confused with vertical transmission where ingested virus is passed directly from parents to offspring or eggs usually following sublethal infection of parents.

Virus may also persist on pupal cuticle (Doane, 1975), larval cadavers, and tent material of silk spinning insects. Kukan (1996) experimentally tested the persistence of NPV on tents of tent caterpillars contaminated with infected caterpillars. PCR analysis and bioassays of extracts of the tent material confirmed the persistence of active virus for at least 4 months in the field.

2. Vectors

Mechanical transfer of viruses between insect hosts can occur via vectors but generally without virus multiplication on or within the vector (Andreadis, 1987), as occurs in some insect vector-borne diseases of mammals. Wasp and fly parasitoids that develop within or oviposit in infected hosts may transmit virus during subsequent oviposition either through the ovipositor or by contamination of the host or hosts' environment (Stairs, 1965; Irabagon and Brooks, 1974; Beegle and Oatman 1975; Vail, 1981; Young and Yearian, 1990; Hochburg, 1991a). Birds and mammals may also act as vectors for insect viruses following predation on infected insects (Entwistle and Adams, 1978; Lautenschlager *et al.*, 1980). Wind and rainsplash also disperse insect viruses (Entwistle *et al.*, 1983; D'Amico and Elkinton, 1995).

B. Vertical Transmission

Vertical transmission from parent to offspring may occur by contamination of the egg surface (transovum) or within the ovary (transovarian), and is important in more benign pathogens that usually kill only young infected hosts. Transovum transmission is important amongst CPVs of Lepidoptera. Adults may contaminate eggs from reproductive organs, meconia (waste material released following pupal emergence), and feces or regurgitate, resulting in infection of progeny (Bullock *et al.*, 1969; Vail and Gough, 1970; Sikorowski *et al.*, 1973; Payne, 1982; Sikorowski and Lawrence, 1994). In *Heliothis virescens* (Lepidoptera), larvae can be infected when they consume CPV-contaminated chorion when hatching from the egg (Sikorwoski *et al.*, 1973). By contrast, transovarian transmission may be important for viruses of flies (Marshall, 1973) such as in *A. taeniarhynchus* infected with an iridescent virus (Linley and Nielson 1968a; Woodard and Chapman, 1968) and sigma virus of *Drosophila melanogaster*, which is transmitted through female and male gametes (Fleuriet and Periquet, 1993).

Polydnaviruses (PVs) of parasitic Hymenoptera (Ichneumonidae and Braconidae) may be vertically transmitted through the germline by integration into wasp chromosomal DNA (Fleming, 1992). PVs are found primarily in female wasps in the lumen of the oviducts and in the epithelial cells of the calyx, where they replicate (Stoltz and Vinson, 1979; Fleming, 1992). This is a fascinating virus–host system, as the interaction between the wasp and virus is mutualistic. PV co-injected with the parasitoid egg alters the physiology and suppresses the immune system of the lepidopteran host of the wasp parasitoid, enabling the development of the parasitoid eggs and larvae (Dover *et al.*, 1987; Fleming, 1992). Because PV replication occurs only in the wasp, the virus benefits from successful reproduction of the wasp, and the wasp benefits by successful parasitization of its caterpillar host (Fleming, 1992).

Though vertical (transovum) transmission is expected to be more important among benign viruses, it can also augment horizontal routes in the usually virulent baculoviruses such as NPVs. Hamm and Young (1974) reported transovum transmission of an NPV of *H. zea* when female moths were fed the virus. Transmission occurred without infection of the adult; the virus passed through the digestive tract and contaminated the tip of the abdomen. Males fed NPV could indirectly transmit the virus to progeny by contaminating females during mating. Similar findings on vertical transmission of NPVs are reported for *Heliothis armigera* (Lepidoptera) (Dhandapani *et al.*, 1993b) and *S. frugiperda* (Fuxa and Richter, 1991; Fuxa *et al.*, 1992). Kukan (1996) reviewed studies that tested for vertical transmission of NPV in Lepidoptera, and it was not uncommon to find low levels of infection (<2%). However, it is always difficult to rule out contamination.

C. Latent Virus

Early studies of insect pathogens featured latent virus as an important route of vertical transmission, leading to the sudden appearance of viral infection in insects following stress (e.g., Steinhaus, 1958). This interpretation was subsequently questioned because environmental contamination was always possible. Hughes *et al.* (1993) more recently demonstrated latent NPV-like virus in a laboratory culture of *M. brassicae*. The virus, which appeared identical to NPV of *M. brassicae*, was activated by feeding larvae with NPVs from other species of moths. PCR amplification demonstrated the latent virus in host eggs, larvae, pupae, and adults. Fat body from caterpillars harboring this latent virus, when fed to control caterpillars, caused an MbNPV-like infection (Hughes *et al.*, 1997). This exciting finding confirms latent virus as a mechanism of vertical transmission for the usually virulent viruses such as NPVs of Lepidoptera. How common latent virus is remains to be discovered.

IV. VIRUSES AND THE HOST INDIVIDUAL

A. Virulence

Virulence is the level of host mortality resulting from parasite infection (May and Anderson, 1983). Virulent pathogens have high rates of replication within the host. They therefore tend to cause more damage to host tissues and are thus more lethal or kill more quickly than less virulent pathogens. The type of host tissue infected may also influence the probability of host mortality (Tanada and Fuxa, 1987).

The virulence of insect viruses depends on the type of virus, the host species infected, the age of the host, and variation in resistance between individuals and populations (see Milks [1997] for an example of interpopulation variation). Virulent pathogens such as the baculoviruses produce systemic infections, while more benign viruses tend to produce infections that are restricted to particular host tissues. For example, infections caused by *Oryctes* virus and the sawfly NPVs and CPVs of Lepidoptera are restricted to the host gut and produce chronic rather than rapidly lethal disease (Bailey, 1973; Kelly, 1981; Payne, 1982). The polydna-viruses replicate in reproductive structures of some female wasps and appear to promote offspring survival rather than harm the host (Fleming, 1992). Sigma virus of *D. melanogaster* multiplies in the cytoplasm of host gametes (Fleuriet and Periquet, 1993), and the only symptom of virus infection is a sensitivity to concentrations of CO_2 (Marshall, 1973). Chronic stunt virus (CSV), an RNA virus of the navel orange worm, *Amyelois transitella* (Lepidoptera) specifically invades granular hemocytes and may take as long as 40 days to kill its host (Kellen and Hoffmann, 1982).

Virulence may depend on the host species or taxon infected. For example, NPVs of Lepidoptera are highly virulent, often killing the host within 10 days or less, while NPVs of some Hymenoptera (sawflies) are restricted to the host gut and usually kill only young larvae (Bailey, 1973). Virulence may also be influenced by host age. Younger individuals tend to be more susceptible to viral infection, and the time taken to die from infection generally increases with host age and/or weight (Linley and Nielson, 1968a,b; Hochburg, 1991b; Sait *et al.*, 1994a, and the references therein; Sikorowski and Lawrence, 1994; Engelhard and Volkman, 1995). Age-related resistance may occur because the virus has insufficient time to become established before host pupation, when the host becomes immune to the virus (Whitlock, 1977; Murray *et al.*, 1991). Engelhard and Volkman (1995) found increasing resistance to NPV in *T. ni* when the virus was administered orally, but not when injected directly into the hemocoel. This suggests that factors mediating resistance operate during infection of the larval midgut. Changes in hormone titer may also play a role in age-related resistance. Ecdysone increases in titer as insect development progresses and is required for the larval–pupal molt (Dover *et al.*, 1987; Park *et al.*, 1993). β-ecdysone delays the onset of pathology and reduces mortality in NPV-infected *Heliothis virescens* (Keeley and Vinson, 1975).

B. Debilitating Effects on Survivors

Finding effective ways to kill pest insects has been a principal focus of much research on insect viruses, and less attention has been given to potential sublethal effects of viral pathogens. Effects of pathogens on hosts that survive infection are often difficult to detect and to distinguish from the effects of other ecological factors, particularly in field situations. Further, because insects are small and very abundant animals, less attention is given to variation in quality among individuals (Wellington, 1957).

Sublethal disease from insect viruses can cause a variety of debilitating symptoms in their hosts, including deformation of adults and reduced rates of development, pupal and adult weights, adult longevity, fecundity, mating success, and egg viability (Vail *et al.*, 1969; Zelazny, 1973; Zelazny *et al.*, 1992; Williamson and von Wechmar, 1995; Hamm *et al.*, 1996; reviewed in Rothman and Myers, 1996a). While sublethal effects are most usually debilitating, a recent lab study of a small RNA virus of *Drosophila* found that infected larvae had reduced survival, but that infected adult females had increased egg production (Thomas, 1996).

Viruses may also increase host susceptibility to other mortality agents. Delays in development could increase hosts' temporal exposure to natural enemies. Infection with one virus can increase susceptibility to other viruses or pathogens (see Gallo *et al.* [1991] and Xu and Hukuhara [1992] and the references therein). Gallo *et al.* (1991) observed that NPV infection in *T. ni* was enhanced by a viral "enhancing factor" extracted from a GV. Xu and Hukuhara (1992) found that spheroids of an EPV enhanced infectivity of an NPV of *Pseudaletia separata* (Lepidoptera). Dual CPV and bacterial infection of *H. zea* acted synergistically to adversely affect growth and development (Bong and Sikorowski, 1991). Debilitating effects of insecticides are also greater when *H. zea* is infected with a CPV (Mohamed *et al.*, 1989).

Sublethal effects may be due to diversion of host energy reserves to support or combat the pathogen (Sikorowski and Thompson, 1979; Wiygul and Sikorowski, 1978, 1991), disruption of oocyte development (Neilson, 1965; Hamm *et al.*, 1996), or hormonal changes induced by the pathogen (O'Reilly and Miller, 1989; Burand and Park, 1992; Park *et al.*, 1993). Wiygul and Sikorowski (1978, 1991) recorded increased oxygen uptake in caterpillars of *H. virescens* infected with a CPV and in *H. zea* infected with an IV. They suggested that viral replication and increased metabolic activity of damaged organs increased the demand for oxygen. Sikorowski and Thompson (1979) speculated that low levels of fatty acids in *H. virescens* infected by a CPV decreased egg production.

NPV of alfalfa the looper *Autographa californica* and of the gypsy moth have an *egt* gene that codes for the enzyme ecdysteroid UDP-glucosyltransferase, which inactivates the molting hormone of infected caterpillars (O'Reilly and Miller, 1989; Park *et al.*, 1993). Production of this enzyme delays molting and could influence the phenotype of individuals that survive infection late in development.

Debilitating effects of sublethal infections are more frequent with more benign pathogens such as CPVs than with the more virulent baculoviruses such as NPV of Lepidoptera (Rothman and Myers, 1996a). However, debilitating effects can occur following treatment with NPV, particularly when host larvae are exposed late in development (Young and Yearian, 1982; Vargas-Osuna and Santiago-Alvarez, 1988; Sait *et al.*, 1994a,b; Rothman and Myers, 1996a). In most studies with NPV, the presence of viral infection has not been confirmed, so that the conclusion that sublethal disease causes the effects remains speculative (Rothman and Myers, 1996a).

V. VIRUSES AND THE HOST POPULATION

A. General theory

Much attention in population biology focuses on the issue of regulation. Regulation is the process whereby a population tends to return to an equilibrium level (Myers and Rothman, 1995b; Sinclair and Pech, 1996). Because population densities always fluctuate to some degree, it is more meaningful to view an equilibrium as a bounded probability distribution of densities rather than as a single point (Turchin, 1995). Regulation occurs when the per-capita growth rate of a population is related to density (i.e., is density dependent) such that the population returns toward an equilibrium density following perturbation. Ecological factors such as intra- and interspecific competition, predation, and disease can regulate a population if their impacts on host mortality, reproduction, or dispersal reduce the per-capita growth rate of the host population at high density. Research in population biology has traditionally focused on whether observed dynamics result from density-dependent or density-independent factors (e.g., weather or weather effects on food quality) (Martinat 1987; White, 1993; Daniel and Myers, 1995). More recently, attention has shifted to the variety of dynamics (e.g., stable, cyclic, chaotic) that can result from density dependence. For example, when per-capita growth rate depends on previous rather than current density (delayed density dependence), cyclic dynamics can occur (Myers and Rothman 1995b; Sinclair and Pech, 1996). Note that a population experiencing delayed density-dependent growth is still regulated, albeit in a more unstable form. Multigeneration population cycles are of particular interest, as they are observed in many species of forest insects in northern temperate regions (Myers, 1988, 1998; Turchin, 1990).

1. Disease, Regulation, and Population Cycles

Transmission of virus is generally considered to be a density-dependent process. High population density increases the probability of contact between susceptible and infected hosts or susceptible hosts and free-living pathogens. Viral epizootics are predicted and often observed following high caterpillar densities (Clark, 1958; Wellington, 1962; Anderson and May, 1981; Tanada and Fuxa, 1987; Myers, 1988; Elkinton and Liebhold, 1990; Woods et al., 1991). Insect viruses have the potential to regulate their host populations if they increase mortality when host population densities are high. In pathogens with long and free-living infectious stages, the environment may become heavily contaminated with viral inoculum following an epizootic at high host density. Thus, virus can continue to infect host populations in generations following an epizootic even though host density has decreased. This is consistent with the findings of Kukan (1996) and Kukan and Myers (1997) for tent caterpillars. NPV infection in caterpillars and NPV contamination of tent material were found most frequently in dense and declining host populations. Density-dependent transmission and time delays introduced by accumulation of virus in the environment can destabilize host insect populations and cause population cycles. Viruses could also interact with other ecological factors to drive host population dynamics. For example, virus could have a destabilizing influence on populations regulated by insect parasitoids (Berryman, 1996; Myers, 1996). Continued mortality from viral infection could drive host populations below densities at which parasitoids are able to persist.

Viruses might also influence population dynamics by reducing host reproduction. Reduced fecundity following high larval densities is observed in several species of forest Lepidoptera and may contribute to population declines (Myers, 1988; see references in Rothman and Myers, 1996b). For example, in western tent caterpillars, fecundity tends to be lower following peak larval densities and remains low for several generations during the population decline (Myers, 1990; Myers and Rothman, 1995b; Myers and Kukan, 1995). As discussed by Myers (1993), sublethal viral disease may play a role in this pattern, and NPV has been found to reduce tent caterpillar fecundity in the laboratory and in the field (Rothman and Myers, 1994; Rothman, 1997).

B. Mathematical Models

The models of Anderson and May (1981) have been influential in the field of population biology, particularly in relation to the study of insect pathogens. These models have been described in a number of publications (see, e.g., Cory et al., 1997). Anderson and May used differential equations to model populations in continuous time and assumed mass action spread of disease. They also assumed exponential growth of the host population without virus (but see model F, Anderson and May, 1981). Mass action is a simplifying assumption with deep roots, stretching back to the beginnings of modern theoretical ecology (Kingsland, 1985). It forms the basis of many population models involving predator–prey, interspecific, and pathogen–host interactions. With mass action, contact between susceptible hosts and infected hosts or pathogen particles (leading to new infections) is considered to be similar to molecules of a gas in a closed container in which the number of collisions between particles of different gases will be proportional to the product of their densities (Kingsland, 1985). Individuals are treated as identical units, and the spatial distribution of the host and virus or variation in host susceptibility are ignored.

The Anderson and May models predict that pathogens, such as viruses, will be maintained within a host population if host density exceeds a certain threshold. Thus, the interaction between host and pathogen is density dependent, and pathogens are capable of regulating their host population, and of causing stable or cyclic dynamics. Regulation is more likely with pathogens that usually kill infected hosts (high virulence) and/or decrease host reproduction. Anderson and May (1981) model G incorporates a free-living pathogen population and predicts that population cycles are more likely to arise in associations where the virus kills infected hosts and produces many long-lived infective stages, such as occurs in NPVs and some GVs. This agrees with predictions of more general theory discussed above (§V.A.). The presence of long-lived infective stages introduces time delays into the pathogen–host system (Hassell and May, 1989), thereby destabilizing host population dynamics. Models also yield a somewhat counterintuitive prediction: prevalence is inversely related to virulence and regulatory potential. Less lethal pathogens usually have low threshold host densities and therefore tend to be more prevalent than more virulent pathogens, but they are less likely to regulate the host population (Anderson and May, 1981).

Anderson and May (1980, 1981) suggested that pathogens may drive population cycles observed in many forest Lepidoptera based on characteristics of many virus–forest insect associations (e.g., high virulence long-lived infective stages) and comparisons between predictions of models and observed dynamics of the larch budmoth *Zeiraphera diniana* (Lepidoptera) and its GV. This work stimulated further development of insect–pathogen models (see Dwyer, 1995). Bowers *et al.* (1993) added density dependence to the host birth rate in the Anderson and May model of forest insects and applied it to the larch budmoth–GV system. They found that, with this more realistic modification, disease prevalence generated by the model peaked at too high a level, compared to its observed prevalence in the field, while peaks in host density in the model were too low. They concluded that GV alone could not drive population cycles in the larch budmoth.

Vezina and Peterman (1985) tested Anderson and May (1981) model G and several variations (which included density-dependent natural [nonvirus] mortality, vertical transmission, and an incubation period for the virus) with data on the Douglas fir tussock moth *O. pseudotsugata* and its NPV. Though cycles could be produced, they found that none of the model's predictions of period, amplitude, or host density matched the observed dynamics when they used realistic parameter values. They concluded that predators, parasitoids, and/or food availability probably also play significant roles in the dynamics. A more recent modification of the Anderson and May (1981) model G by Dwyer (1994) included density-dependent birth rate in the *O. pseudotsugata*–NPV system and predicted the observed cycle period for this insect over a range of parameter values. Dwyer concluded that the addition of greater realism to the model strengthened Anderson and May's original prediction that pathogens could drive population cycles. According to White *et al.* (1996), however, it is not the addition of density dependence that makes cycles more likely in the Dwyer (1994) model, but the removal of reproduction of infected hosts.

Other modifications were made by Hochburg (1989), who included a long-term virus reservoir, which stabilized host–pathogen interactions. Infection could only occur after the pathogen moved from the reservoir (e.g., from the soil) to a transmissible subpopulation where contact between pathogen and hosts could occur (e.g., foliage). Pathogens with long-lived infective stages could now produce stable host dynamics or cycles depending on the rate of movement between the reservoir and the transmissible subpopulation. The population cycles predicted from the above-discussed models all have relatively long periods (i.e., multigenerational). Briggs and Godfray (1995) incorporated stage structure to an insect–pathogen model such that only certain host life stages were susceptible to disease. This led to cycles that were the length of a generation or longer.

Long-term population cycles appear to be a feature of many forest Lepidoptera in temperate regions (Myers, 1988, 1998). The strong seasonality limits population growth to discrete intervals, and generations do not overlap. It is surprising that, until very recently, models of these insects have been framed in continuous rather than discrete time. Briggs and Godfray (1996) explored the dynamics in insect–pathogen systems in seasonal environments by modeling within-season dynamics using continuous (differential) equations and between-season dynamics using discrete (difference) equations. This model was less stable than Anderson and May (1981) model G, and persistence of the insect–pathogen interaction was possible only when pathogen transmission was assumed to be nonlinear, that is, transmission increased with pathogen density, but with a decreasing slope (sublinear transmission). Long-term cycles were generally less likely than in continuous models but were produced by extensions to the models that included vertical transmission or the presence of a pathogen reservoir.

Predictions of models indicate the possibility that viruses could drive the observed dynamics of some insect populations. However, the variety of dynamics possible from the same model with different parameter values, from similar models examined by different authors (even when applied to the same system), and from the various model extensions designed to incorporate greater biological realism, argues for greater emphasis on experimental approaches to explore the role of pathogens in the dynamics of insect populations.

C. Experiments

1. Population Level Experiments

To directly test the role of viral disease in host population dynamics, population level experiments must be performed, with replication and controls, and over several generations. Such large-scale experiments are difficult or impossible to perform in the field but are more easily performed in the laboratory. Studies of the Indian grain moth *Plodia interpunctella* and its GV by Begon and colleagues provide an excellent example. Infected and control populations were maintained in containers to which food (flour) was added on a regular basis. Virus-free populations fluctuated in density with a periodicity of one generation. The addition of GV increased the cycle length slightly and reduced adult density (Sait *et al.*, 1994c). These authors suggested that the greater length of cycles in the virus-infected populations could have resulted from sublethal infection increasing host development time.

Addition of a parasitoid, *Venturia canescens* (Hymenoptera), to virus-free populations also resulted in generation-length cycles, while multigeneration length cycles occurred with both the GVs and parasitoid present. The authors suggested that generation cycles resulted from a concentration of mortality during specific stages of the host life cycle as predicted by Briggs and Godfray's (1995) stage-structured model (see above). The combination of virus and parasitoid served to distribute mortality more evenly throughout the host life cycle and to increase the period of the fluctuations as predicted by models without stage structure (Begon *et al.*, 1996). However, more recent work (Knell *et al.*, 1998) suggests that food quality and quantity are the major determinants of population cycles of lab populations of *Plodia*, and interpretations of these experiments must be made cautiously.

2. Small-Scale Experiments

Small-scale experiments can be used to test hypotheses about specific mecha-nisms that influence population dynamics. For example, to explore the potential role of NPV in population cycles of tent caterpillars, Rothman (1997) introduced NPVs to caterpillars at different densities on individual trees. The proportion of virus-killed caterpillars was higher during the second wave of mortality on trees with more caterpillar colonies (Fig. 1). The following year, significantly greater mortality from NPVs occurred among caterpillars on trees that previously har-bored higher caterpillar densities (Fig. 2). Persistence of NPVs from the first to the second year caused delayed density-dependent mortality.

With a single-factor experiment it is difficult to assess the importance of NPVs relative to other factors that may influence populations. For example, insects feeding on trees that previously experienced heavy defoliation may be negatively affected by induced changes in foliage quality (delayed induced resistance or DIR) (Haukioja, 1990). Negative effects of viral disease or high caterpillar density in the parental generation could be inherited by offspring via maternal effects or genetic selection (Myers, 1993; Rossiter, 1994). Both DIR and inheritance could cause delayed density-dependent population growth. Rothman (1997) simultane-ously explored the effects of NPV persistence, DIR, and inherited effects of virus and high density. No evidence for DIR was found, and inherited effects of previous treatments were weak. Rothman concluded that NPV persistence (Fig. 2) was more likely to contribute to population instability than was DIR or inheritance.

Another approach to examine mechanisms of population change involves trans-planting individuals from source populations to areas with no history of insect infestations. Myers (1990) introduced *M. c. pluviale* egg masses from increasing and peak populations to sites with low or no natural populations. If population declines are caused by factors in the insects' environment (e.g., food shortage and food quality), then transplanted populations should increase and their dynamics should be out of phase with source populations. Myers found that transplanted populations declined in synchrony or at most a year later than source populations.

Though this experiment does not test directly for virus impact, it suggests that the future fate of the population may be carried with the egg masses rather than in the surrounding environment and could be related to contamination of eggs with NPVs or to the quality of eggs. Similar experiments comparing surface-sterilized and -unsterilized egg masses might further our understanding of the role of virus in the dynamics of this host species.

Small-scale experiments can also be used to test specific assumptions or predictions in mathematical models. One fruitful line of research involves testing whether transmission of virus is a mass action process or, in other words, whether transmission is linearly proportional to the density of susceptible hosts, and the density of pathogens or infected hosts. Testing for mass action transmission is important, since it is assumed in most population models of pathogens and their animal hosts (D'Amico *et al.*, 1996), including insects and their viruses, and can have important consequences for host population dynamics (Liu *et al.*, 1986; Hochburg, 1991c; Briggs and Godfray, 1996).

The mass action assumption can be tested by calculating ν, the transmission parameter in small, discrete, or enclosed groups of insects over one infection cycle at various host and pathogen densities. If we assume no addition of virus after the start of the experiment and a virus decay rate of zero during the experimental period, then

$$\nu = -\ln(S_t/S_0)/P_0$$

where P_0 is the initial amount of virus introduced at the beginning of the experiment, S_t is the number of susceptibles remaining at the end of the experiment, and S_0 is the original number of susceptible hosts (D'Amico *et al.*, 1996). If transmission fits a mass action process, ν should remain constant over manipulated levels of P_0 and S_0.

Studies have shown that ν does not remain constant when P_0 and S_0 are varied. D'Amico *et al.* (1996) found greater-than-expected transmission (higher ν) at low densities of susceptibles and infecteds in *L. dispar*. Dwyer and Elkinton (1993) and Dwyer (1995) found that a modified Anderson and May model was reasonably accurate in predicting the pattern and magnitude of within-season mortality from NPV for gypsy moth when the transmission parameter was estimated from small-scale experiments. However, the model underestimated the infection level for low-density host populations. The mass action assumption may be inappropriate for predicting mortality from NPV at low host density. Beisner and Myers (unpublished) found that ν remained constant with density, but actually declined with P_0 in experiments involving colonies of gregarious western tent caterpillars. The transmission parameter ν may also be influenced by the stage of host larvae and the spatial distribution of virus (Dwyer 1991, but see also Goulson *et al.*, 1995). Such interplay between field observation, modeling, and experimentation is rare, but it is necessary to refine theory and improve our ability to predict the effects of viruses on dynamics of host insect populations.

D. Biological Control

Insect viruses are used to control several pest species. These bioinsecticides usually have only short-term effects on the host population, although long-term control has been achieved in a few cases. For example, a severe outbreak of the introduced European spruce sawfly *Gilpinia hercyniae* in northeastern North America from 1930 to 1942 was terminated by an NPV accidentally introduced with parasites from Europe. The virus spread rapidly, causing an epizootic that reduced sawfly populations. The virus persisted and achieved long-term control (Stairs, 1971; Cunningham, 1982). The Rhinocerus beetle *O. rhinoceros* causes serious damage to coconut palms in Southeast Asia and many Pacific islands. *Oryctes* virus has provided successful control following its introduction and rapid spread in the Maldives Islands in 1984 and 1985 (Zelazny *et al.*, 1992). Control resulted from reduced adult life-span and fecundity rather than from virus-induced mortality of larvae (Zelazny *et al.*, 1992). It is interesting to note that these examples of long-term control involve viruses that are less virulent than the NPVs of Lepidoptera. Highly virulent pathogens may be less efficiently transmitted and require high host densities for a virus epizootic. They may only suppress an insect population once outbreak densities have been reached, as in the tussock moth (Otvos *et al.*, 1987). Persistence of more virulent pathogens in the environment can cause cycles in the host population that are not desirable from the viewpoint of biocontrol. NPVs may be more successful when used for short-term (insecticidal) control (Cory *et al.*, 1997). The development of recombinant viruses for insect control has also been reviewed (Bonning and Hammock, 1996).

VI. SYNTHESIS: EVOLUTION OF PATHOGEN–HOST ASSOCIATIONS

The synthesis of the ecological and evolutionary interactions between viruses and their insect hosts is difficult, because so little is known about the long-term persistence and transmission of virus in field populations. The evolution of viruses will be determined by the primary mode of transmission, the rate of replication of the virus, and selection for resistance in the host. We have summarized the expected patterns of virulence and transmission of insect viruses in Myers and Rothman (1995a) and Figure 3. Viruses that rely on horizontal transmission through environmental contamination will need to replicate many times and are

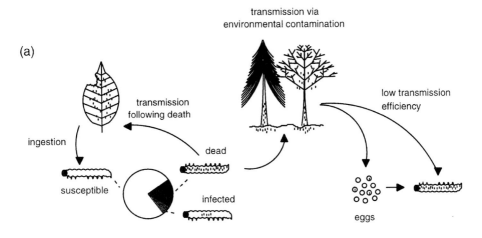

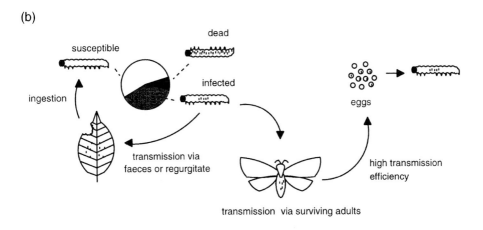

Fig. 3.   Some characteristics of (**a**) virulent and (**b**) benign pathogen–host associations. Circles represent the host population, while shaded areas represent the relative proportions of infected and dead hosts. Reprinted with permission from Myers and Rothman (1995b).

therefore likely to infect all host tissues. Because they employ a less efficient "sit-and-wait" strategy in the hosts' environment, selection will favor production of many (greater virulence) and long-lived infective stages (see Ewald, 1994). These viruses will be more likely to promote cyclic or fluctuating dynamics in their host populations than will benign viruses. Viruses with greater dependence on vertical transmission, and which are transmitted horizontally via live hosts (in feces or regurgitate) will tend to cause infections that are restricted to specific host tissues and cause sublethal effects rather than host death. Selection will favor production of fewer infective stages (reduced virulence) and longer survival of infected hosts.

Genetic variation in virulence, rate of replication, and type of transmission complicates predictions of how viruses will evolve. Lipsitch *et al.* (1996) modeled the evolution of virulence in a pathogen population that included benign vertically transmitted and virulent horizontally transmitted strains. Increased vertical transmission favored the evolution of lower virulence. However, models predicted that increased opportunity for horizontal transmission would actually lower the equilibrium level of virulence.

The degree of genetic variability within a horizontally transmitted virus population could influence selection for virulence. For a virus that is transmitted through environmental contamination, rapid replication may allow more rounds of infection per host generation (Fig. 1). On the other hand, slower replication could increase production of inoculum if infected individuals continue to grow before death. If the initial infection of an individual host involves several genotypic variants of a virus that vary in their rate of replication, then the fastest replicating genotype should produce the most copies and be selected for because of intrahost competition for resources. However, if infection of host individuals involves a single virus genotype, then between-genotype selection will not occur within the infected host individual. The virus that produces more polyhedra by allowing the host to grow could be favored in this case. Therefore, quantifying the "genetic" variation within a virus population is an aid to understanding virus evolution (see Cory *et al.* [1997] for a review of baculovirus examples).

It is still not known whether viruses regulate their host populations, whether this regulation results in stable dynamics or cycles, or whether viruses destabilize host regulation by other density-dependent processes, for example, parasitism by insect parasitoids. Predictions from insect–virus models remain largely untested, and both small-scale experiments examining mechanisms of population change as well as population level experiments are uncommon. Further, most of what we know about the dynamics of viral disease and host populations is based on outbreak species, and almost nothing is known about the role of viruses in low-density hosts in the field. Finally, work at the population level has concentrated on baculoviruses that have DNA genomes. RNA viruses could influence host populations through their debilitating effects and have important interactive effects in the case of mixed infections with baculoviruses. Until studies are undertaken to quantify the existence and persistence of these viruses in host populations, their potential impacts will be overlooked.

REFERENCES

Anderson, R. M., and May, R. M. (1980). Infectious diseases and population cycles of forest insects. *Science* 210, 658–661.

Anderson, R. M., and May, R. M. (1981). The population dynamics of microparasites and their invertebrate hosts. *Philos. Trans. R. Soc. London, Ser. B* 291, 451–524.

Andreadis, T. G. (1987). Transmission. *In* "Epizootiology of Insect Diseases" (J. R. Fuxa and Y. Tanada, eds.), pp. 159–176. Wiley, New York.

Bailey, L. (1973). Viruses of Hymenoptera. *In* "Viruses and Invertebrates" (A. J. Gibbs, ed.), pp. 442–454. North-Holland, Amsterdam.

Beegle, C. C., and Oatman, E. R. (1975). Effect of a nuclear polyhidrosis virus on the relationship between *Trichoplusia ni* (Lepidoptera: Noctuidae) and the parasite *Hyposoter exiguae* (Hymenoptera: Ichneumonidae). *J. Invertebr. Pathol.* 25, 59–71.

Begon, M., Sait, S. M., and Thompson, D. J. (1996). Predator–prey cycles with period shifts between two- and three-species systems. *Nature* 381, 311–315.

Benz, G. (1987). Environment. *In* "Epizootiology of Insect Diseases" (J. R. Fuxa and Y. Tanada, eds.), pp. 177–214. Wiley, New York.

Berryman, A. (1996). What causes population cycles of forest Lepidoptera? *Trends Ecol. Evol.* 11, 28–32.

Bong, C. F. J., and Sikorowski, P. P. (1991). Effects of cytoplasmic polyhidrosis virus and bacterial contamination on growth and development of the corn earworm, *Helicoverpa zea* (Lepidoptera: Noctuidae). *J. Invertebr. Pathol.* 57, 406–412.

Bonning, B. C., and Hammock, B. D. (1996). Development of recombinant baculoviruses for insect control. *Annu. Rev. Entomol.* 41, 191–210.

Bowers, R. G., Begon, M., and Hodgkinson, D. E. (1993). Host–pathogen population cycles in forest insects? Lessons from simple models reconsidered. *Oikos* 67, 529–538.

Briggs, C. J., and Godfray, H. C. J. (1995). The dynamics of insect–pathogen interactions in stage-structured populations. *Am. Nat.* 145, 855–887.

Briggs, C. J., and Godfray, H. C. J. (1996). The dynamics of insect–pathogen interactions in seasonal environments. *Theor. Popul. Biol.* 50, 149–177.

Bullock, H. R., Magnum, C. L., and Guerra, A. A. (1969). Treatment of eggs of the pink bollworm, *Pectinophora gossypiella*, with formaldehyde to prevent infection with a cytoplasmic polyhidrosis virus. *J. Invertebr. Pathol.* 14, 271–273.

Burand, J. P., and Park, E. J. (1992). Effect of nuclear polyhidrosis virus infection on the development and pupation of gypsy moth larvae. *J. Invertebr. Pathol.* 60, 171–175.

Carter, J. B. (1973a). The mode of transmission of *Tipula* iridescent virus, I: Source of infection. *J. Invertebr. Pathol.* 21, 123–130.

Carter, J. B. (1973b). The mode of transmission of *Tipula* iridescent virus, II: Route of infection. *J. Invertebr. Pathol.* **21**, 136–143.

Clark, E. C. (1958). Ecology of the polyhidrosis of tent caterpillars. *Ecology* **39**, 132–139.

Cory, J. S., Hails, R. S., and Sait, S. M. (1997). Baculovirus ecology. *In* "The Baculoviruses" (L. K. Miller, ed.), pp. 301–339. Plenum, New York.

Crawford, A. M., and Kalmakoff, J. (1977). A host–virus interaction in a pasture habitat: *Wiseana* spp. (Lepidoptera: Hepialidae) and its baculoviruses. *J. Invertebr. Pathol.* **29**, 81–87.

Cunningham, J. C. (1982). Field trials with baculoviruses: Control of forest insect pests. *In* "Microbial and Viral Pesticides" (E. Kurstak, ed.), pp. 335–386. Marcel Dekker, New York.

D'Amico, V., and Elkinton, J. S. (1995). Rainfall effects on transmission of gypsy moth (Lepidoptera, Lymantriidae) nuclear polyhidrosis virus. *Environ. Entomol.* **24**, 1144–1149.

D'Amico, V., Elkinton, J. S., Dwyer, G., and Burand, J. P. (1996). Virus transmission in gypsy moths is not a simple mass action process. *Ecology* **77**, 201–206.

Daniel, C. J., and Myers, J. H. (1995). Climate and outbreaks of the forest tent caterpillar. *Ecography* **18**, 353–362.

David, W. A., Gardiner, B. O. C., and Woolner, M. (1968). The effects of sunlight on a purified granulosis virus of *Pieris brassicae* applied to cabbage leaves. *J. Invertebr. Pathol.* **11**, 496–501.

Dhandapani, M., Jayaraj, S., and Rabindra, R. J. (1993a). Cannibalism on nuclear polyhidrosis virus infected larvae by *Heliothis armigera* (Hubn.) and its effect on viral infection. *Insect Sci. Appl.* **14**, 427–430.

Dhandapani, M., Jayaraj, S., and Rabindra, R. J. (1993b). Transmission of a nuclear polyhidrosis of *Heliothis armigera* (Hbn.) to progeny through adult feeding. *Indian J. Exp. Biol.* **31**, 721–722.

Doane, C. C. (1975). Infectious sources of nuclear polyhidrosis virus persisting in natural habitats of the gypsy moth. *Environ. Entomol.* **4**, 392–394.

Dover, B. A., Davies, D. H., Strand, M. R., Gray, R. S., Keeley, L. L., and Vinson, S. B. (1987). Ecdysteroid-titre reduction and developmental arrest of last-instar *Heliothis virescens* larvae by calyx fluid from the parasitoid *Campoletis sonorensis*. *J. Insect Physiol.* **33**, 333–338.

Dwyer, G. (1991). The roles of density, stage, and patchiness in the transmission of an insect virus. *Ecology* **72**, 559–574.

Dwyer, G. (1994). Density dependence and spatial structure in the dynamics of insect pathogens. *Am. Nat.* **143**, 533–562.

Dwyer, G. (1995). Simple models and complex interactions. *In* "Population Dynamics: New Approaches and Synthesis" (N. Cappuccino and P. W. Price, eds.), pp. 209–227. Academic Press, San Diego.

Dwyer, G., and Elkinton, J. S. (1993). Using simple models to predict virus epizootics in gypsy moth populations. *J. Anim. Ecol.* **62**, 1–11.

Elkinton, J. S., and Liebhold, A. M. (1990). Population dynamics of the gypsy moth in North America. *Annu. Rev. Entomol.* **35**, 571–596.

Engelhard, E. K., and Volkman, L. E. (1995). Developmental resistance in fourth instar *Trichoplusia ni* orally inoculated with *Autographa californica* nuclear polyhidrosis virus. *Virology* **209**, 384–389.

Entwistle, P. F., and Adams, P. H. W. (1978). Epizootiology of a nuclear polyhidrosis virus in European spruce sawfly (*Gilpinia hercyniae*): The rate of passage of infective virus through the gut of birds during cage tests. *J. Invertebr. Pathol.* **31**, 307–312.

Entwistle, P. F., and Evans, H. F. (1985). Viral control. *In* "Comprehensive Insect Physiology, Biochemistry, and Pharmacology" (L. I. Gilbert and G. A. Kerkut, eds.), pp. 347–412. Pergamon, Oxford.

Entwistle, P. F., Adams, P. H. W., Evans, H. F., and Rivers, C. F. (1983). Epizootiology of a nuclear polyhidrosis virus (Bacuoloviridae) in European spruce sawfly (*Gilpinia hercyniae*): Spread of disease from small epicentres in comparison with spread of baculovirus diseases in other hosts. *J. Appl. Ecol.* **20**, 473–487.

Ewald, P. W. (1994). "Evolution of Infectious Disease." Oxford University Press, Oxford.

Fleming, J. G. W. (1992). Polydnaviruses: Mutualists and pathogens. *Annu. Rev. Entomol.* **37**, 401–425.

Fleuriet, A., and Periquet, G. (1993). Evolution of the *Drosophila melanogaster*-sigma virus system in natural populations from Languedoc (southern France). *Arch. Virol.* **129**, 131–143.

Fuxa, J. R., and Richter, A. R. (1991). Selection for an increased rate of vertical transmission of *Spodoptera frugiperda* (Lepidoptera: Noctuidae) nuclear polyhidrosis virus. *Environ. Entomol.* **20**, 603–609.

Fuxa, J. R., Weidner, E. H., and Richter, A. R. (1992). Polyhedra without virions in a vertically transmitted nuclear polyhidrosis virus. *J. Invertebr. Pathol.* **60**, 53–58.

Gallo, L. G., Corsaro, B. G., Hughes, P. R., and Granados, R. R. (1991). In vivo enhancement of baculovirus infection by the viral enhancing factor of a granulosis virus of the cabbage looper, *Trichoplusia ni* (Lepidoptera: Noctuidae). *J. Invertebr. Pathol.* **58**, 203–210.

Gildow, F. E., and D'Arcy, C. J. (1988). Barley and oats as reservoirs for an aphid virus and the influence on barley yellow dwarf virus transmission. *Phytopathology* **78**, 811–816.

Goulson, D., Hails, R. S., Williams, T., Hirst, M. L., Vasconcelos, S. D., Green, B. M., Carty, T. M., and Cory, J. S. (1995). Transmission dynamics of a virus in a stage-structured insect population. *Ecology* **76**, 392–401.

Granados, R. R. (1981). Entomopoxvirus infections in insects. In "Pathogenesis of Invertebrate Microbial Diseases" (E. W. Davidson, ed.), pp. 101–126. Allanheld, Osmun.

Hamm, J. J., and Young, J. R. (1974). Mode of transmission of nuclear polyhidrosis virus to progeny of adult *Heliothis zea*. *J. Invertebr. Pathol.* **24**, 70–81.

Hamm, J. J., Carpenter, J. E., and Styer, E. L. (1996). Oviposition day effect on incidence of agonadal progeny of *Helicoverpa zea* (Lepidoptera: Noctuidae) infected with a virus. *Ann. Entomol. Soc. Am.* **89**, 266–275.

Hassell, M. P., and May, R. M. (1989). The population biology of host–parasite and host–parasitoid associations. In "Perspectives in Ecological Theory" (J. Roughgarden, R. M. May, and S. A. Levin, eds.), pp. 319–348. Princeton University Press, Princeton.

Haukioja, E. (1990). Induction of defenses in trees. *Annu. Rev. Entomol.* **36**, 25–42.

Hochburg, M. E. (1989). The potential role of pathogens in biological control. *Nature* **337**, 262–265.

Hochburg, M. E. (1991a). Extra-host interactions between a braconid endoparasitoid, *Apanteles glomeratus*, and a baculovirus for larvae of *Pieris brassicae*. *J. Anim. Ecol.* **60**, 65–77.

Hochburg, M. E. (1991b). Viruses as costs to gregarious feeding behaviour in the Lepidoptera. *Oikos* **61**, 291–296.

Hochburg, M. E. (1991c). Non-linear transmission rates and the dynamics of infectious disease. *J. Theor. Biol.* **153**, 301–321.

Hughes, D. S., Possee, R. D., and King, L. A. (1993). Activation and detection of a latent baculovirus resembling *Mamestra brassicae* nucleopolyhedrosis virus in M. brassicae insects. *Virology* **194**, 608–615.

Hughes, D. S., Possee, R. D., and King, L. A. (1997). Evidence for the presence of a low level, persistent baculovirus infection of *Memestra brassicae* insects. *J. Gen. Virology* **78**, 1801–1805.

Hurpin, B., and Robert, P. H. (1972). Comparison of the activity of certain pathogens of the cockchafer, *Melolontha melolontha*, in plots of natural meadowland. *J. Invertebr. Pathol.* **19**, 291–298.

ICTV (International Committee on Taxonomy of Viruses) (1995). "Virus Taxonomy: Sixth Report of the International Committee on Taxonomy of Viruses" (F. A. Murphy, C. M. Fauquet, D. H. L. Bishop, S. A. Ghabrial, A. W. Jarvis, G. P. Martelli, M. A. Mayo, and M. D. Summers, eds.), pp. 1–507. Springer-Verlag, Vienna.

Ignoffo, C. M., Hostetter, D. L., Sikorowski, P. P., Sutter, G., and Brooks, W. M. (1977). Inactivation of representative species of entomopathogenic viruses, a bacterium, fungus, and protozoan by an ultraviolet light source. *Environ. Entomol.* **6**, 411–415.

Irabogon, T. A., and Brooks, W. M. (1974). Interaction of *Campoletis sonorensis* and a nuclear polyhidrosis virus in larvae of *Heliothis virescens. J. Econ. Entomol.* **67**, 229–231.

Jaques, R. P. (1967). The persistence of a nuclear polyhidrosis virus in the habitat of the host insect, *Trichoplusia ni. Can. Entomol.* **99**, 785–794.

Jaques, R. P. (1972). The inactivation of foliar deposits of viruses of *Trichoplusia ni* (Lepidoptera: Noctuidae) and *Pieris rapae* (Lepidoptera: Pieridae) and tests on protectant additives. *Can. Entomol.* **104**, 1985–1994.

Jaques, R. P. (1975). Persistence, accumulation, and denaturation of nuclear polyhidrosis and granulosis viruses. *In* "Baculoviruses for Insect Pest Control: Safety Considerations" (M. Summers, R. Engler, L. A. Falcon, and P. Vail, eds.), pp. 90–101. American Society for Microbiology, Washington, DC.

Keeley, L. L., and Vinson, S. B. (1975). β-ecdysone effects on the development of nucleopolyhedrosis in *Heliothis* spp. *J. Invertebr. Pathol.* **26**, 121–123.

Kellen, W. R., and Hoffmann, D. F. (1982). Dose–mortality and stunted growth responses of larvae of the navel orangeworm, *Amyelois transitella*, infected by chronic stunt virus. *Environ. Entomol.* **11**, 214–222.

Kellen, W. R., and Hoffmann, D. F. (1983). Longevity and fecundity of adult *Amyelois transitella* (Lepidoptera: Pyralidae) infected by two small RNA viruses. *Environ. Entomol.* **12**, 1542–1546.

Kelly, D. C. (1981). Non-occluded viruses. *In* "Pathogenesis of Invertebrate Microbial Diseases" (E. W. Davidson, ed.), pp. 38–60. Allanheld, Osmun.

Kingsland, S. E. (1985). "Modeling Nature." University of Chicago Press, Chicago.

Knell, R. J., Begon, M., and Thompson, D. J. (1998). Host–pathogen population dynamics, basic reproductive rates and threshold densities. *Oikos* **81**, 299–308.

Kukan, B. (1996). "The Occurrence and Persistence of Nuclear Polyhidrosis Virus in Fluctuating Populations of Tent Caterpillars." PhD dissertation, University of British Columbia, Vancouver.

Kukan, B., and Myers, J. H. (1997). Prevalence and persistence of nuclear polyhidrosis virus in fluctuating populations of forest tent caterpillars (Lepidoptera: Lasiocampidae) in the area of Prince George, British Columbia. *Environ. Entomol.* **26**, 882–887.

Lautenschlager, R. A., Podgwaite, J. D., and Watson, D. E. (1980). Natural occurrence of the nucleopolyhedrosis virus of the gypsy moth, *Lymantria dispar* (Lepidoptera: Lymantriidae) in wild birds and mammals. *Entomophaga* **25**, 261–267.

Lipsitch, M., Siller, S., and Nowak, M. A. (1996). The evolution of virulence in pathogens with vertical and horizontal transmission. *Evolution* **50**, 1729–1741.

Liu, W., Levin, S., and Iwasa, Y. (1986). Influence of nonlinear incidence rates upon the behavior of SIRS epidemiological models. *J. Math. Biol.* **23**, 187–204.

Linley, J. R., and Nielson, H. T. (1968a). Transmission of a mosquito iridescent virus in *Aedes taeniarhynchus*, I: Laboratory experiments. *J. Invertebr. Pathol.* **12**, 7–16.

Linley, J. R., and Nielson, H. T. (1968b). Transmission of a mosquito iridescent virus in *Aedes taeniarhynchus*, II: Experiments related to transmission in nature. *J. Invertebr. Pathol.* **12**, 17–24.

Marshall, I. D. (1973). Viruses of Diptera. *In* "Viruses and Invertebrates" (A. J. Gibbs, ed.), pp. 406–427. North-Holland, Amsterdam.

Martinat, P. J. (1987). The role of climatic variation and weather in forest insect outbreaks. *In* "Insect Outbreaks" (P. Barbosa and J. Schultz, eds.), pp. 241–268. Academic Press, San Diego.

May, R. M., and Anderson, R. M. (1983). Parasite–host coevolution. *In* "Coevolution" (D. J. Futuyama and M. Slatkin, eds.), pp. 186–206. Sinauer, Sunderland.

Milks, M. L. (1997). Comparative biology and susceptibility of cabbage looper (Lepidoptera: Noctuidae) lines to a nuclear polyhidrosis virus. *Environ. Entomol.* **26**, 839–848.

Mohamed, A. K., Yang, J. R., and Nelson, F. R. S. (1989). Influence of insecticides–cytoplasmic polyhidrosis virus combinations on pupal weight and fecundity of tobacco budworm (Lepidoptera: Noctuidae). *J. Entomol. Sci.* **24**, 539–544.

Moore, N. F. (1991). The Nudaurelia β family of insect viruses. *In* "Viruses of Invertebrates" (E. Kurstak, ed.), pp. 277–285. Marcel Dekker, New York.

Murray, K. D., and Elkinton, J. S. (1989). Environmental contamination of egg masses as a major component of transgenerational transmission of gypsy moth nuclear polyhidrosis virus (LdMNPV). *J. Invertebr. Pathol.* **53**, 324–334.

Murray, K. D., Shields, K. S., Burand, J. P., and Elkinton, J. S. (1991). The effect of gypsy moth metamorphosis on the development of nuclear polyhidrosis virus infection. *J. Invertebr. Pathol.* **57**, 352–361.

Myers, J. H. (1988). Can a general hypothesis explain population cycles of forest Lepidoptera? *Adv. Ecol. Res.* **18**, 179–241.

Myers, J. H. (1990). Population cycles of western tent caterpillars: Experimental introductions and synchrony of fluctuations. *Ecology* **71**, 986–995.

Myers, J. H. (1993). Population outbreaks in forest Lepidoptera. *Am. Sci.* **81**, 240–251.

Myers, J. H. (1996). Important may not be enough to explain population cycles. *Trends Ecol. Evol.* **11**, 335–336.

Myers, J. H. (1998). Synchrony in outbreaks of forest Lepidoptera: A possible example of the Moran Effect. *Ecology* **79**, 1111–1117.

Myers, J. H., and Kukan, B. (1995). Changes in the fecundity of tent caterpillars: A correlated character of disease resistance or sublethal effect of disease? *Oecologia* **103**, 475–480.

Myers, J. H., and Rothman, L. D. (1995a). Virulence and transmission of infectious diseases in humans and insects: Evolutionary and demographic patterns. *Trends Ecol. Evol.* **10**, 194–198.

Myers, J. H., and Rothman, L. D. (1995b). Field experiments to study regulation of fluctuating populations. *In* "Population Dynamics: New Approaches and Synthesis" (N. Cappuccino and P. W. Price, eds.), pp. 229–250. Academic Press, San Diego.

Neilson, M. M. (1965). Effects of a cytoplasmic polyhidrosis on adult Lepidoptera. *J. Invertebr. Pathol.* **7**, 306–314.

Olofsson, E. (1988). Dispersal of the nuclear polyhidrosis virus of *Neodiprion sertifer* from soil to pine foliage with dust. *Entomol. Exp. Appl.* **46**, 181–186.

O'Reilly, D. R., and Miller, L. K. (1989). A baculovirus blocks insect molting by producing ecdysteroid UDP-glucosyl transferase. *Science* **245**, 1110–1112.

Otvos, I. S., Cunningham, J. C., and Alfaro, R. I. (1987). Aerial application of nuclear polyhidrosis virus against Douglas-fir tussock moth, *Orgyia pseudotsugata* (McDunnough) (Lepidoptera: Lymantriidae), II: Impact 1 and 2 years after application. *Can. Entomol.* **119**, 707–715.

Park, E. J., Burand, J. P., and Chih-Ming, Y. (1993). The effect of baculovirus infection on ecdysteroid titer in gypsy moth larvae. *J. Insect Physiol.* **39**, 791–796.

Payne, C. C. (1981). Cytoplasmic polyhidrosis virus. *In* "Pathogenesis of Invertebrate Microbial Diseases" (E. W. Davidson, ed.), pp. 61–100. Allanheld, Osmun.

Payne, C. C. (1982). Insect viruses as control agents. *Parasitology* **84**, 35–77.

Pringle, R. L., and Lewis, L. C. (1997). Field persistence of a multiple nucleopolyhedrosis virus of the celery looper, *Anagrapha falciferea*, on sweet corn. *J. Invert. Pathol.* **69**, 282–284.

Roland, J., and Kaupp, W. J. (1995). Reduced transmission of forest tent caterpillar (Lepidoptera: Lasiocampidae) nuclear polyhidrosis virus at the forest edge. *Environ. Entomol.* **24**, 1175–1178.

Rossiter, M. C. (1994). Maternal effects hypothesis of herbivore outbreak. *BioScience* **44**, 752–763.

Rothman, L. D. (1997). Immediate and delayed effects of a viral pathogen and density on tent caterpillar performance. *Ecology,* **78**, 1481–1493.

Rothman, L. D., and Myers, J. H. (1994). Effect of nuclear polyhidrosis virus treatment on reproductive potential of western tent caterpillar (Lepidoptera: Lasiocampidae). *Environ. Entomol.* **23**, 864–869.

Rothman, L. D., and Myers, J. H. (1996a). Debilitating effects of viral diseases on host Lepidoptera. *J. Invertebr. Pathol.* **67**, 1–10.

Rothman, L. D., and Myers, J. H. (1996b). Is fecundity correlated with resistance to viral disease in the western tent caterpillar? *Ecol. Entomol.* **21**, 396–398.

Rothman, L. D., and Roland, J. (1998). Forest fragmentation and colony performance of forest tent caterpillar. *Ecography* **21**, 383–391.

Sait, S. M., Begon, M., and Thompson, D. J. (1994a). The influence of larval age on the response of *Plodia interpunctella* to a granulosis virus. *J. Invertebr. Pathol.* **63**, 107–110.

Sait, S. M., Begon, M., and Thompson, D. J. (1994b). The effects of a sublethal baculovirus infection in the Indian meal moth, *Plodia interpunctella. J. Anim. Ecol.* **63**, 541–550.

Sait, S. M., Begon, M., and Thompson, D. J. (1994c). Long-term population dynamics of the Indian meal moth *Plodia interpunctella* and its granulosis virus. *J. Anim. Ecol.* **63**, 861–870.

Santiago-Alvarez, C., and Vargas-Osuna, E. (1988). Reduction of reproductive capacity of *Spodoptera littoralis* males by a nuclear polyhidrosis virus (NPV). *J. Invertebr. Pathol.* **52**, 142–146.

Sikorowski, P. P., and Lawrence, A. M. (1994). *Heliothis* cytoplasmic polyhidrosis virus and its effect upon microbial contaminant-free *Heliothis virescens. J. Invertebr. Pathol.* **63**, 56–62.

Sikorowski, P. P., and Thompson, A. C. (1979). Effects of cytoplasmic polyhidrosis virus on diapausing *Heliothis virescens. J. Invertebr. Pathol.* **33**, 66–70.

Sikorowski, P. P., Andrews, G. L., and Broome, J. R. (1973). Trans-ovum transmission of a cytoplasmic polyhidrosis virus of *Heliothis virescens* (Lepidoptera: Noctuidae). *J. Invertebr. Pathol.* **21**, 41–45.

Sinclair, A. R. E., and Pech, R. P. (1996). Density dependence, stochasticity, compensation and predator regulation. *Oikos* **75**, 164–173.

Stairs, G. R. (1965). Artificial initiation of virus epizootics in forest tent caterpillar populations. *Can. Entomol.* **97**, 1059–1062.

Stairs, G. R. (1971). Use of viruses for microbial control of insects. *In* "Microbial Control of Insects and Mites" (H. D. Burges and N. W. Hussey, eds.), pp. 97–124. Academic Press, New York.

Steinhaus, E. A. (1958). Crowding as a possible stress factor in insect disease. *Ecology* **39**, 503–514.

Stoltz, D. B., and Vinson, S. B. (1979). Viruses and parasitism in insects. *Adv. Virus Res.* **24**, 125–171.

Tanada, Y., and Fuxa, J. R. (1987). The pathogen population. *In* "Epizootiology of Insect Diseases" (J. R. Fuxa and Y. Tanada, eds.), pp. 113–157. Wiley, New York.

Tatchell, G. M. (1981). The transmission of a granulosis virus following contamination of *Pieris rapae* adults. *J. Invertebr. Pathol.* **37**, 210–213.

Thomas, O. M. (1996). A virus–*Drosophila* association: The first steps towards co-evolution? *Biodivers. Conserv.* **5**, 1015–1021.

Tinsley, T. W., and Kelly, D. C. (1985). Taxonomy and nomenclature of insect pathogenic viruses. *In* "Viral Insecticides for Biological Control" (K. Maramorosch and K. E. Sherman, eds.), pp. 3–25. Academic Press, San Diego.

Turchin, P. (1990). Rarity of density dependence or population regulation with lags? *Nature* **344**, 660–663.

Turchin, P. (1995). Population regulation: Old arguments and a new synthesis. *In* "Population Dynamics: New Approaches and Synthesis" (N. Cappuccino and P. W. Price, eds.), pp. 19–40. Academic Press, San Diego.

Vail, P. V. (1981). Cabbage looper nuclear polyhidrosis virus–parasitoid interaction. *Environ. Entomol.* **10**, 517–520.

Vail, P. V., and Gough, D. (1970). Effects of a cytoplasmic polyhidrosis virus on adult cabbage loopers and their progeny. *J. Invertebr. Pathol.* **15**, 397–400.

Vail, P. V., Hall, I. M., and Gough, D. (1969). Influence of a cytoplasmic polyhidrosis on various developmental stages of the cabbage looper. *J. Invertebr. Pathol.* **14**, 237–244.

Vargas-Osuna, E., and Santiago-Alvarez, C. (1988). Differential response of male and female *Spodopteras littoralis* (Boisduval) (Lepidoptera: Noctuidae) individuals to a nuclear polyhidrosis virus. *J. Appl. Entomol.* **105**, 374–378.

Vasconcelos, S. D. (1996). Alternative routes for the horizontal transmission of a nucleopolyhedrovirus. *J. Invertebr. Pathol.* **68**, 269–274.

Vezina, A., and Peterman, R. M. (1985). Tests of the role of a nuclear polyhidrosis virus in the population dynamics of its host, douglas-fir tussock moth, *Orgyia pseudotsugata* (Lepidoptera: Lymantriidae). *Oecologia* **67**, 260–266.

Wellington, W. G. (1957). Individual differences as a factor in population dynamics: The development of a problem. *Can. J. Zool.* **35**, 293–323.

Wellington, W. G. (1962). Population quality and the maintenance of nuclear polyhidrosis between outbreaks of *Malacosoma pluviale* (Dyar). *J. Invertebr. Pathol.* **4**, 285–305.

White, A., Bower, R. G., and Begon, M. (1996). Host–pathogen cycles in self-regulated forest insect systems: Resolving conflicting predictions. *Am. Nat.* **148**, 220–225.

White, T. C. R. (1993). "The Inadequate Environment: Nitrogen and the Abundance of Animals." Springer-Verlag, Berlin.

Whitlock, V. H. (1977). Effect of larval maturation on mortality induced by nuclear polyhidrosis and granulosis virus infections of *Heliothis armigera*. *J. Invertebr. Pathol.* **30**, 80–86.

Williamson, C., and von Wechmar, M. B. (1995). The effects of two viruses on the metamorphosis, fecundity and longevity of the green stinkbug, *Nezara viridula*. *J. Invertebr. Pathol.* **65**, 174–178.

Wiygul, G., and Sikorowski, P. P. (1978). Oxygen uptake in tobacco budworm larvae (*Heliothis virescens*) infected with cytoplasmic polyhidrosis virus. *J. Invertebr. Pathol.* **32**, 191–195.

Wiygul, G., and Sikorowski, P. P. (1991). Oxygen uptake in larval bollworms, (*Heliothis zea*) infected with iridescent virus. *J. Invertebr. Pathol.* **58**, 252–256.

Woodard, D. B., and Chapman, H. C. (1968). Laboratory studies with the mosquito iridescent virus (MIV). *J. Invertebr. Pathol.* **11**, 296–301.

Woods, S. A., and Elkinton, J. S. (1987). Bimodal patterns of mortality from nuclear polyhidrosis virus in gypsy moth (*Lymantria dispar*) populations. *J. Invertebr. Pathol.* **50**, 151–157.

Woods, S. A., Elkinton, J. S., and Podgwaite, J. D. (1989). Acquisition of nuclear polyhidrosis virus from tree stems by newly emerged gypsy moth (Lepidoptera: Lymantriidae) larvae. *Environ. Entomol.* **18**, 298–301.

Woods, S. A., Elkinton, J. S., Murray, K. D., Liebhold, A. M., Gould, J. R., and Podgwaite, J. D. (1991). Transmission dynamics of a nuclear polyhidrosis virus and predicting mortality in gypsy moth (Lepidoptera: Lymantriidae) populations. *J. Econ. Entomol.* **84**, 423–430.

Xu, J., and Hukuhara, T. (1992). Enhanced infection of a nuclear polyhidrosis virus in larvae of the armyworm, *Pseudaletia separata*, by a factor in the spheroids of an entomopoxvirus. *J. Invertebr. Pathol.* **60**, 259–264.

Young, S. Y., and Yearian, W. C. (1982). Nuclear polyhidrosis virus infection of *Pseudoplusia includens* (Lepidoptera: Noctuidae) larvae: Effect on post larval stages and transmission. *Entomophaga* **27**, 61–66.

Young, S. Y., and Yearian, W. C. (1990). Transmission of nuclear polyhidrosis virus by the parasitoid *Microplitis croceipes* (Hymenoptera: Braconidae) to *Heliothis virescens* (Lepidoptera: Noctuidae) on soybean. *Environ. Entomol.* **19**, 251–256.

Zelazny, B. (1972). Studies on Rhabdionvirus *oryctes*, I: Effect on larvae of *Oryctes rhinoceros* and inactivation of virus. *J. Invertebr. Pathol.* **20**, 235–241.

Zelazny, B. (1973). Studies on Rhabdionvirus *oryctes*, II: Effect on adults of *Oryctes rhinoceros*. *J. Invertebr. Pathol.* **22**, 122–126.

Zelazny, B. (1976). Transmission of a baculovirus in populations of *Oryctes rhinoceros*. *J. Invertebr. Pathol.* **27**, 221–227.

Zelazny, B., and Alfiler, A. R. (1991). Ecology of baculovirus-infected and healthy adults of *Oryctes rhinoceros* (Coleoptera: Scarabaeidae) on coconut palms in the Philippines. *Ecol. Entomol.* **16**, 253–259.

Zelazny, B., Lolong, A., and Pattang, B. (1992). *Oryctes rhinoceros* (Coleoptera: Scarabaeidae) populations suppressed by a baculovirus. *J. Invertebr. Pathol.* **59**, 61–68.

11

Ecology of Viruses of Cold-Blooded Vertebrates

V. GREGORY CHINCHAR

Department of Microbiology
University of Mississippi Medical Center
Jackson, Mississippi 39216

I. INTRODUCTION

Viruses are obligate intracellular parasites whose survival is dependent on how successfully they interact with their host at both the cellular and organismic levels. To complete their "life cycle," viruses must attach to and replicate within permissive cells, spread from the site of primary infection to specific target tissues, evade

413

the host immune response, and successfully infect another host. The mechanisms by which viruses accomplish these diverse tasks comprise the subject of viral ecology.

Viruses infecting warm-blooded animals (homeotherms) and immune responses directed against these agents have been the focus of intense research interest. In contrast, relatively little is known about viruses infecting cold-blooded (ectotherms) vertebrates and about immune responses in these animals. Although this differential is explicable in view of the impact that viruses have on human health and on the health of domesticated animals (chiefly cattle, horses, swine, and poultry), the relative paucity of information concerning viruses of "lower" vertebrates is surprising in terms of the number of species of ectothermic animals. In contrast to 8600 species of birds and 4500 species of mammals, there are 30,000 species of bony fish (class *Osteichthyes*), 4000 species of amphibians, and 7000 species of reptiles (Campbell, 1996). Moreover, in view of the large number of species within class *Osteichthyes*, it is surprising that only about 60 different viruses have been isolated from bony fish (Ahne, 1994). Does this underrepresentation (*vis-à-vis* mammals and their viruses) reflect the relative resistance of bony fish to virus infection, or is it simply due to a lack of comprehensive surveys? Aside from comparative studies, interest in viruses infecting ectotherms appears to be justified on several grounds: the increased culture of fish, especially catfish, trout, and salmon, as a source of human protein, the use of aquatic ectotherms as biomonitors of environmental quality, and human interest in preserving endangered animals such as sea turtles.

To illustrate key features of viral ecology, this review will focus on select viruses infecting teleosts (i.e., bony fish), amphibians (frogs), and reptiles (turtles). Although this approach should facilitate discussion, it must be appreciated that knowledge gained studying one particular virus–host system may not be directly applicable to different virus families infecting animals of diverse taxonomic classes. Before discussing specific host–virus systems, the role of extrinsic (e.g., temperature, stress, and environmental toxins) and intrinsic (e.g., age and genetic composition) factors that affect virus replication and host susceptibility will be considered. General features of virus replication will not be discussed in this review. For the most part, replication strategies of viruses infecting lower vertebrates are similar in broad detail to events occurring with other members of the same taxonomic family. For the interested reader, details of viral macromolecular synthesis can be found elsewhere (Fields *et al.*, 1996).

II. IMMUNE RESPONSE OF COLD-BLOODED VERTEBRATES TO VIRUS INFECTIONS

Because virus replication occurs within cells of an infected organism, the host immune response contributes significantly to the success or failure of a given pathogen. It is generally accepted that too vigorous an immune response will eliminate the virus before it has the opportunity to ensure its efficient transmission, whereas too slow an immune response may lead to death of the host and loss of the species required to sustain its continued existence. Because of its importance in pathogenesis, a synopsis of immune responses in "lower vertebrates" is provided.

A. Organization and Structural Components of Ectothermic Immune Systems

Although the immune response of amphibians and reptiles has been studied in some detail (Du Pasquier, 1993; Du Pasquier *et al.*, 1989), the bulk of what we know about immunity among ectotherms is derived from studies on a few species of commercially important fish such as rainbow trout, salmon, catfish, and carp (Ellis, 1989, 1999; Iwama and Nakanishi, 1996). Thus, piscine immunity will be considered as a model for ectothermic organisms. Broadly speaking, the types of immune responses seen in homeotherms are paralleled by functionally similar responses in ectotherms. Fish possess innate (e.g., complement, acute-phase proteins, interferons, regulatory cytokines, natural killer [NK] cells, and activated macrophages) and acquired (e.g., antigen-specific antibody) immune responses, as well as an array of major histocompatibility (MHC) genes important in antigen presentation (Iwama and Nakanishi, 1996). At the cellular level, fish possess antigen-presenting cells, B cells, and T cells (Clem *et al.*, 1991; Vallejo *et al.*, 1992; Partula *et al.*, 1995). Surprisingly, despite demonstration of the T cell receptor (TCR) and the presence of several correlates of cell-mediated immunity (e.g., mixed lymphocyte reactions, acute skin graft rejection), authentic MHC-restricted, antigen-specific, and cytotoxic T cells (CTLs) have not yet been convincingly demonstrated in fish. This gap in our knowledge is due, for the most part, to the absence of syngeneic fish strains. The development of syngeneic cell lines from gynogenetic carp, trout, and catfish should prove invaluable and make detection of piscine cytotoxic T cells a reality within the near future.

Although fish possess an immune system functionally analogous to those of homeothermic animals, it is not structurally identical. Whereas piscine and amphibian antibody genes are organized along the pattern seen in mice and humans (i.e., multiple copies of variable [V], diversity [D], and joining [J] elements located upstream from constant [C] gene-coding regions), the major serum antibody molecule in fish is a high-molecular-weight tetramer more similar to IgM than to mammalian IgG (Wilson and Warr, 1992; Kattari and Paganelli, 1996). In addition, although fish cytokines and interferons have been functionally identified, few fish cytokine genes and no interferon genes have been cloned and sequenced (Yano,

1996; Manning and Nakanishi, 1996). The absence of interleukin 1 (IL-1), IL-2, and interferon sequence information suggests that fish cytokine genes (like avian interferon genes) are markedly different from their mammalian counterparts. Thus, although fish and mammalian immune systems are functionally similar and share several structural features in common, some elements of the piscine system appear to be quite divergent from their mammalian counterparts.

B. Immune Response of Ectothermic Animals to Virus Infection

1. Temperature Considerations

Because fish are ectothermic animals, their ability to respond to a specific antigen is dependent on temperature. For example, cold-water fish such as trout respond very well at 12°C, but slowly, if at all, at 4°C. In contrast, warm-water fish such as carp respond optimally at 22°C but are essentially immunosuppressed at 12°C (Trust, 1986). The cellular mechanisms underlying this phenomenon have been extensively studied in channel catfish (Miller and Clem, 1984; Clem et al., 1991; Bly and Clem, 1991, 1992; Bly et al., 1997). In the channel catfish, 27–32°C is considered immunologically permissive, whereas 17–22°C is nonpermissive for T-cell responses, and 12°C is nonpermissive for both B- and T-cell functions. Interestingly, although T-cell responses are impaired at 17°C, if fish are first acclimated *in vivo* to low temperatures, T-cell responses are not lost at 17°C. From these and other studies, it was determined that low temperature-induced immune suppression does not involve T suppressor cells or suppressive factors, nor is it due to a defect in accessory cells, antigen processing, or cytokine synthesis. Rather, since low temperature-suppressed fish were able to synthesize antibody during a secondary immune response, it appears as if low temperature inhibits the generation and/or activation of virgin helper T cells. The role that low temperature-induced immune suppression plays in disease processes in cold-blooded animals will be discussed further in Section III.B.

2. Kinetics of the Immune Response

Antiviral immunity is a three-stage process. In mammalian systems, innate responses (e.g., complement, interferon, and NK cells) appear within a few days following infection and limit virus replication. CTL responses develop a few days later, peak between days 7–10, and are thought to be the most important element in resolving an ongoing infection. Finally, antiviral antibodies reach maximum titers later than CTLs, persist for many months to many years, and are thought to provide protection against reinfection (Zinkernagel, 1993; Whitton and Oldstone, 1996). With the exception of CTL responses (which have not yet been convincingly demonstrated), bony fish possess all these elements.

Complement, interferon (both types I and II), and NK cells have been demonstrated in fish (Du Pasquier, 1993; Yano, 1996). Interferon-like responses were induced in several fish species following virus infection or exposure to double-stranded RNA (Oie and Loh, 1971; de Kinkelin and Dorson, 1973; de Sena and Rio, 1975; Eaton, 1990; Renault et al., 1991; Snegaroff, 1993; Rogel-Gaillard et al., 1993). Although this activity is considered to be interferon based on physical characteristics (i.e., pH stability, heat lability, protease sensitivity, and resistance to RNase), the gene(s) encoding fish interferon have not yet been cloned and sequenced. However, the presence of piscine Mx genes (which in mammals are induced by interferon and associated with viral resistance) supports the notion that fish contain interferon genes (Trobridge and Leong, 1995). Induction of interferon-like activity following virus infection or exposure to poly I:C occurred within as little as 24 hr at temperatures greater than 14°C (Snegaroff, 1993; de Sena and Rio, 1975; de Kinkelin and Dorson, 1973) or required as long as 2–3 weeks at 6°C (Eaton, 1990; Renault et al., 1991). The response time at temperatures >14°C is almost as rapid as that seen in mammalian cells and is surprising since fish and fish cell cultures are grown at temperatures 10–20°C below that used for mammals. It is crucial for host survival that the kinetics of the interferon response match the rate of virus replication, because many fish viruses undergo a complete replication cycle within 24 hr. Clearly, the kinetics of interferon induction in fish is such that interferon likely serves as a first line of defense against virus infection.

NK-like cells (also referred to as nonspecific cytotoxic cells [NCCs]) have been identified in several species of fish (Evans and Jaso-Friedmann, 1992). Although their role in controlling virus infection has not been extensively studied, several reports suggest that they recognize and lyse virus-infected allogeneic and autologous cells (Yoshinaga et al., 1994; Hogan et al., 1996). In mammals NK activity is enhanced following virus infection. Although this phenomenon has not been examined in fish systems, Stuge et al. (1997) showed that incubation of catfish peripheral blood lymphocytes (PBLs) with a lethally irradiated catfish allogeneic cell line resulted in marked cell proliferation and a 100-fold increase in NK-like activity. Whether this is an accurate reflection of what occurs in vivo following allograft transfer or virus infection is not known.

Consistent with responses occurring at lower temperatures, fish antibody responses are seen beginning 10–15 days after antigen exposure, reach peak levels at 5–6 weeks, and persist for up to a year (Ellis, 1999; Plumb, 1973; Heartwell, 1975). In contrast to mammals, fish generally show a poor response to secondary immunization, with antibody titers only 10-fold higher than those seen in the primary response (compared to 100-fold higher in mammalian systems) (Ellis,

1999; Kattari and Paganelli, 1996; Wilson and Warr, 1992). Antiviral antibody has been shown to be protective by *in vitro* neutralization tests, as well as by observations that (1) fish surviving virus infection as fry or fingerlings are subsequently resistant to viral disease, and (2) vaccination of fry protects against subsequent challenge with infectious virus (see below).

3. Protection from Disease

The protective nature of piscine immune responses has been demonstrated convincingly by passive antibody transfer experiments and by the ability of several viral vaccines to protect against disease (Leong and Fryer, 1993; Leong, 1993; Wolf, 1988; Kattari and Paganelli, 1996). Protection has been achieved following immunization with inactivated, live attenuated, and DNA vaccines. Whether protection is due solely to induced antibodies or whether there is also a cell-mediated component is not clear, because there is no assay system for the latter. However, the finding that immunization with plasmid DNA-encoding viral structural genes protects trout from challenge with infectious hematopoietic necrosis virus (IHNV) suggests that cell-mediated immunity may be involved in protection (Anderson *et al.*, 1996).

III. EFFECT OF EXTRINSIC AND INTRINSIC FACTORS ON DISEASE DEVELOPMENT AND VIRUS REPLICATION

A. Intrinsic Factors: Age, MHC Haplotype, and Development of Disease

Many viruses that cause disease in cold-blooded vertebrates are more pathogenic in young rather than old animals (Wolf, 1988). For example, inoculation of frog tadpoles with the iridovirus frog virus 3 (FV3) results in a lethal infection, whereas no disease is seen in adults (Tweedell and Granoff, 1968). Among herpesviruses, channel catfish herpesvirus (CCV) results in high morbidity and mortality among young-of-the-year catfish (i.e., animals <10 cm in length) but does not cause disease when adult animals are injected i.p. with up to 10^6 pfu CCV (Plumb, 1973, 1989). Likewise, infection with IHNV, a rhabdovirus, or infectious pancreatic necrosis virus (IPNV), a birnavirus, results in clinical disease primarily in fry and fingerlings (Wolf, 1988). Infection of older fish is rare, and in those cases where IHNV infects smolts (i.e., fish about 1 year of age), disease is not as severe as in younger fish. The reasons why fry and fingerlings are more susceptible to infection are not clear. The ability of fry to be vaccinated and protected from subsequent challenge suggests that young fish are immunocompetent. For example, immunization with an attenuated strain of CCV led to protection against subsequent virus challenge (Wolf, 1988; Zhang and Hanson, 1995). Whereas the resistance of older fish may simply be a reflection of widespread prior virus exposure, it is also possible that older fish are less susceptible for other reasons.

Aside from age, another intrinsic determinant that affects disease susceptibility and progression is the MHC haplotype (Wiegertjes *et al.*, 1996). As with mammals, class I and II major histocompatibility (MHC) genes have been identified in amphibians, fish, and reptiles, and preliminary evidence suggests the existence of multiple loci, each containing multiple alleles (Kaufman *et al.*, 1991; Dixon *et al.*, 1995; Manning and Nakanishi, 1996; Goodwin *et al.*, 1997; Antao *et al.*, 1999). Class I gene products bind peptides derived from intracellular pathogens such as actively replicating viruses and play a key role in the development of antiviral cell-mediated immunity (Hansen *et al.*, 1993). Class II gene products, on the other hand, bind and present antigen derived from extracellular pathogens such as bacteria and extracellular viruses and are crucial to the development of an antibody response. Since MHC gene products differ in their ability to bind specific antigens, members of an outbred population will differ in their response to virus infection. Individuals that bind immunologically important peptides effectively may mount a protective immune response, whereas those that bind key proteins poorly may fail to respond, or generate only a low level of antiviral activity. In addition, the susceptibility of young animals to virus infection may be due to a defect in MHC class I expression. Flajnik *et al.* (1986) showed that *Xenopus* do not express MHC class I proteins until after metamorphosis. However, because tadpoles express class II proteins, they are partially immunocompetent, that is, they synthesize antibody, reject skin grafts, and their splenic lymphocytes proliferate in response to mitogen or allogeneic cells. Thus, tadpoles should be able to combat most bacterial and some viral infections through a humoral response, but they may fail to respond effectively against most viral invaders because they lack the class I proteins necessary for recognition by cytotoxic T cells. Clearly, it remains to be determined whether the susceptibility of lower vertebrates to virus infection is a reflection of MHC haplotype, MHC expression, or other mechanisms.

B. Temperature and Disease Progression: Possible Interplay Between Low Temperature-Induced Immune Suppression and Pathogenesis

In one of the best-studied systems, Bly and her colleagues (Bly and Clem, 1991, 1992; Clem *et al.*, 1991; Bly *et al.*, 1992, 1997) demonstrated that abrupt exposure to low temperature suppressed immune responsiveness in the channel catfish. To date studies linking low temperature-induced immune suppression to enhanced viral disease in catfish (or other fish species) have not been performed. However, low temperature has been associated with a disease phenomenon termed "winter saprolegniosis" (Bly *et al.*, 1992). Illness resulted when fish were subjected to an abrupt temperature drop from 22 to 10°C over 24 hr and challenged with the opportunistic cold-water pathogen *Saprolegnia* spp. In contrast, fish maintained at 22°C, as well as fish acclimated at 10°C for 8 weeks prior to challenge, did not show signs of illness. Although low temperature-induced immune suppression results in B- and T-cell dysfunction, the direct cause of winter saprolegniosis may be a block in mucus cell migration through the dermis, which reduces the thickness of this vital physical barrier (Bly *et al.*, 1997). During the period when the protective mucus coat is absent or reduced, *Saprolegnia* cysts attach and germinate. It is thought that low temperature-induced immune suppression blunts the normally vigorous inflammatory response directed against the developing hyphae and allows invasion of underlying tissues.

Carey and her colleagues (Carey, 1993; Carey and Bryant, 1995; Carey *et al.*, 1996a,b; Maniero and Carey, 1997) attempted to explain progressive die-offs of amphibians from areas where they have been previously abundant by linking low temperature and other sublethal environmental stresses with immune suppression and subsequent infection. She observed that low temperatures did not significantly alter mitogenic responses of toads (*Bufo marinus*), but adversely affected mitogenic responses in frogs (*Rana pipiens*). When frogs maintained for 5 weeks at 5°C were returned to 20°C and then injected with *Aeromonas hydrophilia*, excess morbidity and mortality were not seen. While this result suggests that low temperature-induced immune suppression of *R. pipiens* did not result in enhanced sensitivity to an opportunistic bacterial pathogen, the absence of disease may reflect acclimation to low temperature or an increased level of complement (Maniero and Carey, 1997). Whether low temperature-induced immune suppression plays a role in viral disease in frogs and/or fish remains an open question. Since virus replication is generally more efficient at the higher end of an ectotherm's physiological temperature range (e.g., CCV replicates more rapidly and reaches higher titers at 28–30°C than it does at 20°C), it is likely that an overwhelming virus infection may result if fish, previously maintained at low temperatures, are shifted to higher temperatures and challenged with virus. However, experiments to test this hypothesis have not been performed, and such extreme temperature shifts are unlikely in nature or in an aquaculture setting.

Although low temperature-induced immune suppression is potentially a very damaging phenomenon, its impact on animal health may be limited by several factors. With the notable exception of *Saprolegnia*, temperatures low enough to cause immune suppression are often below the optimum for pathogenic organisms. Clinical disease should only occur within a narrow temporal and thermal window corresponding to the time required for an animal to recover from low temperature-induced immune suppression and a temperature range permissive for pathogen growth. In addition, at low environmental temperatures, elements of the innate response (e.g., phosphorylcholine-reactive protein [Szalai *et al.*, 1994] and NK cells [Le Morvan-Rocher *et al.*, 1996]) have been suggested to offset suppression of specific immunity.

C. Stress, Immune Dysfunction, and Disease Progression

In addition to low temperature-induced immune suppression, other natural and manmade stresses such as changes in salinity or photoperiod, crowding, hormonal changes related to spawning, low oxygen tension, and handling depress immune responsiveness (Schreck, 1996; Tatner, 1996). All stresses are not equal, and low temperature-induced immune suppression and handling stress affect immune function in different ways (Ellsaesser and Clem, 1986; Bly and Clem, 1991). Handling stress is accompanied by elevated levels of glucocorticoids, ACTH, and serum glucose, along with neutrophilia and lymphopenia, whereas none of these occur after low temperature-induced immune suppression. Although the impact of stress on virus replication has not been extensively evaluated, one example will serve to illustrate the interaction between stress and virus replication. A percentage of fingerlings survive IHNV infection and become lifelong "carriers." Although IHNV in adult fish is occult, it reappears during spawning as massive amounts of virus are produced by multiple tissues and released in urine and sex products (Wolf, 1988). It is possible that virus reactivation is a result of weakening or complete loss of immune function. A similar situation may exist for IPNV, where clinical disease is usually confined to fry and fingerlings, but carrier broodfish actively shed virus in sex products during spawning.

D. Toxins, Environmental Pollution, Immune Suppression, and Disease Progression

Increasing concentrations of heavy metals, organic compounds, and other man-made products are known to have an adverse effect on immune function in ectothermic animals (Zelikoff, 1993; Anderson, 1996). In a few studies where responses to viruses were examined, heavy metal exposure was shown to enhance susceptibility to virus replication, presumably as a result of depressing immune responsiveness. For example, exposure to copper increased the susceptibility of rainbow trout fry to IHNV. Exposure to copper elevates serum corticosteroid levels in salmon and results in immune suppression. However, exposure of rainbow trout fry to organic compounds (polychlorinated biphenyls or 2,3,7,8-tetrachlorodibenzo-*p*-dioxin [TCDD]) at concentrations that depress immunity in mammals did not increase the susceptibility of fish to infection with IHNV (Spitsbergen *et al.*, 1988). Furthermore, experience with sea turtle fibropapillomatosis, a disease associated with a putative herpesvirus infection, suggests that both morbidity and mortality are enhanced by exposure to manmade pollutants (Herbst, 1994). Moreover, there is a higher incidence of epidermal papillomas and carcinomas in white suckers and more aggressive hematopoietic tumors in northern pike and muskellunge from polluted waters (Poulet *et al.*, 1994). Interestingly, although environmental toxins are generally thought to adversely affect immune function, low levels of some compounds have been found in some cases to be stimulatory (Zelikoff, 1993).

E. Seasonal Variation in Disease Symptoms

Several viral diseases show marked seasonal variation. These include tumors resulting from infection with Lucké tumor virus (LTV) (Granoff, 1983), walleye dermal sarcoma virus (WDSV) (Poulet *et al.*, 1994), sea turtle fibropapillomatosis (Herbst, 1994), as well as sea turtle gray patch disease (GPD) (Hoff and Hoff, 1984) and channel catfish virus disease (CCVD) (Wolf, 1988). With the three aforementioned "tumors," tumor mass varied with the season. In these cases, the host immune response and the temperature dependence of virus replication, along with other poorly understood factors, may contribute to the seasonal variation. In contrast to tumors associated with WDSV, which regressed during summer months, clinical disease associated with GPD and CCVD was exacerbated by elevated summer temperatures. Although other environmental factors such as low oxygen tension and crowding may contribute to disease in nature, mortality due to CCV infection is reproducibly increased at elevated temperatures. At 18–25°C mortality of infected fingerlings ranges from 40–70%, whereas replicate controls maintained below 18°C show few if any deaths (Leong and Fryer, 1993). It is possible that CCVD may represent a "race" between rapid virus replication at elevated temperature and temperature-enhanced immune function.

IV. ECOLOGY OF SELECTED VIRUS–HOST SYSTEMS

A. Channel Catfish Herpesvirus (CCV)

CCV is a large DNA-containing virus of the family Herpesviridae (Davison, 1992, 1994). Although virions are morphologically similar to other herpesviruses (e.g., human pathogens such as herpes simplex virus type 1, varicella-zoster virus), CCV gene products share little homology to other herpesviruses (Booy *et al.*, 1996), and thus CCV may represent the prototype of a new herpesvirus subfamily.

1. Host Range

CCV possesses a very limited host range (Wolf, 1988; Davison, 1994; Plumb, 1989). Infection occurs primarily in young-of-the-year catfish (<10 cm in length). Adult animals are generally resistant to bath immersion or i.p. injection of virus, although infection and death have been noted after experimental infection (Hedrick *et al.*, 1987). Only channel catfish (*Ictalurus punctatus*) are naturally infected, although fingerling blue catfish (*I. furcatus*) and hybrids between blue catfish and channel catfish can be infected by injection. Brown bullheads (*I. nebulosus*) and yellow catfish (*I. natalis*) are resistant even to injection. Restriction extends to *in vitro* growth as replication occurs only in cells of ictalurid and clariid fish. CCO (channel catfish ovary) and BB (brown bullhead) cells are commonly used to propagate the virus. Virus replication is markedly influenced by temperature, and, although CCV replicates at temperatures between 10–35°C, optimal virus yields are observed at 30°C (Bowser and Plumb, 1980).

2. Pathogenesis

CCVD is best described as severe hemorrhagic viremia (Wolf, 1988). Infected fish show generalized edema and marked necrosis of kidney, liver, gastrointestinal tract, spleen, skeletal muscle, neural tissue, and pancreas. Externally, fingerlings show a distended abdomen and urogenital vent, exophthalmia, and hemorrhages at the base of fins. When water temperatures are 25°C or greater, onset of clinical symptoms is sudden and massive mortality occurs within 7–10 days.

3. Epizootic Cycle

CCVD is a disease primarily of aquaculture, as CCV has not been isolated from feral channel catfish. It is likely that environmental conditions such as dense stocking of fry and fingerlings and poor water quality coupled with elevated summer temperatures predispose farm-raised fish to disease. Fish that survive primary infection are thought to be lifelong carriers and "immune" to reinfection with the same virus. Carriers remain virus-free and asymptomatic until "stress" (i.e., low temperatures, crowding, spawning) leads to virus reactivation and shedding. Young fish may be infected at hatching by virus excreted by adults in urine and sex products. Alternatively, CCV may be transmitted vertically in eggs (Wise *et al.*, 1988). The primary response of non-immune adult fish is rapid, but highly variable (Plumb, 1973; Hedrick *et al.*, 1987; Heartwell, 1975). Moreover, prior infection with CCV does not always correlate with the persistence of detectable antiviral antibody (Plumb *et al.*, 1981). Thus, although antibody indicates prior infection, the absence of detectable anti-CCV antibody does not rule out earlier infection.

Fish-to-fish transmission is thought to occur via free virus in urine without the need for a biological vector. Although an aquatic virus, CCV is not especially stable in pond water. At 25°C CCV survived only 2 days in pond water; at 4°C, however, survival was extended to ~1 month. Pond mud rapidly inactivated virus, and virus was also quickly inactivated by drying. CCV has been widely spread throughout the United States via infected broodstock.

As is characteristic of herpesvirus infections, survivors of CCV infections are thought to harbor latent virus throughout life (Kieff, 1996). Herpesvirus latency is characterized by the presence of the viral genome (either as an unintegrated episome or integrated within the host chromosome) and the absence of infectious viral particles (Ahmed *et al.*, 1996). With human herpes simplex virus types 1 and 2 and varicella-zoster virus (VZV), latent infections are confined to neural tissue. Stress "reactivates" the virus, resulting in replication, seeding of target tissues, and clinical disease. For example, in man primary infection with VZV results in "chickenpox," whereas reactivated virus causes "shingles." Reactivation does not invariably result in clinical disease. For example, primary infection of humans with Epstein–Barr virus (EBV) may or may not result in clinical disease (e.g., mononucleosis), whereas reactivation of EBV is characterized by virus shedding in the saliva without overt pathology. Although it is thought that similar mechanisms occur with CCV, the site of latency and the molecular mechanisms controlling reactivation have not been defined.

Latent infections of catfish with CCV have been detected by cell culture and molecular approaches. Although serological approaches did not invariably identify asymptomatic carriers, carriers were identified by co-cultivation, nucleic acid hybridization, and PCR amplification (Plumb, 1973; Bowser *et al.*, 1985; Wise *et al.*, 1985; Boyle and Blackwell, 1991; Gray *et al.*, 1999). However, because virus and/or viral DNA was detected in numerous tissues, it was not possible to determine the primary site of latency. Since CCV replicates in a variety of catfish lymphoid cells *in vitro* (Chinchar *et al.*, 1993), and because other herpesviruses appear to use lymphoid tissue as a site of latent infection (Kieff, 1996), it is possible that catfish lymphocytes are the site of latent infections. If so, this would explain the detection of virus in multiple organs of carrier fish. However, experiments to systematically examine this possibility have not been conducted.

The proximal trigger for CCV reactivation is unknown. Suggestive of low temperature-induced immune suppression, CCV was isolated from a moribund brood fish maintained at 8°C in a pond (Bowser *et al.*, 1985). However, whether low temperature is the sole modulator of reactivation or other stimuli are involved is not known. Perhaps, as with other fish viruses (e.g., IHNV and IPNV), stress associated with spawning leads to reactivation and release of virus in urine, sperm, and eggs.

Viruses have evolved in a number of ways to evade host immunity (Gooding, 1992; Kotwal, 1996). These include: (1) encoding proteins that block immune responses (e.g., vaccinia virus, adenovirus), (2) downregulating host immunity (e.g., measles, HIV-1), (3) perturbing antigen processing and presentation (e.g., HSV-1, human cytomegalovirus), and (4) undergoing rapid antigenic variation (e.g., human influenza virus and HIV-1). Latent (discussed here and in Section IV.C. [Lucké tumor virus]) and persistent infections (see Section IV.B. [IHNV]) can also be viewed as mechanisms by which virus remains "viable" and protected from host immunity. Because viral antigens are not expressed during latency, antiviral antibody and cell-mediated responses are unable to eliminate latently infected cells. Spawning stress, which is often a proximal trigger for virus reactivation, concomitantly depresses the immune system. Thus, during spawning, virus is reactivated and released with little pressure from the host immune system. Furthermore, virus release is temporally linked to a period when susceptible young are abundant, thus ensuring efficient transmission of the virus to potential targets.

4. Economic and Environmental Importance

CCV is an important virus because of its economic impact on the catfish industry; infection in hatcheries has decimated fingerling populations. However, because the virus has not been isolated in nature, its impact on feral species may be marginal.

5. Control Measures

CCV outbreaks have been significantly curtailed in recent years as fish farmers identified and utilized carrier-free broodstocks. Recent advances in virus detection should make it possible to identify virus-free fish stocks with a high degree of accuracy. Thus, molecular approaches, coupled with good animal husbandry, should markedly reduce the incidence of CCVD. In addition to these preventive measures, ongoing outbreaks can be resolved by destruction of infected fish, as well as by draining and drying of infected ponds. Furthermore, since catfish strains show marked differences in their susceptibility to CCV, it should be possible to select resistant animals for commercial production and breeding (Wolf, 1988). Finally, the ability of live attenuated vaccines to protect fish from challenge with virulent virus offers the promise of protecting young fish from infection (Zhang and Hanson, 1995). If attenuated CCV could be engineered to express antigens for other important catfish pathogens (e.g., the bacterial pathogen *Edwardsiella ictaluri*), the impact of vaccination would be greatly extended.

B. Infectious Hematopoietic Necrosis Virus (IHNV)

IHNV is an enveloped virus containing a single-stranded RNA genome encoding six viral-specific polypeptides (Kuruth and Leong, 1985). It, along with several serologically distinct aquatic viruses (e.g., spring viremia of carp virus, Egtved virus) are members of the family Rhabdoviridae. IHNV resembles other rhabdoviruses (e.g., vesicular stomatitis virus and rabies virus) both morphologically and in terms of genetic organization.

1. Host Range

Salmonids are the only natural host for IHNV (Wolf, 1988). Infection occurs in brown trout (*Salmo trutta*), chinook salmon (*Oncorhynchus tshawytscha*), pink salmon (*O. gorbuscha*), sockeye salmon (*O. nerka*), steelhead salmon (rainbow trout, *O. mykiss*), chum salmon (*O. keta*), omago (*O. rhodurus*), and yamame (*O. masou*). Atlantic salmon (*Salmo salar*) are susceptible after injection. IHNV infection was initially confined to the Pacific Northwest coasts of the United States and Canada. However, it has been inadvertently spread to Japan and several locations within the United States following transfer of infected eggs and fry. Although *in vivo* infections are confined to salmonids, infection *in vitro* occurs in cell lines of both salmonid (e.g., CHSE-214 and RTG-2) and non-salmonid (e.g., EPC and FHM) origin.

2. Pathogenesis

IHNV causes an acute systemic infection that is associated with high morbidity and mortality (Wolf, 1988). Infection is most often seen in fry and fingerlings; smolts (fish about 1 year old) may be infected, but they show signs of less severe disease. Infected fry are exophthalmic, darker than normal, possess swollen abdomens, and often trail fecal pseudocasts. Internally, major degenerative changes are seen in the kidneys, hematopoietic tissues, pancreas, gastrointestinal tract, adrenal cortex, and liver. Among infected fry and fingerlings, cumulative losses during outbreaks approach 100%.

IHNV replicates between 4 and 20°C; 15°C is optimal, whereas 23°C is nonpermissive. Most infections occur at water temperatures below 12°C. Interestingly, mortality was reduced if fish, infected for less than 24 hr, were maintained in water warmed to 18°C for 4–6 days. Although this finding suggests that temperature manipulation is an effective control measure, it is not the complete answer, since water temperatures in the Snake River Valley of Idaho are nearly constant at 15°C yet IHNV is a persistent problem.

3. Epizootic Cycle

IHNV is maintained in the environment as free virus in water, possibly associated with one or more vectors, and as an occult infection in persistently infected carrier fish. IHNV is acid-labile, ether-sensitive, and rapidly inactivated at temperatures greater than 38°C (Gosting and Gould, 1981). However, it is relatively stable in an aquatic environment. Virus added to fresh water at 15°C survives for 25 days, whereas virus in brackish water or seawater survives half as long (Wolf, 1988). Virus may also be concentrated or replicate within invertebrate vectors. In one study, 60–80% of leeches associated with sockeye salmon contained IHNV. In addition, both leeches and fish lice were reported to act as mechanical vectors for a related aquatic rhabdovirus: spring viremia of carp virus (Yamamato et al., 1989). However, whether vectors play a significant role in the maintenance of IHNV in the environment, or in the transmission of the virus during epizootics, is not known.

Since fry and fingerlings are primary targets of IHNV infection, it must be determined how the virus is maintained from year to year. In the absence of long-term stability in water, or replication within a vector or an alternative fish host, Leong and her colleagues suggested that IHNV is maintained in survivors of earlier infections as a persistent infection mediated by defective interfering particles (Drolet et al., 1995; Kim et al., 1999). Defective interfering particles (DIs) are naturally occurring deletion mutants containing a truncated viral genome. Because DIs encode a reduced number of viral proteins and require wild-type (helper) virus to replicate, they interfere with the replication of wild-type virus and prevent the latter from completely destroying the cell culture. In long-term persistent infections in vitro, levels of infectious wild-type virus and DI particles vary in abundance in a cyclical fashion. Although the role of DIs in modulating virus replication in cell culture is well established, their relevance to the establishment and maintenance of persistent infections in vivo has been questioned. The demonstration of DI-like particles in IHNV carrier fish provides the first evidence that DI particles mediate persistence in vivo. Persistence was broken and virus replication re-initiated by spawning stress or exposure to other pathogens. Subsequently, virions became abundant in all organs and fluids, and were released and transmitted to young fish. Although the precise molecular mechanism by which stress triggers the transition from noninfectious carrier to infectious adult is unknown, the change is dramatic. Among Pacific salmon, virus was unable to be isolated from sea-run fish, appeared in 2% of fish when they entered fresh water a month before spawning, increased to 50–80% during spawning, and peaked at 100% in the post-spawn period (Wolf, 1988).

As indicated above, IHNV is maintained in nature by yearly infection of susceptible young in the period following spawning. Three criteria favor an IHNV outbreak: low water temperature (peak losses occur at temperatures <7°C), susceptible young fish, and a source of virus (Wolf, 1988). All three criteria are met at the time of spawning. High concentrations of virus are present in water during spawning and ensure efficient infection of young. In one study, virus titers as high as 1600 pfu/ml were detected downstream of a dense collection of adult sockeye salmon showing 100% virus prevalence. Since in the laboratory, bath immersion of rainbow trout fry in water containing 3.7×10^3 IHNV pfu/ml results in 20% mortality after 30 days, the level of virus seen downstream from the spawning sockeyes may be sufficient to initiate an IHNV outbreak (Drolet *et al.*, 1995). Aside from direct infection of young fish, virus is tightly associated with both sperm and eggs, and this may be important in transmission. Each year a variable number of survivors of acute infection become long-term 2carriers via persistent infection mediated by DI particles. As carrier adults prepare to spawn, persistence is broken, virus replication begins anew, and, along with the release of eggs and sperm, high levels of virus are shed into the environment.

4. Economic and Environmental Importance

Salmon and trout are major food and game fish and are important to both aquaculture and the recreational fishing industry. IHNV has caused devastating epizootics in trout and salmon hatcheries in the Pacific Northwest, Europe, and Asia, with mortality among juvenile fish approaching 100%. The geographic range of IHNV has been expanded by the export of infected eggs and fry. Thus, man has been an unwitting agent in extending the host range of IHNV. Fish farmers and hatchery personnel must be aware of the dangers of shipping fish of unknown viral status.

5. Control Measures

Morbidity and mortality associated with IHNV can be reduced in several ways (Wolf, 1988). In hatcheries, identification and culling of persistently infected broodstock has markedly reduced losses due to IHNV. Inactivation of virus on egg surfaces has been achieved by treatment with iodine products. Exposure of infected fry to higher water temperatures has been shown to significantly reduce mortality in some studies. Finally, vaccination may prove to be an effective way to limit disease in fry and to develop carrier free broodstock. Both inactivated and live attenuated vaccines have proven successful in laboratory studies, but their widespread use in the field is limited by cost (for the inactivated vaccine) and safety concerns (for the live attenuated virus vaccine) (Wolf, 1988). A DNA vaccine has been developed, which, while showing the efficacy of an attenuated vaccine, overcomes many if not all of the safety concerns (Anderson *et al.*, 1996).

C. Lucké Tumor Herpesvirus (LTV)

LTV is a herpesvirus similar in size and virion structure to other members of the family Herpesviridae (Granoff, 1983). Little is known about LTV replication because the virus cannot be propagated in tissue culture. Despite this considerable limitation, LTV is of research interest because it was one of the first viruses causally associated with carcinoma in vertebrates.

Renal carcinoma in *Rana pipiens* was the first neoplasm to be causally linked with a herpesvirus (Granoff, 1983; McKinnel, 1984). Although tumors were first described in leopard frogs in 1905, it was not until 1934 that Lucké properly identified the malignancy as a renal tumor and suggested that a virus was responsible (Lucké, 1934). This suggestion was bolstered in 1952 when a herpesvirus was detected in renal tumors by electron microscopy, and strengthened when Tweedell (1967) demonstrated that virus-containing tumor homogenates were oncogenic. Tumors were more recently shown to arise after injection of purified LTV or ascitic fluid containing virus particles.

1. Host Range

Leopard frogs from the Midwestern United States are susceptible to renal tumors under natural and experimental conditions. However, tumor incidence in leopard frogs from other parts of the United States varies: tumor-bearing frogs were found in Vermont, Minnesota, and Michigan, but not in northwest Montana, North Dakota, or Louisiana (McKinnel, 1984). Reasons for regional differences in tumor incidence are not understood. While *R. pipiens* is the only naturally susceptible frog, other species (*R. palustris*, *R. clamitans*, and *R. utricularia*) could be infected by injection of LTV-containing tumor homogenates.

2. Pathogenesis

LTV leads to tumor development in the mesonephric kidney. Neoplasms probably develop as a transformation of cells within the proximal tubules. Unlike other tumors in experimental animals, metastasis occurs spontaneously and is enhanced by elevated temperature.

3. Epizootic Cycle

There appear to be both yearly and seasonal variations in LTV-induced renal carcinoma. Most tumors were seen prior to (i.e., in the fall) and just after (i.e., in the spring) emergence from hibernation. The absence of tumors in summer may reflect a survival disadvantage imposed on tumor-bearing animals (i.e., they are less able to escape prey, feed effectively, etc.). Thus, animals with large tumor masses in the spring are at higher risk during the summer, whereas animals with small or newly developing tumors survive the summer and become apparent only later in the year. In addition to seasonal variation, there has been, over the last 30 years, dramatic declines in both amphibian numbers and tumor incidence. Specifically, tumor incidence dropped from 8.5% of collected frogs in 1963 to 0% in 1977–1979. Reasons for both declines are not clear, but it is possible that a decrease in frog density is responsible for the decrease in tumor incidence.

Although LTV is the etiological agent of renal carcinoma, during the warm summer months when the tumor is enlarging, virus particles cannot be detected (Granoff, 1983). However, virions are present during hibernation. The cycle of virus-free tumor development at elevated temperatures, followed by virus replication and dissemination at lower temperatures, can be replicated in the laboratory. Frogs maintained at 20–25°C show no nuclear inclusions (i.e., virus). However, after 2 months at 5°C (a condition that mimics hibernation temperature), nuclear inclusions develop, virus replication occurs, and virions can be found in urine, ascitic fluid, and possibly sex products (eggs and sperm). It is unclear why virus replication is temperature dependent. Does the virus simply not replicate at temperatures above 5–10°C, or do lower temperatures suppress a key aspect of immunity that modulates virus replication?

It is likely that LTV is transmitted from tumor-bearing adults to gametes, or immature offspring, while adults are present in breeding ponds. There is no evidence supporting the existence of alternative vertebrate hosts or invertebrate vectors. Prefeeding frog embryos and early larval stages (prior to hind limb development) are susceptible to LTV infection, but older animals are resistant. In the wild, tumor development progresses during the summer months. Virions are not seen in developing tumors, but viral RNA can be detected. By fall (i.e., prior to hibernation), tumors become palpable, but virus is still absent. During winter, virus replication (or virion assembly) begins and proceeds throughout the hibernation period. Early in the spring, tumor-bearing virus-positive frogs emerge from hibernation and excrete infectious virus in addition to, or along with, gametes. The physical state of the viral genome during summer is not known. It is not known whether the LTV genome is present in a true latent state (as with other herpesviruses) or as a persistent infection modulated by DIs or some other mechanism. The decline in frog populations and tumor abundance, and the inability to propagate LTV in tissue culture, have made it extremely difficult to experimentally approach this very interesting system.

4. Economic and Environmental Importance

Frogs are of little direct economic value in the United States. However, they are useful indicators of changes in aquatic environments, and their widespread disappearance from many locations over the past 20–30 years has been the subject of intense speculation and concern (Carey, 1993).

5. Control Measures

No studies designed to limit LTV-induced renal carcinoma of *Rana pipiens* have been attempted or contemplated.

D. Retroviruses Associated with Neoplasia in Fish

Retroviruses are enveloped RNA-containing viruses that replicate their genome through an obligatory DNA intermediate (Coffin, 1996). They are agents of neoplastic disease in vertebrates, with most tumors occurring in rodents and birds (Nevins and Vogt, 1996). In addition, retroviruses have been linked to immunodeficiency, neurological illness, and subclinical infections. Retroviruses are natural carcinogens whose infection leads to tumor development in several ways (Coffin, 1996; Nevins and Vogt, 1996). *Cis*-acting viruses such as avian leukosis virus precipitate tumor development by integrating within or near a protooncogene (i.e., a normal gene controlling cell growth and development) and disrupting the latter's regulatory properties. In contrast, transducing viruses such as Rous sarcoma virus or avian myelocytomatosis virus encode viral oncogenes (i.e., altered versions of cellular protooncogenes that have been introduced into the viral genome by transduction) that when expressed in the appropriate cell type lead to transformation.

In lower vertebrates, retroviruses have been identified in fish, reptiles, and amphibians, but only in fish have they been linked to neoplastic disease (Tristam *et al.*, 1996; Poulet *et al.*, 1994; Bowser and Casey, 1993). The precise role of piscine retroviruses in disease progression is not clear, because in no system has virus been passaged successfully in tissue culture. Therefore, because plaque-purified or cloned virus is not available for transmission studies, it is not possible to ascertain whether the virus detected in tumors is the etiological agent or merely an endogenous virus whose replication is adventitious. More than a dozen retroviruses have been detected in fish. Several viruses (e.g., walleye dermal sarcoma virus, damselfish neurofibromatosis virus, and salmon leukemia virus) have been identified and linked to neoplastic disease by demonstrating that cell-free tumor homogenates transmit the disease. Malignancies induced by piscine retroviruses are seen at frequencies that are among the highest recorded for naturally occurring tumors. Perhaps high-density fish farming has disturbed the host–virus balance and is responsible for these apparently new epizootic diseases. A brief discussion of three representative viruses follows.

1. Walleye Dermal Sarcoma Virus (WDSV)

Infection with WDSV results in skin lesions that appear as nodular tumors (mean diameter = 5 mm) on walleye (*Stizostedion vitreum*) (Poulet *et al.*, 1994; Bowser and Casey, 1993). Tumor incidence is high, with up to 27% of walleyes collected from Lake Oneida (New York) showing evidence of dermal sarcoma. Bowser and Casey (1993) speculate that the virus is transmitted by close fish-to-fish contact during spring spawning. Despite its high prevalence in the walleye population, transmission of WDSV to other fish species has not been detected. Moreover, infection is likely more common than tumor development, since PCR analysis demonstrated that both tumor-bearing and tumor-free fish were infected by WDSV, but that only the former harbor transcriptionally active virus (Poulet *et al.*, 1996).

There is marked seasonal variation in tumor incidence, with prevalence high in the spring and fall, but low in summer. An inflammatory response is associated with lesions suggesting that host immunity may play a role in the development and regression of tumors. As with other piscine retroviruses, passage of WDSV in cell culture has not been achieved. Tumor tissue from infected fish, and gradient fractions enriched in retrovirus-like particles, show reverse transcriptase activity, strengthening the suggestion that a retrovirus is the etiological agent. Infection has been transmitted from infected tumor-bearing fish to normal animals by injection of walleye fingerlings with a cell-free sarcoma homogenate. While this observation strengthens the putative viral etiology of walleye dermal sarcoma, it does not prove that the virus detected within tumors is the causative agent.

Viral RNA was isolated from tumor tissue and subsequently cloned and sequenced (Holzschu *et al.*, 1995). The viral genome is 13.2 kb, making it one of the largest known retroviruses. The genome contains three open reading frames in addition to the standard *gag–pol–env* genes. Although sequence similarity within the reverse transcriptase gene suggests that WDSV is closely related to viruses within the murine leukemia virus group, other structural features place it closer to the spumaviruses or lentiviruses. Thus, WDSV may represent the prototype of a new group of vertebrate retroviruses.

2. Damselfish Neurofibromatosis Virus (DNFV)

Multiple neurofibromas, malignant schwannomas, and hyperpigmented epidermal lesions have been observed in the bicolor damselfish *Pomacentrus partitus* (Poulet *et al.*, 1994; Bowser and Casey, 1993). Tumor incidence varies from reef to reef, but can be as high as 24%. Damselfish are territorial, and it is thought that aggressive behavior facilitates contact transmission among fish (Schmale and Henley, 1988). In the laboratory, DNFV can be transmitted by injection of cell-free tumor homogenates (Schmale and Henley, 1988). It has been shown that tumor cell lines derived from fish with spontaneous or experimentally induced tumors produce retrovirus-like particles (Schmale *et al.*, 1996). Consistent with this observation, particles purified by rate zonal sedimentation possess reverse transcriptase activity with a temperature optimum of 20°C. As with other fish retroviruses, it has not been possible to passage DNFV in tissue culture. Thus, it is unclear whether the retrovirus observed in tumor cell lines is the etiological agent of DNF or merely an endogenous virus of no pathogenic significance.

The role of the immune system in controlling DNF is unresolved. McKinney and Schmale (1994) demonstrated that lymphocytes from tumor-bearing fish lysed DNF tumor cell lines *in vitro*. Surprisingly, lymphocytes from healthy animals lack cytotoxic activity directed against DNF cell lines. Cytotoxic activity was enhanced following injection of animals with purified virus, suggesting that cytotoxic cells recognized "viral" antigens on target cells rather than "tumor" antigens (McKinney *et al.*, 1997). Although these results suggest that immune responses may play a role in the control of DNF, the inability of lymphocytes from tumor-bearing animals to resolve an infection indicates that cell-mediated immunity alone may not be sufficient to block malignant cell growth.

3. Salmon Leukemia Virus (SLV)

In 1987, an apparently new disease appeared in chinook salmon and resulted in up to 50% mortality in market-sized fish in seawater facilities in British Columbia and in 4-month-old freshwater fingerlings in Washington and California (Poulet *et al.*, 1994). Infected adult fish were anemic, lethargic, and darker than normal, with distended abdomens, exophthalmia, severe reno- and splenomegaly, and petechial hemorrhages. Microscopically, the disease was characterized by proliferation of plasmacytoid leukemic cells, which infiltrated visceral organs and the retrobulbar tissue of the eyes (Poulet *et al.*, 1994). A retrovirus, termed salmon leukemia virus (SLV), was isolated from tumor tissues of chinook salmon with plasmacytoid leukemia and shown to be transmitted along with the disease to recipient fish by injection of cell-free homogenates of tumor tissue from diseased fish. Although suggestive, the role of SLV in plasmacytoid leukemia is not clear.

As with other fish retroviruses associated with neoplasia, attempts at propagating SLV in cell lines failed. Eaton *et al.* (1993) developed two cell lines from plasmacytoid-bearing chinook salmon kidney and eye tumor tissue. Eight weeks after plating, primary cell lines underwent syncytium formation and vacuolation. Electron microscopy detected a retrovirus-like particle that was confirmed by demonstration of reverse transcriptase activity in gradient fractions containing virions.

Taken together, study of WDSV, DNFV, and SLV indicates that a retrovirus is most likely responsible for neoplastic disease. However, until virus can be cloned or plaque purified, it will not be possible to demonstrate formally that a retrovirus is the etiological agent. Clearly, further advances in fish tissue culture, that is, development of cell lines that permit propagation of and transformation by these viruses, will be necessary to convincingly link piscine retroviruses and neoplastic disease.

E. Vertebrate Iridoviruses

Iridoviruses are large enveloped DNA viruses. Viral DNA is double-stranded, encodes ~100 polypeptides, and, characteristic of vertebrate iridoviruses, is highly methylated, terminally redundant, and circularly permutated (Willis *et al.*, 1985; Williams, 1996; Tidona and Darai, 1997). Pathognomonic for iridovirus infection is the presence of paracrystalline arrays of icosahedral nucleocapsids (200 ± 50 nm in diameter) within the cytoplasm of infected cells.

Iridoviruses are classified into four genera. Two genera contain viruses found only in invertebrates, whereas the remaining two (*lymphocystivirus* and *ranavirus*) contain viruses that infect only fish, reptiles, and amphibians (Goorha, 1995). The genus *lymphocystivirus* contains two viruses — lymphocystis disease virus type 1 (LCDV-1) and LCDV-2 — that infect marine and freshwater fish (Anders, 1989). Frog virus 3 (FV3) is the prototype of genus *ranavirus*. Although it was originally thought (and reflected in the genus name) that ranaviruses infected only amphibians, recent work indicates that the genus contains viruses infecting fish and reptiles, as well as amphibians (Mao *et al.*, 1997).

1. Lymphocystis Disease Virus (LCDV)

LCDV is the etiological agent of a chronic benign infection affecting nearly 100 different species of fresh- and saltwater fish representing a wide variety of advanced orders (Anders, 1989; Weissenberg, 1965). Despite the large number of diverse fish species susceptible to infection, it appears as if only two distinct viral species (LCDV-1 and LCDV-2) are involved. Molecular studies indicate that

LCDV-1 is usually associated with infection of European flounder and plaice, whereas LCDV-2 is found in the dab (Wolf, 1988).

Lymphocystis disease results from terminal infection of individual connective tissue cells. The effect on the host is usually minor, and although large numbers of fish may be infected, mortality is extremely low (Wolf, 1988). Externally, the disease is characterized by the appearance on the skin of irregularly shaped elevated masses with a pebbled texture. Internally similar masses sometimes occur. The wart-like lesions characteristic of lymphocystis disease represent hypertrophied individual cells that have undergone a 50,000- to 100,000-fold increase in cell volume. Virus is abundant within lesions and is readily released by mechanical homogenization. In nature, contact transmission appears to be the means by which infection spreads from fish to fish. Experimental transmission has been achieved by contact, implantation, or application of infected material to gills or damaged skin. Transmission does not appear to involve the oral route. As expected, factors such as high population density and external trauma, enhance transmission. Infection rates as high as 100% have been reported in an intensively cultivated freshwater pond. The duration of a typical infection cycle varies with fish species and temperature. With cold-water fish, infections are lengthy and may last for a year, whereas in warm-water fish the cycle is usually complete within several weeks. Although many different fish species are susceptible to infection, in mixed natural populations there is a tendency for infection to be confined to only one or two species. However, under cohabitation conditions multiple species can be infected. Thus, it is unclear whether apparent host selectivity reflects properties inherent within the virus and/or poorly understood environmental factors.

As with other viruses, primary infection usually occurs among young fish, and resistance, characterized by humoral (antibody) and cell-mediated responses, is thought to protect against reinfection and clinical disease in adults. Because lesions are generally located on the surface of fish, at the periphery of the vascular system, and protected by a hyaline capsule, antigen recognition is inhibited. Thus, immune responses appear to be delayed until late in the course of the disease, when infected cells rupture and antigen is released.

The mechanism(s) maintaining LCDV within a population are not known. Virions of the related iridovirus FV3 are remarkably stable (Granoff et al., 1966), suggesting that free virus may remain viable in an aquatic environment for extended periods. There is no convincing evidence for biological vectors, or for persistent or latent infection. In a setting with a large number of susceptible fish, virus may simply be maintained by continuous fish-to-fish passage. Furthermore, the marked temperature sensitivity of virus replication and disease progression suggests that infections initiated late in the year may overwinter and provide a source of virus for a new round of infections in the spring. Although this scenario is plausible, isolation of FV3 and related iridoviruses from ostensibly healthy adult frogs (Granoff *et al.*, 1966) suggests that some form of persistent, chronic, or subclinical infection may occur. However, this possibility has not been systematically investigated.

Lymphocystis disease is not life threatening to fish and has only marginal impact on human economic interests. However, lesions present on infected fish are unsightly and depress the marketability of diseased animals. Specific control measures are not available, although culling of infected fish may be useful. The absence of significant adverse economic impact, coupled with the difficulty of growing the virus in tissue culture, has limited study of this widespread pathogen.

2. *Genus* Ranavirus

Frog virus 3, the prototype of the genus *ranavirus*, was identified by Granoff *et al.* (1965) following cultivation of kidney tissue from a leopard frog with renal carcinoma. At first, it was thought that there was an association between tumor development and FV3 infection, but later it became apparent — after several serologically related viruses were isolated from the kidneys of healthy frogs — that FV3 was not etiologically associated with renal carcinoma. Because FV3 can be isolated from apparently healthy frogs (Granoff *et al.*, 1966), the role of FV3 in amphibian disease is not clear. FV3 replicates in various tissues of adult Fowler's toads and produces glomerular nephritis leading to death (Goorha and Granoff, 1979). In contrast, while leopard frog tadpoles could be lethally infected with as little as 900 pfu of FV3, adults were resistant to infection with as many as 10^8 pfu (Tweedell and Granoff, 1968). However, it is not clear whether tadpoles succumbed to a progressive systemic infection, or whether death was the result of a generalized toxic phenomenon as seen following injection of inactivated FV3 into mice (Kirn *et al.*, 1972). A related virus, tadpole edema virus (TEV), was isolated from both moribund and healthy bullfrogs (*R. catesbeiana*) and was shown to cause illness and death after experimental inoculation into three species of adult toads and 8/20 metamorphosing bullfrogs (Wolf *et al.*, 1968). Virus was also transmitted by mouth, and two of four toads died following oral exposure. However, as with some insect iridoviruses (Williams, 1996), oral infection was not efficient as direct injection of the virus. Although infection occurred in several anuran species, newly hatched salamanders, bluegill fry, and juvenile brook trout were refractory to infection with TEV.

As a virus with seemingly little economic and environmental impact, FV3 might have been easily bypassed had it not possessed a number of interesting molecular features: a highly methylated, circularly permutated, and terminal redundant genome; a virus-encoded restriction-modification system; the ability of heat- or UV-inactivated virus to rapidly block host cell macromolecular synthesis; its use of transcriptional and translational mechanisms to modulate viral gene expression; a viral transcriptase that is able, in contrast to other eukaryotic systems, to transcribe methylated DNA; and a two-stage system of DNA replication in which early rounds of viral DNA synthesis take place in the nucleus, whereas later rounds of viral DNA synthesis, packaging, and virion assembly occur in the cytoplasm within morphologically distinct viral assembly sites (Willis *et al.*, 1985; Williams, 1996). Moreover, FV3 (unlike LCDV) is readily grown in tissue culture, making study of these features possible.

3. Host Range

Until the mid-1980s, ranaviruses were viewed as interesting models, but not as important pathogens. However, this benign situation changed when Langdon and Humphrey (1987) reported the isolation of an iridovirus-like agent from redfin perch (*Perca fluviatilis*) in Australia (Langdon *et al.*, 1988). Infection with this agent led to an acute systemic illness characterized by necrosis of renal hematopoietic tissue, liver, and spleen, and variable involvement of the pancreas. Reflecting the clinical presentation, the virus was designated "epizootic hematopoietic necrosis virus" (EHNV) and was shown to be associated with infections of rainbow trout and several other fish species (Hetrick and Hedrick, 1993). Iridoviruses were subsequently detected in catfish (*I. melas*) in France (Pozet *et al.*, 1992), sheatfish (*Siluris glanis*) in an aquaculture facility in Germany (Ahne *et al.*, 1989), white sturgeon (*Acipenser transmontanus*) in the Western United States (Hedrick *et al.*, 1992a), guppies (*Poecilia reticulata*) and doctor fish (*Labroides dimidatus*) imported into the United States from Southeast Asia (Hedrick and McDowell, 1995), largemouth bass (*Micropterus salmoides*) from the Southeast United States (Plumb *et al.*, 1996; Mao *et al.*, 1999b), Red Sea bream in South Asia (Hetrick and Hedrick, 1993), and tiger salamanders in North America (Jancovich *et al.*, 1997; Bollinger *et al.*, 1999). Bohle iridovirus was detected in the ornate burrowing frog *Limnodynastes ornatus* and the green tree frog *Litoria cauarulea* in Australia and was experimentally transmitted to an amphibian (*Bufo marinus*) and barramundi fish (*Lates calcarifer*) (Moody and Owens, 1994; Cullen *et al.*, 1995). Recently, iridoviruses linked to systemic disease have been isolated from turtles (*Terrapene c. carolina* and *Testudo horsfieldi*) in the Western United States (Mao *et al.*, 1997) and from frogs (*R. aurora*) and stickleback fish (*Gasterostelus aculeatus*) isolated from the same location in Northern California (Mao *et al.*, 1999a). Finally, Cunningham *et al.* (1996) have suggested that primary iridovirus infection, with or without secondary bacterial infection, may be responsible for natural outbreaks of "red-leg" in frogs.

As indicated above, the geographic range of iridoviruses associated with systemic clinical disease is worldwide. However, it is not clear whether a single isolate is responsible for global disease, or whether distinct viral "species" are responsible for regional outbreaks. Hedrick *et al.* (1992b) showed that viruses isolated from redfin perch, sheatfish, and catfish were similar by electron microscopy of virus particles, SDS-polyacrylamide gel analysis of virion polypeptides, and indirect immunofluorescence assays. More recently, Mao *et al.* (1997) analyzed nine iridovirus isolates from fish, turtles, and frogs and found that viruses isolated from the same geographic region were similar by SDS-polyacrylamide gel electrophoresis of infected cell proteins, by restriction endonuclease (REN) digestion profiles of viral DNA, and by partial sequence analysis of the gene encoding the viral major capsid protein (MCP). For sequence analysis, Mao *et al.* (1997) designed a pair of oligonucleotide primers specific for two conserved regions within the 5' end of the gene encoding the MCP (Mao *et al.*, 1996). Using these primers, they successfully amplified (and subsequently cloned and sequenced) the 5' terminal portion of the MCP gene from these nine isolates and found that all were more similar to the ranavirus FV3 than they were to the fish virus LCDV-1. In addition, sequence analysis of two other viral genes, one encoding the 18-kDa early protein (Mao *et al.*, 1997) and another encoding the viral methyltransferase (Mao *et al.*, 1999b), confirmed that newly isolated piscine iridoviruses were more closely related to the ranavirus FV3 than to LCDV-1, and demonstrated that the similarity detected by analysis of the MCP gene was not an artifact of the gene chosen.

The host range of ranaviruses infecting fish, frogs, and reptiles is not known. Whereas some studies support a relatively broad host range, other work suggests that infection is confined to one or a few species. For example, in an outbreak among sheatfish, eel and carp in the same aquaculture facility were not affected (Ahne *et al.*, 1989; Hetrick and Hedrick, 1993), and in a pond outbreak in which catfish were lethally infected other fish appeared to be spared (Pozet *et al.*, 1992). Whether this represents an absolute restriction in host range or whether the apparent selectivity under natural conditions indicates that some species are more susceptible than others is not known. However, with other ranaviruses the host range appears to be quite broad. EHNV infects redfin perch, rainbow trout, and several other fish species (Hetrick and Hedrick, 1993; Langdon, 1989), and Bohle iridovirus infects three species of amphibians as well as one species of fish after experimental inoculation (Moody and Owens, 1994; Cullen *et al.*, 1995). We have obtained evidence that the same virus (as judged by SDS-polyacrylamide gel analysis of protein synthesis in infected cells, REN profiles, and partial sequence analysis of the MCP gene) infects and is associated with clinical disease in sympatric *R. aurora* and the three-spine stickleback *Gastrostelus aculeatus* (Mao *et al.*, 1999a). This last observation appears to confirm what other investigators have suggested: that an individual iridovirus species may be capable of infecting both amphibians and fish, and that members of one taxonomic class (e.g., amphibians) can serve as reservoirs for viruses infecting another taxonomic class (e.g., fish). However, it is not clear whether fish harbor iridoviruses that "normally" infect frogs or vice versa, or whether each species, depending on environmental and other factors, is equally susceptible to infection. Given the growing importance of aquaculture and the apparent increased incidence of amphibian "die-offs," it will be necessary to establish the host–reservoir relationship for iridoviruses affecting lower vertebrates.

V. OVERALL SUMMARY

Viruses infecting lower vertebrates were viewed as agents that, for the most part, had a minimal impact on human endeavors. The growing importance of aquaculture, coupled with concerns about environmental degradation (as reflected in amphibian extinctions) have refocused attention on these heretofore little-studied pathogens. Continued study of these viruses may provide insight into human and animal disease and fundamental biochemical processes, as well as clarify the conflicting taxonomic organization of these agents (especially iridoviruses) and their role in disease.

REFERENCES

Ahmed, R., Morrison, L. A., and Knipe, D. M. (1996). Persistence of viruses. In "Fundamental Virology" (B. N. Fields, D. M. Knipe, and P. M. Howley, eds.), pp. 207–238. Lippincott-Raven, Philadelphia.

Ahne, W. (1994). Viral infection of aquatic animals with special reference to Asian aquaculture. *Annu. Rev. Fish Dis.* 4, 375–388.

Ahne, W., Schlotfeldt, H. J., and Thomsen, I. (1989). Fish viruses: Isolation of an icosahedral cytoplasmic deoxyribovirus from sheatfish (*Silurus glanis*). *J. Vet. Med. B* **36**, 333–336.

Anders, K. (1989). Lymphocystis disease of fishes. In "Viruses of Lower Vertebrates" (W. Ahne and E. Krustak, eds.), pp. 141–160. Springer-Verlag, New York.

Anderson, D. P. (1996). Environmental factors in fish health: Immunological aspects. In "The Fish Immune System: Organism, Pathogen, and Environment" (G. Iwama and T. Nakanishi, eds.), pp. 289–310. Academic Press, San Diego.

Anderson, E. D., Mourich, D. V., Fahrenkrug, S. C., LaPatra, S., Sheperd, J., and Leong, J. C. (1996). Genetic immunization of rainbow trout (*Oncorhynchus mykiss*) against infectious hematopoietic necrosis virus. *Molec. Marine Biol. Biotechnol.* **5**, 114–122.

Antao, A., Chinchar, V. G., McConnell, T. J., Miller, N. W., Clem, L. W., and Wilson, M. R. (1999). MHC class I genes of the channel catfish: Sequence analysis and expression. *Immunogenetics* **49**, 303–311.

Bly, J. E., and Clem, L. W. (1991). Temperature-mediated processes in teleost immunity: *In vitro* immunosuppression induced by *in vivo* low temperature in channel catfish. *Vet. Immunol. Immunopathol.* **28**, 365–377.

Bly, J. E., and Clem, L. W. (1992). Temperature and teleost immune functions. *Fish Shellfish Immunol.* **2**, 159–171.

Bly, J. E., Lawson, L. A., Dale, D. J., Szalai, A. J., Durborow, R. M., and Clem, L. W. (1992). Winter saprolegniosis in channel catfish. *Dis. Aquatic Org.* **13**, 155–164.

Bly, J. E., Quiniou, S., and Clem, L. W. (1997). Environmental effects on fish immune mechanisms. *In* "Fish Vaccinology" (R. Gudding, A. Lillehaug, P. Midtlyng, and F. Brown, eds.), Vol. 90, pp. 33–43. Karger, Basel.

Bollinger, T. K., Mao, J., Schock, D., Brigham, R. M., and Chinchar, V. G. (1999). Pathology, isolation, and preliminary molecular characterization of a novel iridovirus from tiger salamanders in Saskatchewan. *J. Wildlife Dis.* **35**, 413–429.

Booy, F. P., Trus, B. L., Davison, A. J., and Steven, A. C. (1996). The capsid architecture of channel catfish virus: An evolutionarily distant herpesvirus is largely conserved in the absence of discernible sequence homology with herpes simplex virus. *Virology* **215**, 134–141.

Bowser, P. R., and Casey, J. W. (1993). Retroviruses of fish. *Ann. Rev. Fish Dis.*, pp. 209–224.

Bowser, P. R., and Plumb, J. A. (1980). Growth rates of a new cell line from channel catfish ovary and channel catfish virus replication at different temperatures. *Can. J. Fish. Aquat. Sci.* **37**, 871–873.

Bowser, P. R., Munson, A. D., Jarboe, H. H., Francis-Floyd, R., and Waterstrat, P. R. (1985). Isolation of channel catfish virus from channel catfish, *Ictalurus punctatus* (Rafinesque), broodstock. *J. Fish Dis.* **8**, 557–561.

Boyle, J., and Blackwell, J. (1991). Use of polymerase chain reaction to detect latent channel catfish virus. *Am. J. Vet. Res.* **52**, 1965–1968.

Campbell, N. A. (1996). "Biology." Benjamin/Cummings, New York.

Carey, C. (1993). Hypothesis concerning the causes of the disappearance of boreal toads from the mountains of Colorado. *Conserv. Biol.* **7**, 355–362.

Carey, C., and Bryant, C. J. (1995). Possible interrelations among environmental toxicants, amphibian development, and decline of amphibian populations. *Environ. Health Persp.* **103** (suppl. 4), 13–17.

Carey, C., Maniero, G. D., and Stinn, J. F. (1996). Effect of cold on immune function and susceptibility to bacterial infection in toads (*Bufo marinus*). *In* "Adaptations to the Cold: Tenth International Hibernation Symposium" (G. F. Hulbert and S. C. Nichol, eds.), pp. 123–129. University of New England Press, Armidale, ME.

Carey, C., Maniero, G., Harper, C. W., and Synder, G. (1996). Measurements of several aspects of immune function in toads (*Bufo marinus*) after exposure to low pH. *In* "Modulators of Immune Responses: The Evolutionary Trail" (J. S. Stolen, T. C. Fletcher, C. J. Bayne, C. J. Secombes, and J. T. Zelikoff, L. E. Twerdok, and D. P. Anderson, eds.), pp. 565–577, SOS Publications, Fair Haven, NJ.

Chinchar, V. G., Rycyzyn, M. R., Clem, L. W., and Miller, N. W. (1993). Productive infection of continuous lines of channel catfish leukocytes by channel catfish virus. *Virology* **193**, 989–992.

Clem, L. W., Miller, N. W., and Bly, J. E. (1991). Evolution of lymphocyte subpopulations, their interactions, and temperature sensitivities. *In* "Phylogenies of Immune Functions" (G. Warr and N. Cohen, eds.), pp. 191–213. CRC Press, Boca Raton, FL.

Coffin, J. M. (1996). Retroviridae: The viruses and their replication. *In* "Fundamental Virology" (B. N. Fields, D. M. Knipe, and P. M. Howley, eds.), pp. 763–844. Lippincott-Raven, Philadelphia.

Cullen, B. R., Owens, L., and Whittington, R. J. (1995). Experimental infection of Australian anurans (*Limnodynastes terraereginae* and *Litoria latopalmata*) with Bohle iridovirus. *Dis. Aquat. Org.* **23**, 83–92.

Cunningham, A. A., Langton, T. E. S., Bennett, P. M., Lewin, J. F., Drury, S. E. N., Gough, R. E., and MacGregor, S. K. (1996). Pathological and microbiological findings from incidents of unusual mortality of the common frog (*Rana temporaria*). *Phil. Trans. R. Soc. Lond. B* **351**, 1539–1557.

Davison, A. J. (1992). Channel catfish virus: A new type of herpesvirus. *Virology* **186**, 9–14.

Davison, A. J. (1994). Fish herpesviruses. *In* "Encyclopedia of Virology" (R. G. Webster and A. Granoff, eds.), pp. 470–480. Academic Press, San Diego.

de Kinkelin, P., and Dorson, M. (1973). Interferon production in rainbow trout (*Salmo gairdneri*) experimentally infected with Egtved virus. *J. Gen. Virol.* **19**, 125–127.

de Sena, J., and Rio, G. J. (1975). Partial purification and characterization of RTG-2 fish cell interferon. *Inf. Immun.* **11**, 815–822.

Dixon, B., van Erp, S. H. M., Rodrigues, P. N. S., Egberts, E., and Stet, R. J. M. (1995). Fish major histocompatibility complex genes: An expansion. *Develop. Compar. Immunol.* **19**, 109–133.

Drolet, B. S., Chiou, P., Heidel, J., and Leong, J. C. (1995). Detection of truncated virus particles in a persistent RNA virus infection *in vivo. J. Virol.* **69**, 2140–2147.

Du Pasquier, L. (1993). Evolution of the immune system. *In* "Fundamental Immunology" (W. E. Paul, ed.), pp. 199–234. Raven, New York.

Du Pasquier, L., Schwager, J., and Flajnik, M. F. (1989). The immune system of *Xenopus laevis. Annu. Rev. Immunol.* **7**, 251–275.

Eaton, W. D. (1990). Anti-viral activity in four species of salmonids following exposure to polyinosinic:cytidylic acid. *Dis. Aquat. Org.* **9**, 193–198.

Eaton, W. D., Folkins, B., Bagshaw, J., Traxler, G., and Kent, M. L. (1993). Isolation of a retrovirus from two fish cell lines developed from chinook salmon (*Oncorhynchus tshawytscha*) with plasmacytoid leukaemia. *J. Gen. Virol.* **74**, 2299–2302.

Ellis, A. E. (1989). The immunology of teleosts. *In* "Fish Pathology," 2nd ed. (R. J. Roberts, ed.), pp. 135–152. Bailliere-Tindall, London.

Ellis, A. E. (1999). The fish immune system. *In* "Encyclopedia of Immunology" (I. M. Roitt, ed.), pp. 920–926. Academic Press, San Diego.

Ellsaesser, C. F., and Clem, L. W. (1986). Hematological and immunological changes in channel catfish stressed by handling and transport. *J. Fish Biol.* **28**, 511–521.

Evans, D. L., and Jaso-Friedmann, L. (1992). Nonspecific cytotoxic cells as effectors of immunity in fish. *Annu. Rev. Fish Dis.* **2**, 109–122.

Fields, B. N., Knipe, D. M., and Howley, P. M., eds. (1996). "Fundamental Virology," 3rd ed. Lippincott-Raven, Philadelphia.

Flajnik, M. F., Kaufman, J. F., Hsu, E., Manes, M., Parisot, R., and Du Pasquier, L. (1986). Major histocompatibility complex-encoded class I molecules are absent in immunologically competent *Xenopus* before metamorphosis. *J. Immunol.* **137**, 3891–3899.

Gooding, L. R. (1992). Virus proteins that counteract host immune defenses. *Cell* **71**, 5–7.

Goodwin, U. B., Antao, A., Wilson, M. W., Chinchar, V. G., Miller, N. W., Clem, L. W., and McConnell, T. J. (1997). MHC class II *B* genes in the channel catfish (*Ictalurus punctatus*). *Dev. Compar. Immunol.* **21**, 13–23.

Gosting, L. H., and Gould, R. W. (1981). Thermal inactivation of infectious hematopoietic necrosis and infectious pancreatic necrosis viruses. *Appl. Environ. Microbiol.* **41**, 1081–1082.

Goorha, R. (1995). Iridoviruses. *Arch. Virol. Suppl.* **10**, 95–99.

Goorha, R., and Granoff, A. (1979). Icosahedral cytoplasmic deoxyriboviruses. *In* "Comprehensive Virology," Vol. 14 (H. Fraenkel-Conrat and R. R. Wagner, eds.), pp. 347–399. Plenum, New York.

Granoff, A. (1983). Amphibian herpesviruses. *In* "The Herpesviruses" (B. Roizman, ed.), Vol. 2, pp. 367–384. Plenum, New York.

Granoff, A., Came, P. E., and Rafferty, K. A. (1965). The isolation and properties of viruses from *Rana pipiens*: The possible relationship to renal carcinoma. *Annu. N. Y. Acad. Sci.* **126**, 237–255.

Granoff, A., Came, P. E., and Breeze, D. C. (1966). Virus and renal carcinoma of *Rana pipiens*, I: The isolation and properties of virus from normal and tumor tissue. *Virology* **29**, 133–148.

Gray, W. L., Williams, R. J., Jordan, R. L., and Griffin, B. R. (1999). Detection of channel catfish virus DNA in latently infected catfish. *J. Gen. Virol.* **80**, 1817–1822.

Hansen, T. H., Carreno, B. M., and Sachs, D. H. (1993). The major histocompatibility complex. *In* "Fundamental Immunology" (W. E. Paul, ed.), pp. 577–628. Raven, New York.

Heartwell, C. M. (1975). "Immune Response and Antibody Characterization of the Channel Catfish to a Naturally Pathogenic Bacterium and Virus." U.S. Fish and Wildlife Service Technical Paper No. 85.

Hedrick, R. P., and McDowell, T. S. (1995). Properties of iridoviruses from ornamental fish. *Vet. Res.* **26**, 423–427.

Hedrick, R. P., Groff, J. M., and McDowell, T. S. (1987). Response of adult channel catfish to waterborne exposures of channel catfish virus. *Prog. Fish-Culturist* **49**, 181–187.

Hedrick, R. P., McDowell, T. S., Groff, J. M., Yun, S., and Wingfield, W. H. (1992a). Isolation and some properties of an iridovirus-like agent from white sturgeon *Acipenser transmontanus*. *Dis. Aquatic Org.* **12**, 75–81.

Hedrick, R. P., McDowell, T. S., Ahne, W., Torhy, C., and de Kinkelin, P. (1992b). Properties of three iridovirus-like agents associated with systemic infections of fish. *Dis. Aquat. Org.* **13**, 203–209.

Herbst, L. H. (1994). Fibropapillomatosis of marine turtles. *Annu. Rev. Fish Dis.* **4**, 389–425.

Hetrick, F. M., and Hedrick, R. P. (1993). New viruses described in finfish from 1988 to 1992. *Annu. Rev. Fish Dis.* **3**, 187–207.

Hoff, G. L., and Hoff, D. M. (1984). Herpesviruses of reptiles. *In* "Diseases of Amphibians and Reptiles" (G. L. Hoff, F. L. Frye, and E. R. Jacobson, eds.), pp. 159–167. Plenum, New York.

Hogan, R. J., Stuge, T. B., Clem, L. W., Miller, N. W., and Chinchar, V. G. (1996). Anti-viral cytotoxic cells in the channel catfish (*Ictalurus punctatus*). *Dev. Compar. Immunol.* **20**, 115–127.

Holzschu, D. L., Martineau, D., Fodor, S. K., Vogt, V. M., Bowser, P. R., and Casey, J. W. (1995). Nucleotide sequence and protein analysis of a complex piscine retrovirus, Walleye Dermal Sarcoma Virus. *J. Virol.* **69**, 5320–5331.

Iwama, G., and Nakanishi, T., eds. (1996). "The Fish Immune System: Organism, Pathogen, and Environment." Academic Press, San Diego.

Jancovich, J. K., Davidson, E. W., Morado, J. F., Jacobs, B. L., and Collins, J. P. (1997). Isolation of a lethal virus from the endangered tiger salamander *Ambystoma tigrinum* stebbinsi. *Dis. Aquat. Org.* **31**, 161–167.

Kattari, S. L., and Piganelli, J. D. (1996). The specific immune system: Humoral defenses. *In* "The Fish Immune System" (G. Iwama and T. Nakanishi, eds.), pp. 207–254. Academic Press, San Diego.

Kaufmann, J., Flajnik, M., and DuPasquier, L. (1991). The MHC molecules of ectothermic vertebrates. *In* "Phylogenies of Immune Function" (G. Warr and N. Cohen, eds.), pp. 125–150. CRC Press, Boca Raton, FL.

Kieff, E. (1996). Epstein–Barr virus and its replication. *In* "Fundamental Virology" (B. N. Fields, D. M. Knipe, and P. M. Howley, eds.), pp. 1109–1162. Lippincott-Raven, Philadelphia.

Kim, C. H., Dummer, D. M., Chiou, P. P., and Leong, J.-A. (1999). Truncated particles produced in fish surviving infections of hematopoietic necrosis virus infection: Mediators of persistence. *J. Virol.* **73**, 843–849.

Kirn, A., Gut, J. P., Bingen, A., and Hirth, C. (1972). Acute hepatitis produced by FV3 in mice. *Arch. Gesamte Virusforsch.* **36**, 394.

Kotwal, G. J. (1996). The great escape: Immune evasion by pathogens. *The Immunologist* **4/5**, 157–164.

Kuruth, G., and Leong, J. C. (1985). Characterization of IHNV mRNA species reveals a non-virion rhabdovirus protein. *J. Virol.* **53**, 462–468.

Langdon, J. S. (1989). Experimental transmission and pathogenicity of epizootic haematopoietic necrosis virus (EHNV) in redfin perch, *Perca fluviatius* L., and eleven other teleosts. *J. Fish Dis.* **12**, 295–310.

Langdon, J. S., and Humphrey, J. D. (1987). Epizootic haematopoietic necrosis, a new viral disease in redfin perch, *Perca fluviatilis* L., in Australia. *J. Fish Dis.* **10**, 289–297.

Langdon, J. S., Humphrey, J. D., and Williams, L. M. (1988). Outbreaks of an EHNV-like iridovirus in cultured rainbow trout in Australia. *J. Fish Dis.* **11**, 93–96.

Le Morvan-Rocher, C., Troutaud, D., and Deschaux, P. (1995). Effects of temperature on carp leukocyte mitogen-induced proliferation and nonspecific cytotoxic activity. *Develop. Compar. Immunol.* **19**, 87–95.

Leong, J. C. (1993). Molecular and biotechnological approaches to fish vaccines. *Curr. Opin. Biotechnol.* **4**, 286–293.

Leong, J. C., and Fryer, J. L. (1993). Viral vaccines for aquaculture. *Annu. Rev. Fish Dis.* **3**, 225–240.

Lucké, B. (1934). A neoplastic disease of the kidney of the frog, *Rana pipiens. Am. J. Cancer* **20**, 352–379.

Maniero, G. D., and Carey, C. (1997). Changes in selected aspects of immune function in the leopard frog, *Rana pipiens*, associated with exposure to cold. *J. Comp. Physiol. B* **167**, 256–263.

Manning, M. J., and Nakanishi, T. (1996). The specific immune system: Cellular defenses. *In* "The Fish Immune System" (G. Iwama and T. Nakanishi, eds.), pp. 160–206. Academic Press, San Diego.

Mao, J., Tham, T. N., Gentry, G. A., Aubertin, A.-M., and Chinchar, V. G. (1996). Cloning, sequence analysis, and expression of the major capsid protein of the iridovirus frog virus 3. *Virology* **216**, 431–436.

Mao, J., Hedrick, R. P., and Chinchar, V. G. (1997). Molecular characterization, sequence analysis, and taxonomic position of newly isolated fish iridoviruses. *Virology* **229**, 212–220.

Mao, J., Green, D. E., Fellers, G., and Chinchar, V. G. (1999a). Molecular characterization of iridoviruses isolated from sympatric amphibians and fish. *Virus Res.* **63**, 45–52.

Mao, J., Wang, J., Chinchar, G. D., and Chinchar, V. G. (1999b). Molecular characterization of a ranavirus isolated from largemouth bass *Micropterus salmoides. Dis. Aquat. Org.* **37**, 107–114.

McKinnel, R. G. (1984). Lucké tumor of frogs. *In* "Diseases of Amphibians and Reptiles" (G. L. Hoff, F. L. Frye, and E. R. Jacobson, eds.), pp. 581–605. Plenum, New York.

McKinney, E. C., and Schmale, M. C. (1994). Damselfish with neurofibromatosis exhibit cytotoxicity toward tumor targets. *Develop. Compar. Immunol.* **18**, 305–313.

McKinney, E. C., Kamper, S., and Schmale, M. C. (1997). Damselfish with neurofibromatosis exhibit cytotoxicity toward retrovirus-infected cells. *Develop. Compar. Immunol.* **21**, 219.

Miller, N. W., and Clem, L. W. (1984). Temperature-mediated processes in teleost immunity: Differential effects of temperature on catfish *in vitro* primary and secondary responses to thymus-dependent and thymus-independent antigens. *J. Immunol.* **133**, 2356–2359.

Moody, N. J. G., and Owens, L. (1994). Experimental demonstration of the pathogenicity of a frog virus, Bohle iridovirus, for a fish species, barramundi *Lates calcarifer. Dis. Aquat. Org.* **18**, 95–102.

Nevins, J. R., and Vogt, P. K. (1996). Cell transformation by viruses. *In* "Fundamental Virology" (B. N. Fields, D. M. Knipe, and P. M. Howley, eds.), pp. 267–309. Lippincott-Raven, Philadelphia.

Oie, H. K., and Loh, P. C. (1971). Reovirus type 2: Induction of viral resistance and interferon production in fathead minnow cells. *Proc. Soc. Exp. Biol. Med.* **136**, 369–373.

Partula, S., de Guerre, A., Fellah, J. S., and Charlemagne, J. (1995). Structure and diversity of the T cell antigen receptor β chain in a teleost fish. *J. Immunol.* **155**, 699–706.

Plumb, J. A. (1973). Neutralization of channel catfish virus by serum of channel catfish. *J. Wildlife Dis.* **9**, 324–330.

Plumb, J. A. (1989). Channel catfish herpesvirus. *In* "Viruses of Lower Vertebrates" (W. Ahne and E. Krustak, eds.), pp. 198–216. Springer-Verlag, New York.

Plumb, J. A., Thune, R. L., and Klesius, P. H. (1981). Detection of channel catfish virus in adult fish. *Develop. Biol. Stand.* **49**, 29–34.

Plumb, J. A., Grizzle, J. M., Young, H. E., and Noyes, A. D. (1996). An iridovirus isolated from wild largemouth bass. *J. Aquat. Anim. Health* **8**, 265–270.

Poulet, F. M., Bowser, P. R., and Casey, J. W. (1994). Retroviruses of fish, reptiles, and molluscs. *In* "The Retroviridae" (J. A. Levy, ed.), Vol. 3, pp. 1–38. Plenum, New York.

Poulet, F. M., Bowser, P. R., and Casey, J. W. (1996). PCR and RT–PCR analysis of infection and transcriptional activity of walleye dermal sarcoma virus (WDSV) in organs of adult walleyes (*Stizostedion vitreum*). *Vet. Pathol.* **33**, 66–73.

Pozet, F., Morand, M., Moussa, A., Torhy, C., and de Kinkelin, P. (1992). Isolation and preliminary characterization of a pathogenic icosahedral deoxyribovirus from the catfish *Ictalurus melas*. *Dis. Aquat. Org.* **14**, 35–42.

Renault, T., Torchy, C., and de Kinkelin, P. (1991). Spectrophotometric method for titration of trout interferon and its application to rainbow trout fry experimentally infected with viral hemorrhagic septicemia virus. *Dis. Aquat. Org.* **10**, 23–29.

Rogel-Gaillard, C., Chilmonczyk, S., and de Kinkelin, P. (1993). *In vitro* induction of interferon-like activity from rainbow trout leucocytes stimulated by Egtved virus. *Fish Shellfish Immunol.* **3**, 383–394.

Schmale, M. C., and Henley, G. T. (1988). Transmissibility of a neurofibromatosis-like disease in bicolor damselfish. *Cancer Res.* **48**, 3828–3833.

Schmale, M. C., Aman, M. R., and Gill, K. A. (1996). A retrovirus isolated from cell lines derived from neurofibromas in bicolor damselfish (*Pomacentrus partitus*). *J. Gen. Virol.* **77**, 1181–1187.

Schreck, C. B. (1996). Immunomodulation: Endogenous factors. *In* "The Fish Immune System: Organism, Pathogen, and Environment" (G. Iwama and T. Nakanishi, eds.), pp. 311–338. Academic Press, San Diego.

Snegaroff, J. (1993). Induction of interferon synthesis in rainbow trout leucocytes by various homeotherm viruses. *Fish Shellfish Immunol.* **3**, 191–198.

Spitsbergen, J. M., Schat, K. A., Kleeman, J. M., and Peterson, R. E. (1988). Effects of 2,3,7,8-tetrachlorodibenzo-*p*-dioxin (TCDD) or Aroclor 1254 on the resistance of rainbow trout, *Salmo gairdneri* Richardson, to infectious haematopoietic necrosis virus. *J. Fish Dis.* **11**, 73–83.

Stuge, T. B., Yoshida, S. H., Chinchar, V. G., Miller, N. W., and Clem, L. W. (1997). Cytotoxic activity generated from channel catfish peripheral blood leukocytes: Mixed lymphocyte cultures. *Cell. Immunol.* **177**, 154–161.

Szalai, A. J., Bly, J. E., and Clem, L. W. (1994). Changes in serum concentrations of channel catfish phosphorylcholine-reactive protein (PrP) in response to inflammatory agents, low temperature shock, and infection by the fungus *Saprolegnia* spp. *Fish Shellfish Immunol.* **4**, 323–336.

Tatner, M. F. (1996). Natural changes in the immune system of fish. *In* "The Fish Immune System" (G. Iwama and T. Nakanishi, eds.), pp. 255–287. Academic Press, San Diego.

Tidona, C. A., and Darai, G. (1997). The complete DNA sequence of lymphocystis disease virus. *Virology* **230**, 207–216.

Tristem, M., Herniou, E., Summers, K., and Cook, J. (1996). Three retroviral sequences in amphibians are distinct from those in mammals. *J. Virol.* **70**, 4864–4870.

Trobridge, G. D., and Leong, J. C. (1995). Characterization of a rainbow trout Mx gene. *J. Interferon Cytokine Res.* **15**, 691–702.

Trust, T. J. (1986). Pathogenesis of infectious diseases of fish. *Annu. Rev. Microbiol.* **40**, 479–502.

Tweedell, K., and Granoff, A. (1968). Virus and renal carcinoma of *Rana pipiens*, V: Effect of frog virus 3 on developing frog embryos and larvae. *J. Natl. Cancer Inst.* **40**, 407–410.

Tweedell, K. (1967). Induced oncogenesis in developing frog kidney cells. *Cancer Res.* **27**, 2042–2052.

Vallejo, A. N., Miller, N. W., and Clem, L. W. (1992). Antigen processing and presentation in teleost immune responses. *Annu. Rev. Fish Dis.* **2**, 73–89.

Weissenberg, R. (1965). Fifty years of research into the lymphocystis virus disease of fish (1914–1964). *Annu. N. Y. Acad. Sci.* **126**, 362–374.

Whitton, J. L., and Oldstone, M. B. A. (1996). Immune response to viruses. In "Fundamental Virology" (B. N. Fields, D. M. Knipe, and P. M. Howley, eds.), pp. 311–340. Lippincott-Raven, Philadelphia.

Wiegertjes, G. F., Stet, R. J. M., Parmentier, H. K., and van Muiswinkel, W. B. (1996). Immunogenetics of disease resistance in fish: A comparative approach. *Develop. Compar. Immunol.* **20**, 365–381.

Williams, T. (1996). The iridoviruses. *Adv. Virus Res.* **41**, 345–412.

Willis, D. B., Goorha, R., and Chinchar, V. G. (1985). Macromolecular synthesis in cells infected by frog virus 3. *Curr. Topics Microbiol. Immunol.* **116**, 77–106.

Wilson, M. R., and Warr, G. W. (1992). Fish immunoglobulins and the genes that encode them. *Annu. Rev. Fish Dis.* **2**, 201–221.

Wise, J. A., Bowser, P. R., and Boyle, J. A. (1985). Detection of channel catfish virus in asymptomatic adult channel catfish, *Ictalurus punctatus* (Rafinesque). *J. Fish Dis.* **8**, 485–493.

Wise, J. A., Harrell, S. F., Busch, R. L., and Boyle, J. A. (1988). Vertical transmission of channel catfish virus. *Am. J. Vet. Res.* **49**, 1506–1509.

Wolf, K. (1983). Biology and properties of fish and reptilian herpesviruses. In "The Herpesviruses" (B. Roizman, ed.), Vol. 2, pp. 319–366. Plenum, New York.

Wolf, K. (1988). "Fish Viruses and Fish Viral Diseases." Cornell University Press, Ithaca, NY.

Wolf, K., Bullock, G. L., Dunbar, C. E., and Quimby, M. C. (1968). Tadpole edema virus: A viscerotropic pathogen for anuran amphibians. *J. Infect. Dis.* **118**, 253–262.

Yamamato, T., Arakawa, C. K., Batts, W. N., and Winton, J. R. (1989). Comparison of IHNV in natural and experimental infections of spawning salmonids by infectivity and immunochemistry. In "Viruses of Lower Vertebrates" ((W. Ahne and E. Krustak, eds.), pp. 411–429. Springer-Verlag, New York.

Yano, T. (1996). The non-specific immune system: Humoral defenses. In "The Fish Immune System" (G. Iwama and T. Nakanishi, eds.), pp. 106–159. Academic Press, San Diego.

Yoshinaga, K., Okamoto, N., Kurata, L., and Ikeda, Y. (1994). Individual variations of natural killer activity of rainbow trout leucocytes against IPN virus-infected and uninfected RTG-2 cells. *Fish Pathol.* **29**, 1–4.

Zelikoff, J. T. (1993). Metal pollution-induced immunomodulation in fish. *Annu. Rev. Fish Dis.* **3**, 305–315.

Zinkernagel, R. M. (1993). Immunity to viruses. In "Fundamental Immunology" (W. E. Paul, ed.), pp. 1211–1250. Raven, New York.

Zhang, H. G., and Hanson, L. A. (1995). Deletion of thymidine kinase gene attenuates channel catfish herpesvirus while maintaining infectivity. *Virology* **209**, 658–663.

12

Virus Cycles in Aquatic Mammals, Poikilotherms, and Invertebrates

ALVIN W. SMITH

College of Veterinary Medicine
Oregon State University
Corvallis, Oregon 97331

VIRAL ECOLOGY

I. INTRODUCTION

Viruses can be transported across the land and through the air, but it is in a water substrate that they replicate. Their release from host tissues into bodies of water frequently provides them with a highly effective means of transport and survival outside the host cell.

A. Viruses in a Water Substrate

Reports of zoonotic viruses emerging from ocean reservoirs signaled the importance of this ecosystem as a spawning ground for new diseases (Smith *et al.*, 1998a,b). To gain an understanding of virus movements in and out of aquatic and terrestrial animals, similarities and differences between cycles in these differing ecologic niches need to be examined. In both systems, host animals can be viewed as islands that must be sought out and colonized by viruses if viruses are to survive.

Two differing mechanisms meet this need. One requires very high viral replicative numbers to ensure by mathematical odds that contact with susceptible hosts will occur. Hazards for the virus are dilution combined with decay rates over time. The second mechanism functions with relatively fewer virus numbers and is a "smart-bomb" approach. The virus is carried directly from one compliant host to another. The disadvantages that the second holds for the virus include reliance on intermediate delivery systems.

Having established these two extremes, it is understood that most virus cycles depend on chance distribution in concert with some delivery systems, and that both will involve a water substrate in either a micro- or macroenvironment. Thus, viruses of both aquatic and terrestrial animals are water dependent for their replication and movements at the molecular and macroscopic levels. Aquatic substrates are irrevocably linked to viruses and are the lifeblood of every ecologic niche where water is up 70 to 80% of the mass of living matter, both plant and animal. Without water, replicative function, growth, and life for both viruses and their hosts ceases.

Water not only provides a system through which both organic and inorganic matter is transported but also the means to move elements, molecules, compounds, and particulates using the push-and-tug of van der Waals forces, ions, dipoles, solubility, buoyancy, and hydrophobicity/hydrophilicity. Under these influences and bathed in water inside living cells, viruses are spawned, then released into a water medium for export to another susceptible host cell. This general theme holds true for the viruses infecting aquatic mammals, poikilotherms, and invertebrates, even when moving between these animal groups or when they emerge from their aquatic host to parasitize terrestrial species including humans.

For terrestrial species, viral dispersion can be and is often airborne. Successful host targeting will have much to do with infective particle buildup in the vicinity of a susceptible host, and then portal-of-entry contact. For example, arthropods as true vectors or mechanical carriers can facilitate these movements, including oral–fecal spread among terrestrial hosts. By contrast, airborne viral transmission and cycles involving flying and biting insets or ticks would be minimal for aquatic species. Instead, aquatic animals live and eat bathed in their own excreta or that of other animals, and viral packaging and delivery would occur through metazoan parasite-vectored cycles or food and water intake.

When aquatic animals and terrestrial species share the same foods, fish for example, and if these fish are the vector for a viral disease of aquatic animals, then that same disease may be transmitted to terrestrial animals eating these fish. This chapter will focus on the Caliciviridae and explore several ocean cycles for caliciviruses of aquatic animals, then discuss the movement of these viruses from the sea into terrestrial species, including humans.

B. Calicivirus Structure and Taxonomy

Members of the family Caliciviridae are naked single-stranded positive-sense RNA icosahedral viruses having but one major capsid polypeptide and distinctive cuplike surface depressions as visualized using negative-stain electron microscopy (Fig. 1). It is from this feature that they derive their name (Cooper et al., 1978) and are easily recognized as distinct from all other families of virus. The Caliciviridae are currently divided into five genera based on tentative genomic organization (Fig. 2) (Berke et al., 1997). Their diameter ranges from 27 to 36 nm, depending in part on whether the capsid protein has undergone cleavage. Genomic length is usually less than 8 kilobases, and four of the five genera are resistant to in vitro propagation (Clarke and Lambden, 1997).

Three of these four genera of so-called noncultivatable caliciviruses are the Norwalk-like groups, infectious for humans and some other primates and causing gastroenteritis; the Sapporo group also causes gastroenteritis, particularly in Japanese children (Jiang et al., 1997); and hepatitis E, which occurs primarily in third world countries, where lethal hepatitis and hemorrhage are seen in up to 25% of the pregnant women who contract the disease. All three have fecal–oral transmission cycles, with contaminated water, shellfish, or other foods being implicated as the usual sources of infection (Kapikian and Chanock, 1990). Although zoonotic reservoirs have been considered, none have been shown, with the possible exception of hepatitis E (Meng et al., 1997).

The fourth virus in the noncultivatible group is the calicivirus associated with rabbit hemorrhagic disease (RHD). From time of exposure, hemorrhagic death can

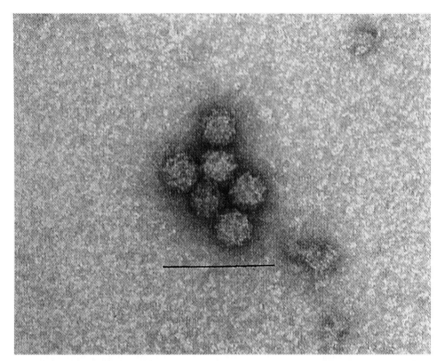

Fig. 1. Electron photomicrograph of Cetacean calicivirus showing typical morphology. Negative stain. Bar equals 100 nm. Photo by Douglas E. Skilling, Laboratory for Calicivirus Studies, Oregon State University.

occur within 24 hours in up to 95% of exposed rabbits (Xu and Chen, 1989), making this one of the most deadly known viruses. Rabbit hemorrhagic disease differs from the other four genera in that it is the only member of the Caliciviridae not yet proved to cause disease in humans. Although this may no longer be true (Smith *et al.*, 1998a), the RHD virus, as with other caliciviruses, can survive a variety of environmental challenges; thus, water, food, fomites, and other mechanical modes of transmission including insects and perhaps birds as true intermediate vectors can all be expected to spread the disease (Poet *et al.*, 1996).

The fifth genus of the Caliciviridae is separated into three subgroups based on the species, time, or place of isolation rather than any antigenic or genomic relationship. Prior to 1956, all calicivirus isolates were from swine and are referred to as vesicular exanthema of swine virus (VESV). Beginning in 1948, each of these swine serotypes was named alphabetically A through H followed by the subscript of the year of isolation. Thus, VESV-A_{48} became the prototype strain for the family Caliciviridae (Smith and Boyt, 1990). After 1956, all the caliciviruses isolated from cats were called feline calicivirus (FCV). All of these formed a single neutralizing serotype. Beginning in 1972, and continuing through today, nearly all

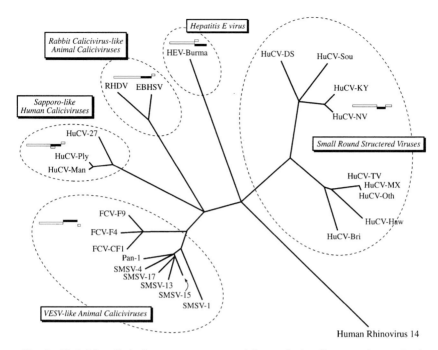

Fig. 2. Fitch–Margoliash distance tree constructed from pairwise distance estimates for the hypervariable capsid region. Percentage divergence is represented by the total linear distance between terminal tips. Branches with bootstraps greater than 95% are shown. The comparable region of one picornavirus (human rhinovirus 14) is included as an outgroup. Circles were applied to indicate groups of strains (five proposed genera) sharing similar genome organization.

calicivirus isolates, which total more than 150 strains divided into 27 neutralizing serotypes (Smith *et al.*, 1998a), have been recovered from 25 animal species — including shellfish, teleosts, seals, cetaceans, and humans — and are loosely grouped as San Miguel sea lion viruses (SMSVs). Occasionally, new types have been named for the animal group yielding the first isolate of that particular serotype. Examples are reptilian calicivirus (RCV), first isolated from snakes, then amphibians, and recently from three species of marine mammals (Barlough *et al.*, 1998), primate calicivirus (PCV) (Smith *et al.*, 1983c), and bovine calicivirus (BCV) (Smith *et al.*, 1983a). Within each of the SMSV examples, including the VESV and FCV groupings, multiple host species are known or suspected for each serotype, and a diversity of genomic and antigenic variants are the rule. The concept of quasispecies is clearly demonstrated by this diversity, yet all members of these three groups are present in ocean reservoirs.

The remainder of this chapter will discuss endemic and epidemic movements of caliciviruses through aquatic ecosystems and their dispersion onto land. It will provide a calicivirocentric look at how these events occur and document the emergence of these zoonotic viruses from their ocean reservoirs to cause disease in humans. This is a paradigm shift for viral diseases (Smith *et al.*, 1998b).

II. CALICIVIRUS CYCLES IN PINNIPEDS

Studies of aquatic mammals and their virus parasites provide an unusually rich source of information if one wishes to examine the aquatic viral disease reservoirs and the ecologic bridges between them and their terrestrial hosts.

A. Northern Fur Seal (*Calorhinus ursinus*): Epidemics and Disease

Caliciviruses were first isolated from Northern fur seal in 1972 from two dead newborn pups examined on St. Paul Island (Pribilof Island Group), Alaska (Fig. 3). This virus type was designated SMSV type 1 and was the same as that isolated 4 months earlier from two aborting California sea lions (*Zalophus californianus*) on San Miguel Island, California, 1600 km to the south (Smith *et al.*, 1973). Serum samples collected from 200 juvenile males 3 to 5 years old that year and the following year and examined using SMSV-1 virus neutralizing tests were all negative. The virus was not seen to cause blistering disease and, based on serology, did not move through the fur seal herd on the rookeries. In contrast, these same

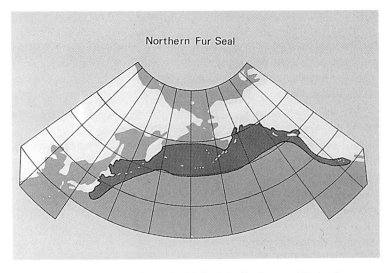

Fig. 3.   Northern fur seal range in the North Pacific Ocean and Bering Sea.

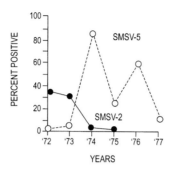

Fig. 4. SMSV-1 was isolated from the fur seals in 1972, but neutralizing antibodies were not found. SMSV-2 was not isolated; however, antibodies present from previous exposure declined over time. The SMSV-5 antibody profiles document the start of the epidemic in 1973, and the 1975 serologic tests reflect the fact that most animals sampled that year had been at sea as yearlings when the epizootic occurred.

sera showed a declining incidence of antibody for SMSV-2. The third year, using the same procedure for sampling and testing, all SMSV-2 sero-positive animals had disappeared (Fig. 4). This suggests that sometime prior to 1972 SMSV-2 had been introduced into the Pribilof Island colony, and had moved through the herd. By 1974 it had disappeared without observable effect. By contrast, SMSV-1 had been introduced into the herd but with no detectable movement through the herd other than perhaps an occasional late-term abortion or neonatal death; these two viruses seemed to differ in their host relationships. The first documented virus epidemic in seals began in 1973, and the epidemiologic history of this virus may help explain these differences. A blister-causing calicivirus of unknown type was isolated from diseased seals in the Pribilof colony (Fig. 5; see color insert). This new type was named SMSV-5 (Smith *et al.*, 1977).

Extensive blistering was observed on the flippers of the subadult male seals harvested for their pelts under an international agreement among the United States, Canada, Japan, and Russia. Within 2 weeks, the blistering condition was observed in harem bulls (10–13 years old, breeding males), breeding females (6–24 years old), and young pups. Thus, in 1973 all age classes except for the yearling seals at sea (fur seal pups spend 18 months at sea during their first pelagic cycle) were exposed to SMSV-5 and showing disease. For the following years, little or no disease was observed, but retrospective serology testing for type-specific antibodies to SMSV-5 over six seasons provided an excellent epidemiologic record (see Fig. 4) of this first documented viral epidemic among marine mammals (Smith *et al.*, 1980b).

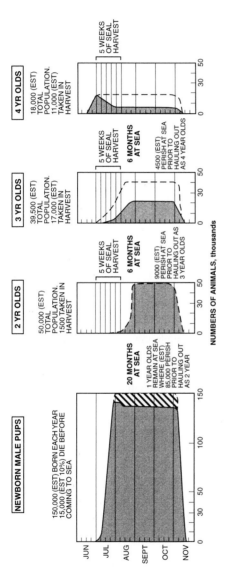

AGE COMPOSITION OF THE BACHELOR MALE
Callorhinus ursinus POPULATION ON ST PAUL
ISLAND FROM BIRTH THROUGH 4 YEARS

Fig. 6. The reduction in numbers of male fur seals each year from their birth (approximately 150,000) through age 4 years. The approximate number lost at sea each year can be seen, as can the weekly drop in numbers during 5 weeks of commercial seal harvesting (now discontinued, with only subsistence harvest occurring). Thus, about 5000 4-year-old males were left each year, and the harem breeding bulls (10–13 years old) will be drawn from this group as it ages.

Each year, 200 seals were bled. Their age distribution was about 5% 2 year olds, 55% 3 year olds, 30% 4 year olds, and 10% 5 year olds (Fig. 6). The net survival and yearly composition of the male/female ratio of fur seals can be seen in Figure 7. Beginning in 1972, year 1 (Fig. 4), there were no antibodies to SMSV-5; by year 2 (the year of the epidemic), a few animals had been exposed and developed antibodies by when they were sampled. By year 3, there was an increased incidence in seropositive seals (85% of sample), which could be explained by an abundance of animals sampled as 3 and 4 year olds. These had been exposed as 2 and 3 year olds the previous year on the rookery. Year 4 saw a very sharp decline in seropositive animals (25%), which coincided with the sample being predominantly 3 year olds from the 1972 age class that had been at sea during the epidemic. The year 5 sample peaked again (58%), showing most animals to be seropositive, and this is explained by the 3-year-old group being the age class born and nursed on the rookeries during the height of the epidemic. Year 6 showed marked decline (less than 10%), and year 7 (not shown) had virtually no antibody-positive samples, as the last of the animals born during the epidemic year were sampled as 5 year olds.

These data document the occurrence and failure of SMSV-1 to spread, and the serologic evidence of a previous epidemic of SMSV-2 and the introduction and epidemic spread of SMSV-5 followed by population resistance and disappearance of that virus from the population. Fur seal populations were generally negative for antibodies to most caliciviruses. This differs markedly from California sea lions, where every adult female tests positive often to numerous (up to 18 of 22)

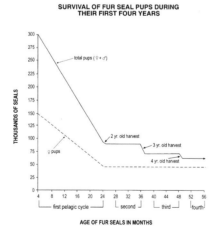

Fig. 7. On St. Paul Island, Alaska, the annual seal harvest reduces the total number of each age class dramatically by reducing the number of males. After age 5, when male seals are no longer harvested, the total year-class recruitment will be about 60,000 total with ¾ of this made up of females entering the breeding herd at age 6.

serotypes (Table I). These species differences raised questions of calicivirus survival and movement in the ocean ecosystems. How was a virus common off Southern California introduced into a Bering Sea colony, and why was the virus not sustained from year to year in that community?

California gray whales (*Eschrichtius robustus*) migrate each year from the Sea of Cortez through the Bering Sea and Bering Strait. They are known to become infected with multiple calicivirus types (Smith and Latham, 1978) and could therefore transport the San Miguel caliciviruses northward. However, they seem to have no specific Northern fur seal interaction. Female and juvenile fur seals also migrate as far south as Mexico, and then return to the Bering Sea. To be carriers, these animals would need to become persistently infected, which did not appear to be occurring. Therefore, other means of virus transport were sought, and one involving a parasitic cycle received considerable attention.

Northern fur seals can become infected with lungworms thought to be similar or identical to the California sea lion lungworm *Parafilaroides decorus*. Infective larvae were collected in mucus from the throat of a heavily parasitized California sea lion pup. These were washed free of mucus, bathed in tissue culture media containing 10^8 caliciviruses (SMSV-5), then rinsed twice and fed to their intermediate host, the opaleye perch (*Girella nigricans*). Alternatively, two additional groups of opaleye perch were infected with calicivirus by intraperitoneal injection, and one group was also fed untreated larvae. At 32 days postinfection (the infective lungworm larvae encyst in the gut wall by day 30), the fish were killed, minced, assayed for SMSV-5, and fed to three groups, each made up of three colostrum-deprived fur seal pups on St. Paul Island. All samples of fish contained infective virus, and all nine seal pups became infected, shed virus, and yielded virus isolates at time of necropsy. One of three pups receiving virus-contaminated larvae developed a blister on the flipper 21 days postinfection, and SMSV-5 was isolated from this lesion. This individual and several others also developed hepatitis and encephalitis (Smith *et al.*, 1980b).

This work demonstrated that SMSV-5 carried by lungworm larvae could infect fish and remain infective for 32 days in the fish in saltwater, then be ingested by a fur seal, resulting in classic blistering disease caused by the SMSV-5 virus 28 days later. The entire series of events spanned 60 days, plenty of time for fur seal movements between San Miguel and St. Paul.

Although the fur seal epidemics suggested that calicivirus overwintering in northern waters did not occur, two subsequent discoveries suggest that this does occur in walrus and bowhead whales. Both examples will be discussed in Sections II.C and III.A. Because some calicivirus types seem not to overwinter in fur seals, but instead cause epidemics followed by disease die-out in fur seals, this phenomena came under further investigation.

In 1968, a Northern fur seal breeding herd began to colonize coastal areas in Southern California on San Miguel Island. This provided an opportunity for

Fig. 5. Blisters on the flippers of Northern fur seals with partially eroded coverings reveal shallow ulcers 1–2 cm across. SMSV-5 was first isolated from these lesions.

Fig. 13. Stillborn sea lion pup delivered about 30 days premature. Not all but some of these stillborns proved to be infected with caliciviruses. Other seals in the picture are sleeping yearling Northern elephant seals.

Fig. 14. Subdural hemorrhage in a stillborn pup often associated with the perinatal hemorrhagic syndrome described in four species of North Pacific seals. Caliciviruses have been isolated from such animals.

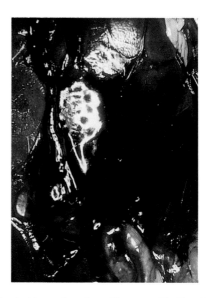

Fig. 15. Massive abdominal hemorrhage in a stillborn pup. The liver (mass in the upper right side of figure, which is pale and friable) and a loop of gut (middle bottom of figure) are identifiable in a pool of blood. Caliciviruses have been isolated from such animals.

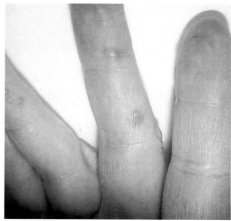

Fig. 22. Opaleye taken from Southern California tidal pools and retained in a saltwater aquarium for calicivirus infectivity studies.

Fig. 30. San Miguel sea lion virus type 5 infection in humans. The infected individual complained of flu-like symptoms prior to several dozen blisters appearing on the fingers, toes, palms of the hands, and soles of the feet. Such spread is consistent with a generalized blood-borne infection.

TABLE I

San Miguel Island Breeding Female *Zalophus californianus*[a,b]

Viruses	F1	F2	F3	F4	F5	F6	F7	F8	F9	F10	F12	F13	F14	F15	F16	F17	F18	F19	F20	F21
SMSV-1	1:80	1:30	1:20	1:80	1:60	1:15	1:10	1:15	1:80	1:60	1:10	1:15	1:60	1:80	1:10	<1:10	1:60	1:80	1:60	1:10
SMSV-2	1:80	1:20	1:60	1:10	1:10	1:20	1:80	1:20	1:10	1:40	1:15	1:20	1:70	1:60	1:100	1:80	1:80	1:120	1:15	1:15
SMSV-3	1:60	1:60	1:60	1:70	1:70	1:10	1:10	1:20	1:20	1:15	1:60	1:15	1:50	1:80	1:60	1:10	1:50	1:80	1:100	1:20
SMSV-4	1:10	<1:10	1:50	1:80	<1:10	<1:10	1:15	1:10	<1:10	1:80	1:20	1:20	<1:10	<1:10	<1:10	<1:10	<1:10	<1:10	1:60	<1:10
SMSV-5	1:20	1:30	1:40	1:160	1:20	1:10	1:20	1:20	1:20	1:40	1:15	1:15	1:40	1:40	1:840	1:60	1:280	1:80	1:80	1:80
SMSV-6	1:80	1:10	1:50	1:70	1:80	1:10	1:40	1:40	1:10	1:60	1:50	1:20	<1:10	1:80	1:15	1:15	<1:10	1:60	1:20	<1:10
SMSV-7	1:70	1:60	1:80	1:100	1:50	1:10	<1:10	1:15	<1:10	1:15	1:60	1:15	1:10	1:20	1:80	1:15	1:70	1:60	1:20	1:50
SMSV-8	<1:10	<1:10	<1:10	<1:10	<1:10	<1:10	<1:10	<1:10	<1:10	<1:10	<1:10	<1:10	<1:10	1:10	<1:10	<1:10	<1:10	<1:10	<1:10	<1:10
SMSV-9	1:80	1:120	1:40	1:50	1:15	<1:10	1:10	1:20	1:10	1:20	1:60	1:10	1:40	1:10	1:80	<1:10	1:60	1:80	<1:10	1:60
SMSV-10	<1:10	1:20	1:10	<1:10	<1:10	<1:10	1:10	1:10	<1:10	<1:10	1:40	1:30	<1:10	1:10	1:20	<1:10	<1:10	<1:10	1:10	<1:10
SMSV-11	1:40	1:15	1:15	1:40	1:20	<1:10	1:20	1:15	<1:10	1:20	1:20	<1:10	<1:10	1:10	<1:10	<1:10	1:30	1:20	1:15	<1:10
SMSV-12	1:10	1:10	1:20	1:15	1:10	<1:10	1:10	<1:10	<1:10	<1:10	1:10	1:10	1:20	<1:10	1:10	1:10	1:15	1:15	1:20	1:20
SMSV-13	1:10	1:10	<1:10	<1:10	1:10	<1:10	1:10	<1:10	<1:10	1:10	1:10	<1:10	<1:10	<1:10	<1:10	1:10	1:10	1:10	1:10	<1:10
SMSV-14	<1:10	1:10	1:10	1:10	<1:10	<1:10	1:10	<1:10	<1:10	1:10	<1:10	1:10	<1:10	<1:10	<1:10	1:10	1:10	1:10	1:10	<1:10
SMSV-15	<1:10	1:10	1:10	1:60	1:15	1:10	1:20	1:10	<1:10	1:20	1:15	1:60	1:10	1:15	1:20	1:15	1:15	1:80	1:20	1:10
SMSV-16	1:80	1:20	1:15	1:60	1:10	<1:10	1:15	<1:10	<1:10	1:20	1:60	1:10	1:60	<1:10	1:20	1:10	1:20	1:80	1:10	1:80
041	1:320	1:320	1:220	1:380	1:240	1:50	1:50	1:160	<1:10	1:160	1:60	1:10	1:15	<1:10	1:90	1:10	1:160	1:240	1:80	1:90
7420	<1:10	<1:10	<1:10	<1:10	<1:10	<1:10	<1:10	<1:10	<1:10	<1:10	<1:10	<1:10	<1:10	<1:10	<1:10	<1:10	<1:10	<1:10	<1:10	<1:10
W-6	1:10	<1:10	<1:10	1:80	1:40	<1:10	1:50	1:70	1:60	1:10	1:70	1:120	1:50	1:20	1:10	1:50	1:10	1:10	1:60	1:80
F-9	1:20	1:15	1:20	1:120	1:20	1:60	1:60	1:15	1:30	1:20	1:180	1:60	<1:10	1:50	1:12	1:160	1:220	<1:10	<1:10	1:80
Reptile Cv	1:160	1:220	1:90	1:280	1:220	1:50	1:15	1:320	1:80	1:240	1:100	1:280	1:20	1:100	1:320	<1:10	1:160	1:320	1:240	1:50
CaCv	<1:10	<1:10	<1:10	<1:10	<1:10	<1:10	<1:10	<1:10	<1:10	<1:10	<1:10	<1:10	<1:10	<1:10	<1:10	<1:10	<1:10	<1:10	<1:10	<1:10

[a] Serum neutralization 50%, endpoint vs. 100 $TCID_{50}$ calicivirus.

[b] Twenty adult female California sea lions were screened for type-specific neutralizing antibody to 22 different calicivirus serotypes. All were positive to 8 or more of 21 serotypes, and 17 were positive (titers up to 1:220) for feline calicivirus, whereas none showed antibody to a canine serotype and walrus calicivirus. SMSV = San Miguel sea lion virus; 041 = cetacean calicivirus, 7420 = walrus calicivirus, W-6 = gray whale isolate typed as vesicular exanthema of swine virus A_{48} (the prototype virus for the Caliciviridae family and the foreign animal disease vesicular exanthema of swine), F9 = vaccine strain of feline calicivirus, CaCv = canine calicivirus.

comparison of caliciviral introduction into this herd and the Northern herds. On San Miguel Island, Northern fur seals, California sea lions, Northern elephant seals, harbor seals, and occasionally Steller sea lions all haul out and establish dense rookeries in close proximity. The fur seals and sea lions give birth and breed at approximately the same time of year. Therefore, if direct contact and near-shore phenomena are major events for calicivirus transmission in these two species, one would expect the San Miguel population of Northern fur seals to present calicivirus antibody profiles resembling those of the California sea lions. This does not occur (Smith and Latham, 1978).

Instead, the San Miguel colonies of fur seals show calicivirus antibody profiles very similar to those seen in the Pribilof colonies, yet fur seals are seemingly just as susceptible to calicivirus infections as the California sea lions. The answer seems to lie in the feeding habit differences between the two species. Fur seals feed on fish species 150–200 km offshore near the deep-water margins of the continental shelf, while leaving their nursing pups behind on the rookeries. Thus, after 5–10 days they return to their pups on the rookeries, having digested all of their food and discharged their feces at sea. This reduces the very important fecal–oral transmission cycles of the virus. If epidemics occur as those seen in the Pribilof Islands, they are likely to involve pathotypes well adapted to direct contact transmission. This would happen with the blistering and release of virus from skin erosions, as occurred with the epizootic in the Pribilof colony involving SMSV-5 (Smith *et al.*, 1977).

Calicivirus infections have been associated with reproductive failure in at least four pinniped species. A so-called perinatal hemorrhagic syndrome has been seen in California sea lions, Northern elephant seals, northern fur seals, and Steller sea lions. Pups were found dead or dying a few weeks or days before full term, and frequently these have what biologists term "red-eye." This condition resulted from hemorrhage into the anterior chamber of the eye, which usually denotes massive hemorrhages of the brain, subperiosteal hemorrhage of the parietal bones, and pooled blood in the abdomen from ruptures of a very friable liver and generalized bleeding from other major organs (see §II.C and Figs. 14 and 15). This manifestation appears to be more sporadic than epidemic in all species. In the case of fur seals, because of their pelagic life right up to within a few days of giving birth, exposure and presumably some aborting of pups is believed to occur at sea rather than at breeding rookeries. Further evidence of this is the epidemiologic observation that these occurrences diminish as the breeding season progresses.

It had been documented and first reported in 1978 that the Northern fur seal colonies on St. Paul Island, Alaska, had declined (Smith, 1978). This observation was based on studies attempting to link population dynamics with the occurrence of the SMSV-5 epidemic (Smith *et al.*, 1980b). Beginning in 1964 and continuing on through 1978, the number of seals returning for harvest had dwindled by half (Figs. 8 and 9). This was eventually manifested by a reduced number of breeding females and reduced pup counts.

SEAL HARVEST BY YEAR CLASS

Fig. 8.   The year class is made up of the number of animals born any given year, while the harvest will have components of four year classes (2, 3, 4, and 5 year olds).

Those same trends have continued into the late 1990s, even though by 1980 commercial harvests were dramatically reduced such that only the pelts of seals killed for Alaskan-native subsistence could be taken. Seemingly, from a calicivirocentric perspective, the Northern fur seal is neither an important nor a natural reservoir for virus. Epidemics, though rare, can occur, and yet endemnicity seems to be lacking. The major impact on this species may be the added reproductive burden associated with spontaneous abortion.

There has been much conjecture regarding the causes of this 30-year decline in fur seal stocks, but the impact of the abortogenic caliciviruses, which can actively cause ongoing reproductive failure in this and other seal species, has not yet been added into the dynamic that has driven this once robust marine mammal population into protracted decline (Fig. 9). Something of this same decline has been seen in Steller sea lions (*Eumetopias jubatus*).

B. Steller Sea Lion: Epidemic and Endemic Disease

Large compartments of the general Steller sea lion population have undergone protracted decline through the 1980s and 1990s. Although there have not been well-defined calicivirus epidemics passing through the Steller herd as occurred with fur seals, calicivirus presence has been documented to be far more common (Barlough *et al.*, 1987, 1988). Herd dynamics, population, distribution, and feeding habits can account for this ongoing caliciviral presence.

Breeding populations of Steller sea lions have been reported as far south as San Miguel Island (this population disappeared in the 1970s). Permanent populations

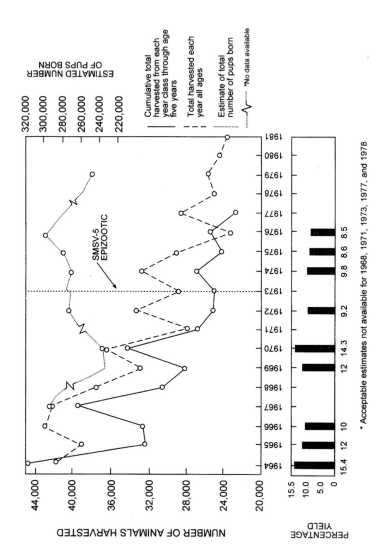

Fig. 9. Decline by total harvest and by year class from 1964 to 1981. This reflects an overall decline that was eventually felt in pup production declines brought on by reduced numbers of breeding-age female seals. The year class born during the SMSV-5 epizootic (1973) seemed not to show declining harvest numbers, suggesting a "non-effect" for survival, but such an effect if it occurred could be overshadowed by alternate factors degrading survivability.

remain spotted up the coasts of California (Farallon Islands), Oregon (Rogue Reef), Washington, Canada, and Alaska, then east along the Aleutian Island chain and onto the Pribilof Islands (Fig. 10). These populations have local movements and migrations, with feeding cycles that occur near shore that can be daily. Therefore, there is much rookery contact all year around, and fecal discharge following feeding is near shore and on haul-out sites rather than 100 to 200 km offshore, as occurs with Northern fur seals. These factors provide insight into the caliciviral disease dynamic that seems to be occurring.

Infected Steller sea lions can discharge feces contaminated with caliciviruses, and virus transmission can occur either directly or indirectly by intermediate or prey species in the tidal pools and kelp beds. In this way, a much more intense mixing of infected seals and potentially infected reservoirs along the near-shore feeding zones can take place, resulting in a high incidence of exposure for all susceptible individuals in the population regardless of sex or age class. Such a dynamic should, and apparently does, sustain caliciviruses within this species (Barlough *et al.*, 1987, 1988) far more effectively than occurs in fur seals, where only seasonal direct contact occurs and feeding and fecal discharge are predominantly offshore in deep water. Given these background notes, what might the impacts of caliciviral infection be on this declining species?

The first evidence of protracted change in the composition of the Steller sea lion herd that suggested an infectious disease component was presented at a National Marine Fisheries Service workshop on the Steller sea lion population decline

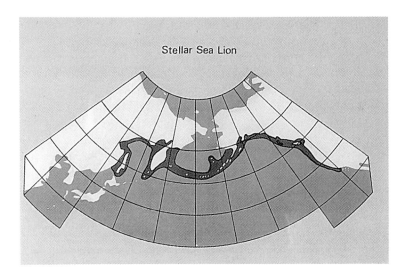

Fig. 10. Northern sea lion range extending along the North American Coast, the Bering Sea, and into Asia.

(Smith, 1986). Morphometric studies of beached animals showed a marked decrease in body size, and weight between animals of the same age when 1980s data
were compared to similar 1970s data (Fig. 11). This difference ran through all year
groups, from yearling to age 11 years; then from 11 years to terminal age (20 years)
no size and weight differences were detectable. These data could be seen to
indicate that whatever occurred was initiated early in life and resulted in "stunting"
during the first year, and this overall event had been an annual occurrence for 10
years. The event seemed not to be food deprivation involving fish availability,
because the body condition of the older animals (11 to 20) was not affected
(Caulkins, 1986).

A second clue pointing to an infectious component came out of a statistical
report of loss in fecundity and reduced birth rates over a protracted period (York,
1986). In combination, the factors of reproductive failure — including abortion,
decreased numbers of surviving pups, and poor first-year growth performance —
were all operating to erode the Steller sea lion population base. Presumably, the
cause for such broad-based effects would be multifactorial, and this may be true

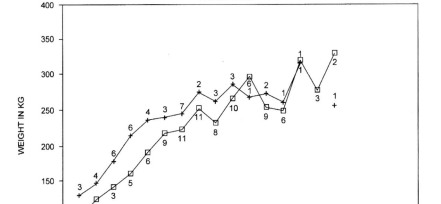

Fig. 11. The same animals shown in this figure were also measured for length and girth. These
additional measurements revealed the same differences shown with total weight (Calkins, 1986). Given
that other unknown compounding factors were not at work, these data suggest that during gestation or
the first year of life something caused stunting of growth that was never overcome. This was an annual
occurrence from the mid-1970s through the mid-1980s, and possibly beyond. Numbers shown at each
data point are the number of animals sampled of each age class.

for the Steller sea lion herd. However, calicivirus activity among reproducing females was seen (Barlough *et al.*, 1987). Fetuses presumably infected near full term in 1985 had neutralizing antibodies against caliciviruses, as did their dams. This pattern was observed for four virus serotypes (Barlough *et al.*, 1987). Although sample size precluded definitive conclusions, there was a suggestion that new calicivirus types were entering the herd and impacting reproduction on a temporal basis (Barlough *et al.*, 1987). The disease effects of abortion, stunting of offspring, and agalactia (lack of or dramatically reduced milk production) are well documented for calicivirus infections in some species (Smith and Boyt, 1990), and these same effects would be expected in Steller sea lions, where their occurrence could contribute substantially to population decline.

Thus, from a calicivirocentric view, virus survival in Steller sea lions occurs where endemic infections are sustained, multiple serotypes become active concurrently, and the Steller sea lion habitat provides the means for effective virus transmission and long-term survival once the virus is introduced into a population naive for that pathotype. Because of these dynamics and the very long south-to-north and warm-to-cold-water distribution of the Steller range and the lack of long-distance migration cycles, certain discrete or local colonies and herds have been unequally affected. This also suggests that caliciviral disease may have somewhat local effects within the overall population and that the virus can be sustained year-round in these locales. This would require a reservoir species such as fish or some lower life form or persistently infected seals that shed virus in infective doses. Either one or perhaps all three mechanisms of virus survival are operating within the ecosystem inhabited by Steller sea lions and may have contributed to the protracted decline of some Steller sea lion populations, which began subsequent to the mid-1970s.

C. California Sea Lion (*Zalophus californianus*)

The first virus isolates recovered from any pinniped species were from aborting California sea lions and their fetuses on San Miguel Island in California (Fig. 12) (Smith *et al.*, 1973). These were caliciviruses of multiple serum neutralizing types. Over time, numerous additional serotypes were isolated, and retrospective serology demonstrated a sea lion population that showed evidence of exposure to multiple calicivirus serotypes by the time they reached 4 months of age (Smith and Latham, 1978). Furthermore, every adult female was antibody-positive to many calicivirus serotypes (Table I).

Epidemiologically, this presented a perplexing picture. Historically, in domestic animals, infection with caliciviruses resulted in type-specific immunity, but sus-

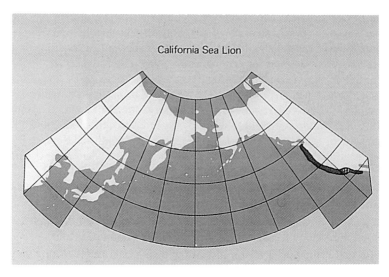

Fig. 12. Male California sea lion range from Baja to Vancouver Island, B.C. Breeding does not occur north of the Santa Barbara Channel Islands (shown by 5 dots off the Southern California Coast), and females do not migrate to the north of these islands.

ceptibility to new serotypes remained. There was no cross-protection between the virus serotypes.

This concept was applied to sea lions, and, as this picture emerged to include 17 distinct serotypes, it was no longer feasible to accept the California sea lions as the primary reservoir host for these agents. To do so would require an explanation of how 17 distinctly different viruses could each remain active in a total population of about 100,000 individuals, all seemingly exposed to and resistant to the plurality of caliciviruses that were active in the ecosystem. Recruitment from new pups (5,000 to 10,000 per year) did not provide a significant pool of new susceptible animals. It was concluded that the sea lion was an incidental host and that a much larger population base was needed to sustain this highly complex system of calicivirus types (Smith *et al.*, 1980b, 1981a). Having concluded this and knowing from a historical perspective that these same viruses had been repeatedly introduced into swine from some unknown reservoir when the pigs were fed raw garbage from coastal California cities, some species of marine fish resident to the California coast became suspect.

This hypothesis was explored by setting the following four criteria and then testing each indigenous fish species along the California Coast against these criteria. Fish species were sought that were (1) taken in the sport fishery, (2) taken in the commercial fishery, (3) had an intimate relationship with California sea lions, and (4) were resident to the Southern California region but not the colder northern waters, where raw garbage from port city restaurants had not been implicated in introducing the virus into swine. Only one fish species was found that comfortably fit all four criteria (see §IV.B and Figs. 21 and 22; see color insert) (Smith, 1975).

That species, the opaleye perch (*Girella nigricans*) was considered a game fish and could be caught by rod and line in tidal pools or kelp beds using peas or mussels as bait. When taken in commercial catches, opaleye were simply filleted as perch. Opaleye would be one of the first prey species eaten by young sea lions and are the intermediate host for the California sea lion lungworm (*Parafloroides decorus*). Infections with the parasite are almost universal among young sea lions. The opaleye spend their juvenile phase in tidal pools and adulthood in the kelp beds and are found in the southern waters from Point Conception, California south.

Juvenile opaleye on St. Nicholas Island were caught, and several dozen were processed for virus isolation. Three isolates were recovered, two of which had previously been isolated from marine mammals, and one a new serotype (Smith *et al.*, 1981b). This established the principal of a marine teleost serving as the reservoir host for viruses infecting marine mammals (see §IV.B). Further testing in swine extended the concept to include terrestrial mammals (Smith *et al.*, 1980a) (see §VI.A).

Opaleye ingest primarily ribbon grass, digest the surface slime off the grass, and defecate the grass back into the water, with the grass looking unaltered. This provides a mechanism for ingesting large quantities of grasses, thereby increasing the probability of eating grasses contaminated with sea lion feces, lungworm larvae, and caliciviruses (see §II.A). The lungworm larvae encyst in the fish gut wall, where they are presumed to remain viable for several years. If there is a nematode link with the virus, this would provide a much needed answer to the ecology of caliciviruses. Such linkage would result in opaleye fish accumulating calicivirus laden larvae that would encyst and survive over time, thereby allowing the same fish to accumulate additional serotypes of caliciviruses. In this way a single opaleye could become a repository for long-term accumulation and storage of multiple calicivirus serotypes and could then deliver these all into a single sea lion when it ate fish (Smith *et al.*, 1980b).

This cycle would nicely explain the early and near universal exposure of California sea lions to multiple virus types. It would also explain why the Southern California Coast is a hotbed of caliciviruses of new types and variants. Specific factors involved would be the recombinant events that can occur during simultaneous infections with genomic variants of RNA virus having nonsegmented genomes. This, as well as the innate error associated with RNA replication and the immune pressures from infected hosts, are all factors that may explain why the opaleye↔California sea lion cycle is an unusually effective incubator for producing new calicivirus pathotypes and serotypes. These factors all converge in the

Southern California coastal ecosystem formed by the Santa Barbara Channel Islands.

The impact of endemic caliciviral infections on the California sea lion herd appears to be absorbed without driving the population into a negative balance. In this respect, the marine calicivirus and California sea lions have attained some measure of mutual adaptation. The virus is associated with reproductive losses and can be isolated routinely from California sea lions. Occasionally, transient vesicular disease can be seen, and infected pups occasionally have massive hemorrhages (Figs. 13–15; see color insert), suggesting that this very dangerous virulence factor is active sporadically. At any time, about 3% of the animals (nose and rectal) swabbed for virus isolation can be shown to be shedding viable caliciviruses of various types (Smith *et al.*, 1981b). This serves the virus by providing a continuous flush of multiple calicivirus types into the tidal pools.

California sea lions may provide an added dimension to caliciviral spread. The annual migration of adult and subadult males takes them up to 1400 km north each winter to Vancouver Island, British Columbia (Canada). In this way, persistently infected individuals, as would be expected with endemic viral disease agents, can shed new virus types along the California, Oregon, Washington, and Canadian coasts, where intermixing with Steller sea lions and fish species resident to colder water can take place.

D. Walrus (*Odobenus rosmarus*): Endemic Calicivirus

The same serotype of calicivirus (walrus calicivirus) was first isolated from three walrus fecal samples taken in the Chukchi Sea in 1977 (Smith *et al.*, 1983b), and again 10 years later from two additional animals from the Eastern Siberian Sea (Fig. 16). Walrus populations remain predominantly north of the Aleutian Island chain, so that mixing with southern species as occurs with fur seals is unlikely. The walrus virus also appears to have one important dynamic that can dramatically alter its survival in a host species. Serologic surveys show very little neutralizing antibody activity against this serotype virus (Smith *et al.*, 1983b). Even more perplexing, when given experimentally to three groups of swine (three per group), all nine shed virus and developed hepatitis but did not develop neutralizing antibody (Smith and Boyt, 1990; Smith *et al.*, 1988). This provides yet another insight into the host–parasite dynamic of caliciviruses. The above observations suggest some mechanism of immunotolerance involving the walrus calicivirus whereby overt disease manifestations were not observed, even though some animals became infected and shed a virus that seemed not to induce neutralizing antibody in the host. These same walrus populations did show modest seroreactivity to other calicivirus virus types, so the immunotolerance-like observation with walrus calicivirus appears to be a result of special adaptation of the virus, not the host.

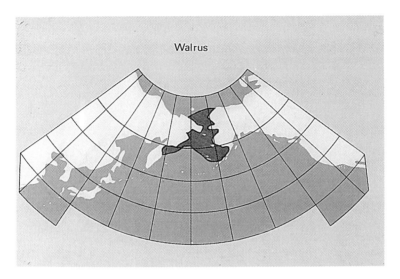

Fig. 16. Walrus range in the Bering Sea through the Bering Strait into the Chukchi and Eastern Siberian seas.

E. Northern Elephant Seal (*Mirounga augustirostris*): Endemic Infection

Elephant seals provide an important glimpse of viral ecology. Recall that, by the turn of the century, this species was believed to be extinct. The current population descended from just a few breeding animals discovered about 50 years ago (Fig. 17). Certain diseases require a large host population base, and severe host population bottlenecks can result in extinction of these diseases. Presumably, that happened with some elephant seal diseases, and this works to the advantage of depleted populations during recovery. Some portion of the complements of natural diseases that normally stabilize population numbers will be missing, allowing rapid population growth. That has happened to the elephant seals, and, presumably at some time in the future, new diseases will be introduced, become endemic, and play a role in stabilizing population growth. Elephant seals and calicivirus inter- actions now seem sporadic. Occasionally, pups have been born with hemorrhage syndrome, and caliciviruses have been isolated from these. Weaned pups that share the rookeries with sea lions occasionally shed caliciviruses, and there will be a low level of calicivirus antibody background among the more mature animals. All this seemingly incidental calicivirus involvement could be accounted for by sea lion contact and near-shore tidal pool activities that do not involve feeding. Whereas elephant seals share haul-out sites, and splash zones with sea lions, they feed in long cycles involving prey species caught at depths down to 3000 feet. It

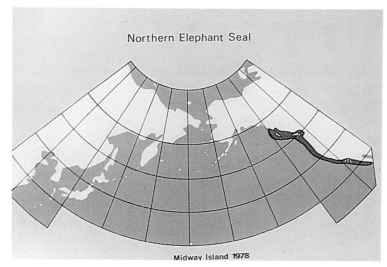

Fig. 17. Northern elephant sea range along the North American Coast. Individuals tagged in California have been spotted on Midway Island.

seems obvious that calicivirus cycles involving deep-water offshore prey species and elephant seals either have not yet developed or are inadequate to infect elephant seals with caliciviruses to the extent observed in their warm-water neighbors, the California sea lions. Again, the divergent ecologic niches occupied by two marine mammal species sharing the same haul-out sites has resulted in very different calicivirus infection and disease profiles.

III. OCEAN CYCLES IN CETACEANS

Cetaceans, and specifically the baleen whales, provide an especially intriguing glimpse into a rather indiscriminate feed mechanism (filter feeding) that serves the smallest (protozoa) as well as the largest animals on the planet (baleen whales). Filter feeding is notorious for scavenging and concentrating particulates, in this case viruses from water columns into discrete packages (e.g., plankton, bivalves, and whales).

A. Bowhead Whales (*Balaena mysticetus*)

Bowhead whales remain in relative proximity to the margins of pack-ice the year around (Fig. 18), yet as they too have type-specific antibodies to multiple

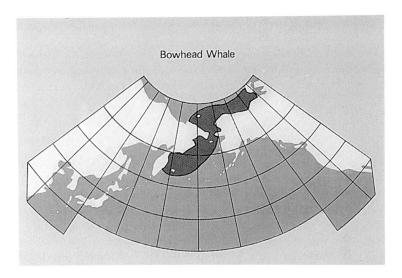

Fig. 18. Bowhead whale range generally associated with the margins of the pack ice.

calicivirus types, including some having high prevalence along the Southern California Coast (Smith *et al.*, 1981c). Even more interesting, animals examined in the 1980s and 1990s have type-specific antibodies to one serotype that has not been re-isolated since its discovery in California swine in 1934, and two other serotypes that were isolated only once previously, and that was from swine in Secaucus, New Jersey (O'Hara *et al.*, 1998; Smith *et al.*, 1981c, 1987). It is apparent that the different calicivirus serotypes can be carried great distances and infect new host species in geographically remote areas. The impact of calicivirus, if any, on the population dynamic of bowhead whales is not known, nor is there evidence suggesting the presence of bowhead whale and prey species (krill and small fish) cycles. California gray whales or migratory pelagic birds may provide the means for delivering new calicivirus types into this most northern cetacean population.

B. California Gray Whale (*Eschrichtius robustus*)

California gray whales migrate from breeding lagoons in the Sea of Cortez into the Arctic Ocean and return again each year (Fig. 19). Retrospective serologic surveys have shown this species to be infected with numerous calicivirus types (Smith and Latham, 1978). This is not surprising. Their migrations each spring with their young calves take them through the California sea lion range, so that

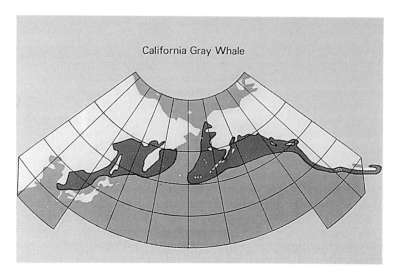

Fig. 19. California gray whale range. Thought to have Asian and North American populations. Migration can be from breeding lagoons in Baja to feeding grounds in the Arctic Ocean.

they feed near kelp beds and in shallow water, where they can ingest and come in contact with the same virus sources as sea lions (Smith and Latham, 1978).

Caliciviral disease effect, if it exists for gray whales, is unknown, but they do become infected and can greatly impact the dynamic of viral transport. For example, one gray whale fecal sample is known to contain 10^6 caliciviruses per gram, which would calculate out as 10^{13} caliciviruses passed in the feces daily; thus, gray whales migrating offshore can discharge tremendous numbers of caliciviruses (Smith *et al.*, 1998a).

If gray whales can become persistently infected, and one could assume that this happens because the individual shedding 10^6 virus per gram of feces had serum neutralizing antibody titers of 1:80 for the specific virus type that was continuing to be shed (Smith and Latham, 1978), then caliciviruses from Mexico and Southern California can be distributed up the North Pacific Coast into the Bering and Chukchi seas, where all species resident to these regions can become exposed. Perhaps this explains the bowhead whale phenomenon.

The gray whale is of particular interest because the individual whale described above was the first marine mammal of any species to yield a virus isolate using *in vitro* propagation. This occurred in 1968, and the virus was thought to be an enterovirus (Watkins *et al.*, 1969). Twenty years later, the isolate was reexamined and shown by electron microscopy to be a calicivirus. Furthermore, the virus serotype was vesicular exanthema of swine virus A_{48}, which was first isolated from swine in Fontana, California, in 1948. This virus, VESV-A$_{48}$, had been designated the type species for the family Caliciviridae and then twenty years later was recovered from a marine mammal (Watkins *et al.*, 1969). Of further interest and to be discussed later (§VI.A), the gray whale VESV-A$_{48}$ isolate caused fulminating vesicular disease in exposed swine (Smith *et al.*, 1998c).

C. Ocean Presence in Miscellaneous Whale Species

Type-specific antibodies to various SMSV/VESV serotypes have been seen in other whale species sampled in the North Pacific. This includes Sei whales (*Balaenoptera borealis*), fin whales (*Balaenoptera physalus*), and sperm whales (*Physter catodon*) (Smith and Latham, 1978). Two factors indicate global distribution for the ocean caliciviruses. First, the migration of sperm whales is not well understood, but their distribution is into all the major oceans, with females ranging 40° north and south of the equator and males extending to 55° north and south (Fig. 20). Second, Sei whales taken in the Antarctic also have type-specific antibodies to caliciviruses, suggesting probable reservoirs in the Southern Hemisphere. Pacific bottle-nose dolphins (*Tursiops gilli*) captured adjacent to the Hawaiian Islands also had serologic profiles indicating previous calicivirus exposure and probable infection. Caliciviral infection in several captive *Tursiops* and one *Delphinus* has resulted in a blistering and erosive skin disease.

Toothed whales (sperm whales and dolphins) and all four baleen whale species tested were positive, suggesting calicivirus presence within the ocean ecosystems traversed by all these diverse host species.

IV. POIKILOTHERM AND AVIAN HOSTS

Viruses such as caliciviruses that can survive and replicate in both poikilotherms and avian hosts seem to reside in the best of two worlds. Poikilothermic hosts provide massive host numbers and cooler temperatures where conditions for virus survival seem idealized, while avian hosts provide the means to distribute these viruses globally into new and diverse ecologic settings.

A. Reptilian Calicivirus (RCV)

The first calicivirus isolates recovered from reptiles were from dead hatchlings of an endangered Aruba Island rattlesnake (*Crotalus unicolar*) held in a zoologic collection in San Diego, California. Eventually two other reptilian species and an amphibian, the Bell's horned frog (*Ceratophyrs orata*), were found dead and

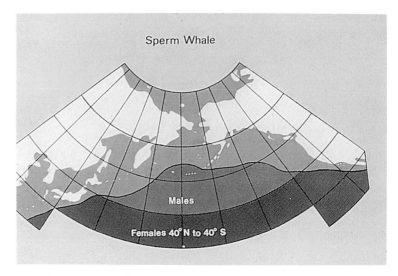

Fig. 20.   Sperm whale range is essentially worldwide from the middle latitudes to both sides of the equator.

shown to be infected with reptilian calicivirus (Smith *et al.*, 1986). Ten years later, RCV was recovered from three species of pinnipeds (California sea lions, Northern fur seals, and Steller sea lions) ranging along the California and Oregon coasts (Barlough *et al.*, 1998).

It cannot be determined whether RCVs moved from land to the sea or moved from sea to land. However, the land involvements were in a port city among captive animals where there would be considerable opportunity for virus spread within these confines. In contrast, and the ocean data show wide geographic presence of the virus where reptilian calicivirus appears to be very comfortable, in ocean ecosystems involving the warmer southern California bite (the coastal region between Point Conception, California to the north, and Baja, Mexico to the South) and the much colder Oregon Coast. The finding of infections in four terrestrial poikilotherm species, including an amphibian, suggest that RCVs could likely infect marine poikilotherms and maintain an ocean presence.

B. Calicivirus Reservoir in Ocean Fish

The finding of calicivirus types infecting both fish (opaleye perch) and mammals was mentioned previously (§II.C). Samples from opaleye (Figs. 21 and 22; see color insert) were examined each year for 2 years and resulted in three virus

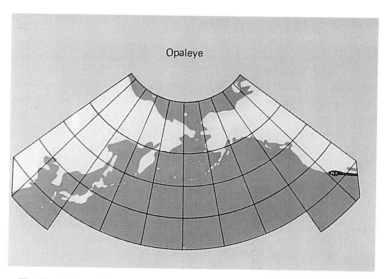

Fig. 21. Opaleye fish range extends along the near-shore of Southern California.

isolates. One was from the spleen of an opaleye and was a known serotype (SMSV-6) previously isolated from two California sea lion pups and a fur seal pup on San Miguel Island (Smith *et al.*, 1979). The two additional isolates were recovered on San Nicholas Island and were designated SMSV-7 (Smith *et al.*, 1980a) or, in today's nomenclature, SMSV-7 *Girella*-1 (SMSV-7-Gir-1), the prototype strain for the SMSV-7 serotype.

Other isolates of SMSV-7 were made from the nasal swabs of three elephant seal pups sampled early in their 6-week pre-weaning pre-aquatic phase on San Miguel Island, and another from a lung fluke (*Zalophatrema* species unknown) collected from a California sea lion dying of verminous pneumonia in San Diego (Smith *et al.*, 1980a).

These findings were pivotal in formulating workable hypotheses to answer longstanding questions relevant to caliciviral ecology. The presence of calicivirus in a metazoan parasite that passes through this teleost intermediate suggested mechanisms for virus movement that are treated in more detail in the section on fur seals (§II.A). The intense involvement of California sea lions but not their co-inhabitants on the same haul-out sites (Northern elephant seals and Northern fur seals) is also explained by this finding of a calicivirus-infected, near-shore, primary-prey species for the California sea lions.

Experimentally infecting opaleye fish with SMSV-5 added three more pieces of information (Smith *et al.*, 1981a). First, caliciviruses can retain viability when exposed to seawater for 14 days at 15°C. Second, the virus can replicate to high titers (10^7 infective viruses per gram) in opaleye spleen, without overt signs of disease occurring; third, compliant cell lines (*Vero*) that are usually lysed by SMSV-5 within 24 hours postinfection at 37°C were permissive for viral replication for 14 days at 15°C, but lysis was not observed even though viral titers reach 10^6 tissue culture infective doses. These findings suggest that the primary lytic cycle of caliciviruses may occur during replication at warm temperatures. Thus, mammals would not be the natural reservoir hosts (the lytic cycle with rapid cell death could suggest poor adaptation to such host species), whereas cold-temperature hosts may better support virus survival through nonlytic cycling.

One of the important answers in viral ecology derived from this has been resolution of the origins of devastating disease in swine that affected California and then the entire nation (see §VI.A). Fish scraps have been shown to be a source (probably the only important source) of calicivirus introduction into this major animal industry (Smith *et al.*, 1980a). A new paradigm in viral ecology was established with this first demonstration of a virus from natural ocean reservoirs spilling into land animals and then spreading among them to cause not just sporadic disease but a nationwide epidemic.

C. Fish and Migratory Birds

Calicivirus movements cannot always be explained on the basis of fish and marine mammal migrations. Some seabirds have the longest migratory routes of any animal, but these species have been only superficially investigated for their calicivirus profiles. One such investigation from the mid-Pacific (French Frigate Shoals) demonstrated retrospectively that several species of reef fish (small fingerlings) netted to feed a white tern hatchling were infected with caliciviruses (Table II), as shown by cDNA and monoclonal antibody probe. The tern developed blisters on its feet, clear fluid from which was shown by electron microscopy to contain calicivirus. The calicivirus from both the fish and the tern sample were resistant to *in vitro* cultivation, but this study (Poet *et al.*, 1996) demonstrated for the first time that migratory sea birds could acquire caliciviral disease from the ocean.

V. INVERTEBRATE CALICIVIRUS HOSTS

Invertebrates introduce an interesting aspect of viral ecology where large numbers of infectious virus can be scavenged from the water column by molluscan shellfish and then retained for long periods of time. In a sense, this duplicates persistent infections in higher animals where the "carriers" provide a means of

TABLE II

Summary of Calicivirus cDNA Probe Analysis,[a] Monoclonal Antibody Reactions, and Direct
Electron Microscopy of Diverse Samplings of Ocean Sources

Sample and source	cDNA probe	Monoclonal antibody	EM
22 mussels	22/22	22/22	4/4[b]
Gut/gill and residual water from each mussel	62/66	63/66	4/4
19 Hawaiian monk seals[c]	13/19[a]	19/19	5/5[b]
Throat and rectal swabs and WBC buffy coat from each seal	22/57[a]	54/57	5/5
5 fish, shad mackerals[c] and Hawaiian forktail	1/5[a]	3/5	–
1 white tern chick	1/1	–	1/1[d]

[a]cDNA probe analysis of seal and fish samples should be updated. These results were accumulated
prior to improving the cDNA probe methods (compare with the results of mussel assays).

[b]Only a few samples were spot-checked by direct EM.

[c]19 monk seals, 5 fish, and the white tern chick with a blister on its foot were all sampled on
French Frigate Shoals of the Hawaiian chain.

[d]Direct EM of vesicular fluid revealed many antibody-coated caliciviruses, and this could account for the negative Mab test.

storing a virus through those periods of time when susceptible hosts are unavail-
able but then serve as a source of infection for renewed periods of viral transmission, replication, and amplification when compliant hosts once again do become
available.

A. Bivalve Molluscan Shellfish

Some outbreaks of caliciviral-induced vomiting and diarrhea in humans have
been traced directly to contaminated shellfish eaten either raw or lightly steamed
(McDonnell et al., 1997; Stafford et al., 1997). The usual calicivirus group
involved is the Norwalk-like small round structured viruses. Although fecal–oral
transmission is well documented for this group, not all shellfish beds implicated
in such disease are shown to be contaminated with fecal wastes. This raises the
possibility that caliciviruses of this class could originate from nonhuman sources
and be scavenged from the water column by filter feeders or that shellfish may be

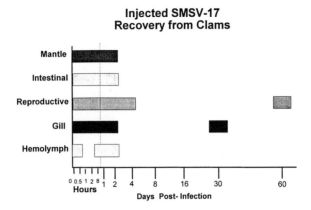

Fig. 23. San Miguel sea lion virus type (SMSV-17) recovery from soft-shelled clams (*Mya arenaria*) experimentally infected by injection. Virus was recovered by *in vitro* isolation in Vero cells (African green monkey kidney) incubated at 37°C. Infected clams were held under conditions of depuration with flow-through sterile seawater.

replicating the virus. Investigative efforts addressing this resulted in a calicivirus (SMSV-17) being isolated *in vitro* from Pacific mussels (*Mytilus californianus*) collected adjacent to a pinniped haul-out area (Poet *et al.*, 1994). Mussels, oysters (*Crassostrea virginica*), and soft-shelled clams (*Mya arenaria*) were then experimentally infected with two calicivirus types and held for 60 days under conditions of depuration with flow-through sterile seawater. Periodic sampling showed that some retained virus for 60 days (Smith *et al.*, 1994) (Fig. 23). The question of viral replication was not resolved. However, this question became somewhat irrelevant based on shellfish retention of tissue culture-infective quantities of virus for at least 60 days.

Other studies demonstrated that some soft-shelled clams in an Oregon estuary were also positive for caliciviruses by direct electron microscopy and cDNA probe. When placed into depuration conditions for 60 days, these clams remained strongly positive for caliciviruses (Fig. 24).

Such findings prompted investigation for calicivirus contamination of various commercially important shellfish beds and species on the Atlantic Coast from Maine into the Gulf of Mexico and the Pacific Coast from California to Alaska. Both open and closed beds involving multiple shellfish samples on both seaboards were found to be positive (see Table III). This has resulted in an unresolved dilemma. Shellfish stocks are readily identifiable that are contaminated with caliciviruses, but the calicivirus pathotypes and their threat to human health cannot be tested with today's existing diagnostic reagents.

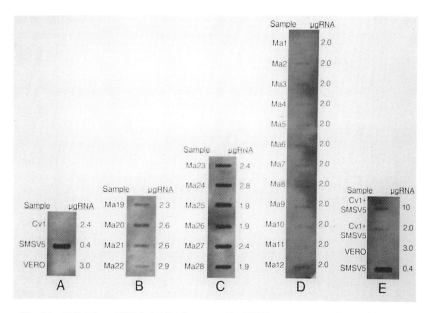

Fig. 24. Calicivirus cDNA hybridization assay. Total RNA was extracted from soft-shelled clam (*Mya arenaria*) tissue samples. Assessment of calicivirus persistence in clam tissue from Alsea Bay, Oregon. Initial collection: 6/22/93. Second RNA extraction from 6/22/93 collection group: 7/20/93. Second collection: 7/20/93. Collected individuals were maintained in pathogen-free aquaria at Hatfield Marine Science Center, Newport, Oregon. The amount of RNA applied to the nylon membrane, in micrograms, is indicated for each sample. The hybridization conditions for each panel were identical. (**A**) Control panel for **B** and **C**. VERO = Vero cell RNA-negative control; SMSV = San Miguel sea lion virus type 5-positive control, purified RNA; Cv1 = eastern oyster whole animal calicivirus-negative RNA pool. (**B**) Ma19–Ma22 = pooled tissue RNA from individual soft-shelled clams collected 7/20/93. (**C**) Ma23–Ma28 = RNA from individual soft-shelled clams collected 6/22/93 and held in pathogen-free aquaria until 7/20/93. (**D**) Ma1–Ma12 = pooled tissue RNA from pools of five individual soft-shelled clams collected 6/22/93. (**E**) Control panel for **D**. SMSV5 = San Miguel sea lion virus type 5-positive control, purified RNA; VERO = Vero cell RNA-negative control; Cv1 + SMSV5 = eastern oyster pooled tissue calicivirus-negative RNA pool with approximately 9,000 (2 μg) or 45,000 (10 μg) $TCID_{50}$ SMSV5 particles added before extraction.

Very little can be said of the ecologic relationships involving the invertebrates. All caliciviruses associated with invertebrates, with the exception of SMSV-17, have defied efforts for *in vitro* propagation. One could speculate that bivalve feeders such as walruses in Alaska may contribute to movements and amplification of these caliciviruses, and that the walrus calicivirus (Smith *et al.*, 1983b) has some attributes that set it apart from most other SMSV types (see §II.D). However, other than the above-mentioned food safety issues, little has been accomplished to define the ecology of vertebrate–calicivirus relationships.

TABLE III

Survey of Atlantic, Gulf, and Pacific Coast Shellfish of Economic Importance for Calicivirus Presence Using *in vitro* Isolation, cDNA Hybridization, and Monoclonal Immunoblot Assays in Both Open and Closed Beds

Species[a]	Location	State	Date collected	Virus isolation	Immuno-blot	Hybrid-ization
Crassostrea gigas	Samish Bay, closed beds	WA	04/27/94	neg	pos	neg
Crassostrea virginica	Galveston Bay, closed beds	TX	08/17/94	neg	pos	neg
Crassostrea virginica	Galveston Bay, closed beds	TX	08/17/94	neg	pos	neg
Crassostrea virginica	Morehead City, Swan Quarter Bay, open beds	NC	08/23/94	neg	pos	neg
Mercinaria mercinaria	Indian River, variable quality beds	FL	08/26/94	neg	pos	pos
Crassostrea virginica	Severin River, closed beds	MD	09/29/94	neg	pos	pos
Mytilus californianus	San Miguel Island, near pinnipeds	CA	10/05/94	neg	pos	pos
Mytilus californianus	San Miguel Island, away from pinnipeds	CA	10/06/95	neg	neg	pos/neg
Crassostrea virginica	Chester River, open beds	MD	10/14/94	neg	neg	neg
Mercinaria mercinaria	St. Martin's River, closed beds	MD	10/26/94	neg	pos	pos
Mercinaria mercinaria	Assaoman Bay, open beds	MD	11/07/94	neg	neg	pos
Mya arenaria	Lamoine Bay, open beds	ME	11/07/94	neg	pos	pos
Mya arenaria	Bar Harbor, closed beds	ME	11/07/94	neg	neg	pos

[a] *Crassostrea gigas* = Pacific oyster; *Crassostrea virginica* = Atlantic or eastern oyster; *Mercinaria mercinaria* = cherry clams; *Mytilus californianus* = Pacific mussel; *Mya arenaria* = soft-shelled clams.

VI. TERRESTRIAL HOSTS OF MARINE CALICIVIRUSES

Although a concept of viruses having their primary cycles in the sea and then spilling into and cycling through terrestrial species does seem paradoxical, the ocean is, in a sense, an effective calicivirus generator, pumping out new and variant forms of the virus that infect and impact numerous terrestrial species, including humans.

A. Swine Epizootics

Beginning in 1932 and ending in 1956, there were repeated epidemics of a foot-and-mouth-like viral disease of swine called vesicular exanthema of swine. Most of these occurrences were confined to California, but in the early 1950s they became pandemic in all of the swine-growing areas of the United States. Eradication measures eliminated the disease, and all 13 serotypes of the virus (VESV) that had been described were eventually classified as "foreign animal disease (FAD) agents" by the U.S. Department of Agriculture. The origins of the disease were unknown, and swine were said to be the only naturally infected host. The virus was reported to not infect humans (Bankowski, 1965; Barlough et al., 1986a,b), although zoonotic potential was suspect (Smith et al., 1978; Smith and Boyt, 1990). The first isolates of San Miguel sea lion virus in 1972 were shown to produce classic clinical vesicular exanthema in swine (Smith et al., 1973).

With that, the mystery of the origins of VESV was solved, and for regulatory purposes the SMSV viruses were all declared to be viruses indistinguishable from VESV. Retrospective studies showed that all the early VESV outbreaks could be accounted for by introduction of the virus into swine from raw garbage. Now with the ecology of caliciviruses better defined, the fish-to-swine cycle that began by feeding garbage containing raw fish is well established (Smith et al., 1980a). In recognition of this, the federal regulations governing swine feeding require that fish or fish scraps associated with the Pacific Coast be cooked, as is required of garbage, before being fed to swine (CFR, 1990). Furthermore, a calicivirus isolate identified as VESV-A$_{48}$, the prototype for the family Caliciviridae, was isolated from a California grey whale 20 years after the original isolation from diseased swine, yet this virus biotype retained full virulence for swine (Smith et al., 1998c).

Although the vesicular disease first seen in swine has not reappeared, and the virus has been officially declared eradicated, it does remain endemic in U.S. swine and has been isolated from domestic herds in California, Minnesota, and Pennsylvania (Smith and Boyt, 1990; Smith et al., 1992) (see Table IV and Fig. 25). One of these isolates is of the walrus calicivirus serotype, another is SMSV-4, and a third, which causes abortion and neonatal death of piglets, is not typeable with

TABLE IV

Calicivirus Isolations from Livestock Since VESV Was Declared an FAD

No. of isolates	Location	Species	Year	Virus type	Associated disease
3	Sonoma County, California	Swine	1976	SMSV-4[a1]	Lameness, brucella
3	Tillamook, Oregon	Cattle	1982	Bovine[a2]	Respiratory
1	Shedd, Oregon	Cattle	1989	SMSV-5[a3]	Diarrhea
4	Sleepy Eye, Minnesota	Swine	1990	MPDLP[a4] 7420[a5]	PRRS
3	Denver, Pennsylvania	Swine	1991	MPDLP[a4] 7420[a5]	PRRS
1	Hawaii	Cattle	1995	New serotype	Diarrhea

[a]Notes on isolation and histories and experimental data:

1. SMSV-4 was first isolated in 1973 from an aborted sea lion. Experimental infection in swine causes vesicular exanthema, which spreads horizontally.

2. Bovine calicivirus is also found in marine mammals. Experimentally, it causes a persistent infection in cattle and classic vesicular exanthema in swine.

3. SMSV-5 was first isolated from fur seals in Alaska. Experimentally, it infects fish, seals, swine (causing vesicular exanthema), monkeys, and causes (by an unknown route) a blistering disease in man.

4. MPDLP is an new serotype of calicivirus that has only been isolated from stillborn and porcine reproductive respiratory syndrome (PRRS) piglets. Experimental oral inoculation of 95-day pregnant sows (3) causes 70% piglet mortality involving premature births, stillborn piglets and neonatal deaths.

5. 7420 calicivirus was first isolated from walruses in the Chukchi Sea in 1978, and has only been isolated from walruses and PRRS piglets. Experimental oral inoculation of 95-day pregnant sows (3) did not cause disease. Oral and intradermal inoculation of weanling piglets caused moderate to marked hepatocellular degeneration (hepatitis), which spread to pen-contact piglets.

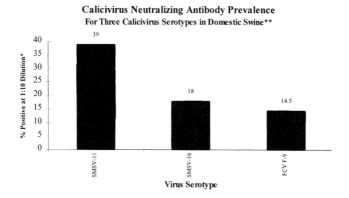

Fig. 25. Two (San Miguel sea lion virus types 11 and 16) out of three caliciviruses testing positive for neutralizing antibody in aborting domestic swine are indistinguishable from the foreign animal disease agent vesicular exanthema of swine virus. The third calicivirus, feline calicivirus, designated FCV F-9, is a vaccine strain of live virus for domestic cats. Presence of neutralizing antibody suggests infectivity occurring in swine, possibly from a live-virus feline vaccine reservoir or similar feline calicivirus. *Positive serum dilutions ranged from 1:10 up to 1:40. **28 sera were drawn from three different swine farms in Minnesota that were experiencing porcine reproductive and respiratory syndrome.

current typing sera. These data support a hypothesis that the marine caliciviruses continue to be introduced into swine, presumably through inadequately heat-treated fish lysate protein or fish meal.

The use of fish protein for animal food supplements may also be involved in transmitting caliciviruses to other terrestrial species.

As an aside, once VESV was introduced into swine, the cycle was greatly amplified and a nationwide epidemic occurred in swine by feeding raw calicivirus-infected pork scraps in garbage to susceptible pigs. This historic event is not unlike the more recent bovine spongiform encephalopathy (BSE) epidemic in Britain resulting from a similar form of cannibalism (Westaway *et al.*, 1995).

B. Cattle: Endemic Infections

One of the differential diagnostic tests used for more than 50 years was to inoculate cattle with unknown viruses that produce vesicles in infected species. If disease occurred in cattle, the virus was either foot-and-mouth virus or vesicular stomatitis virus, but not VESV. Cattle were said to be absolutely resistant to calicivirus disease. In 1982, caliciviruses were isolated from dairy calves with pneumonia in Tillamook, Oregon, and these isolates produced severe vesicular

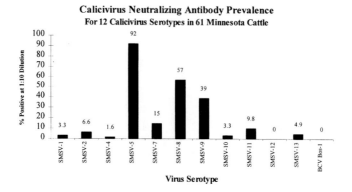

Fig. 26.   San Miguel sea lion virus types SMSV-1 through SMSV-13 are all of ocean (pinniped) origin except SMSV-7, which was first isolated from a fish. BCV Bos-1 was isolated from three calves having respiratory problems and is known to be present in ocean communities and to cause vesicular (blistering) disease in swine.

exanthema in swine but little or no discernable disease in cattle (Smith *et al.*, 1983a). Later, a very virulent virus for seals and swine (SMSV-13) was tested in cattle, where it induced a vesicular disease that spread by contact to other calves (Smith and Boyt, 1990). These findings led to an extensive serologic survey of cattle, and some herds were over 90% positive for certain calicivirus serotypes (Fig. 26). All herds tested showed antibody to one or several of the VESV/SMSV calicivirus types, including VESV-A_{48}, the prototype strain for family Caliciviridae (Figs. 27 and 28). Three different serotypes of calicivirus have been isolated

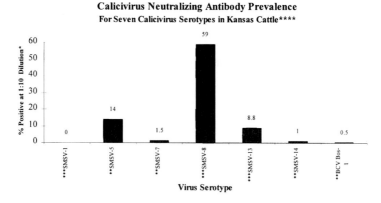

Fig. 27.   *Positive serum dilutions ranged from 1:10 up to 1:1280. **1046 sera tested. ***158 sera tested. ****Test sera drawn randomly from a county-by-county National Animal Health Monitoring System sampling of both dairy and beef cattle.

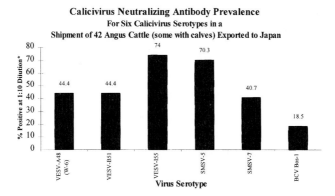

Fig. 28.   Demonstrates potential for viral traffic (caliciviruses) using intercontinental shipping of terrestrial animals having presumably been infected with caliciviruses of ocean origin. *Positive serum dilutions ranged from 1:10 up to 1:110.

from cattle showing diarrhea and/or respiratory disease. There is serologic linkage between abortion and caliciviruses in dairy cattle.

Disease manifestations are poorly defined in cattle, but there is no doubt that domestic cattle in the United States are being repeatedly exposed to a variety of serotypes (Fig. 29), and all of these have been shown to have an oceanic presence.

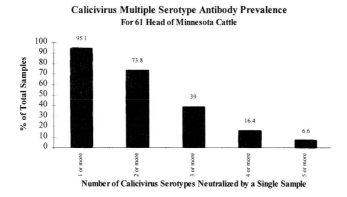

Fig. 29.   In Minnesota, nearly all of a randomly selected sample (61 head) of cattle tested positive for one or more serotypes of calicivirus indistinguishable from the foreign animal disease agent vesicular exanthema of swine virus. Because there is no cross-neutralization of virus types, with the test used this result shows exposure and presumably subsequent infection with multiple caliciviruses of putative "foreign animal disease agent" classification. That so many individual cattle are being exposed and seemingly infected with multiple virus types points to very active viral movements into and among domestic cattle herds.

Again, a host species, the bovine, said not to be susceptible to infection with VESV/SMSV calicivirus, is highly susceptible. Has the virus changed, or have previous observations and conclusions been poorly founded?

C. Miscellaneous Animal Infection

The most recently documented terrestrial host for caliciviruses of the SMSV group is the skunk (Seal et al., 1995). Several other infected species (e.g., the mink, canine, and feline) have potential links to fish protein food supplements. In none of these species does the virus appear to be species specific. The mink virus is readily shown by type-specific serologic tests to have an ocean presence (Smith and Boyt, 1990) (Table IV). Antibodies to the canine calicivirus have been detected in marine mammals; but, in contrast, the feline calicivirus, in addition to infecting dogs, coyotes (Evermann et al., 1984), and all members of family Felidae, also infects adult California sea lions (Table I). As recently as 1996, the feline calicivirus mistakenly was being reported by some to be host specific (Westbury et al., 1994), yet feline calicivirus has a wide host range.

The last terrestrial species to be mentioned will be the donkey, because this species may illustrate yet another aspect of caliciviral spread and survival. An insular population of donkeys was found to have type-specific antibodies to SMSV (Smith and Latham, 1978). This was perplexing, because this herd did not frequent the beaches, nor were they ever observed there. Instead, they grazed the high ground and drank at freshwater springs that were never frequented by pinnipeds. One conclusion could be that the caliciviruses were carried inland as aerosols. The basis for this is that bubbles are very efficient pumps for carrying particulates from the water into the air. A bubble 1 mm in diameter will scavenge particulates from a water column of like diameter as it rises. At the surface, the bubble bursts and surface tension pops a droplet 15 cm into the air, where it becomes airborne on prevailing winds and can carry particulates, in this case caliciviruses, inland, and then deposit them on forages or in waterways. The high concentration of life forms carrying caliciviruses and inhabiting the tidelands combined with surf action along the shoreline make this an especially attractive theoretical mechanism for virus transport along coastal zones.

VII. ZOONOTIC EXPRESSION OF CALICIVIRUSES OF OCEAN ORIGIN

The ease with which caliciviruses of ocean origin could be replicated *in vitro* in primate and human cell lines and cause disease in primates following experimental infection (Smith et al., 1978) gave early warning of probable human health implications for these viruses.

A. Primate Caliciviral Diseases

A single serotype of calicivirus designated primate calicivirus or PCV Pan-1 was isolated from five species of primates in a zoologic collection in San Diego, California. Phylogenetically, these were from Old and New World primate groups (Smith *et al.*, 1983c, 1985a,b).

The prototype strain for PCV Pan-1 was isolated from a herpesvirus-like lesion on the lip of a pygmy chimpanzee (*Pan paniscus*) and re-isolated 6 months later from a throat swab taken from the same animal. A second isolate came from the spleen of a lowland gorilla (*Gorilla gorilla*) who had died of a generalized fungal infection. A third isolate was recovered from a gingival swab form a spider monkey (*Ateles fusciceps*) suffering from severe gingivitis. The fourth isolate was recovered from the throat of a normal silver leaf langur (*Presbytis cristata*), and the fifth isolate was recovered from the brain of a baby douc langur (*Pygathrix nemaeus*) dying with encephalitis (Smith *et al.*, 1983c, 1985a,b).

These findings are remarkable for several reasons. They represent the first isolation of calicivirus from any naturally infected primate. In three of the five cases, clinical disease was associated with the tissues where calicivirus was recovered, and a persistent infection of at least 6 months duration was established and demonstrated in one species (Smith *et al.*, 1983b). Finally, retrospective serology of California sea lions demonstrated an ocean presence for PCV Pan-1.

This account demonstrates much the same lesson on calicivirus ecology that comes from studying the RCVs (see §IV.A). Once caliciviruses of ocean origin are introduced into new species, they very often adapt and move through the new host populations as endemic and/or epidemic disease agents. A second lesson that should not be lost based on these findings is that caliciviruses of ocean origin would be expected to cause disease in humans.

B. Human Caliciviral Disease

The Caliciviridae are divided into five tentative genera (Berke *et al.*, 1997). Three of these — the Sapporo virus, Norwalk, and hepatitis E — are considered primary human pathogens with no other important reservoirs recognized except possibly for hepatitis E, which may also occur in swine (Meng *et al.*, 1997; Clayson *et al.*, 1996). A fourth member, rabbit hemorrhagic disease virus, may or may not cause disease in humans (Smith *et al.*, 1998a). The fifth member, the VESV/SMSV group, was for over 50 years dogmatically said not to infect people. We now know that it does (Smith *et al.*, 1998b).

The caliciviruses of ocean origin will cause a blistering disease in humans (Fig. 30; see color insert), but infections more often probably go undetected. The seroprevalence in humans of antibodies that react to these viruses is about 20% (Smith *et al.*, 1998a,b). In other species, the same or similar serotypes that have infected people can cause an array of diseases, including abortion, pneumonia, diarrhea, hepatitis, myocarditis, and encephalitis. Theoretically, instances of these conditions, including abortion (Table V), which go without etiologic diagnosis among human populations, could be the result of calicivirus infection (Smith *et al.*, 1998b). The necessary reagents are simply not available to test such a hypothesis.

The modes of enzootic calicivirus transmission into humans remain obscure but could involve eating poorly cooked fish, seaside activities, and contact with infected animals. In any event, the avenues whereby viruses from the sea can spill

TABLE V

Human Antibody to Calicivirus Antigen[a]

Sample[b]	Dilu-tion	Species	Antigen reactivity pool[c]					
			A	B	C	D	E	F
OHS-1	1:5	Human	++	+	+++	+++	+++	+++
OHS-2	1:5	Human	++	+	+++	+++	+++	+++
OHS-3	1:5	Human	+	+++	+	+	+/−	++
OHS-4	1:5	Human	+/−	+	+	+	+/−	+
OHS-5	1:5	Human	+/−	+/−	+/−	+	+	+
OHS-6	1:5	Human	+	+	+	+	+	+
OHS-7	1:5	Human	+	+	+	+	+	+
OHS-8	1:5	Human	+++	++	+++	+++	+++	+++
OHS-9	1:5	Human	++	+/−	++	+++	+	+++
OHS-10	1:5	Human	+++	++	+++	+++	+++	+++

[a]Preliminary test: spontaneous abortion sera tested against various calicivirus antigens using an immunoblot technique.

[b]Known negative and positive human serum is not available, nor is the prevalence of such antibodies in the normal population using this test. The very strong positives (sera 1, 2, 8, 9, 10, and 3B) as opposed to the minimal reactivity (sera 4, 5, 6 and 7) suggests a working test but fails to prove abortogenic association in this small sampling. When these antigen pools were run against their respective typing sera using this assay, all tested strongly positive (3+ at a 1:200 dilution of typing sera).

[c]Calicivirus antigen pools (whole virus): A contains SMSV-1, SMSV-2, SMSV-3, SMSV-4; B contains SMSV-5, SMSV-6, SMSV-7, SMSV-8; C contains SMSV-9, SMSV-10, SMSV-11, SMSV-12; D contains SMSV-13, SMSV-14, SMSV-15, SMSV-16; E contains cetacean, bovine, reptile, and walrus; F contains all of the above 20 serotypes.

over into humans and other land species appear to be open and functioning quite well. For the first time, documented human disease has been caused by a zoonotic virus contained in primary ocean reservoirs. This is a paradigm shift (Smith *et al.*, 1998b).

VIII. CONCLUSIONS

The ecologic picture of the caliciviruses that have an ocean presence shows them to be viruses of unusual adaptation that have colonized an unexpectedly broad range of host species. The hallmark of these agents has been the ease with which a single pathotype can move into a new host species and then spread horizontally through that species. In all instances where detailed studies have been possible, caliciviruses having ocean reservoirs have been seen as potent pathogens, and, although asymptomatic infections are not uncommon, the virus does carry cytopathic virulence factors that can result in disease manifestation at some point during the host life cycle. Additional virulence factors are hinted at by the perinatal hemorrhagic syndrome seen in pinnipeds, the purpura hemorrhagica seen in baby pigs, and the tissue trophisms that have resulted in abortion, hepatitis, pneumonia, gastroenteritis, skin blistering, myocarditis, joint pain, and encephalitis.

This chapter has reviewed the survival mechanisms and movements of ocean caliciviruses through diverse ecosystems and hosts. They have been shown to impact animal health while also receiving a high biological security classification as a foreign animal disease (FAD) virus. These agents move through the environment in water and food, so that concerns for water quality and the public health aspects of food safety arise. The depleted or declining marine mammal populations will not be fully understood without first defining the impact of endemic and epidemic caliciviral disease on these populations, and this has not been done. Public health connection has now been well documented with the report of emerging zoonotic infections and human disease, but the overall threat in humans remains poorly defined due to a lack of data (Smith *et al.*, 1998a,b).

The caliciviruses of ocean origin provide a classic model of RNA virus quasispecies movements played out across the global reaches of our oceans, our land masses, and the air we breathe.

ACKNOWLEDGMENTS

Much of this information has been compiled through personnel and funding support provided by the U.S. Air Force, Office of Naval Research, Naval Bureau of Medicine and Surgery, Naval Ocean Systems Center, San Diego Zoologic Society, U.S. Department of Agriculture, Agriculture Research Service and Animal Plant Health Inspection Service, the Oregon State University Agriculture Experiment Station and College of Veterinary Medicine, the National Marine Fisheries Service, the Federal Drug Administration through an Oregon Sea Grant, the Center for Pediatric Research at the East Virginia Medical School, and the Alaska Department of Fish and Game. Most pinniped samples were collected and processed in cooperation with Drs. Robert DeLong, William Gilmartin, Bud Antonellis, Tom Laughlin, Francis Fay, Enid Goodwin, and Robin Brown. Most cetacean samples were collected and processed in cooperation with Drs. H. M. S. Watkins, Tom Albert, Murray Daily, and Sam Ridgeway. Drs. Jeffrey Barlough, Peter Boyt, Neylan Vedros, and Steven Poet were especially helpful in the areas of immunology and assay development. Dr. David Matson collaborated on virus characterization and taxonomy. Henry Bray, Catherine Prato, and Arthur Latham provided field and laboratory support. Douglas Skilling was central to all aspects of the work, including collection, virus isolation and characterization, and all technical assays, including electron microscopy. Christine Robinette provided technical, editorial, and production assistance.

REFERENCES

Bankowski, R. A. (1965). Vesicular exanthema. *Adv. Vet. Sci.* **10**, 23–64.

Barlough, J. E., Berry, E. S., Skillling, D. E., and Smith, A. W. (1986a). The marine calicivirus story, part I. *Compend. Cont. Ed. Prac. Vet.* **8**, F75–F82.

Barlough, J. E., Berry, E. S., Skilling, D. E., and Smith, A. W. (1986b). The marine calicivirus story, part II. *Compend. Cont. Ed. Prac. Vet.* **8**, F5–F14.

Barlough, J. E., Berry, E. S., Goodwin, E. A., Brown, R. F., DeLong, R. L., and Smith, A. W. (1987). Antibodies to marine caliciviruses in the Steller sea lion (*Eumetopias jubatus* Schreber). *J. Wildl. Dis.* **23**(1), 34–44.

Barlough, J. E., Berry, E. S., Skilling, D. E., and Smith, A. W. (1988). Prevalence and distribution of serum neutralizing antibodies to San Miguel sea lion virus types 6 and 7 in selected populations of marine mammals. *Dis. Aquat. Org.* **5**, 75–80.

Barlough, J. E., Matson, D. O., Skilling, D. E., Berke, T., Berry, E. S., Brown, R. F., and Smith, A. W. (1998). Isolation of reptilian calicivirus Crotalus Type I from feral pinnipeds. *J. Wildl. Dis.* **34**(3), 451–456.

Berke, T., Golding, B., Jiang, X., Cubitt, D. W., Wolfaardt, M., Smith, A. W., and Matson, D. O. (1997). Phylogenetic analysis of the Caliciviruses. *J. Med. Virol.* **52**(4), 419–424.

Caulkins, D. G. (1986). "Morphometric Comparisons of Steller Sea Lions Prior to and During a Population Decline." Preliminary Data, Northern Sea Lion Workshop, National Marine Mammal Laboratory, National Marine Fisheries Service, National Oceanic and Atmospheric Administration, Seattle, WA, 8–11 December.

CFR (Code of Federal Regulations) (1990). Subchapter K: Swine health protection. Part 166.2: General Restrictions. Fish exempt from cooking includes, fish from the Atlantic Ocean within 200 miles of the Continental United States or Canada or fish from inland waters of the United States or Canada which do not flow into the Pacific Ocean. 9 CFR, chap. 1, 1-1-90 ed.

Clarke, I. N., and Lambden, P. R. (1997). The molecular biology of caliciviruses. *J. Gen. Virol.* **78** (pt. 2), 291–301.

Clayson, E. T., Snitbhan, R., Ngarmpochana, M., Vaughn, D. W., and Shrestha, M. P. (1996). Evidence that the hepatitis E virus (HEV) is a zoonotic virus: Detection of natural infections among swine, rats, and chickens in an area endemic for human disease. *In* "Enterically Transmitted Hepatitis Viruses" (Y. Buisson, P. Coursaget, and M. Kane, eds.), pp. 329–335. La Simarre, Joue-les Tours, France.

Cooper, P. D., Angol, V. I., Bachrach, H. L., Brown, F., Ghendon, V., Gibbs, A. J., Gillespie, J. H., Holm, K., Mandel, B., Melnick, J. L., Mohanty, S. B., Porey, R. C., Rueckert, R. R., Schaffer, F. L., and Tyrell, D. A. (1978). Picornaviridae: Second report. *Intervirolgy* **10**, 165.

Evermann, J. F., McKieman, A. J., Smith, A. W., Skilling, D. E., Ott, R. L. (1984). Isolation and identification of caliciviruses from dogs with enteric infections. *Am. J. Vet. Res.* **46**(1), 218–220.

Jiang, X., Cubitt, D. W., Berke, T., Zhong, W., Dai, X., Nakata, S., Pickering, L. K., and Matson, D. O. (1997). Sapporo-like human caliciviruses are genetically and antigenically diverse. *Arch. Virol.* **142**, 1813–1827.

Genetic and antigenic diversity of human caliciviruses (HuCVs) using RT-PCR and new EISs. *Arch. Virol. Suppl.* **12**, 251–262.

Kapikian, A. Z., and Chanock, R. M. (1990). Norwalk group of viruses. *In* "Fields Virology" (B. N. Fields, D. M. Knipe, R. M. Chanock, M. S. Hirsch, J. L. Melnick, T. P. Monath, and B. Roizman, eds.), Vol. 1, pp. 671–693. Raven Press, New York.

McDonnell, S., Kirkland, K. B., *et al.* (1997). Failure of cooking to prevent shellfish-associated viral gastroenteritis. *Arch. Intern. Med.* **157**(1), 111–116.

Meng, X.-J., Purcell, R. H., Halbur, P. G.,, Lehman, J. R., Webb, D. M., Tsareva, T. S., Haynes, J. S., Thacker, B. J., and Emerson, S. U. (1997). A novel virus in swine is closely related to the human hepatitis E virus. *Proc. Natl. Acad. Sci. U.S.A.* **94**, 9860–9865.

O'Hara, T. M., House, C., House, J. A., Suydam, R. S., and George, J. C. (1998). Viral serologic survey of bowhead whales in Alaska. *J. Wildl. Dis.* **34**(1), 39–46.

Poet, S. E., Skilling, D. E., DeLong, R. L., and Smith, A. W. (1994). Detection and isolation of a calicivirus from a bivalve mollusk (*Mytilus californianus*) collected from rocks adjacent to pinniped rookeries. *In* "Proceedings of the 25th Annual Meeting of the International Association of Aquatic Animal Medicine, Vallejo, CA."

Poet, S. E., Skilling, D. E., Megyesi, J. L., Gilmartin W. G., and Smith, A. W. (1996). Detection of a non-cultivatable calicivirus from the white tern. *J. Wildl. Dis.* **32**(3), 461–467.

Seal B. S., Lutze-Wallace, C., Kreutz, L. C., Sapp, T., Dulac, G. C., and O'Neill, J. D. (1995). Isolation of caliciviruses from skunks that are antigenically and genotypically related to San Miguel sea lion virus. *Virus Res.* **37**(1), 1–12.

Smith, A. W. (1975). "Characteristics and Marine Reservoirs of Caliciviruses Currently Indistinguishable from Vesicular Exanthema of Swine Virus." PhD Dissertation, University of California, Berkeley.

Smith, A. W. (1978). Marine caliciviruses. *In* "Proceedings of the Ninth Annual Conference and Workshop, International Association for Aquatic Animal Medicine, San Diego, California."

Smith, A. W. (1986). "Report on Infectious Disease, Northern Sea Lion Workshop." National Marine Mammal Laboratory, National Marine Fisheries Service, National Oceanic and Atmospheric Administration, Seattle, WA, 8-11, December.

Smith, A. W., and Boyt, P. M. (1990). Caliciviruses of ocean origin: A review. *J. Zool. Wild. Med.* **12**, 2–23.

Smith, A. W., and Latham, A. B. (1978). Prevalence of vesicular exanthema of swine antibodies among feral mammals associated with the southern California coastal zones. *Am. J. Vet. Res.* **39**(2), 291–296.

Smith, A. W., Akers, T. G., Madiu, S. H., and Vedros, N. A. (1973). San Miguel sea lion virus isolation, preliminary characterization and relationship to vesicular exanthema of swine virus. *Nature* **244**(5411), 108–110.

Smith, A. W., Prato, C. M., and Skilling, D. E. (1977). Characterization of two new serotypes of San Miguel sea lion virus. *Intervirology* **8**(1), 30–36.

Smith, A. W., Prato, C. M., and Skilling, D. E. (1978). Caliciviruses infecting monkeys and possibly man. *Am. J. Vet. Res.* **39**, 287–289.

Smith, A. W., Akers, T. G., Latham, A. B., Skilling, D. E., and Bray, H. L. (1979). A new calicivirus isolated from a marine mammal. *Arch. Virol.* **61**, 255–259.

Smith, A. W., Skilling, D. E., Dardiri, A. H., and Latham, A. B. (1980a). Calicivirus pathogenic for swine: A new serotype from Opaleye *Girella nigricans*, an ocean fish. *Science* **209**(4459), 940–941.

Smith, A. W., Skilling, D. E., and Brown, R. J. (1980b). Preliminary investigation of a possible lung worm (*Parafilaroides decorus*), fish (*Girella nigricans*), and marine mammal (*Callorhinus ursine*) cycle for San Miguel sea lion virus type 5. *Am. J. Vet. Res.* **41**(11), 1846–1850.

Smith, A. W., Skilling, D. E., Prato, C. M., and Bray, H. L. (1981a). Calicivirus (SMSV-5) infection in experimentally inoculated opaleye fish (*Girella nigricans*). *Arch. Virol.* **67**, 165–168.

Smith, A. W., Skilling, D. E., and Latham, A. B. (1981b). Isolation and identification of five new serotypes of calicivirus from marine mammals. *Am. J. Vet. Res.* **42**(4), 693–694.

Smith, A. W., Skilling, D. E., and Benirschke, K. (1981c). Investigations of the serum antibodies and viruses of the bowhead whale, Balaena mysticetus. *Tissue Structural Studies, and Other Investigations on the Biology of Endangered Whales in the Beaufort Sea, U.S. Department of the Interior, Bureau of Land Management, Alaska OCS Office, Anchorage, Alaska* **1**, 233–254.

Smith, A. W., Mattson, D. E., Skilling, D. E., and Schmitz, J. A. (1983a). Isolation and partial characterization of a calicivirus from calves. *Am. J. Vet. Res.* **44**(5), 851–855.

Smith, A. W., Ritter, G. C., Ray, B., Skilling, D. E., and Wartzok, D. (1983b). A new calicivirus isolate from walrus feces (*Odobenus rosmarus*). *J. Wildl. Dis.* **19**, 86–89.

Smith, A. W., Skilling, D. E., Ensley, P. K., Lester, T. L., and Benirschke, K. (1983c). Calicivirus isolation and persistence in a Pygmy chimpanzee (*Pan paniscus*). *Science* **221**, 79–81.

Smith, A. W., Skilling, D. E., and Benirschke, K. (1985a). Calicivirus isolated from three species of primates: An incidental finding. *JAVMA* **46**, 2197–2199.

Smith, A. W., Skilling, D. E., Anderson, M. P., and Benirschke, K. (1985b). Isolation of primate calicivirus *Pan paniscus* Type I from a douc langur (*Pygathrix nemaeus* Linne). *J. Wildl. Dis.* **21**, 426–428.

Smith, A. W., Anderson, M. P., and Benirschke, K. (1986). First isolation of calicivirus from reptiles and amphibians. *Am. J. Vet. Res.* **47**(8), 1718–21.

Smith, A. W., Skilling, D. E., Benirschke, K., and Albert, Y. F. (1987). Serology and virology of the bowhead whale (*Balaena mysticetus* L.). *J. Wildl. Dis.* **23**(1), 92–8.

Smith, A. W., Lassen, D. E., Hedstrom, O., and Skilling, D. E. (1988). Observed hepatitis in swine infected experimentally with calicivirus of ocean origin. "Procedings of the Ninth Annual Food Animal Disease Research Conference, Moscow, Idaho."

Smith, A. W., Skilling, D. E., Applegate, G. L., Trayor, T. R., Lola, T. J., and Poet, S. E. (1992). Calicivirus isolates from stillborn piglets in two outbreaks of swine infertility and respiratory syndrome (SIRS). *In* "Proceedings of the 12th International Symposium of the World Association of Veterinary Microbiologists, Immunologists, and Specialists in Infectious Disease, School of Veterinary Medicine, University of California at Davis."

Smith, A. W., Reno, P., Poet, S. E., Skilling, D. E., and Stafford, C. (1994). Retention of an ocean-origin calicivirus in bivalve mollusks maintained under experimental depuration conditions. *In* "Proceedings of the 25th Annual Meeting of the International Association of Aquatic Animal Medicine, Vallejo, CA."

Smith, A. W., Skilling, D. E., Cherry, N., Mead, J. H., and Matson, D. O. (1998a). Calicivirus emergence from ocean reservoirs: Zoonotic and interspecies movements. *Emerg. Infect. Dis.* **4**(1), 13–20.

Smith, A. W., Berry, E. S., Skilling, D. E., Barlough, J. E., Poet, S. E., Berke, T., Mead, J., and Matson, D. O. (1998b). In vitro isolation and characterization of a calicivirus causing a vesicular disease of the hands and feet. *Clin. Infect. Dis.* **26**(2), 43–49.

Smith, A. W., Skilling, D. E., and House, J. A. (1998c). Re-isolation and reclassification of a whale enterovirus as a calicivirus serotype VESV A_{487}. Unpublished data from the Laboratory for Calicivirus Studies, Oregon State University, Corvallis, and the Plum Island National Animal Disease Laboratory, Plum Island, New York.

Stafford, R., Strain, D., *et al.* (1997). An outbreak of Norwalk virus gastroenteritis following consumption of oysters. *Commun. Dis. Intell.* **21**(21), 317–320.

Watkins, H. M. S., Worthington, G. R. L., and Latham, A. B. (1969). Isolation of enterovirus from the California gray whale (*Eschrictius gibbosus*). *Bacteriol. Proc. Am. Soc. Microbiol.* **69**, 180.

Westaway, B., Carlson, G. A., and Pruisner, S. B. (1995). On safari with PrP: Prion disease of animals. *Trends Microbiol.* **3**, 141–147.

Westbury, H., Lenghaus, C., and Munro, R. (1994). A review of the scientific literature relating to RHD. *In* "Rabbit Haemorrhagic Disease: Issues in Assessment for Biological Control" (R. K. Munro and R. T. Williams, eds.), pp. 91–103. Bureau of Resource Sciences, Australian Government Publishing Services, Canberra.

York, A. (1986). "A Statistical Analysis of Reproduction." Northern Sea Lion Workshop, National Marine Mammal Laboratory, National Marine Fisheries Service, National Oceanic and Atmospheric Administration, Seattle, WA, 8–11, December.

Xu, Z. J., and Chen, W.-X. (1989). Viral haemorrhagic disease in rabbits: A review. *Vet. Res. Commun.* **13**(3), 205–212.

13

Macroecology and Microecology of Viruses of Terrestrial Mammals

CHARLES H. CALISHER* and FRANK J. FENNER[†]

*Arthropod-Borne and Infectious Diseases Laboratory
Colorado State University
Fort Collins, Colorado 80523

[†]The John Curtin School of Medical Research
The Australian National University
Canberra, ACT 0200, Australia

> Viruses may be mistletoe on the Tree of Life.
>
> *Duncan J. McGeoch*

I. INTRODUCTION

The remarkable diversity of viruses is apparent not only from their sizes, shapes, physical and chemical characteristics, and gene sequences, but by their hosts, by the diseases they cause, and by their extraordinary ability to adapt to novel

493

surroundings. Because viruses are transmitted by various means, their genes must contain messages providing for proficiency even under external pressures, such as drying, temperature contrasts, pH and other chemical disparities, antibodies, and host cell metabolic divergences. For the most part, they appear to be able to adapt to all these potential vicissitudes. Were these adaptations not so usual for viruses, they would be considered extraordinary. Given that all viruses have such compensatory attributes, they are not at all extraordinary. In this chapter, we will focus on the ecologies and concomitant adaptations of a few viruses; we use the term "microecology" to denote the ecology of a virus within a particular host animal and the term "macroecology" to describe its ecology within populations of animal hosts, which may include different species of vertebrates and arthropods as well. Constraints of space demand some unavoidable superficiality in regard to the details outlined.

II. TAXONOMY OF VIRUSES OF TERRESTRIAL VERTEBRATES

As of 1995, 51 virus families and an additional 25 "floating" genera (no family yet established) were recognized by the International Committee for Taxonomy of Viruses (ICTV, 1995). Of these, 22 families and two "floating" genera include viruses that infect terrestrial mammals. These are (with nucleic acid type and strandedness and at least a single example of each provided) as follows: single-stranded DNA: Circoviridae (chicken anemia virus), Parvoviridae (canine parvovirus); double-stranded DNA: Poxviridae (myxoma virus), Hepadnaviridae (hepatitis B virus), Herpesviridae (human herpesvirus 1), Papovaviridae (murine polyomavirus), Adenoviridae (human adenovirus 2), *African swine fever-like viruses* (African swine fever virus); single-stranded RNA: Picornaviridae (foot-and-mouth disease virus), Flaviviridae (yellow fever virus), Togaviridae (western equine encephalitis virus), Bunyaviridae (La Crosse virus, Crimean–Congo hemorrhagic fever virus, Rift Valley fever virus), Rhabdoviridae (rabies virus), Orthomyxoviridae (influenza A virus), Paramyxoviridae (mumps virus), Arenaviridae (lymphocytic choriomeningitis virus), Retroviridae (human immunodeficiency 1 virus), Filoviridae (Ebola virus), Coronaviridae (avian infectious bronchitis virus), Caliciviridae (vesicular exanthema of swine virus, rabbit hemorrhagic disease virus), Astroviridae (human astrovirus 1), Arteriviridae (equine arteritis virus); double-stranded RNA: Birnaviridae (infectious bursal disease virus), Reoviridae (simian rotavirus SA11, bluetongue viruses).

III. RELATION OF TAXONOMIC PLACEMENT AND VIRAL ECOLOGY

It seems probable that every species of organism, from bacterium to mammal, is host to at least one species of virus, and those that are best studied (*Escherichia coli*, tobacco plants, mice, humans) to many more than one virus, several of which viruses are species specific. All mammals tested for viruses have been shown to be infected by at least one. This is not to say that all mammals are suffused with viruses, but it is mentioned merely to indicate that all mammals are susceptible to virus infections and that many individuals of all species of mammals are infected at any given time. Most viruses do no serious harm; many infections, including those caused by viruses that under other circumstances may cause dramatic illnesses in their vertebrate hosts, are subclinical or cause only minor illnesses or transient episodes. Thus, the apparent virus infections, the dramatic or exotic ones, are the tip of the proverbial iceberg. Obviously, there is not sufficient space in this chapter to list and describe the ecology of all viruses of terrestrial mammals. Instead, we will focus on a few selected viruses, using their natural cycles as representative of certain mechanisms, and trust these will be instructive.

Placement of viruses in taxa is based on (a) whether the genome is DNA or RNA, (b) the strandedness of this nucleic acid, (c) the morphologic features, (d) possession of an envelope, and many other, less important features. Whether coincidentally or not, this scheme roughly parallels the ecology of viruses and can be considered in relation to their evolutions.

A. General Principles of Transmission: Gene Expression

All cells of a blue-eyed human contain genes for blue-eyedness, yet only the pupils of that person manifest the coloration. This coloration is a consequence of the genetic expression of a set of genes that together produce proteins that give the pupils their blue color. Why are toes, fingers, knees, and other parts of the body not blue when the cells that compose them have the same genes as do the eyes? The answer lies in the polypeptides expressed by the nucleic acids composing the genes. Under certain circumstances (states) gene expression is permitted, and under other circumstances it is not. The conditions under which these occur are poorly understood, but it is clear that an understanding of these conditions would be fundamental to an overall understanding of the expression of genes, including viral genes associated with pathogenicity and virulence.

One must attribute a virus–host relationship to characteristics of both the virus and the host. The pathogenicity of a virus must be assumed to be a consequence of superimposition of virus genetic information on host susceptibility. In isolation, that is, in a glass vial, by definition a virus is not and cannot be pathogenic. It is only when it infects a host that the effects of a virus are manifested. Ultimately, differences between viruses can be attributed to the genetics of the virus and its gene sequences, in concert with the genes of the host.

Elegant studies of pathogenicity and pathophysiology, such as that done by Monath *et al.* (1981) with yellow fever virus in monkeys, begin to answer the questions of "what" and "how." To understand the microecology (pathophysiology) of a virus within a macroecosystem, some small indications might be obtained from the organ, tissue, and even specific cell predilection of a virus. Some signs and symptoms (e.g., fever, rash, hepatitis, headache, and encephalitis) occur in many viral and nonviral infections. Others are relatively specific for a particular infection or disease. For example, edema of the supraorbital fossa is characteristic of African horse sickness. Similarly, corneal edema in canine adenovirus infection, distended parotid glands in mumps virus infection, inflammation of the coronal band of the equid hoof in vesicular stomatitis, Koplik's spots in measles, and other signs and symptoms with other viral infections are at least suggestive of, if not pathognomonic for, those illnesses. The relative specificity of such features suggests a particular adaptation of virus and host. Indeed, were we to be able to grasp the pathogenetic significance of these distinctive indications of particular infections, we might have a basis on which to understand the pathogenetic mechanisms of many virus infections. At present, we have no such insights. In the future, pathognomonic signs may be helpful to this end as we accumulate more information about the interactions of viral genes and host cells.

B. Viruses with Mammals as Primary or Secondary Hosts

Mutation rates are the proportion of misincorporated nucleotides that occur during nucleic acid synthesis, and they can be expressed as the number of substitutions per nucleotide per round of template copying. Mutation frequency is defined as the proportion of mutants in a given population and is expressed as substitutions per nucleotide (Domingo and Holland, 1994). It is impossible here to delve deeply into the details of such mechanisms, but they have been covered elsewhere (e.g., Gibbs *et al.*, 1995).

The rate of generation of virus mutants is dependent on a number of factors, including type of nucleic acid (DNA, RNA, single-stranded, double-stranded), Darwinian pressures brought to bear with regard to fitness, and chance. Double-stranded DNA viruses have an advantage over RNA viruses with respect to stability. Because they cannot replicate without two complementary nucleic acid strands and because such physical complementarity is the *sine qua non* of completeness, small differences (errors) in replication do not lead to poor integrity or inaccurate translation; they lead to no replication at all. This mechanism is not possessed by single-stranded RNA viruses, which have no method of correcting such errors, that is, lack a mechanism to ensure fidelity of replication. When a small error (nucleotide substitution, nucleotide insertion, nucleotide deletion) occurs during the replication cycle of a single-stranded RNA virus, the error is reproduced. Whether that error yields progeny that is more or less fit than the parent virus is another matter, one that is dependent on other factors, such as infectivity for the host cell, ability to replicate, and ability to elude the host immune system. However, the bottom line is that heterogeneous viral populations (quasi- species) are the result of erroneous replication.

Table I presents a selected list of viruses, showing their modes of transmission and their microecologies (their terrestrial mammal hosts). Of course, many more examples could be given for each diverse category of transmission, for each virus, and for each host.

1. Example 1: A hantavirus

Not only Sin Nombre virus, but all hantaviruses ostensibly are transmitted between rodents, and many are transmitted from rodents to humans by inhalation of aerosols of infected rodent urine, feces, or saliva. Nonetheless, different hantaviruses cause different diseases in their human hosts (*vide infra*). No doubt this difference is related to the genetics of the virus, not the genetics of the human host, virus virulence not being a characteristic of the virus but of the virus–host interaction.

2. Example 2: A paramyxovirus

A recently discovered paramyxovirus was isolated from tissues collected post-mortem from a human who had died in acute respiratory distress. Infection with this virus apparently was contracted from infectious saliva of a sick horse that had also died in acute respiratory distress caused by this virus. Further studies demonstrated that bats of the genus *Pteropus* are a, if not the, vertebrate host of this virus (Young *et al.*, 1996). The mechanism of transmission of virus between these hosts is not yet clear.

3. Example 3: Other paramyxoviruses

Paramyxoviruses in the genus *Morbillivirus* include measles virus of humans as well as canine distemper, dolphin distemper, peste-des-petits-ruminants, phocine (seal) distemper, porpoise distemper, and rinderpest (of ruminants in Africa and Asia) viruses. In spite of the apparent divergence in host specificity of these closely related viruses, only a single serotype of canine distemper virus has been recognized, yet the host range of this virus includes not only dogs but other canines (wolf, fox, dingo, coyote, and jackal), raccoons and pandas, and weasels, ferrets, mink, skunk, badgers, martens, and otters.

TABLE I

A Few Examples of Modes of Transmission of Certain Viruses and Their Usual (Natural) Terrestrial Mammal Hosts

Ostensible mode of transmission	Virus	Usual terrestrial mammal host
Aerosols or urine feces, saliva, nasal discharge	Sin Nombre	rodents
	pseudorabies	pigs
	bovine virus diarrhea	cattle
	equine arteritis	equids
	influenza A	human
	swine influenza	pigs
	equine influenza	equids
	equine morbillivirus	bats
	canine distemper	dogs
	lymphocytic choriomeningitis	rodents
	foot-and-mouth disease	cattle
	vesicular exanthema	pigs
	rabbit hemorrhagic disease	rabbits
	rabies	dog
	encephalomyocarditis	rodents
	Nipah	pigs
Direct contact	rabies	wild animals
	canine parvovirus	dogs
	foot-and-mouth disease	cattle
	human papillomaviruses	humans
	deer papillomavirus	deer
	European elk papillomavirus	elk
	elephant papillomavirus	elephants
	reindeer	reindeer
	rabbit oral papillomavirus	rabbits
	canine oral papillomavirus	dogs
	rhesus monkey	rhesus monkeys
	sheep	sheep
	bovine papillomaviruses	cattle
	equine papillomavirus	equids
Fecal–oral	rotaviruses	humans
	adenoviruses	humans
Venereal	human immunodeficiency 1	human
	herpesviruses (many)	(many)
Mechanical	myxoma	rabbits
Arthropod-borne	St. Louis encephalitis	rodents?[a]
	West Nile	unknown[a]
	equine infectious anemia	equids
	western equine encephalitis	none[a]
	(bluetongue viruses)	none[b]
	Colorado tick fever	rodents
	Congo–Crimean hemorrhagic fever	rodents
	dengue (four viruses)	humans

[a]Natural cycle involves birds.

[b]Restricted to biting flies as reservoir host.

In rare and unfortunate instances, measles virus can persist in the brains of infected, then clinically recovered, humans, only to later cause subacute sclerosing panencephalitis; this appears to be due to slow continuing virus replication. Similarly, canine distemper virus has been known to cause "old dog encephalitis" in dogs that have recovered from distemper. It is assumed that the mechanisms of these late-onset illnesses are similar, suggesting similar pathogenetic mechanisms in addition to the similar virologic characteristics that have caused them to be placed in the same taxon. This seems an excellent example of a peculiar viral microecology.

4. Example 4: Arenaviruses

As with hantaviruses, transmission of viruses of the family Arenaviridae is between susceptible mammals, usually rodents, by fighting, or saliva exchange, or by random contact of a susceptible rodent with infected body fluids from an infected rodent. One arenavirus, lymphocytic choriomeningitis virus, has a novel means of persistence and transmission. The virus causes a persistent and tolerated infection in the host, usually a wild or laboratory-reared mouse. The virus is transmitted *in utero* (vertically), and the host generally appears normal throughout its life, although some transitory metabolic changes are noted. Infected mice have persistent viremia and viruria, but circulating antibody cannot be detected. Nonetheless, these animals are not completely immunologically tolerant. They have circulating virus–IgG antibody–complement complexes that are infectious. Inbred but not wild mice manifest signs of illness later in life, principally due to the effect of immune complexes on renal glomeruli. Disease caused by this virus may provide a model for study of Machupo, Junin, Guanarito, and other arenaviruses that also cause inapparent infections in their rodent hosts.

5. Example 5: Rabies-related viruses

Rabies, foot-and-mouth disease, various parvoviruses, and many other viruses also are transmitted either by direct contact of infected body fluids or exudates of infected animals with mucus membranes of susceptible mammals. Rabies virus can be transmitted from an infected animal that appears "furious" or "dumb" by the infected host either biting (puncturing) the skin or other parts, or licking, and thereby introducing saliva containing virus.

An interesting aspect of rabies virus is that there are many viruses related to it genetically and antigenically. Not only rabies virus but Duvenhage (bat), European bat 1, European bat 2, Lagos bat, and a recently identified virus from *Pteropus* sp. bats in Australia as well as Mokola (from *Crocidura* sp. = shrew) in Nigeria, kotonkon (from *Culicoides* sp. = midge) in Nigeria, obodhiang (from *Mansonia uniformis* = mosquito) in Sudan, and possibly Rochambeau (from *Coqillettidia albicosta* = mosquito) are genetically and antigenically related to standard ("street") rabies virus. Further, elegant molecular and monoclonal antibody studies of isolates of rabies virus from throughout the world have shown that what once appeared to be a rather simple situation is in reality quite complex (Smith *et al.*, 1986). When a virus shares a majority of its gene sequences with another virus, how does one distinguish between the two, and is there an epidemiologic significance to such a difference? This depends on the definition of "majority." Is 60% sufficient? 80%? 95%? 99%? If a virus is essentially identical genetically to rabies virus and causes a rabies-like illness in humans or other animals, is it not rabies virus or a variant thereof? The further apart two viruses are in terms of their gene sequences, the less inclined one would be to call them by the same name. Mokola virus could not be confused virologically with rabies virus in terms of macroecology (Africa only), microecology (limited in its mammalian host range), genetics, or antigenicity, yet Mokola virus causes a rabies-like illness in humans. The term "rabies-related" was coined to comprise these viruses. Whether the microecologic differences evident between these viruses is a reflection of a series of divergent evolutionary events or simply a dispersal into various ecosystems as a result of such divergence remains for discussion and study. Nonetheless, in this instance, the clinical term "rabies" cannot be used to identify the etiologic agent.

It is now generally accepted that one should evaluate the polypeptide differences (amino acid substitutions) that result from nucleic acid substitutions. That is, some nucleic acid substitutions do not result in amino acid changes, so that they do not specify polypeptides different from the unchanged sequence. In sum, even if the nucleic acid sequences of two viruses differ significantly (any definition of "significant"), if those differences do not specify meaningful protein changes, the genetic differences are considered biologically minimized. Likewise, if the nucleic acid sequences of two viruses do not diverge in terms of apparent differences but those small differences are translated to biologically meaningful proteins, the genetic differences are considered important.

In the few examples we have presented, sundry hosts, strategies of transmission, and remarkably diverse environments not only contribute to but provide unique macroecologic essentials for continued replication of virus lineages in particular ecologic niches, prerequisites for perpetuation of a virus species (Van Regenmortel *et al.*, 1991).

6. Example 6: Two respiratory viruses

That various viruses classified as influenzaviruses (family Orthomyxoviridae, genus *Influenza virus A*) can be transmitted between humans and can cause serious illness is well known. Influenza B virus is a specifically human infection. However, the more common influenza A viruses are primarily asymptomatic infections of aquatic birds, but may infect other birds (chickens, turkeys) and certain mammal species (humans, swine, equids), in which they may cause serious disease. In birds, these viruses are transmitted by the fecal–oral route, and in mammals by the respiratory route. The recent experience in Hong Kong with influenzavirus A H5N1 in chickens is a relevant example of the difficulties such viruses may have in crossing species barriers and the seriousness with which the human community takes such a threat (ProMED-mail, 1997–1998).

Although the first hantaviruses (family Bunyaviridae, genus *Hantavirus*) to be discovered were associated with hemorrhagic fever with renal syndrome, the recently discovered hantaviruses of the Americas are the etiologic agents of pulmonary syndromes. Under natural conditions, hantaviruses infect but cause no apparent illness in their principal rodent hosts, indicating a long-term evolutionary or coevolutionary relationship between them. When hantaviruses are transmitted to humans by excreta from infected rodents, humans can become seriously ill, suggesting that the hantavirus–human association is an unnatural one and a recent occurrence.

a. *Macroecology and Microecology of Influenza A Virus.* Influenzaviruses are unusual among animal viruses in that their genome consists of several pieces of RNA. This leads to two different kinds of genetic changes that cause what are called "antigenic drift" and "antigenic shift." Sequential replication of an influenza A virus in the lungs of millions of people yields not only immense populations of progeny but an enormous number of mutants, produced by chance errors of replication. Under the pressure of host immune systems, most of these mutants will be eliminated from the virus population. However, certain quasispecies will have found an appropriate niche in which they can elude host defenses. By this process of genetic (and antigenic) drift, these influenzavirus populations may then outcompete and supersede populations that hitherto had predominated and become the Virus of the Year. Another process that occurs at intervals of decades can cause pandemics, that is, worldwide epidemics. This is believed to happen when pigs are co-infected with the circulating human virus and a bird virus, usually from a duck. Reassortment of the RNA segments occurs, and some of the reassortants are able to infect humans who tend pigs. If such viruses are able to replicate in humans, they encounter a wholly susceptible population and quickly spread around the world. The World Health Organization and its collaborating international centers follow the changes of antigenic drift and advise on strains to be used for vaccines each year. They also try to recognize new viruses produced by reassortment (antigenic shift) so as to produce vaccines to protect against potential pandemic strains.

b. *Macroecology and Microecology of Sin Nombre Virus.* Five genera have been described within the family Bunyaviridae: *Bunyavirus, Nairovirus, Phlebovirus, Tospovirus,* and *Hantavirus*. Viruses in the first four genera are transmitted by arthropods; hantaviruses are not. Most of the bunyaviruses are mosquito-borne, most of the nairoviruses are tick-borne, most of the phleboviruses are phlebotomine fly-borne, and all of the tospoviruses, which are plant viruses, are transmitted by thrips. Thus, not only have viruses of this family evolved unique patterns regarding their reservoir hosts, they clearly are transmitted by disparate vectors.

Depending on how one looks at the hantaviruses, they are either the most or least primitive of the viruses of this family in that they have either never had a requirement for arthropod transmission or have lost this requirement. In any event, they are transmitted from vertebrate to vertebrate by direct contact or by contact with saliva, urine, or feces from an infected animal.

Hantaviruses use rodents as their natural reservoirs. The hantaviruses are found wherever there are rodents, essentially worldwide. Hantaan virus, the prototype of this antigenic group and genus causes hemorrhagic fever with renal syndrome (also known as epidemic hemorrhagic fever and Korean hemorrhagic fever) in Asia and parts of Europe. Puumala virus causes a milder form of this disease, known as "nephropathia epidemica," in northern Europe and parts of Asia. Many other members of the genus have been isolated from various genera and species of mice, rats, voles, shrew, and bandicoot. One fascinating feature of these viruses is the close association of each with a specific rodent species (Table II). This is not to say that they do not infect rodents of species different from the principal host, but only that most infections have been associated with a single rodent species per virus.

In 1993, an outbreak of adult respiratory distress syndrome was recognized in humans of the southwestern United States in the area where New Mexico, Colorado, Utah, and Arizona meet. It came as a considerable surprise when laboratory testing indicated that the etiologic agent, Sin Nombre virus, was a hantavirus, because all previous hantavirus infections of humans had renal involvement as a primary manifestation. Subsequent elegant molecular and antigenic analyses showed that Sin Nombre was one that had not been recognized and was being transmitted from its principal rodent host, the deer mouse *Peromyscus maniculatus* to humans (Nichol *et al.*, 1993).

Patients with hantavirus pulmonary syndrome may have abrupt onset of acute respiratory distress associated with interstitial pulmonary edema due to infection of capillary endothelium. In this way, the pathophysiology of hantavirus pulmonary syndrome may be similar to that of hemorrhagic fever with renal syndrome, caused by Hantaan, Puumala, and Seoul viruses, in which the analogous principally effected organ is the kidney.

Ongoing studies on the prevalence of antibody to Sin Nombre virus and the mechanisms of trans-seasonal maintenance are providing information about the endemicity and epidemiology of the virus as well as important information about the dynamics of the rodent host populations. Apparently, Sin Nombre virus is

TABLE II

Viruses of the Genus *Hantavirus*, Family Bunyaviridae, Listed by Geographic Distribution of the Rodent Virus and Whether Associated with Human Illness

Virus	Rodent host	Geographic distribution	Disease
Hantaan	*Apodemus agrarius*	Asia, Europe	HFRS[b]
Seoul	*Rattus norvegicus, R. rattus*	Asia, Europe, the Americas	HFRS
Thai-749	*Bandicota indica*	Asia	unknown
Dobrava	*A. flavicollis*	Europe, Middle East	HFRS
Puumala	*Clethrionomys glareolus*	Europe, Asia	HFRS/NE[c]
Prospect Hill	*Microtus pennsylvanicus*	North America	unknown
Tula	*M. arvalis*	Europe	unknown
Khabarovsk	*M. fortis*	Asia	unknown
Isla Vista	*M. californicus*	North America	unknown
Bloodland Lake	*M. ochrogaster*	North America	unknown
El Moro Canyon	*Reithrodontomys megalotis*	North America	unknown
Rio Segundo	*R. mexicanus*	Mexico, Central and South America	unknown
Black Creek Canal	*Sigmodon hispidus*	United States	HPS
Muleshoe	*S. hispidus*	United States	unknown
Bayou	*Oryzomys palustris*	United States	HPS
Sin Nombre	*Peromyscus maniculatus*	North America	HPS[d]
New York	*P. leucopus*	North America	HPS
Thottapalayam	*Suncus murinus*	Asia	unknown
Rio Mamore	*Oligoryzomys microtis*	South America	unknown
Andes	*Oligoryzomys longicaudatus*	South America	HPS
Topografov	*Lemmus sibiricus*	Russia, Asia, Alaska, Canada	unknown
Lechiguanas	*Oligoryzomys flavescens*	Argentina	HPS
Juquituba	unknown	Brazil	HPS
Laguna Negra	*Calomys laucha*	Paraguay	HPS
Pergamino	*Akodon azarae*	Argentina	unknown
Maciel	*Bolomys obscurus*	Argentina	unknown
Oscar	?	Argentina	unknown
Cano Delgadito	*Sigmodon alstoni*	Venezuela	unknown

transmitted between rodents and may persist in them for life. Long-lived and persistently infected animals provide ample opportunity for the virus to persist through periods conducive to low rates of transmission, such as severe cold, excessive heat, drought, flooding, and, as a result of any of these environmental conditions, low rodent population densities. When such populations are high, intraspecific aggressive behavior, prompted by the need to acquire and retain a mate, protect territory, or defend food sources, may result in transmission of virus through biting or scratching; a greater proportion of males than females are infected with Sin Nombre and with at least one other hantavirus. Thus, microecologically, Sin Nombre virus may have two mechanisms of transmission: a trans-seasonal maintenance mechanism and an epizootic amplification mechanism. In either situation, the virus appears to be in equilibrium with the rodent reservoir host, and infection of humans is only an accident of nature, one that is of no benefit to the virus.

C. Viruses with Arthropod Vectors as Hosts: Delivery Systems

Although some viruses may be transmitted mechanically by arthropods, others, called arboviruses (*ar*thropod-*bo*rne *viruses*), replicate in and are transmitted by arthropods that have fed on a virus-infected host. This is a remarkable capacity in that the virus must microecologically adapt to the widely variant ecology, biological condition, metabolism, and enzymes of each host, sometimes a human and a tick, a rodent and a culicoid fly, a kangaroo and a mosquito, or any of a large number of other exotic pairings.

The biological transmission cycle of arboviruses begins and ends with an infected arthropod. This ensues when an uninfected, susceptible, and competent hematophagous arthropod takes a blood meal from a viremic vertebrate. Ingested virus then replicates in the arthropod, in the posterior midgut and, if there is no "mesenteronal escape barrier," is disseminated elsewhere (Kramer *et al.*, 1981); a mesenteronal escape barrier exists for epizootic, but not enzootic strains of Central American Venezuelan equine encephalitis virus (Weaver *et al.*, 1984), indicating how specific and how complicated these mechanisms are.

Ludwig *et al.* (1989) contended that, because arboviruses are exposed to two notably different environments, vertebrate cells and arthropod cells, enzyme processing might play a role in the capacity of the virus to adjust to such disparate conditions. They then showed that exposure of a bunyavirus to proteolytic enzymes, such as those in the mosquito midgut, increased virus affinity for mosquito cells. The enzymes they used removed the G1 glycoprotein but left intact the G2 glycoprotein. They concluded that processing of the viral glycoproteins in the mosquito midgut "may be necessary to expose attachment proteins on the virion surface before attachment to, and infection of, midgut cells can occur," and suggested that this model may answer questions about the molecular basis for midgut infection barriers and susceptibility to arbovirus infections.

Irrespective of the mechanism by which virus reaches the mosquito hemocoel, it is then transported in the hemolymph or along nerves to various organs and tissues, including the salivary gland, where it replicates. Then, if there is no "salivary gland escape barrier," within a few days of itself becoming infected the arthropod can transmit virus to a vertebrate host on which the arthropod subsequently feeds. In areas ecologically unusual for a particular virus, or during drought or other conditions incompatible with long-term persistence of an arbovirus, an atypical arthropod species may be employed. However, such arthropods may be incompetent transmitters and only peripherally involved in viral transmission. Tabachnick (1991a) has made an excellent case for continuing evolutionary studies of arboviruses and their vectors, suggesting that such studies may provide methods for understanding vector–host interactions.

Mosquitoes, fleas, biting flies, houseflies, true bugs of many species, and many other insects have been shown to carry virus on their mouth parts from an infected to an uninfected vertebrate host. Such mechanical transmission of some viruses (e.g., parvoviruses, encephalomyocarditis virus, and polioviruses) is more an anomaly than a concern. However, mechanical transmission by arthropods is of crucial importance for myxoma and fowlpox viruses, in which the mouth parts become contaminated during probing through skin lesions, and equine infectious anemia virus, in which stable flies and tabanids contaminate their mouth parts during blood feeds.

1. Macroecology and Microecology of African Swine Fever Virus (no family placement; genus "African swine fever-like viruses")

The sylvatic cycle of African swine fever virus in warthogs (*Phacochoerus athiopicus*) contrasts with the domestic cycle in swine. In both instances, soft ticks of the genus *Ornithodoros* (*moubata* in warthog burrows and domestic pigpens in sub-Saharan Africa and *erraticus* in the pigpens of Portugal and southern Spain), live in the burrows of warthogs and serve as biological hosts and vectors. The virus is transmitted trans-stadially, trans-ovarially, and sexually (venereally) in these ticks. The infected warthog remains asymptomatic, but European wild pigs and domestic swine herds are devastated. Subsequent transmission between pigs can be by contact with infected aerosols, by infection from the bite of an infected tick, by transmission through fomites, mechanically by biting flies in which the virus does not replicate, or by feeding on scraps of infected meat. "Virulent" and "less virulent" isolates are recognized.

2. Macroecology and Microecology of Western Equine Encephalitis Virus (family Togaviridae, genus Alphavirus)

During an extensive epizootic of equine encephalitis in the San Joaquin Valley of California in 1930, Meyer and colleagues isolated western equine encephalitis (WEE) virus from the brain of a horse (Meyer *et al.*, 1931). The next summer, the disease reappeared in epizootic form in the San Joaquin Valley and spread to other parts of California. In the summer of 1932, the disease again appeared, extending to most of the state and most of the other Western states. In each of those years, the summer outbreaks disappeared with the onset of cool weather, and viremias were notably high, leading to suspicion that the virus was arthropod-borne. Association of a similar illness in humans who were in contact with sick horses gave further impetus to studies of the virus.

WEE virus subsequently was shown to replicate in and be transmitted by various mosquitoes, including naturally infected *Culex tarsalis*, and to infect birds and mammals in the wild. In the usual cycle, WEE virus is transmitted between passerine birds by *Cx. tarsalis* mosquitoes and only rarely infects horses and humans, infections of the latter being a consequence of enormous populations of mosquitoes with high virus prevalence. Both of these mammals are dead-end hosts for WEE virus, not mounting sufficient viremia to serve as a source of virus for subsequently feeding mosquitoes.

As noted, the basic transmission cycle involves *Cx. tarsalis* as the primary vector mosquito species and house finches (*Carpodacus mexicanus*) and house sparrows (*Passer domesticus*) as the primary amplifying hosts. Secondary amplifying hosts, on which *Cx. tarsalis* frequently feed, include other passerine species, chickens, and possibly pheasants in areas where they are abundant. Another transmission cycle that most likely is initiated from the *Cx. tarsalis*–wild bird cycle involves *Aedes melanimon* and the blacktail jackrabbit *Lepus californicus*. Humans, horses, squirrels, and a few other wild mammal species are involved tangentially in the WEE virus cycle but do not contribute significantly to virus amplification (Hardy, 1987).

Not only is better irrigation management leading to better management of mosquito populations, a correlation between increased use of air conditioning and television during peak mosquito activity periods and decreased attack rates of WEE and other arboviruses has been noted in California (Gahlinger *et al.*, 1986). Apparently, these modern appliances promote remaining indoors during summer evenings, the period when infected *Cx. tarsalis* mosquitoes feed and usually transmit these viruses. It is possible that expanded education programs may change the macroecosystem of WEE virus by eliciting other changes in human behavioral patterns and may help protect humans from infection with this virus. This would complement existing vector control programs, which in any event are intended to alter the macroecosystem.

In addition to these characteristics of the transmission cycle, the epidemiology, in effect the macroecology, of WEE virus is distinct in terms of geographic distribution. The virus has been isolated only in the Western Hemisphere, from Argentina to Western Canada, and only in ecologic areas favoring a sympatric and competent mosquito vector. Alphaviruses closely related to WEE virus are known. Indeed, a genetic and antigenic complex of WEE viruses has been recognized, most of which are transmitted by mosquitoes, but at least two, Fort Morgan and Buggy Creek viruses, are transmitted by bugs (*Oeciacus vicarius*) that parasitize nestling birds. However, wherever WEE virus has been found it is transmitted between birds by bird-feeding mosquitoes. Birds likely spread progenitor WEE viruses north and south and from continent to continent, probably accounting for its widespread geographic distribution and its antigenic and genetic subtypes and variants. Throughout its range, WEE virus is an equine pathogen, but human disease in South America occurs less frequently and with less severity than in North America, perhaps an indication of modest genetic differences that translate to substantial epidemiologic differences.

Isolates of WEE virus have been obtained from naturally infected snakes, frogs, and tortoises (Gebhardt and Hill, 1960; Burton *et al.*, 1966; Sudia *et al.*, 1975), and experimentally infected garter snakes yielded viremias that persisted through hibernation and were of sufficient titer to infect subsequently feeding mosquitoes (Thomas and Eklund, 1960). Further efforts to extend these findings in field and laboratory studies have had mixed results, and studies to determine the possibility that WEE virus overwinters in poikilotherms have not provided convincing evidence (Reeves, 1990).

WEE virus is an RNA virus, so genetic variation is common and increased opportunities exist for selection of more fit genotypes. Comparisons of WEE complex viruses from many areas of the world indicate parallels between geographic distribution and antigenic relatedness. Viruses of the WEE complex with lesser antigenic differences may develop in discrete ecologic conditions (Calisher *et al.*, 1988). Under such circumstances it is not surprising to find genotypic, and therefore phenotypic, variations. What is surprising is that there are not more recognized variations. This must be due to selective pressures brought to bear by both ecologic circumstances and characteristics of the natural cycles of these viruses.

Hahn and coworkers (1988) sequenced the 3′-terminal 4288 nucleotides of the RNA of WEE virus. The sequences of the capsid protein and of the (untranslated) 3′-terminal 80 nucleotides of WEE were shown to be closely related to the corresponding sequences of a New World alphavirus, eastern equine encephalitis virus, whereas the sequences of glycoproteins E2 and E1 of WEE were determined to be more closely related to those of an Old World alphavirus, Sindbis virus. From their determinations and the results of elegant comparisons of molecular sequences of several other alphavirus genomes, they concluded that alphaviruses have descended from a common ancestor by divergent evolution. They deduced that WEE virus is a recombinant of a predecessor of Sindbis virus and a predecessor of EEE virus and that WEE virus has the encephalogenic properties of EEE virus and the antigenic specificity of Sindbis virus.

3. Macroecology and Microecology of Yellow Fever, Dengue, and St. Louis Encephalitis Viruses (family Flaviviridae, genus Flavivirus)

The more than 60 recognized flaviviruses comprise an ecologically disparate set of viruses having similar replicative mechanisms and similar responses to host immunity. The best known is yellow fever virus, which continues to cause periodic outbreaks and epidemics of yellow fever in Africa and South America. The so-called jungle cycle of this virus involves monkey-to-monkey mosquito transmission. The urban cycle involves human-to-human mosquito transmission. In remote areas of the rain forest where monkeys live, the virus cycles without human notice. Unimmunized humans entering the forest at such a time may become infected and carry the virus back to urban areas, where they then become ill. The relatively long incubation period for this virus puts it at an epidemiologic advantage in that a truly ill person would not be able to travel but one who was simply viremic and asymptomatic could lead a normal life until becoming ill. Nonetheless, it does not appear to be evolutionarily advantageous to yellow fever virus to expand its host range from monkeys to humans, because humans often respond with vaccine programs.

Other flaviviruses are zoonotic and cycle between wild animals and only rarely infect humans, or are not know to infect them. Many of these viruses are known only from single or a few isolations from rodents, bats, and other small mammals and have not been isolated from arthropods. Flaviviruses that cause disease in humans and are arthropod-borne are transmitted by mosquitoes (which transmit yellow fever, dengue, and dengue hemorrhagic fever [dengue 1, dengue 2, dengue 3, dengue 4], Murray Valley encephalitis, Japanese encephalitis, St. Louis encephalitis, and West Nile viruses) or ticks (Kyasanur Forest disease, louping ill, Omsk hemorrhagic fever, Powassan, tick-borne encephalitis [Central European encephalitis and Russian Spring-Summer encephalitis]).

RNA sequencing of many of these viruses has revealed large areas of conserved sequences, suggesting a common origin. Although different parts of the flaviviral genomes have different functions, specifying different polypeptide products, such as enzymes, in general, there is a correlation between flavivirus genome motif, location (structure), and function. When the conserved RNA sequence portion of a flavivirus specifies a protein, that protein is common to all other flaviviruses. Should such a protein serve as an antigen and participate in antigen–antibody reactions, we observe what is known as group specificity. When a nonconserved

RNA sequence portion specifies a protein, that protein likely is type specific, that is, is not shared by other flaviviruses; therefore, genotype is reflected in phenotype. Thus, superinfection by the same flavivirus is not known to occur, but sequential infections with flaviviruses do occur, and may have serious clinical consequences.

It is thought that the basis for the pathogenetic mechanism causing dengue hemorrhagic fever/dengue shock syndrome is based on the occurrence of sequential infections with different dengue viruses, which may cause uncomplicated dengue fever in primary infections and then shock syndrome in subsequent infections.

Halstead (1988) has shown that antibody-dependent enhancement of flavivirus replication occurs in cells with Fc receptors. It is thought that increased absorption of virions to mammalian host cell membranes is mediated by antibody present in quantities too low to neutralize. This has been named "antibody-dependent enhancement," has been shown also to occur in infections with yellow fever and Japanese encephalitis viruses (Gould and Buckley, 1989), and may provide an explanation for the pathogenesis of dengue hemorrhagic fever/dengue shock syndrome and neurovirulence caused by other flaviviruses.

The flavivirus St. Louis encephalitis (SLE) virus was discovered in the United States but is now known to occur, at least focally but usually over large areas, from Canada to Argentina wherever grain, corn, and other foods can be obtained by the usual vertebrate host, birds. There are genetic and concomitant antigenic and virulence differences between strains, but most have been isolated from mosquitoes or birds; a few isolates have been recovered from tissues of rodents. Microecologically, SLE viruses differ with respect to the species of mosquito vector and therefore differ with respect to geographic distribution. In Florida, SLE virus is found in a *Culex nigripalpus*–bird cycle; in the Ohio–Mississippi River basin in the *Culex pipiens*– and *Culex quinquefasciatus*–bird cycles; and in the Western United States and Canada in a *Culex tarsalis*–bird cycle. Molecular correlates of these differing natural cycles have been documented (Trent *et al.*, 1980). These data may provide evidence for maintenance (trans-seasonal transmission) of SLE viruses in local reservoirs. Indeed, SLE virus has been isolated more than a month after infection, suggesting the occurrence of a mechanism for SLE virus persistence.

4. Macroecology and Microecology of Bluetongue Viruses (family Reoviridae, genus Orbivirus)

More than 250 viruses have been placed in the family Reoviridae, 150 of which are considered members of the genus *Orbivirus*. Among the orbiviruses are African horse sickness, bluetongue, epizootic hemorrhagic disease, equine encephalosis, Orungo, and others, some of these being livestock or human pathogens of considerable economic and aesthetic importance. Basically, orbiviruses have 10 RNA genome segments and are transmitted by arthropods of one sort or another, including culicoid midges, phlebotomine flies, ticks, and mosquitoes. Orbiviruses have been isolated essentially worldwide.

Having multiple RNA segments, each one apparently responsible for production of a different protein, each protein playing a role in virus structure or replication, segment reassortment among orbiviruses in dually infected mammalian hosts and arthropod vectors and nucleotide substitution, insertions, and deletions assures the generation of variants with altered capacity to infect, altered pathogenicity, and altered antigenicity. The latter guarantees complex immunologic relationships among these viruses as well as confounding attempts to produce effective vaccines against them.

5. Macroecology and Microecology of Crimean–Congo Hemorrhagic Fever Virus (family Bunyaviridae, genus Nairovirus)

The virus family Bunyaviridae is comprised of more than 300 viruses, making it the largest family of vertebrate viruses. Viruses within each genus have in common the different 3′ and 5′ termini of each of their three RNA species. However, viruses of different genera have different 3′ and different 5′ termini (sequences of 3′ termini of bunyaviruses, hantaviruses, nairoviruses, and phleboviruses are shown in Table III). These characteristics serve as the basis for genus placement within the family Bunyaviridae, but, ultimately, they reflect divergent evolutionary events, a result of chance mutations and adaptation to defined hosts by selection of the winners.

Among viruses of the genus *Bunyavirus* are LaCrosse and Akabane viruses, the etiologic agents of important diseases of humans and livestock, respectively, and at least 165 others; these are transmitted by mosquitoes or culicoid flies. The genus *Phlebovirus* includes Rift Valley fever and the sandfly fever viruses, etiologic agents of human, livestock, and wild animal diseases of considerable importance, as well as 50 others; these are transmitted by biting flies (the phleboviruses) and ticks (the uukuviruses). Thirty-four viruses comprise the genus *Nairovirus*, to which belong Congo–Crimean hemorrhagic fever and Nairobi sheep disease viruses, the etiologic agents of diseases of livestock and of humans; these are transmitted by ticks. Tomato spotted wilt virus, a plant pathogen, and three other viruses, comprise the genus *Tospovirus*; these are transmitted by thrips. Another genus of the family is *Hantavirus*, comprising 28 viruses. Consequent to the great numbers of viruses, their wide geographic distributions, their capacity to replicate in a wide variety of hosts, and their remarkable genomic plasticity, the family Bunyaviridae must be considered one of the most remarkable of all RNA virus taxa.

TABLE III

3′ Nucleotide Sequences of L, M, and S RNA Segments of Representative Viruses of Genera within the Family Bunyaviridae

Genus	Virus	RNA segment	3′ terminus[a]
Bunyavirus	LaCrosse	S	3′ UCAUCACAUGAGGUG
		M	3′ UCAUCACAUGAUGGU
		L	3′ UCAUCACAUGAGGAU
Hantavirus	Hantaan	S	3′ AUCAUCAUCUGAGGG
		M	3′ AUCAUCAUCUGAGGC
		L	3′ AUCAUCAUCUGAGGG
Nairovirus	Qalyub	S	3′ AGAGAUUCUGCCUGC
		M	3′ AGAGAUUCUUUAUGA
		L	3′ AGAGAUUCUUUAAUU
Phlebovirus	Rift valley fever	S	3′ UGUGUUUCGG
		M	3′ UGUGUUUCUGCCACGU
		L	3′ UGUGUUUCUG

[a]Sequences identical on all three genome segments are underlined.

An example of the extraordinary capacity of these viruses is Congo–Crimean hemorrhagic fever (C–CHF) virus. Isolated in 1956 from a human with fever, headache, nausea, vomiting, backache, joint pains, and photophobia in the Democratic Republic of Congo (the former Zaire), the virus has been recovered from humans, livestock, and many species of ticks. In 1967 it was shown that the etiologic agent of Crimean hemorrhagic fever was a virus named after that disease in Asia. When it was realized that the two viruses (Congo virus and Crimean hemorrhagic fever virus) were the same, the names and data were merged. C–CHF virus is now recognized to occur from Africa to Asia, including portions of southern Europe. C–CHF virus may occur focally, with human and livestock disease unrecognized, or may be recognized as a persistent zoonosis with periodic outbreaks.

The principal natural vertebrate hosts of C–CHF virus are hares and hedgehogs (hosts for the immature stages of the tick vectors) and livestock (hosts for the adult stages of the tick vectors). Birds may serve as hosts for some of these tick species and may provide a means for transportation of C–CHF virus over wide geographic areas.

Human disease caused by C–CHF virus occurs in rural areas, usually related to infections of cattle infested with virus-infected ticks. Human infections usually occur in livestock handlers slaughtering cattle and bitten by infected ticks, but secondary human infections occur when hospital personnel handle virus-contaminated blood from hemorrhaging patients. Thus, the ecology of this virus involves a variety of vertebrate and invertebrate hosts in defined but focal situations.

6. Macroecology and Microecology of Myxoma Virus (family Poxviridae, genus Leporipoxvirus)

Myxoma virus is one of the few viruses (others are rabbit fibroma and rabbit papilloma, fowlpox, and equine infectious anemia virus) that depend on mechanical transmission by arthropods for its spread from one animal to another. In its natural hosts in South America (*Sylvilagus brasiliensis*) and California (*S. bachmani*), it is a benign infection, producing localized fibromas through which mosquitoes (and other biting arthropods) can contaminate their mouth parts with virus. By biological accident, the only other susceptible animal is the European rabbit (*Oryctolagus cuniculus*), in which the virus from *Sylvilagus* causes a very lethal generalized disease (Fenner and Ross, 1994). Transmission from one European rabbit to another depends on transfer by biting insects that probe through the skin lesions and then bite another rabbit. Myxomatosis was deliberately introduced into wild European rabbits in Australia in 1950 and spread quickly over the continent, causing enormous mortality. Highly and moderately lethal viruses, but not highly attenuated strains, produced enough virus in the superficial layers of the skin over the lesions to contaminate the mouth parts of probing mosquitoes and fleas. Because mosquitoes (the major vector in Australia) became rare over the winter, viruses that were infective for mosquitoes but killed slowly were selected, and variants that killed only 90% of rabbits soon appeared and came to dominate the picture. The 10% of survivors were those rabbits that were genetically more resistant, and the strong selection pressure soon led to most rabbits becoming highly resistant. However, the moderately virulent virus behaved in rabbits like attenuated viruses in unselected rabbits, and it was not transmitted. Instead, new mutants were selected that produced transmissible lesions in resistant rabbits but were highly virulent for unselected rabbits, a classical example of coevolution driven by the need to produce transmissible lesions in the skins of rabbits, whether these were unselected or highly resistant.

D. Viruses with No Known Vector and No Known Reservoir (They Must Live Somewhere)

One of the fascinating enigmas of the last 20 years has been the lack of identification of a natural reservoir for the Ebola and Marburg viruses (family

Filoviridae, genus *Filovirus*). In 1967, simultaneous outbreaks of a fatal hemorrhagic fever among laboratory workers in Germany and Yugoslavia led to isolation of Marburg virus; of 31 cases, 7 died. These workers had been processing kidneys from African green monkeys (*Cercopithecus aethiops*) to make cell cultures. The virus was isolated from both the patients' and monkey tissues. Many of the monkeys had originated in Uganda, and many of them died. However, subsequent serologic studies of nonhuman primates in Uganda did not reveal the presence of antibody to Marburg virus, suggesting that they were not the vertebrate reservoir of the virus. Other studies provided no information useful in identifying such a reservoir.

In the 1970s, Ebola virus caused outbreaks of a fatal hemorrhagic disease in the Democratic Republic of Congo and the Sudan. Later, outbreaks in the United States in monkeys imported from Southeast Asia and a human infection in Ivory Coast allowed isolation of other Ebola-like viruses.

Although many able investigators with considerable resources have attempted to determine the reservoir of these viruses, all such efforts, including inoculation of plants, have failed (Swanepoel *et al.*, 1996).

There is no doubt that Marburg and Ebola viruses are pathogenic for monkeys, yet monkey virulence is considered the prime reason to not consider these animals natural reservoirs because a host that dies after infection with a virus is not likely to be a suitable reservoir. Until an elegant piece of work is done, a remarkably insightful hypothesis is presented, or a fortuitous convergence of circumstances occur, the macroecology of the filoviruses remains unknown.

Until recently, another lethal virus of European rabbits, rabbit hemorrhagic disease virus (family Caliciviridae, genus *Calicivirus*) fell into this category. It appeared in domestic rabbits shortly after they had been imported from Germany into China in 1984. It was then spread from China to Europe, Mexico, and many other countries via contaminated rabbit meat, and caused consternation among rabbit breeders (OIE, 1991). In commercial rabbitries, between which there has always been extensive trading of live rabbits, it appeared to be spread by contact or by the fecal–oral route. Unlike myxoma virus, which originated in a different genus of leporids on another continent, rabbit hemorrhagic disease virus appears to have arisen by mutation from a previously unrecognized calicivirus that caused an inapparent infection among European rabbits in Europe (Capucci *et al.*, 1996). Because rabbits were Australia's most serious agricultural and environmental pest, rabbit hemorrhagic disease virus was released in Australia in 1996 for rabbit control. The virus spread widely and caused very heavy mortalities in many areas. Among Australian wild rabbits, it appears to be spread mainly by flies of various species that contaminate their mouth parts when feeding on the internal organs of dead rabbits that have been ripped open by scavenging mammals or birds. Under suitable meteorological conditions, such contaminated flies can be carried for hundreds of kilometers, setting up new foci of infection. Speculatively, if this is indeed the principal mechanism of maintenance of the virus in Australia, selection may favor highly virulent strains that kill rabbits when there are high concentrations of virus in their livers. The rabbits are unlikely to be eradicated because immature rabbits are physiologically more resistant than mature rabbits, so some will survive to breed, but they will not have been selected for genetic resistance.

IV. EFFECTS OF VIRUSES ON VERTEBRATE AND ARTHROPOD POPULATIONS

If viruses have evolved as "renegade genes" in certain cells, it would be no mystery as to why they are so well adapted to those cells and to the general milieu in which those cells are found, their microecology. If viruses arose extracellularly, or at least not from the cells in which they are found, and have simply adapted as parasites of those cells, again it is no wonder they are found there. Whether by parallel evolution or coevolution in which, respectively, the virus and host happen to evolve together or evolve interdependently, the fact of virus evolution is clear.

Pathogen evolution likely has been influenced by the internal (and by inference the external) environment of the vector, certainly the arthropod vector in the case of arboviruses. Whether obvious or not, mammalian biology must be influenced by the presence of a virus, whether as a pathogen or nonpathogen (e.g., mammalian host behavior, body temperature, antibody production, or for vaccine use).

Arthropod hosts may be preferred sites for arboviruses to evolve. Persistent infection is common in competent arthropods, and such persistence is conducive to the accumulation of intramolecular genomic alterations and the accumulation of spontaneous mutations (Beaty et al., 1997). It would seem likely that the longer a virus–host relationship occurs, the less incompatible the relationship.

Most virus infections of mammals lead to subclinical infections. This seems to be the case with virus infections of arthropods, but even slight changes due to virus infection can have significant consequences. For example, virus infection may alter the capacity of the arthropod to transmit the virus, and such infections may affect arthropod feeding behavior, survival, and fertility (Platt et al., 1997).

V. CONCLUSIONS

No doubt viruses have sublime fitness for their natural hosts. How that fitness has evolved is a field of study in itself and can only be postulated, not determined. For coevolution of virus and host to occur there must be a precise and co-dependent reciprocal interaction between the vector and the virus (Tabachnick, 1991b). Specific changes in the microecology of the virus likely are caused by parallel changes in the vector, and changes in the virus then are caused by further recipro-cal changes in the vector (Tabachnick, 1998). Whether such congruent reciprocity leads to parallel speciation and subspeciation or not, the remarkable and widely various adaptations evidenced by viruses will continue to provide fascinating and dangerous epidemiologic circumstances as human populations increase and eco-logic variation decreases. As we continue to change the ecologic balance of nature, we will see concomitant and continuing viral evolutionary events, leading to the emergence of virus diseases caused by these opportunists.

REFERENCES

Beaty, B. J., Borucki, M., Farfan, J., and White, D. (1997). Arbovirus-vector interactions: Determinants of arbovirus evolution. *In* "Factors in the Emergence of Arbovirus Diseases" (J. F. Saluzzo and B. Dodet, eds.), pp. 23–35. Elsevier, Paris.

Burton, A. N., McLintock, J., and Rempel, J. G. (1966). Western equine encephalitis virus in Sas-katchewan garter snakes and leopard frogs. *Science* 154, 1029–1031.

Calisher, C. H., Karabatsos, N., Lazuick, J. S., Monath, T. P., and Wolff, K. L. (1988). Reevaluation of the western equine encephalitis antigenic complex of alphaviruses (family Togaviridae) as determined by neutralization tests. *Am. J. Trop. Med. Hyg.* 38, 447–452.

Capucci, L., Fusi, P., Lavazza, A., Pacciarini, M. L., and Rossi, C. (1996). Detection and preliminary characterization of a new rabbit calicivirus related to rabbit hemorrhagic disease virus. *J. Virol.* 70, 8614–8623.

Domingo, E., and Holland, J. J. (1994). Mutation rates and rapid evolution of RNA viruses. *In* "Evolutionary Biology of Viruses" (S. S. Morse, ed.), pp. 161–184. Raven, New York.

Fenner, F., and Ross. J. (1994). Myxomatosis. *In* "The European Rabbit" (H. V. Thompson and C. M. King, eds.), pp. 205–239. Oxford University Press, Oxford.

Gahlinger, P. M., Reeves, W. C., and Milby, M. M. (1986). Air conditioning and television as protec-tive factors in arboviral encephalitis risk. *Am. J. Trop. Med. Hyg.* 35, 601–610.

Gebhardt, L. P., and Hill, D. W. (1960). Overwintering of western equine encephalitis virus. *Proc. Soc. Exp. Biol. Med.* 104, 695–698.

Gibbs, A. J., Calisher, C. H., and Garcia-Arenal, F., eds. (1995). "Molecular Basis of Virus Evolution." Cambridge University Press, Cambridge.

Gould, E. A., and Buckley, A. (1989). Antibody-dependent enhancement of yellow fever and Japanese encephalitis virus neurovirulence. *J. Gen. Virol.* 70, 1605–1608.

Hahn, C. S., Lustig, S., Strauss, E. G., and Strauss, J. H. (1988). Western equine encephalitis virus is a recombinant virus. *Proc. Natl. Acad. Sci. U.S.A.* 85, 5997–6001.

Halstead, S. B. (1988). Pathogenesis of dengue: Challenges to molecular biology. *Science* 239, 476–481.

Hardy, J. L. (1987). The ecology of western equine encephalomyelitis virus in the Central Valley of California, 1945–1985. *Am. J. Trop. Med. Hyg.* 37 (suppl.), 18S–32S.

ICTV (International Committee on Taxonomy of Viruses) (1995). "Virus Taxonomy: Sixth Report of the International Committee on Taxonomy of Viruses" (F. A. Murphy, C. M. Fauquet, D. H. L. Bishop, S. A. Ghabrial, A. W. Jarvis, G. P. Martelli, M. A. Mayo, and M. D. Summers, eds.). Springer-Verlag, Vienna.

Kramer, L. D., Hardy, J. L., Presser, S. B., and Houk, E. J. (1981). Dissemination barriers for western equine encephalomyelitis virus in *Culex tarsalis* infected after ingestion of low viral doses. *Am. J. Trop. Med. Hyg.* **30**, 190–197.

Ludwig, G. V., Christensen, B. M., Yuill, T. M., and Schultz, K. T. (1989). Enzyme processing of La Crosse virus glycoprotein G1: A bunyavirus-vector infection model. *Virology* **171**, 108–113.

Meyer, K. F., Haring, C. M., and Howitt, B. (1931). The etiology of epizootic encephalomyelitis in horses in the San Joaquin Valley, 1930. *Science* **74**, 227–228.

Monath, T. P., Brinker, K. R., Chandler, F. W., Kemp, G. E., and Cropp, C. B. (1981). Pathophysiologic correlations in a rhesus monkey model of yellow fever. *Am. J. Trop. Med. Hyg.* **30**, 431–443.

Nichol, S. T., Spiropoulou, C. F., Morzunov, S., Rollin, P. E., Ksiazek, T. G., Feldmann, H., Sanchez, A., Childs, J., Zaki, S., and Peters, C. J. (1993). Genetic identification of a novel hantavirus associated with an outbreak of acute respiratory illness in the southwestern United States. *Science* **262**, 914–917.

OIE (1991). Viral haemorrhagic disease viruses. *Rev. Sci. Tech. Off. Int. Epizooties (Paris)* **10**.

Platt, K. B., Linthicum, K. J., Myint, K. S., Innis, B. L., Lerdthusnee, K., and Vaugn, D. W. (1997). Impact of dengue virus infection on feeding behavior of *Aedes aegypti. Am. J. Trop. Med. Hyg.* **57**, 119–125.

ProMED mail (1997–1998). Influenza, bird-to-human — China (Hong Kong) [online]. Available from: URL http://www.healthnet.org/programs.

Reeves, W. C. (1990). "Epidemiology and Control of Mosquito-Borne Arboviruses in California, 1943–1987." California Mosquito and Vector Control Association, Sacramento.

Smith, J. S., Reid-Sanden, F. L., Roumillat, L. F., Trimarchi, C., Clark, K., Baer, G. M., and Winkler, W. G. (1986). Demonstration of antigenic variation among rabies virus isolates by using monoclonal antibodies to nucleocapsid proteins. *J. Clin. Microbiol.* **24**, 573–580.

Sudia, W. D., McLean, R. G., Newhouse, V. F., Johnston, J. G., Miller, D. L., Bowen, G. S., and Sather, G. (1975). Epidemic Venezuelan equine encephalitis in North America in 1971. *Am. J. Epidemiol.* **101**, 36–50.

Swanepoel, R., Leman, P. A., Burt, F. J., Zachariades, N. A., Braack, L. E. O., Ksiazek, T. G., Rollin, P. E., Zaki, S. R., and Peters, C. J. (1996). Experimental inoculation of plants and animals with Ebola virus. *Emerg. Infect. Dis.* **2**, 321–325.

Tabachnick, W. J. (1991a). The evolutionary relationships among arboviruses and the evolutionary relationships of their vectors provides a method for understanding vector–host interactions. *J. Med. Entomol.* **28**, 297–298.

Tabachnick, W. J. (1991b). Reappraisal of the consequences of evolutionary relationships among California serogroup viruses (Bunyaviridae) and *Aedes* mosquitoes (Diptera:Culicidae). *J. Med. Entomol.* **28**, 297–298.

Tabachnick, W. J. (1998). Arthropod-borne pathogens: Issues for understanding emerging infectious diseases. *In* "Emerging Diseases" (Richard Krause, ed.). Academic Press, New York, In press.

Thomas, L. A., and Eklund, C. M. (1960). Overwintering of western equine encephalomyelitis virus in experimentally infected garter snakes and transmission to mosquitoes. *Proc. Soc. Exp. Biol. Med.* **105**, 52–55.

Trent, D. W., Monath, T. P., Bowen, G. S., *et al.* (1980). Variation among strains of St. Louis encephalitis virus: Basis for a genetic, pathogenetic, and epidemiologic classification. *Ann. N. Y. Acad. Sci.* **354**, 219–237.

Van Regenmortel, M. H., Maniloff, J., and Calisher, C. H. (1991). The concept of virus species. *Arch. Virol.* **120**, 313–314.

Weaver, S. C., Scherer, W. F., Cupp, E. W., and Castello, D. A. (1984). Barriers to dissemination of Venezuelan encephalitis viruses in the Middle American enzootic vector mosquito, *Culex* (*Melanoconion*) *taeniopus. Am. J. Trop. Med. Hyg.* **33**, 953–960.

Young, P. L., Halpin, K., Selleck, P. W., Field, H., Gravel, J. L., Kelly, M. A., and Mackenzie, J. S. (1996). Serologic evidence for the presence in pteropus bats of a paramyxovirus related to equine morbillivirus. *Emerg. Infect. Dis.* **2**, 239–240.

14

Relationship Between Humans and Their Viruses

CHRISTON J. HURST and NOREEN J. ADCOCK

United States Environmental Protection Agency
26 Martin Luther King Drive West
Cincinnati, Ohio 45268

I. INTRODUCTION

Many of the viruses that can infect humans should not be considered as viruses of humans, but rather as zoonotic. Zoonotic viruses are those viruses of animals that can cross boundaries such that they occasionally infect humans. Some examples of diseases induced in humans by zoonotic viruses are dengue and Ebola fever, the equine encephalitids (i.e., Eastern, St. Louis, Venezuelan, and Western), hantavirus pneumonia, Lassa fever, Marburg fever, rabies, and yellow fever. Additionally, it should be noted that the zoonotic category includes most, if not all, of the human illnesses induced either by arboviruses (viruses whose transmission is vectored by arthropods) or by the hemorrhagic fever viruses. With respect to the zoonotic viruses humans are, at best, alternate hosts. Humans do in fact usually represent dead-end hosts for these zoonotic viruses, meaning that subsequent transmission of those viruses either to new humans or back to the natural host is not sustained.

There is a subgroup of zoonotic viruses that, although principally remaining viruses of animals, seem to have adapted themselves to use humans as natural hosts. This adaptation is indicated by the fact that these viruses have demonstrated an ability to sustain a chain of transmission among humans. Examples of zoonotic viruses that have shown this ability to adapt themselves to become viruses of humans are those members of the family Flaviviridae (genus *Flavivirus*) that induce the human diseases known as dengue and yellow fever.

All of the above-mentioned zoonotic viruses contrast with the viral agents that clearly are known by their nature to be viruses of humans. Examples of viruses of humans are those that induce the diseases known as acquired immunodeficiency syndrome, fever blisters, measles, mumps, polio, rubella, smallpox, T-cell leukemia, T-cell lymphoma, and type A influenza. The aim of this chapter is to address those viruses that are considered to be viruses of humans. Those viruses of terrestrial mammals that are considered to be zoonotic are addressed by Calisher and Fenner in Chapter 13.

Every virus species needs to have a successful overall approach for sustaining its existence. That overall approach must enable the virus to attain its two principal goals, namely, that the virus be able to reproduce itself within a host and that the virus then be transmitted onward to a new host. Those mechanisms that any given virus species employs for achieving its sustainment, of course, have been developed through a process that involved an initiation of events by random chance followed by evolutionary selection. The most successful overall approaches may be those that subsequently evolve into the types of relationships between a virus and its host species that will allow the virus to persist without eliminating the host population. This latter point is very important, because the virus may in turn become extinct if it kills off the host population. It is for this reason that excessive virulence will be detrimental to the virus, and an interesting side point may be that, if an individual host cannot successfully surmount the infection, then the death of that individual host may be seen as an altruistic defense mechanism for the host population as a whole.

II. ACHIEVING THE GOAL OF VIRAL REPRODUCTION

Achieving self-reproduction is the first principal goal of the virus. The processes involved can be divided into three aspects. The first aspect is the virus's overall approach with regard to the course of the infection it establishes within the host. This involves the time course of the infection and the extent to which infectious progeny viral particles are produced during the course of the infection. The second aspect is the replication strategy employed by the virus. This involves the issues of where the virus begins its march through the host body and the physical trajectory followed until the virus exits in search of a new host. The third aspect involves which approaches, if any, the virus uses to avoid the host defensive mechanisms.

A. Strategy of the Infection Course

As mentioned in Chapter 1, the goal of establishing an effective course of infection is an aspect of viral reproduction that can be attained in many ways. We can summarize the strategies that viruses use as following four basic patterns.

1. Productive Infections

Three out of the four patterns of the course of viral infection are considered to be productive. The infection will be acquired in the form of infectious viral particles, and subsequently produced progeny viral particles then serve to infect future hosts. Productive, in this context, means that the number of progeny viral particles produced during the course of an infection is sufficient for the particles to transmit the infection to a new host with some reasonable probability. The productive approach involves an evolutionary decision as to whether there will be either a very short initial but highly productive course of infection (short term-initial), a course of infection that is prolonged but only intermittently productive (recurrent), or a productive though very slow course of infection whose severity progressively increases with time to reach a dramatic end-stage (increasing to end-stage). These three patterns can be described as follows.

a. *Short Term-Initial.* Viral production only occurs during a short time course near the time of initiation of the infection, which then abruptly ends. The human host may or may not survive beyond the course of this short infection. Host survival depends on the type of virus involved, the extent to which the involved virus and humans have had time to coevolve as species, and whether or not the ancestral humans of that particular subgroup of the human host population previously have had contact with the causative virus. Coevolution usually will tend to make the outcome of this pattern of viral infection sufficiently mild as to be associated with a fairly low incidence of mortality in an otherwise healthy population of human hosts. Some examples of this pattern would be the infections caused by the human caliciviruses (family Caliciviridae), human influenza viruses (family Orthomyxoviridae), human polioviruses and human rhinoviruses (family Picornaviridae), and the human rotaviruses (family Reoviridae).

b. *Recurrent.* Viral production, often very pronounced initially, is recurrent when the virus persists in a latent state within the host body and viral particle production periodically recurs but is not life threatening. Some examples of this pattern would be the infections caused by the human herpesviruses (family Herpesviridae) and the human papillomaviruses (family Papovaviridae).

c. *Increasing to End-Stage.* Viral infection normally is associated with a slow, almost innocuous, start, followed by a gradual progression, associated with an increasing level of viral production and eventual host death. Death of the host may relate to destruction of host immunological defense systems, which then results in death by secondary infections. This pattern of infection may take from 10 to 40 years to kill a human host. The infections caused by the human immunodeficiency viruses and the human T-lymphotropic viruses (all belonging to the family Retroviridae) represent examples of this pattern.

2. Nonproductive Infections

The fourth basic pattern of viral infection is considered to be nonproductive. A nonproductive infection is one in which the production of infectious virus particles is so limited that the virus must transmit itself through other means, usually by transferring a copy of only the nucleic acid genome of the virus. In these instances, the viral infection is normally acquired by direct transfer of the viral genetic material from the human parents to their developing fetuses, such transfer occurring via the egg and sperm cells. There may be no apparent health effects associated with such an infection. An example of this pattern is the infections caused by the endogenous retroviruses (family Retroviridae), whose genomes are incorporated into the chromosomal material of every cell in the human body (Villareal,

1997). The nonproductive pattern of infection seems to suggest the highest degree of coevolution between a virus and its host, since a nonproductive virus has no means of transmitting itself to a new host without some very active, albeit perhaps unwitting, participation on the part of the present host.

B. Strategy of Viral Replication

This section addresses the questions of where and how the virus begins its march through the host body, and how the virus then continues the course of this attack, leading ultimately to the concept of viral reproduction strategies at the host population level.

1. Cellular Metabolic Level

Discussion of the strategy of viral replication within the body of a host organism begins at the most basic level, which is attachment of the virus to a particular molecule present on the surface of host cells. Such a molecule is said to be the "virus's receptor," and will be some cellular protein or lipid component naturally produced by those cells. The virus's choice of receptor is a product of viral evolution. After binding to its receptor, the virus gains entrance to the interior of the cell and viral replication begins. Viruses whose genomes are composed of DNA generally replicate mainly within the nucleus. In contrast, those viruses with RNA genomes generally focus their center of replication in the cytoplasm. During the course of replication, the virus must decide which cellular systems and machinery it will employ. Some large viruses carry the genomic coding capacity for many of their own enzymes, while others may rely almost completely on the enzymatic machinery possessed by the host cell. Many viruses, including those belonging to the genus *Enterovirus* of the family Picornaviridae, are said to be highly cytopathogenic, meaning that they usually quickly kill the host cell as a result of infection. Other viruses, such as those that occupy the genus *Rubivirus* of the family Togaviridae, may cause prolonged and severe crippling of the cell rather than killing it outright. A further discussion of these issues can be found in Chapter 3.

2. Tissue and Organ Tropism Level

Viruses vary greatly with respect to the tissues they tend to target for infection. On a larger scale, this then leads to identification of those organs that the viruses are affecting. This selective targeting is referred to as a "tropism." Viral tropisms can be divided into those considered primary and those considered secondary. Primary tropisms will be associated with production of those viral particles that subsequently contribute to transmission of the viral infection to a new host. As such, the primary tropisms tend to be related to those sites (termed "portals") through which the virus either enters or exits the host body. Secondary tropisms may represent accidents. Some of these accidents may come about as a result of the molecule that a virus uses as its receptor, existing on the cells of tissues that are unrelated to those that the virus must employ in order to achieve transmission. Nevertheless, secondary tropisms may contribute greatly to the types and severity of the illnesses associated with infection of humans by any particular virus.

3. Host Population Level

When considered at the host population level, the strategy of viral replication includes the ease with which or likelihood that a virus is transmitted to new hosts plus the severity of infection and accompanying likelihood of death (including the age-related likelihood of death) for any given host individual.

C. Strategies for Evading Host Defensive Mechanisms

Through the course of evolution, many viruses have developed mechanisms for either countering or evading the human immune and non-immune defenses as a means for enhancing the probability of viral success.

1. Avoiding the Host Immune Defenses

The human immune system includes both humoral (antibody-mediated) and cellular components. The cellular components can include granulomatous reactions, which play a role in defense against protozoans, though their possible role in antiviral defenses has been incompletely explored. Those mechanisms that viruses use to either avoid or minimize attack by the host immune system can be divided into the four following groups. The use of these types of mechanisms seems particularly critical in association with those viral infections that persist within a human host for very long periods of time, often up to decades.

a. *Antigenic Mimicry.* The produced antigens are similar to those of the host, as with prions.

b. *Rapid Viral Mutation.* This mechanism includes both antigenic drifting and shifting. Some viral types demonstrate rapid viral mutation during the course of an infection, as occurs with the human immunodeficiency viruses of the family Retroviridae. Other virus types, such as the influenza viruses of the family Orthomyxoviridae, demonstrate rapid viral mutation between reinfections of the same host.

c. *Low Antigenicity.* Some viruses inherently seem to provoke little, if any, immune response. This often occurs because the virus persists in a latent state within host cells, during which time either little or no viral antigenic material is produced. Examples include the endogenous retroviruses of the family Retroviridae and the human herpesviruses of the family Herpesviridae.

d. *Infect the Immune Cells!* The most direct attack may be the most effective. Exceptionally notorious examples of this approach are the genus *Rubivirus* of the family Togaviridae and the genus *Lentivirus* of Retroviridae.

Aside from the above groupings, some viruses such as the Norwalk virus of the family Caliciviridae seem to be antigenic but provoke an immune response that is minimally effective.

2. Avoiding the Host Non-Immune Defenses

The body has non-immune defense mechanisms that help to protect against viral infections. These mechanisms are associated with the portals through which viruses can enter the body of a host. Examples of non-immune defenses include the enzymes secreted as a part of pancreatic fluid, saliva, and tears. Various glands associated with mucosal tissues secrete antimicrobial compounds into the mucus produced by those tissues. Some mucosal tissues also possess cilia whose movement helps to expel both the mucus and any foreign materials, including pathogens, that become entrapped within the mucus. An additional example of a non-immune defense is the stomach acid produced to aid digestion of organic compounds. Many gastroenteritis viruses, such as the rotaviruses of the family Reoviridae and the astroviruses of Astroviridae, have evolved such an effective resistance to attack by proteolytic enzymes that those viruses virtually need partial proteolysis to facilitate their infectivity. The influenza viruses of the family Orthomyxoviridae are known for their ability to paralyze the activity of the mucosal cilia located within the respiratory tract. One of the defining characteristics for the enteroviruses of the family Picornaviridae is their resistance to acidic exposure.

III. ACHIEVING THE GOAL OF VIRAL TRANSMISSION
   BETWEEN HOSTS

The task of achieving viral transmission between hosting individuals involves two aspects. The first is the type of infectious material in which a virus will leave its present host, while the second involves the route by which the virus can encounter its proximate host.

A. Type of Infectious Bodily Material in Which Virus Is Released
   from the Host

The types of bodily materials within which viruses can be released include substances that exit during the course of normal body functions. Among these substances are feces and a variety of liquids, including menstrual blood, respiratory secretions of the upper and lower tracts, saliva, semen, tears, urine, and vaginal fluid. Sweat is another fluid that is naturally released from the body; however, it is not known to contain viruses. Viruses can also be found in blood released from wounds in the skin; from blood acquired by blood-consuming parasitic insects, among which are the fleas and several groups of flies, ticks, and mosquitoes; and blood leaked from swollen or ruptured capillaries into mucosal tissue and skin pores.

B. Route of Transmission between Hosts

Those natural routes by which viruses are transferred to and between humans are the same routes associated with all surface-dwelling terrestrial vertebrates. These routes are tightly associated with the portals of entry and exit that any particular virus family uses as it tries to survive and find its way from one host to the next. Viral transmission routes can be subdivided into two broad groups. The first is transmission by direct contact (also known as direct transfer) between two members of those species that host the virus. This includes both the possibility of transmission between two members of the principal host species and the possibility of transmission between a member of that principal host species and some alternate host species, and the latter may represent a vectoring species. The second group is transmission by indirect contact (also known as indirect transfer). These routes have been described in detail by Hurst and Murphy (1996) and are represented in Figures 9 and 12 of Chapter 1 herein. As explained in Chapter 1, there are some routes of viral transmission that are considered unnatural vehicular routes. Such routes represent the use of unnatural vehicles as a means to evade the host defenses associated with natural portals of entry. These unnatural routes involve invasive medical devices (such as syringes, endoscopes, and other surgical instruments) and transplanted tissues, including transfused blood and blood products. The remainder of this section describes the natural routes of viral transmission between hosts.

1. Direct Contact

The direct contact approach offers one major advantage and one major drawback to the virus. The advantage is that those viruses transmitted by direct contact need not be stable when exposed to ambient environments. The drawback is that the number of new hosts to which they have potential access may be smaller than for viruses transmitted by indirect contact. Viruses that are endogenous by their nature will survive for as long as the host survives. Although endogenous viruses can only be transmitted to host progeny, these viruses neither have to adapt themselves to nor coevolve with any other hosting species. An example of this type of endogenous agent would be the endogenous retroviruses of the family Retroviridae. Viruses that are venereal in nature, that is, transmitted in semen and vaginal secretions during sexual activity, have a somewhat greater potential for contacting new hosts. Represented among the venereal viruses of humans are some species of the genera *Simplexvirus* (family Herpesviridae) and *Papilloma-virus* (family Papovaviridae). Once these venereal viruses infect a host, they remain associated with the host for the rest of the host's life in the form of a permanent infection. Thus, although the frequency with which endogenous and venereal viruses can find a new host is restricted, the viruses compensate for this to some degree by remaining with the host for a very long time. The next step on the scale of host access is represented by those viruses transmitted via direct contact with insect vectors. These viruses have greatly increased access to new hosts and tend not to remain with their present host for the remainder of the host's life. Those viruses transmitted by biting insects are commonly referred to as "arboviruses," which is an abbreviation of "*ar*thropod-*bo*rne *viruses*." Included among those arboviruses that infect humans are members of the genera *Alphavirus* (family Togaviridae); *Bunyavirus, Nairovirus,* and *Phlebovirus* (all of the family Bunyaviridae); and *Flavivirus* (family Flaviviridae).

Viruses transmitted by way of saliva may be perceived as bridging the categories of direct and indirect contact. If any particular type of virus that is secreted into saliva has either no stability when exposed to ambient environments in oral secretions or only limited stability under those conditions, then the virus will have to be transmitted by saliva that is transferred during oral contact between hosts. Conversely, if the particular virus type has good stability when exposed to ambient environments in oral secretions, then that virus can be transmitted on shared food or in association with fomites. Some of the viruses transmitted in saliva do remain associated with the host as a permanent infection, and these often are the viruses that possess limited stability in ambient environments, such as members of the family Herpesviridae. Many of the viruses that are secreted into saliva and can be transferred to a new host in association with fomites do not remain associated with the host as a permanent infection, such as members of the family Picornaviridae. In general, those viruses transmitted by indirect contact between hosting individuals tend to produce only transient infections of their individual hosts rather than remain associated with the individual host as a permanent infection. These several latter points suggest that there may be some evolutionary relationship between ease of viral transmission to a new potential host individual or the frequency of opportunities for viral transmission, and the length of time that a virus must be capable of remaining with its present host to have a reasonable chance of achieving eventual transmission.

2. Indirect Contact (Vehicle-Borne)

The indirect contact approach also offers one major advantage and one major drawback to a virus. Viruses transmitted by indirect contact have the advantage of potential access to a far greater number of hosting individuals than is the case for viruses transmitted by direct contact. The drawback is that the viruses must have evolved stability when they are exposed to the ambient environment. The vehicles that viruses utilize to achieve transmission between hosting individuals by indirect contact are subdivided into the following four categories: food, water, air (in actuality, this refers to aerosols), and fomites. Transmission by any of these four categories of vehicles will usually be associated with some specific physical activity on the part of the present host, and will always be associated with some physical activity on the part of the proximate host.

Of course, foods are items intentionally ingested for their caloric or nutritional value. Food contamination can occur by way of the food being a virally infected animal consumed by the proximate host. In such cases, there is no specific physical activity on the part of the present host (the one being eaten) that can be identified as having caused the proximate host to be ingesting contaminated food (indeed, perhaps it is a lack of physical activity on the part of the present host that is to blame). Otherwise, viral contamination of foods can result from fecal material being transferred via contact with unwashed hands or if contaminated aerosols fall into the food. A particularly notable example of a virus of humans that is transmitted via foods is the hepatitis A virus of the genus *Hepatovirus* (family Picornaviridae).

Water usually serves as a vehicle after it has been contaminated with fecal material. Acquisition of a viral infection from water usually results from the proximate host ingesting contaminated water. Physical contact of the skin or mucosa of the proximate host with contaminated water, as may occur during recreational activities or body washing, can result in acquisition of infection. Notable examples of viruses transmitted by these waterborne routes are those belonging to the viral families Astroviridae and Caliciviridae, and the genera *Enterovirus* and *Hepatovirus* of the family Picornaviridae.

Viral contamination of air can occur by two principal mechanisms. The first, and most significant, involves release of aerosols that contain droplets of respiratory secretions (i.e., nasal, oral, or pulmonary mucus). This type of transmission route is referred to as being the route of droplet aerosols. Notable examples of viruses transmitted in this manner are those belonging to the viral families Coronaviridae, Orthomyxoviridae, and Paramyxoviridae, and members of the genus *Rhinovirus* of the family Picornaviridae. The second mechanism is by way of particulate aerosols. This involves generation of aerosols composed of soil particles coated with dried urine or feces. Notable examples of viruses transmitted by this route belong to the genera *Arenavirus* (family Arenaviridae) or *Hantavirus* (family Bunyaviridae).

Viral contamination of fomites (defined as solid environmental surfaces that can serve in transmission of infections) can occur in many ways. The variety of things that represent fomites include items used to conserve warmth (e.g., blankets, clothing), items used while eating (e.g., cups, dinner plates, and utensils), tables on which diapers are changed, doorknobs, medical devices, toilet seats, and toys. The ways by which these environmental items become contaminated include projection of droplet aerosols onto environmental objects during sneezing or coughing, aerosols falling onto objects, and unintended surface contamination (including children's clothing, blankets and toys) by blood, feces, fluid from skin lesions (rashes), nasal secretions, saliva, or urine. The task of achieving viral transmission via this route is accomplished when these objects are subsequently handled or used by a potential proximate host. Among the viral genera whose members can be transmitted via fomites are *Orthopoxvirus* (family Poxviridae) and *Rhinovirus* (family Picornaviridae).

IV. SUMMARY OF VIRAL FAMILIES THAT AFFLICT HUMANS

Twenty of the viral families contain members capable of infecting humans. Together, they cause a broad range of illnesses in humans. The terminology used in describing these illnesses is given in Table I. The rest of this section summarizes the ecology of these viral families. Figure 1 schematizes the manner in which the different aspects of the ecology of viral infection fit together. The literature sources used for to compile this summary include Hurst and Murphy (1996), ICTV (1995), Evans and Kaslow (1997), and White and Fenner (1994).

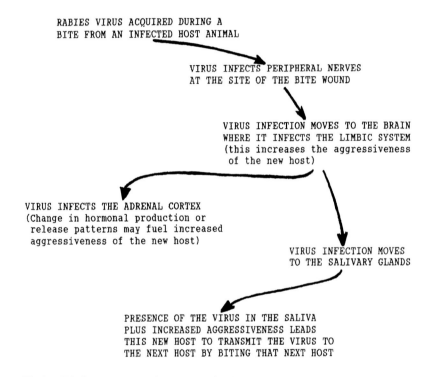

Fig. 1.   This figure shows the viral ecology of rabies virus (genus *Lyssavirus*, family Rhabdoviridae) in association with a natural host. Transmission of this virus between hosts occurs when an infected animal bites an uninfected animal, with the virus being transferred by saliva into the bite wound. Subsequent movement of viral infection into the nervous system and salivary glands of the newly bitten host animal is considered to represent primary tropisms, as infection at these sites is directly related to movement of virus into the body of the current host and subsequent transfer of virus to the next host. Infection of the adrenal cortex is considered a secondary tropism, since those viruses produced in the adrenal cortex will not be transferred to a subsequent host animal. However, infection of the adrenal cortex may play a role in the viral ecology by augmenting the aggressiveness of this newly infected host, thereby increasing the likelihood that this animal will then bite other potential host animals.

TABLE I

Terminology of Human Illnesses Induced by Viruses

Term	Definition
Acquired immunodeficiency syndrome	Syndrome of frequent and chronic infections caused by opportunistic pathogens resulting from immunodepletion
Adenopathy	Physiological changes associated with degeneration of adenoid tissues
Anemia	Low iron level in the blood
Arthralgia	Pain in skeletal joints
Arthritis	Syndrome associated with pain and swelling in skeletal joints
Auditory	Related to hearing
Broncheolitis (bronchitis)	Syndrome associated with swelling or related dysfunction of bronchiole tissues
Carcinoma	Tumor of epithelial tissues
Cardiological	Relating to the heart
Carditis	Syndrome associated with swelling or the related dysfunction of heart tissues
Conjunctivitis	Swelling and reddening of the conjunctiva (the mucosal membranes that line the eyelids)
Coryza	Watery discharge from eyes and nose
Diabetes	Syndrome associated with underproduction or dysfunctional use of insulin
Diabetic	A person who has diabetes or some health characteristic related to diabetes
Encephalitis	Syndrome associated with swelling or the related dysfunction of brain tissues
Encephalomyelitis	Syndrome associated with swelling or the related dysfunction of brain and spinal cord tissues
Encephalopathy	Physiological changes associated with degeneration of brain tissues
Encephalopathy, demyelinating	Encephalopathy associated with the myelin tissue that surrounds nerve cells
Edema	Swelling
Encephalitic	Relating to the brain
Enteritis	Syndrome associated with swelling or the related dysfunction of intestinal tissues, frequently evidenced as diarrhea
Erythema	Form of macular rash showing diffused redness of the skin

continued

Term	Definition
Exanthema	Skin rash
Facial	Relating to the face
Fetal	Relating to the fetus
Fetal developmental abnormalities	Defects occurring during development of the fetus
Fetal loss	Spontaneous abortion
Gastritis	Syndrome associated with swelling or the related dysfunction of stomach tissues, frequently evidenced as vomiting
Gastroenteritis	Combined syndrome of gastritis and enteritis, frequently evidenced as vomiting and diarrhea
Hematemesis	Vomiting of blood
Hematuria	Blood in the urine
Hemolytic uremia	Syndrome consisting of hemolytic anemia (anemia due to blood cell lysis), a reduced level of thrombocytes, and acute degeneration of kidney tissues
Hemorrhagic fever	Syndrome consisting of massive hemorrhage and high fever
Hepatic	Relating to the liver
Hepatitis	Syndrome associated with swelling or the related dysfunction of liver tissues
Hepatomegaly	Enlargement of the liver
Immunodepletion	Reduced level of circulating immune cells
Immunosuppression	Suppressed functioning of the immune system
Keratoconjunctivitis	Combined syndrome associated with swelling or the related dysfunction of both the cornea and conjunctiva tissues
Leukemia	Cancerous syndrome associated with an extremely high level of circulating white cells in the blood
Lymphoma	Solid tumor of the immune system
Malaise	Syndrome characterized by an extremely low level of motivational energy
Malignancy	A spreading cancer
Melena	Bleeding into the lumen of the intestines evidenced by the voiding of tarlike fecal material

continued

Term	Definition
Meningitis	Syndrome associated with swelling or the related dysfunction of meningeal tissues (membranes that enclose the brain and spinal cord)
Meningoencephalitis	Combined syndrome of encephalitis with meningitis
Mucosa	Tissues that secrete mucus
Myalgia	Tenderness or pain in the muscles
Myelopathy	Physiological changes associated with degeneration of the spinal cord
Myocarditis	Syndrome associated with swelling or the related dysfunction of heart tissues
Myositis	Syndrome associated with swelling or the related dysfunction of muscle tissues
Nasopharyngitis	Combined syndrome associated with swelling or the related dysfunction of both the nasal passage and pharynx tissues
Necrosis	Death of tissue cells
Necrotic lesion	Focal area of tissue cell death
Nerve deafness	Loss of hearing resulting from reduced functioning of nerve cells
Neuralgia	Severe sharp pain along the course of a nerve
Neuronal	Relating to the nerve cells
Nodule	A small knotlike protuberance or swelling of tissue
Orchitis	Syndrome associated with swelling or the related dysfunction of testicular tissues
Otitis media	Syndrome associated with swelling or the related dysfunction of middle ear tissues
Paralysis	Loss of mobility
Pericarditis	Syndrome associated with swelling or the related dysfunction of pericardial tissues (fibrous membrane sack that enfolds the heart)
Pharyngitis	Syndrome associated with swelling or the related dysfunction of the pharynx tissues
Pharyngoconjunctival fever	Combined syndrome of conjunctivitis and pharyngitis with fever
Pneumonia	Syndrome associated with swelling or the related dysfunction of lung tissues (by definition, this term indicates that the swelling was induced by an infection)

continued

Term	Definition
Pneumonia, hemorrhagic	Pneumonia accompanied by bleeding into the lungs
Pneumonitis	Syndrome associated with swelling or the related dysfunction of lung tissues (by definition, this term indicates that the swelling was induced by an unknown irritation)
Rash, hemorrhagic (petechial)	Bleeding within the skin evidenced by purple spots termed petechiae
Rash, macular	Discolored spots of various size and shape on the skin (or on the mucosa) that are neither elevated nor depressed
Rash, maculopapular	The presence of both macules and papules on the skin or mucosa
Rash, papular	Small, red circular elevated solid areas on the skin (or on mucosa) that may progress to become either vesicles (filled with clear liquid) or pustules (filled with pus), or may first fill with clear liquid and then with pus
Renal dysfunction	Dysfunction of the kidneys
Retinitis	Syndrome associated with swelling or the related dysfunction of the retina
Retroocular pain	Pain centered behind the eyeball
Salivary glands	Glands that produce saliva
Sarcoma	Cancer arising from connective tissue such as muscle or bone
Sinusitis	Syndrome associated with swelling or the related dysfunction of nasal sinus tissues
Splenomegaly	Enlargement of the spleen
Tracheobronchitis	Syndrome associated with swelling or the related dysfunction of both trachea and bronchiole tissues
Trimester	One third of a time period; usually refers to either the first, second, or third 3-month segment of the 9-month human fetal gestation period
Tumor	Solid abnormal growth of cells
Visual	Relating to vision

A. Viral Family Adenoviridae

Genus affecting humans: *Mastadenovirus.*

Familial nature with respect to members affecting humans: Viruses of humans.

Alternate hosts: Species affecting humans seem naturally limited to humans.

Types of illnesses induced in humans: Adenopathy (the origin of the family name), conjunctivitis, coryza, encephalitis, gastroenteritis, keratoconjunctivitis, pharyngitis, pharyngoconjunctival fever, and pneumonia.

Familial strategies:

Infection course — productive, both short term-initial and recurrent.

Viral replication — at the individual host level, the primary tissue and organ tropisms are toward the cervix, conjunctiva, pharynx, small intestine, and urethra; the secondary tissue and organ tropisms are toward the brain, kidney, lungs, and lymph nodes; at the host population level, these viruses generally are endemic and initially acquired at a very early age, with the infections very often asymptomatic in young children.

Evasion of host defenses — uncertain.

Predominant routes of transmission between hosts: Direct contact via host-to-host and indirect (vehicle-borne) contact via fecally contaminated water, food, fomites, and fomites contaminated by respiratory secretions.

B. Viral Family Arenaviridae

Genus affecting humans: *Arenavirus.*

Familial nature with respect to members affecting humans: Zoonotic.

Natural hosts: Rodents, including commensal voles and mice, as well as commercial colonies of hamsters and nude mice.

Types of illnesses induced in humans: Arthralgia, carditis, encephalomyelitis, encephalopathy, facial edema, fetal loss, focal necrosis of liver, gastritis, hemorrhagic fever, hepatitis, inhibition of platelet function (causes the fatal bleeding associated with this virus family), malaise, meningitis, myalgia, nerve deafness, and pneumonia.

Familial strategies:

Infection course — productive, short term-initial.

Viral replication — at the individual host level, the primary tissue and organ tropisms presumably are toward the liver and lungs; the secondary tissue and

organ tropisms are toward the brain, fetus, heart, joints, and nerves; at the host population level, these viruses can be extremely devastating to individual hosts but are poorly transferred between humans, which usually represent dead-end hosts.
Evasion of host defenses — uncertain.

Predominant routes of transmission between hosts: Indirect (vehicle-borne) contact via inhalation of particulate aerosols bearing dried rodent urine or acquisition of infectious materials through skin abrasion (a form of surface contact).

C. Viral Family Astroviridae

Genus affecting humans: *Astrovirus*.

Familial nature with respect to members affecting humans: Viruses of humans.

Alternate hosts: Species affecting humans seem naturally limited to humans.

Types of illnesses induced in humans: Enteritis, gastroenteritis.

Familial strategies:
Infection course — productive, short term-initial.
Viral replication — at the individual host level, the primary tissue and organ tropisms are toward the small intestine; the secondary tissue and organ tropisms presently are unknown; at the host population level, these viruses are endemic, principally causing a mild enteritis seen in young adults.
Evasion of host defenses — avoids host non-immune defenses by resistance to proteolytic attack (their infectivity is actually increased by proteolytic attack).

Predominant routes of transmission between hosts: Indirect (vehicle-borne) contact via fecally contaminated water, food, and fomites.

D. Viral Family Bunyaviridae

Genera affecting humans: *Bunyavirus, Hantavirus, Nairovirus, Phlebovirus*.

Familial nature with respect to members affecting humans: Zoonotic.

Natural hosts: Largely rodents, but also hares and rabbits, and some ungulates.

Types of illnesses induced in humans: Arthralgia, encephalitis, hematemesis, hematuria, hemolytic uremia, hemorrhagic fever, hemorrhagic pneumonia, hemorrhagic (petechial) skin rash, hepatitis, melena, myalgia, pneumonia, renal dysfunction, retinitis, retroocular pain.

Familial strategies:

Infection course — productive, short term-initial.

Viral replication — at the individual host level, the primary tissue and organ tropisms are toward the kidneys, liver, and lungs; the secondary tissue and organ tropisms are toward the brain and eyes; at the host population level, these viruses are not well sustained within human populations, and humans usually represent dead-end hosts.

Evasion of host defenses — uncertain, but may include avoiding host immune defenses by infecting immune cells.

Predominant routes of transmission between hosts: Direct host-to-vector contact by gnats, midges, mosquitoes, sandflies, and ticks for the genera *Bunyavirus*, *Nairovirus*, and *Phlebovirus*, and indirect (vehicle-borne) contact via particulate aerosols containing dried rodent urine, or contact with rodent excreta or contaminated fomites for the genus *Hantavirus*.

E. Viral Family Caliciviridae

Genus affecting humans: *Calicivirus.*

Familial nature with respect to members affecting humans: Viruses of humans and zoonotic.

Natural or alternate hosts: Fish, terrestrial as well as marine mammals (especially swine).

Types of illnesses induced in humans: Gastroenteritis, hepatitis (nonprogressive, but extraordinarily high fatality rate — 10 to >25% — for women if contracted during the third trimester of pregnancy), myalgia.

Familial strategies:

Infection course — productive, short term-initial.

Viral replication — at the individual host level, primary tissue and organ tropisms are toward the small intestine; secondary tissue and organ tropisms are toward the liver; at the host population level, these tend to be epidemic within human populations; for the hepatitis E virus it seems that acquisition occurs from swine, with the result being epidemics (often very widespread) of human disease; some acquisition from animals may come from eating infected animals; subsequent transmission of all caliciviruses within human populations is by fecally contaminated waste and thus can be very widespread.

Evasion of host defenses — avoids host non-immune defenses by resistance to proteolytic attack.

Predominant routes of transmission between hosts: Indirect (vehicle-borne) contact via fecally contaminated water, food, and fomites.

F. Viral Family Coronaviridae

Genera affecting humans: *Coronavirus, Torovirus.*

Familial nature with respect to members affecting humans: Viruses of humans.

Alternate hosts: Some terrestrial ungulates and carnivores.

Types of illnesses induced in humans: Coryza, gastroenteritis.

Familial strategies:
Infection course — productive, short term-initial.
Viral replication — at the individual host level, the primary tissue and organ tropisms are toward the intestines, lungs (possibly), nasopharynx, and sinuses; at the host population level, these viruses are very widespread and essentially nonfatal.
Evasion of host defenses — avoids host immune defenses by viral mutation and recombination.

Predominant routes of transmission between hosts: Indirect (vehicle-borne) contact via fecally contaminated water, food, and fomites.

G. Viral Family Filoviridae

Genus affecting humans: *Filovirus.*

Familial nature with respect to members affecting humans: Generally zoonotic.

Natural hosts: Unknown, but may include bats and rodents, with primates serving as intermediary hosts leading to human exposure.

Types of illnesses induced in humans: Conjunctivitis, hemorrhagic fever (frequently fatal, death possibly resulting from extreme inflammatory response), hepatic necrosis, myalgia, pharyngitis.

Familial strategies:
Infection course — productive, short term-initial.
Viral replication — at the individual host level, primary tissue and organ tropisms are toward the immune cells and liver (possibly); secondary tissue and organ tropisms are toward the adrenal glands, kidneys, liver, and spleen; at the host population level, these viruses are transferred between humans but seem unable to be sustained in human populations; humans usually represent dead-end hosts.

Evasion of host defenses — avoids host immune defenses by infecting immune cells.

Predominant routes of transmission between hosts: Direct contact via host-to-host transfer of contaminated bodily fluids.

H. Viral Family Flaviviridae

Genera affecting humans: *Flavivirus, Hepatitis C-like viruses.*

Familial nature with respect to members affecting humans: Viruses of humans and zoonotic.

Natural or alternate hosts: Members of the genus *Flavivirus* cross-infect a variety of birds and terrestrial mammals via mosquitoes or ticks (depending on the viral species) and most clearly are zoonotic, although those that cause yellow fever and the four that cause dengue may have become viruses of humans without time to coevolve; one species of the genus *Hepatitis C-like viruses* affects humans and seems naturally limited to humans.

Types of illnesses induced in humans: Arthritis with rash, encephalitis, hemorrhagic fever, hepatitis (chronic, which may lead to hepatocellular carcinoma).

Familial strategies:
Infection course — short term-initial for the genus *Flavivirus*, increasing to end-stage for the hepatitis C virus.
Viral replication — at the individual host level, primary tissue and organ tropisms are toward the immune cells (principally monocytes and macrophages) and liver; secondary tissue and organ tropisms are toward the brain and liver; at the host population level, most of these viruses are zoonotic, with humans representing dead-end hosts; however, some can be sustained within human populations and occasionally have high lethality rates.
Evasion of host defenses — avoids host immune defenses by infecting immune cells.

Predominant routes of transmission between hosts: For flaviviruses, direct host-to-vector contact; for hepatitis C virus, presumably direct contact via host-to-host transfer of contaminated bodily fluids.

I. Viral Family Hepadnaviridae (and Genus *Deltavirus*)

Genera affecting humans: *Orthohepadnavirus.* The hepatitis D virus (HDV) is a member of the floating genus *Deltavirus*; it is a defective satellite virus which can coinfect humans, but only in association with the hepatitis B virus (HBV) because HDV encapsidates itself with proteins encoded by the genome of the coinfecting HBV.

Familial nature with respect to members affecting humans: Viruses of humans.

Alternate hosts: One species of viral family Hepadnaviridae (hepatitis B virus) is known to infect humans, and it seems naturally limited to humans.

Type of illness induced in humans: Hepatitis, which may become chronic in adults.

Familial strategies:
   Infection course — productive, short term-initial, and increasing to end-stage.
   Viral replication — at the individual host level, primary tissue and organ tropisms are toward the liver; secondary tissue and organ tropisms are toward the bile duct epithelium, circulating immune cells, and pancreatic acinar cells; at the host population level, when acquired by adults and older children, these viruses generally cause an acute but short-term illness that sometimes can be fulminant; when acquired by neonates or younger children, initially tends to be subclinical but becomes chronic, and the tendency to be chronic can be racially associated (Chinese, possibly also Black African).
   Evasion of host defenses — avoids host immune defenses by infecting immune cells.

Predominant routes of transmission between hosts: Direct contact via host-to-host transfer of contaminated bodily fluids and perinatally from contaminated maternal blood.

J. Viral Family Herpesviridae

Genera affecting humans: *Cytomegalovirus, Lymphocryptovirus, Roseolovirus, Simplexvirus, Varicellovirus.*

Familial nature with respect to members affecting humans: Viruses of humans.

Alternate hosts: Species affecting humans seem naturally limited to humans, but may pass to primates.

Types of illnesses induced in humans: Carcinoma, carditis, chronic gastrointestinal infection, encephalitis, hepatomegaly, keratoconjunctivitis, lymphoma, myelitis, neuralgia, papular rash of skin and mucosa, paralysis, retinitis, splenomegaly.

Familial strategies:
   Infection course — productive, recurrent.
   Viral replication — at the individual host level, primary tissue and organ tropisms are toward the genital and oral mucosa, pharynx, and salivary glands; secondary tissue and organ tropisms are toward the eyes, kidneys, liver, lymph nodes, nervous system including brain, and spleen; at the host population level, these viruses are ubiquitous, tend to be acquired in childhood or early adulthood, and seldom directly result in host death.
   Evasion of host defenses — avoids host immune defenses by infecting immune cells.

Predominant routes of transmission between hosts: Direct contact via host-to-host transfer of fluid from viral-induced lesions of skin or mucosa and by saliva contaminated by chronically infected salivary glands; plus transmission to offspring either transplacentally, intrapartum (during the birth process), or via breast milk.

K. Viral Family Orthomyxoviridae

Genera affecting humans: *Influenzavirus A*, *Influenzavirus B*, and *Influenzavirus C*.

Familial nature with respect to members affecting humans: Generally viruses of humans.

Alternate hosts: Birds (possibly), swine.

Types of illnesses induced in humans: Coryza, malaise, myalgia, nasopharyngitis, pneumonia, retroocular pain, tracheobronchitis.

Familial strategies:
Infection course — productive, short term-initial.
Viral replication — at the individual host level, primary tissue and organ tropisms are toward the ciliated columnar epithelium of the respiratory tract (the exact tissue tropism is directly related to the virus hemagglutinin [HA] serotype); at the host population level, these viruses constantly undergo antigenic drift and antigenic shift and cause wide-scale seasonal epidemics in humans, although infection-related fatality is usually limited to humans aged 65 or older (most notably, age 75 or older).
Evasion of host defenses — avoids host immune defenses by antigenic mimicry and by rapid viral mutation.

Predominant routes of transmission between hosts: Indirect (vehicle-borne) contact via droplet aerosols (from sneezing and coughing) and aerosol-contaminated fomites.

L. Viral Family Papovaviridae

Genera affecting humans: *Papillomavirus, Polyomavirus.*

Familial nature with respect to members affecting humans: Viruses of humans.

Alternate hosts: Species affecting humans seem naturally limited to humans.

Types of illnesses induced in humans: Benign tumors of skin and mucosa that may progress to malignancy, progressive demyelinating encephalopathy.

Familial strategies:
Infection course — productive, recurrent.

Viral replication — at the individual host level, primary tissue and organ tropisms are toward the mucosa and skin (genus *Papillomavirus*) and toward the upper respiratory tract (genus *Polyomavirus*); secondary tissue and organ tropisms are toward the brain and kidneys (genus *Polyomavirus*); at the host population level, these viruses are ubiquitous and almost never directly responsible for host death.

Evasion of host defenses — avoids host immune defenses by antigenic mimicry.

Predominant routes of transmission between hosts: Presumably direct contact via host-to-host or indirect (vehicle-borne) contact by way of fomites (genus *Papillomavirus*); indirect (vehicle-borne) contact via aerosols (genus *Polyomavirus*).

M. Viral Family Paramyxoviridae

Genera affecting humans: *Morbillivirus, Paramyxovirus, Pneumovirus, Rubulavirus.*

Familial nature with respect to members affecting humans: Viruses of humans.

Alternate hosts: Species affecting humans seem naturally limited to humans.

Types of illnesses induced in humans: Bronchiolitis, conjunctivitis, coryza, encephalitis, glandular enlargement (especially salivary glands), immunosuppression (*Morbillivirus* causes an immunosuppression that is temporary, but which is arguably the most severe induced by a virus of humans, and can result in death by other coinfecting pathogens, such as enteric protozoans, that normally would not cause fatality), macular rash, nerve deafness, orchitis, pneumonitis.

Familial strategies:
Infection course — productive, short term-initial.
Viral replication — at the individual host level, primary tissue and organ tropisms are toward the epidermis and mucosa (including conjunctival, oral and respiratory); secondary tissue and organ tropisms are toward the brain, breasts, circulating immune cells, and testicles; at the host population level, these viruses tend to be acquired at a young age and are almost never directly responsible for host death, although severe sequelae can result if acquired beyond early childhood.
Evasion of host defenses — avoids host immune defenses by infecting immune cells.

Predominant routes of transmission between hosts: Indirect (vehicle-borne) contact via aerosols.

N. Viral Family Parvoviridae

Genus affecting humans: *Parvovirus*.

Familial nature with respect to members affecting humans: Viruses of humans.

Alternate hosts: Species affecting humans seem naturally limited to humans.

Types of illnesses induced in humans: Anemia, arthralgia, erythema, myalgia.

Familial strategies:
Infection course — productive, short term-initial.
Viral replication — at the individual host level, primary tissue and organ tropisms are toward the throat; secondary tissue and organ tropisms are toward the circulatory system, erythrocyte precursor cells in bone marrow, possibly reticulocytes in blood, and skin; at the host population level, these viruses usually cause a disease of childhood; parvoviral disease is either mild or self-limiting in otherwise healthy children or adults.
Evasion of host defenses — uncertain.

Predominant routes of transmission between hosts: Uncertain, but potentially direct host-to-host contact, including transplacental, and indirect (vehicle-borne) contact via aerosols and fecally contaminated water, food, and fomites.

O. Viral Family Picornaviridae

Genera affecting humans: *Enterovirus, Hepatovirus, Rhinovirus*.

Familial nature with respect to members affecting humans: Viruses of humans.

Alternate hosts: Species affecting humans seem naturally limited to humans, but may pass to primates and canines.

Types of illnesses induced in humans: Diabetes, encephalitis, macular and maculopapular rashes of skin and mucosa, meningitis, myocarditis, otitis media, paralysis of skeletal muscles (occasionally including the diaphragm), pericarditis, retroocular pain, sinusitis (genus *Enterovirus*); hepatitis (nonprogressive) (genus *Hepatovirus*); coryza (genus *Rhinovirus*).

Familial strategies:
Infection course — productive, short term-initial.
Viral replication — at the individual host level, primary tissue and organ tropisms are toward the nasopharynx and small intestine; secondary tissue and organ tropisms are very genus and species specific and toward the beta cells of the pancreas, conjunctiva, liver, meninges, muscles (including the heart), neurons (including those of the central nervous system), and skin; at the host population level, infections caused by members of the genus *Enterovirus* usually are nonfatal, and both *Enterovirus* and *Hepatovirus* tend to result in asymptomatic

infections if acquired in infancy, though the likelihood of severe symptomatology increases with age at acquisition; infections caused by members of the genus *Rhinovirus* generally are symptomatic but essentially nonfatal regardless of host age.

Evasion of host defenses — members of genus *Enterovirus* avoid host non-immune defenses by resistance to low pH (resistant to stomach acid) and to moderate alkalinity.

Predominant routes of transmission between hosts: Indirect (vehicle-borne) contact via aerosols and fecally contaminated water, food, and fomites.

P. Viral Family Poxviridae

Genera affecting humans: *Molluscipoxvirus, Orthopoxvirus, Parapoxvirus.*

Familial nature with respect to members affecting humans: Viruses of humans and zoonotic.

Alternate hosts: One species affecting humans (smallpox) seems naturally limited to humans; monkeypox is a very notable but rare zoonotic exception and is presumably acquired from monkeys; several other species may cycle with domesticated bovines and ovines appearing as lesions on teats and udder.

Types of illnesses induced in humans: Necrotic lesions of abdominal organs and skin, nodules and tumors in skin, papular rash.

Familial strategies:
Infection course — productive, short term-initial.
Viral replication — at the individual host level, primary tissue and organ tropisms are toward the skin; secondary tissue and organ tropisms are toward the internal organs and lymph nodes; at the host population level, these viruses have very low transmissibility but have prolonged survivability on fomites due to extreme resistance to desiccation.
Evasion of host defenses — avoids host immune defenses by antigenic mimicry.

Predominant routes of transmission between hosts: Direct contact via host-to-host contact with skin lesions and indirect (vehicle-borne) contact via lesion-contaminated fomites (very notably blankets and other bedding items).

Q. Viral Family Reoviridae

Genera affecting humans: *Coltivirus, Orthoreovirus, Rotavirus.*

Familial nature with respect to members affecting humans: Viruses of humans and zoonotic.

Natural or alternate hosts: Those species of the genus *Coltivirus* infecting humans seem zoonotic with terrestrial mammals (notably rodents and squirrels) serving as their natural hosts; species of the genus *Orthoreovirus* cross-infect nearly all known terrestrial mammals (especially rodents); those species of the genus *Rotavirus* affecting humans seem naturally limited to humans.

Types of illnesses induced in humans: Hemorrhagic fever, meningoencephalitis (genus *Coltivirus*); upper respiratory symptoms (possibly associated with genus *Orthoreovirus*); gastroenteritis (genus *Rotavirus*).

Familial strategies:
Infection course — productive, short term-initial.
Viral replication — at the individual host level, primary tissue and organ tropisms are highly genus specific, and are toward the immune cells (genus *Coltavirus*), possibly upper respiratory area (genus *Orthoreovirus*), and the small intestine (genus *Rotavirus*); secondary tissue and organ tropisms are toward the brain and meninges; at the host population level, these viruses have high transmissibility, especially among newborns, for whom they usually produce asymptomatic infections; in older children and adults, these viruses likewise have a tendency to produce asymptomatic infections; although rarely fatal in well-nourished children, members of the genus *Rotavirus* are estimated to cause a million deaths every year in undernourished children.
Evasion of host defenses — avoids host immune defenses by infecting immune cells (genus *Coltivirus*), avoids host non-immune defenses by resistance to heat, low pH, and proteolytic attack (infectivity actually increased by proteolytic attack) (members of the genera *Orthoreovirus* and *Rotavirus*).

Predominant routes of transmission between hosts: Indirect (vehicle-borne) contact via fecally contaminated water, food, and fomites with the orthoreovirus possibly also being spread by aerosols.

R. Viral Family Retroviridae

Genera affecting humans: *BLV-HTLV retroviruses, Lentivirus, Spumavirus*.

Familial nature with respect to members affecting humans: Viruses of humans.

Alternate hosts: Species affecting humans seem naturally limited to humans.

Types of illnesses induced in humans: Carcinoma, encephalitis, leukemia (adult T-cell), lymphoma (adult T-cell), progressive chronic immunosuppression and immunodepletion (including acquired immunodeficiency syndrome), progressive myelopathy, sarcoma.

Familial strategies:
Infection course — productive, short term-initial, often followed by increasing to end-stage; also may seem nonproductive in the case of some endogenous retroviruses.

Viral replication — at the individual host level, primary tissue and organ tropisms are toward the immune cells (largely T-cell populations); secondary tissue and organ tropisms are toward the brain and intestines; at the host population level, those viruses considered transmissible (i.e., excluding endogenous retroviruses) have a very low transmissibility rate, produce infections whose incubation times are very long (10–40 years), and may pass through breast milk; the endogenous retroviruses are permanently integrated into the human genome and are passed genetically to all offspring.

Evasion of host defenses — avoids host immune defenses by rapid viral mutation and by infecting immune cells.

Predominant routes of transmission between hosts: Direct contact via host-to-host transfer of contaminated bodily fluids.

S. Viral Family Rhabdoviridae

Genera affecting humans: *Lyssavirus, Vesiculovirus*.

Familial nature with respect to members affecting humans: Zoonotic.

Natural hosts: Foxes, skunks, and vampire bats (genus *Lyssavirus*), cattle and horses (genus *Vesiculovirus*).

Types of illnesses induced in humans: Neuronal infections leading to encephalitis which appears invariably fatal (genus *Lyssavirus*); myalgia (genus *Vesiculovirus*).

Familial strategies:
Infection course — productive, short term-initial.
Viral replication — at the individual host level, primary tissue and organ tropisms are toward the neurons, including those in the spinal cord and limbic system of the brain, and the salivary glands (genus *Lyssavirus*); and toward either the muscles or nerves (genus *Vesiculovirus*); secondary tissue and organ tropisms are toward the adrenal cortex and pancreas (genus *Lyssavirus*); at the host population level, these viruses are essentially nontransmissible.
Evasion of host defenses — avoids host immune defenses by limited antigenic exposure within the host because the virus largely remains within neuronal cells until near end-stage (genus *Lyssavirus*).

Predominant routes of transmission between hosts: Direct contact via host-to-host contact associated with deposition of contaminated saliva into a bite wound and possibly associated with contamination of skin or mucosal wounds by other types of bodily fluids; in the case of the genus *Vesiculovirus*, vesicular fluids.

T. Viral Family Togaviridae

Genera affecting humans: *Alphavirus, Rubivirus.*

Familial nature with respect to members affecting humans: Viruses of human and zoonotic.

Natural or alternate hosts: Species of the genus *Alphavirus* cross-infect a wide variety of terrestrial vertebrates, mostly via mosquitoes and ticks; one species of the genus *Rubivirus* affects humans and it seems restricted to humans.

Types of illnesses induced in humans: Arthralgia, arthritis, diabetes, encephalitis, fetal developmental abnormalities (cardiological, diabetic, and neurological — including auditory, encephalitic, and visual — caused by *Rubivirus* if contracted during the first trimester of pregnancy), macular rash of skin, myalgia, myositis.

Familial strategies:
Infection course — productive, short term-initial.
Viral replication — at the individual host level, primary tissue and organ tropisms are toward the immune cells (specifically monocytes and macrophages in bone marrow, liver, lymph nodes, and spleen) and oropharynx; secondary tissue and organ tropisms are toward the beta cells of the pancreas, muscles, neurons of the central nervous system including the brain, skin, and synovial cells of joints; at the host population level, most members of the genus *Alphavirus* seem poorly transmitted between humans, and humans probably represent a dead-end host; infection by the genus *Rubivirus* is seldom fatal but highly transmissible via aerosols and usually causes a trivial exanthema of childhood or mild symptoms in adults, although infection during the first trimester of pregnancy can result in extremely severe developmental abnormalities.
Evasion of host defenses — avoids host immune defenses by infecting immune cells.

Predominant routes of transmission between hosts: Direct contact via either host-to-vector (genus *Alphavirus*), host-to-host (genus *Rubivirus*), or indirect (vehicle-borne) contact via aerosols (genus *Rubivirus*).

V. CONCLUSIONS

There are many types of viruses that afflict humans. We have managed to coevolve with some of these to lessen our misery. The struggle will continue as new viruses appear and as the existing ones reshuffle their genes or change their antigenicity by mutation. In the end, the contest is a struggle of biology versus biology, and the basic biology of the viruses is the same as ours.

REFERENCES

Evans, A. S., and Kaslow, R. A., eds. (1997). "Viral Infections of Humans: Epidemiology and Control," 4th ed. Plenum, New York.

Hurst, C. J., and Murphy, P. A. (1996). The transmission and prevention of infectious disease. *In* "Modeling Disease Transmission and Its Prevention by Disinfection" (C. J. Hurst, ed.), pp. 3–54. Cambridge University Press, Cambridge.

ICTV (International Committee on Taxonomy of Viruses) (1995). "Virus Taxonomy: Sixth Report of the International Committee on Taxonomy of Viruses" (F. A. Murphy, C. M. Fauquet, D. H. L. Bishop, S. A. Ghabrial, A. W. Jarvis, G. P. Martelli, M. A. Mayo, and M. D. Summers, eds.), pp. 1–507. Springer-Verlag, Vienna.

Villareal, L. P. (1997). On viruses, sex and motherhood. *J. Virol.* **71**, 859–865.

White, D. O., and Fenner, F. J. (1994). "Medical Virology," 4th ed. Academic Press, San Diego.

15

Impact of Avian Viruses

MICHAEL L. PERDUE and BRUCE S. SEAL

Southeast Poultry Research Laboratory
Agriculture Research Service
United States Department of Agriculture
Athens, Georgia 30605

VIRAL ECOLOGY

I. THE AVIAN HOST

A. Comparison of Class Aves with Other Vertebrates

In considering the ecology of the avian viruses and their impact on life on earth, it may be useful to first consider the host itself. The class Aves first diverged from the reptiles between 150 and 300 million years ago, depending on which current paleontological interpretations one accepts. Of the vertebrate classes, they are most often compared with the reptiles from which they evolved and with mammals because of their common warm-blooded nature. This shared feature with mammals is probably the most influential with respect to ecology, since a virus adapted to warm-blooded physiology would not fare well in the cold-blooded world, and vice versa. Along these lines, our recent experience is that some virologists think that a bird is a bird, and that if a given virus replicates in one, it will replicate in them all. This is, of course, far from the truth and perhaps should be a starting point for discussing avian virus ecology.

According to fossil records, the class Aves emerged from the extinctions of the Late Cretaceous period 65,000,000 years ago, somewhat bottlenecked, as did the class Mammalia, but since that time they have undergone parallel evolution with mammals and are equally diverse in their own right. Certainly some viruses, such as avian paramyxovirus 1 (of which Newcastle disease virus is the prototype), are infectious for numerous orders of birds. There are also other important avian virus strains such as the gallid herpesvirus 1 (infectious laryngotracheitis) and the Avihepadnaviridae, which appear to be exclusively confined to a single bird family or even genus. It is clear that birds, because of their close association with powered flight, do present a more homogeneous anatomy and physiology than do the class Mammalia (Feduccia, 1996). Whether this affects or specifies the molecular nature of viruses that infect birds is not really known. This feature, however, almost certainly uniquely impacts the natural distribution and ecology of the viruses that inhabit the flying birds. A virus infecting or persisting in an arctic tern could potentially be translocated up to 12,000 miles in a few weeks. Viruses such as some avian orthomyxovirus or avian paramyxovirus strains, which may exhibit subclinical infections, might be shared among a migrating flock of waterfowl and persist for indefinite periods as the flocks move from lake to lake. Along the way the virus might be shared with other birds crossing flight paths. So birds do present unique environments for transmission of viruses.

The commercial practices of humans have further provided unique opportunities for transmission not normally seen for birds. The order Galliformes in particular, which includes domestic and wild fowl; pheasants, quail, and turkeys, has been unquestionably affected. It would be quite safe to say that humans have both determined and upset the ecological balance among members of this order and the

viruses that affect them. Due to continuous breeding practices, live-virus vaccination regimes, and by housing tens of thousands of birds in a single enclosure,
situations never encountered in natural settings are created. Whether this has
affected the virus–host ecological balance in other orders of undomesticated birds
as well is not known; but it would seem highly likely.

One of the most important features of a host–parasite relationship, of course, is
the host's immune defense. While there are similarities shared with respect to the
immune system, particularly the dual (humoral and cell-mediated) nature, there
are marked differences between cold- and warm-blooded vertebrates and additional significant differences between birds and mammals (Eerola *et al.*, 1987).
The discovery of processing and maturation of immunoglobulin-producing lymphocytes was made as a result of characterization of an avian-specific organ, the
Bursa of Fabricius (Ratcliffe, 1989). Although functional equivalents exist in
mammals, the bursa is a distinct and wholly avian-specific organ. Since several
viruses are known to affect this organ specifically, it should be considered a unique
ecological niche.

A second significant difference in the immune system of birds is in the apparent
genetic content responsible for specifying the avian major histocompatibility
complex (MHC). The chicken MHC appears considerably more simple (providing
the oxymoron: a simpler complex) than the mammalian MHC (Kaufman and
Wallny, 1996). Several alleles have arisen and considerable recombination documented in the mammalian MHCs that thus far have been studied. These are
responsible for producing a great variety of class I-, II-, and III-type proteins used
in recognition and presentation of antigen. In the chicken, some MHC haplotypes
produce only one type of class I protein and there is no evidence for recombination
at all (Kaufman and Wallny, 1996). This has led to speculation that the relationship
with the avian pathogens has evolved significantly differently from the mammals.
One result of this difference may be the occurrence and frequency of either
resistance or sensitivity to specific viral infections encountered in chickens (see
below). Thus, the class Aves presents several unique features that might ultimately
affect the ecology of viruses that infect them.

B. Elements of Avian Systematics

The classification of birds has presented a significant challenge to systematists.
The number of species has actually decreased over the years because of the clearer
genetic relationships that have emerged. Conversely, additional species have been
defined as recognition of convergent evolution has become clearer. Currently, new
species are identified at a rate of about two per year. While there remains some
disagreement among specialists, most accept the current classification of 30

orders, 174 families, 2044 genera, and 9020+ species (Gill, 1990). Molecular analysis of avian genes for phylogenetic studies is still in its early stages. Restriction fragment length polymorphism (RFLP) analysis and sequence analysis of 12S mitochondria DNA has yielded some molecular phylogenetic information but not enough to gain any insight regarding the viruses of birds (Hedges *et al.*, 1996; Cooper and Penny, 1997; Mindell *et al.*, 1997). Roughly speaking, the flightless orders represented by ostriches, rheas, cassowaries, and kiwis are thought to be the most ancient, while the bewildering array of members of the order Passeriformes (representing 60% of known species and 40% of known families) are thought to be the most recent (Gill, 1990). Trying to determine how long viruses have been associated with various avian groups is of course as impossible as it is with virus–host relationships in any other classes. Orthomyxoviruses, paramyxoviruses, and coronaviruses have recently been isolated from ostriches, rheas, and emus, indicating no absolute barriers in these more ancient birds.

C. Geographic Distribution of the Avian Host

The class Aves is of course distributed worldwide. In addition, there are several hundred species that migrate, sometimes in spectacular fashion. Biogeographers have divided the earth into six distinct faunal regions, which correspond somewhat with the major continental areas (Welty, 1982). These include the Nearctic (North America and Greenland), Palearctic (Asia, Europe, and North Africa), Ethiopian (Central and Southern Africa), Oriental (India, Southeast Asia), Australasian (Australia, New Guinea), and Neotropical (South and Central America). Each area contains its own characteristic birds. In the Northern Hemisphere (Nearctic and Palearctic), most species are migratory, which is not the case in the other areas. The Neotropical has the richest and most abundant bird life, and the Southern Hemisphere has by far the most families represented, as well as the most families peculiar to a given region.

Each year, billions of land birds and waterfowl in North America and Asia head south to South America and Africa, respectively, carrying their viruses with them. The size and scale of these geographic relocations are unmatched by any other land vertebrates. The sea mammals are the only other comparable migrating vertebrates, and they surely cross paths with the birds. In the most interesting putative contacts, purely avian-origin type A orthomyxovirus of at least two different subtypes were isolated from dead and dying seals off the coast of New England in 1980 and 1982–1983 (Webster *et al.*, 1981b; Hinshaw *et al.*, 1984). This represents viral ecology at its most forceful, being effected between two warm-blooded vertebrate orders during their natural migration and geographic interaction.

II. REPLICATION AND PERSISTENCE OF VIRUSES IN THE AVIAN HOST

A. The Embryonated Egg

If one takes a broad look at the virus families associated with avian hosts (Table I), it is actually easier to list the families of viruses that infect vertebrates but that do *not* yet have a clear avian member. These are the Iridoviridae, Arenoviridae, *African swine fever-like* viruses, and Filoviridae. A similar comparison of families that do not contain a mammalian member yields only the Birnaviridae. So, in one sense, the mammalian host might be considered more virus-friendly, rather than the alternative. Still, many investigators think of bird tissues, in particular the embryo, as being an ideal medium for identifying new viruses. If just from an experimental and practical, rather than a natural, point of view, the avian host has played a tremendous role in our understanding of viral ecology. Many of the most important findings in virology have been made utilizing the chicken embryo and in chicken cell cultures. The egg also continues to provide an abundant and important substrate for the production of veterinary and human vaccines.

If we think of ecology as the study of organisms and their relationship with their environment to include all other organisms, we most certainly encounter a unique relationship with viruses. Unlike higher pathogens, viruses really do not mate, communicate, or colonize (except in the very broadest stretches of the imagination); they simply parasitize and replicate. We try and make them more familiar by defining higher organism-type genetic alterations as "evolution," when in truth the viruses may simply be adapting to the evolutionary pressures encountered by, or within, its host. Obviously, the host is central to the viruses' lifestyle and must be considered a major part of virus ecology. In this sense, information gathered on various viruses as they replicate in embryos or embryo cells, in the absence of immune pressure, should only be considered "natural" for those viruses that are transmitted vertically. The embryo or cell culture then can only provide a window on the natural ecology of the organism in which the immune system is not a factor. In ecological terms, only the natural state of the virus–host relationship becomes important. Viruses being parasites, these relationships more often than not eventually result in a disease state.

B. Unique Tissues and Viral Replication

It is not possible to distinguish the molecular biology of avian viruses from that of non-avian viruses on the basis of features unique to the avian system. However,

TABLE I

Families of Viruses Infecting Birds[a]

Virus family (1)	Representative member(s) (2)	Primary host(s) isolation (3)	Range of hosts (4)	Pathogenesis (5)	Transmission (6)	Ecological/economic impact (7)
dsDNA						
Poxviridae Genus: *Avipoxvirus*	Fowlpox, Canarypox	Domestic poultry, turkeys, pigeons	>60 species, and 20 families of birds; distribution worldwide	Mild cutaneous form and more severe diphtheric form; mortality may reach 100% in canaries	Common in domestic poultry; slowly spreading cutaneous form; Respiratory, mechanical — insects	Low to moderate in poultry; can be devastating in aviaries
Herpesviridae	Infectious laryngotracheitis (ILT); Marek's disease virus (MDV)	Domestic poultry: chickens, turkeys, ducks, pigeons	Several strains of herpes in many orders, including falcons, cormorants, psittacines, cranes, quail	ILT: Acute respiratory, & milder enzootic MDV: classical and acute forms — oncogenic; DPV: acute enteric disease in ducks, geese and swans	Lateral — respiratory and ocular; virus may be reactivated from latency and transmitted; MDV can spread from feather follicles	Quite high; MDV and ILT two of the major viral diseases of commercial poultry; virulence increasing in field isolates of MDV
	Duck plague virus (DPV)	Ducks	Limited			
Adenoviridae Genus: *Aviadenovirus*	Group I — CELO/Phelps strain; Group II — HE & MSD viruses; Group III — egg drop syndrome virus	Chickens, turkeys, pheasants	I. Widespread in fowl; II. Primarily Galliformes; III. Ducks, geese, laying chickens	I. Low: poultry; high: quail. II. Variable in turkeys, chicks. III. Low	Vertical transmission very important mode	Low to moderate in poultry; different in different countries
Papovaviridae	Fringilla papilloma virus; Budgerigar fledgling disease (polyoma-like)	Finches Fledgling budgerigars	Finches; perhaps African green parrots; Budgerigars	Typical papillomas on legs and feet	Assumed same as mammalian papillomas	None or very low; occasional problems in aviaries

continued

1	2	3	4	5	6	7
ssDNA						
Parvoviridae	Goose parvovirus (GPV); Muscovy duck parvovirus	Geese, Muscovy ducks	Only Geese and ducks thus far	Derzy's disease; age dependent; high mortality in young birds; only seroconversion in adults	Vertical transmission to goslings most serious; horizontal also important	Restricted thus far to commercial geese, duck operations
Circoviridae	Chick anemia agent virus; Psittacine beak and feather disease virus / Pigeon circovirus (PCV)	Chickens, pigeons, psittacines / (mostly EM evidence)	Chickens, pigeons, psittacines, doves, canaries, finches	All strains immunosuppressive thus far; low mortality in chickens; high in pigeons and psittacines	Horizontal and vertical; mixing of racing pigeons important mechanism for PCV	Difficult to assess because of immunosuppressive effect and secondary infections
Reverse-transcribing DNA/RNA viruses						
Hepadnaviridae Genus: *Avihepadnavirus*	Duck hepatitis B virus (DHBV); heron hepatitis B virus	Ducks, herons	Limited to ducks and herons	Very low; no known association with hepatocarcinomas	Primarily vertical	Essentially none
Retroviridae Genus: *Avian type C retroviruses*	Lymphoid leukosis virus (LLV)/sarcoma virus group / Reticuloendotheliosis virus	Chickens, pheasants / Turkeys, ducks	Other Galliformes / Turkeys, ducks, chickens, quail, pheasants	Lymphoid leukosis; various other neoplasms	Vertical and horizontal; present in large proportion of flocks	Variable: sporadic significant problems yield significant economic loss
dsRNA						
Reoviridae	Avian orthoreoviruses, avian rotaviruses, avian-associated orbiviruses	Chickens, turkeys	Numerous strains isolated form turkeys and chickens; numerous isolations — other orders	Variable, generally low in commercial birds; predominantly arthritis/tenosynovitis	Vertical transmission important as "seed" to promote horizontal transmission	Significant although usually low-performance rather than high-mortality
Birnaviridae	Infectious bursal disease virus (IBDV)	Chickens, turkeys	Antibody and other surveillance suggest widespread distribution	Variable: inapparent to acute; highly pathogenic strains in Europe; immunosuppressive	Horizontal, highly contagious; "natural" hosts unknown because of widespread vaccine use	Worldwide distribution; significant poultry pathogen

continued

1	2	3	4	5	6	7
sRNA (–)						
Orthomyxoviridae	Avian influenza virus (AIV) subtypes H1-H15; N1-N9	Ducks, shorebirds, turkeys, chickens	Strain dependent; can include diverse array, e.g., ratites, but not unlimited	Subclinical to highly lethal; apparently subtype dependent; HP = systemic infection	Horizontal only, highly contagious, oral and respiratory; documented transmission to mammals and humans	Low path form varies from insignificant to mildly significant; HP form: high impact and potential as OIE list A disease
Paramyxoviridae	Newcastle disease virus; avian paramyxoviruses (PMV) 2 and 3; Turkey rhinotracheitis virus, avian pneumovirus	Waterfowl; chickens; pigeons; exotic birds, turkeys	Very widespread, many orders of birds susceptible	Three broadly defined pathotypes of NDV: mild to highly lethal	Horizontal only, highly contagious, oral/resp.; documented transmission to humans	Very significant world-wide; sporadic in nature; vaccination has influenced isolates; lethal form: high impact, OIE list A disease
Rhabdoviridae	No natural avian isolates; vesicular stomatitis virus, rabies	Antibodies in a few species including wild turkeys, chickens	Pathogens of mammals that may be transmitted to birds	None except under experimental conditions	None, appears to only be acquired from infected mammals during heavy outbreaks	None
Bunyaviridae	Too numerous to list	Several	Over 50 members of the family isolated from more than 20 bird species	No characteristic pathogenesis; many infections detected by Ab	All arthropod-borne viruses; various invertebrate vectors	Not well understood if any impact at all
ssRNA (+)						
Coronaviridae	Infectious bronchitis virus (IBV); Turkey coronavirus	Chicken	Very limited; some evidence in other Galliformes; worldwide	Variable; some can yield 10–20% mortality; respiratory symptoms predominant	One of the most rapidly transmitted poultry respiratory viruses	Very significant; vaccination has generated new strains; new antigenic variants continue to arise
		Turkeys	Turkeys exclusively	Enteric disease	Low serum titers against IBV in human workers	Virulence increasing?

continued

1	2	3	4	5	6	7
Togaviridae	Eastern equine encephalitis virus (EEEV)	Horse; pheasants, turkeys, ducks, pigeons, passerine birds	Arbovirus; *Culiseta melanura* main vector; wild Passeriformes	Neuropathic; high mortality in pheasants and young poultry	Arbovirus vector; birds appear to only transmit to insects, except EEE, which can transmit bird to bird	Minor, except in affected game-bird operations
	Western EEV and 10 others, from naturally infected wild birds	Various wild birds	Wide	Insignificant		
Flaviviridae	Turkey meningoencephalitis	Turkeys	Unknown but limited	Neuropathic; variable mortality; egg production drop returns to normal following disease	No known arthropod vector	Minor
	16 or more isolated from naturally infected wild birds					
Astroviridae	None isolated	None isolated	Unknown but seen by EM only in turkey poults	Moderate enteric disease associated with weight loss	Unknown; detectable only by EM thus far	Unknown but minor
Picornaviridae	Avian encephalomyelitis virus (AEV)	Chickens, pheasants, quail, turkeys	Limited, not found in wild birds	Encephalomyelitis	Oral, horizontal; vertical also important; natural isolate enteric	Significant in commercial chickens, only in chicks of unvaccinated stocks
	Avian nephritis virus (ANV)	Chickens, turkeys	Chickens, turkeys	Enteric disease and nephritis	Oral, not completely understood	Insignificant
	Duck hepatitis virus (DHV)	Domestic ducks	Other hosts; experimentally no natural	Gross hepatitis and opisthotonos; high mortality in young birds	Respiratory; contagious; becomes endemic in affected flocks	Only significant in commercial duck populations
Caliciviridae (tentative assignment)	Chicken calicivirus	Chickens	Unknown	Enteric disease shown using purified virus preparation	Unknown	As member of enteric disease complex, holds some importance

[a]Information collected primarily from three sources: Calnek *et al.* (1997), McFerran and McNulty (1993), and ICTV (1995).

there are some important distinctions with which viruses must deal. One dramatic difference between birds and mammals is body temperature. On average, birds operate at 3–4°C higher than mammals, and most can readily regulate their temperature, some dropping body temperature as much as 10°C during the night (Gill, 1990). While this certainly could influence such things as rates of viral enzyme activity, polymerase fidelity, or protein stability, there are no situations documented in which a particular virus group is restricted to class Aves solely on the basis of body temperature.

It is possible to distinguish some important receptor-specific distinctions unique to avian systems. One example is found in the influenza type A viruses. The glycosidic linkages associated with sialic acid residues on the avian versus mammalian cell surface serves to restrict the strains of viruses able to replicate. Influenza virus hemagglutinin (HA) proteins bind to sialic acid residues on the surface of the host cell. Although there is no absolute barrier, those viruses that replicate well in avian cells have a receptor binding pocket on the surface of the HA that has a preference for the α-2,3 siallyl–sugar linkages abundant in bird tissues. Those viruses that replicate well in mammalian cells have a preference for α-2,6 linkages more prevalent in mammalian cells (Rogers and Paulson, 1983; Murphy and Webster, 1996). There are probably many more receptor specificities associated with other avian virus infections, particularly among the herpesviruses, where host range is narrowly dictated.

Another example of virus-specified tissue tropism unique to birds would be infectious bursal disease virus (IBDV; genus *Avibirnavirus*), which has "evolved" an affinity for the bursa, a bird-specific organ. The other two known genera of Birnaviridae infect fish and invertebrates. Since there are as yet no mammalian members, it is tempting to speculate that IBDV is bird specific because of its evolved affinity for the bursa.

Air sacs are also uniquely avian structures. The bird lung is a fixed tissue incapable of expansion like mammalian lungs. The air sacs are a series of extensions of the respiratory system that expand with the musculature of the body cavity and allow the large-scale rapid oxygen transfer needed for powered flight. The air sac system is extensive throughout the bird, even encroaching into bone tissue; some birds have as many as nine distinct air sacs. Many viruses replicate and cause disease in this unique organ system, and there are examples of apparent preferences by some viruses for air sac tissue over other respiratory tissue. Finally, feathers, the most notable and prominent distinguishing feature of the class Aves, provide a unique niche for at least two three viruses — avipox, Marek's disease, and psittacine beak-and-feather disease virus — which can replicate in and are spread from feather follicles (Biggs, 1985; Tripathy and Reed, 1997).

C. Effects of the Immune System on Virus Replication and Spread in Birds

As mentioned previously, the immune system of the intact bird is a critical feature in the establishment of ecological relationships between many viruses and hosts. As the immune system plays a major role in the relationships, they will be covered, but there are some extraordinary examples that deserve their own mention. Marek's disease, caused by an alphaherpesvirus, is most commonly associated with lymphomas developing relatively early in the life of a chicken. The initial infection in MD is a respiratory infection probably initiated in macrophages. A typical field infection would then progress through the following phases in a bird:

$$\text{Initial lytic infection} \Rightarrow \text{Latency} \Rightarrow \text{2nd lytic infection} \Rightarrow \text{Oncogenesis}$$

Latency is established primarily in lymphoid cells, and mostly in activated T lymphocytes. In susceptible birds, the latent state progresses to a second round of lytic infections at multiple sites in the bird. Interestingly, at this point the feather follicle epithelium is the only site where complete virus replication occurs and becomes a significant source of environmental infectious virus (Calnek and Witter, 1997). Concomitant and permanent immunosuppression occurs in the affected bird. The disease then progresses into a lymphoproliferative phase in susceptible birds that can range in severity depending on the virus strain and breed of chickens. Some breeds have shown natural resistance to this progression, and, although the mechanism of resistance is not completely understood, it involves primarily lymphoid tissue and is dictated mostly by genes involved in the immune response (Venugopal and Payne, 1995). It should be clear from this scenario that the immune system in this host–parasite relationship plays a critical role in the ecology of Marek's disease virus.

Infectious bursal disease virus, as mentioned earlier, is a virus that replicates exclusively in lymphoid tissue. The virus can be detected replicating in the bursa of Fabricius and within circulating lymphocytes as early as 4 hours after infection of a chicken. It causes acute degeneration of various lymphoid tissues within the first day of infection and results in a severe, albeit age-dependent, depression in the humoral immune response, being most dramatic in very young birds. Interestingly, the infection does not suppress B-cell responses to the viral antigens themselves; in fact, there is stimulation of proliferation of B cells committed to anti-IBDV antibody production (McFerran, 1993). One may only speculate as to what role, if any, this may play in the viral ecology or replicative cycle, but the immune system is once again a major participant in this host–parasite relationship.

Finally, the effects of vaccination programs on the ecology of avian viruses cannot be overemphasized. In one sense, we have an ongoing experiment where humans control the type of viruses to which certain species of birds are exposed. Since live viruses generally yield much better immune responses, they are employed most often. In the case of the single-stranded RNA viruses, noted for their ability to rapidly mutate and avoid the immune system, this has the effect of artificially challenging the immune system, creating selection pressure between host and parasite that would not normally occur. This may be effective in the short run, protecting against disease, but the long-term effects are unknown.

III. THE MAJOR GROUPS OF VIRUSES THAT INFECT BIRDS

Summarized in Table I are some of the most important features of relationships of various viruses with their avian hosts as well as the wide variety of relationships that exist. It is not feasible to list all the interesting attributes for each virus–host relationship, but the table presents a variety of relationships that will be covered in more detail in the following sections. As such, the table should not be taken as the final word on each member virus. For example, in the case of turkey meningoencephalitis virus, a *Flavivirus*, reduced egg production is listed under pathogenesis. Reduction in egg production is a common feature in many virus infections of poultry, and while listing it under pathogenesis somewhat stretches the meaning of that word, this clinical sign is one of the most important aspects of that disease in turkeys.

Although there is variation in the economic or ecological impact of various viral groups from year to year and among geographic sites, the "Top Ten" list of virus groups exhibiting routine significant impact on commercial poultry worldwide (not necessarily in order of impact) are paramyxoviruses (Newcastle disease); coronaviruses (infectious bronchitis); herpesviruses (infectious laryngotracheitis; Marek's disease; duck enteritis); reoviruses (viral arthritis); picornaviruses (avian encephalomyelitis); adenoviruses (egg drop syndrome); retroviruses (lymphoid leukosis); orthomyxoviruses (avian influenza); poxviruses (fowlpox); and birnaviruses (infectious bursal disease). The circoviruses (chick anemia) could likely be included in the above list, except it is not yet known to what extent the viral infection alone influences morbidity and mortality (see below).

What may not be obvious from Table I is that if one searches for viruses in a given avian species one will likely find them. In some of the virus families listed, investigators were forced to clearly separate the disease-causing virus in question from accompanying "contaminant" viruses, which may or may not have influenced the original disease manifestation. There appear to be many viruses of birds that in certain ecological conditions and in certain species may be considered "normal flora" and are not associated with disease. These, obviously, are less interesting to any funding agencies and consequently do not receive much research attention. No one really knows to what extent their transmission and persistence in avian populations affect their own ecology or that of their hosts.

The Office International des Epizooties (OIE), the principal world organization for animal health, provides listings of the most serious infectious diseases of animals (OIE, 1996) and divides them into two groups: list A diseases, which "includes those diseases that spread rapidly, the scope of which extends beyond national borders" and "have particularly serious socioeconomic or public health consequences"; and list B diseases, which include those "that are considered to be of socioeconomic and/or public health importance within countries." Of the avian viral diseases listed above, only highly pathogenic avian influenza and velogenic Newcastle disease are in list A; Marek's disease, infectious bursal disease, infectious bronchitis, duck enteritis, and infectious laryngotracheitis are in list B. The only avian disease in OIE's list and not on our top 10 list is duck hepatitis, which is a complex of diseases caused by at least three virus families, and generally limited to country-specific origins.

IV. IMPACT OF VIRAL TRANSMISSION AMONG MEMBERS OF THE CLASS AVES

A. Commercial Poultry Production

The ecological impact of viruses of birds ultimately interests us as *Homo sapiens*, perhaps only to the extent that we are affected. In this respect, there is no question that the major impact thus far has been on raising birds as a food source for our species. Ecologically speaking, this impact could have very significant consequences when one considers that poultry provide the most widely used protein source in the world. Thus, we will consider for the most part how viruses affect this food source. The most significant and widespread infections of wild birds will be discussed as they are encountered relative to commercial and domestic birds.

The most imminent and significant human public health concerns with regard to bird viruses appear to be twofold:

1. **The potential relationships with type A orthomyxoviruses that have become at least partially adapted in a totally nonpathogenic state to some avian orders.** There is compelling evidence that these viruses may also replicate in pigs and re-assort with pig and/or human strains of influ-

enza, yielding new variants capable of replicating and causing disease in humans. They also find their way into commercial poultry, sometimes with devastating consequences. With the recent documented transmission of a lethal avian influenza virus from commercial poultry to humans, these ecological relationships take on new significance.

2. **The presence of a large reservoir of arboviruses in wild birds, some of which, when transmitted by invertebrate vectors to mammals, cause disease.**

Beyond these two examples, other relationships are less directly important to human public health. Other avian-origin viruses are capable of replicating in and sometimes causing mild disease in humans, but there is obviously not room in one chapter to cover each in detail.

B. Infectious Laryngotracheitis and Other Nononcogenic Herpesvirus Infections

1. Infectious Laryngotracheitis

Infectious laryngotracheitis (ILT) is a respiratory disease almost exclusively of chickens. Infections in turkeys and pheasants have been reported, but surveys have yielded no wild bird reservoir or other domestic poultry reservoir (Cranshaw and Boycott, 1982). Based on this and the knowledge that ILT apparently exhibits little antigenic heterogeneity, it has been proposed that through proper husbandry practices and appropriate vaccination techniques the disease could be eliminated from commercial poultry (Bagust and Johnson, 1998). The virus is a member of the Alphaherpesvirinae subfamily and is identified taxonomically as gallid herpesvirus I. The disease is almost exclusively respiratory, with no systemic involvement. The severity of disease can vary from significant mortality (70%) in young birds to an inapparent infection of adult birds. There are age-dependent effects on the pathognomonic signs, and there do appear to be strain-specific virulence differences. However, different isolates do not exhibit sufficient genetic heterogeneity thus far to identify specific virulence factors (Bagust and Guy, 1997). The most interesting aspect of ILT is its capacity for persistence in infected birds and flocks showing no disease signs. This persistence is most likely due to establishment of the latent state and recrudescence. Numerous studies have demonstrated re-isolation of virus many months after initial infection, and one more recent study demonstrated reactivation of latent virus due to stress factors (Hughes *et al.*, 1989). This latency achieved by herpesviruses could certainly be considered a unique ecological state, and most herpesvirus infections in birds are associated with its establishment.

2. Duck Viral Enteritis

Duck plague, also known as duck viral enteritis (DVE), is caused by an alpha-herpesvirus classified as anatid herpesvirus-1, which infects free-living and domestic ducks, geese, and swans (Sandhu and Leibovitz, 1997). The disease is acute and often associated with high morbidity and mortality. The virus has a worldwide distribution and has caused numerous documented outbreaks in free-living ana-tids. The first documented North American outbreak was in commercial ducks on Long Island, New York in 1967 (Leibovitz and Hwang, 1968), and since that time sporadic reappearance of the virus in commercial and wild populations has occurred. Major outbreaks in free-living birds along the Mississippi flyway and a large epornitic in South Dakota in 1973 killed tens of thousands of wild ducks and geese (Brand, 1987). Vertical transmission and recrudescence of latent virus has been established experimentally in mallard ducks but has not been demonstrated in wild waterfowl. Species susceptibility may vary among various waterfowl, although more than 30 species have been shown to be naturally or experimentally infected and virulence differences among DVE strains have been demonstrated.

3. Herpesvirus Infections of Pigeons and Wild Birds

The other notable avian herpesvirus infection occurs in pigeons. The virus is taxonomically designated as columbid herpesvirus I and is antigenically indistin-guishable from natural isolates taken from wild falcons and owls (Vindevogel and Duchatel, 1997). The causative virus is antigenically distinct from ILT, MDV, herpesvirus of turkeys (HVT), and the anatid herpesvirus I. The disease associated with infection by columbid herpesvirus I is a major cause of growth retardation and bad performance in homing pigeons, though mortality is generally low. This virus, like ILT, becomes latent, reappears, and is shed in asymptomatic birds and flocks. Other antigenically distinct herpesviruses have also been isolated from cormorants, quail, and storks. Kaleta (1990) has proposed the division of these various herpesviruses into eight antigenic groups. The host specificity for each group varies, but in general they appear to be strictly adapted to the host of origin (as exemplified by the gallid herpesvirus I). Gallid herpesvirus II and HVT will be discussed in the following section.

C. Avian Tumor Viruses

1. Herpesviruses

The impact of transmissible neoplastic diseases of poultry has been quite variable over the years. Prior to vaccination, losses to Marek's disease were often devastating, and even in marketable flocks condemnations due to lymphomas at processing plants could reach 2%. Lymphomas caused by MDV and retroviruses are still the most common viral neoplastic diseases of poultry, and a recent increase in mortality and evolution of more virulent MDV strains indicates that the impact of these viruses will continue to be felt (Witter, 1996). Marek's disease is caused by a herpesvirus that has two very similar relatives: a second nononcogenic serotype and the herpesvirus of turkeys (HVT). These are sometimes classified as serotypes 1–3, respectively, of the gallid herpesvirus II strain. They share common antigens that distinguish them from the nononcogenic herpesviruses but can be distinguished on the basis of antigenic differences (Calnek and Witter, 1997). Only the oncogenic MDVs (serotype 1) cause significant problems in commercial poultry. The herpesvirus of turkeys is ubiquitous among commercial flocks and quite prevalent in wild turkeys but has not been directly associated with disease. The serotype 2 strains were originally thought to be nononcogenic apathogenic MDV isolates until they were serologically distinguished from the original isolates. The type 2 serotypes are associated with subclinical infections in chickens, although not as prevalent as turkey strains.

The HVT isolate is a very interesting and important isolate, as it was used (and is still used) in vaccine formulations against Marek's disease in chicken flocks. It has been very effective at protecting flocks against lymphomas and thus correctly billed as the first anticancer vaccine. Marek's disease is relatively well controlled by using mixtures of primarily serotype 2 and 3 isolates in vaccine formulations. However, this vaccination program may ultimately extract a price. Shown in Figure 1 is a diagrammatic representation of the evolution of virulence in MDV strains associated with changing vaccine formulations. The extent to which these formulations have influenced the evolution of virulence is not proven, but certainly the association is undeniable. The 1990s have brought an increase in incidence of Marek's disease cases, and the presence of these acutely virulent strains raises concerns for the future. What is clear is the role of humans in the generation of these strains. There is evidence that the commercial housing practices developed in the 1950s and 1960s resulted in generation of strains that were oncogenic and that this rapid evolution of virulence in the last 40 years is probably due to human control of commercial bird populations.

2. Retroviruses

The avian retroviruses have one of the most interesting histories of all of the avian viruses. The first transmissible lymphomas were demonstrated in 1908 by Ellermann and Bang (1908) and the first cell-free transmissible solid tumor by Peyton Rous three years later (Rous, 1911). The etiologic agents of both of these diseases were later shown to be members of what is now known as the avian leukosis virus–avian sarcoma virus complex (ALV–ASV) of related retroviruses.

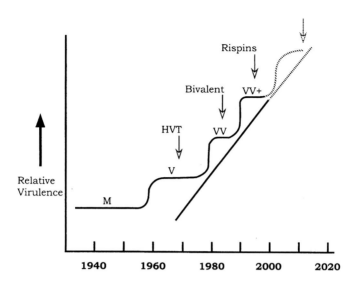

Fig. 1. Representation of field and laboratory observations in the United States on the development of virulence in Marek's disease (MD) isolates. The turkey herpesvirus (HVT) vaccine was first introduced in 1970 to protect against MD. Then in 1983 a bivalent vaccine was introduced, followed in 1995 by the Rispens strain vaccine. There is an apparent relationship between introduction of these vaccines and development of virulent phenotypes among recovered field outbreak isolates. M = moderate, V = virulent, VV = very virulent, VV+ = very virulent plus. Graciously provided by Dr. Robert Silva, Avian Disease and Oncology Laboratory, USDA, East Lansing, Michigan.

Scientists investigating this interesting group of viruses have garnered more Nobel Prizes (six) than with any other group. Because of their relative simplicity in genetic content and their close association with the genetic character of their host, they have provided a bountiful model for the study of oncogenesis. These are RNA-containing viruses that replicate via an intermediate DNA stage that is most often incorporated into the host genome. The integrated viral genomes serve as templates for production of new progeny genomic RNA molecules and the mRNAs needed to make new viral proteins (Coffin, 1996). In the process, these viruses transform the host cell into a tumor cell. There is a significant array of viral subtypes and relationships that exist between the host and avian retroviruses and numerous reviews that may be consulted (Crittenden, 1981; Swanstrom and Vogt, 1990). There are several viral subgroups, based on antigenic differences alone, in the surface envelope glycoproteins. Viral subtypes have also been grouped on the basis of whether the viruses rapidly induce neoplasia because of the presence of a viral oncogene, or whether they induce slow development of tumors due to their integration into the host genome and subsequent activation of cellular oncogenes.

In commercial poultry, the predominant problems are caused by the latter, the slow-inducing lymphoid leukosis viruses (LLVs). The sarcoma-, myeloblastosis-, and erythroblastosis-inducing strains cause only sporadic problems. Additionally, it has been calculated that losses due to LLV because of poultry performance decreases are actually greater than those due to the lymphomas. Still, these viruses continue to cause condemnations at slaughter and must be distinguished from the more rapidly forming MDV-induced lymphomas.

Two other oncogenic avian retroviruses pose significant problems and possess much wider host ranges than the LLV–sarcoma complex. The relatively new subgroup "J" has been characterized in Europe as causing significant myelocytoma and endothelioma. In contrast to the LLVs, subgroup J viruses have a high tropism for cells of the myelomonocytic series but a very low tropism for bursal cells (Arshad et al., 1997). A related subgroup J virus has been identified in broiler–breeder flocks in the United States as well (Smith et al., 1997). Additionally, of note are those retroviruses of the reticuloendotheliosis virus (REV) group. Infection with members of this group only rarely results in disease, and then disease most commonly occurs only in the hosts from which REV was first isolated: turkeys. Still the REV group is predicted to have significant potential for causing problems in poultry (Witter and Johnson, 1985), and the wide host range of the virus group has prompted speculation regarding potential as a public health risk. There are, in fact, data demonstrating antibodies to the virus in human and other mammalian sera (Johnson et al., 1998).

D. Coronavirus Infections

1. Chickens

Coronaviruses contain a large positive-sense RNA genome of approximately 30 kb. Members of this virus family infect both mammals and birds (Cavanagh et al., 1994). Infectious bronchitis, caused by a *Coronavirus*, infectious bronchitis virus (IBV), is one of the major poultry viral diseases of worldwide economic importance. It is also one of the most rapidly spreading avian respiratory diseases known (McMartin, 1993). The virus is antigenically variable owing to a high mutation rate in the surface glycoprotein S gene. New antigenic variants of IBV continue to be isolated from various geographic regions. Thus far, it has not been possible to determine the number of distinct IBV subtypes, but there are at least six (Siddell, 1995). Strains can differ in their virulence and tissue tropism, but in general the disease is a rapid-onset respiratory distress that can cause significant mortality in young birds. In established flocks, the infection is often associated with growth retardation and reduction in egg production that may be exacerbated by other respiratory pathogens. Estimates of a 10 to 20% loss in market value have been made for a typical outbreak in a flock. Vaccination is employed to control the disease, and an interesting related feature is the demonstration that vaccine strains can undergo recombination with wild-type strains. Consequently, new IBV antigenic variants of the S1 gene emerge with characteristics of both viruses (Cavanagh *et al.*, 1992; Wang *et al.*, 1993). IBV is also an excellent example of a virus exquisitely adapted to its host, as it is not found naturally in any other reservoir. Chickens seem to be the only reservoir, and the virus is capable of persisting in some manner within populations and later being transmitted to naive flocks. Evidence suggests the mode is primarily airborne transmission. In controlled studies, birds in houses 115 feet away from an experimentally "seeded" house were infected via aerosol transmission (McMartin, 1993), and circumstantial data from natural outbreaks suggest transmission over distances of several hundred yards (Cumming, 1970).

2. Turkeys

Coronaviral enteritis of turkeys, also known as bluecomb disease, may also be species specific since chickens, pheasants, and quail do not exhibit any disease following inoculation with a strain of virus that is virulent for turkeys (Hofstad *et al.*, 1970; Larsen, 1979). Coronaviruses have been isolated from ratites with enteric disease in zoological collections, but the relationship of these isolates with the turkey coronavirus (TCV) has not been investigated (Frank and Carpenter, 1992; Kennedy and Brenneman, 1995). Transmission of TCV appears to be restricted to the fecal–oral route, although cell-free lysates of the bursa from infected birds can be used to transmit the agent orally. Among turkey flocks, the primary mode of transmission of coronaviral enteritis is via contaminated personnel and equipment. However, since TCV is excreted in fecal material and is very stable, it is conceivable that wild birds may serve as vectors for the agent. Morbidity is usually as high as 100%, with mortality becoming as high as 50% in an infected turkey flock. Mortality among poults may be much higher with increased numbers of secondary bacterial gastrointestinal infections contributing to severity of disease.

Turkey coronavirus also shares features reported for mammalian hemagglutinating coronaviruses and is closely related by sequence and antigenic crossreactivity to bovine coronavirus and a human coronavirus isolate (Verbeek and Tijssen, 1991). This close relationship suggests recent interspecies transmission of the coronaviruses. While no etiologic agent has been pinpointed, a bovine origin coronavirus has been implicated in an emerging enteric disease of turkeys called poult enteritis mortality syndrome (PEMS) (Barnes and Guy, 1997).

E. Arbovirus Infections

The arthropod-borne viruses (arboviruses) present a unique challenge to evaluating ecological virus–host relationships. This grouping includes members from several virus families, six of which have members isolated from birds: Togaviridae, Flaviviridae, Reoviridae, Arenoviridae, Bunyaviridae, and Rhabdoviridae. Of these, only the Togaviridae and Flaviviridae have strains that have caused documented disease in commercial poultry and game birds. Although difficult to assess in feral populations, isolates from the remaining virus families as well as several isolates from the Togaviridae and Flaviviridae are not associated with any pathology in birds.

Of the Togaviridae, the genus *Alphavirus* contains the encephalitic strains causing eastern equine encephalitis (EEE), western equine encephalitis (WEE), and a closely related wild bird isolate — the Highlands J virus (HJV). Pheasants have been the primary targets of significant outbreaks of EEE. Signs of infection are primarily but not exclusively neural, and mortality rates have reached 80% in some natural outbreaks (Ficken *et al.*, 1993; Eleazer and Hill, 1994). Economically significant outbreaks of disease due to EEE virus in turkeys have occurred in Wisconsin, with severity of the disease decreasing with increasing bird age. Significant outbreaks have also been recorded in chukar partridges, ducks, and chickens. WEE has been only rarely associated with disease in avian species, and the closely related HJ virus appears to be the Eastern United States equivalent to WEE. HJV has caused severe neuropathogenic outbreaks in chukar partridges and has been associated with infections in turkey flocks resulting in acute reduction in egg production (Wages *et al.*, 1993). Of course, these encephalitic viruses also cause disease in humans and horses.

Another interesting alphavirus, Ockelbo virus, related to Sindbis virus, has been implicated in causing arthralgia and rash in humans following its isolation from mosquitos collected during the outbreak. These viruses are transmitted by mosquitos among bird populations, which may act as the vector for transmission between humans and avian species. Ockelbo virus, therefore, is apparently maintained in an enzootic cycle involving birds and mosquitos with transmission to other hosts such as humans. Antibodies to Ockelbo virus, either experimentally or naturally infected, have been detected in Passeriformes, Galliformes, and Anseriformes (Lundstrom *et al.*, 1992; Lundstrom and Niklasson, 1996). Viremia in the absence of disease resulting from infection with Ockelbo has also been demonstrated in these bird groups. Given the widespread occurrence of antibodies to these encephalitic alphaviruses, it seems logical to conclude that birds in many cases act as "natural" and reservoir hosts.

The only flavivirus thus far associated with disease in birds is the Israel turkey meningoencephalitis virus. Infected birds exhibit neurological dysfunction and occasional significant mortality. This virus has also been identified in turkeys in South Africa.

F. Infections with Double-Stranded RNA Viruses

1. Reoviruses

Double-stranded segmented RNA viruses are unique and intriguing. The distinguished virologist Dr. Wolfgang K. Joklik, when asked what in the world he thought they meant in the grand scheme of biology, replied (and I paraphrase), "they represent an evolutionary step forward in the establishment of the ideal genetic material." The reoviruses have fascinated virologists for some time. They have an unusual and complex replication strategy, are able to undergo reassortment of their genes, and are curious in that they retain their infecting subviral cores as their RNA replication template. Reoviruses are also quite stable outside the host, remaining viable for up to a year at room temperature (Nibert *et al.*, 1996). The mammalian strains have a wide host range but are found without any associated clinical signs, the origin of the name REO (*r*espiratory and *e*nteric *o*rphan) telling most of the story. In poultry, several antigenic subtypes have been identified, and the virus can be classified based on serotyping and virulence, but there is no unified typing scheme as yet (Kawamura and Tsubahara, 1960; Robertson and Wilcox, 1986). The most severe and common pathology associated with reoviruses is arthritis–tenosynovitis, although associations with other clinical syndromes including respiratory and enteric disorders have been described (Rosenberger and Olson, 1997). Often, the clinical states are influenced by the presence of other pathogens. Arthritis is a significant problem in birds but primarily only in young chicks and turkey poults. Other Reoviridae in birds include several members of the genera *Orbivirus* (arboviruses; see below) and *Rotavirus*, which have been associated with a runting–stunting syndrome in chickens.

2. Infectious Bursal Disease Virus

The Birnaviridae are a relatively recently characterized family of viruses that contain two double-stranded segments of RNA and have no mammalian members thus far (ICTV, 1995). They were difficult to classify for many years because of their cell-associated nature and slow replication in cell culture systems. Currently, two serotypes (1 and 2) are accepted, with significant antigenic variation and numerous proposed subtypes within each serotype (McFerran *et al.*, 1980; Jackwood and Saif, 1987). Chickens are the only animals known to develop disease and lesions when naturally infected by avibirnaviruses. Both serotypes are distributed worldwide, but serotype 1 strains are the only ones associated with pathogenicity and immunosuppression. The bursa is the primary target organ, and strains of differing virulence have been identified. The disease is most significantly manifested when young birds are infected and permanent immunosuppression results. This sets the stage for subsequent severe viral and bacterial infections later in life. Interestingly, IBD is an example where maternal antibodies derived by either vaccinations or natural infections provide significant immune protection. Highly virulent strains of these viruses exist in various countries worldwide, and there are indications that new virulent variants do arise in the face of immune pressure (Chettle *et al.*, 1989b).

G. Adenovirus Infections

Avian adenoviruses are double-stranded DNA viruses containing 30–40 genes. These viruses can be loosely divided into three groups based on antigenic relationships of internal proteins (McFerran, 1997), although the three groups have not been officially recognized by the ICTV. The avian group I adenoviruses include at least 12 serologically distinct types, all of which have been isolated from mildly or asymptomatic poultry. The type species is known as the CELO (chick embryo lethal orphan), or Phelps strain, or F1 (fowl) strain. There are also numerous isolates from other avian species plus electron microscopy and immunological evidence that the type I aviadenoviruses are widely distributed in birds. By themselves, these viruses present no particular clinical problems in commercial poultry, but they are thought to cause significant problems in mixed infections with immunosuppressive viruses such as IBDV and chick infectious anemia virus (CIAV). A virus virtually indistinguishable from the CELO strain known as quail bronchitis virus (QBV) can be devastating in commercial quail operations, causing as high as 80% mortality (DuBose and Grumbles, 1959; Montreal, 1992). Virus isolations and significant antibody levels in wild quail suggest that the virus may present disease problems in nature (King *et al.*, 1981), although there are no documented cases or epizootics.

The group II aviadenoviruses include turkey hemorrhagic enteritis virus (HEV), pheasant marble spleen disease virus (MSDV), and the avian adenovirus splenomegaly (AAS) of chickens. These three agents are serologically indistinguishable but induce differing clinical manifestations in the different species. Infections caused by this virus group appear to target lymphoid tissue, often resulting in immunosuppression, and there are strain-specific differences in virulence among the various group II isolates (Pierson and Domermuth, 1997). HEV reached epidemic proportions in the 1960s and still causes significant problems in turkey-producing states in the United States and elsewhere. MSDV is a significant pathogen in confinement pheasant operations, causing significant economic losses. AAS is virtually ubiquitous among chicken flocks in the United States, in which case mild respiratory signs and splenomegaly are occasionally associated. The syndrome is similar to the disease state observed in pheasants, but in general AAS virus causes no significant problems among chickens. Evidence indicates that these group II avian adenoviruses are limited to the order Galliformes, and that wild bird populations (even wild turkeys) are unaffected.

The more interesting group III adenoviruses first appeared in 1976 associated with a distinct clinical manifestation called egg drop syndrome (EDS), in which egg production decreases and thin-shelled or shell-less eggs are produced. The disease has caused significant egg production losses mostly in the Eurasian and Australian–Pacific poultry markets. A virus named EDS76, which has been found associated with this syndrome, may have been originally introduced via a contaminated vaccine, but it seems clear now that sporadic cases are initiated by introduction of the causative virus from domestic and wild waterfowl, mostly ducks and geese. When endemic in flocks, the virus is transmitted vertically and exhibits a latent phase, which is reactivated in laying hens, usually after egg production begins. The virus initially appears to replicate in lymphoid tissues but rapidly moves to the oviduct, where replication causes inflammation and production of aberrant eggs. Unlike other avian adenoviruses, the group III strains do not replicate in the intestinal mucosa. It is likely that EDS76 virus or very similar strains are present ubiquitously in wild ducks and geese, but rarely appear to be associated with disease (McFerran, 1997).

H. Poxvirus Infections

The avipoxviruses are widely distributed throughout the class Aves, having been isolated from some 60 species representing 20 avian taxonomic families. The avipoxviruses are responsible for economically important disease problems in commercial poultry and aviaries (Tripathy, 1993). They may cause a slowly developing cutaneous disease with low mortality or, conversely, significant mortality, and generalized infections when in the diphtheritic form on mucosal surfaces of the respiratory tract and associated areas. These large DNA viruses replicate in the cytoplasm of the cell, where they form characteristic inclusion bodies within rapidly proliferating nodular lesions. The poxviruses may be transmitted by mechanical means such as introduction from poultry workers into abrasions in the skin of uninfected poultry. There is also real evidence for transmission of the disease by mosquitos and other vectors such as mites in close conditions, where the number of diseased birds is high. The avipoxviruses can also apparently establish a latent state and be naturally or chemically induced to reactivate (Kirmse, 1967b). Cutaneous lesions containing infectious virus persistent for more than a year have also been documented in wild birds (Kirmse, 1967a). Variant strains of avipoxviruses have since been isolated from previously vaccinated flocks that were experiencing significant mortality from the diphtheritic form of the disease (Tripathy and Reed, 1997). This suggests that, similar to Marek's disease, vaccination against poxviruses may ultimately lead to escape of more virulent forms of the viruses. The avipoxviruses are also considered one of the most promising DNA virus vectors for delivery of recombinant poultry vaccines, and poxvirus-vectored vaccines against Newcastle disease and avian influenza have been licensed.

I. Avian Encephalomyelitis and Other Avian Picornavirus Infections

Avian encephalomyelitis is primarily a disease of young chicks caused by a picornavirus. The disease was quite economically important prior to initiation of live-virus vaccination. The host range is very limited (order Galliformes), and there is only one virus serotype. The natural isolate is enterotropic and is transmitted horizontally (orally) and vertically. There is a gradient of pathology dependent on the age of infection of young chicks. Pre-immune chicks from non-immune parents will generally die if infected within 1 to 2 days of hatch. If infected between 2 and 8 days of age, they may live but exhibit significant nervous involvement (encephalomyelitis). If infected beyond that age, they may exhibit enteric pathology but not neural signs. At adulthood or full immunocompetence, they may be infected but refractory to any clinical signs. So the immune status of the host in this case is very important in affecting the course of the viral replication (Calnek, 1997).

Another avian picornavirus, distinct from AEV, is the avian nephritis virus (Shirai *et al.*, 1991). This virus is similar to AEV in that it is primarily a problem in very young birds and has a very limited host range. As the name implies, a distinct clinical syndrome is associated with the agents, but it is has not yet been determined how extensive infections are in commercial poultry. Finally, another picornavirus, duck hepatitis 1 (DHV-1), has caused significant problems in commercial duck operations. There are still unknowns with regard to the overall importance of DHV-1, and there are additional virus-induced types of hepatitis disease with at least two others often encountered.

J. Other Avian Enteric Infections

The most significant cost involved in raising birds for food sources is feeding them. Consequently, diseases that affect the feed–protein conversion ratio will directly affect the cost of marketable birds. As such, enteric disease, even when nonlethal or mild in nature, can affect the performance of food-source birds. In recent years, investigators have identified rotaviruses in chickens, astroviruses in turkeys, and enterovirus-like particles in several avian species associated with enteric diseases. Many of these have been associated with significant pathogenesis, particularly in mixed infections where immunosuppression occurs, with subsequent decrease in marketability due to weight loss (Barnes, 1997).

K. Chicken Infectious Anemia

The virus designated chicken infectious anemia virus (CIAV) is a single-stranded circular DNA-containing virus, tentatively classified with two similar agents in a new family: Circoviridae. The agent has only recently been recognized, purified, and characterized, and its unequivocal role in economically important disease is not yet fully established. Purified virus inoculated into young chicks results in severe anemia and immunosuppression, but inoculation of 2- to 3-week old birds results in little if any pathology (Yuasa and Imai, 1986). The virus has been associated with adult anemic conditions, however, in conjunction with other viral and bacterial infections and may thus play a significant role. The agent is widespread in commercial poultry flocks worldwide, and infection with the virus has been statistically related to a decrease in overall growth and performance (Chettle *et al.*, 1989a). The extent to which two other tentative members of the family Circoviridae — psittacine beak-and-feather disease virus (BFDV) and porcine circovirus — are related to CIAV is under question, and these stated taxonomic relationships may differ in the future. BFDV has proven to be a significant pathogen in aviaries and commercial in psittacine birds to the extent that vaccines are now routinely administered. Recent evidence also suggests that *Circovirus* particles can be detected by EM and isolated from several other wild and captive bird species. Consequently, the Circoviridae may emerge as a more significant disease-causing agent in the near future.

L. Caliciviruses

The Caliciviridae is a family of small single-strand positive-sense RNA viruses with a polyadenylated genome. Virus-like particles resembling caliciviruses have been isolated from a variety of wild-bird and captive-raised species. A chicken calicivirus has been replicated in cell culture and caused apparent gastrointestinal disease in specific pathogen-free day-old chicks (Cubitt and Barrett, 1985). Intestinal contents of goldfinches with hemorrhagic enteritis have been found to contain calicivirus-like particles and gastrointestinal disease associated with caliciviruses has been reported in guinea fowl and pheasants (Gough et al., 1985). Caliciviruses were originally isolated from marine mammals and have caused disease in both domestic and feral swine due to consumption of uncooked garbage containing seafood. Interestingly, caliciviruses have been detected in pelagic birds, such as the white tern (Poet et al., 1994). Calicivirus isolations such as these lead to the speculation that wild sea birds may be important in the transmission of these agents across large areas or to other animals.

M. Parvoviruses

Parvoviruses are the smallest of the DNA-containing viruses, carrying a single-stranded genome of about 5000 nucleotides. They cause or are associated with three infections in birds (Kisary, 1993). In wild and domestic geese and Muscovy ducks, Derzy's disease refers to a syndrome that had been variously called goose influenza, goose hepatitis, gosling enteritis, and infectious myocarditis in different countries. This collection of names gives some indication of the variety of signs associated with Derzy's disease. The virus, which can be transmitted horizontally and vertically, induces differing pathologies depending on the age of the bird and, in the case of hatchlings, the level of maternal antibodies. An interesting pathology often associated with infection of older birds is the virtually complete loss of feathers. The extreme stability of the parvoviruses makes control of this disease difficult in commercial operations worldwide, and the disease can result in 100% mortality in young hatchlings.

In chickens and young turkeys, parvoviruses have been associated with runting stunting syndrome (RSS), as have other agents. Experimental inoculations have indicated that the associated viruses, which are distinct from the goose viruses, can cause significant pathology, but the extent to which these viruses participate in the RSS is not yet established (Trampel et al., 1983; Decaesstecker et al., 1986). Another parvovirus, the avian adeno-associated virus is almost always found associated with adenovirus infections in poultry. The avian adeno-associated viruses are grouped with their mammalian counterparts in the *Dependovirus* genus, although they are serologically unrelated. The avian dependovirus appears to contribute to disease only in the sense that it can affect multiplication of the associated adenoviruses (Yates and Piela, 1993).

N. Newcastle Disease and Other Paramyxovirus Infections

Newcastle disease virus (NDV), classified as avian paramyxovirus-1, contains a single-strand negative-sense RNA genome of 15 kb containing coding sequences for six genes. The virus infects all bird species tested to date, including free-living and domestic species. One panzootic outbreak of the disease is thought to have originated in Asia, with subsequent spread to Europe, with outbreaks of disease first reported in poultry during the 1920s in Java, Indonesia, and Newcastle-upon-Tyne, England. Worldwide dissemination of the disease, particularly during the 1960s and 1970s, has been attributed to increased international trade of both commercial poultry and pet birds. This led to development of inactivated and live-virus vaccines for commercial poultry. The transmission of infectious NDV may occur by either ingestion or inhalation, and this knowledge is the basis for mass-application vaccination procedures during poultry production.

Isolates of NDV are grouped into three main pathotypes depending on the severity of disease that they cause (Alexander and Parsons, 1974). Mildly virulent "lentogenic" viruses may cause unnoticeable infections in adult chickens or a mild respiratory distress and are used extensively as live-virus vaccines. "Mesogenic" NDVs are of intermediate virulence and cause respiratory distress, with mild infections of various organs detectable only by histopathology. Highly virulent viruses that cause severe morbidity and mortality are termed "velogenic." Velogenic viruses can manifest themselves as neurotropic or viscerotropic forms of Newcastle disease with extensive systemic replication throughout a bird. Virulent forms of NDV can replicate within cultures of most avian and mammalian cell types without the addition of trypsin, while lentogens require added proteases for replication in cell culture. The presence of dibasic amino acids at the proteolytic cleavage site (PCS) of the viral fusion protein and the ability of cellular proteases to cleave the fusion protein of various pathotypes specify the molecular basis for NDV virulence. Fewer basic amino acids are present in the fusion protein cleavage site of lentogenic NDV than is the case for mesogenic or velogenic isolates. The presence of the increased number of dibasic amino acids in the fusion protein PCS sequence of NDV allows for systemic replication of these more pathogenic viruses in the host (Nagai *et al.*, 1976).

Although the principal concern with NDV is its effect on poultry production, it may have severe consequences in other free-living avian species. Major outbreaks of Newcastle disease have occurred in North American cormorants during the summers of 1990 and 1992, and again in 1997. Outbreaks during 1990 and 1992 occurred in the north central United States and south central Canada, while in 1997

Newcastle disease occurred among cormorants of the western United States and Canada. Mortality in young nestlings in some areas was as high as 80 to 90%, and the affected birds had characteristic neurotropic lesions (Heckert *et al.*, 1996). During the 1992 outbreak in cormorants, an unvaccinated North Dakota turkey flock became infected with NDV, resulting in high mortality. Using nucleotide sequence analysis, the virus causing disease in turkeys was proven to be the same virus isolated from afflicted cormorants in the central United States and Canada. The cormorant virus was also phylogenetically related to other known NDV isolates of psittacine origin (Seal *et al.*, 1995) that had caused a major outbreak in Southern California poultry during the early 1970s that resulted in a depopulation of two million chickens (Utterback and Schwartz, 1973).

Illegal importation of pet bird species into the United States continues to be a source of highly virulent NDV and certainly may play a role in spread of viruses that threaten commercial poultry worldwide (Panigrahy *et al.*, 1993). What role free-living birds play in spread of NDV is unclear. Although persistent infections of chickens by NDV do not appear to occur, the virus' persistence in poultry flocks may result from virus reintroduced from wild populations. Also, different bird species vary in their level of susceptibility to NDV. Ducks, geese, and certain psittacine birds do not exhibit signs of disease when infected with highly virulent NDV, while other psittacine species may have high mortality. Persistent infections have been demonstrated in various psittacine birds, with virus isolations noted up to a year following experimental infection of parrots with velogenic NDV chicken isolates (Erickson *et al.*, 1978). Although psittacine birds have been directly linked as a source of NDV highly virulent for gallinaceous birds, such as chickens and pheasants, no studies have shown direct isolation of NDV from feral psittacines.

Newcastle disease occurred in racing and show pigeons during the 1980s in both western Europe and North America. The outbreaks in western Europe were linked to contaminated feed, and disease was controlled in both areas via vaccination with mildly virulent NDV commonly used for poultry. The virus isolated from these birds varied somewhat from traditionally virulent NDV in having only one set of dibasic amino acids in the fusion protein cleavage site. Increased virulence for poultry subsequently occurred only after passage in chickens (Collins *et al.*, 1994). These examples demonstrate the fact that variant forms of NDV may arise in different bird species and that free-living birds may be a constant source of viruses that affect both domestic and wild birds.

Vaccination programs against NDV for the most part have been effective at controlling the virus in commercial poultry. However, all of the vaccine strains available today, while effective against lethal disease, share less sequence identity with currently circulating disease strains. This is effectively illustrated in Figure 2, which may also be considered a useful example for other RNA-containing viruses. The phylogenetic tree demonstrates the relationship between current

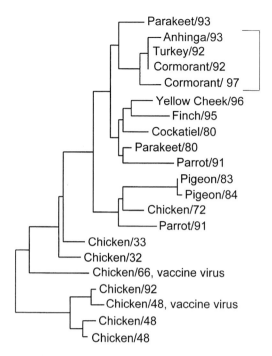

Fig. 2. Phylogenetic relationships among Newcastle disease virus isolates. Nucleotide sequences from the fusion protein gene coding for the cleavage site from several Newcastle disease virus isolates were aligned and analyzed by parsimony analysis. All the viruses listed are intermediately virulent mesogens or highly virulent velogenic isolates when inoculated into chickens, with the exception of the two vaccine viruses and the chicken/92 field isolate. The data demonstrate that field isolates from chickens related to vaccine virus are isolated from poultry. The chicken/72 virus caused the last major outbreak of Newcastle disease in U.S. poultry and was epidemiologically linked to a pet bird. It is genetically related to a parrot isolate and viruses that caused disease in pigeons during the 1980s. Viruses isolated from cormorants in 1992 infected an unvaccinated turkey flock. These viruses appear to be circulating among cormorants, since closely related viruses were isolated again in 1997. No viruses related to the virulent chicken viruses from 1948 have been isolated since that time, indicating these virus types may no longer be circulating among birds. All the recent virulent Newcastle disease viruses were isolated from pet or exotic birds and chickens infected by free-living birds. These viruses are related to virulent viruses originally isolated during the 1930s.

strains and the vaccine strains based on the sequence of the fusion protein. It shows that vaccine strains are relatively far removed from the recent isolates coming from wild birds. The future consequences of such heterologous vaccination approaches are unknown. The same scenario might unfold for NDV, as has occurred in the attempt to control Marek's disease where the immune status of vaccinated flocks promotes generation of novel antigenic variants with new virulence characteristics.

O. Avian Influenza

Avian influenza (AI) presents one of the most interesting ecological relationships between birds and their viruses. Its ecology has been reviewed in depth (Webster *et al.*, 1992; Hinshaw *et al.*, 1980), but AI presents such a complete picture of a virus with multiple impacts on several species that it must be included in this chapter. The type A orthomyxoviruses are essentially bird viruses. The infections in birds can range from clinically inapparent with minimal serologic response, to a devastating systemic disease that can result in 100% mortality within a matter of days. There are 15 identified subtypes of AI viruses based on the hemagglutinin (HA) protein antigenic structure and nine subtypes based on the neuraminidase (NA) structure. These two genes code for the predominant surface glycoproteins, which are embedded in the lipid bilayer of the viral envelope. These are only 2 of 10 genes coded for by the virus. The virus' genetic material is contained on eight separate negative-sensed single-stranded RNA segments ranging from 890 to 2450 nucleotides in length.

All of the HA and NA viral subtypes have been identified in feral waterbirds, and it has been proposed that these birds act as a "natural" reservoir for the virus (Webster *et al.*, 1992; Slemons and Easterday, 1977). Determination of the phylogenetic relationships of several genes from several subtypes of AIV collected from wild birds indicated that the general rate of evolution in the avian reservoir is low compared to the rate of evolution observed in human and other mammalian strains of type A orthomyxoviruses (Gorman *et al.*, 1992). Other recent studies, however, have shown that the mutation rate of the HA gene and the NS gene even in wild birds approaches that seen in human strains (Garcia *et al.*, 1997; Suarez and Perdue *et al.*, 1998). Measurement of clinical signs among infected migrating waterbirds is, of course, a difficult task. Thus far, only a single instance of severe clinical signs has been associated with free-living birds — a lethal outbreak in terns in South Africa in 1961 caused by an H5N1 strain (Becker, 1966). Given the recently measured mutation rates, and the appearance of this severe lethality in a wild bird population, it must be considered that AIVs do continue to evolve in feral birds (indeed 15 subtypes have arisen already!). There is no reason to believe that the present count of 15 subtypes will be the final tally.

While some have proposed that the AIV strain in wild birds causes no disease problems as a result of "evolutionary stasis," without question the virus has dramatic effects when it leaves this proposed reservoir. In addition to the fixed influenza populations in pigs and horses, avian strains can directly infect mammals and have been identified in whales, seals, and mink (Lang *et al.*, 1981; Berg *et al.*, 1990). The dramatic association with seals in 1980 and 1982 mentioned earlier clearly points out the potential impact of AIVs. This avian reservoir then serves as a sort of base for the viral "biological invasions" that are a part of interspecies transmission. Ecologically and economically, the transmission of these AIV sub-types to commercial poultry has perhaps yielded the most important impact. The situation is most clearly played out in Minnesota each year. Due to the large number of lakes in the state, migratory waterfowl, primarily Anseriformes (ducks, geese), frequent the area in large numbers seasonably. Nearby commercial turkey operations suffer infections nearly every year, particularly in the cooler fall months, caused by different influenza subtypes of different virulence (Halvorson *et al.*, 1985). In severe years, the associated costs due to turkey mortality, as well as performance and egg production losses, can reach several millions of dollars (Poss and Halvorson, 1986).

In chickens, introduction of influenza viruses from waterfowl has been more devastating. In the northeastern United States, the H5 subtype caused the first large-scale outbreak of highly pathogenic AI in numerous flocks in 1983. Twenty-three million birds were destroyed at a cost of more than $65,000,000 in order to contain the infection. Clinically, the disease was indistinguishable from the disease originally described as "fowl plague" in 1927 in Europe (Eckroade and Silverman, 1986). Today we know that all of the classical "fowl plague"-type outbreaks have been associated exclusively with only two subtypes: H5 and H7. Evolution to virulence in these subtypes has been closely associated with accumulation of basic amino acids at the proteolytic cleavage site of the hemagglutinin protein (Bosch *et al.*, 1981; Webster and Rott, 1987). In order for AIV strains to be infectious, the HA protein must be cleaved into HA1 and HA2 subunits, which subsequently allows structural rearrangement and exposure of a protein sequence needed to fuse the viral envelope with the plasma membrane of the target cell (Klenk *et al.*, 1975; Skehel *et al.*, 1995). Isolates of low pathogenicity lack multiple basic amino acids and are only cleaved by trypsin-like extracellular proteases. These proteases are abundant on the mucosal surfaces of the respiratory tract and gut, so the viruses replicate unrestricted at these sites. The highly pathogenic (HP) forms, however, have additional basic amino acids, which are recognized by the intracellular furin-like proteases ubiquitous in bird tissues (Rott *et al.*, 1995). Thus, in birds, these viruses are able to escape the respiratory tract and infect a wider variety of internal tissues and organs. This is one of the most important of the virulence factors (Webster and Rott, 1987). Only two subtypes have been shown to accumu-late these basic amino acids: H7 and H5. These subtypes appear to do so by either base substitution or by insertion events (Wood *et al.*, 1993; Perdue *et al.*, 1997). Nucleotide insertion events appear to be the most common mechanism, and Table II and Figure 3 illustrate what is presently known about occurrence of these HP virus isolates in commercial poultry. The HP isolates, with only two exceptions, have additional inserted basic amino acids within the well-conserved region surrounding the cleavage site. From the data in Table II and Figure 3, it may be easily surmised that both the number of HP isolates and the number of basic amino acids at the cleavage site are increasing.

TABLE II

Cleavage-Site Sequence Conservation among Type A Influenza
Hemagglutinin Proteins and Insertions in HP Strains[a]

Subtype	HA1		HA2
1	GLRNVPSIQSR	⇓	GLFGAIAGFIE
2	GLRNVPQIESR		GLFGAIAGFIE
3	GMRNVPEKQTR		GLFGAIAGFIE
4	GMRNIPEKATR		GLFGAIAGFIE
5	GMRNVPQRETR		GLFGAIAGFIE
6	GLRNVPQIETR		GLFGAIAGFIE
7	GMRNVPENPKTR		GLFGAIAGFIE
8	GLRNTPSVEPR		GLFGAIAGFIE
9	GLRNVPAVSSR		GLFGAIAGFIE
10	GMRNVPEVVQGR		GLFGAIAGFIE
11	GPRNVPAIASR		GLFGAIAGFIE
12	GLRNVPQVQDR		GLFGAIAGFIE
13	GLRNVPAISNR		GLFGAIAGFIE
14	GMRNIPGKQAK		GLFGAIAGFIE
15	GMKNVPEKIRTR		GLFGAIAGFIE

Subtype	Selected HP Isolates
H5	
A/Tk/Ireland/83:	PQRKRKKR ⇓ GLF...
A/Ck/Puebla/94:	PQRKRKTR ⇓ GLF...
A/Ck/Queretaro-20/95:	PQRKRKRKTR ⇓ GLF...
A/Hong Kong/157/97:	PQRERRRKKR ⇓ GLF...
H7	
A/Tk/England/199/79:	PEIPKKREKR ⇓ GLF...
A/Ck/Queensland/95:	PEIPRKRKR ⇓ GLF...
A/Ck/Pakistan/95:	PEIPKRKRKR ⇓ GLF...

[a]The amino acids surrounding the hemagglutinin protein cleavage sites
of a representative member of each of the 15 subtypes of avian influenza
viruses are shown above. Subtypes 7, 10, and 15, which cluster together
in a phylogenetic tree, all have five amino acids between the conserved
proline and arginine; the remaining subtypes have four. Highly pathogenic
isolates (subtypes H5 and H7) more often than not, have insertions of
arginines and lysines, increasing the length of the cleavage site and
making the hemagglutinin accessible to ubiquitous proteases. This has the
effect of greatly increasing the tissue distribution and virulence of the
virus.

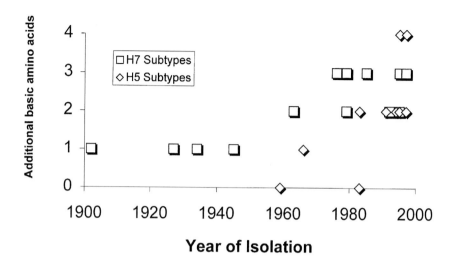

Fig. 3. Accumulation of basic amino acids at the HA protein proteolytic cleavage site in highly pathogenic avian influenza isolates. The year of isolation of all known isolates of HP viruses from commercial poultry is shown and compared with the number of inserted basic amino acids at the HA cleavage site of the isolate. See Table II for further explanation.

The avian influenza viruses are distributed worldwide. In addition to the United States, outbreaks of highly pathogenic AI have occurred in Mexico, Canada, Europe, Asia, and Australia. In most cases, there is some sort of connection between the affected flocks and nearby waterfowl. The extent to which the reservoir of viruses exists in waterfowl is not really known. Surveys have shown that, in addition to the viruses being widespread, the various subtypes increase and decrease in prevalence over successive years. The order Charadriiformes (shorebirds and allies), in addition to Anseriiformes, has been shown to carry AIV strains (Kawaoka *et al.*, 1988), and the extent to which AIVs are carried in other orders has really only been superficially explored. Molecular and phylogenetic evidence have clearly shown that these segmented viruses share various genes back and forth within the populations of circulating virus strains (Garcia *et al.*, 1997). They are also capable of donating genes through genetic reassortment to viruses in swine and those genes ultimately end up in humans (Webster *et al.*, 1993). Add this capacity to the known (albeit rare) transmission of purely avian influenza strains directly into mammals and one creates a potential ecological bonanza for the virus, and potential nightmares for new hosts.

VI. IMPACT OF TRANSMISSION OF AVIAN VIRUSES OUTSIDE THE CLASS AVES

A. The Variety of Unknowns

As advanced as our study of viruses has become, there are still a variety of unknowns with regard to the extent to which avian viruses infect other vertebrates. It would not be unreasonable to suggest that, because of their mobility and longevity of existence, that the class Aves may have played a critical role in distribution of viruses to other classes. In the case of the avian orthomyxoviruses, phylogenetic evidence points to recent introduction of these viruses from birds to mammals. Following introduction, the natural mutation and evolution of the virus produced strains that now appear to be adapted to their new hosts. There are swine, equine, and human strains of type A influenza viruses that do not replicate well outside their respective host species. If birds are indeed responsible for us coming down with the flu today, we know of no reason why they might not have participated over the eons in the establishment of other purely mammalian diseases. Of course, one could easily argue the opposite relationship, but, intuitively, birds do present an ideal mobile vector, particularly for the enterically transmitted viruses.

In addition to the diseases discussed here, there is a growing group of emerging diseases or diseases of as yet unknown etiology in commercial poultry (Saif, 1997). Many of them have virus particles clearly associated with the syndrome, but only limited research and characterization has been carried out thus far. It is probably safe to assume that the future will continue to bring new avian diseases and bring to light new or undiscovered avian viruses.

B. Birds, Viruses, and Insects

The relationships existing among birds, insects, and viruses is one area that should be explored in more detail by increased research efforts. In free-living birds, the large number of apparently innocuous infections with Bunyaviridae, Togaviridae, and Flaviviridae suggest that birds may act as an ecological reservoir for maintenance of these virus populations. From a virocentric point of view (if there is such a thing), having a potentially highly mobile population of susceptible hosts on which transmission vectors may feed would be advantageous. It would provide a mechanism by which to transfer viruses over large geographical distances where they could then be transmitted by new related or unrelated vectors, thus increasing host range. Of course, one runs the risk here of assuming that the virus would be ecologically improved by extending its replicative capacity in new hosts. This may not be a valid assumption. For example, if variola virus or human poliovirus type 1 had been exploiting this strategy, they made a bad choice in extending their host range into humans!

From one human vantage point, birds provide a useful monitor for detection of arboviral diseases in a given area. Sentinel chickens are actually in use today along the East Coast of the United States to gauge the extent to which some mosquito populations are carrying eastern equine encephalitis and St. Louis encephalitis viruses. The future is perhaps the most important aspect to consider. What viruses other than the orthomyxoviruses may someday establish a similar relationship where transmission to a mammalian intermediate or even direct transmission may allow introduction into the human population?

C. Transmission to Mammals and Humans

One fundamental unanswered question is whether birds, acting as reservoirs for arboviruses, should be considered an important ecological niche for those viruses. There is no evidence of which we are aware to indicate either yes or no. There have been no documented transmissions of arboviruses from birds to mammals without the insect vector in nature, however, so at this point wild birds could be considered dead-end hosts for the arboviruses. In commercial poultry during outbreaks of the alphavirus EEEV, there is no evidence of infection of workers in close contact with infected birds. If, however, a bird provides the *necessary* reservoir of EEEV that infected the human, then this bird becomes very important. But there are no data that as yet identify the bird as the required reservoir for disease transmission to human.

Newcastle disease virus provides an interesting new disease problem in humans that has only arisen as a result of increased vaccination of birds. A live-virus vaccine is available for control of NDV that is administered by aerosol spray. In a small number of cases where precautions have not been taken, vaccinators can contract a conjunctivitis caused by the vaccine. This outcome has been seen in the heavy poultry-producing areas of northern Georgia, on a common enough basis now, to be easily recognized by ophthalmologists. There is usually no seroconversion to the virus in these cases, and they are not particularly serious infections. There is also suggestive evidence in the form of measured seroconversion that poultry workers occasionally may be subclinically infected with avian coronaviruses and avian retroviruses.

The evidence for direct infection of humans with avian influenza viruses is growing. Human strains have classically included only three subtypes (H1–3). Until recently, the only other strains shown to cause disease in humans were of the H7 subtype. Conjunctivitis caused by purely avian H7 subtype viruses has been reported on two occasions (Kurtz *et al.*, 1996; Webster *et al.*, 1981a). The most disturbing set of events, however, was the more recent highly publicized influenza outbreak in Hong Kong. A purely avian H5 subtype virus was isolated from some 18 patients, 6 of whom died, from May through December of 1997. The index case, a 3-year-old child, in May followed a highly pathogenic outbreak of avian influenza that had just previously occurred in chickens in Hong Kong and its environs. The child had been exposed to ill birds at a day care center, and it was clear that the recovered virus was essentially the same as that recovered during the chicken outbreak. Considerable efforts were made to determine whether the virus was a contaminant and whether it was replicating in the child. The evidence strongly suggested that it was indeed replicating (Subbarao *et al.*, 1998). Subsequent to this event, in November–December the remaining confirmed cases were reported and an intense effort was mounted to determine exactly how this virus was transmitted. It is clear now that the H5N1 viruses infecting humans in Hong Kong 1997 were all of avian origin (Suarez *et al.*, 1998). There is limited serologic suggestion of human-to-human transmission but no genetic evidence of adaptation to humans. This case was touted as a premier test run for the next predicted influenza pandemic. Whether this outbreak represents a simple dead-end zoonotic transmission or whether these purely avian influenza viruses can become fixed in the human population is a matter for conjecture. The avian H5 and H7 influenza strains have been the only natural influenza strains thus far to unequivocally cause systemic lethal infection in any species owing to the unique structure of their HA protein (see Fig. 3). A major fear is that by adding such HA subtypes to a population of mammalian viruses a whole new class of virulent strains might be created. Our experience in working with numerous highly pathogenic and nonpathogenic strains has been one of very little or no evidence for human infection in the 25 years that scientists and technicians have spent working with hundreds of different stains. But our experience has also been that, when one attempts to confine influenza viruses by setting biological rules for them, they usually find a way to break them.

VI. CONCLUSIONS AND FUTURE CONSIDERATIONS

If ecology is the study of the relationship of organisms to their environment, for bird viruses, the active, pertinent, environment is always within the bird and its tissues. While one cannot discount the effects of the environment outside the host, there are no positive effects on an avian virus life cycle outside the host of which we are aware. That is, there are no activation events outside the avian host of which we are aware; only events of inactivation. Certainly, avian viruses, as all viruses, have evolved survival strategies to allow passage from host to host, and some are much more refractory to environmental inactivation than others. But without really knowing which, if any, "evolutionary direction" viruses are taking, it is impossible to determine whether a large herpesvirus of 150 genes is ecologically more fit than the enterovirus with five genes!

The relationship of birds and insects in the transmission and life cycle of the arboviruses is poorly understood. Until this recent transmission of H5N1 influenza strains to humans, this relationship appeared to be the only significant one in which there is interaction among a genetically unaltered virus, an avian host, and a non-avian host. As mentioned earlier, other documented cases of avian-to-mammalian viral transmission are Newcastle disease infections of humans and influenza A infections of mammals and humans. There is also serological evidence for avian retrovirus and avian coronavirus infections. In the case of Newcastle disease, the documented infections are mostly minor conjunctivitis. In the case of influenza, discussed earlier, the avian viruses most often acquire RNA segments from other sources before they are established in a new host. Thus, in general, the impact of avian viruses on public health might currently be considered small, but clearly the potential exists for significant future impact.

A major interesting question arises when evaluating virulence of avian-origin viruses. Conventional wisdom seems to be suggesting that virus entry into naive host populations is more likely to result in a general decrease in virulence as the virus adapts to the new host (Morse, 1994). This may be true under truly "natural" conditions; but what happens under human-dictated conditions such as vaccination and by defining the host population through breeding and housing? In the case of Marek's disease, we see an example of virulence increasing in the face of continued vaccination against the virus (Fig. 1). Whether this is occurring as a result of the vaccination or a combination of factors, the fact is that the population of Marek's disease herpesviruses in nature is becoming more virulent. The vaccination programs for Newcastle disease utilize vaccines whose genetic compositions are becoming farther and farther removed from the virulent strains circulating in wild fowl (Fig. 2). This may not bode well for future control of virulent strains in commercial poultry.

In the case of avian influenza viruses, we have seen the number of highly virulent outbreaks increase in recent years, often following prior circulation of a nonpathogenic precursor, and we have seen concomitant changes in genetic structure related to virulence (Fig. 3). Again there are unknowns in this system, but the fact remains that we are seeing more virulence as the nonpathogenic AIV subtypes replicate in commercial poultry settings. Thus, this naive commercial population, unlike the waterfowl hosts to which the virus has adapted, appears to allow generation of heretofore unencountered virulence phenotypes. Similar scenarios appear to be playing out as more virulent infectious bronchitis strains and infectious bursal disease strains appear worldwide. One must be cautious, then, in working strictly under the previously mentioned virulence-decrease paradigms for mammalian viruses. These have been suggested mostly by the experiences with artificial introduction of rabbit papilloma viruses in Australia or where new introductions occur as human encroachment ensues (such as in the recent cases of Ebola virus or monkey poxvirus infections). In these cases, it was proposed that virus transmission results first in high virulence for the host followed by subsequent adaptation and attenuation.

Finally, it may be dangerous to attempt to confine avian viruses to some of our more "logical" inferences based on Darwinian evolution. One paraphrased definition of evolution is "moving from a prior more primitive (or less fit) state to a current more advanced (or more fit) state." It is difficult, for us at least, to say that any avian viruses are following such a progression. Additionally, other than use as molecular biological vectors, viruses in general have not been shown to provide anything positive to the ecology of any other organisms. The future may yet reveal such relationships, but, for now, ridding ourselves of these fascinating parasites by continuing such policies as those resulting in the eradication of smallpox and polio still appears to be our most prudent course.

REFERENCES

Alexander, D. J., and Parsons, G. (1974). Newcastle disease virus pathotypes. *Avian Pathol.* **3**, 269–278.

Arshad, S. S., Howes, K., Barron, G. S., Smith, L. M., Russell, P. H., and Payne, L. N. (1997). Tissue tropism of the HPRS-103 strain of J subgroup avian leukosis virus and of a derivative acutely transforming virus. *Vet. Path.* **34**, 127–137.

Bagust, T. G., and Guy, J. S. (1997). Laryngotracheitis. In "Diseases of Poultry," 10th ed. (B. W. Calnek, R. E. Luginbuhl, and C. F. Helmboldt, eds.), pp. 527–540. Iowa State University Press, Ames.

Bagust, T. G., and Johnson, M. A. (1998). Avian infectious laryngotracheitis: Virus–host interactions in relation to prospects for eradication. *Avian Path.* **24**, 373–391.

Barnes, H. J. (1997). Viral enteric infections. "Diseases of Poultry," 10th ed. (B. W. Calnek, R. E. Luginbuhl, and C. F. Helmboldt, eds.), pp. 685–710. Iowa State University Press, Ames.

Barnes, H. J., and Guy, J. S. (1997). Poult enteritis syndrome of turkeys — mortality. "Diseases of Poultry," 10th ed. (B. W. Calnek, R. E. Luginbuhl, and C. F. Helmboldt, eds.), pp. 1025–1031. Iowa State University Press, Ames.

Becker, W. B. (1966). The isolation and classification of tern virus influenza virus A/tern/South Africa/1961. *J. Hyg.* **64**, 309–320.

Berg, M., Englund, L., Abusugra, I. A., Klingeborn, B., and Linne, T. (1990). Close relationship between mink influenza (H10N4) and concomitantly circulating avian influenza viruses. *Arch. Virol.* **113**, 61–71.

Biggs, P. M. (1985). Spread of Marek's disease. In "Marek's Disease" (L. N. Payne, ed.), pp. 329–340. Martinus Nijhoff, Boston.

Bosch, F. X., Garten, W., Klenk, H. D., and Rott, R. (1981). Proteolytic cleavage of influenza virus hemagglutinins: Primary structure of the connecting peptide between HA1 and HA2 determines proteolytic cleavability and pathogenicity of Avian influenza viruses. *Virology* **113**, 725–735.

Brand, C. J. (1987). Duck plague. *In* "Field Guide to Wildlife Diseases" (M. Friend, ed.). U.S. Department of Interior, Fish and Wildlife Services Resource Publication No. 167, Washington, DC.

Calnek, B. W. (1997). Avian encephalomyelitis. *In* "Diseases of Poultry," 10th ed. (B. W. Calnek, R. E. Luginbuhl, and C. F. Helmboldt, eds.), pp. 571–582. Iowa State University Press, Ames.

Calnek, B. W., and Witter, R. L. (1997). Marek's disease. *In* "Diseases of Poultry," 10th ed. (B. W. Calnek, R. E. Luginbuhl, and C. F. Helmboldt, eds.), pp. 369–413. Iowa State University Press, Ames.

Calnek, B. W., Luginbuhl, R. E., and Helmboldt, C. F., eds. (1997). "Diseases of Poultry," 10th ed. Iowa State University Press, Ames.

Cavanagh, D., Davis, P. J., and Cook, J. K. A. (1992). Infectious bronchitis virus: Evidence for recombination within the Massachusetts serotype. *Avian Path.* **21**, 401–408.

Cavanagh, D., Brian, D. A., Brinton, M. A., Enjuanes, L., Holmes, K. V., Horzinek, M. C., Lai, M. M. C., Laude, H., Plagemann, P. G. W., Siddell, S. G., Spaan, W., Taguchi, F., and Talbot, P. J. (1994). Revision of the taxonomy of the Coronavirus, Torovirus and Arterivirus genera. *Arch. Virol.* **135**, 227–237.

Chettle, N., Eddy, R. K., Wyeth, P. J., and Lister, S. A. (1989a). An outbreak of disease due to chicken anemia agent in broiler chickens in England. *Vet. Rec.* **124**, 211–215.

Chettle, N., Stuart, J. C., and Wyeth, P. J. (1989b). Outbreak of virulent infectious bursal disease in East Anglia. *Vet. Rec.* **125**, 271–272.

Coffin, J. M. (1996). Retroviridae: The viruses and their replication. *In* "Fields Virology," 3rd ed. (B. N. Fields, D. M. Knipe, and P. M. Howley, eds.), pp. 1767–1847. Lippincott-Raven, Philadelphia.

Collins, M. S., Strong, I., and Alexander, D. J. (1994). Evaluation of the molecular basis of pathogenicity of the variant Newcastle disease viruses termed "pigeon PMV-1" viruses. *Arch. Virol.* **134**, 403–411.

Cooper, A., and Penny, D. (1997). Mass survival of birds across the Cretaceous–Tertiary boundary: Molecular evidence. *Science* **275**, 1109–1113.

Cranshaw, G. J., and Boycott, B. R. (1982). Infectious laryngotracheitis in peafowl and pheasants. *Avian Dis.* **26**, 397–401.

Crittenden, L. B. (1981). Exogenous and endogenous leukosis virus genes: A review. *Avian Path.* **10**, 101–112.

Cubitt, W. D., and Barrett, D. T. (1985). Propagation and preliminary characterization of a chicken candidate calicivirus. *J. Gen. Virol.* **66**, 1431–1438.

Cumming, R. B. (1970). Studies on Australian infectious bronchitis virus, IV: Apparent farm-to-farm airborne transmission of infectious bronchitis. *Avian Dis.* **14**, 191–195.

Decaesstecker, M., Charlier, G., and Meullemans, G. (1986). Significance of parvoviruses, entero-like viruses and reoviruses in the aetiology of the chicken maladsorption syndrome. *Avian Path.* **15**, 769–782.

DuBose, R. T., and Grumbles, L. C. (1959). The relationship between quail bronchitis virus and chicken embryo lethal orphan virus. *Avian Dis.* **3**, 321–344.

Eckroade, R. J., and Silverman, L. A. (1986). Avian influenza in Pennsylvania: The beginning. *Proc. 2nd Int. Symp. Avian Influenza*, pp. 22–32.

Eerola, E., Veromaa, T., and Toivanen, P. (1987). Special features in the structural organization of the avian lymphoid system. *In* "Avian Immunology: Basis and Practice" (A. Toivanen and P. Toivanen, eds.), pp. 9–22. CRC, Boca Raton, FL.

Eleazer, T. H., and Hill, J. E. (1994). Highland J virus-associated mortality in chukar partridges. *J. Vet. Diag. Invest. 6, 98–99.*

Ellermann, V., and Bang, O. (1908). Experimentelle Leukamie bei Huhnern. *Zentralbl. Bakteriol.* **46**, 595–605.

Erickson, G. A., Mare, C. J., Gustafson, G. A., Miller, L. D., Proctor, S. J., and Carbrey, E. A. (1978). Interactions between viscerotropic velogenic Newcastle disease virus and pet birds of six species, I: Clinical and serological responses, and viral excretion. *Avian Dis.* **21**, 642–654.

Feduccia, A. (1996). "The Origin and Evolution of Birds." Yale University Press, New Haven.

Ficken, M. D., Wages, D. P., Guy, J. S., Quinn, J. A., and Emory, W. H. (1993). High mortality of domestic turkeys associated with Highlands J virus and eastern equine encephalitis virus infections. *Avian Dis.* **37**, 585–590.

Frank, R. K., and Carpenter, J. W. (1992). Coronaviral enteritis in an ostrich (*Struthio camelus*) chick. *J. Zoo Wildl. Med.* **23**, 103–107.

Garcia, M., Suarez, D. L., Crawford, J. M., Latimer, J. W., Slemons, R. D., Swayne, D. E., and Perdue, M. L. (1997). Evolution of H5 subtype avian influenza A viruses in North America. *Virus Res.* **51**, 115–124.

Gill, F. B. (1990). "Ornithology," 2nd ed. W. H. Freeman and Company, New York.

Gorman, O. T., Bean, W. J., and Webster, R. G. (1992). Evolutionary processes in influenza viruses: Divergence, rapid evolution and stasis. *Curr. Topics Microbiol. Immunol.* **176**, 75–97.

Gough, H. B., Drury, S. E. D., Bygrave, A. C., and Mechie, S. C. (1985). Detection of caliciviruses from pheasants with enteritis. *Vet. Rec.* **131**, 290–291.

Halvorson, D. A., Kelleher, C. J., and Senne, D. A. (1985). Epizootiology of avian influenza: Effect of season on incidence in sentinel ducks and domestic turkeys in Minnesota. *Appl. Environ. Microbiol.* **49**, 914–919.

Heckert, R. A., Collins, M. S., Manvell, R. J., Strong, I., Pearson, J. E., and Alexander, D. J. (1996). Comparison of Newcastle disease viruses isolated from cormorants in Canada and the USA in 1975, 1990 and 1992. *Can. J. Vet. Res.* **60**, 50–64.

Hedges, S. B., Parker, P. H., Sibley, C. G., and Kumar, S. (1996). Continental breakup and the ordinal diversification of birds and mammals. *Nature* **381**, 226–229.

Hinshaw, V. S., Webster, R. G., Bean, W. J., and Sriram, G. (1980). The ecology of influenza viruses in ducks and analysis of influenza viruses with monoclonal antibodies. *Comp. Immunol. Microbiol. Infect. Dis.* **3**, 155–164.

Hinshaw, V. S., Bean, W. J., Webster, R. G., Rehg, J. E., Fiorelli, P., Early, G., Geraci, J. R., and St. Aubin, D. J. (1984). Are seals frequently infected with avian influenza viruses? *J. Virol.* **51**, 863–865.

Hofstad, M. S., Adams, N., and Frey, M. L. (1970). Studies of filterable agent associated with infectious enteritis (bluecomb) of turkeys. *Avian Dis.* **13**, 386–393.

Hughes, C. S., Gaskell, R. M., Jones, R. C., Bradbury, J. M., and Jordan, F. T. W. (1989). Effects of certain stress factors on the re-excretion of infectious laryngotracheitis virus from latently infected carrier birds. *Res. Vet. Sci.* **46**, 247–276.

ICTV (International Committee on Taxonomy of Viruses) (1995). "Virus Taxonomy: Sixth Report of the International Committee on Taxonomy of Viruses" (F. A. Murphy, C. M. Fauquet, D. H. L. Bishop, S. A. Ghabrial, A. W. Jarvis, G. P. Martelli, M. A. Mayo, and M. D. Summers, eds.), pp. 240–244. Springer-Verlag, Vienna.

Jackwood, D. J., and Saif, Y. M. (1987). Antigenic diversity of infectious bursal disease viruses. *Avian Dis.* **31**, 766–770.

Johnson, E. S., Nicholson, L. G., and Durack, D. T. (1998). Detection of antibodies to avian leukosis/sarcoma/viruses (ASLV) and reticuloendotheliosis viruses (REV) in humans by Elisa. *Cancer Det. Prev.* **19**, 394–404.

Kaleta, E. F. (1990). Herpesviruses of birds: A review. *Avian Path.* **19**, 193–211.

Kaufman, J., and Wallny, H. J. (1996). The chicken MHC complex. *In* "Immunology and Developmental Biology of the Chicken" (O. Vaino and B. A. Imhof, eds.), pp. 129–141. Springer-Verlag, New York.

Kawamura, H., and Tsubahara, H. (1960). Common antigenicity of avian reoviruses. *Natl. Inst. Anim. Health Quart. (Tokyo)* **6**, 187–193.

Kawaoka, Y., Chambers, T. M., Sladen, W. L., and Webster, R. G. (1988). Is the gene pool of influenza viruses in shorebirds and gulls different from that in wild ducks? *Virology* **163**, 247–250.

Kennedy, M. A., and Brenneman, K. A. (1995). Enteritis associated with a coronavirus-like agent in a rhea (*Rhea americana*) chick. *J. Avian Med. Surg.* **9**, 138–140.

King, D. J., Pursglove, S. R., and Davidson, W. R. (1981). Adenovirus isolation and serology from wild bobwhite quail (*Colinus virginianus*). *Avian Dis.* **25**, 678–682.

Kirmse, P. (1967a). Host specificity and long persistence of pox infection in the flicker (*Colaptes auratus*). *Bull. Wildl. Dis. Assoc.* **3**, 14–20.

Kirmse, P. (1967b). Pox in wild birds: An annotated bibliography. *Wildl. Dis.* **49**, 1–10.

Kisary, J. (1993). Parvovirus infections of chickens. *In* "Virus Infections of Birds" (J. B. McFerran and M. McNulty, eds.), pp. 153–162. Elsevier, Amsterdam.

Klenk, H. D., Rott, R., Orlich, M., and Blodorn, J. (1975). Activation of influenza A viruses by trypsin treatment. *Virology* **68**, 426–439.

Kurtz, J., Manvell, R. J., and Banks, J. (1996). Avian influenza virus isolated from a woman with conjunctivitis [letter]. *Lancet* **348**, 901–902.

Lang, G., Gagnon, A., and Geraci, J. R. (1981). Isolation of an influenza A virus from seals. *Arch. Virol.* **68**, 189–195.

Larsen, C. T. (1979). "The Etiology of Bluecomb Disease of Turkeys." PhD thesis, University of Minnesota, Minneapolis–St. Paul.

Leibovitz, L., and Hwang, J. (1968). Duck plague on the American continent. *Avian Dis.* **12**, 361–378.

Lundstrom, J. O., and Niklasson, B. (1996). Ockelbovirus (Togaviridae: *Alphavirus*) neutralizing antibodies in experimentally infected Swedish birds. *J. Wildl. Dis.* **32**, 87–93.

Lundstrom, J. O., Turell, M. J., and Niklasson, B. (1992). Antibodies to Ockelbo virus in three orders of birds (Anseriformes, Galliformes, and Passeriformes) in Sweden. *J. Wildl. Dis.* **28**, 144–147.

McFerran, J. B. (1993). Infectious bursal disease. *In* "Virus Infections of Birds" (J. B. McFerran and M. McNulty, eds.), pp. 213–228. Elsevier, Amsterdam.

McFerran, J. B. (1997). Adenovirus infections. *In* "Diseases of Poultry," 10th ed. (B. W. Calnek, R. E. Luginbuhl, and C. F. Helmboldt, eds.), pp. 607–642. Iowa State University Press, Ames.

McFerran, J. B., and McNulty, M. S., eds. (1993). "Virus Infections of Birds." Elsevier, Amsterdam.

McFerran, J. B., McNulty, M. S., McKillop, E. R., Connor, T. J., McCracken, R. M., Collins, M. S., and Allan, G. M. (1980). Isolation and serological studies with infectious bursal disease viruses from fowl, turkey, and duck: Demonstration of a second serotype. *Avian Path.* **9**, 395–404.

McMartin, D. A. (1993). Infectious bronchitis viridae. *In* "Virus Infections of Birds" (J. B. McFerran and M. McNulty, eds.), pp. 249–268. Elsevier, Amsterdam.

Mindell, D. F., Sorenson, M. S., Huddleston, C. J., Miranda, H. C., Knight, A., Sawchuk, S. J., and Yuri, T. (1997). Phylogenetic relationships among and within select avian orders based on mitochondrial DNA. *In* "Avian Molecular Evolution and Systematics" (D. F. Mindell, ed.), pp. 211–245. Academic Press, New York.

Montreal, G. (1992). Adenoviruses and adeno-associated viruses of poultry. *Poultry Sci. Rev.* **4**, 1–27.

Morse, S. (1994). The viruses of the future? Emerging viruses and evolution. *In* "The Evolutionary Biology of Viruses" (S. Morse, ed.), pp. 325–335. Raven, New York.

Murphy, B. R., and Webster, R. G. (1996). Orthomyxoviruses. *In* "Fields Virology," 3rd ed. (B. N. Fields, D. M. Knipe, and P. M. Howley, eds.), pp. 1397–1446. Lippincott-Raven, Philadelphia.

Nagai, Y., Klenk, H. D., and Rott, R. (1976). Proteolytic cleavage of the viral glycoproteins and its significance for the virulence of Newcastle disease virus. *Virology* **72**, 494–508.

Nibert, M. L., Schiff, L. A., and Fields, B. N. (1995). Reoviruses and their replication. *In* "Fields Virology," 3rd ed. (B. N. Fields, D. M. Knipe, and P. M. Howley, eds.), pp. 1557–1596. Lippincott-Raven, Philadelphia.

OIE (Office International des Epizooties) (1996). "Manual of Standards for Diagnostic Tests and Vaccines," 3rd ed. Office International des Epizooties, Paris.

Panigrahy, B., Senne, D. A., Pearson, J. E., Mixson, M. A., and Cassidy, D. R. (1993). Occurrence of velogenic viscerotropic Newcastle disease in pet and exotic birds in 1991. *Avian Dis.* **37**, 254–258.

Perdue, M. L., Garcia, M., Senne, D., and Fraire, M. (1997). Virulence-associated sequence duplication at the hemagglutinin cleavage site of avian influenza viruses. *Virus Res.* **49**, 173–186.

Pierson, F. W., and Domermuth, C. H. (1997). Hemorrhagic enteritis, marble spleen disease, and related infections. *In* "Diseases of Poultry," 10th ed. (B. W. Calnek, R. E. Luginbuhl, and C. F. Helmboldt, eds.), pp. 607–642. Iowa State University Press, Ames.

Poet, S. E., Skilling, D. E., Megyesi, J. L., Gilmartin, W. G., and Smith, A. W. (1994). Detection of non-cultivatable calicivirus from the white tern (*Gygis alba Rothschild*). *J. Wildl. Dis.* **32**, 461–467.

Poss, P. E., and Halvorson, D. A. (1986). The nature of avian influenza in turkeys in Minnesota. *Proc. 2nd Int. Symp. Avian Influenza, Athens, GA*, pp. 112–117. U.S. Animal Health Association.

Ratcliffe, M. J. H. (1989). Development of the avian B lymphocyte lineage. *Crit. Rev. Poultry Biol.* **2**, 207–234.

Robertson, M. D., and Wilcox, G. E. (1986). Avian reoviruses. *Vet. Bull.* **56**, 155–174.

Rogers, G. N., and Paulson, J. C. (1983). Receptor determinants of human and animal influenza virus isolates: Differences in receptor specificity of the H3 hemagglutinin based on species of origin. *Virology* **127**, 361–373.

Rosenberger, J. K., and Olson, N. O. (1997). Viral arthritis. *In* "Diseases of Poultry," 10th ed. (B. W. Calnek, R. E. Luginbuhl, and C. F. Helmboldt, eds.), pp. 711–719. Iowa State University Press, Ames.

Rott, R., Klenk, H. D., Nagai, Y., and Tashiro, M. (1995). Influenza viruses, cell enzymes, and pathogenicity. *Am. J. Respir. Crit. Care Med.* **152**, S16-9.

Rous, P. (1911). A sarcoma of the fowl transmissible by an agent separable from the tumor cells. *J. Exp. Med.* **13**, 397–411.

Saif, Y. M. (1997). Emerging diseases and diseases of complex or unknown etiology. *In* "Diseases of Poultry," 10th ed. (B. W. Calnek, R. E. Luginbuhl, and C. F. Helmboldt, eds.), pp. 1007–1053. Iowa State University Press, Ames.

Sandhu, T. S., and Leibovitz, L. (1997). Duck viral enteritis (duck plague). *In* "Diseases of Poultry," 10th ed. (B. W. Calnek, R. E. Luginbuhl, and C. F. Helmboldt, eds.), pp. 675–683. Iowa State University Press, Ames.

Seal, B. S., King, D. J., and Bennett, J. D. (1995). Characterization of Newcastle disease virus isolates by reverse transcription PCR coupled to direct nucleotide sequencing and development of a sequence database for pathotype prediction and molecular epidemiological analysis. *J. Clin. Microbiol.* **33**, 2624–2630.

Shirai, J. K., Nakamura, K., Shinohara, K., and Kawamura, H. (1991). Pathogenicity and antigenicity of avian nephritis isolates. *Avian Dis.* **35**, 49–54.

Siddell, S. G., ed. (1995). "The Coronaviridae." Plenum, New York.

Skehel, J. J., Bizebard, T., Bullough, P. A., Hughson, F. M., Knossow, M., Steinhauer, D. A., Wharton, S. A., and Wiley, D. C. (1995). Membrane fusion by influenza hemagglutinin. *Cold Spring Harb. Symp. Quant. Biol.* **60**, 573–580.

Slemons, R. D., and Easterday, B. C. (1977). Type-A influenza viruses in the feces of migratory waterfowl. *J. Am. Vet. Med. Assoc.* **171**, 947–948.

Smith, E. J., Williams, S. M., and Fadly, A. M. (1997). Detection of avian leukosis virus subgroup J using the polymerase chain reaction. *Avian Dis.* **42**, 375–380.

Suarez, D. L., and Perdue, M. L. (1998). Multiple alignment comparison of the non-structural genes of influenza A viruses. *Virus Res.* **54**, 59–69.

Suarez, D. L., Perdue, M. L., Cox, N., Rowe, T., Bender, C., Huang, J., and Swayne, D. E. (1998). Comparisons of highly virulent H5N1 influenza A viruses isolated from humans and chickens from Hong Kong. *J. Virol.* **72**, 6678–6688.

Subbarao, K., Klimov, A., Katz, J. M., Regnery, H., Lim, W., Hall, H., Perdue, M. L., Swayne, D. E., Bender, C., Huang, J., Hemphill, M., Rowe, T., Shaw, M., Xu, X., Fukuda, K., and Cox, N. (1998). Characterization of an avian influenza A (H5N1) virus isolated from a child with a fatal respiratory illness. *Science* **279**, 393–396.

Swanstrom, R., and Vogt, P. K. (1990). "Retroviruses: Strategies of Replication." Springer-Verlag, Berlin.

Trampel, D. W., Kinden, D. A., Solorzano, R. F., and Stogsdill, P. L. (1983). Parvovirus-like enteropathy in Missouri turkeys. *Avian Dis.* **27**, 49–54.

Tripathy, D. N. (1993). Apoxviruses. *In* "Virus Infections of Birds" (J. B. McFerran and M. McNulty, eds.), pp. 5–18. Elsevier, Amsterdam.

Tripathy, D. N., and Reed, W. M. (1997). Pox. *In* "Diseases of Poultry," 10th ed. (B. W. Calnek, R. E. Luginbuhl, and C. F. Helmboldt, eds.), pp. 643–659. Iowa State University Press, Ames.

Utterback, W. W., and Schwartz, J. H. (1973). Epizootology of velogenic viscerotropic Newcastle disease in Southern California. *J. Am. Vet. Med. Assoc.* **163**, 1080–1090.

Venugopal, K., and Payne, L. N. (1995). Molecular pathogenesis of Marek's disease — recent developments. *Avian Path.* **24**, 597–609.

Verbeek, A., and Tijssen, P. (1991). Sequence analysis of the turkey enteric coronavirus nucleocapsid and membrane protein genes: A close genomic relationship with bovine coronaviruses. *J. Gen. Virol.* **72**, 1659–1666.

Vindevogel, H., and Duchatel, J. P. (1997). Miscellaneous herpesvirus infections. *In* "Diseases of Poultry," 10th ed. (B. W. Calnek, R. E. Luginbuhl, and C. F. Helmboldt, eds.), pp. 757–760. Iowa State University Press, Ames.

Wages, D. P., Ficken, M. D., Guy, J. S., Cummings, T. S., and Jennings, S. R. (1993). Egg production drop in turkeys associated with alphaviruses: Eastern equine encephalitis virus and Highlands J virus. *Avian Dis.* **37**, 1163–1166.

Wang, L., Junker, D., and Collisson, E. W. (1993). Evidence of natural recombination within the S1 gene of infectious bronchitis virus. *Virology* **192**, 710–716.

Webster, R. G., and Rott, R. (1987). Influenza virus A pathogenicity: The pivotal role of hemagglutinin. *Virology* **50**, 665–666.

Webster, R. G., Geraci, J., Petursson, G., and Skirmisson, K. (1981a). Conjunctivitis in human beings caused by influenza A virus of seals. *New Engl. J. Med.* **304**, 911–916.

Webster, R. G., Hinshaw, V. S., Bean, W. J., Van Wyke, K. L., Geraci, J. R., St. Aubin, D. J., and Petursson, G. (1981b). Characterization of an influenza A virus from seals. *Virology* **113**, 712–724.

Webster, R. G., Bean, W. J., Gorman, O. T., Chambers, T. M., and Kawaoka, Y. (1992). Evolution and ecology of influenza A viruses. *Microbiol. Rev.* **56**, 152–179.

Webster, R. G., Wright, S. M., Castrucci, M. R., Bean, W. J., and Kawaoka, Y. (1993). Influenza: A model of an emerging virus disease. *Intervirology* **35**, 16–25.

Welty, J. C., ed. (1982). "The Life of Birds," 3rd ed. Saunders College Publishing, Philadelphia.

Witter, R. L. (1997). Increased virulence of Marek's disease virus field isolates. *Avian Dis.* **41**, 149–163.

Witter, R. L., and Johnson, D. C. (1985). Epidemiology of reticuloendotheliosis virus in broiler breeder flocks. *Avian Dis.* **29**, 1140–1154.

Wood, G. W., McCauley, J. W., Bashiruddin, J. B., and Alexander, D. J. (1993). Deduced amino acid sequences at the haemagglutinin cleavage site of avian influenza A viruses of H5 and H7 subtypes. *Arch. Virol.* **130**, 209–217.

Yates, V. J., and Piela, T. H. (1993). Avian Adenovirus-associated virus. *In* "Virus Infections of Birds" (J. B. McFerran and M. S. McNulty, eds.), pp. 163–165. Elsevier, Amsterdam.

Yuasa, N., and Imai, K. (1986). Pathogenicity and antigenicity of eleven isolates of chicken anemia agent (CAA). *Avian Path.* **15**, 639–645.

16

Prion Diseases of Humans and Animals

GLENN C. TELLING

Prion Pathogenesis Group
Sanders–Brown Center on Aging
University of Kentucky
Lexington, Kentucky 40536-0230

I. INTRODUCTION

For many years, the unusual physical and biological properties of the infectious agents causing scrapie in sheep and kuru and Creutzfeldt–Jakob disease (CJD) in humans puzzled researchers. While these diseases share many features in common with conventional viral diseases (for example, the infectious agent is filterable and exists as multiple strains), the scrapie agent exhibited unusual properties that were

593

not shared by conventional viruses (Alper *et al.*, 1966; Hunter, 1972). To distinguish these "proteinaceous infectious" agents from viruses and viroids, Prusiner (1982) introduced the term "prion." The unifying hallmark of prion diseases is the aberrant metabolism of the prion protein (PrP) that exists in at least two conformational states with different physicochemical properties. The normal form of the protein, referred to as PrP^C, is expressed in the presence and absence of disease, while the disease-associated isoform, referred to as PrP^{Sc}, is associated exclusively with infection. Although the molecular structure of the prion still eludes definitive identification, considerable evidence argues that prions lack nucleic acid and are composed largely, if not exclusively, of PrP^{Sc} molecules and that prion diseases are disorders of protein conformation.

II. THE PRION DISEASES

The prion disorders of animals and humans, which are often referred to as transmissible spongiform encephalopathies (TSEs), share a number of common features (Table I). They all have long incubation periods, ranging from months to years, and are invariably fatal once clinical symptoms have appeared. Clinical symptoms are variable but are a consequence of degeneration of the central nervous system (CNS). While the brains of patients or animals with prion diseases frequently show no recognizable abnormalities on gross examination, light-microscopic examination of the CNS reveals characteristic neuropathological features consisting of neuronal degeneration and neuronal loss, giving the cerebral gray matter a microvacuolated or "spongiform" appearance, severe astrocytic gliosis, which is often out of proportion to the degree of nerve cell loss, and, while by no means a constant feature, deposition of amyloid plaques composed of insoluble aggregates of prion protein in some examples of prion disease.

A. Scrapie

Scrapie has been recognized as a fatal neurodegenerative disorder of sheep and goats for more than 250 years. Disease is endemic in sheep populations in many parts of the world, and affected animals exhibit behavioral changes that are characterized by anxiousness and hypersensitivity followed by intense pruritus. In fact, the term *scrapie* derives from the tendency of affected animals to rub themselves against walls and fences, thus removing much of their fleece. Later stages of the disease are characterized by unsteady gait, eventually resulting in

TABLE I

The Animal and Human Prion Diseases

Disease	Host	Etiology
A. Animal prion diseases		
Scrapie	Sheep, goats	Infectious disease of uncertain etiology
Transmissible mink encephalopathy (TME)	Captive mink	Probably food-borne
Chronic wasting disease (CWD)	Captive and free-range mule deer and elk	Uncertain etiology
Bovine spongiform encephalopathy (BSE)	Cattle	Prion-contaminated meat and bone meal (MBM)
Feline spongiform encephalopathy (FSE)	Domestic cats Cheetah Puma Tiger Ocelot	Exposure to BSE prions via the food chain
Exotic ungulate encephalopathy	Kudu Gemsbok Nyala Onyx Eland Ankole cow	Prion-contaminated meat and bone meal (MBM)

Disease	Mani-festa-tion	Distribution	Etiology
B. Human prion diseases			
Creutzfeldt–Jakob disease (CJD)	Sporadic	1 case per million population	Unknown
Familial CJD; Gerstmann–Sträussler–Scheinker disease (GSS); fatal familial insomnia (FFI)	Inherited	10–20% of all human prion diseases are inherited	Autosomal dominant transmission of a germline *PRNP* mutation
Kuru	Acquired	Fore people, Papua New Guinea	Cannibalism
Iatrogenic CJD		About 80 cases	Accidental exposure to prions during medical procedure
New variant CJD		~50 young adults and teenagers in the UK and France	Exposure to BSE prions

severe ataxia, with animals becoming recumbent in the terminal phase of the disease.

Several British and French investigators attempted to transmit disease from affected animals to healthy sheep by injection of various tissues and body fluids. Transmission was convincingly demonstrated under experimental conditions by intraocular injection of scrapie-free animals with spinal cord derived from scrapie-affected animals (Cuillé and Chelle, 1939). Iatrogenic transmission was also demonstrated after more than 1500 sheep developed scrapie as a result of vaccination against looping ill virus with a formalin-treated extract of sheep lymphoid tissue apparently contaminated with scrapie prions (Gordon, 1946). This incident also provided one of the first indications of the unusual resistance of the infectious agent to certain chemical treatments.

While experimental transmission of scrapie eventually became well established, the etiology of natural scrapie was the subject of intense debate for many years. Parry was convinced that scrapie was a purely genetic disorder in which contagious spread played little or no part, arguing that proper breeding measures would ultimately eradicate the disease (Parry, 1962). Other studies suggested that scrapie was an endemic viral-like illness (Dickinson et al., 1974) in which susceptibility was modulated by host genes (Dickinson et al., 1965). Ultimately, both interpretations proved to be at least partly correct. The successful transmission of scrapie to mice (Chandler, 1961) led to identification of the *Sinc* gene, an autosomal locus that controls scrapie incubation time (Dickinson et al., 1968).

B. Bovine Spongiform Encephalopathy and Related Diseases

Bovine spongiform encephalopathy (BSE), often referred to as "mad cow disease," was first diagnosed in the United Kingdom in April 1985. The clinical features are now well described and include temperament changes, postural abnormalities, coordination problems, and terminal recumbence. By 1986, pathologists identified lesions in the brains of affected cattle that were similar to the characteristic neuropathological hallmarks of scrapie in sheep (Wells et al., 1987). Experimental transmission of BSE to mice (Fraser et al., 1988) and cattle (Dawson et al., 1990a) confirmed that BSE was a prion disease. Experimental strain-typing studies in panels of inbred mice demonstrated that different sources of BSE are biologically invariable, strongly suggesting that BSE represents a single major strain that is different from the 20 distinct strains of scrapie defined by similar means (Bruce et al., 1994, 1991; Bruce, 1996).

Epidemiological studies point to contaminated offal that was used in the manufacture of meat and bone meal (MBM) and fed to cattle as a protein supplement as the source of prions responsible for BSE (Wilesmith *et al.*, 1988). It has been hypothesized that a change in the late 1970s in the price of tallow, which is a byproduct of the rendering process, resulted in the diminished use of a hydrocarbon solvent extraction stage during the production of MBM that allowed prions to persist during the modified rendering process (Wilesmith, 1991). Since Great Britain has a relatively large sheep population and 0.5–1% of that population is affected annually with scrapie, it was hypothesized that scrapie-contaminated sheep offal was the initial source of BSE, which was subsequently amplified by recycling of infected cattle with subclinical disease. However, the spontaneous occurrence of prion disease in cattle cannot be ruled out as the original source of BSE prions. Since 1988, the practice of feeding ruminant-derived protein to ruminants has been banned in the United Kingdom, but not before BSE reached epidemic proportions, affecting over 170,000 cattle during the past decade (Anderson *et al.*, 1996). It is estimated that up to 1,000,000 cattle were infected with BSE, with an incubation period of 5 years, although this may be an underestimation of the extent of exposure since most cattle were slaughtered between 2 to 3 years of age (Stekel *et al.*, 1996).

Given the potential for human exposure to BSE prions via the food chain, the emergence of the BSE epidemic prompted concerns over the safety of BSE-contaminated foodstuffs. Based on pathogenesis studies of natural scrapie in goats and sheep (Hadlow *et al.*, 1982, 1980), the specified bovine offal (SBO) ban was introduced in the United Kingdom in 1989 to prevent inclusion in the human food chain of bovine tissues from the lymphoreticular system (LRS) and CNS — namely, brain, spinal cord, tonsil, thymus, spleen, and intestine. Attempts to establish an infectious link between sporadic CJD and scrapie had been unrewarding, and this observation convinced many scientists that a similar protective barrier would prevent passage of BSE from cattle to humans. However, the unusually wide host-range of BSE has been appreciated since the early days of the epidemic, and BSE-contaminated foodstuffs have caused disease in several other species in Great Britain, including feline spongiform encephalopathy (FSE) in approximately 80 domestic cats (Wyatt *et al.*, 1993) and captive wild cats in zoos from five different species (Kirkwood and Cunningham, 1994; A. Cunningham, personal communication). FSE has been experimentally transmitted to laboratory mice (Pearson *et al.*, 1992). Bovidae in zoos from eight other species have also been affected (Kirkwood and Cunningham, 1994; A. Cunningham, personal communication), and brain extracts from a nyala and greater kudu have also transmitted disease to mice (Cunningham *et al.*, 1993). In addition to these "natural" infections, BSE has been experimentally transmitted to a variety of species including mice (Fraser *et al.*, 1988), cattle (Dawson *et al.*, 1990a), sheep and goats (Foster *et al.*, 1993), pigs (Dawson *et al.*, 1990b), and mink (Robinson *et al.*,

1994). Of particular concern was the transmission of BSE to the nonhuman primate marmoset (Baker *et al.*, 1993).

C. Chronic Wasting Disease and Transmissible Mink Encephalopathy

Chronic wasting disease (CWD) was first recognized in captive mule deer in north central Colorado in 1967 (Williams and Young, 1980). Signs in clinically affected deer and elk include weight loss, behavioral alterations, apparent ruminal atony, and salivary defluxion in late-stage disease. Like other prion diseases, pathognomonic lesions are confined to the CNS and consist of intraneuronal vacuolation, neuropil spongiosis, astrocytic hypertrophy and hyperplasia, and occasionally amyloid plaques (Williams and Young, 1993). The origins of the disease are unknown, but since its identification CWD has transmitted with an efficiency that appears unparalleled in other prion diseases. This remarkable transference of disease is documented at a Colorado Division of Wildlife facility wherein 90% of deer (*n* = 60 animals) resident for 2 years or longer developed CWD between 1970 and 1981 (Williams and Young, 1992).

Surveillance of hunter-killed deer has revealed that the disease is no longer restricted to captive animals. In 1996–97, the incidence of CWD infection in free-ranging animals in the northern Colorado endemic area was 5–6% in deer and <1% in elk. Moreover, CWD is no longer confined to the original northern Colorado endemic area. In 1998, 39 of 179 (22%) of game farm elk in South Dakota were infected with CWD. CWD has also been diagnosed in captive elk in Nebraska, Oklahoma, and Canada (M. Miller, Colorado Division of Wildlife, personal communication). While the most plausible natural route of CWD transmission is via ingestion of forage or water contaminated by secretions, excretions, or other sources (e.g. placenta, carcasses), the natural route of CWD transmission remains unknown.

Transmissible mink encephalopathy (TME) is a rare disease of ranch-raised mink in the United States that are fed a variety of protein-based diets. Experimental transmission studies suggest that scrapie is not the source of TME. However, mink are readily infected by oral exposure to BSE and cattle-passaged TME (Robinson *et al.*, 1994, 1995), and it has been suggested that an outbreak of TME in 1985 was due to exposure to a sporadic case of BSE (Marsh *et al.*, 1991).

D. Human Prion Diseases

Scrapie remained an obscure veterinary disorder until the late 1950s and the discovery of an exotic neurodegenerative illness called kuru, or "laughing death," occurring among the Fore highlanders in Papua New Guinea (Gajdusek and Zigas, 1957). Kuru, which is thought to have been spread by cannibalism (Gajdusek, 1977), had its origins at the beginning of this century and reached epidemic proportions in midcentury, killing more than 3000 people in the affected population of 30,000. Since the cessation of cannibalistic practices, the disease has all but died out, with only a handful of cases currently occurring in older individuals who were presumably exposed to kuru as young children.

The clinical features of kuru are similar to scrapie, being a progressive cerebellar ataxia accompanied by dementia in the later stages and death occurring usually within 9 months. However, it was the neuropathological features of the two diseases that strongly suggested that they might be closely related. Hadlow recognized that the spongiform vacuolation and plaque lesions in kuru patients were similar to the neuropathological profile in the brains of scrapie-affected sheep. Since scrapie had been demonstrated to be experimentally transmissible, Hadlow speculated that kuru brain extracts might also be transmissible by inoculation of nonhuman primates (Hadlow, 1959). Seven years later, Gajdusek, Gibbs and Alpers demonstrated experimental transmission when brain extracts from kuru-affected patients produced a progressive neurodegenerative condition in inoculated chimpanzees with clinical and pathological features similar to kuru after a prolonged incubation period of 18 to 21 months (Gajdusek *et al.*, 1966). The similar neuropathology of kuru and CJD (Klatzo *et al.*, 1959) prompted additional transmission experiments with chimpanzees, and transmissions of CJD were subsequently reported (Gibbs *et al.*, 1968).

For many years, the transmission of human prion diseases was studied largely with apes and monkeys, where >90% of cases are thought to be transmissible (Brown *et al.*, 1994b). The scarcity of primates, the expense of their long-term care and the increasing ethical objections to such experiments has limited subsequent investigations. While the most reliable transmission data has been said to emanate from studies using nonhuman primates, some cases of human prion disease have been transmitted to mice, rats, and hamsters but with much more variable results (Gibbs *et al.*, 1979; Manuelidis *et al.*, 1978, 1976; Prusiner, 1987; Tateishi and Kitamoto, 1995; Tateishi *et al.*, 1983; Telling *et al.*, 1994).

The human prion diseases are unique in biology in that they are manifest in three forms:

1. Sporadic forms of prion disease comprise most (~80%) cases of CJD (Hsiao *et al.*, 1989; Masters *et al.*, 1978), which occurs at a rate of roughly one per million population across the planet (Masters and Richardson, 1978). Sporadic CJD is predominately a disease of old age, with a peak onset between 60 and 65 years. The etiology of sporadic CJD is unknown, but hypotheses include horizontal transmission from humans or animals (Gajdusek, 1977), somatic mutation of the

PrP gene, and spontaneous conversion of PrP^C into PrP^{Sc} as a rare stochastic event (Hsiao *et al.*, 1991; Prusiner, 1989).

2. The human PrP gene (referred to as *PRNP*) is located on the short arm of chromosome 20 (Robakis *et al.*, 1986). Between 10 and 20% of human prion diseases occur in families with an autosomal dominant mode of inheritance. The discovery of genetic linkage between the PrP gene and scrapie incubation times in mice (Carlson *et al.*, 1986) suggested a similar basis for the inherited human prion diseases. The first human prion disease mutation to be identified was a 144-base-pair insertion in the amino-terminal coding sequence of *PRNP* that normally encodes a tandem array of five octapeptide repeats (Owen *et al.*, 1989). Genetic linkage was established for a second nonconservative missense mutation occurring at codon 102 changing a proline to a leucine in two families with Gestmann–Sträussler–Scheinker (GSS) disease (Hsiao *et al.*, 1989), and this mutation has been found in many families in numerous countries including the original GSS family (Doh-ura *et al.*, 1989; Goldgaber *et al.*, 1989; Kretzschmar *et al.*, 1991). To date, 20 different nonconservative missense and octarepeat insertion mutations have been identified in the coding sequence of *PRNP* that segregate with dominantly inherited neurodegenerative disorders (see Fig. 1). Five of these mutations have been genetically linked to loci controlling familial CJD, GSS, and fatal familial insomnia (FFI), all of which are inherited human prion

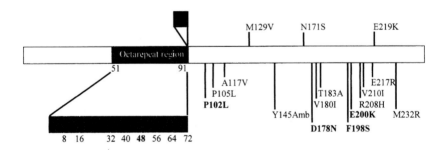

Fig. 1. Pathogenic mutations and polymorphisms in the human prion protein gene. The 253-amino-acid-residue open reading frame (ORF) of *PRNP* is represented as an extended rectangle. In wild-type *PRNP*, the amino-terminal sequence between codons 51 and 91 encodes a tandem array of five octapeptide repeats represented by black boxes within the ORF. The pathogenic mutations associated with human prion disease are shown below the ORF. These consist of 8, 16, 32, 40, 48, 56, 64, and 72 amino acid insertions within the octarepeat region and point mutations causing nonconservative missense amino acid substitutions. Point mutations are designated by the wild-type amino acid preceding the codon number, followed by the mutant residue, using the single letter amino acid conventions. The *PRNP* mutations for which genetic linkage with familial CJD, GSS, and FFI has been established are shown in bold type. Deletion of a single octapeptide repeat is polymorphic and is not associated with disease. Other polymorphisms found at codons 129, 171 and 219 are also shown above the ORF.

diseases that can be transmitted to experimental animals (Dlouhy *et al.*, 1992; Gabizon *et al.*, 1993; Hsiao *et al.*, 1989; Petersen *et al.*, 1992; Poulter *et al.*, 1992).

In addition to the disease-associated mutations, there are a number of benign polymorphisms of *PRNP*, including one at codon 129 that encodes either methionine (Met) or valine (Val) (Owen *et al.*, 1990). Homozygosity for Met or Val at codon 129 has been reported to predispose individuals to development of sporadic CJD (Palmer *et al.*, 1991) and iatrogenic CJD (Brown *et al.*, 1994a; Collinge *et al.*, 1991; Deslys *et al.*, 1994). The interplay between residue 129 and the pathogenic mutations is illustrated dramatically by the clinical and pathological phenotypes of FFI and familial CJD. Mutation at position 178, changing aspartic acid to asparagine (D178N), is linked to FFI or a subtype of familial CJD depending on whether Met or Val is encoded at position 129 on the mutant allele (Goldfarb *et al.*, 1992).

3. While all forms of human prion disease are experimentally transmissible, the naturally acquired forms have been confined to rare and unusual situations, for example, kuru and iatrogenic cases of CJD caused by accidental human prion transmission during medical procedures. Iatrogenic CJD has been caused by human growth hormone (HuGH) and gonadotrophin derived from cadaveric pituitaries as well as dura mater and corneal grafting (Brown *et al.*, 1994a, 1992). However, in 1995 two cases of CJD were reported in teenagers in the United Kingdom (Bateman *et al.*, 1995; Britton *et al.*, 1995). Since sporadic CJD is extremely rare in individuals under 30 years of age (only four patients under 20 years of age had ever previously been diagnosed with CJD, and none of these were in the United Kingdom [Collinge, 1997]), these cases caused considerable concern that they might be linked to BSE. By March 1996, additional CJD cases in young adults and teenagers were reported (Will *et al.*, 1996), and it became clear that the unusual neuropathology in these new cases, consisting of amyloid plaques reminiscent of kuru pathology, was remarkably consistent. These cases were termed "new variant" CJD (vCJD). In the past 4 years, ~50 cases of vCJD have been reported in teenagers and young adults in the United Kingdom as well as a single case of vCJD in France (Chazot *et al.*, 1996). All reported cases of vCJD have occurred in individuals who were homozygous (Met/Met) at codon 129 (Collinge *et al.*, 1996a; Zeidler *et al.*, 1997)

Transmission of BSE to three macaques has been reported to produce disease with neuropathological features similar to those reported for cases of vCJD in humans (Lasmézas *et al.*, 1996). Recent strain-typing experiments have demonstrated that vCJD is indistinguishable from BSE (Bruce *et al.*, 1997; Collinge *et al.*, 1996a; Hill *et al.*, 1997a), providing the strongest proof so far for a causal relationship between BSE and vCJD.

III. CHARACTERIZING THE INFECTIOUS AGENT

Because prions can only be assayed in experimental animals, and the time between inoculation and disease is extremely long, characterization of the infectious agent has been enormously difficult. Experimental transmission studies consist of intracerebral or intraperitoneal inoculation of brain homogenate into a set of recipient animals; the time to appearance of a characteristic set of clinical symptoms (the incubation period) is measured and CNS pathology is characterized. Early investigations were hampered by the intractability of sheep as experimental models, so successful transmission of scrapie to mice (Chandler, 1961) and Syrian hamsters (Marsh and Hanson, 1977) offered more convenient systems for studying prion diseases.

Early attempts to characterize the scrapie agent suggested a structure that was different from conventional pathogens. Alper and colleagues demonstrated in UV irradiation studies that the target size of the agent was considerably smaller than a virus and suggested that a nucleic acid might not be involved (Alper *et al.*, 1966). In 1967 Griffith proposed a number of models to account for the unusual characteristics of the infectious agent, one being the alteration of a normal protein into a pathological protein (Griffith, 1967), although it must be said that these ideas were met with considerable skepticism.

Fifteen years later, Prusiner and colleagues succeeded in purifying scrapie infectivity from hamster brains and demonstrated characteristics that were typical of proteins and inconsistent with the agent containing nucleic acid (Prusiner, 1982; Prusiner *et al.*, 1982). Subsequent thorough attempts to locate nucleic acids in highly purified scrapie preparations have consistently failed (Kellings *et al.*, 1992). To distinguish these "proteinaceous infectious" agents from conventional pathogens, Prusiner coined the term *prion*, defined as small proteinaceous infectious particles that resist inactivation by procedures that modify nucleic acids (Prusiner, 1982). A partially protease-resistant protein of relative molecular mass in SDS-PAGE of 27 to 30 kD, designated PrP27-30, which is highly insoluble in non-ionic detergents, was enriched in purified fractions of infectious activity (Bolton *et al.*, 1982; McKinley *et al.*, 1983).

The determination of the amino-terminal amino acid sequence of purified PrP27-30 (Prusiner *et al.*, 1984) led to the recovery of PrP-encoding cDNA clones from libraries derived from the brains of scrapie-infected Syrian hamster (Oesch *et al.*, 1985) and mouse (Chesebro *et al.*, 1985). Surprisingly, PrP mRNA proved to be the product of a host cell gene that was expressed at the same levels in the brains of infected and uninfected animals (Chesebro *et al.*, 1985; Oesch *et al.*, 1985), and PrP primary structure was shown to be identical in normal and infected animals (Basler *et al.*, 1986). These studies demonstrated that PrP exists in two varieties with different physicochemical properties. The normal form of the protein found in the brains of infected and uninfected animals, which is sensitive to protease treatment and soluble in detergents, is referred to as PrP^C (Meyer *et al.*, 1986; Oesch *et al.*, 1985), while the disease-associated isoform found only in the brains of infected animals is partially resistant to protease treatment, insoluble in detergents, and forms aggregates, and is referred to as PrP^{Sc}. Protease cleavage of the N-terminal 66, or so, amino acids of PrP^{Sc} gives rise to a protease-resistant core represented by PrP27–30.

While the molecular structure of the prion still eludes definitive identification, considerable evidence argues that prions are composed largely, if not exclusively, of PrP^{Sc} molecules. During the disease process, PrP^C is converted to PrP^{Sc} by a posttranslational process (Borchelt *et al.*, 1990). Whereas no covalent modifications have been found that distinguish PrP^C from PrP^{Sc} (Stahl *et al.*, 1993), considerable evidence argues that prion diseases are disorders of protein conformation. The so-called "protein-only" model of prion propagation contends that prion replication results from conformational conversion of the benign host-encoded PrP^C into pathogenic PrP^{Sc}. During prion replication, PrP^{Sc} is thought to act as a template by imposing its conformation on PrP^C (Cohen *et al.*, 1994; Huang *et al.*, 1996). Support for this comes from the results of low-resolution structural studies showing that PrP^C has a high α-helical content and is virtually devoid of β-sheets, and that PrP^{Sc} has a high β-sheet content (Caughey *et al.*, 1991; Pan *et al.*, 1993; Pergami *et al.*, 1996; Prusiner *et al.*, 1983; Safar *et al.*, 1993). The three-dimensional structures of bacterially expressed recombinant mouse PrP (residues 121–231 and 23–231) (Riek *et al.*, 1996, 1997) and Syrian hamster PrP (residues 29–231) (James *et al.*, 1997) have been determined by nuclear magnetic resonance (NMR) spectroscopy. The structured globular carboxy-terminal portion consists of three α-helices interspersed with two short sections that form a short β-pleated sheet, while the amino-terminal region is largely unstructured (Fig. 2).

IV. THE ROLE OF PrP^C IN THE DISEASE PROCESS

The first indication that PrP^C was involved in the disease process came from studies to determine the molecular basis of scrapie incubation times in mice, which demonstrated that a single autosomal gene, referred to as *Sip*, determined incubation time in sheep (Dickinson, 1976). Studies with inbred strains of mice demonstrated that a similar incubation time locus in mice, referred to as *Sinc* (Dickinson *et al.*, 1968) or *Prn-i*, was genetically linked to the mouse PrP gene, referred to as *Prnp* (Carlson *et al.*, 1986). Differences at two coding positions within *Prnp* were found in strains of mice with long and short mouse scrapie incubation times (Westaway *et al.*, 1987).

Fig. 2. Structural features of the prion protein. The 254-amino-acid-residue translation product is processed by removal of an amino-terminal signal peptide of 22 amino acids and a carboxyl-terminal hydrophobic peptide of 23 amino acids, both represented by boxes with square dots. After cleavage of the carboxyl-terminal peptide, a glycosyl phosphatidylinositol (GPI) anchor is added (Stahl *et al.*, 1987). The locations of the three α-helices (designated H1, H2, and H3) and two short sections forming a β-pleated sheet (represented as arrows), which have been identified by nuclear magnetic resonance (NMR) spectroscopy of recombinant PrP are also indicated. H1 extends from amino acid residues 143 to 153, H2 from amino acid residues 175 to 192, and H3 from amino acid residues 199 to 218, while the two regions making up the β-pleated sheet extend from amino acid residues 127 to 130 and amino acid residues 160 to 163 (mouse PrP numbering). The amino-terminal portion of the molecule is largely unstructured and contains a tandem array of five octapeptide repeats between codons 51 and 90, which are represented by single-letter amino-acid code. Asn-linked oligosaccharides are attached at residues 180 and 196, and H2 and H3 are connected by a disulfide bond that joins codons 178 and 213. Protease cleavage of the amino-terminal 66, or so, amino acids of PrP^{Sc} gives rise to the protease-resistant core represented by PrP27–30, which is shown in gray.

Perhaps the most compelling evidence to date for the protein-only model of prion replication has come from two different experimental approaches involving transgenic mice. Since PrP^C is the source of PrP^{Sc}, the model predicts that elimination of PrP^C would result in abolition of prion replication. To test this, the *Prnp* locus was disrupted by homologous recombination in embryonic stem cells and knockout mice were established (Büeler *et al.*, 1992; Manson *et al.*, 1994a; Sakaguchi *et al.*, 1995). Unlike wild-type mice, the resultant homozygous null mice (*Prnp*$^{0/0}$), which express no PrP^C, fail to develop the characteristic clinical and neuropathological symptoms of scrapie after inoculation with mouse prions and do not propagate prion infectivity (Büeler *et al.*, 1993; Manson *et al.*, 1994b; Prusiner *et al.*, 1993; Sakaguchi *et al.*, 1995).

In a second approach, transgenic mice overexpressing a proline-to-leucine mutation at codon 101 of mouse PrP, equivalent to the human GSS P102L mutation, referred to as Tg(MoPrP–P101L), spontaneously developed clinical and neuropathological symptoms similar to mouse scrapie between 150 and 300 days of age (Hsiao *et al.*, 1990; Telling *et al.*, 1996a). In contrast, transgenic mice overexpressing wild-type mouse PrP at equivalent levels did not develop disease (Telling *et al.*, 1996a). Importantly, the *de novo* generation of prions in the brains of spontaneously sick mice was demonstrated by serial propagation of infectivity in mice expressing low levels of mutant protein that otherwise do not get sick (Hsiao *et al.*, 1994; Telling *et al.*, 1996a). These findings suggest that the disease-associated mutations destabilize the structure of PrP^C, causing it to spontaneously adopt the pathologic PrP^{Sc} conformation and explain how prion diseases can be both genetic and infectious. Prion infectivity from brain extracts of humans expressing the P102L GSS mutation was also propagated in transgenic mice expressing a chimeric mouse–human PrP gene with the P101L mutation (Telling *et al.*, 1995). Other inherited human prion diseases have also been transmitted to transgenic mice expressing human and chimeric mouse–human PrP (Telling *et al.*, 1995, 1996b; Collinge *et al.*, 1995b) and nonhuman primates (Brown *et al.*, 1994b).

V. THE NORMAL ROLE OF PrP^C

Two *Prnp*$^{0/0}$ lines in which the PrP coding sequence was disrupted were independently generated in Zurich and Edinburgh. Contrary to expectation, these mice developed normally and suffered no gross phenotypic defects (Büeler *et al.*, 1992; Manson *et al.*, 1994a). These results raised the possibility that adaptive changes occur during the development of *Prnp*$^{0/0}$ mice that compensate for the loss of PrP^C function. To test this hypothesis, transgenic mice were produced in which expression of transgene-expressed PrP^C could be controlled at will. Using the tetracycline gene-response system, mice were produced that co-express a tetracycline-responsive transactivator, referred to as tTA, and a tTA-responsive promoter that drives PrP expression (Tremblay *et al.*, 1998). The tTA consists of the tetracycline repressor fused to the transactivation domain of herpes simplex virus VP16 and binds specifically and with high affinity to the tetracycline operator (tetO). Binding of tTA normally activates transcription of the PrP gene, but binding of doxycycline, a tetracycline analogue, to tTA prevents the tTA protein from binding to tetO, which in turn prevents PrP gene expression. Thus, PrP^C is expressed in the absence of doxycycline but not in its presence. Repressing PrP^C expression by oral administration of doxycycline was not deleterious to adult mice. However, since doxycycline treatment did not completely inhibit PrP^C expression in these mice, it is not clear whether this residual expression masks the true phenotype of *Prnp*$^{0/0}$ mice.

A third line of gene-targeted *Prnp*$^{0/0}$ mice generated in Nagasaki also developed normally, but, unlike the *Prnp*$^{0/0}$ mice created in Zurich and Edinburgh, they showed progressive ataxia and cerebellar Purkinje cell degeneration at about 70 weeks (Sakaguchi *et al.*, 1996). In addition to the PrP coding sequence, ~900 nucleotides from the second intron and 450 nucleotides of the 3' noncoding sequence of *Prnp* is deleted in these mice. Because the extent of the gene ablation differs among these *Prnp*$^{0/0}$ lines, there are several possible explanations of these discrepant phenotypes. Since the Zurich and Edinburgh $Prnp^{0/0}$ lines express a PrP/neo fusion transcript, this raises the possibility that any expressed product may be at least partially functional in masking the true phenotype of the PrP gene knockout. The possibility of an intron control element that is eliminated in the Nagasaki $Prnp^{0/0}$ line is relevant in light of studies using the so-called "half-genomic" construct. In this transgenic expression vector, referred to as phgPrP, the larger of the two $Prnp$ introns is completely absent (Fischer $et\ al.$, 1996). Interestingly, PrP RNA is not detected in cerebellar Purkinje cells of transgenic mice produced from phgPrP, supporting the idea that this $Prnp$ intron contains one or more Purkinje cell-specific enhancers controlling expression of non-PrP gene(s). Resolving the molecular basis of the phenotypic differences in these $Prnp^{0/0}$ lines is important, because the results may have widespread applications for understanding of general mechanisms of neurodegeneration.

Several other phenotypic defects are also being investigated in $Prnp^{0/0}$ mice, including altered circadian rhythms and sleep patterns (Tobler $et\ al.$, 1996), alterations in superoxide dismutase activity (SOD-1) (Brown $et\ al.$, 1997a) and defects in copper metabolism (Brown $et\ al.$, 1997b). Electrophysiological studies have demonstrated that $GABA_A$ receptor-mediated fast inhibition and long-term potentiation are impaired in hippocampal slices from $Prnp^{0/0}$ mice (Whittington $et\ al.$, 1995).

VI. SPECIES BARRIERS AND PROTEIN X

The species barrier was first described by Pattison in his studies of passage of the scrapie agent between sheep and rodents (Pattison, 1965). In those investigations, the initial passage of prions between species was associated with a prolonged incubation time, with only a few animals developing illness. Subsequent passage in the same species was characterized by all the animals becoming ill after greatly shortened incubation times.

Experiments designed to probe the molecular basis of the species barrier have also provided important clues about the mechanism of prion propagation involving association and conformational conversion between PrP^C and PrP^{Sc}. The seminal experiments involved transgenic mice that were engineered to express Syrian hamster PrP. As a result of the species barrier, wild-type mice are normally resistant to infection with Syrian hamster prions. However, expression of Syrian hamster PrP^C in Tg(SHaPrP) mice rendered mice susceptible to Syrian hamster prion infectivity and produced CNS pathology similar to that found in scrapie-sick Syrian hamsters (Scott $et\ al.$, 1989). These and subsequent studies (Prusiner $et\ al.$,

1990) indicated that disease transmission is facilitated when elements of the primary structure of host-encoded PrP^C and PrP^{Sc} in the inoculum are identical.

Transgenic mice expressing chimeric mouse–hamster PrP and mouse–human PrP gene constructs have been valuable in mapping functional domains of PrP (Scott *et al.*, 1993; Telling *et al.*, 1994, 1995) and point to the first α-helix of PrP^C as a region where homology between PrP^C and PrP^{Sc} is important, presumably as the site for interaction between the two isoforms during prion propagation. These *in vivo* results are supported by *in vitro* PrP conversion studies (Kocisko *et al.*, 1995).

The infrequent transmission of human prion disease to rodents has also been cited as an example of the species barrier. Based on the results with Tg(SHaPrP) mice, it was expected that the species barrier to human prion propagation would be abrogated in transgenic mice expressing human PrP. However, transmission of human prion disease was generally no more efficient in Tg(HuPrP) mice trans-genic mice expressing human PrP^C on a wild-type background than in nontrans-genic mice (Telling *et al.*, 1994), with one notable exception (Telling *et al.*, 1994; Collinge *et al.*, 1995a). The barrier to CJD transmission in Tg(HuPrP) mice was abolished by expressing HuPrP on the null background (Telling *et al.*, 1995). In contrast, propagation of human prions was highly efficient in transgenic mice expressing a chimeric mouse–human PrP gene, referred to as Tg(MHu2M) (Tell-ing *et al.*, 1994, 1995). These findings argue that MoPrP inhibited the transmission of prions to Tg mice expressing HuPrP but not to those expressing chimeric PrP. To explain these and other data, it was proposed that the most likely mediator of this inhibition is an auxiliary non-PrP molecule, provisionally designated protein X, that participates in the formation of prions by interacting with the carboxy-ter-minal region of PrP^C to facilitate conversion to PrP^{Sc} (Telling *et al.*, 1995). The ability of one certain human prion inoculum to override the inhibitory effects of mouse PrP^C (Telling *et al.*, 1994; Collinge *et al.*, 1995a) suggests that strain effects are also important. More recent *in vitro* experiments and structural studies suggest that two separate clusters of amino acid residues are juxtaposed in the three-di-mensional structure of PrP^C to form the putative protein X binding site (Billeter *et al.*, 1997; James *et al.*, 1997; Kaneko *et al.*, 1997). While protein X has been postulated from genetic arguments, factors that interact with PrP^C involved in its conversion to PrP^{Sc} await identification and characterization. To date, no less than 12 proteins have been identified as potential PrP^C ligands (Edenhofer *et al.*, 1996; Kurschner and Morgan, 1995; Martins *et al.*, 1997; Oesch and Prusiner, 1992; Rieger *et al.*, 1997; Yehiely *et al.*, 1997). In no case has physiological relevance been confirmed, although other studies suggest that protein X may function as a molecular chaperone during PrP^{Sc} formation (Tatzelt *et al.*, 1995).

VII. PRION STRAINS

Like viruses, prions exist as different strains with well-defined, heritable properties, which are defined in terms of incubation time and CNS pathology in inbred mice (Bruce and Dickinson, 1987; Dickinson et al., 1968). For many years, the issue of prion strains was largely of academic interest and was used to argue that prion diseases are caused by viruses (Bruce and Dickinson, 1987). In the face of convincing evidence that vCJD is caused by human exposure to BSE, the issue of prion strains has acquired new importance. Recent developments indicate that strain specificity is linked to conformational differences in PrP^{Sc}, providing a mechanism for the propagation of prion strains.

Studies of different strains of TME suggested that they might be represented by different conformational states of PrP^{Sc} (Bessen and Marsh, 1992). Evidence supporting the concept that PrP^{Sc} conformation encrypts strain information has emerged from transmission studies of human prion diseases in transgenic mice. A point mutation of the PrP gene at codon 178 is the cause of FFI (Goldfarb et al., 1992; Medori et al., 1992), while mutation of *PRNP* at codon 200, changing a glutamate to a lysine residue, referred to as fCJD(E200K), segregates with a different subtype of fCJD originally found in Libyan Jews with CJD (Hsiao et al., 1991). Expression of these different mutant prion proteins in patients with FFI and fCJD(E200K) results in variations in PrP conformation, reflected in altered proteinase K cleavage sites that generate PrP^{Sc} molecules with molecular weights of 19 kD in FFI and 21 kD in fCJD(E200K) (Monari et al., 1994). Extracts from the brains of FFI and fCJD(E200K) patients transmit disease to Tg(MHu2M) mice and induce the formation of the 19 kD PrP^{Sc} and 21 kD PrP^{Sc}, respectively (Telling et al., 1996b). On subsequent passage, these characteristic molecular sizes remain constant, but the incubation times for FFI and fCJD(E200K) prions diverge (Telling and Prusiner, unpublished observations). These studies suggest that mutant prion proteins with different primary structures produce distinct prion strains and argue that different tertiary structures of PrP^{Sc} can be imparted upon a single PrP^C by conformational templating.

Other studies have shown that different sporadic and iatrogenic CJD cases associated with specific codon 129 genotypes can be typed according to PrP^{Sc} fragment sizes following proteinase K treatment and western blotting of brain extracts (Collinge et al., 1996b; Parchi et al., 1996). These banding patterns are also maintained on experimental transmission, except when type 1 CJD (which are all of genotype Met/Met) is transmitted to Tg(HuPrP)/*Prnp*$^{0/0}$ mice that encode HuPrP with Val at 129. Of particular interest is the type 4 banding pattern of PrP^{Sc} found in vCJD patients and BSE-infected animals that has a distinct signature of protease-resistant mono-, di-, and unglygosylated PrP forms that distinguishes it from the patterns observed in classical CJD (Collinge et al., 1996b; Hill et al.,

1997a). Because glycosylation is thought to be a co-translational process, PrP conformation may be the primary determinant of strain type; however, a preferential affinity of particular PrP^{Sc} glycoforms to cells expressing similarly glycosylated proteins could be the mechanism by which different prion strains target distinct regions of the brain (Aguzzi and Weissmann, 1997; Collinge, 1997; Collinge et al., 1996b; Hecker et al., 1992).

While studies with transgenic mice have provided crucial information about the importance of PrP primary structure in determining prion species barriers, the ability of BSE prions to naturally affect a wide range of different species in addition to cattle forces the conclusion that the species barrier involves a complex interplay between PrP^{C}: PrP^{Sc} compatibility dictated by the primary structure and strain information dictated by PrP tertiary structure and/or glycosylation.

VIII. PERIPHERAL PATHOGENESIS

Although the pathological consequences of prion infections occur in the CNS and experimental transmission of these diseases is most efficiently accomplished by intracerebral inoculation, most natural infections occur through the peripheral route, for example, kuru by the oral route (Gajdusek, 1977) and iatrogenic CJD by intramuscular injection with prion-contaminated HuGH (Brown et al., 1992). Many studies over the years have established that prion replication in the lymphoreticular system (LRS) plays an essential role in the pathogenesis of prion diseases. Of particular relevance is the observation that type 4 PrP^{Sc} accumulates in the lymphoid tissue of tonsils in patients with vCJD (Hill et al., 1997b). Recent studies have established that differentiated B lymphocytes are crucial for neuroinvasion (Klein et al., 1997). Future studies on the peripheral pathogenesis of prion disease should identify additional crucial steps in the spread of prions from the periphery that may be amenable to pharmacological intervention.

IX. CONCLUSIONS AND FUTURE PROSPECTS

What were once esoteric studies of rare sheep and human diseases have been transformed by the recognition that prions are novel subcellular pathogens that represent an entirely new biological paradigm. In the 15 years since it was first proposed, the prion hypothesis has gone from being a heretical notion that contradicted the central dogma of molecular biology to being almost universally accepted. In the past decade, studies of the prion diseases have taken on a new significance with the emergence of the BSE epidemic that devastated the beef farming industry in the United Kingdom and other European countries because of concerns that BSE could infect humans via the food chain. The emergence of vCJD in the United Kingdom suggests that these fears were justified and that our understanding of these infectious pathogens is far from complete. While recent structural and molecular biological studies have clearly advanced our understanding of the prion diseases, the exact nature of the conformational changes involved in prion propagation remain to be determined.

The mechanism by which information is transmitted by conformational templating of proteins without the aid of nucleic acids is not unique to the mammalian prion diseases. It has become clear over the past few years that epigenetic transmission of prion-like cytoplasmic traits occurs in yeast, allowing them to switch physiological states and thereby adapt to new environments (see Lindquist [1997] and Masison *et al.* [1997] for reviews). It seems likely that similar prion-like modes of information transfer will turn out to be even more ubiquitous in nature.

REFERENCES

Aguzzi, A., and Weissmann, C. (1997). Prion research: The next frontiers. *Nature* **389**, 795–798.
Alper, T., Haig, D. A., and Clarke, M. C. (1966). The exceptionally small size of the scrapie agent. *Biochem. Biophys. Res. Commun.* **22**, 278–284.
Anderson, R. M., Donnelly, C. A., Ferguson, N. M., Woolhouse, M. E. J., Watt, C. J., Udy, H. J., MacWhinney, S., Dunstan, S. P., Southwood, T. R. E., Wilesmith, J. W., Ryan, J. B. M., Hoinville, L. J., Hillerton, J. E., Austin, A. R., and Wells, G. A. H. (1996). Transmission dynamics and epidemiology of BSE in British cattle. *Nature* **382**, 779–788.
Baker, H. F., Ridley, R. M., and Wells, G. A. H. (1993). Experimental transmission of BSE and scrapie to the common marmoset. *Vet. Rec.* **132**, 403–406.
Bamborough, P., Wille, H., Telling, G. C., Yehiely, F., Prusiner, S. B., and Cohen, F. E. (1996). Prion protein structure and scrapie replication: Theoretical, spectroscopic and genetic investigations. *Cold Spring Harb. Symp. Quant. Biol.* **61**, 495–509.
Basler, K., Oesch, B., Scott, M., Westaway, D., Wälchli, M., Groth, D. F., McKinley, M. P., Prusiner, S. B., and Weissmann, C. (1986). Scrapie and cellular PrP isoforms are encoded by the same chromosomal gene. *Cell* **46**, 417–428.
Bateman, D., Hilton, D., Love, S., Zeidler, M., Beck, J., and Collinge, J. (1995). Sporadic Creutzfeldt–Jakob disease in a 18-year-old in the UK [letter]. *Lancet* **346**, 1155–1156.
Bessen, R. A., and Marsh, R. F. (1992). Identification of two biologically distinct strains of transmissible mink encephalopathy in hamsters. *J. Gen. Virol.* **73**, 329–334.
Billeter, M., Riek, R., Wider, G., Hornemann, S., Glockshuber, R., and Wüthrich, K. (1997). Prion protein NMR structure and species barrier for prion diseases. *Proc. Natl. Acad. Sci. U.S.A.* **94**, 7281–7285.
Bolton, D. C., McKinley, M. P., and Prusiner, S. B. (1982). Identification of a protein that purifies with the scrapie prion. *Science* **218**, 1309–1311.

Borchelt, D. R., Scott, M., Taraboulos, A., Stahl, N., and Prusiner, S. B. (1990). Scrapie and cellular prion proteins differ in their kinetics of synthesis and topology in cultured cells. *J. Cell Biol.* **110**, 743–752.

Britton, T. C., Al-Sarraj, S., Shaw, C., Campbell, T., and Collinge, J. (1995). Sporadic Creutzfeldt–Jakob disease in a 16-year-old in the UK [letter]. *Lancet* 346, 1155.

Brown D. R., Schulz-Schaeffer, W. J., Schmidt B., and Kretzschmar H. A. (1997a) Prion protein-deficient cells show altered response to oxidative stress due to decreased SOD-1 activity. *Exp. Neurol.* **146**, 104–112.

Brown D. R., Qin K., Herms, J. W., Madlung, A., Manson, J., Strome, R., Fraser, P. E., Kruck, T., von Bohlen A., Schulz-Schaeffer, W., Giese, A., Westaway, D., and Kretzschmar, H. (1997b). The cellular prion protein binds copper *in vivo. Nature* **390**, 684–687.

Brown, P., Preece, M. A., and Will, R. G. (1992). "Friendly fire" in medicine: Hormones, homografts, and Creutzfeldt–Jakob disease. *Lancet* **340**, 24–27.

Brown, P., Cervenáková, L., Goldfarb, L. G., McCombie, W. R., Rubenstein, R., Will, R. G., Pocchiari, M., Martinez-Lage, J. F., Scalici, C., Masullo, C., Graupera, G., Ligan, J., and Gajdusek, D. C. (1994a). Iatrogenic Creutzfeldt–Jakob disease: An example of the interplay between ancient genes and modern medicine. *Neurology* **44**, 291–293.

Brown, P., Gibbs Jr., C. J., Rodgers-Johnson, P., Asher, D. M., Sulima, M. P., Bacote, A., Goldfarb, L. G., and Gajdusek, D. C. (1994b). Human spongiform encephalopathy: The National Institutes of Health series of 300 cases of experimentally transmitted disease. *Ann. Neurol.* **35**, 513–529.

Bruce, M. E. (1996). Strain typing studies of scrapie and BSE. *In* "Methods in Molecular Medicine: Prion Diseases" (H. F. Baker and R. M. Ridley, eds.), pp. 223–236. Humana Press, Totowa, NJ.

Bruce, M. E., and Dickinson, A. G. (1987). Biological evidence that the scrapie agent has an independent genome. *J. Gen. Virol.* **68**, 79–89.

Bruce, M. E., McConnell, I., Fraser, H., and Dickinson, A. G. (1991). The disease characteristics of different strains of scrapie in *Sinc* congenic mouse lines: Implications for the nature of the agent and host control of pathogenesis. *J. Gen. Virol.* **72**, 595–603.

Bruce, M. [E.], Chree, A., McConnell, I., Foster, J., Pearson, G., and Fraser, H. (1994). Transmission of bovine spongiform encephalopathy and scrapie to mice: Strain variation and the species barrier. *Phil. Trans. R. Soc. Lond. B* **343**, 405–411.

Bruce, M. E., Will, R. G., Ironside, J. W., McConnell, I., Drummond, D., Suttie, A., McCardle, L., Chree, A., Hope, J., Birkett, C., Sousens, S., Fraser, H., and Bostock, C. J. (1997). Transmissions to mice indicate that "new variant" CJD is caused by the BSE agent. *Nature* **389**, 498–501.

Büeler, H., Fischer, M., Lang, Y., Bluethmann, H., Lipp, H.-P., DeArmond, S. J., Prusiner, S. B., Aguet, M., and Weissmann, C. (1992). Normal development and behaviour of mice lacking the neuronal cell-surface PrP protein. *Nature* **356**, 577–582.

Büeler, H., Aguzzi, A., Sailer, A., Greiner, R.-A., Autenried, P., Aguet, M., and Weissmann, C. (1993). Mice devoid of PrP are resistant to scrapie. *Cell* **73**, 1339–1347.

Carlson, G. A., Kingsbury, D. T., Goodman, P. A., Coleman, S., Marshall, S. T., DeArmond, S. J., Westaway, D., and Prusiner, S. B. (1986). Linkage of prion protein and scrapie incubation time genes. *Cell* **46**, 503–511.

Caughey, B. W., Dong, A., Bhat, K. S., Ernst, D., Hayes, S. F., and Caughey, W. S. (1991). Secondary structure analysis of the scrapie-associated protein PrP 27-30 in water by infrared spectroscopy. *Biochemistry* **30**, 7672–7680.

Chandler, R. L. (1961). Encephalopathy in mice produced by inoculation with scrapie brain material. *Lancet* **1**, 1378–1379.

Chazot, G., Broussolle, E., Lapras, C. I., Blättler, T., Aguzzi, A., and Kopp, N. (1996). New variant of Creutzfeldt–Jakob disease in a 26-year-old French man. *Lancet* **347**, 1181.

Chesebro, B., Race, R., Wehrly, K., Nishio, J., Bloom, M., Lechner, D., Bergstrom, S., Robbins, K., Mayer, L., Keith, J. M., Garon, C., and Haase, A. (1985). Identification of scrapie prion protein-specific mRNA in scrapie-infected and uninfected brain. *Nature* **315**, 331–333.

Cohen, F. E., Pan, K.-M., Huang, Z., Baldwin, M., Fletterick, R. J., and Prusiner, S. B. (1994). Structural clues to prion replication. *Science* **264**, 530–531.

Collinge, J. (1997). Human prion diseases and bovine spongiform encephalopathy (BSE). *Hum. Mol. Genet.* **6**, 1699–1705.

Collinge, J., Palmer, M. S., and Dryden, A. J. (1991). Genetic predisposition to iatrogenic Creutzfeldt–Jakob disease. *Lancet* **337**, 1441–1442.

Collinge, J., Whittington, M. A., Sidle, K. C., Smith, C. J., Palmer, M. S., Clarke, A. R., and Jefferys, J. G. R. (1994). Prion protein is necessary for normal synaptic function. *Nature* **370**, 295–297.

Collinge, J., Palmer, M. S., Sidle, K. C., Hill, A. F., Gowland, I., Meads, J., Asante, E., Bradley, R., Doey, L. J., and Lantos, P. L. (1995a). Unaltered susceptibility to BSE in transgenic mice expressing human prion protein. *Nature* **378**, 779–783.

Collinge, J., Palmer, M. S., Sidle, K. C. L., Gowland, I., Medori, R., Ironside, J., and Lantos, P. (1995b). Transmission of fatal familial insomnia to laboratory animals [letter]. *Lancet* **346**, 569–570.

Collinge, J., Beck, J., Campbell, T., Estibeiro, K., and Will, R. G. (1996a). Prion protein gene analysis in new variant cases of Creutzfeldt–Jakob disease. *Lancet* **348**, 56.

Collinge, J., Sidle, K. C. L., Meads, J., Ironside, J., and Hill, A. F. (1996b). Molecular analysis of prion strain variation and the aetiology of "new variant" CJD. *Nature* **383**, 685–690.

Cuillé, J., and Chelle, P. L. (1939). Experimental transmission of trembling to the goat. *C. R. Seances Acad. Sci.* **208**, 1058–1060.

Cunningham, A. A., Wells, G. A. H., Scott, A. C., Kirkwood, J. K., and Barnett, J. E. F. (1993). Transmissible spongiform encephalopathy in greater kudu (*Tragelaphus strepsiceros*). *Vet. Rec.* **132**, 68.

Dawson, M., Wells, G. A. H., and Parker, B. N. J. (1990a). Preliminary evidence of the experimental transmissibility of bovine spongiform encephalopathy to cattle. *Vet. Rec.* **126**, 112–113.

Dawson, M., Wells, G. A. H., Parker, B. N. J., and Scott, A. C. (1990b). Primary parenteral transmission of bovine spongiform encephalopathy to the pig. *Vet. Rec.* **127**, 338.

Deslys, J.-P., Marcé, D., and Dormont, D. (1994). Similar genetic susceptibility in iatrogenic and sporadic Creutzfeldt–Jakob disease. *J. Gen. Virol.* **75**, 23–27.

Dickinson, A. G. (1976). Scrapie in sheep and goats. *In* "Slow Virus Diseases of Animals and Man" (R. H. Kimberlin, ed.), pp. 209–241. North-Holland, Amsterdam.

Dickinson, A. G., Young, G. B., Stamp, J. T., and Renwick, C. C. (1965). An analysis of natural scrapie in Suffolk sheep. *Heredity* **20**, 485–503.

Dickinson, A. G., Meikle, V. M. H., and Fraser, H. (1968). Identification of a gene which controls the incubation period of some strains of scrapie agent in mice. *J. Comp. Pathol.* **78**, 293–299.

Dickinson, A. G., Stamp, J. T., and Renwick, C. C. (1974). Maternal and lateral transmission of scrapie in sheep. *J. Comp. Pathol.* **84**, 19–25.

Dlouhy, S. R., Hsiao, K., Farlow, M. R., Foroud, T., Conneally, P. M., Johnson, P., Prusiner, S. B., Hodes, M. E., and Ghetti, B. (1992). Linkage of the Indiana kindred of Gerstmann–Sträussler–Scheinker disease to the prion protein gene. *Nat. Genet.* **1**, 64–67.

Doh-ura, K., Tateishi, J., Sasaki, H., Kitamoto, T., and Sakaki, Y. (1989). Pro→Leu change at position 102 of prion protein is the most common but not the sole mutation related to Gerstmann–Sträussler syndrome. *Biochem. Biophys. Res. Commun.* **163**, 974–979.

Edenhofer, F., Rieger, R., Famulok, M., Wendler, W., Weiss, S., and Winnacker, E.-L. (1996). Prion protein PrP^C interacts with molecular chaperons of the Hsp60 family. *J. Virol.* **70**, 4724–4728.

Fischer, M., Rulicke, T., Raeber, A., Sailer, A., Moser, M., Oesch, B., Brandner, S., Aguzzi, A., and Weissmann, C. (1996). Prion protein (PrP) with amino terminal deletions restoring susceptibility of PrP knockout mice to scrapie. *EMBO J.* **15**, 1255–1264.

Foster, J. D., Hope, J., and Fraser, H. (1993). Transmission of bovine spongiform encephalopathy to sheep and goats. *Vet. Rec.* **133**, 339–341.

Fraser, H., McConnell, I., Wells, G. A. H., and Dawson, M. (1988). Transmission of bovine spongiform encephalopathy to mice. *Vet. Rec.* **123**, 472.

Gabizon, R., Rosenmann, H., Meiner, Z., Kahana, I., Kahana, E., Shugart, Y., Ott, J., and Prusiner, S. B. (1993). Mutation and polymorphism of the prion protein gene in Libyan Jews with Creutzfeldt–Jakob disease. *Am. J. Hum. Genet.* **33**, 828–835.

Gajdusek, D. C. (1977). Unconventional viruses and the origin and disappearance of kuru. *Science* **197**, 943–960.

Gajdusek, D. C., and Zigas, V. (1957). Degenerative disease of the central nervous system in New Guinea: The endemic occurrence of "kuru" in the native population. *New Engl. J. Med.* **257**, 974–978.

Gajdusek, D. C., Gibbs Jr., C. J., and Alpers, M. (1966). Experimental transmission of a kuru-like syndrome to chimpanzees. *Nature* **209**, 794–796.

Gibbs Jr., C. J., Gajdusek, D. C., Asher, D. M., Alpers, M. P., Beck, E., Daniel, P. M., and Matthews, W. B. (1968). Creutzfeldt–Jakob disease (spongiform encephalopathy): Transmission to the chimpanzee. *Science* **161**, 388–389.

Gibbs Jr., C. J., Gajdusek, D. C., and Amyx, H. (1979). Strain variation in the viruses of Creutzfeldt–Jakob disease and kuru. *In* "Slow Transmissible Diseases of the Nervous System," Vol. 2 (S. B. Prusiner and W. J. Hadlow, eds.), pp. 87–110. Academic Press, New York.

Goldfarb, L. G., Petersen, R. B., Tabaton, M., Brown, P., LeBlanc, A. C., Montagna, P., Cortelli, P., Julien, J., Vital, C., Pendelbury, W. W., Haltia, M., Wills, P. R., Hauw, J. J., McKeever, P. E., Monari, L., Schrank, B., Swergold, G. D., Autilio-Gambetti, L., Gajdusek, D. C., Lugaresi, E., and Gambetti, P. (1992). Fatal familial insomnia and familial Creutzfeldt–Jakob disease: Disease phenotype determined by a DNA polymorphism. *Science* **258**, 806–808.

Goldgaber, D., Goldfarb, L. G., Brown, P., Asher, D. M., Brown, W. T., Lin, S., Teener, J. W., Feinstone, S. M., Rubenstein, R., Kascsak, R. J., Boellaard, J. W., and Gajdusek, D. C. (1989). Mutations in familial Creutzfeldt–Jakob disease and Gerstmann–Sträussler–Scheinker's syndrome. *Exp. Neurol.* **106**, 204–206.

Gordon, W. S. (1946). Advances in veterinary research. *Vet. Res.* **58**, 516–520.

Griffith, J. S. (1967). Self-replication and scrapie. *Nature* **215**, 1043–1044.

Hadlow, W. J. (1959). Scrapie and kuru. *Lancet* **2**, 289–290.

Hadlow, W. J., Kennedy, R. C., Race, R. E., and Eklund, C. M. (1980). Virologic and neurohistologic findings in dairy goats affected with natural scrapie. *Vet. Pathol.* **17**, 187–199.

Hadlow, W. J., Kennedy, R. C., and Race, R. E. (1982). Natural infection of Suffolk sheep with scrapie virus. *J. Infect. Dis.* **146**, 657–664.

Hecker, R., Taraboulos, A., Scott, M., Pan, K.-M., Torchia, M., Jendroska, K., DeArmond, S. J., and Prusiner, S. B. (1992). Replication of distinct prion isolates is region specific in brains of transgenic mice and hamsters. *Genes Dev.* **6**, 1213–1228.

Hill, A. F., Desbrulais, M., Joiner, S., Sidle, K. D. L., Gowland, I., Collinge, J., Doey, L. J., and Lantos, P. (1997a). The same strain causes vCJD and BSE. *Nature* **389**, 448–450.

Hill, A. F., Zeidler, M., Ironside, J., and Collinge, J. (1997b). Diagnosis of new variant Creutzfeldt–Jakob disease by tonsil biopsy. *Lancet* **349**, 99–100.

Hsiao, K., Baker, H. F., Crow, T. J., Poulter, M., Owen, F., Terwilliger, J. D., Westaway, D., Ott, J., and Prusiner, S. B. (1989). Linkage of a prion protein missense variant to Gerstmann–Sträussler syndrome. *Nature* **338**, 342–345.

Hsiao, K. K., Scott, M., Foster, D., Groth, D. F., DeArmond, S. J., and Prusiner, S. B. (1990). Spontaneous neurodegeneration in transgenic mice with mutant prion protein. *Science* **250**, 1587–1590.

Hsiao, K., Meiner, Z., Kahana, E., Cass, C., Kahana, I., Avrahami, D., Scarlato, G., Abramsky, O., Prusiner, S. B., and Gabizon, R. (1991). Mutation of the prion protein in Libyan Jews with Creutzfeldt–Jakob disease. *New Engl. J. Med.* **324**, 1091–1097.

Hsiao, K. K., Groth, D., Scott, M., Yang, S.-L., Serban, H., Rapp, D., Foster, D., Torchia, M., DeArmond, S. J., and Prusiner, S. B. (1994). Serial transmission in rodents of neurodegeneration from transgenic mice expressing mutant prion protein. *Proc. Natl. Acad. Sci. U.S.A.* **91**, 9126–9130.

Huang, Z., Prusiner, S. B., and Cohen, F. E. (1996). Scrapie prions: A three-dimensional model of an infectious fragment. *Folding Design* **1**, 13–19.

Hunter, G. D. (1972). Scrapie: A prototype slow infection. *J. Infect. Dis.* **125**, 427–440.

James, T. L., Liu, H., Ulyanov, N. B., Farr-Jones, S., Zhang, H., Donne, D. G., Kaneko, K., Groth, D., Mehlhorn, I., Prusiner, S. B., and Cohen, F. E. (1997). Solution structure of a 142-residue recombinant prion protein corresponding to the infectious fragment of the scrapie isoform. *Proc. Natl. Acad. Sci. U.S.A.* **94**, 10086–10091.

Kaneko, K., Zulianello, L., Scott, M., Cooper, C. M., Wallace, A. C., James, T. L., Cohen, F. E., and Prusiner, S. B. (1997). Evidence for protein X binding to a discontinuous epitope on the cellular prion protein during scrapie prion propagation. *Proc. Natl. Acad. Sci. U.S.A.* **94**, 10069–10074.

Kellings, K., Meyer, N., Mirenda, C., Prusiner, S. B., and Riesner, D. (1992). Further analysis of nucleic acids in purified scrapie prion preparations by improved return refocussing gel electrophoresis (RRGE). *J. Gen. Virol.* **73**, 1025–1029.

Kirkwood, J. K., and Cunningham, A. A. (1994). Epidemiological observations on spongiform encephalopathies in captive wild animals in the British Isles. *Vet. Rec.* **135**, 296–303.

Klatzo, I., Gajdusek, D. C., and Zigas, V. (1959). Pathology of kuru. *Lab. Invest.* **8**, 799–847.

Klein, M. A., Frigg, R., Flechsig, E., Raeber, A. J., Kalinke, U., Bluethmann, H., Bootz, F., Suter, M., Zinkernagel, R. M., and Aguzzi, A. (1997). A crucial role for B cells in neuroinvasive scrapie. *Nature* **390**, 687–690.

Kocisko, D. A., Priola, S. A., Raymond, G. J., Chesebro, B., Lansbury Jr., P. T., and Caughey, B. (1995). Species specificity in the cell-free conversion of prion protein to protease-resistant forms: A model for the scrapie species barrier. *Proc. Natl. Acad. Sci. U.S.A.* **92**, 3923–3927.

Kretzschmar, H. A., Honold, G., Seitelberger, F., Feucht, M., Wessely, P., Mehraein, P., and Budka, H. (1991). Prion protein mutation in family first reported by Gerstmann, Sträussler, and Scheinker. *Lancet* **337**, 1160.

Kurschner, C., and Morgan, J. I. (1995). The cellular prion protein (PrP) selectively binds to Bcl-2 in the yeast two-hybrid system. *Mol. Brain Res.* **30**, 165–168.

Lasmézas, C. I., Deslys, J.-P., Demaimay, R., Adjou, K. T., Lamoury, F., Dormont, D., Robain, O., Ironside, J., and Hauw, J.-J. (1996). BSE transmission to macaques. *Nature* **381**, 743–744.

Lindquist, S. (1997). Mad cows meet psi-chotic yeast: The expansion of the prion hypothesis. *Cell* **89**, 495–498.

Manson, J. C., Clarke, A. R., Hooper, M. L., Aitchison, L., McConnell, I., and Hope, J. (1994a). 129/Ola mice carrying a null mutation in PrP that abolishes mRNA production are developmentally normal. *Mol. Neurobiol.* **8**, 121–127.

Manson, J. C., Clarke, A. R., McBride, P. A., McConnell, I., and Hope, J. (1994b). PrP gene dosage determines the timing but not the final intensity or distribution of lesions in scrapie pathology. *Neurodegeneration* **3**, 331–340.

Manuelidis, E., Kim, J., Angelo, J., and Manuelidis, L. (1976). Serial propagation of Creutzfeldt–Jakob disease in guinea pigs. *Proc. Natl. Acad. Sci. U.S.A.* **73**, 223–227.

Manuelidis, E., Gorgacz, E. J., and Manuelidis, L. (1978). Interspecies transmission of Creutzfeldt–Jakob disease to Syrian hamsters with reference to clinical syndromes and strains of agent. *Proc. Natl. Acad. Sci. U.S.A.* **75**, 3422–3436.

Marsh, R. F., and Hanson, R. P. (1977). The Syrian hamster as a model for the study of slow virus diseases caused by unconventional agents. *Fed. Proc.* **37**, 2076–2078.

Marsh, R. F., Bessen, R. A., Lehmann, S., and Hartsough, G. R. (1991). Epidemiological and experimental studies on a new incident of transmissible mink encephalopathy. *J. Gen. Virol.* **72**, 589–594.

Martins, V. R., Graner, E., Garcia-Abreu, J., De Souza, S. J., Mercadante, A. F., Veiga, S. S., Zantana, S. M., Moura Neto, V., and Brantani, R. R. (1997). Complementary hydropathy identifies a cellular prion protein receptor. *Nature Med.* **3**, 1376–1382.

Masison, D., Maddelein, M.-L., and Wickner, R. B. (1997). The prion model for [URE3] of yeast: Spontaneous generation and requirements for propagation. *Proc. Natl. Acad. Sci. U.S.A.* **94**, 12503–12508.

Masters, C. L., and Richardson Jr., E. P. (1978). Subacute spongiform encephalopathy Creutzfeldt–Jakob disease: The nature and progression of spongiform change. *Brain* **101**, 333–344.

Masters, C. L., Harris, J. O., Gajdusek, D. C., Gibbs Jr., C. J., Bernouilli, C., and Asher, D. M. (1978). Creutzfeldt–Jakob disease: Patterns of worldwide occurrence and the significance of familial and sporadic clustering. *Ann. Neurol.* **5**, 177–188.

McKinley, M. P., Bolton, D. C., and Prusiner, S. B. (1983). A protease-resistant protein is a structural component of the scrapie prion. *Cell* **35**, 57–62.

Medori, R., Tritschler, H.-J., LeBlanc, A., Villare, F., Manetto, V., Chen, H. Y., Xue, R., Leal, S., Montagna, P., Cortelli, P., Tinuper, P., Avoni, P., Mochi, M., Baruzzi, A., Hauw, J. J., Ott, J., Lugaresi, E., Autilio-Gambetti, L., and Gambetti, P. (1992). Fatal familial insomnia, a prion disease with a mutation at codon 178 of the prion protein gene. *New Engl. J. Med.* **326**, 444–449.

Meyer, R. K., McKinley, M. P., Bowman, K. A., Braunfeld, M. B., Barry, R. A., and Prusiner, S. B. (1986). Separation and properties of cellular and scrapie prion proteins. *Proc. Natl. Acad. Sci. U.S.A.* **83**, 2310–2314.

Monari, L., Chen, S. G., Brown, P., Parchi, P., Petersen, R. B., Mikol, J., Gray, F., Cortelli, Montagna, P., Ghetti, B., Goldfarb, L. G., Gajdusek, D. C., Lugaresi, E., Gambetti, P., and Autilio-Gambetti, L. (1994). Fatal familial insomnia and familial Creutzfeldt–Jakob disease: Different prion proteins determined by a DNA polymorphism. *Proc. Natl. Acad. Sci. U.S.A.* **91**, 2839–2842.

Oesch, B., and Prusiner, S. B. (1992). Interaction of the prion protein with cellular proteins. *In* "Prion Diseases of Humans and Animals" (S. B. Prusiner, J. Collinge, J. Powell, and B. Anderton, eds.), pp. 398–406. Ellis Horwood, London.

Oesch, B., Westaway, D., Wälchli, M., McKinley, M. P., Kent, S. B. H., Aebersold, R., Barry, R. A., Tempst, P., Teplow, D. B., Hood, L. E., Prusiner, S. B., and Weissmann, C. (1985). A cellular gene encodes scrapie PrP 27-30 protein. *Cell* **40**, 735–746.

Owen, F., Poulter, M., Lofthouse, R., Collinge, J., Crow, T. J., Risby, D., Baker, H. F., Ridley, R. M., Hsiao, K., and Prusiner, S. B. (1989). Insertion in prion protein gene in familial Creutzfeldt–Jakob disease. *Lancet* **1**, 51–52.

Owen, F., Poulter, M., Collinge, J., and Crow, T. J. (1990). Codon 129 changes in the prion protein gene in Caucasians. *Am. J. Hum. Genet.* **46**, 1215–1216.

Palmer, M. S., Dryden, A. J., Hughes, J. T., and Collinge, J. (1991). Homozygous prion protein genotype predisposes to sporadic Creutzfeldt–Jakob disease. *Nature* **352**, 340–342.

Pan, K.-M., Baldwin, M., Nguyen, J., Gasset, M., Serban, A., Groth, D., Mehlhorn, I., Huang, Z., Fletterick, R. J., Cohen, F. E., and Prusiner, S. B. (1993). Conversion of α-helices into β-sheets features in the formation of the scrapie prion proteins. *Proc. Natl. Acad. Sci. U.S.A.* **90**, 10962–10966.

Parchi, P., Castellani, R., Capellari, S., Ghetti, B., Young, K., Chen, S. G., Farlow, M., Dickson, D. W., Sima, A. A. F., Trojanowski, J. Q., Petersen, R. B., and Gambetti, P. (1996). Molecular basis of phenotypic variability in sporadic Creutzfeldt–Jakob disease. *Ann. Neurol.* **39**, 767–778.

Parry, H. B. (1962). Scrapie: A transmissible and hereditary disease of sheep. *Heredity* **17**, 75–105.

Pattison, I. H. (1965). Experiments with scrapie with special reference to the nature of the agent and the pathology of the disease. *In* "Slow, Latent and Temperate Virus Infections," NINDB Monograph 2 (D. C. Gajdusek, C. J. Gibbs Jr., and M. P. Alpers, eds.), pp. 249–257. U.S. Government Printing, Washington, DC.

Pearson, G. R., Wyatt, J. M., Gruffydd-Jones, T. J., Hope, J., Chong, A., Higgins, R. J., Scott, A. C., and Wells, G. A. H. (1992). Feline spongiform encephalopathy: Fibril and PrP studies. *Vet. Rec.* **131**, 307–310.

Pergami, P., Jaffe, H., and Safar, J. (1996). Semipreparative chromatographic method to purify the normal cellular isoform of the prion protein in nondenatured form. *Anal. Biochem.* **236**, 63–73.

Petersen, R. B., Tabaton, M., Berg, L., Schrank, B., Torack, R. M., Leal, S., Julien, J., Vital, C., Deleplanque, B., Pendlebury, W. W., Drachman, D., Smith, T. W., Martin, J. J., Oda, M., Montagna, P., Ott, J., Autilio-Gambetti, L., Lugaresi, E., and Gambetti, P. (1992). Analysis of the prion protein gene in thalamic dementia. *Neurology* **42**, 1859–1863.

Poulter, M., Baker, H. F., Frith, C. D., Leach, M., Lofthouse, R., Ridley, R. M., Shah, T., Owen, F., Collinge, J., Brown, G., Hardy, J., Mullan, M. J., Harding, A. E., Bennett, C., Doshi, R., and Crow, T. J. (1992). Inherited prion disease with 144 base pair gene insertion, 1: Genealogical and molecular studies. *Brain* **115**, 675–685.

Prusiner, S. B. (1982). Novel proteinaceous infectious particles cause scrapie. *Science* **216**, 136–144.

Prusiner, S. B. (1987). The biology of prion transmission and replication. *In* "Prions: Novel Infectious Pathogens Causing Scrapie and Creutzfeldt–Jakob Disease" (S. B. Prusiner and M. P. McKinley, eds.), pp. 83–112. Academic Press, Orlando.

Prusiner, S. B. (1989). Scrapie prions. *Annu. Rev. Microbiol.* **43**, 345–374.

Prusiner, S. B., Bolton, D. C., Groth, D. F., Bowman, K. A., Cochran, S. P., and McKinley, M. P. (1982). Further purification and characterization of scrapie prions. *Biochemistry* **21**, 6942–6950.

Prusiner, S. B., McKinley, M. P., Bowman, K. A., Bolton, D. C., Bendheim, P. E., Groth, D. F., and Glenner, G. G. (1983). Scrapie prions aggregate to form amyloid-like birefringent rods. *Cell* **35**, 349–358.

Prusiner, S. B., Groth, D. F., Bolton, D. C., Kent, S. B., and Hood, L. E. (1984). Purification and structural studies of a major scrapie prion protein. *Cell* **38**, 127–134.

Prusiner, S. B., Scott, M., Foster, D., Pan, K.-M., Groth, D., Mirenda, C., Torchia, M., Yang, S.-L., Serban, D., Carlson, G. A., Hoppe, P. C., Westaway, D., and DeArmond, S. J. (1990). Transgenetic studies implicate interactions between homologous PrP isoforms in scrapie prion replication. *Cell* **63**, 673–686.

Prusiner, S. B., Groth, D., Serban, A., Koehler, R., Foster, D., Torchia, M., Burton, D., Yang, S.-L., and DeArmond, S. J. (1993). Ablation of the prion protein (PrP) gene in mice prevents scrapie and facilitates production of anti-PrP antibodies. *Proc. Natl. Acad. Sci. U.S.A.* **90**, 10608–10612.

Rieger, R., Edenhofer, F., Lasmezas, C. I., and Weiss, S. (1997). The human 37-kDa laminin receptor precursor interacts with the prion protein in eukaryotic cells. *Nature Med.* **3**, 1383–1388.

Riek, R., Hornemann, S., Wider, G., Billeter, M., Glockshuber, R., and Wüthrich, K. (1996). NMR structure of the mouse prion protein domain PrP(121–231). *Nature* **382**, 180–182.

Riek, R., Hornemann, S., Wider, G., Glockshuber, R., and Wüthrich, K. (1997). NMR characterization of the full-length recombinant murine prion protein mPrP(23–231). *FEBS Lett.* **413**, 282–288.

Robakis, N. K., Devine-Gage, E. A., Kascsak, R. J., Brown, W. T., Krawczun, C., and Silverman, W. P. (1986). Localization of a human gene homologous to the PrP gene on the p arm of chromosome 20 and detection of PrP-related antigens in normal human brain. *Biochem. Biophys. Res. Commun.* **140**, 758–765.

Robinson, M. M., Hadlow, W. J., Huff, T. P., Wells, G. A. H., Dawson, M., Marsh, R. F., and Gorham, J. R. (1994). Experimental infection of mink with bovine spongiform encephalopathy. *J. Gen. Virol.* **75**, 2151–2155.

Robinson, M. M., Hadlow, W. J., Knowles, D. P., Huff, T. P., Lacy, P. A., Marsh, R. F., and Gorham, J. R. (1995). Experimental infection of cattle with the agents of transmissible mink encephalopathy and scrapie. *J. Comp. Path.* **113**, 241–251.

Safar, J., Roller, P. P., Gajdusek, D. C., and Gibbs Jr., C. J., (1993). Conformational transitions, dissociation, and unfolding of scrapie amyloid (prion) protein. *J. Biol. Chem.* **268**, 20276–20284.

Sakaguchi, S., Katamine, S., Shigematsu, K., Nakatani, A., Moriuchi, R., Nishida, N., Kurokawa, K., Nakaoke, R., Sato, H., Jishage, K., Kuno, J., Noda, T., and Miyamoto, T. (1995). Accumulation of proteinase K-resistant prion protein (PrP) is restricted by the expression level of normal PrP in mice inoculated with a mouse-adapted strain of the Creutzfeldt–Jakob disease agent. *J. Virol.* **69**, 7586–7592.

Sakaguchi, S., Katamine, S., Nishida, N., Moriuchi, R., Shigematsu, K., Sugimoto, T., Nakatani, A., Kataoka, Y., Houtani, T., Shirabe, S., Okada, H., Hasegawa, S., Miyamoto, T., and Noda, T. (1996). Loss of cerebellar Purkinje cells in aged mice homozygous for a disrupted PrP gene. *Nature* **380**, 528–531.

Scott, M., Foster, D., Mirenda, C., Serban, D., Coufal, F., Wälchli, M., Torchia, M., Groth, D., Carlson, G., DeArmond, S. J., Westaway, D., and Prusiner, S. B. (1989). Transgenic mice expressing hamster prion protein produce species-specific scrapie infectivity and amyloid plaques. *Cell* **59**, 847–857.

Scott, M., Groth, D., Foster, D., Torchia, M., Yang, S.-L., DeArmond, S. J., and Prusiner, S. B. (1993). Propagation of prions with artificial properties in transgenic mice expressing chimeric PrP genes. *Cell* **73**, 979–988.

Stahl, N., Borchelt, D. R., Hsiao, K., and Prusiner, S. B. (1987). Scrapie prion protein contains a phosphatidylinositol glycolipid. *Cell* **51**, 229–240.

Stahl, N., Baldwin, M. A., Teplow, D. B., Hood, L., Gibson, B. W., Burlingame, A. L., and Prusiner, S. B. (1993). Structural analysis of the scrapie prion protein using mass spectrometry and amino acid sequencing. *Biochemistry* **32**, 1991–2002.

Stekel, D. J., Nowak, M. A., and Southwood, T. R. E. (1996). Prediction of future BSE spread. *Nature* **381**, 119.

Tateishi, J., and Kitamoto, T. (1995). Inherited prion diseases and transmission to rodents. *Brain Pathol.* **5**, 53–59.

Tateishi, J., Sato, Y., and Ohta, M. (1983). Creutzfeldt–Jakob disease in humans and laboratory animals. *In* "Progress in Neuropathology," Vol. 5 (H. M. Zimmerman, ed.), pp. 195–221. Raven, New York.

Tatzelt, J., Zuo, J., Voellmy, R., Scott, M., Hartl, U., Prusiner, S. B., and Welch, W. J. (1995). Scrapie prions selectively modify the stress response in neuroblastoma cells. *Proc. Natl. Acad. Sci. U.S.A.* **92**, 2944–2948.

Telling, G., and Prusiner, S. B. Unpublished observations.

Telling, G. C., Scott, M., Hsiao, K. K., Foster, D., Yang, S.-L., Torchia, M., Sidle, K. C. L., Collinge, J., DeArmond, S. J., and Prusiner, S. B. (1994). Transmission of Creutzfeldt–Jakob disease from humans to transgenic mice expressing chimeric human-mouse prion protein. *Proc. Natl. Acad. Sci. U.S.A.* **91**, 9936–9940.

Telling, G. C., Scott, M., Mastrianni, J., Gabizon, R., Torchia, M., Cohen, F. E., DeArmond, S. J., and Prusiner, S. B. (1995). Prion propagation in mice expressing human and chimeric PrP transgenes implicates the interaction of cellular PrP with another protein. *Cell* **83**, 79–90.

Telling, G. C., Haga, T., Torchia, M., Tremblay, P., DeArmond, S. J., and Prusiner, S. B. (1996a). Interactions between wild-type and mutant prion proteins modulate neurodegeneration in transgenic mice. *Genes Dev.* **10**, 1736–1750.

Telling, G. C., Parchi, P., DeArmond, S. J., Cortelli, P., Montagna, P., Gabizon, R., Mastrianni, J., Lugaresi, E., Gambetti, P., and Prusiner, S. B. (1996b). Evidence for the conformation of the pathologic isoform of the prion protein enciphering and propagating prion diversity. *Science* **274**, 2079–2082.

Telling, G. C., Tremblay, P., Torchia, M., DeArmond, S. J., Cohen, F. E., and Prusiner, S. B. (1997). N-terminally tagged prion protein supports prion propagation in transgenic mice. *Protein Sci.* **6**, 825–833.

Tobler, I., Gaus, S. E., Deboer, T., Achermann, P., Fischer, M., Rülicke, T., Moser, M., Oesch, B., McBride, P. A., and Manson, J. C. (1996). Altered circadian activity rhythms and sleep in mice devoid of prion protein. *Nature* **380**, 639–642.

Tremblay, P., Meiner, Z., Galou, M., Heinrich, C., Petromilli, C., Lisse, T., Cayetano, J., Torchia, M., Mobley, W., Bujard, H., DeArmond, S. J., and Prusiner, S. B. (1998). Doxycycline control of prion protein transgene expression modulates prion disease in mice *Proc. Natl. Acad. Sci. U.S.A.* **95**, 12580–12585.

Wells, G. A. H., Scott, A. C., Johnson, C. T., Gunning, R. F., Hancock, R. D., Jeffrey, M., Dawson, M., and Bradley, R. (1987). A novel progressive spongiform encephalopathy in cattle. *Vet. Rec.* **121**, 419–420.

Westaway, D., Goodman, P. A., Mirenda, C. A., McKinley, M. P., Carlson, G. A., and Prusiner, S. B. (1987). Distinct prion proteins in short and long scrapie incubation period mice. *Cell* **51**, 651–662.

Whittington, M. A., Sidle, K. C. L., Gowland, I., Meads, J., Hill, A. F., Palmer, M. S., Jefferys, J. G. R., and Collinge, J. (1995). Rescue of neurophysiological phenotype seen in PrP null mice by transgene encoding human prion protein. *Nat. Genet.* **9**, 197–201.

Wilesmith, J. W. (1991). The epidemiology of bovine spongiform encephalopathy. *Sem. Virol.* **2**, 239–245.

Wilesmith, J. W., Wells, G. A. H., Cranwell, M. P., and Ryan, J. B. M. (1988). Bovine spongiform encephalopathy: Epidemiological studies. *Vet. Rec.* **123**, 638–644.

Will, R. G., Ironside, J. W., Zeidler, M., Cousens, S. N., Estibeiro, K., Alperovitch, A., Poser, S., Pocchiari, M., Hofman, A., and Smith, P. G. (1996). A new variant of Creutzfeldt–Jakob disease in the UK. *Lancet* **347**, 921–925.

Williams, E. S., and Young, S. (1980). Chronic wasting disease of captive mule deer: A spongiform encephalopathy. *J. Wildl. Dis.* **16**, 89–98.

Williams, E. S., and Young, S. (1992). Spongiform encephalopathies in Cervidae. *Rev. Sci. Tech. Off. Int. Epiz.* **11**, 551–567.

Williams, E. S., and Young, S. (1993). Neuropathology of chronic wasting disease of mule deer (*Odocoileus hemionus*) and elk (*Cervus elaphus* nelsoni). *Vet. Pathol.* **30**, 36–45.

Wyatt, J. M., Pearson, G. R., and Gruffydd-Jones, T. J. (1993). Feline spongiform encephalopathy. *Feline Practice* **21**, 7–9.

Yehiely, F., Bamborough, P., Da Costa, M., Perry, B. J., Thinakaran, G., Cohen, F. E., Carlson, G. A., and Prusiner, S. B. (1997). Identification of candidate proteins binding to prion protein. *Neurobiol. Dis.* **3**, 339–355.

Zeidler, M., Stewart, G., Cousens, S. N., Estibeiro, K., and Will, R. G. (1997). Codon 129 genotype and new variant CJD. *Lancet* **350**, 668.

Index